T0100435

Numerical Methods for Reliability and Safety Assessment

Seifedine Kadry • Abdelkhalak El Hami
Editors

Numerical Methods for Reliability and Safety Assessment

Multiscale and Multiphysics Systems

 Springer

Editors
Seifedine Kadry
American University of the Middle East
Al-Ahmadi, Egaila
Kuwait

Abdelkhalak El Hami
National Institute of Applied Sciences
INSA de Rouen Laboratoire d' Optimisation
Saint Etienne de Rouvray
France

ISBN 978-3-319-07166-4   ISBN 978-3-319-07167-1 (eBook)
DOI 10.1007/978-3-319-07167-1
Springer Cham Heidelberg New York Dordrecht London

Library of Congress Control Number: 2014948196

Printed on acid-free paper

Springer is part of Springer Science+Business Media (www.springer.com)

Foreword

The recent advances in materials, sensors, and computational methods have resulted in a much higher reliability and safety expectations of infrastructures, products, and services. This has been translated into expected longer lives for non-repairable products such as satellites, longer warranty periods for both repairable and non-repairable products such as automobiles, and longer residual lives of infrastructures such as bridges, dams, and high-rising buildings. In order to accomplish these expectations, the designers, engineers, and analysts need to incorporate the system configuration, physics of failure of its components, and the scale and complexity of the system. Therefore, testing begins at the components levels and subsystems. Reliability and safety analyses are conducted at all levels considering different failure modes of the components and subsystems under different operating conditions. Different numerical approaches are required at every aspect and step in the design and implementation processes.

The chapters of this book cover three topics related to different aspects of reliability and safety of complex systems. The first set of topics deals with generic methods and approaches which include theoretical developments and quantification of uncertainties which have effects on the expected lives and performance of the products and structures, approaches for risk assessments due to environmental conditions, methods for conducting and analyzing accelerated life testing, and use of advanced design of experiments methods such as Latin Hypercube for estimating the optimum parameters levels for reliability-based designs. The second set of topics deals with applications and use of reliability as a criterion in the design of civil engineering infrastructures such as blast wall structures, road pavements operating under different environmental conditions and different traffic loads, and other applications. The third set of topics is devoted to mechanical systems, their designs and reliability modeling. They include optimum inspection periods for aircraft structures subject to fatigue loadings and optimum repairs for mechatronics systems.

The book is an excellent reference for the design of systems, structures, and products for reliability and safety. The chapters provide coverage of the use of reliability methods in a wide range of engineering applications.

Piscataway, NJ   E.A. Elsayed

Preface

Reliability and safety analyses are important applications of modern probabilistic methods and stochastic concept (reliability of systems, probability of failure, statistics, and random variables/processes). These fields create a wide range of problems but due to their practical importance, it gave rise to development of new probabilistic methods and can contain interesting and fruitful mathematical settings. The reliability of a structure is traditionally achieved by deterministic methods using safety factors calculated generally under conservative estimators of influent parameters. Structural reliability analysis methods use probabilistic approaches for assessing safety factors or for optimizing maintenance and inspection programs. These methods become essential in the frame of long-term maintenance or life extension.

The main focus of this book is numerical methods for multiscale and multiphysics in reliability and safety. Multiphysics problems are problems involving two or more equations describing different physical phenomena that are coupled together via the equations. Multiscale problems on the other hand are problems on large scales that experience fine scale behavior, which makes them hard to solve using standard methods. Instead of solving the entire problem at once the problem is rewritten into many smaller subproblems that are coupled from each other.

This book includes 29 chapters, contributed by worldwide researchers and practitioners from 16 countries, of innovative concepts, theories, techniques, and engineering applications in various fields. It is designed to assist practicing engineers, students, and researchers in the areas of reliability engineering, safety and risk analysis.

Egaila, Kuwait   Seifedine Kadry
Rouen, France   Abdelkhalak El Hami

Acknowledgments

We would like to thank the Springer editor and the book reviewers for their valuable suggestions to make this quality book project happen and appear to service the public.

Contents

Part III Reliability and Risk Analysis

Contributors

Amir Alani School of Engineering, University of Greenwich, Chatham Maritime, UK

Ciprian Alecsandru Concordia University, Montreal, Canada

Md. Shohel Reza Amin Department of Building, Civil and Environmental Engineering, Concordia University, Montreal, QC, Canada

Marcus Arenius Fachbereich Maschinenbau Arbeits- und Organisationspsychologie, Universität Kassel, Kassel, Germany

Shaham Atashband Civil Engineering Department, Islamic Azad University, Markazi, Iran

Seyed Ahmad Ayatollahi Department of Industrial Engineering and Management Systems, Amirkabir University of Technology (Tehran Polytechnic), Tehran, Iran

A.S. Balu Department of Civil Engineering, National Institute of Technology Karnataka, Mangalore, Karnataka, India

Piotr Breitkopf Roberval Laboratory, University of Technology of Compiegne, Compiegne, France

Luc Chouinard Department of Civil Engineering and Applied Mechanics, McGill University, Montreal, Canada

Sophie Collong University of Technology of Belfort-Montbéliard, Belfort Cedex, France

Abdelkhalak El Hami Laboratoire d'Optimisation et Fiabilité en Mécanique des Structures, INSA Rouen, Saint Etienne de Rouvray, France

Antonio Federico Politecnico di Bari, Faculty of Engineering, Taranto, Italy

Vicente Gonzalez-Prida University of Seville, Seville, Spain

Nupur Goyal Department of Mathematics, Graphic Era University, Dehradun, Uttarakhand, India

John Harrigan Lloyd's Register Foundation (LRF) Centre for Safety and Reliability Engineering, School of Engineering, University of Aberdeen, Aberdeen, UK

Mohammad H. Hedayati School of Engineering, University of Aberdeen, Aberdeen, UK

Taha-Hossein Hejazi Department of Industrial Engineering and Management Systems, Amirkabir University of Technology (Tehran Polytechnic), Tehran, Iran

Chao Hu University of Maryland College Park (Currently at Medtronic, Inc.), Brooklyn Center, MN, USA

Therrar Kadri Beirut Arab University, Beirut, Lebanon

Seifedine Kadry American University of the Middle East, Egaila, Kuwait

Catherine Knopf-Lenoir-Vayssade Roberval Laboratory, University of Technology of Compiegne, Compiegne, France

Yuriy V. Kostyuchenko Scientific Centre for Aerospace Research of the Earth, National Academy of Sciences of Ukraine, Kiev, Ukraine

Department of Earth Sciences and Geomorphology, Faculty of Geography, Taras Shevchenko National University of Kiev, Kiev, Ukraine

Raed Kouta University of Technology of Belfort-Montbéliard, Belfort Cedex, France

Dagang G. Lu School of Civil Engineering, Harbin Institute of Technology, Harbin, China

Sankaran Mahadevan Department of Civil and Environmental Engineering, Vanderbilt University, Nashville, TN, USA

Mojtaba Mahmoodian School of Engineering, University of Greenwich, Chatham Maritime, UK

Abderahman Makhloufi Laboratoire d'Optimisation et Fiabilité en Mécanique des Structures, INSA Rouen, Saint Etienne de Rouvray, France

Carmen Martin Mechanics, Materials, Structure and Process Division, ENIT-INPT, Toulouse University, Tarbes Cedex, France

V.A. Matsagar Department of Civil Engineering, Indian Institute of Technology (IIT), New Delhi, India

Ankush Mittal Department of Computer Science and Engineering, Graphic Era University, Dehradun, Uttarakhand, India

Nicholas A. Nechval University of Latvia, Riga, Latvia

Konstantin N. Nechval Transport and Telecommunication Institute, Riga, Latvia

Richard D. Neilson School of Engineering, University of Aberdeen, Aberdeen, UK

Abayomi Obisesan Lloyd's Register Foundation (LRF) Centre for Safety and Reliability Engineering, School of Engineering, University of Aberdeen, Aberdeen, UK

Agnès Peeters Institut Superieur Industriel de Bruxelles (ISIB)—Haute Ecole Paul-Henri Spaak, Bruxelles, Belgium

François Pérès Decision making and Cognitive System Division, ENIT-INPT, Toulouse University, Tarbes Cedex, France

Daniel Play INSA of Lyon, Villeurbanne Cedex, France

Guy Pluvinage Ecole Nationale d'Ingénieurs de Metz, METZ Cedex, France

Mihail E. Popescu Illinois Institute of Technology, Chicago, IL, USA

Mangey Ram Department of Mathematics, Graphic Era University, Dehradun, Uttarakhand, India

B.N. Rao Department of Civil Engineering, Indian Institute of Technology Madras, Chennai, India

S.K. Saha Department of Civil Engineering, Indian Institute of Technology (IIT), New Delhi, India

Shankar Sankararaman SGT, Inc., NASA Ames Research Center, Moffett Field, CA, USA

Christian Schmitt Ecole Nationale d'Ingénieurs de Metz, METZ Cedex, France

Mirmehdi Seyyed-Esfahani Department of Industrial Engineering and Management Systems, Amirkabir University of Technology (Tehran Polytechnic), Tehran, Iran

Khaled Smaili Lebanese University, Zahle, Lebanon

Iman Soleiman-Meigooni Department of Industrial Engineering and Management Systems, Amirkabir University of Technology (Tehran Polytechnic), Tehran, Iran

Srinivas Sriramula Lloyd's Register Foundation (LRF) Centre for Safety and Reliability Engineering, School of Engineering, University of Aberdeen, Aberdeen, UK

Oliver Straeter Fachbereich Maschinenbau Arbeits- und Organisationspsychologie, Universität Kassel, Kassel, Germany

Umma Tamima Department of Civil Engineering and Applied Mechanics, McGill University, Montreal, Canada

Aurelian C. Trandafir Fugro GeoConsulting, Inc., Houston, TX, USA

Pingfeng Wang Industrial and Manufacturing Engineering Department, Wichita State University, Wichita, KS, USA

Byeng D. Youn Seoul National University, Seoul, South Korea

Xiaohui Yu School of Civil Engineering, Harbin Institute of Technology, Harbin, China

Peipei Zhang School of Mechatronics Engineering, University of Electronic Science and Technology of China, Chengdu, China

H.T. Zhu State Key Laboratory of Hydraulic Engineering Simulation and Safety, Tianjin University, Tianjin, China

Part I
Reliability Education

Mechanical System Lifetime

Raed Kouta, Sophie Collong, and Daniel Play

Abstract We present, in three parts, the approaches for the random loading analysis in order to complete methods of lifetime calculation.

First part is about the analysis methods. Second part considers modeling of random loadings. A loading, or the combination of several loadings, is known as the leading cause of the dwindling of the mechanical component strength. Third part will deal with the methods taking into account the consequences of a random loading on lifetime of a mechanical component.

The motivations of the present document are based on the observation that operating too many simplifications on a random loading lost much of its content and, therefore, may lose the right information from the actual conditions of use. The analysis of a random loading occurs in several ways and in several approaches, with the aim of later evaluate the uncertain nature of the lifetime of a mechanical component.

Statistical analysis and frequency analysis are two complementary approaches. Statistical analyses have the advantage of leading to probabilistic models (Demoulin B (1990a) Processus aléatoires [R 210]. Base documentaire « Mesures. Généralités ». (*)) provide opportunities for modeling the natural dispersion of studied loadings and their consequences (cracking, fatigue, damage, lifetime, etc.). The disadvantage of these statistical analyses is that they ignore the history of events.

On the other hand, the frequency analyses try to remedy this drawback, using connections between, firstly, the frequencies contained in the loading under consideration and, secondly, whether the measured average amplitudes (studied with

R. Kouta (✉) • S. Collong
University of Technology of Belfort-Montbéliard, 90010 Belfort Cedex, France
e-mail: raed.kouta@utbm.fr

D. Play
INSA of Lyon, 69621 Villeurbanne Cedex, France

S. Kadry and A. El Hami (eds.), *Numerical Methods for Reliability and Safety Assessment: Multiscale and Multiphysics Systems*, DOI 10.1007/978-3-319-07167-1_1,
© Springer International Publishing Switzerland 2015

the Fourier transform, FT) or their dispersions (studied with the power spectral density, PSD) (Kunt M (1981) Traitement numérique des signaux. Éditions Dunod; Demoulin B (1990b) Fonctions aléatoires [R 220]. Base documentaire « Mesures. Généralités ». (*)). The disadvantage of frequency analyses is the need to issue a lot of assumptions and simplifications for use in models of lifetime calculation (e.g., limited to a system with one degree of freedom using probabilistic models simplified for the envelope of the loading).

A combination of the two analyses is possible and allows a good fit between the two approaches. This combination requires a visual interpretation of the appearance frequency. Thus, a random loading is considered a random process to be studied at the level of the amplitude of the signal, its speed, and its acceleration.

1 Random Loadings Analysis

1.1 Usual Conditions of a Mechanical System

Mechanical systems and mechanical components provide functions for action more or less complicated. These actions are performed and controlled by one or more users in a variety of conditions (Schütz 1989). The diversity of uses leads to a large number of load situations. The challenge for designers of mechanical systems and mechanical components integrates these actual conditions of use (Heuler and Klätschke 2005). More generally, the challenge is to take into account the possibly nondeclared or explicit wishes of the users. Practically, it is to consider the diversity of loads and stresses applied to mechanical components. This condition is added to the geometric optimization requirements and conditions of material strength (Pluvinage and Sapunov 2006). It requires the development of a calculation tool suitable for both to obtain a representative model of loads and to carry out design calculations (Weber 1999).

Taking into account the actual conditions of use become a technological and economic challenge. But it causes a profound change in attitude since the causes are considered a probable way from assumptions used by a significant segment of the population. And of course, the calculation of the effects will be presented in terms of probability of strength and reliability (Lannoy 2004). This approach is possible because the two parts of the modeling are now well understood. Firstly, the effects of various loads applied to the components can be analyzed and calculated in terms of dynamic loads (Savard 2004; Bedard 2000). Then, the physical behavior of materials subjected to repeated stress is better known (Lu 2002; Rabbe and Galtier 2000). The design engineer can then develop methods for calculations reconciling best current knowledge and objectives he must achieve. Upstream of the approach, the loads from the conditions imposed by the users must be known. And downstream of the approach, it is necessary to calculate the consequences of such loads.

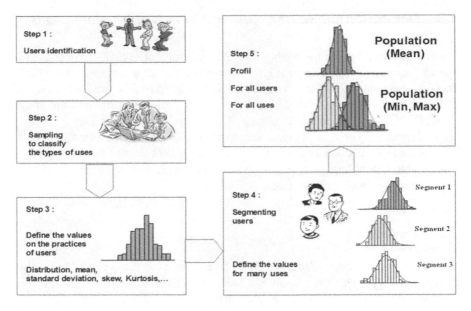

Fig. 1 Taking into account conditions of use

The variety of conditions is the major difficulty encountered in the integration of real condition of use when designing a mechanical component. For example, an equipments model is designed as a response to the needs of a user class (Fig. 1). A user class or class of use (Heuler and Klätschke 2005; Lallet et al. 1995; Leluan 1992a; Ten Have 1989) is often determined based on a profile of life confirmed by a market investigation. Despite the definition of multiple use classes, constructors seek as much as possible on the operating parts of equipments, to make the offer more overall that is to say, to find integrators resemblances between different classes use. The simplest presentation of a class of use or a life profile in the field of transport is by example to determine the number of kilometers traveled by an average user will during a specified period. This number of kilometers is presented as a sum weighted of a set of types of severity often called mission profiles or slices of life (good road, bad road, roundabout, mountain, city, different climatic conditions, etc.). Even if these simplified configurations, predictive calculations of resistance and lifetime require a statement of simplifications and assumptions that lead to the use of safety factors (Clausen et al. 2006) to reduce the risk of defects. Indeed, a class of use (or life profile) is considered by the designer of mechanical components, such as a homogeneous whole. Nevertheless, this homogeneity is accompanied by uncertainties that require consideration in terms of random information. Indeed, it is now proved (Osgood 1982) that a random mechanical stress leads to a lifetime smaller than alternating stress which seems broadly similar.

1.2   Statistical Analysis or Counting Methods of Random Loads

We are interested here in the methods of interpretation of the characteristic parameters of time series (or a discrete graphics representation) to obtain a distribution
law of these parameters (Brozzetti and Chabrolin 1986a). From the viewpoint of
checking the fatigue of the mechanical components, the extent of variation of the
variable load is an essential parameter of the same value as the average stress.
Variable loads may come as external actions as internal stresses. In what follows, we
shall make no distinction knowing that it is possible to determine the stresses from
the variable actions applied to a structure or component, making either quasi-static
or dynamic mechanical calculations.

1.2.1   Load Event

The term "load event" (Grubisic 1994) gives rise to a history of stress (also
called trajectory). This load event is a load state of service, characteristic of the
mechanical system and generating within each component considered, a variation
of stimulations.

Examples Included in the transport sector are the following cases:

- Bridge-road: the passage of a vehicle characterized by mass, number of axles,
  the speed, producing a bias at a particular point of the structure. The passage of
  the vehicle being a function of several parameters (the surface irregularities of
  the coating, the transverse position of the vehicle on the item, the weight of the
  rolling load, speed, etc.).
- Road-chassis: the stresses on the chassis of a vehicle on a road section.
- Marine platform: the action of water depending on the status of storm characterized by the duration, the height, the period, the average direction of waves.

Know the statistical distribution of load events during the intended use of the
system or the mechanical component, then leads to the establishment of a statistical
distribution law given by the average number of occurrences of each type of event.
For a bridge, this distribution is that of the expected traffic; for the chassis of a
vehicle, are the driving conditions; and for a marine structure, it will be a weather
data on the frequency of storms.

When each load event is characterized by one or more parameters, the long-
term distribution is in the form of a histogram, easily representable for one or two
parameters (Rabbe et al. 2000a).

In some cases, experience and theoretical modeling used to have this distribution
analytically. The histogram obtained is then replaced by a continuous distribution
law. The majority of mechanical systems and mechanical components from simple
to more complicated are subject to loads distributions often represented by Weibull
laws (Chapouille 1980).

For example in the field of land transport, this distribution may relate to severe stresses in a car chassis such as:

$$p\left(C_s > c^*\right) = \exp\left[-\left(\frac{c^*}{c_0}\right)^\gamma\right]$$

and

$$p\left(c^* \leq C_s < c^* + dc^*\right) = \frac{\gamma}{c_0}\left(\frac{c^*}{c_0}\right)^{\gamma-1} \exp\left[-\left(\frac{c^*}{c_0}\right)^\gamma\right] dc^*$$

$p(C_s > c^*)$ represents the probability of exceeding a threshold c^*; C_s is the random variable representing a severe stress event which can be here a stress due to the passage on a road in poor condition and shown in a significant stress c_0 ; $p(c^* \leq C_s < c^* + dc^*)$ represents the probability of being located around a threshold. It is thus possible to assess the probability that this stress is between c^* and $c^* + dc^*$. For this example, the statistical knowledge of the total number of sections of bad road then used to define a number of instances is to be associated with a given state of stress.

The difficulty of estimating a statistical distribution load event is that any statistical prediction as it relates to natural events (wind, wave, current, etc.) or in-service use of a considered mechanical system or considered mechanical component (traffic on a bridge, resistance of a frame, etc.). This prediction on the probability distribution of load events can be challenged by the emergence of exceptional causes.

Example We may not have anticipated increased traffic on a bridge for special seasonal reasons. Similarly another example, it is always difficult to extrapolate over the long term, the extreme value of a wave height, based on statistical values of wave heights measured in a few months.

1.2.2 Load Spectrum (Grubisic 1994), Histogram

Example Acceleration recorded on the axle as it passes on a test section, the speed of a gust of wind during a given period, etc.

From this trajectory, the problem is to obtain the information necessary to have at a histogram, or a distribution law of stresses that is called the spectrum of loads or stresses (Grubisic 1994), which is in reality only approximate representation of all charges stresses applied. We also note that obtaining load spectrum reduced information, in the sense that you lose the timing of the cycles of load variations. Therefore, the subsequent calculation of the damage (presented in the third part) may not consider any interaction between successive cycles of stress variations due to these loads. It may, however, admit that many events are largely random and it is unrealistic, at the stage of predicting the behavior of mechanical systems and components claim to have any knowledge of the precise order of appearance, e.g., values of variations ranges of stresses. It thus focuses on the study of the statistical distribution of variations ranges of stresses. And for some applications, the average stress of each cycle is sometimes used. Assume in the following presentation, the overall average stress is zero on the duration of the path.

Statement of Characteristic Data of a Random Loading

Except for some special cases of process (periodic sinusoidal path, stationary narrow band Gaussian process, that is to say with few excitation characteristic frequencies), it is generally difficult to combine with a variations range of stresses of one cycle (Fig. 2). In the case of a very irregular loading path, such as that of Fig. 2, the secondary peaks are problematic. And any a priori definition of how to count variations ranges of stresses may lead to differences in prediction compared to reality, if it is not supported by experimental verification.

The laws of damage based on more or less simple models (Duprat 1997), and the only way to tell if an identification of damaging cycles method is better than another is to correlate the results with those of the studied model of experience where it is possible to achieve (time scale and/or compatible cost, etc.). In fact, the existing methods give results fairly dispersed compared to published (Chang and Hudson 1981) results. For these reasons, the extraction of information from a random stress is to be performed with care. Different types of information to be extracted may occur in the following three forms:

Global analysis: where all the amplitudes of the stress are considered regardless the geometrical shape of the path (amplitude extreme, positive or negative slope, curvature upwards or downwards). This analysis is done using the histogram of the stress or by tracking specific amplitudes in the stress studied.

Local analysis: through the study of extreme values according to the geometrical shape of the path. In this case, the extreme values are separated into four statistical groups: the positive peaks, the trough positive, peaks negative, and trough negative. Amplitudes that do not have a change of direction in the digitized signal are not studied.

Analyses of stress tracts and/or of cycles: When the random stress is considered a constraint, it is useful to think in tract or stress cycle. This definition is consistent with what is done during fatigue tests under sinusoidal stresses that life is measured by the number of cycles. In the case of a sinusoidal stress, a tract concerns only half a cycle. In the case of random stress $S(t)$, defining a cycle is less easy:

- *Definition of a peak and a trough of stress*

    A stress peak S_M (or a trough stress S_m) is defined as the value of a local maximum (or local minimum) of the function $S(t)$. This peak (or the trough) can be positive or negative.

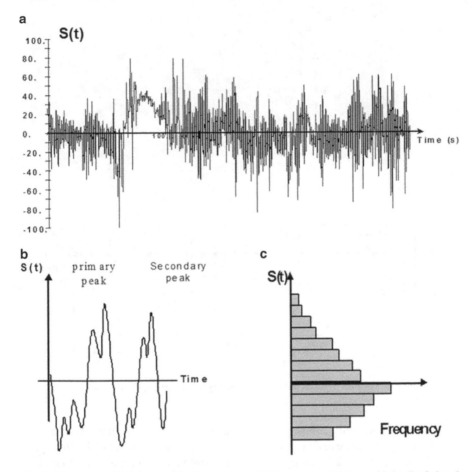

Fig. 2 Viewing a solicitation provided actual use. (**a**) Temporal solicitation, (**b**) detail of signal, (**c**) histogram

- *Definition of a half cycle or one cycle of tract stress variation*

A tract half-cycle variation of stress is defined by the time between two successive local extreme values of S_M and S_m (the tract of the variation of stress is defined by $S = S_M - S_m$) (Fig. 3a). A tract cycle of stress variation is defined as the time between two successive local maxima whose value is the first S_M and the second is S'_M (Fig. 3a), intermediate local minimum with a value S_m. The extent of variation of stress associated with this cycle is not unique in this case, since it may be taken as

$$S = |S_M - S_m| \quad \text{or} \quad S' = |S'_M - S_m|.$$

Another way to define a cycle, and that this is not linked to the counting of the peaks and troughs of a path, is related to the time interval between two zero crossings and by increasing value (or decreasing values) of the path (Fig. 3b).

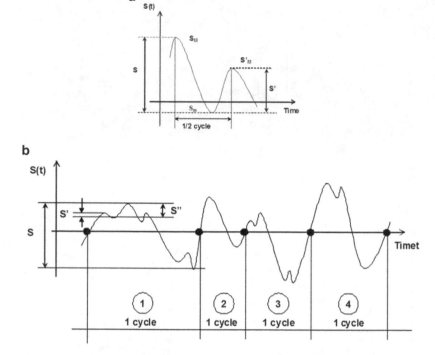

Fig. 3 Definition of characteristics of stress. (**a**) Half cycle definition, (**b**) series of cycles

The example of Fig. 3b with local peaks and local troughs shows the difficulty in defining one cycle and the tract of variation of stress associated with this cycle. Only the cycle no. 2 in the figure is used to define a single tract of variation of stress associated with this cycle.

In summary, three pieces of information must be seen from a random loading: the amplitudes that have imposed load considered (overall analysis), specific amplitudes observed by zooming effect and that reflect the severity of loading (local analysis), and finally tract or extracts cycles of loads studied. And counting methods (Lalanne 1999a) can be divided into three groups: global methods, local methods, and the methods of counting matrix.

Counting Global Methods

The main global counting methods (Lalanne 1999a) are: counting by class and method count overruns levels. For each counting method, an application will be presented around the stress shown in Fig. 4a.

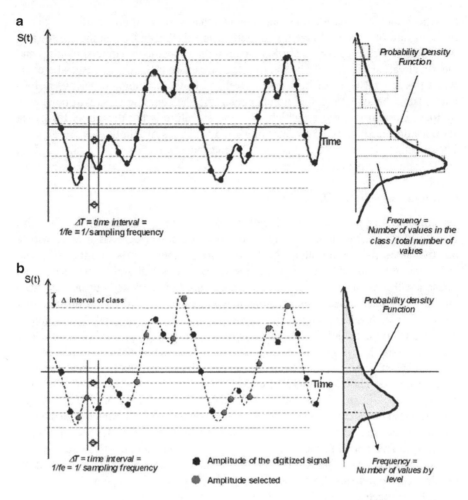

Fig. 4 Descriptive statistics of stress. (**a**) Realization of histogram, (**b**) definition of probability density function

Histogram or Holding Time in a Class of Amplitudes

This method considers the digital signal recorded as a statistical sample not knowing the temporal aspect. The sample is grouped into classes of amplitudes (Fig. 4a, dashed horizontal lines). In this case, no distinction is made between the extreme values and others. The advantages of this method reside in the immediate possibility of statistical modeling and propose a model of probability density (right side of Fig. 4a). Since between two successive points, there is a not predefined time by the method of measurement, the number of points recorded in a class when multiplied by the time step gives the total holding time of the stress studied in this class of amplitudes. This counting method should be reserved only for homogeneous stress (or whose source is considered homogeneous) that is to say, if no significant change in the nature of loading. Indeed, this method is very dependent on the speed (first derivative of the curve considered on a path with a specific dynamic signature) and the acceleration (second derivative) of the stress studied. In the case where the stress has several types of information (related to the braking, cornering, various loads, etc.), the signal loses its homogeneity and the counting class will be altered by these class different uses which dynamic signature is not the same. Figure 4a shows a digitized stress where 28 points and 9 classes of amplitudes are defined. The counting class (from the class below) leads successively to 2, 7, 3, 5, 1, 2, 5, 1, and 2 amplitudes per class.

Counting the Number of Level Crossing

This method, like the previous one, calls for predefined amplitude classes (Fig. 4b). Counting, for a given level, is triggered when the signal exceeds a level with a positive slope (hence the name level crossing). A count of the number of given level crossing is only relevant if an attitude selection of small oscillations is defined. These small oscillations can provide loads staffing (number) without interest from the point of view of calculating the damage and calculating life. And for counting the number of level crossing increment signal noted Δ is defined. It is often in the case of a mechanical component, interval stress below a fatigue limit set to a Wöhler curve (Leybold and Neumann 1963). This increment is considered Δ a threshold reset in the counting process. Historically, several counting methods have been proposed. The most interesting method has only one level if the stress has already gone through at least once this threshold Δ, irrespective of nature of slope. It should also be remembered that the counting is done with a digitized signal, and a proximity rule should be implemented to count very close to the levels determined amplitude. This counting method allows—as the previous one—to build a model of probability density. The application of this method focuses on the levels defined by the class boundaries. Count per level (with the solicitation of Fig. 4b, from low level) leads successively to 0, 0, 2, 1, 0, 1, 1, and 2 level overruns.

Counting Local Methods (Local Events)

Extreme values (peaks and troughs) of a random stress occur from four different families by:

- Positive maximum values preceded by a positive slope (peak > 0)
- Negative minimum values preceded by a negative slope (trough < 0)
- Negative maximum values preceded by a positive slope (peak < 0)
- Positive minimum values preceded by a negative slope (trough > 0)

Figure 5 illustrates these four families of amplitudes. As for the global analysis, grouping into classes for each family gives the possibility to build a model of probability density by type of extreme values. The result of this counting method around the stress shown in Fig. 5 leads, starting from the low class of the nine amplitude classes, to the results in Table 1.

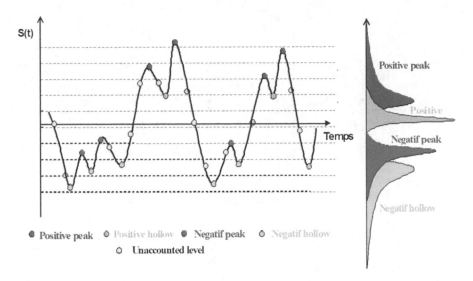

Fig. 5 Descriptive statistics with local events

Table 1 Local counting method

Classes	Trough < 0	Peak < 0	Trough > 0	Peak > 0
1[a]	1	0	0	0
2	4	0	0	0
3	2	2	0	0
4	0	1	0	0
5	0	0	0	0
6	0	0	0	0
7	0	0	2	1
8	0	0	0	1
9	0	0	0	1

[a]The lowest class

Counting Matrix Methods

Matrix counting methods take into account the evolution of the mean stress and/or involve the concept of stress cycle.

Counting Tract Between Peaks and Troughs

A tract (Fig. 6a) is defined as the difference between the given maximum and minimum (negative range), or between a given minimum and maximum (positive range). This method has half cycles. A variant of this method consists in associating with each cycle, the average of the positive and negative tract (to create tract average

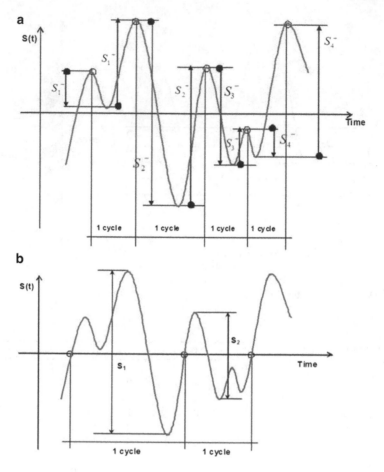

Fig. 6 Counting tracts of stress. (**a**) Counting between peaks and troughs, (**b**) counting means cycles

counts). The results of this counting method for the solicitation studied in this paragraph are shown in Fig. 6a. Indeed, the values S_1^-, S_1^+, S_2^-, S_2^+, S_3^-, S_3^+, S_4^-, et S_4^+ are the only tracts identified in this solicitation.

Counting Means Cycles

A cycle is defined by means the time between two zero crossings by increasing value (positive derivative). The tract of variation of stress S for this mean cycle is defined by the maximum of the local maxima and minimum of local minima inside this mean cycle (Fig. 6b). Just as the previous method, the counting results of this method are shown in Fig. 6b. Indeed, this counting method locates two identified cycles by S_1 and S_2 tracts, respectively, corresponding to the values S_2^- and S_3^- of the previous method.

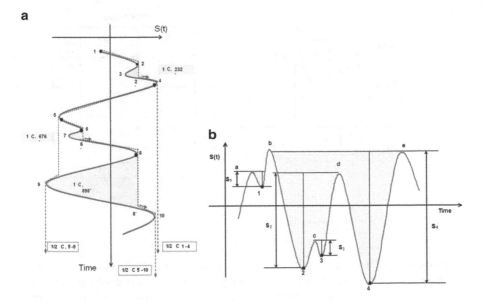

Fig. 7 Global cycle count of stress. (**a**) Rainflow method, (**b**) tanks method

"Rainflow" Method

This method takes into account all sequences of stress and in particular, it has all the secondary stress variations of the trajectory. It takes into account the half cycles and cycles by different, from that which was described in Sect. 2.3.3.1, way. Its name comes from how to identify cycles by the flow of a drop of water that slides along the path from top to bottom (Fig. 7a, vertical time axis to image representation). Whenever a drop of water leaves the path, a new cycle is initiated from the next summit. Counting cycle stops when the drop is blocked (the drop of water cannot follow the same path a path previously followed by another drop). If the end of the path is reached without blocking, there is only one half cycle. This method is cumbersome to implement and it does not lend itself easily to statistical mathematical modeling and the use of statistical properties of the trajectories of a random process. In the case of Fig. 7a, three cycles are observed ($232'$, $676'$, and $898'$) and three half-rings ($1–4$, $5–10$, $4–9$).

Tank Method

This counting method is similar to the previous one. It also has the advantage of taking into account all sequences tracts of stress variation. This method takes its name from the hydraulic analogy it presents with emptying a tank. One can imagine that all the pockets the water path is filled (Fig. 7b, now the time axis is shown horizontally). Was identified by decreasing values all troughs and imagine that each of these troughs has a drain valve. The tap of the lowest trough 1 is open, the tank

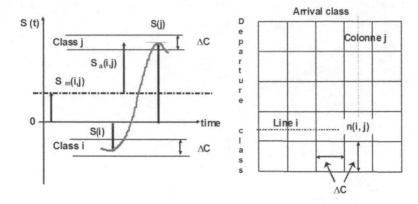

Fig. 8 Construction of MARKOV matrix

empties, leaving the other pockets filled with water. It does so by opening the valve of the next lowest trough and so on until you have emptied every pocket. You identify for each difference in height of water, a tract of variation of stress. A tract of variation of stress is cycle. For example, four reservoirs are detected on the stress of Fig. 7b. Each tank defines a stress cycle. The detected tanks are:

- The tank defined by the level (a) and the trough (1)
- The tank defined by the level (d) and the trough (2)
- The tank defined by the level (c) and the trough (3)
- The tank defined by the level (e) and the trough (4)

This method gives results similar to those obtained with the previous method. From the point of view of its mathematical formulation, it is easier to implement, but it nevertheless has the same disadvantages as the previous one.

Method of Transition Matrix or MARKOV Matrix

This method facilitates the counting of half cycles of stress. It is easier to automate. It leads to the construction of a transition matrix called MARKOV matrix (Lieurade 1980a; Kouta and Play 2006). The latter is obtained by the following steps (Fig. 8):

1. Temporal stress studied is divided into classes of amplitudes width ΔC. This width is considered as a threshold for filtering because all fluctuations within a class of amplitudes are ignored. The class width is often, in the case of a mechanical component, in a range below a stress fatigue limit set to a Wöhler curve (Leybold and Neumann 1963) (similar to the threshold Δ defined in Sect. 2.3.1.2).
2. Pour démarrer le comptage, la classe d'appartenance de la première amplitude observée, est détectée. Cette classe est considérée comme une classe de départ. La classe d'arrivée est déterminée quand le signal atteint un extremum.

Fig. 9 Subdivision of
MARKOV matrix

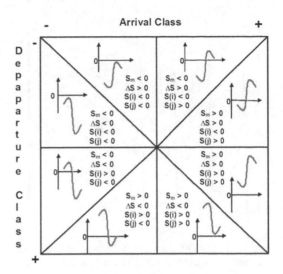

3. Thus, the horizontal dimension (row i) of the transition matrix (right part of Fig. 8) shows the starting classes and the vertical dimension has the arrival classes (column j). Box crossing the line i and column j is used to indicate the number $n(i, j)$ times the transition stress was observed.
4. The MARKOV matrix is thus transitions to an extremum of another $(S(i) \to S(j))$. The alternating amplitude $S_a(i, j)$, the average amplitude $S_m(i, j)$ of the transition, and tract $\Delta S(i, j) = S(j) - S(i)$ can be calculated.

$$S_a = \frac{S(j) - |S(i)|}{2} - S_m(ij)$$

$$S_m(i, j) = \frac{S(j) + |S(i)|}{2}$$

Figure 9 explains the transitions detected in the various subparts of a Markov matrix. This particular figure shows that most of the severity of a stress (with nonzero average value) is located in the top-left quadrant and the lower right quarter. Figure 10 shows a MARKOV matrix obtained for the deflection of a right front axle of the industrial vehicle. This stress is observed upon passage of the vehicle on a specific test section with five concrete bumps.

Conclusion on the Counting Methods

The different counting methods give different results, by the type of path considered. The method of counting of cycles means produces the same results for the three cases of Fig. 11a. The method of counting the peaks gives the same results for the

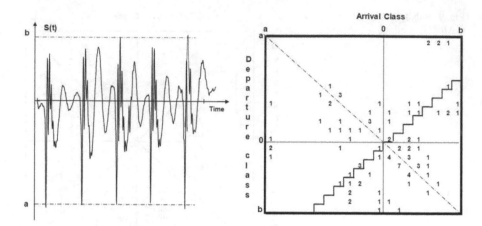

Fig. 10 MARKOV matrix for the deflection of the front right axle of a motor vehicle (Measured on a section consists of five bumps in concrete)

case (2) and (3) of Fig. 11a. Drop of water counting method or tank counting method and tracts between peaks and troughs give very different results for the two paths in Fig. 11b.

Currently, we do not have a number of published experimental tests enough on material fatigue under random stresses actual reasonably to conclude on the choice of method. However, some laboratory studies (Société française de métallurgie (Commission fatigue) 1981; Gregoire 1981; Olagnon 1994) show that in specific cases the method of counting the peaks and the counting method of the drop of water (counting method similar to the tank) give satisfactory results. So many regulations on fatigue verification of structures, systems, and mechanical components currently recommend, take the method of counting the drop of water. But it should take precautions to extrapolation of complex industrial cases. Remember that it is not enough to properly count the events, but it is also essential to develop approaches to mathematical modeling, probability, and statistics. These models offer several opportunities for engineers, for example:

- Objective classification of different types of stresses
- Possibility (after classification) build rules correlations between sources of stresses
- Production significant bench tests and numerical calculations of fatigue representative from actual conditions of use

Similarly, these models will be needed to establish predictive methods of calculating lifetime or to perform reliability analyses.

The difficulties in this type of modeling with some counting methods (counting of the drop of water or tanks, for example) lead to consideration of other methods such as the level crossing or the matrix MARKOV (Klemenc and Fajdiga 2000; Rychlik 1996; Doob 1953). When comparing different methods for counting a given

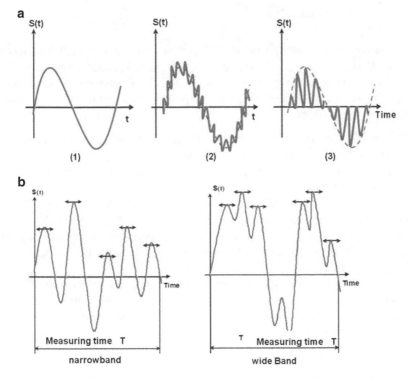

Fig. 11 Different scenarios of stress. (**a**) Comparison of counting methods, (**b**) solicitations with narrow or wide band

historical of use, it must examine the sensitivity of the methods taking into account the following parameters:

- Number of cycles detected.
- Detection of major cycles.
- Consideration of small tract of stress variation.
- Taking into account the average values of each cycle.

Finally, in the context of a particular business, you must choose a representative method load considered and that is consistent with the results of the experiment. But also the choice of the method will be conditioned by the calculation methods implemented in the future to perform the calculations lifetime. Representative path stress event is similar to a random process which is not known a priori, the form that its realization. The mathematical tools needed to solve these problems relate to the theory of probability and statistical properties related to the trajectories of random processes (Parzen 1962; Preumont 1990).

1.3   Frequency Analysis

The frequency analysis (Arquès et al. 2000) provides the notion of Power Spectral Density (PSD). Before developing the concept of PSD, some precautions are necessary for the representation of developments. These precautions are about of nature of stress to study. It is assumed that the path on the stress process $X(t)$ satisfies the following assumptions:

- The process $X(t)$ is continuous over the observation time chosen equal to a time unit that is referred to as the reference period.
- First $\dot{X}(t)$ and second derivatives $\ddot{X}(t)$ of the path $X(t)$ are continuous processes on the observation time.
- The process $X(t)$ is centered, that is to say the average $M(X(t))$ is zero. If $X(t)$ is not centered, the transformation is performed: $X(t) \rightarrow (X(t) - M(X(t)))$.
- The process $X(t)$ has independent statistical properties of time. The process is stationary. Statistical properties by means of spectral moments (defined later in the text) of $(X(t))$ of order 0, 2, and 4.
- The process $X(t)$ is Gaussian, that is to say that the random variable $(X(t_1)$ and $X(t_2), X(t_3), \ldots, X(t_n))$ on $t_1, t_2, t_3, \ldots, t_n$ follows a Gaussian distribution law.
- The statistical characteristics of the random variable at a time are the same as those of the time stress.

1.3.1   Power Spectral Density

PSD is determined according to the frequencies that appear in the stress, following a well-known process (Fig. 12a) (Plusquellec 1991; Borello 2006). This gives a power spectrum. In addition, the quadratic mean of the amplitudes (or variance) is defined. The latter is often called the intensity of the stress or RMS (Root Mean Square). The determination of the PSD is considered material if it is known that stress for a component at a given observation point is caused not only by external forces but also by the interaction effects of the component studied with the different elements system to which it belongs. For example, Fig. 12b (Buxbaum and Svenson 1973) shows the PSD of a record on a mechanical system in automotive signal. Two characteristic frequencies of the frame and of the suspension are identified.

The determination of the power spectrum of a random stress observed in the actual conditions of use of a mechanical component provides two types of information:

- Distinction between the different sources of stress measured.
- Contribution of each frequency to the intensity of the stress, defined rms.

According to WIENER–KHINTCHINE (Max 1989), PSD of a stationary process, as noted, is defined as the frequency distribution of the average energy of a process $X(t)$ where t represents time. $\varnothing_{xx}(\Omega)$ is connected to its autocorrelation function $R_{xx}(\tau)$.

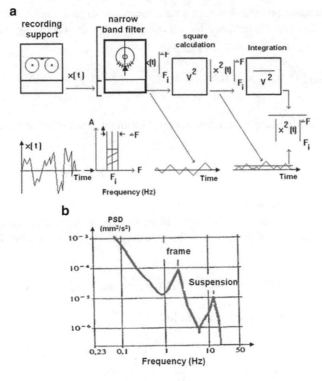

Fig. 12 Obtaining the Power Spectral Density (PSD). (**a**) Numerical procedures, (**b**) PSD example (Max 1989)

For a stationary process, we have:

$$R_{xx}(\tau) = \int_{0}^{+\infty} x(t) \times x(t - \tau) \, dt \qquad (1)$$

Thus,

$$\emptyset_{xx}(\Omega) = \frac{1}{2\pi} \int_{-\infty}^{+\infty} R_{xx}(\tau) . e^{-j\Omega\tau} \, d\tau \qquad (2)$$

and

$$R_{xx}(\tau) = \int_{-\infty}^{+\infty} \emptyset_{xx}(\Omega) . e^{j\Omega\tau} \, d\Omega \qquad (3)$$

So for the same process, but with a zero mean, the expression of the variance (or the RMS in this case) is obtained by imposing $\tau = o$ in the following relationship:

$$R_{xx}(o) = \int_{-\infty}^{+\infty} \varnothing_{xx}(\Omega) \, d\Omega = \sigma_X^2 = E\left[X^2(t)\right] = \text{RMS} = m_0 \qquad (4)$$

This last result is appointed first spectral moment. Two other spectral moments are defined by

$$m_2 = -\ddot{R}(0) \quad \text{and} \quad m_4 = R^{(4)}(0)$$

$R^{(4)}$ is the fourth derivative.

In the case of the process discussed here, m_0, m_2, and m_4 are, respectively, the same as the variance of the stress, of its first derivative and its second derivative as follows:

$$m_0 = V(X(t)), \quad m_2 = V\left(\dot{X}(t)\right) \quad \text{and} \quad m_4 = V\left(\ddot{X}(t)\right)$$

In practice, the PSD is written as follows:

$$\varnothing_{xx}(\Omega) = \lim_{T \to \infty} \frac{1}{2\pi T} \left[\left| \int_{-T/2}^{T/2} X(t).e^{-j\Omega t} \, dt \right|^2 \right] \qquad (5)$$

This relationship provides an interesting interpretation especially in the case of a single measurement. In practice, the signal strength (stress) is determined by bands of frequencies (the most selective possible). This intensity is represented by the area bounded by the time signal curve for each frequency. In general, this is characterized by intensity variations of the curvature of the signal-time trajectory. Figure 12a diagrammatically shows the determination of the PSD. In this case for a real signal, the autocorrelation function is real and even, it is the same to the PSD $\varnothing_{xx}(\Omega)$. Relations Eqs. (2) and (3) can be written:

$$\varnothing_{xx}(\Omega) = \frac{1}{\pi} \int_0^\infty R_{xx}(\tau) \cos(\Omega\tau) \, d\tau$$

$$R_{xx}(\tau) = 2 \int_0^\infty \varnothing_{xx}(\Omega) \cos(\Omega\tau) \, d\Omega \qquad (6)$$

Figure 13a (Rice et al. 1964) shows various forms of random stress and their associated PSD. The stress 1 is almost sinusoidal and the PSD is centered around

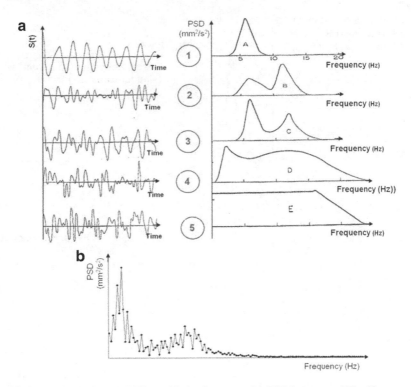

Fig. 13 Example of PSD. (**a**) Different kinds of stresses, (**b**) PSD for stress of Fig. 10

of the main frequency, it is a very uniform stress. The two well-identified peaks of DSPs of stress 2 and 3 are well explained by the presence of two frequencies that appear on the requests presented. The highest amplitudes of DSPs 2 and 3 indicate that these two frequencies are not the same energy signature for each stress. For four and five solicitations, all observed frequencies have almost the same importance. The stress 4 shows a frequency that differs from the other in the low frequency range. The stress 5 has a practically constant energy level whatever the frequency and this frequency range on a wide. And PSDs provide information on the statistical nature of the studied stresses and are indispensable when it comes to a classification of different mission profiles or classes of use of a component or a mechanical system. Its importance is even greater than it is thanks to PSDs that testing laboratories to reproduce test benches solicitations recorded in real conditions of use (Tustin 2001).

Figure 13b shows the PSD on the measurement of a movement, shown in Fig. 10. We see that this is a scenario loading close enough stress 4 shown in Fig. 13a. Indeed, despite the presence of a significant noise energy level, there is a frequency that is distinct from the other and has a high energy level. Physically, this is explained by the five shocks obtained when the vehicle passes over bumps of concrete.

Fig. 14 Example of extrem
response spectrum (ERS)
(**a** sinusoidal stress, **b** stress
Fig. 10)

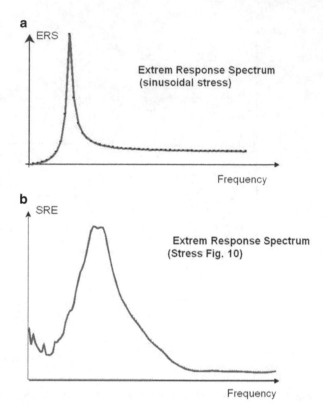

1.3.2 Extrem Response Spectrum

Extrem response (ER) (Lalanne 1999b) is characterized by the quantity

$$ER = \left(2.\pi.f_p\right)^2.Z_m$$

Z_m is the maximum response once met, the probabilistic sense, during the
duration of the random excitation. The Extrem Response Spectrum (ERS) is the
graph of changes in ER based on the natural frequency of the system with one degree
of freedom f_p, for a given damping factor ξ. The extreme response is representative
of the largest stress S_m suffered by the component assuming $S_m = K.Z_m$ (K = spring
constant).

Figure 14a shows spectrum of the extrem response for a sinusoidal stress and
Fig. 14b shows the stress of the ERS shown in Fig. 10.

Obtaining spectrum extreme response requires the definition of the statistical
distribution of extreme values of the stress studied. This is the subject of a paragraph
of Part 2. The process of obtaining the ERS and applications are subject to Annex C
of Part 2.

1.4 Conclusion

Statistical analysis is mainly represented by the counting methods. Frequency analysis is mainly determined by consideration of the PSD and extreme spectral response. The combination of the two analyzes can be considered a stress studied as a random process. Indeed, the uptake of random stress a random path will allow providing probabilistic modeling of a number of methods of numeric counts. These probabilistic models can also include elements of the frequency analysis and in particular PSD of the stress studied.

2 Random Loadings Modeling

The ultimate goal of these three parts is to develop essential methods and essential tools to predicting the lifetime of mechanical components. In the first part, statistical analysis and frequency analysis allowed to have a representation of the stresses applied to a system or a mechanical component. This will allow the second part to move from these representations for models with as ulterior objective to calculate the lifetime. Two types of models will be discussed in this article. The first is based on statistical and probabilistic approaches. The second incorporates an interaction between statistical analysis and frequency analysis.

2.1 Probabilistic Modeling of the Histogram of a Random Stress

The transition from the statistical analysis to a random stress probabilistic modeling with mathematical analysis is always tricky. Indeed, the purpose of modeling random loads is to determine a theoretical probability of occurrence of events, from the observation of a series of values $(x_i)_N$ of a random variable (X). If descriptive statistics seeks to pass judgment on the random variable with respect to preselected values, mathematical statistics seeks to make judgments about the probability of different values. The histogram (Fig. 15a) becomes a "density estimator" and accumulations diagram (Fig. 15b) is "probability estimator" (distribution function). Whichever method of synthesizing a random stress, the information collected can be probabilistic modeling (Saporta 1990). In the remainder of this section, the modeling of the conventional histogram is presented. This can also apply to all methods of unidimensional digital counting. Two-dimensional models are not discussed in this article. Thus for stress studied, the probability of overrun for any level is obtained. Similarly, after modeling the probability density $f_X(x)$ and distribution function $F_X(x)$, all the basic statistical techniques may be applied (including laws minimum and maximum).

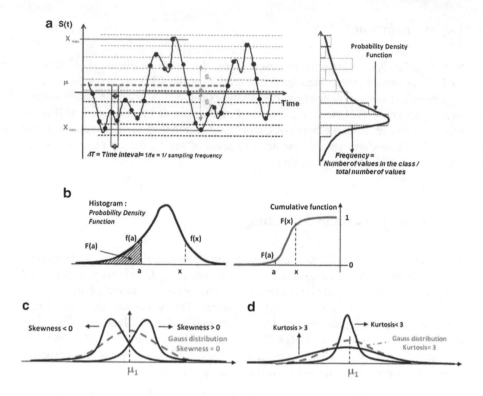

Fig. 15 Introducing the stress reference. (**a**) Example of stress, (**b**) definitions, (**c**) skewness, (**d**) kurtosis

$$\text{Probability } (X < x) = F_X(x) = \int_{-\infty}^{x} f_X(u)\, du \text{ with } \int_{-\infty}^{+\infty} f_x(x)\, dx = 1 \quad (7)$$

In practice, the distribution of X is continuous, the density $f_X(x)$ is obtained from the observed series $X_1, \ldots, x_N$. Call f_N the function that maps x real, the number $f_N(x)(=f_i)$ equal to the height of the rectangle which is relative to the class containing x (Fig. 15a). The purpose of modeling is to say that for all x, the two values $f_N(x)$ and $f_X(X)$ are "close." Some fit tests are used with wariness in the interpretation of a statistical test (χ^2, Kolmogorov tests (Fauchon J Probabilités et statistiques INSA-Lyon)). Given the variety of histograms encountered during the processing of random stress, two modeling approaches can be taken: modeling on the model of *K*. Pearson (Guldberg 1920; Kendall and Stuart 1969) or modeling on the model of Gram-Charlier and Edgeworth (Edgeworth 1916; Charlier 1914). The data to start for these two approaches are shape indicators of stress studied. These indicators are the minimum, maximum, average, standard deviation, asymmetry, and kurtosis stress studied (Fig. 15c, d).

2.1.1 Shape Indicators of a Random Loading

Figure 15a shows schematically a random stress. x_{min} and x_{max} are, respectively, the minimum and maximum values recorded:

$$x_i = x(t_i),$$

The amplitude x_i is measured time t_i knowing that:

$$t_{i+1} = t_i + \Delta t = t_i + \frac{1}{f_e}$$

With Δt time step; $f_e = \frac{1}{\Delta t}$ sampling frequency.

All of these values are the stress studied. The considered average is the arithmetic average:

$$\bar{x} = \mu_1 = \frac{1}{n} \sum_{i=1}^{n} x_i \tag{8}$$

It is the center of gravity of digitized values. The considered standard deviation is the mean of squared difference:

$$s_x = \sqrt{\mu_2} = \sqrt{\frac{1}{n} \sum_{i=1}^{n} (x_i - \bar{x})^2} \tag{9}$$

This is a calculation of the inertia of the stress dispersion studied about the central tendency or its center of gravity. The asymmetry and kurtosis are defined with respect to the law of Gauss.

The asymmetry factor for the law of Gauss is zero; the kurtosis factor is equal to 3. Figure 15c shows a Gaussian distribution and the consequences of a nonzero asymmetry. If the statistical distribution of random stress does not follow a Gaussian distribution, the asymmetry factor indicates the spreading in the right or left of the mean of the distribution.

The kurtosis factor indicates the crushing of the distribution or concentration around a given amplitude (Fig. 15d). Table 2 presents the expressions of asymmetry and kurtosis (Johnson and Kotz 1969).

Centered moment of order r: μ_r, is obtained differently if the evaluation is done from a random stress or a probability distribution. Table 3 presents the two methods of calculation.

In the case of a random stress, a new form factor is defined and is called irregularity factor I. This factor is the ratio of the number of zero crossings N_0 of the focus signal and the number of contained extrema N_e in the studied stress.

Table 2 Expression of asymmetry (skewness) and kurtosis

Parameter	Skewness	Kurtosis
English literature	$\beta_1 = \mu_3^2 / \mu_2^3$	$\beta_2 = \mu_4 / \mu_2^2$
French literature	$G_1 = \dfrac{\mu_3}{\mu_2^{3/2}} = \text{signe}\,(\mu_3)\,(\beta_1)^{1/2}$	$G_2 = \beta_2 - 3$

Table 3 Calculation of central moment of order r

Parameter	From a random stress	From a probability distribution $f_x(x)$
Central moment of order r (μ_r)	$\mu_r = \dfrac{1}{N}\sum_1^N (x_i - \overline{x})^r$	$\mu_r = \int (x - E(X))^r f_x(x)\, dx^{\text{a}}$

[a] $E(X) = \int x f_x(x)\, dx$ represents the expectation of the random variable X (or average x). Note that these integrations are performed on the domain of definition of x

$$I = \frac{N_0}{N_e} \quad I \text{ is between 0 and 1}$$

More I nears 1, the more regular of the stress is important. Thus, almost each extremum is always followed by a zero crossing (Fig. 16a). In this situation, the measure in question is described as "stress narrowband." When I nears 0, the regularity of the stress becomes lower. Thus, between two zero crossings level, the signal goes through many extrema on the same sign (Fig. 16b). In this situation, the measure in question is described as "stress broadband." The irregularity factor gives an important indication of the temporal evolution of stress, what other form factors completely unaware. In Appendix 4, factors of irregularity are defined. They will have a liaison role between statistical modeling and frequency analysis of a random stress.

2.1.2 Pearson Approach for Obtaining the Probability Density

In the study of shapes distributions frequently observed, normal or Gaussian distribution occupies the place of "generator" function. Pearson (Kendall and Stuart 1969; Johnson and Kotz 1969) proposes a differential equation which includes in its solutions a set of curves of probability density.

$$\frac{1}{f_X(x)} \frac{df_X(x)}{dx} = \frac{x + a_0}{b_0 + b_1 x + b_2 x^2} \tag{10}$$

With $f_x(x)$ probability density function of (denoted as p.d.f), a_0, b_0, b_1, and b_2 are parameters to be estimated and are dependent indicators shape of stress studied (mean, standard deviation, asymmetry, and kurtosis).

Fig. 16 Characterization
stresses depending on the
irregularity factor *I*. (**a**) Stress
with an irregularity factor
$I \cong 1$, (**b**) stress with an
irregularity factor $I \cong 0$

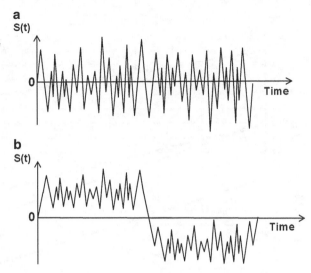

Different values of these give a series of "types" of solution (Jeffreys 1961) that
generate a set of distribution curves. Several solutions of Eq. (10) correspond to
conventional distribution curves with particular values of β_1 and β_2. Other solutions
take into account family values β_1 and β_2.

2.1.3 Introduction of the Pearson System Solutions

If a random variable X has an average m_X and a standard deviation s_X,
$Y = (X - m_X)/s_X$ is centered reduced random variable associated with X. The
average of Y is zero; the standard deviation of Y is 1.

For a random variable centered reduced, the Pearson system leads to seven types
of theoretical solutions. These solutions are defined in terms of asymmetry and of
kurtosis.

Figure 17 shows the chart of system solutions Pearson. The lines D1 and D4 give
the limits of the solutions obtained. Each limited area between two lines corresponds
to a probability density model. The lines D2, D3, D5, and D6 correspond to
particular probability laws. For $\beta_1 = 0$ and $\beta_2 = 3$, we find the Gaussian (or normal
distribution). The solutions obtained on straight D2, D3, D5, D6, or a particular
point N are cases very difficult to obtain figures in reality. Indeed, it is almost
impossible to find numerical values of β_1 and β_2 which are exactly the equations
straight. For this reason and in practice, the solutions of the Pearson system can
be reduced to three characteristics solutions: Beta1 law, Beta2 law, and Gamma
distribution.

The Beta1 law in the area between the straight lines D1 and D2. The Beta2 law in
the area between the lines D3 and D4. The Gamma distribution in the intermediate
zone between areas treated with Beta1 and Beta2 laws.

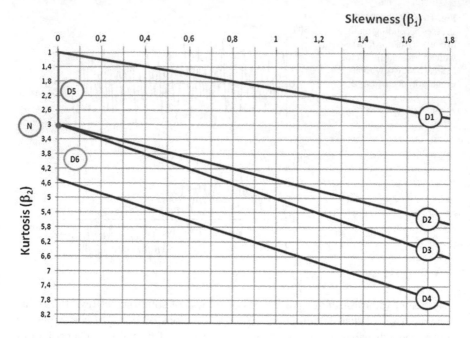

Fig. 17 PEARSON abacus

Appendix 1 presents main solutions of the Pearson system and the laws of probability alternatives that are easier to handle and which estimate their parameters is easier.

Synthesis and Conclusion of the Pearson System

The Pearson system has the advantage of specifying that a random stress can be modeled by three laws presented earlier. In the case where several random stresses from the same environment use must be considered, it is easier to have to provide a single model probability density. Thus, other modeling means are developed with a representation of the distribution curves by the series expansion technique. Gram-Charlier and Edgeworth (Kendall and Stuart 1969; Johnson and Kotz 1969) have developed an expansion method which allows to obtain functions of arbitrary distributions based on the technique of the series expansion, around a continue and known distribution called generator function. The expansion of the normal distribution using a series expansion of Taylor series category.

2.1.4 Gram-Charlier–Edgeworth Approach for Obtaining the Probability Density

Recall that if $f(x)$ is a probability density function, a series function (Eq. (11)) defines a probability density if it justifies the two necessary conditions:

$$g(x) \geq 0.0 \quad \text{and} \quad \int_{-\infty}^{+\infty} g(x)\, dx = 1.$$

In this case, $g(x)$ has some pluralitings $k_1, k_2, \ldots$ (Kendall and Stuart 1969):

$$g(x) = \exp\left[\sum_{j=1}^{\infty} \varepsilon_j \frac{(-1)^j}{j!} D^j f(x) \right]. \tag{11}$$

In practice, the representation of the serial function is taken as a series with few terms. So, the model adopted is as follows:

$$g(x) \approx \alpha(x) \left(1 + \frac{k_3}{6} H_3(x) + \frac{k_4}{24} H_4(x) + \frac{k_3^2}{72} H_6(x) + \cdots \right)$$

$$G(x) = \int_{-\infty}^{x} g(t)\, dt = \varnothing(x) - \left[\frac{k_3}{6} H_2(x) + \frac{k_4}{24} H_3(x) + \frac{k_3^2}{72} H_5(x) + \cdots \right] \alpha(x) \tag{12}$$

With: x between $-\infty$ and $t + \infty$, $k_3 = (\beta_1)^{1/2}$; $k_4 = \beta_2 - 3$;

$$\alpha(x) = \frac{1}{\sqrt{2\pi}} \exp\left(-\frac{x^2}{2} \right) \quad \text{and} \quad \varnothing(x) = \int_{-\infty}^{x} \alpha(u)\, du$$

$$H_2 = x^2 - 1; \quad H_3 = x^3 - 3x; \quad H_4 = x^4 - 6x^2 + 3;$$

$$H_5 = x^5 - 10x^3 + 15x; \quad \text{and} \quad H_6 = x^6 - 15x^4 + 45x^2 - 15$$

This model must verify the following conditions:

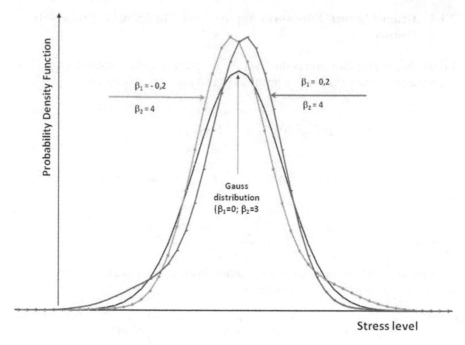

Fig. 18 Results obtained with the GRAM-CHARLIER–EDGEWORTH model

$$\int_{-\infty}^{+\infty} g(x)\,dx = 1 \quad \text{knowing that} \quad \int_{-\infty}^{+\infty} \alpha(x)H_j(x)\,dx = 0 \quad \text{and} \quad g(x) \rightarrow 0 \quad \text{if} \quad x \rightarrow \pm \infty$$

The function $f(x)$ taken into account is that of the Gaussian $\alpha(x)$. $H_r(x)$ are the Hermite polynomials. The coefficients k_j depend on the asymmetry and kurtosis which are expressed in terms of different moments μ_r. When $r > 4$, these are known for their unstable nature. For this reason, in practice, the order of development does not exceed the term H_6, and takes into account the terms μ_3 and μ_4. The disadvantage of this model lies in the fact that the model is always defined between $-\infty$ and $+\infty$. For a random variable (Y) which is defined with a mean (m_Y) and a standard deviation (s_Y), any X is obtained by a reduction and centering such that:

$$X = (Y - m_Y)/s_Y. \quad \text{Thus} \quad m_X = 0 \quad \text{and} \quad s_X = 1.$$

Figure 18 shows different probability densities obtained according to the development of GRAM, CHARLIER, and EDGEWORTH.

2.2 Theoretical Modeling of the Local Events

The statistical models presented in the previous section unknown the intermediate changes in random stress studied. Also to take into account a stress will be represented as a series of ascents and descents. The succession of these changes between peak and trough in the evolution of stresses is another approach to describe the severity of excitatory stress systems and components studied. Thus, modeling is focused on the most representative local events. These are the four families of events (positive peaks, positive trough, negative peaks, and negative troughs) presented in "Counting Local Methods (Local Events)" section. For random stress observed under conditions of use, each type of local events forms a homogeneous sample.

The amplitudes of these four types of events are defined from zero. Thus, the probabilistic modeling of the statistical distribution of these four types of events should consider the following:

- Theoretical model must be defined from zero.
- Upper limit of variation can be considered as infinite because highest amplitudes are much higher than zero.

The statistical distribution model of WEIBULL (Kouta and Play 1999) is used for this kind of phenomenon as the events represent extrema. Appendix 2 presents the technique for estimating the parameters of the WEIBULL distribution for each type of local events. Statistical tests should be performed to ensure the adequacy of proposed laws observed with histograms. Thus, four models of probability density for the four types of local events are obtained. Figure 5 illustrates these four laws of probability.

2.3 Probabilistic Modeling of the Level Crossing

Let $X(t) = \alpha$ level amplitude given, N_α is the number of times this level was exceeded for a small unit of time dt (Rice 1944). $N_\alpha.dt$ is interpreted as the probability that α is exceeded in a time dt, knowing that α may be exceeded if:

$$\alpha - \left| x'(t) \right| dt < x(t) < \alpha \quad x'(t) > 0$$

or

$$\alpha < x(t) < \alpha + \left| x'(t) \right| dt \quad x'(t) < 0 \tag{13}$$

Let $g_{XX'}(u, v)$ the probability density of two random variables $x(t)$ and $x'(t)$ with realizations u and v and

$$g_{XX'}(u, v) \, du \, dv = \text{Prob} \left[(x(t) \in u; u + du) \cap (x'(t) \in v; v + dv) \right].$$

α crossing probability of is then:

$$N_\alpha\, dt = \int\limits_{-\infty}^{0} dv \int\limits_{\alpha}^{\alpha+|v|dt} g_{xx'}\,(u,v)\,\,dv + \int\limits_{0}^{\infty} dv \int\limits_{\alpha-|v|dt}^{\alpha} g_{xx'}\,(u,v)\,\,du \qquad (14)$$

After integration u and considering infinitesimal dt, then:

$$N_\alpha = \int\limits_{-\infty}^{+\infty} |v| g_{xx'}\,(\alpha,v)\,\,dv \qquad (15)$$

If we write $g_{XX'}\,(u,v) = g_{x'}\big/_x\,\,(v/x = u)\,g_X(u)$ with $g_{x'}\big/_x\,\,(v/x = u)$ the conditional probability $x'(t)$, knowing $x(t) = u$ and $g_X(u)$ the probability density of $x(t)$:

$$N_\alpha = g_x\,(\alpha)\,.\int\limits_{-\infty}^{+\infty} |v|\, g_{x'}\big/_x\,\,(v/x = \alpha)\,\,dv = g_x\,(\alpha)\,.E\left\{|x'|\,/x = \alpha\right\} \qquad (16)$$

In the expression $E\{|x'|/x = \alpha\}$, slope $(|x'(t)|)$ taken into account are those level $x(t) = \alpha$. If this mathematical expectation is calculated for $\alpha = 0$, the number N_0 level 0 crossing is obtained:

$$N_0 = g_x(0).E\left\{|x'|\,/x = 0\right\} \qquad (17)$$

Similarly, the number of extreme values cannot be modeled but its determination requires the study of curves $(x''(t))$ and zero slopes.

Let $g_{X'X''}\,(v,w)$, the probability density for two random variables $x'(t)$ and $x'(t)$. Similarly Eqs. (16) and (17)

$$N_e = \int\limits_{-\infty}^{+\infty} |w| g_{x'x''}\,(0,w) = g_{x'}(0).E\left\{|x''|\,/x' = 0\right\} \qquad (18)$$

For a stationary process differentiable (as a differentiable Gaussian process), the cross-correlation function between two successive derivatives of a random variable $x^{(k)}(t)$ and $x^{(k+1)}(t)$ is zero. This has been verified in practice for the majority of studied signals (Osgood 1982).

$$E\left\{x^{(k)}(t)x^{(k+1)}(t)\right\} = E\left\{\frac{d^k}{dt^k}x(t).\frac{d^{k+1}}{dt^{k+1}}x(t)\right\} = E\left\{\frac{d}{dt}\frac{1}{2}\big[x^{(k)}(t)\big]^2\right\}$$

$$= \frac{d}{dt}\frac{1}{2}E\left\{\big[x^k(t)\big]^2\right\} = \frac{d}{dt}\,\text{(constant)} = 0.$$

Thus, Eqs. (16), (17), and (18) take a simple form

$$N_\alpha = g_x(\alpha) . E\left\{|x'|\right\} \tag{19}$$

$$N_0 = g_x(0) . E\left\{|x'|\right\} \tag{20}$$

$$N_e = g_{x'}(0) . E\left\{|x''|\right\} \tag{21}$$

These three relationships give theoretical estimates of the number (N_α) of level α crossing, the number of level 0 crossing (N_0), and the number of extreme values (N_e). This evaluation depends on the probability density model adopted for the studied stress on N_α and N_0 and the first derivative of the N_e. These probability densities are weighted by the average slope $E\{|X'|\}$ for N_α and N_0 and the mean curvatures $E\{|x''|\}$ for N_e.

Pearson model (Sect. 2.1.2) or its replacement laws as well as the model of Gram-Charlier–Edgeworth (Sect. 2.1.3) fit well enough to model the probability densities of X and of X' ($g_X(x), g_X(0), g_{X'}(0)$). $E\{|X'|\}$ and $E\{|x''|\}$ are global estimates obtained from a theoretical and only for a Gaussian stress. To make the evaluation of these elements closest to the specificity of each stress studied (Kouta 1994), we must calculate the average slopes and curvatures for each amplitude histogram studied class.

Thus, a more global modeling is defined. It allows an improvement of counting the number of crossings of either a level, either extreme values, or the number of level 0 crossing:

$$N_\alpha(c_k) = g_x(\alpha) . E\left\{|x'| / x \in \left[\alpha - \frac{\Delta c}{2}; \alpha + \frac{\Delta c}{2}\right]\right\} \tag{22}$$

$$N_0(c_{k*}) = g_x(0) . E\left\{|x'| / x \in \left[-\frac{\Delta c}{2}; \frac{\Delta c}{2}\right]\right\} \tag{23}$$

$$N_e(c_k) = g_{x'}(0) . E\left\{|x''| / x' = 0x \in \left[\alpha - \frac{\Delta c}{2}; \alpha + \frac{\Delta c}{2}\right]\right\} \tag{24}$$

With c_k: class k; c_{k*}: class of center containing the amplitude 0; Δc: class width.

Appendix 4 presents the theoretical modeling of the level crossing in the case of a Gaussian stress. The relations Eqs. (22), (23), and (24) are obtained from the probability densities of theoretical (g_X or $g_{X'}$) and the information observed in the studied random stress (average slopes or curvatures amplitude class). Figure 19 (for stress shown in Fig. 10) shows the difference between the counting results of

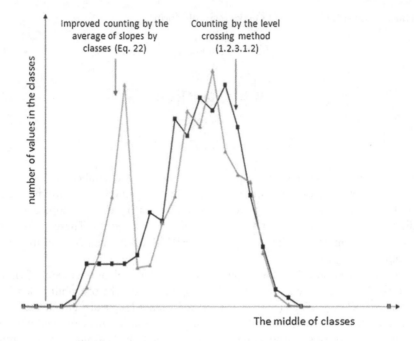

Fig. 19 Comparison of results obtained with different counting methods

numerical level crossings and the results obtained with Eq. (22). In classes of the most negative amplitudes, there are little effective digital values. However, these amplitudes reflect the intensity of the impact when passing over bumps of concrete in this particular case. From the point of view of fatigue of mechanical components, these amplitudes have a significant importance. The product of the number of these amplitudes by the average of their slopes can take into account the effect of these amplitudes.

2.4 Envelope Modeling of a Random Loading

The model of extreme value distribution defines statistics of each extreme group without taking into account their links with the rest of the signal. The objective of this section is to define the probability of exceeding extreme of any level taking into account their statistical distribution and contribution to the dynamics of the signal by taking into account the slopes and curvatures, in the same spirit to that adopted for the development of model exceeded level (2.3).

$f_M(\alpha)d\alpha$ of the probability of occurrence of a maximum at time t with $\alpha < x(t) < \alpha + d\alpha$, this depends on the probability distribution triple $g_{xx'x''}(u, v, w)$ connecting the stress (x), its first derivative (x'), and its second derivative (x''):

$$f_M(\alpha) \ d\alpha . N_M \ dt = \int\limits_{-\infty}^{0} \int\limits_{0}^{|w|dt\alpha+d\alpha} \int\limits_{\alpha} g_{xx'x''}(\alpha, v, w) \ du \ dv \ dw \qquad (25)$$

With N_M: number of maximum values.

This formulation relates positive extremes. Similarly, the probability of appearance of a minimum (negative extreme) between α and $\alpha + d\alpha$ during dt is $f_m(\alpha)$ $\{=f_M(-\alpha)\}$, an extreme appears in the following conditions:

$$\alpha < x(t) < \alpha + d\alpha, \quad 0 < x'(t) < |x''(t)|\,dt \quad \text{and} \quad x''(t) < 0 \text{ (pics)}$$

Then:

$$f_M(\alpha) = N_M . \int\limits_{-\infty}^{0} |w| \, g_{xx'x''}(\alpha, 0, w) \ dw \qquad (26)$$

and

$$f_m(\alpha) = N_m \int\limits_{0}^{\infty} |w| g_{xx'x''}(\alpha, 0, w) \ dw \qquad (27)$$

Appendix 5 presents the modeling of envelope for a Gaussian stress.

2.5 Counting Matrix Methods

These methods include information on slopes and curves observed in a random stress (Rice 1944).

The improved count of level exceeded by the consideration of slope provides a global read of the nature of the severity that the studied mechanical component undergoes during its use. The extreme spectrum response (or strength) presents energy levels depending on the frequencies that constitute the stress studied. The combination of these two types of analysis is done using the two-dimensional representations of transition matrices (Markov, expanses-slopes, and extreme-curvatures). These show the relationship between the local and global behavior. The transition matrix (or Markov) presented in Sect. 1.2.3.2.5 is one of these representations. This representation shows the evolution of the average of expanses in stress studied. For example in the case where the stress is studied stress, the evolution of the average stress influences, in an important way, on the life. Matrices are presented here:

- The matrix expanses-slopes. It gives the number of a given extended for a given slope.
- The matrix extreme-curvatures. It gives the number of an extremum for curvature.

These matrices are built from a clean signal (after grouping into classes of stress). The expanses-slopes matrix is to group different slopes under which a transition occurred between two extreme values. A frequency can be associated with each term of the matrix. Thus, this matrix shows the frequency distribution by type of transition. The matrix connecting the second derivative extrema is to bring together the various extreme amplitudes according to the second derivative or curvature. Figure 20a, b show diagrams of the construction of two matrices. For expanses-slope matrix, the cells of the upper right and the lower ones left are empty because no negative transition can have a positive slope as well as a positive transition can have a negative slope. The top left quarter of the extended-slope matrix (Fig. 20a) shows, from left, the distribution of negative slopes of the largest observed to the lowest. Quarter of the bottom right indicates, always starting from the left, the distribution of positive slopes of the lower to the higher. Thus, the part is on the far left quarter of the top left and the one found on the far right of the bottom right quarter largely explain the severity of stress. Figure 21a shows the matrix for expanses slope the stress of Fig. 10 (Sect. 1). The matrix of extrema-second derivatives (Fig. 21b) provides guidance concerning the regularity or irregularity of the stress studied. Indeed, more the effective of the upper left quarter and the lower ones on the right are important, more the stress is irregular and it contains transitions with nonzero average. A stress has considerable regularity in the case where the top and bottom right quarters of the left have large numbers. Note that the regularity of a stress is synonymous with severity. Stress is regular when it has little or no fluctuation between two successive extrema of opposite signs. Figure 21b shows the matrix of extrema-second derivatives for the stress of Fig. 10, number 5 $(2 + 2 + 1)$, on the top right of the matrix reflects the five major shocks received by the vehicle during its passage over five bumps of concrete.

2.6 Conclusion

Modeling random loads is based on statistical approaches and their interaction with the dynamic behavior of the studied stresses. Statistical approaches include the global approach and local approach. The global statistical modeling leads to a probability distribution that assesses the risk of reach (or exceed) a given amplitude. Local statistical modeling focuses on extreme events of stress studied. Deemed of a random stress to a random path leads probabilistic models which include information and the frequency analysis, in particular, data concerning the speed and acceleration of the stress studied. Now, probabilistic models will help address the fatigue design calculations as well as the definition of laboratory test conditions based on actual conditions of use (Part 3).

Fig. 20 Construction scheme
(**a**) range-slopes matrix and
(**b**) extreme-curvatures matrix

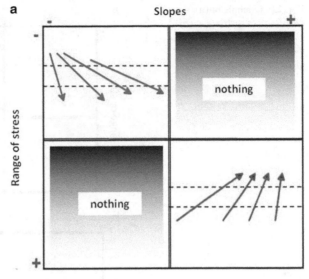

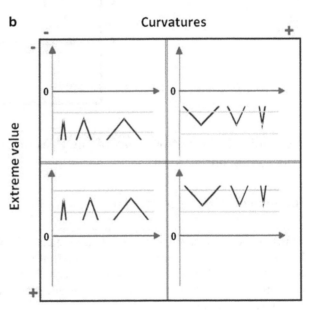

Fig. 21 Example of (**a**) range-slopes matrix and (**b**) extreme-curvatures matrix (stress Fig. 10)

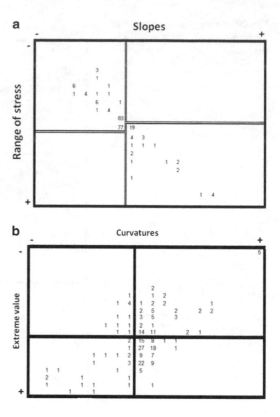

Appendix 1: Main Solutions of the Pearson System and Replacement Laws

Introduction of the Bêta 1 Law

This law takes into account all possible asymmetries between 0 and 1.8. Regarding kurtosis, it takes into account the kurtosis between 1 and at most 5.8. This physically means that this law is limited in its horizontal extension. In fact, this law is theoretically defined on an interval. The standard form of the probability distribution is expressed as follows:

$$f_X(x) = \frac{\gamma(p+q)}{\gamma(p)\gamma(q)} x^{p-1}(1-x)^{q-1} \tag{28}$$

with: x between 0 and 1, $\gamma(t) = \displaystyle\int_0^\infty e^{-u}u^{t-1}\,du$, p and q are the shape parameters of the probability distribution. They are expressed in terms of the mean (m_X) and standard deviation (s_X) of the random variable X.

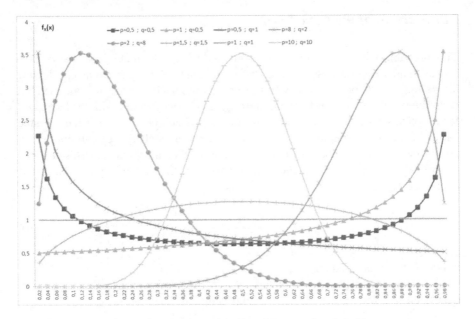

Fig. 22 Probability density function $f_x(x)$ of the Beta 1 law as a function of its parameters

$$p = -m_x + \frac{1 - m_x}{v_x^2} \tag{29}$$

$$q = \frac{m_x - 1}{m_x} \cdot \left(m_x + \frac{m_x - 1}{v_x^2} \right)$$

with $v_x = s_x/m_x$ coefficient of variation.

For a random variable which is defined between any two terminals (a and b), X is obtained by the following change of variable:

$$X = (Y - a) / (b - a)$$

Thus, $m_X = (m_Y - a)/(b - a)$ and $s_X = s_Y/(b - a)$.

Figure 22 shows different shapes of the probability law in function of p and q. This type of probability distribution is used for random stress that is physically defined between two terminals. Both terminals must be of the same order of magnitude.

Introduction of the Bêta 2 Law

This probability law takes into account the medium and high kurtosis. It also takes into account all possible asymmetries between 0 and 1.8. For these reasons, the law favors Beta 2 promotes kurtosis versus asymmetry. This law is theoretically defined from a low left δ (named offset) and infinity. It is easier to work with the "standard form" of this law. The "standard form" for a probability law which the variable varies between a threshold and infinity is obtained by a change of variable which leads to work with a probability density function which the variable varies between 0 (instead of δ) and infinity. The standard form of the probability distribution is expressed as follows:

$$f_X(x) = \frac{\gamma(p+q)}{\gamma(p)\gamma(q)} \frac{x^{p-1}}{(x+1)^{p+q}} \tag{30}$$

with: x is between 0 and ∞.

p and q are the shape parameters of the probability distribution. They are expressed in terms of the mean (m_X) and standard deviation (s_X) of the random variable X.

$$p = m_x + \frac{m_x + 1}{v_x{}^2}$$

and

$$q = 2 + \frac{m_x + 1}{m_x.v_x{}^2} \tag{31}$$

with $v_x = s_x/m_x$.

Figure 23 shows various shapes of the probability law in function of p and q. For this chart, p is 3 and q is between 1 and 6. The variation of p leads only to homotheties on the curves of probability density. This kind of probability distribution is used for random stress which we observe (or we think) that physically dispersion phenomena or kurtosis are more predominant than the phenomena of asymmetry.

In the event that difficulties are encountered when handling this probability law, then it can be replaced by a Log-Normal law which is easier to operate. The standard form of the probability distribution is expressed as follows:

$$f_X(x) = \frac{1}{\sigma\sqrt{2\pi}} \frac{1}{x} . \exp\left(-\frac{1}{2}\left(\frac{\ln(x) - m}{\sigma}\right)^2\right) \tag{32}$$

with: x is between 0 and ∞.

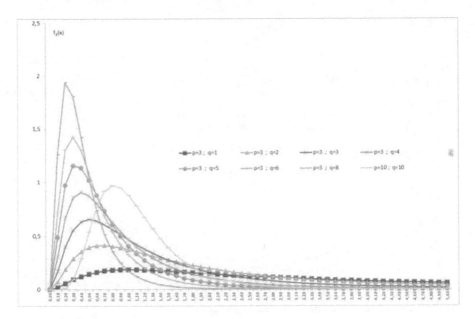

Fig. 23 Probability density function $f_X(x)$ of the Beta 2 law as a function of its parameters

$$m = \ln\left(\frac{m_x}{\sqrt{1 + v_x{}^2}}\right) \quad \text{and} \quad \sigma = \sqrt{\ln(1 + v_x{}^2)} \tag{33}$$

For a random variable which is set between the infinite and δ, X is obtained by the following change of variables: $X = (Y - \delta)$. Thus, $m_X = (m_Y - \delta)$ and $s_X = s_Y$. Figure 24 shows the good approximation of the Beta 2 law by Log-Normal law.

Introduction of the Gamma law

According to the abacus of Pearson, the probability distribution is defined in the intermediate zone between Beta 1 law and Beta 2 law. Thus, its contribution is located mainly in consideration of asymmetry and less kurtosis. For this reason, the Gamma law promotes asymmetry versus kurtosis. This law is theoretically defined between a threshold δ and infinity. The standard form of the probability distribution is expressed as follows:

$$f_X(x) = \frac{a^p}{\gamma(p)} x^{p-1} . \exp(-ax) \tag{34}$$

with: x is between 0 and ∞.

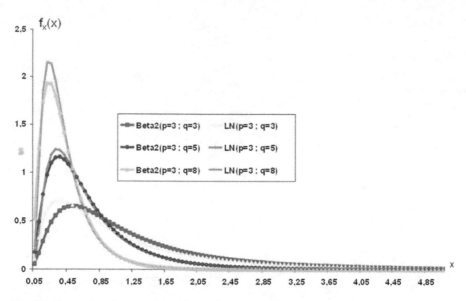

Fig. 24 Comparison between Beta 2 law and Log-Normal

p and q are the shape parameters of the probability distribution. They are expressed in terms of the mean (m_X) and standard deviation (s_X) of the random variable X.

$$p = \frac{1}{v_x{}^2} \quad \text{and} \quad a = \frac{1}{m_x . v_x{}^2} \tag{35}$$

with $v_x = s_x/m_x$.

Figure 25 shows different forms of the probability law function of a and p. For this graph, a is set to 1 and p is between 0.5 and 5. Variation of a led only to homotheties on the probability density curves. This kind of probability law is used for random stress which is observed (or is thought) that physically, the important phenomena of displacement of the medium or of asymmetry that are more predominant kurtosis phenomena.

As for the Beta 2 law, the gamma distribution can be replaced by another law easier to handle and well known is the Weibull distribution. The standard form of this law is expressed as follows:

$$f_X(x) = \frac{\beta}{\eta} \left(\frac{x}{\eta} \right)^{\beta-1} \exp \left\{ -\left(\frac{x}{\eta} \right)^{\beta} \right\} \tag{36}$$

with: x is between 0 and ∞.

With the expectation (or mean) and standard deviation, which are expressed as follows:

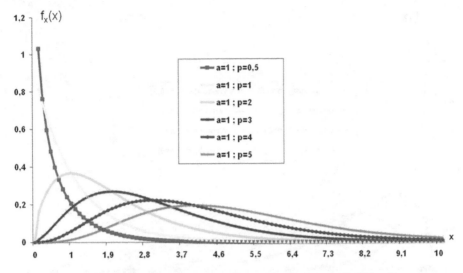

Fig. 25 Probability density function $f_x(x)$ of the Gamma distribution as a function of its parameters

$$E\{X\} = m_X = \eta\gamma\left(1 + \frac{1}{\beta}\right)$$

and

$$s_X = \eta\sqrt{\gamma\left(1 + \frac{2}{\beta}\right) - \left(\gamma\left(1 + \frac{1}{\beta}\right)\right)^2} \tag{37}$$

The β factor is often called form factor η and the scale factor. For a random variable which is set between the infinite and δ, X is obtained by the following change of variable:

$$X = (Y - \delta). \text{ Thus } m_X = (m_Y - \delta) \quad \text{and} \quad s_X = s_Y.$$

Figure 26 shows the good approximation of Gamma law with a WEIBULL law. Thus, WEIBULL law is the most appropriate model for modeling the random dispersion of mechanical stresses that are defined from a given threshold and that their asymmetry and kurtosis them put in the intermediate zone between the law of the Beta 1 and the Beta 2. In addition, WEIBULL law is easy to handle and estimation of its parameters methods is now well known.

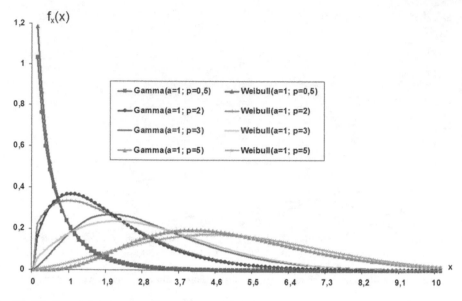

Fig. 26 Comparison between Gamma law and WEILBULL law

Appendix 2: Method of Estimate of the Weibull Law Parameters Around the Four Types of Local Events

The separation of the four types of local events in four samples used to calculate indicators for each sample form: mean $= \overline{x}$, standard deviation $= s_X$, asymmetry $= G_1$ (or β_1), and kurtosis $= G_2$ or (β_2). X, a random variable with Weibull W $(\beta, \eta; \delta = 0)$. The formulation of the probability density function of X, expectation (or its average) and standard deviation are given by the relations Eq. (37) of "Introduction of the Gamma law" of Annex A. Table below gives the theoretical expressions of the dispersion coefficient (v_X), of the asymmetry G_1 and of the kurtosis G_2 the Weibull distribution. These formulas are expressed exclusively in terms of the form factor of the distribution (β).

$$v_x = \frac{s_x}{m_x} = \frac{\left(\gamma_2 - (\gamma_1)^2\right)^{1/2}}{(\gamma_1)}$$

$$G_1 = \frac{\gamma_3 - 3\gamma_2\gamma_1 + 2(\gamma_1)^3}{\left(\gamma_2 - (\gamma_1)^2\right)^{3/2}}$$

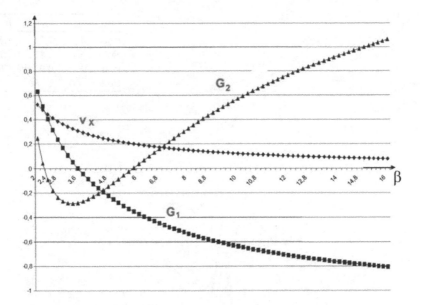

Fig. 27 Evolution of v_X, G_1 and G_2 of the law the Weibull based on shape factor parameters β

Fig. 28 Viewing the input torque to the front stabilizer bar of a motor vehicle

$$G_2 = \frac{\gamma_4 - 3\gamma_3\gamma_1 - 3(\gamma_2)^2 + 12\gamma_2(\gamma_1)^2 - 6(\gamma_1)^4}{\left(\gamma_2 - (\gamma_1)^2\right)^2} \tag{38}$$

with $\gamma_r = \gamma\left(1 + \frac{r}{\beta}\right)$.

Thus, the shape factor may be obtained β by inverting one of three functions v_x, G_1, or G_2. Figure 27 shows the theoretical evolution of v_x, G_1, and G_2. According to β which the $v_X = F_1(\beta)$, $G_1 = F_2(\beta)$, and $G_2 = F_3(\beta)$. Given the shape of the last three functions, the easiest way is to reverse (a numerical method) function G_1. Figure 28 shows a random stress representing the couple in an industrial vehicle stabilizer bar when passing over a poorly maintained road. The recording time is 5 min.

Table 4 Parameters of
Weibull laws

Parameter	G_1	β
Peak > 0	1.25	1.36
Hollow < 0	0.812	1.76
Peak < 0	0.925	1.63
Hollow > 0	1.686	1.12

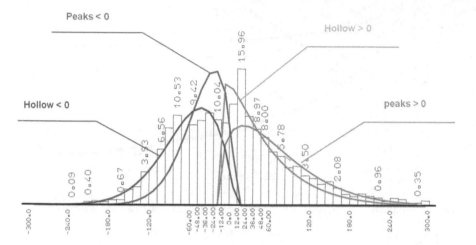

Fig. 29 Histograms peaks > 0 and hollow < 0 and probability distributions of four local events

Table 4 presents the values obtained for the different parameters of Weibull laws for the four types of local events. Figure 29 shows the adequacy of the proposed relationship with histograms experimentally observed laws.

Appendix 3: Extrem Response Spectrum for a Random Loading

Extrem response ER is calculated from the maximum met once during the excitation time response. For recording a duration T, the total number of peaks above $z0$ is given by

$$N = N_e T F_M (u_0) \tag{39}$$

With $F_M(u) =$ cumulative probability function given in E3 of Annex E and N_e is the number of extreme values.

The largest peak for the duration T (on average) approximately corresponds to u_0 level that is exceeded only once, hence:

$$F_M(u_0) = 1/(N_e T) \tag{40}$$

The level u_0 is determined by successive iterations. Distribution $F_M(u)$ is a decreasing function of u. We consider two values u, such that:

$$F_M(u_1) < F_M(u_0) < F_M(u_2) \tag{41}$$

and, at each iteration, the interval is reduced (u_1, u_2) until, for example:

$$\frac{F_M(u_1) - F_M(u_2)}{F_M(u_0)} < 10^{-2} \tag{42}$$

Hence extrapolation:

$$z_m \approx z_0 = (V(X))^{1/2} \left\{ [u_2 - u_1] \frac{F_M(u_1) - F_M(u_2)}{F_M(u_0)} + u_1 \right\} \tag{43}$$

and $ER = (2\pi f_0)^2 Z_m$.

With the same assumptions, the average number of threshold exceedances $z = \alpha$ response with positive slope during a recording period T for stress a Gaussian is given by the relationship:

$$N_\alpha^+ = TN_\alpha = TN_0 \exp\left(-\frac{\alpha^2}{2\sqrt{V(X)}}\right) \quad \text{and} \quad N_0 = \frac{1}{\pi}\sqrt{\frac{V(x')}{V(x)}} \tag{44}$$

Considering that the threshold α is exceeded only once, is obtained by $N_\alpha^+ = 1$

$$\alpha = \sqrt{2V(X)\ln(N_0 T)}$$

Whence

$$ER = 4\pi^2 f_0^2 \sqrt{2V(X)\ln(N_0 T)} \tag{45}$$

In the case of a random stress whose PSD is represented by several levels $(G_i = \Phi$ cf. Sect. 1) each of which is defined between two frequencies f_i et f_{i+1}, the various components of ER are obtained as follows:

$$V(X) = \int_0^\infty G(f).df = \frac{1}{(2\pi)^4.f_0^3} \frac{\pi}{4.\xi} \sum_i G_i.[I_0(h_i + 1) - I_0(h_i)] \tag{46}$$

$$V\left(\dot{X}\right) = (2\pi)^2 . \int_0^\infty f^2 . G(f) . df = \frac{1}{(2\pi)^2 . f_0} \frac{\pi}{4.\xi} \sum_i G_i . [I_2\,(h_i + 1) - I_2\,(h_i)]$$

(47)

$$V\left(\ddot{X}\right) = (2\pi)^4 . \int_0^\infty f^4 . G(f) . df = f_0 . \frac{\pi}{4.\xi} \sum_i G_i . [I_4\,(h_i + 1) - I_4\,(h_i)] \quad (48)$$

with $G(f) = |H(f)|^2 . G_{\ddot{x}}(f)$, $h_i = f_i / f_0$ and $h_{i+1} = f_{i+1} / f_0$.

$$I_0 = \frac{\xi}{\pi.\alpha} . \ln\left(\frac{h^2 + \alpha h + 1}{h^2 - \alpha h + 1}\right) + \frac{1}{\pi} . \left[\mathrm{Arctan}\left(\frac{2.h + \alpha}{2.\xi}\right) + \mathrm{Arctan}\left(\frac{2.h - \alpha}{2.\xi}\right)\right]$$

(49)

$$I_2 = -\frac{\xi}{\pi.\alpha} . \ln\left(\frac{h^2 + \alpha h + 1}{h^2 - \alpha h + 1}\right) + \frac{1}{\pi} . \left[\mathrm{Arctan}\left(\frac{2.h + \alpha}{2.\xi}\right) + \mathrm{Arctan}\left(\frac{2.h - \alpha}{2.\xi}\right)\right]$$

(50)

$$I_4 = \frac{4.\xi}{\pi} . h + \beta . I_2 - I_0$$

(51)

$$\alpha = 2.\sqrt{1 - \xi^2}; \quad \beta = 2\left(1 - 2.\xi^2\right); \quad Q = 1/(2.\xi)$$

(52)

with ξ = damping factor

Thus:

$$N_e = \frac{1}{\pi} \sqrt{\frac{V(x'')}{V(x')}}, \quad N_0 = \frac{1}{\pi} \sqrt{\frac{V(x')}{V(x)}} \quad \text{and} \quad I = \frac{N_0}{N_e}$$

(53)

In cases where the statistical distribution of peaks (trough respectively) is modeled by a Weibull law $W(\beta, \eta, \delta)$ with β: shape factor, η; scale parameter and δ: shift (often close to zero), we seeks probability $\pi_M(u_0)$ exceeds a maximum for any level u_0 of the amplitude of the response. $\pi_M(u_0) = 1 - F_M(u_0)$ with:

$$F_M\left(U_0\right) = \int_{-\infty}^{u_0} p(u)\, du$$

and over a period of observation T, the mean number of peaks is greater than u_0 always $N = n_p^+ . T\pi(u_0)$. When $N = 1$, the corresponding level is defined by $\pi(u_0) = 1/(n_p^+ . T)$.

with

$$\pi(u_0) = \exp\left(-\left(\frac{u_0}{\eta}\right)^\beta\right)$$

and knowing the value of n_p^+, it has

$$u_0 = \eta.\left[\ln\left(\frac{\ddot{Z}_{\text{eff}}.T}{2\pi.\dot{Z}_{\text{eff}}}\right)\right]^{1/\beta} \tag{54}$$

While the extrem response becomes

$$\text{ER} = (2\pi.f_0)^2.\eta.\left[\ln\left(\frac{\ddot{Z}_{\text{eff}}.T}{2\pi.\dot{Z}_{\text{eff}}}\right)\right]^{1/\beta} \tag{55}$$

Appendix 4: Theoretical Modeling of the Overrun of Level for a Gaussian Loading

In the Gaussian case, $x(t)$, $x'(t)$, and $x''(t)$ are Gaussian random variables with three densities as successive probabilities:

$$g_x(u) = \frac{1}{\sqrt{2\pi V(x)}}.\exp\left\{-\frac{u^2}{2V(x)}\right\} \tag{56}$$

$$g_{x'}(v) = \frac{1}{\sqrt{2\pi V(x')}}.\exp\left\{-\frac{v^2}{2V(x')}\right\} \tag{57}$$

$$g_{x''}(w) = \frac{1}{\sqrt{2\pi V(x'')}}.\exp\left\{-\frac{w^2}{2V(x'')}\right\} \tag{58}$$

Thus, we have:

$$E\{|x'|\} = \int_{-\infty}^{+\infty} |v| \, g_{x'}(v) \, dv = \sqrt{\frac{2}{\pi}V(x')} \tag{59}$$

$$E\left\{\left|x''\right|\right\} = \int_{-\infty}^{+\infty} |w| g_{x''}(w)\, dw = \sqrt{\frac{2}{\pi} V\left(x''\right)} \tag{60}$$

Thus, in the case of a Gaussian process, Eqs. (19), (20), and (21) can be written

$$N_\alpha = \frac{1}{\pi} \sqrt{\frac{V\left(x'\right)}{V(x)}} \cdot \exp\left\{-\frac{1}{2}\frac{\alpha^2}{V(x)}\right\} \tag{61}$$

$$N_0 = \frac{1}{\pi} \sqrt{\frac{V\left(x'\right)}{V(x)}} \tag{62}$$

$$N_e = \frac{1}{\pi} \sqrt{\frac{V\left(x''\right)}{V\left(x'\right)}} \tag{63}$$

Hence expression of Gaussian irregularity factor

$$I_g = \frac{N_0}{N_e} = \frac{V\left(x'\right)}{\sqrt{V(x)V\left(x''\right)}} \tag{64}$$

From this factor, the factor ε is defined as the bandwidth of the studied process. The last (as I and I_g) is between 0 and 1.

Bandwidth is expressed by the number of zero crossings by increasing values N_0 and the number of local maxima N_e (positive or negative) of the observation time. We put:

$$\varepsilon = \sqrt{1 - I_g^2} \tag{65}$$

Notes:

1. The relation Eq. (65) is a definition and simply reflects the fact that, when a process narrowband $N_0 = N_e$ (see Fig. 11a) and therefore $\varepsilon = 0$, and when a broadband process (Fig. 11b) $N_e > N_0$, where $N_e \gg N_0$ $\varepsilon \to 1$.
2. When $\varepsilon = 0$, in the case of a trajectory narrowband cycle thinking has a precise meaning, since $N_0 = N_{e.}$
3. When $\varepsilon = 1$, the notion of cycle is likely to various interpretations, the definition of "number of cycles" depends on the counting method.

Appendix 5: Envelope Modeling for a Gaussian Loading

For a Gaussian trajectory, correlation coefficient between the signal $x(t)$ and its first derivative $x'(t)$ is zero. As follows:

$$g_{XX'X''}(u, v, w) = g_{X'}(v)\, g_{XX''}(u, w)$$

Dual distribution $g_{xx''}(u, w)$ between the signal and its second derivative is expressed in a bi-normal distribution:

$$g_{xx''}(u, w) = \frac{1}{2\pi \sqrt{k}} \exp\left[-\frac{1}{2k}\left\{V\left(x''\right)u^2 + 2V\left(x'\right)uw + V(x)w^2\right\}\right] \quad (66)$$

with $k = V(x)V(x'') - [V(x')]^2$

The probability density peaks (Eq. (26)) to be written:

$$f_M(\alpha) = \mathrm{Prob}\,(\alpha < \mathrm{Max} \le \alpha + d\alpha) =$$

$$= \frac{\sqrt{1-I_g^2}}{\sqrt{2\pi}} \cdot \exp\left(-\frac{u^2}{2(1-I_g^2)}\right) + \frac{I_g u}{2}\exp\left(-\frac{u^2}{2}\right)\left[1 + \mathrm{erf}\left(\frac{I_g u}{\sqrt{2(1-I_g^2)}}\right)\right] \quad (67)$$

And the cumulative probability function is:

$$F_M(\alpha) = \mathrm{Prob}\,(\mathrm{Max} \le \alpha)$$

$$= \frac{1}{2}\left\{1 - \mathrm{erf}\left(\frac{u}{\sqrt{2(1-I_g^2)}}\right)\right\} + \frac{I_g}{2}\exp\left(-\frac{u^2}{2}\right)\left\{1 + \mathrm{erf}\left(\frac{I_g u}{\sqrt{2(1-I_g^2)}}\right)\right\} \quad (68)$$

with $I_g = \dfrac{V(X')}{\sqrt{V(X)V(X'')}}$ Gaussian irregularity factor,

$$u = \frac{\alpha}{\sqrt{V(X)}} \quad \text{and} \quad \mathrm{erf}(x) = \frac{2}{\sqrt{\pi}}\int_0^x e^{-\lambda^2/2}\, d\lambda.$$

For $I_g \approx 1$ ($\varepsilon = 0$, stress narrowband), probability density becomes: $f_M(\alpha; I_g) \approx u.\exp(-u^2/2)$. This approximation corresponds to the Rayleigh distribution. Thus, we find that for a narrowband random stress, the distribution of maximum values follow a Weibull distribution with shape factors of individual $\beta = 2$; $\eta = \sqrt{2V(x)}$; $\delta \ge 0$.

The influence of three main variances $V(x)V(x')$ and $V(x'')$ is studied on the distribution of extreme values. This influence is observed through the irregularity factor I_g. The difficulty of this study lies in the fact that evolutions of x, x', and x'' are interdependent for a random trajectory.

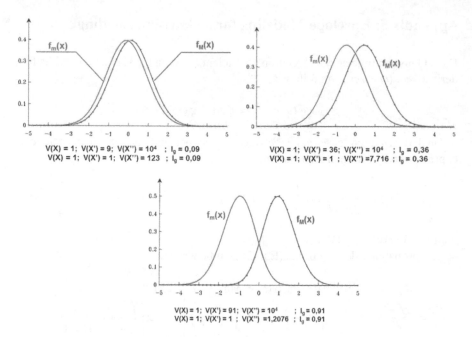

Fig. 30 Influence of the variance of the first derived and second derived on the modeling of the envelope

To simplify the presentation, consider the assumption of a variance $V(x) = 1$. Figure 30 shows the evolution of the extreme value distribution based on the irregularity factor I_g. The latter is directly proportional to $V(x')$. Increasing this variance is due to the appearance of a variety of slopes increasingly important in the stress. This results in a greater probability of observing extreme values t_o higher levels of amplitudes and conversely lower probability amplitudes of fluctuations around the middle of the path. Thus, the distribution of positive extreme and negative extreme deviate from each other in the growth of the irregularity or the variance factor of the second derivative. Significant dispersion of the variance of the curves or the second derivative $(V(x''))$ indicates an occurrence of a wide variety of shapes for the peaks and ridges on the random path. This dispersion occurs in all classes of amplitudes, where a large irregularity (I_g decreases). I_g is inversely proportional to $V(x'')$.

Study of the Random Loading Impact

Predictive calculating of the lifetime of a system or component mechanical, operating in real conditions of use, is made from the fatigue of the materials of the components studied.

Material fatigue occurs whenever the efforts and stress vary over time. These random stresses take very different looks (see Sects. 1 and 2). The rupture may well occur with relatively low stress, sometimes less than a conventional limit called "endurance limit" S_D.

Material fatigue is approached in two ways.

The first is based on a global approach where the material is considered as a homogeneous medium at a macroscopic scale. The mechanical properties of the material are presented by the curves of fatigue; the most famous is the "Wöhler." The critical points of the components are defined by the points of the most damaging stress, and lifetime calculations are made at these points.

The second approach to material fatigue based on a local approach which characteristics are considered potential material defects (cracks). In these areas, the stresses lead to the definition of a cracking speed and is obtained when the out of crack length limit is reached.

These two approaches to the calculation of the fatigue lifetime of the systems or mechanical components use the same distributions of random loads.

The presentation of this part will be limited to the global approach because this approach for the strength of materials is widespread and based on a broad competence in the industry. A significant gain in the quality of forecasts is then expected with the inclusion of random stress. Note, however, that the second approach is not excluded and that all developments presented here can be extrapolated.

Principle of Predictive Calculating

Predictive calculating of the lifetime of a system or component mechanical, operating in real conditions of use, is made from four elements:

- Knowledge of load (or stress) it undergoes (Fig. 31a, b). Figure 31a shows the definition of measurement points over time (with predefined pace of time constant Δt). Figure 31b can recall the quantities considered thereafter.
- An endurance law (often based on the Wöhler) (Fig. 31c). Tests conducted for an average stress S_m^0 zero and a stress S_a give results with different numbers of cycles to failure. The distribution is represented by a sequence of points. The different values of stresses are considered. Thereafter, the three curves can be derived with, respectively, $p_r = 50\ \%$, p_0, and $1 - p_0$ of failure probability. Typically, p_0 is between 1 and 10 %.
- A fatigue requirement that sets the limit resistance (Fig. 31d). When the mean stress is not zero, the fatigue limit S_D depends on the value of the mean stress S_m, different models can then be used according to the stress S_m between 0 and R_E or R_m.
- A law of accumulation of damage to account for all the stresses applied to the system or component. Generally, it is the law that is used Miner.

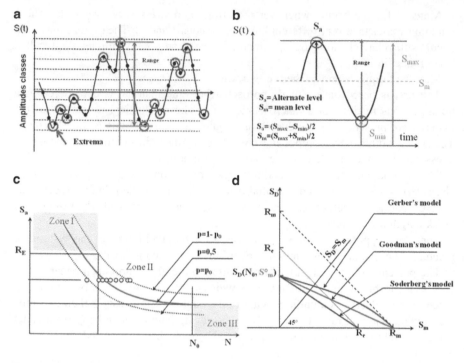

Fig. 31 The four steps of predicting of the lifetime. (**a**) Random stress, (**b**) after counting, (**c**) probabilized Wöhler's curve, (**d**) fatigue criteria

Fatigue and Damage in the Lifetime Validation of a Mechanical Component

The results of the analysis of the stresses measured under actual conditions of use should be exploitable by the steps of the method of calculating the lifetime. Thus, we must find (Sect. 1) counting method for identifying a "statistical load event" in a random loading history. For example, the event may consist of amplitudes of grouped into classes, the extrema, stretches or "cycles" of the stress studied. Often, they are extended or stress cycles which are operated to perform the calculation of a lifetime.

To strengthen the capacity of prediction, calculation of lifetime must have a probabilistic approach to announce reliability. Rather than events or extended cycle type, the load events are grouped into classes where these are the extrema which are better suited to probabilistic modeling (Sect. 2).

Endurance Law or Wöhler Curve

A Wöhler curve (Fig. 31c) represents a probability p to rupture, the magnitude of the cyclic stress ((S_a) considered average for a given stress S_m) based on the lifetime N. Also known as curve S–N (Stress–Number of cycles).

The test conditions must be fully specified in both the test environment and the types of stress applied. Rupture occurs for a number of cycles increases when the stress decreases.

There are *several models of these* curves according to an endurance limit implicitly or explicitly appears. Indeed, when a material is subjected to low-amplitude cyclic stress, fatigue damage can occur for large numbers of cycles. Points of the curve of Wöhler are obtained with the results of fatigue tests on specimens subjected to cyclic loading of constant amplitude.

The curve is determined from which each batch of sample is subjected to a periodic stress maximum amplitude S_a and constant frequency, the failure occurring after a number of cycles N. Each sample corresponds to a point in the plane (S_a, N).

The *results of the fatigue tests* are randomly distributed, such that one can define curves corresponding to given failure probabilities according to the amplitude and the number of stress cycles.

This finding requires the construction of a model to the median ($p_r = 0.5$, 50 % of failures) observed lifetimes, curves at p_0 and at $(1 - p_0)$ are then deducted p_0 is often taken equal to 0.01 (1 % of failures). These curves are called the Wöhler curves probabilized.

Wöhler curves are generally broken down *into three distinct areas* (Fig. 31c):

- Area I: Area of *oligocyclic plastic fatigue*, which corresponds to the higher stresses above the elastic limit R_E of the material. Breaking occurs after a small number of cycles typically varying from one cycle to about 10^5 cycles.
- Area II: Area *of fatigue or limited endurance*, where the break is reached after a limited number of cycles (between 10^5 and 10^7 cycles).
- Area III: Area *of endurance unlimited or safety area* under low stress, for which the break does not occur after a given number of cycles (10^7 and even 10^{10}), higher the lifetime considered for the part.

In many cases, *an asymptotic horizontal branch* to the curve of Wöhler can be traced: the asymptote is called endurance limit or fatigue limit and denoted S_D. The latter is defined at zero mean stress ($S_m^0 = 0$) and corresponds to a lifetime N_0 ($N_0 = 10^7$ cycles often).

In other cases, a conventional endurance limit may be set for example to 10^7 cycles.

Generally, you must use a model that approximates the curves to perform further calculations. Various expressions have been proposed to account for the shape of the curve of Wöhler (Lieurade 1980b). The most practical was proposed by Basquin and she wrote as follows:

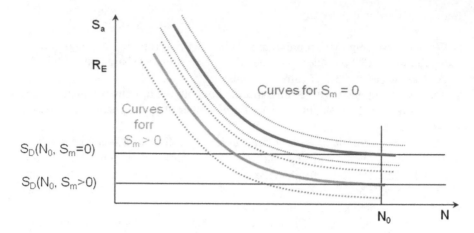

Fig. 32 Influence of mean stress on the network Wöhler curves

$$NS^b = C_1 \tag{69}$$

and

$$\log N + b \log S = C \ (C = \log(C_1)) \tag{70}$$

With N = number of cycles, S = amplitude of the stress, b = slope of the line depends on the material, C = constant dependent right material and the average of the alternating stress fatigue test performed.

In the following text, the coefficient C will be used.

Fatigue tests are usually long term and it is rare to have the results of experiments conducted with a nonzero mean stress.

The increase in mean stress causes a reduction of the lifetime: a network of Wöhler curves can be thought (Fig. 32). Thus, the endurance limit for a chosen number of cycles (e.g., to $N_0 = 10^7$ cycles) decreases. Then, for each nonzero average stress, an endurance limit should be determined. Lacking often experimental results, the calculation is based on a relationship called "fatigue requirement."

Fatigue Requirements

A fatigue test is a threshold defined by a mathematical expression for a fixed lifetime (N_0) and a given material. The threshold separates the state where the part is operating in the state where it is damaged by fatigue.

In general, a fatigue test was developed for cyclic loading with constant amplitude.

It is a relationship between the evolution of the Wöhler curve and the average stress considered. Figure 31d shows the main criteria of fatigue. These graphs are built with a lifetime N_0 (often given as 10^7 cycles) and a given probability of failure ($p_r = 0.5, 0.1$, and 0.01). These different models are (Rabbe et al. 2000b):

$$\text{Goodman model} \quad S_D = S_D^* \cdot \left(1 - \frac{S_m}{R_m}\right) \tag{71}$$

$$\text{Soderberg model} \quad S_D = S_D^* \cdot \left(1 - \frac{S_m}{R_E}\right) \tag{72}$$

$$\text{Gerbermodel} \quad S_D = S_D^* \cdot \left(1 - \left(\frac{S_m}{R_m}\right)^2\right) \tag{73}$$

With $S_D^* = S_D(S_m^0 = 0; N_0)$ endurance limit defined for zero mean stress and number of cycles N_0 selected.

Then, for a given cycle (amplitude, a range) having a nonzero mean, these relationships allow to obtain approximate the endurance limit for the average stress.

Calculation of Fatigue Damage, of Damage Accumulation, and of Lifetime

The concept of damage represents the state of degradation of the material in question. This condition results in a quantitative representation of the endurance of materials subjected to various loading histories.

A law of accumulative damage is a rule to accumulate damage variable D (also called "damage D"), itself defined by a law of damage.

As for the lifetime is defined by a number of cycles N which leads to breakage.

Thus, the application of n cycles ($n < N$) causes a partial deterioration of the treated piece to the calculation. The assessment of damage at a given time is crucial to assess the remaining capacity of lifetime.

Fatemi and Yang (Fatemi and Yang 1998) identified in the literature more than 50 laws of accumulated damage. The most commonly used today is the law of linear cumulative damage to Palmgren–Miner remains the best compromise between ease of implementation and the quality of predictions for large lifetimes (Banvillet 2001). Miner's rule is as follows:

$$d_i = \frac{n_i}{N_i} \tag{74}$$

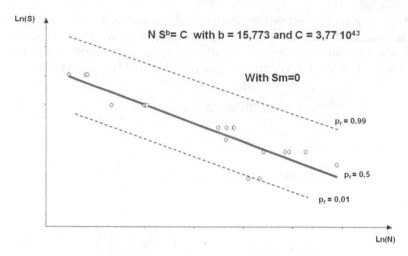

Fig. 33 Basquin Model of the Wöhler curve for Steel A42FP (from Leluan (1992b))

With n_i = the number of repetitions of a given cycle (with amplitude or extended), N_i = the number of repetitions of the same cycle necessary to declare the failed component ($n_i < N_i$).

For different cycles of random stress studied, the global damage is obtained by linear addition of the elementary damage:

$$D = \sum_i d_i = \sum_i \frac{n_i}{N_i} \qquad (75)$$

Fracture occurs when D is 1. The lifetime is equal to $1/D$.

Introductory Examples

The different steps of the calculation will be presented with a A42FP case (Leluan 1992b) steel. Figure 33 shows the model Basquin concerning fatigue tests on this steel with alternating stresses whose average is zero. Figure 34 shows the various criteria of fatigue such as steel. An endurance limit calculated from the right Soderberg is lower than those obtained with the right Goodman and the parabola of Gerber.

This difference between endurance limits calculated using criteria of fatigue leads to some significant differences on the calculated life.

Example 1 The first question in this introductory example is how to calculate the damage suffered by the steel A42FP subjected to a stress of 10^{75} cycles, the alternating stress (S_a) is 180 MPa, and mean stress (S_m) of 100 MPa, calculating with a failure probability of 0.5. The second question is to determine the number of repetitions of this stress must be applied for failure:

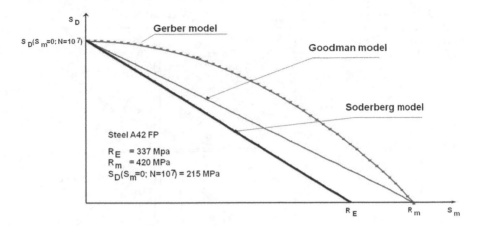

Fig. 34 Criteria of fatigue for Steel A42FP

Step 1: Calculation of the new endurance limit, according to each criteria of fatigue:

Criteria of fatigue	S_D ($S_m = 100$ MPa; $N = 10^7$)
Gerber parabola	203 MPa
Goodman right	164 MPa
Soderberg right	151 MPa

Step 2: Calculation of the new constant C in the model Basquin:

Criteria of fatigue	$C(S_m = 100$ MPa; $S_D) = N \times (S_D)^b = 10^7 (S_D)^b$
Gerber parabola	2.45×10^{43}
Goodman right	8.45×10^{41}
Soderberg right	2.39×10^{41}

Step 3: Calculation of the number of cycles to failure for the stress studied according to the criteria of fatigue. In this case, the calculation is performed at 50 % of probability of failure:

Criteria of fatigue	$N = C \times (S_a)^b$
Gerber parabola	9.15×10^{78}
Goodman right	3.16×10^{77}
Soderberg right	8.93×10^{76}

Step 4: Calculation of damage due to studied stress and the lifetime expectancy. Determination of the number of repetitions to reach failure

Criteria of fatigue	$D = n/N = 10^{75}/N$	Lifetime[a] $= 1/D$
Gerber parabola	1.093×10^{-4}	9,150 times
Goodman right	3.16×10^{-3}	316
Soderberg right	1.12×10^{-2}	89

[a]This result means that the stress will be able to repeat before reaching failure = number of repetitions.

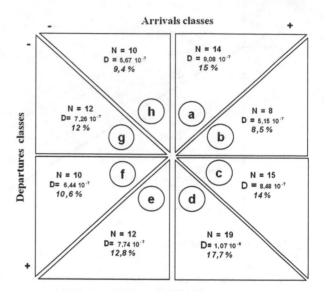

Fig. 35 Distribution of damage per area of the Markov matrix (stress shown in Fig. 10, part 1)

In the case of multilevel constraints, these four steps of calculation (Example 1) are applied for each level and the total damage is obtained according to the rules of linear cumulative damage Miner.

Example 2 It involves the stress of an axle front right of the vehicle. Figure 35 shows the distribution of damage per area of the Markov matrix for stress shown in Fig. 10 (Kouta and Play 2007a). The total damage is 6.05 E-06 and the number of repetitions of this stress before reaching failure is 165,273 times. The separation of Fig. 35 in eight zones ((a)–(h)) allows to visualize the contribution of the elementary stresses.

The nature of the stresses together in each area specified in Fig. 9. Fifteen percent of damage in the area (a) of Fig. 35 can be explained mainly by effective stress that are at the end of the Markov matrix (top right of Fig. 10).

Among a workforce of $N = 15$, a workforce of 5 $(2 + 2 + 1)$ corresponds to the five largest observed this stress extended. These result from extensive shocks when moving the vehicle in five concrete bumps. A total of 17.7 % of the damage area of the figure reflects the severity of the extended amplitude whose average is positive. They reflect the effect of compression after passing on concrete bumps.

Contributions and Impact of a Statistical Modeling for Random Loading in the Calculation of Lifetime

Principle of Damage Calculating

Data around which theoretical models of probability density are of three kinds (part 1):

- The amplitudes grouped into classes.
- Extreme amplitudes.
- Envelope of the stress.

Thus, three models of calculation will be presented in the following paragraphs. The evolution of damage is always calculated from the model Miner. We write an element of damage by fatigue dD is due to a stress element dn. Under the law of Miner

$$dD = \frac{dn}{N(S)} \tag{76}$$

with dn number of amplitudes solicitation studied at the level between S and $S + ds$.

- For a given duration T,

$$dn = N_T \times f_S(s)\, ds,$$

with N_T the total number of stresses and $f_S(s)$ probability density of the time stress whose level is equal to S.
- To stress measured with a time step $\Delta t = 1/f_e$ (f_e sampling frequency), the total number of stresses is equal to $N_T = f_e \times T$. Thus, the global damage is the sum of the partial damage for all values f S:

$$D = N_T \times \int_\Delta \frac{f_S(s)}{N(S)} \times ds \tag{77}$$

with $\Delta =$ domain of definition of S.
- Where the Wöhler is regarded as the model Basquin:
  In this case, ($NS^b = C$, C is a constant depending on the level of stress), Eq. (76) becomes:

$$D = \frac{N_T}{C} \times \int_\Delta s^b \times f_S(s) \times ds \tag{78}$$

If we set

$$u = \frac{s}{\sqrt{V(S)}}$$

with $V(S)$ variance S. Relation Eq. (78) is written:

$$D = [V(S)]^{b/2}\frac{N_T}{C} \times \int_\Delta u^b \times f_U(u) \times du \tag{79}$$

Table 5 Calculation of damage according to the Pearson system (see Fig. 36)

Laws of probability[a]	Damage D	
Beta 1 (Fig. 36)	$D = [V(S)]^{b/2} \frac{N_T}{C} \frac{\gamma(p+q)\gamma(b+p)}{\gamma(p)\gamma(b+p+q)}$	(80)
Beta 2 (Fig. 36)	$D = [V(S)]^{b/2} \frac{N_T}{C} \frac{\gamma(q-b)\gamma(b+p)}{\gamma(p)\gamma(q)}$	(81)
Log-Normal (Fig. 36)	$D = [V(S)]^{b/2} \frac{N_T}{C} \exp\left(bm + \frac{b^2 ff^2}{2}\right)$	(82)
Gamma (Fig. 36)	$D = [V(S)]^{b/a} \frac{N_T}{C} \frac{\gamma(b+p)}{a^b \gamma(p)}$	(83)
Weibull (Fig. 36)	$D = [V(S)]^{b/2} \frac{N_T}{C} \eta^b \gamma\left(1 + \frac{b}{\beta}\right)$	(84)
Normale (Fig. 36)	$D = [V(S)]^{b/2} \frac{N_T}{C\sqrt{2\pi}} 2^{\frac{b}{2}-1} \gamma\left(\frac{b+1}{2}\right)$	(85)

[a] See also Appendix 1

Thus, the damage is D calculated for each amplitude family around which a probabilistic modeling is proposed (Sect. 1). Firstly, the calculation can be done according to the system of Pearson (Beta 1 law, Beta 2 law and Log-Normal law, Gamma Law and Weibull law, Normal (or Gaussian) law. Calculation can also be done using either the law of Gram-Charlier–Edgeworth or the law of the envelope (or Rice).

Damage Calculating on the Basis of Laws Obtained by Pearson System

Table 5 gives the result of the relationship Eq. (79), obtained by the Pearson system for each probability law or each equivalent law. The laws of probability so that their parameters are given in Appendix 1.

Figure 36 shows the influence of parameters that define the laws of probability on the global damage. It is clear that the damage D decreases with increasing exponent b because when b increases (for a given C), the lifetime increases with the model Basquin.

Damage Calculating on the Basis of the Gram-Charlier–Edgeworth Law

In this case, the calculation from Eq. (79) is performed with the probability density function expressed with Eq. (11). Equation (79) leads to the following Eq. (86):

$$D = [V(S)]^{b/2} \frac{N_T}{C\sqrt{2\pi}} \left\{ \begin{array}{l} \gamma\left(\frac{(1+b)}{2}\right)\left(1+(-1)^b\right) \cdot \left\{ \begin{array}{l} 2^{\left(\frac{b-1}{2}\right)} + 2^{\left(\frac{b+1}{2}\right)} \\ \times \left[\left(\frac{k_4^2 b}{18}\right) + \left(\frac{k_3^2 b^2}{14}\right) + \left(\frac{k_4 b^2}{48}\right) - \left(\frac{k_4 b + k_3^2 b^2}{24}\right)\right] \end{array} \right\} \\ + \gamma\left(\frac{b}{2} + 1\right) 2^{\frac{b}{2}} k_3 \left(\frac{b-1}{6}\right)\left(1-(-1)^b\right) \end{array} \right\}$$

(86)

Fig. 36 Evolution of damage
depending on the settings of
each probability distribution.
(**a**) Damage with Beta 1,
(**b**) damage with Beta 2,
(**c**) damage with gamma,
(**d**) damage with Weibull,
(**e**) damage with Log Normal,
(**f**) damage with Gauss

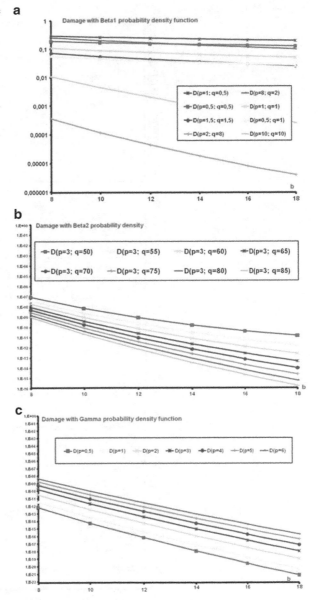

Figure 37 shows the evolution of the damage function of b, the skewness β_1, and
β_2 kurtosis of the stress distribution. All curves decrease with the increase of the
parameter b. For a given β_1 parameter when kurtosis β_2 increases, that is to say, the
dispersion of the stress distribution and damage increases, when the skewness ratio
increases, the damage also.

Fig. 36 (continued)

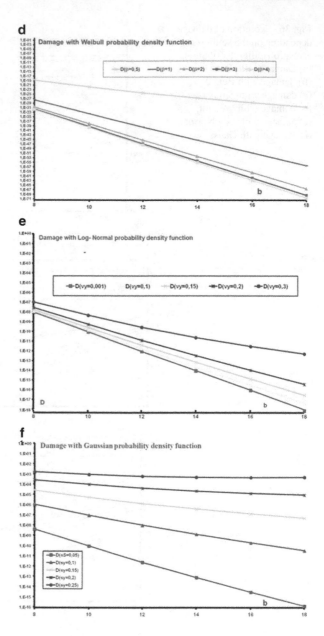

Damage Calculating on the Basis of the Rice Law on the Envelope

The presented model of the envelope in Sect. 2.4 and Appendix 5 relates extrem amplitudes, so the total number of stresses is equal to the number N_T extrem amplitudes N_e. In this case, the damage is determined by the relationship

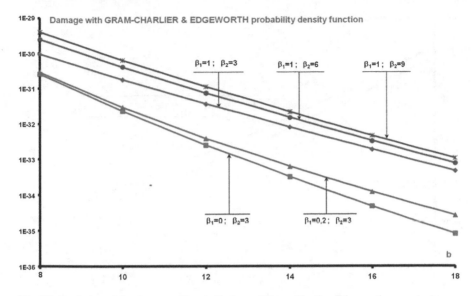

Fig. 37 Evolution of damage according to the law of Gram-Charlier–Edgeworth

Table 6 Calculation of damage to individual cases of irregularity factor I

$I = 0$ (stress broadband)	$I = 1$ (stress narrowband)
$D = [V(S)]^{\frac{b+1}{2}} \frac{N_e}{C} \dfrac{2^{\frac{b}{2}-1}\gamma\left(\frac{b+1}{2}\right)}{\sqrt{\pi}}$ (88)	$D = [V(S)]^{\frac{b+1}{2}} \frac{N_e}{C} 2^{\frac{b}{2}-1}\gamma\left(\frac{b}{2}+1\right)$ (89)

$$D \approx \frac{[V(S)]^{\frac{b+1}{2}} N_e}{C} \left\{ \begin{array}{l} \dfrac{2^{\frac{b}{2}-1}\left(1-I^2\right)^{\frac{b}{2}+1}\gamma\left(\frac{b+1}{2}\right)}{\sqrt{\pi}} + 2^{\frac{b}{2}-1}I\gamma\left(\frac{b}{2}+1\right) \\[2ex] + \dfrac{2^{\frac{b+1}{2}}I^3}{(1-I^2)\sqrt{(b+2)\pi}}\Phi\left(\frac{2I^2}{1-I^2};1;[b+2]\right) \end{array} \right\} \qquad (87)$$

With $\Phi\,(x, 1, [b+2])$ is the cumulative probability function of Fischer–Snedecor of x in the degrees of freedom with one and the integer part of $(b+2)$.

Note $\Phi(0, 1, [b+2]) = 1$ and $\Phi(\infty, 1, [b+2]) = 0$. The results for the particular case of irregularity factor I are presented in Table 6:

The damage curves (Fig. 38) decrease as the parameter b is increased (slopes substantially equal to the slopes Fig. 37). When the irregularity factor changes from 0 (broadband) to a value of 1 (narrowband), damage increases. This effect is due to the increased number of cycles with a narrowband signal.

Applications (Brozzetti and Chabrolin 1986b)

Fatigue strength of a node type T "jacket" of an "offshore" structure must be checked. Available records concerning the height of the waves beat structure; values are based on the frequency of occurrence (Table 7). These records are related to a reference time of 1 year.

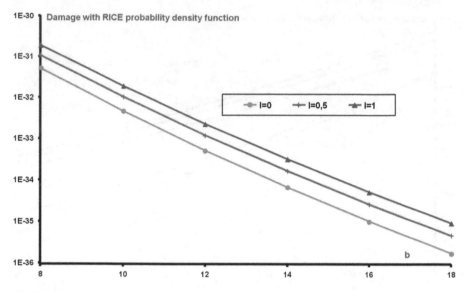

Fig. 38 Evolution of damage according to the law of the envelope of Rice

Table 7 Frequency of occurrence of solicitation for a T structure

No.	Wave height H (m)	Stress sizing $\Delta\sigma$	Annual number of waves $= n_i$
1	0	0	4,482,356
2	1.52	6.45	2,014,741
3	3.05	14.87	759,383
4	4.57	24.15	281,009
5	6.10	34.16	104,059
6	7.62	44.61	38,765
7	9.14	55.49	14,558
8	10.67	66.81	5,516
9	12.19	78.39	2,108
10	13.72	90.34	812
11	15.24	102.48	314
12	16.76	114.87	122
13	18.29	127.56	47
14	19.81	140.39	18
15	21.34	153.50	7
16	22.86	166.71	3
17	24.38	180.00	1
18	25.91	193.74	...

Fig. 39 Endurance curve (for T type node jacket) of an offshore structure

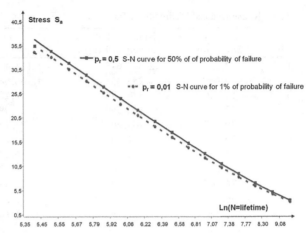

At each of these wave heights, the calculation of the structure leads to a stress on the design of the T node ($\Delta\sigma$, in MPa, Table 7). This constraint sizing takes into account a coefficient of stress concentration T node considered.

Fatigue curve at 50 % of the considered node failures is given by the expression (Basquin deviation) as follows:

$$N\Delta\sigma^{10} = 10^{10} \quad \text{or} \quad \log(N) = -3\log(\Delta\sigma) + 10 \tag{90}$$

Fatigue curve 1 % failure is as follows: $N\Delta\sigma^{2.9} = 10^{9.8}$. These curves are presented in Fig. 39.

Calculation of cumulative damage to the node as well as its lifetime can be made using the methods presented in the previous sections.

Calculation of Damage and Lifetime According to the Classical Method

Table 8 presents the calculation of cumulative damage by applying Miner's rule ("Calculation of fatigue damage, of damage accumulation and of lifetime" and "Introductory examples" in Appendix E). The accumulated damage of the fatigue curve with 50 % of failures provides a value $D_{(0.5)} = \Sigma(n_i/N_i) = 0.0147$ and the lifetime of the node $= 1/D = 68$ years. The calculation with the fatigue curve gives 1 % damage $D_{(0.01)} = 0.0193$ so a minimum lifetime of 52 years.

This approach to calculation is based on the observed numbers (n_i) and the lifetime limit estimated (N_i) for each level of the stress concerned. The statistical nature of the distribution of staff (or the probability amplitudes of the stress studied) is not taken into account. Moreover, this calculation of the global lifetime of endurance depends on the model given in Eq. (90).

Table 8 Damage and lifetime for a T structure

No	Stress (MPa)	Number of waves of height H (m)	N_i	n/N_i ($\times 10^{-4}$)
1	3, 225	2, 467, 615	42, 112, 482, 458	0.5860
2	10, 660	1, 255, 358	1, 166, 080, 106	10.7656
3	19, 510	478, 374	190, 207, 743	25.1501
4	29, 155	176, 950	56, 998, 163	31.0449
5	39, 385	65, 294	23, 121, 045	28.2401
6	50, 050	24, 207	11, 266, 467	21.4859
7	61, 150	9, 042	6, 177, 470	14.6371
8	72, 600	3, 408	3, 691, 392	9.2323
9	84, 365	1, 296	2, 352, 410	5.5092
10	96, 410	498	1, 576, 281	3.1593
11	108, 675	192	1, 100, 553	1.7446
12	121, 215	75	793, 105	0.9456
13	133, 975	29	587, 393	0.4937
14	146, 945	11	445, 180	0.2471
15	160, 105	4	344, 180	0.1162
16	173, 355	2	271, 138	0.0738
17	186, 870	1	216, 462	0.0462
Sum				0.0153

Calculation of Damage and Lifetime by Statistical Modeling

The process of calculation, in this paragraph, incorporates the statistical nature of the distribution of staff (or the probability amplitudes of the stress studied) and uses the same model of endurance expressed by Eq. (90). In this case, the damage is assessed from the following four elements:

- Law of distribution of wave heights in the long term
  In this example, distribution of wave heights corresponds to a Weibull law:

$$H = H_0 + p_{cte} \log(n). \tag{91}$$

  With H_0 is the maximum wave height recorded in the reference period (within 1 year); n is the number of wave heights greater than H_0, p_{cte} is a constant.
- Damage law (or curve of fatigue or S–N curve)

$$\text{Let}: \ N\Delta\sigma^b = C, \ \text{let}: N = C\Delta\sigma^{-b} \text{ (same Eq. (90))} \tag{92}$$

  With b = slope of the S–N curve, N = number of cycles associated with the variation of the stress sizing $\Delta\sigma$, $\Delta\sigma$ extent of stress variation in the node in T of the structure.

- Relationship between wave height H to the stress variation $\Delta\sigma$
  The stress variation $\Delta\sigma$ is laid as:

$$\Delta\sigma = \alpha.H^\beta \tag{93}$$

with α and β two coefficients that are obtainable by smoothing spots $(\Delta\sigma, H)$.

If the value of $\Delta\sigma$ given by the expression Eq. (93) in Eq. (92) is postponed, we obtain:

$$N = C\alpha.\left(H^\beta\right)^{-b} \tag{94}$$

Transformation of expression Eq. (91) gives:

$$n = n_0.\exp\left(2.3026H/p_{cte}\right) \tag{95}$$

With $n_0 = 10^{-H_0/p_{cte}}$.

- Rule of damage cumulation

$$\Delta D = dn/N \tag{96}$$

Thus,

$$dn = (2.3026/p_{cte})\,n_0\exp\left(2.3026H/p_{cte}\right)\,dH \tag{97}$$

We obtain:

$$\Delta D = dn/N = \alpha^b\,(2.3026/C.p_{cte})\,n_0\exp\left(2.3026H/p_{cte}\right)H^{b\beta}\,dH \tag{98}$$

Let:

$$D = \frac{2.3026 n_0 \alpha^b}{C.p_{cte}}\int_0^\infty \exp\left(\frac{2.3026\,H}{p_{cte}}\right)H^{b\beta}\,dH \tag{99}$$

We recall that : $\int_0^\infty t^u \exp\left(-vt^w\right)dt = \frac{\gamma[(u+1)/w]}{wv^{(u+1)/w}}$; $\gamma(t)$ is the gamma function (see Eq. (28)) with $u = \beta b$ $v = 2.3026/p_{cte}$ and $w = 1$, we obtain

$$D = \frac{2.3026 n_0 \alpha^{-b}}{C.p_{cte}}\,\frac{\gamma\,(\beta b + 1)}{\left(\frac{2.3026}{p_{cte}}\right)^{(\beta b + 1)}} \tag{100}$$

Example 3 From the data in Table 7, we obtain by smoothing:

$$H = 6.1 + 0.913\log(n)\quad (H\text{ in meters})\quad \Delta\sigma = 3.9H^{1.2}.$$

Table 9 Summary of results for structure T application

Summary of results for structure T application	Traditional approach	Statistical modeling
Material distribution	Yes	Yes
Load distribution	No	No
50 % failure probability (reliability 0.5)	LT = 68 years	LT = 73 years
1 % failure probability (reliability 0.99)	LT = 52 years	LT = 49 years

Let: $\alpha = 3.9$; $\beta = 1.2$; $b = 3$; $p_{cte} = 0/913$; $C = 10^{10}$; $H_0 = 6.1$; $n_0 = 10^{6.1/0.913}$.

Hence: $\gamma(b\beta + 1) = \gamma(4,6) = 13.38$; $\alpha^b = 59.32$; $2.3026 n_0 = 11{,}053{,}226$; p_{cte}.
$C = 0.913 \times 10^{10}$.

Thus, damage: $D = 0.0137$.

So the lifetime is $1/D = 73$ years.

With the model of endurance to 1 % failure ($N \Delta\sigma^{2.9} = 10^{9.8}$), the lifetime is equal to 49 years.

In summary for this application (see Table 9), the transition from a traditional approach to statistical modeling provides a better prediction of behavior (7 % saving). It is also noted that the increase of reliability of from 0.5 to 0.99 decreases calculated lifetime to about 28 %.

Calculation of Damage and Lifetime on the Basis of Power Spectral Density (Lalanne 1999b)

A system or a mechanical component has a response to applied stresses (PSD). This response can be either measured on a prototype under tested qualification, either calculated from a global mechanical model.

The expression Eq. (79) shows that the damage depends on the variance of the amplitudes of the stress. The relation Eq. (46) gives the expression for the variance of a random stress according to its PSD. It must be previously represented by several levels (G_i). In this case, the relation Eq. (79) is then expressed as follows:

$$D = \frac{N_T}{C}[V(S)]^{b/2} \int_\Delta u^b \, f_U(u) \, du = \frac{N_T}{C} \cdot \left[\frac{\frac{\pi}{4\xi} \sum_{i=1}^{n} G_i \left[I_0 \left(h_{i+1} \right) - I_0 \left(h_i \right) \right]}{(2.\pi)^4 f_0^3} \right]^{b/2} \int_\Delta u^b \, f_U(u) \, du$$

(101)

• This expression is called "Fatigue Damage Spectrum" (Lalanne 1999b). The terms $\int_\Delta u^b f_U(u) \, du$ and $\frac{N_T}{C}$ are two multiplicative constants of the term expressed in the brackets of the expression Eq. (101). This term represents the evolution of the variance of the stress according to the frequencies.

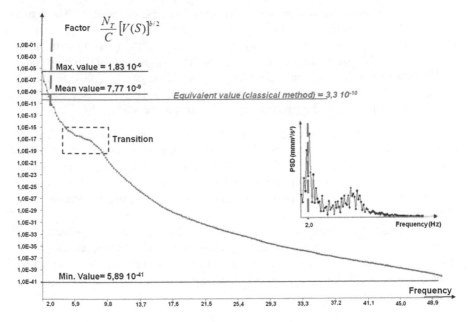

Fig. 40 Fatigue damage spectrum depending on the frequency

- Figure 40 shows the next term $\frac{N_T}{C}[V(S)]^{b/2}$ of the Fatigue Damage Spectrum on the stress shown in Fig. 10. This graph is obtained for $C = 10^7$ and $b = 12$, the value of b is given by (Lambert 1980) as an average value of the slope of the line for Basquin steels. The failure probability is 1 %. In Fig. 40, there is not a particular representative value to summarize the shape of the curve. However, an average value may be deducted, equal to 7.77×10^{-9} (the minimum is 5.89×10^{-41} and the maximum is 1.83×10^{-06}).

- The value of the term $\int_\Delta u^b f_U(u)\, du$ depends on the model probability density which is adopted for the stress studied. In the case of a Gaussian distribution, this sum is $\frac{2^{\frac{b}{2}}}{\sqrt{2\pi}} \gamma\left(\frac{b+1}{2}\right)$ (see relation Eq. (85)). The value here is equal to 18,425 (for $b = 12$). Thus, the average damage D_{av} is 1.43×10^{-4} (with a minimum damage equal to 1.09×10^{-36} D_{min} and D_{max} maximum damage equal to 3.37×10^{-2}). The lifetime (or the number of repetitions to failure $1/D_{av}$) is equal to 6,993.

- This calculation of the lifetime by the PSD as takes into account all the contributions of the stress on the frequency domain of stresses. The shape of the Fatigue Damage Spectrum is thus brought into relationship with the shape of the PSD (Power Spectral Density). The PSD of the stress studied (Fig. 13b) is outlined in Fig. 40. Thus, the almost flat spectrum observed in the fatigue damage in the area of transition nature (framed part in Fig. 40) is due to the transition between the two major contributions observed on the PSD. These two are the two main modes, the stress observed studied.

- Note that the damage calculation by conventional method (described in "Calculation of fatigue damage, of damage accumulation and of lifetime" and "Introductory examples" in Appendix 5) gives damage $D_{\text{conventional}}$ equal to 6.05×10^{-6}. This level of damage is equivalent value $\frac{N_T}{C}[V(S)]^{b/2} = 3.3 \times 10^{-10}$, value that is located in the terminals and mini max damage (Fig. 40).

However, conventional method gives a value of damage to a single class of frequency (around 2 Hz) corresponding to the largest amplitudes of the PSD. Lifetime (or the number of repetitions before failure $1/D_{\text{conventional}}$) is then equal to 165,273 repetitions. The difference between the result of the lifetime by PSD and that obtained by the conventional method is now 95 %, results become very different.

Result obtained with Fatigue Damage Spectrum allows differentiation the role of each frequency band to obtain the damage. Damage results consequences of any stresses represented by the spectral density PSD. Damage obtained by this method takes into account all frequency levels that appear with their stress levels and their respective numbers.

Plan of Validation Tests

The different models presented to assess the damage and the lifetime in real use conditions based on different types of analysis of stresses measured on a system or a mechanical component. These models also allow defining test environments that must cause similar effects to those observed in the actual conditions of use.

- The experimental means perfect more and more to reproduce in laboratory the stresses recorded in real conditions of use. These types of tests are designed to validate models of mechanical calculation and are not intended to validate the lifetime estimated by the models presented earlier. Such validation would require too much time. Indeed, the need to reduce more and more the validation period of a product required to do as much as possible, test duration shorter and shorter and more severe than the stress level observed in the actual conditions of use. Naturally, these severe tests shall show equivalent to those observed in real use conditions consequences.

Damage is an equivalence indicator.

- Figure 41 illustrates the principle of equivalent damage. In the case of alternating stresses and with a material whose curve of resistance to fatigue (Wöhler curve) is known, the limit lifetime is expressed in terms of alternating stress of solicitation. If we assume that the actual conditions of use are summarized in n_{real} repetitions of a sinusoidal stress ($S_a = S_{\text{real}}$) which is N_{real} the limit lifetime in these conditions of use ($n_{\text{real}} < N_{\text{real}}$). In this case (according to Miner), damage under real conditions of use is: $d_{\text{real}} = \frac{n_{\text{real}}}{N_{\text{real}}}$.

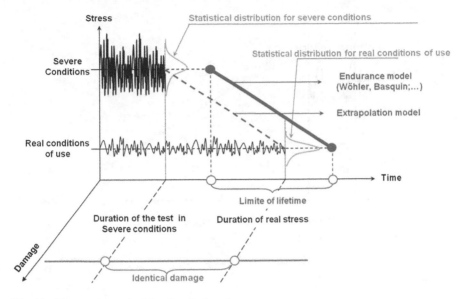

Fig. 41 Schema of the principle of equivalent damage

- Conducting a test in more severe conditions and leads to the same level of damage d_{real} is to define the number of times (n_{severe}) to be repeated a sinusoidal stress ($S'_a = S_{\text{severe}}$) so that the damage d_{severe} is equal to d_{real}.
- For a mechanical component, the level ($S'_a = S_{\text{severe}}$) is determined depending on the material which constitutes the studied and knowledge about limits behavior component. The equivalence between damage ($d_{\text{severe}} = d_{\text{real}}$) leads to the calculation of n_{severe}. As follows:

$$n_{\text{severe}} = \frac{N_{\text{severe}}}{N_{\text{real}}} n_{\text{real}} \tag{102}$$

According to the model Basquin $N = CS^{-b}$, then

$$n_{\text{severe}} = \left(\frac{S_{\text{severe}}}{S_{\text{real}}}\right)^{-b} n_{\text{real}} \tag{103}$$

Figure 42 illustrates the process of calculating the number of cycles required to produce, under severe conditions, damage equivalent to that observed under actual use conditions.

- A detailed analysis of behavior in endurance (long period) should help define tests (short term) to perform in laboratory or on simulation bench. In this case, it is advisable to retain only the critical stresses. Two selection criteria are possible: take only the stresses that have contributed from a given percentage of the global damage or take the stresses whose amplitude alternating (S_a, Fig. 31b) is greater than the limit endurance.

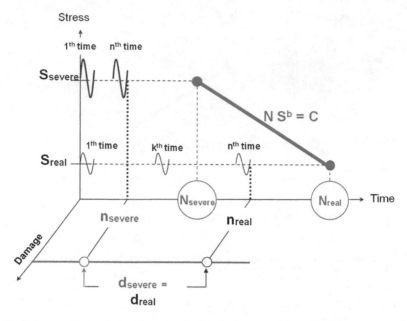

Fig. 42 Schema of the equivalent damage principle in the case of sinusoidal stresses

Fig. 43 Transition matrix of the stress with distinction parts of the alternating stresses which amplitude is less than the endurance limits (see Fig. 10)

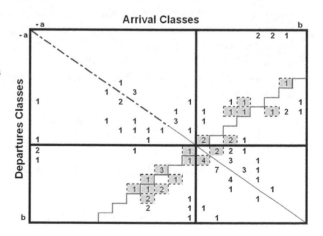

For the examples presented in the following, this is the second criterion that is used.

- Figure 43 shows the transition matrix corresponding to the stress of Fig. 40 by specifying the boxes (with gray background) in which the associated alternating stress is below the endurance limit. If these stresses are not considered, the cumulative damage remaining 72 % of the global damage.

- The remaining part of the transition matrix can be used for testing in the laboratory. Of course, it is only the nonempty boxes of the matrix are taken into account.
- In the case of this stress, the information obtained by the extended-slope matrix (Fig. 21a) and the extreme-curves matrix (Fig. 21b) contributes to the definition of the dynamic sinusoidal stresses considered for each nonempty cell of the matrix of transition.
- In the top right quarter of the matrix linking extreme amplitudes and curvatures (Fig. 21b), there is five number which summarizes the five extreme amplitudes (negative) observed following the passage of the five concrete bumps. Based on the transition matrix (Fig. 10), these five amplitudes are the starting points of which types of stress do not have exactly the same end point. But the information provided by the extreme-curvature matrix confirms that these five amplitudes the frequency signature is identical.
- The frequency according to which occur these five amplitudes is calculated as:

$$\text{frequence} = \frac{1}{2\pi} \sqrt{\left| \frac{y''}{y} \right|} \tag{104}$$

with y'' is the second derivative of the signal y of load plotted against time.
- This relationship can also calculate all the frequencies of stresses that must be applied during laboratory tests. These are the quarters on up and right and down left of the extremes-curvatures matrix which are favored because they plot stresses whose sign of the amplitude of departure is different from the amplitude of arrival, reflecting extent of high amplitude. The quarters up left and down right plot intermediate fluctuations which are not important as long as they remain concentrated around the center of the matrix.

• The transition matrix of Fig. 44a shows a stress history near a fillet weld on a transport vehicle chassis (Kouta et al. 2002) in an actual path of usage of 464 km. Figure 44b shows this same matrix in the form of Iso-curves (or level curves). This second representation gives iso-level curves of log of numbers, for an easier viewing. In red (center of the butterfly), there are more effective but they are the least damaging. By against in blue, the lowest numbers are shown but the most damaging. For this case of loading, figure is symmetrical with respect to the first diagonal, and is offset slightly downwardly. Note also the existence of two small separate areas from the main figure.

- A first failure on the ground was observed on this path of 464 km after 1E + 06 km (2,155 rehearsals). Endurance characteristics of the material (Wöhler curve) were determined through laboratory tests on this type of component. The calculation of lifetime through the approach presented earlier ("Calculation of fatigue damage, of damage accumulation and of lifetime" and "Introductory examples" in Appendix 5) leads to a calculated lifetime of 1.2E + 06 km, which was satisfactory.

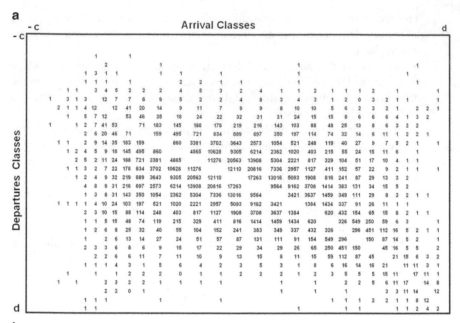

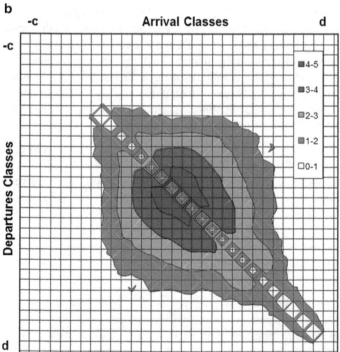

Fig. 44 Two representations (**a** stress history and **b** iso-curves) of a transition matrix for a fillet weld of the bogie chassis (according to (Kouta et al. 2002))

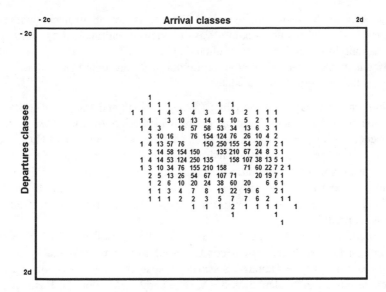

Fig. 45 Transition matrix corresponding to an endurance test obtained with a ratio of severity equal to 2 from the matrix in Fig. 43 (Matrix obtained under the rule of equivalent damage, see Eq. (102))

- The conditions of an endurance test of short-term representative of the actual damage were defined from Eq. (103). Figure 45 shows the transition matrix corresponding to an endurance test which is obtained with a ratio of severity equal to 2 $\{(S_{\text{severe}}/S_{\text{real}}) = 2$, amplitude $-2c$, $+2d\}$. This matrix provides the same damage as observed by considering the matrix in Fig. 44 but with a test duration which corresponds to a distance of 5 km with a rehearsal. Thus, this new matrix can be used to achieve a short-term endurance test in laboratory whose consequences (analysis of cracks, probable failure, etc.) are representative of a real long-term use.

Conclusion

The approach presented in three parts was motivated by the challenges of quality requirements for mechanical systems and components. And more specifically, lifetime and reliability (probability of no break) are two main features. All of the tools can be naturally applied to all cases of random stresses in mechanical. Engineers of design and development in mechanical have to know in detail:

- Actual conditions of use from users' behaviors.
- Actual conditions of use from users' behaviors.
- Behaviors (or responses) of studied systems.
- Models of material degradation.

Thus achieve the provision of values lifetime is not an insurmountable task. However, numerical results should always be viewed and analyzed the same time as the assumptions that lead to their obtaining.

Different approaches for calculating lifetime were presented as well as intermediate hypotheses required to calculate:

- Approach with statistical modeling ("Contributions and Impact of a Statistical Modelling for Random Loading in the Calculation of Lifetime" section in Appendix 5) can now relay conventional approach ("Calculation of Fatigue Damage, of Damage Accumulation and of Lifetime" and "Introductory Examples" sections in Appendix 5) traditionally used in design departments. This approach with statistical modeling provides a global indicator of damage, lifetime, and reliability which remains easy to obtain while having improved the quality of behavior prediction.
- Approach Fatigue Damage Spectrum with the use of the PSD is however in conjunction with the actual conditions of use of mechanical systems and components. Fatigue Damage Spectrum whose the method of obtaining has been developed in "Calculation of Damage and Lifetime on the Basis of Power Spectral Density (Lalanne 1999b)" in Appendix 5, allows to connect lifetime (or damage) of a mechanical component with dynamic characteristics of stresses sustained by the material. The use of PSD in this calculation allows taking into account the dynamic behavior of the system which belongs to the mechanical component studied. In addition, this approach allows taking into account the nature of the statistical distribution of the solicitation.
- Validation of predictive calculations in systems and mechanical components design also calls for the establishment of specific test procedures. Naturally, market pressure requires test times which are compatible with the terms of product development. In general, these periods have nothing to do with expected lifetime in service. The obtained results allow constructing simple environments of bench testing of similar severity to that observed in tests of endurance.

References

Arquès P-Y, Thirion-Moreau N, Moreau E (2000) Les représentations temps-fréquence en traitement du signal [R 308]. Base documentaire « Mesures et tests électroniques ». (*)

Banvillet A (2001) Prévision de durée de vie en fatigue multiaxiale sous chargements réels : vers des essais accélérés. Thèse de doctorat no 2001-17, ENSAM Bordeaux, France (274 pp)

Bedard S (2000) Comportement des structures de signalisation aérienne en aluminium soumises à des sollicitations cycliques. Mémoire de maîtrise ès sciences appliquées, École polytechnique de Montréal, 0-612-60885-9

Borello G (2006) Analyse statistique énergétique SEA [R 6 215]. Base documentaire « Mesures mécaniques et dimensionnelles ». (*)

Brozzetti J, Chabrolin B (1986a) Méthodes de comptage des charges de fatigue. Revue Construction Métallique (1)

Brozzetti J, Chabrolin B (1986b) Méthodes de comptage des charges de fatigue. Revue Construction Métallique, no 1

Buxbaum O, Svenson O (1973) Zur Beschreibung von Betriebsbeanspruchungen mit hilfe statistischer Kenngrößen. ATZ Automobiltechnische Zeitschrift 75(6):208–215

Chang JB, Hudson CM (1981) Methods and models for predicting fatigue crack growth under random loadings. ASTM, STP 748, Philadelphia, PA

Chapouille P (1980) Fiabilité. Maintenabilité [T 4 300]. Bases documentaires « Conception et Production » et « Maintenance ». (*)

Charlier CVL (1914) Contributions to the mathematical theory of statistics. Arkio für Matematik (Astronomi och Fysik) 9(25):1–18

Clausen J, Hansson JSO, Nilsson F (2006) Generalizing the safety factor approach. Reliab Eng Syst Safety 91:964–973

Doob JL (1953) Stochastic process. Wiley, New York, 654 pp

Duprat D (1997) Fatigue et mécanique de la rupture des pièces en alliage léger [BM 5 052]. Base documentaire « Fonctions et composants mécaniques ». (*)

Edgeworth FY (1916) On the mathematical representation of statistical data. J Roy Stat Soc A, 79, Section I, pp 457–482; section II, pp 482–500; A, 80, section III, pp 65–83; section IV, pp 266–288; section V, pp 411–437 (1917)

Fatemi A, Yang L (1998) Cumulative fatigue damage and life prediction theories: a survey of the state of the art for homogeneous materials. Int J Fatigue 20(1):9–34

Fauchon J Probabilités et statistiques INSA-Lyon, cours polycopiés, 319 pp

Gregoire R (1981) La fatigue sous charge programmée, prise en compte des sollicitations de service dans les essais de simulation. Note technique du CETIM no 20, Senlis

Grubisic V (1994) Determination of load spectra for design and testing. Int J Vehicle Des 15(1/2):8–26

Guldberg S (1920) Applications des polynômes d'Hermite à un problème de statistique. In: Proceedings of the International Congress of mathematicians, Strasbourg, pp 552–560

Heuler P, Klätschke H (2005) Generation and use of standardized load spectra and load-time histories. Int J Fatigue 27:974–990

Jeffreys H (1961) Theory of probability, 3rd edn. Clarendon, Oxford

Johnson NL, Kotz S (1969) Distributions in statistics: continuous unvaried distributions—1. Houg. Mifflin Company, Boston, 300 pp

Kendall MG, Stuart A (1969) The advanced theory of statistics, vol 1. Charles Griffin and Company Limited, London, 439 pp

Klemenc J, Fajdiga M (2000) Description of statistical dependencies of parameters of random load states (dependency of random load parameters). Int J Fatigue 22:357–367

Kouta R (1994) Méthodes de corrélation par les sollicitations pour pistes d'essais de véhicules. Thèse de doctorat, Institut national des sciences appliquées de Lyon, no 94ISAL0096

Kouta R, Play D (1999) Correlation procedures for fatigue determination. Trans ASME J Mech Des 121:289–296

Kouta R, Play D (2006) Definition of correlations between automotive test environments through mechanical fatigue damage approaches. In: Proceedings of IME Part D: Journal of Automobile Engineering, vol 220, pp 1691–1709

Kouta R, Play D (2007a) Durée de vie d'un système mécanique—Analyse de chargements aléatoires. [BM 5 030]. (*)

Kouta R, Guingand M, Play D (2002) Design of welded structures working under random loading. In: Proceedings of the IMechE Part K: journal of multi-body dynamics, vol 216, no 2, pp 191–201

Lalanne C (1999a) Vibrations et chocs mécaniques—tome 4: Dommage par fatigue. Hermes Science Publications, Paris

Lalanne C (1999b) Vibrations et chocs mécaniques—tome 5: Élaboration des spécifications. Hermes Science Publications, Paris

Lallet P, Pineau M, Viet J-J (1995) Analyse des sollicitations de service et du comportement en fatigue du matériel ferroviaire de la SNCF. Revue générale des chemins de fer, Gauthier-Villars éditions, pp 19–29

Lambert RG (1980) Criteria for accelerated random vibration tests. Thèses. Proceedings of IES, pp 71–75

Lannoy A (2004) Introduction à la fiabilité des structures [SE 2 070]. Base documentaire « Sécurité et gestion des risques ». (*)

Leluan A (1992) Méthodes d'essais de fatigue et modèles d'endommagement pour les structures de véhicules ferroviaires. Revue générale des chemins de fer, Gauthier-Villars éditions, pp 21–27

Leybold HA, Neumann EC (1963) A study of fatigue life under random loading. Proc Am Soc Test Mater 63:717–734, Reprint no 70—B

Lieurade H.-P. (1980a) Les essais de fatigue sous sollicitations d'amplitude variable—La fatigue des matériaux et des structures. Collection université de Compiègne (éditeurs scientifiques Bathias C, Bailon JP). Les presses de l'Université de Montréal

Lieurade H-P (1980b) Estimation des caractéristiques de résistance et d'endurance en fatigue— La fatigue des matériaux et des structures. Collection université de Compiègne, Éditeurs scientifiques: Bathias C, Bailon JP Les presses de l'université de Montréal

Lu J (2002) Fatigue des alliages ferreux. Définitions et diagrammes [BM 5 042]. Base documentaire « Fonctions et composants mécaniques ». (*)

Max J (1989) Méthodes et techniques de traitement du signal et applications aux mesures physiques, vol 1 & 2, 4th edn. Masson, Paris

Olagnon M (1994) Practical computation of statistical properties of rain flow counts. Fatigue 16:306–314

Osgood CC (1982) Fatigue design. Pergamon, New York

Parzen E (1962) Stochastic process. Holdenday, San Francisco, 324 pp

Plusquellec J (1991) Vibrations [A 410]. Base documentaire « Physique - Chimie ». (*)

Pluvinage G, Sapunov VT (2006) Prévision statistique de la résistance, du fluage et de la résistance durable des matériaux de construction. ISBN: 2-85428-735-5, Cepadues éditions

Preumont A (1990) Vibrations aléatoires et analyse spectrale. Presses polytechniques et universitaires romandes, Lausanne, 343 pp

Rabbe P, Galtier A. (2000) Essais de fatigue. Partie I [M 4 170]. Base documentaire « Étude et propriétés des métaux ». (*)

Rabbe P, Lieurade H-P, Galtier A (2000a) Essais de fatigue. Partie 2 [M 4 171]. Base documentaire « Étude et propriétés des métaux ». (*)

Rabbe P, Lieurade HP, Galtier A (2000b) Essais de fatigue—Partie I. [M 4 170]. (*)

Rice SO (1944) Mathematical analysis of random noise. Bell Syst Tech J 23:282–332, 24:46–156 (1945)

Rice J.R., Beer F.P., Paris P.C. (1964) On the prediction of some random loading characteristics relevant to fatigue. In: Acoustical fatigue in aerospace structures: proceedings of the second international conference, Dayton, Ohio, avril 29 - mai, pp 121–144

Rychlik I (1996) Extremes, rain flow cycles and damage functional in continuous random processes. Stoch Processes Appl 63:97–116

Saporta G (1990) Probabilités. Analyse des données et statistiques. Éd. Technip, 495 pp

Savard M (2004) Étude de la sensibilité d'un pont routier aux effets dynamiques induits par la circulation routière. 11e colloque sur la progression de la recherche québécoise des ouvrages d'art, université Laval, Québec

Schütz W (1989) Standardized stress-time histories: an overview. In: Potter JM, Watanabe RT (eds) Development of fatigue loading spectra. American Society for Testing Materials, ASTM STP 1006, Philadelphia, PA, pp 3–16

Société française de métallurgie (Commission fatigue) (1981) Méthodes d'analyse et de simulation en laboratoire des sollicitations de service. Groupe de travail IV « Fatigue à programme », document de travail, Senlis

Ten Have A (1989) European approaches in standard spectrum development. In: Potter JM, Watanabe RT (eds) Development of fatigue loading spectra. ASTM STP 1006, Philadelphia, PA, pp 17–35

Tustin W (2001) Vibration and shock inputs identify some failure modes. In: Chan A, Englert J (eds) Accelerated stress testing handbook: guide for achieving quality products. IEEE, New York

Weber B (1999) Fatigue multiaxiale des structures industrielles sous chargement quelconque. Thèse de doctorat, INSA-Lyon, 99 ISAL0056

(*)In Technical editions of the engineer (French publications).

Part II
Uncertainty Quantification and Uncertainty Propagation Analysis

Likelihood-Based Approach for Uncertainty Quantification in Multi-Physics Systems

Shankar Sankararaman and Sankaran Mahadevan

Abstract This chapter presents a computational methodology for uncertainty quantification in multi-physics systems that require iterative analysis between models corresponding to each discipline of physics. This methodology is based on computing the probability of satisfying the inter-disciplinary compatibility equations, conditioned on specific values of the coupling (or feedback) variables, and this information is used to estimate the probability distributions of the coupling variables. The estimation of the coupling variables is analogous to likelihood-based parameter estimation in statistics and thus leads to the likelihood approach for multi-disciplinary analysis (LAMDA). Using the distributions of the feedback variables, the coupling can be removed in any one direction without loss of generality, while still preserving the mathematical relationship between the coupling variables. The calculation of the probability distributions of the coupling variables is theoretically exact and does not require a fully coupled system analysis. The LAMDA methodology is first illustrated using a mathematical example and then applied to the analysis of a fire detection satellite.

1 Introduction

Multi-physics or multi-disciplinary systems analysis and optimization is an extensive area of research, and numerous studies have dealt with the various aspects of coupled multi-disciplinary analysis (MDA) in several engineering

S. Sankararaman (✉)
SGT Inc., NASA Ames Research Center, Moffett Field, CA 94035, USA
e-mail: shankar.sankararaman@gmail.com

S. Mahadevan
Department of Civil and Environmental Engineering, Vanderbilt University,
Nashville, TN 37235, USA

S. Kadry and A. El Hami (eds.), *Numerical Methods for Reliability and Safety*   87
Assessment: Multiscale and Multiphysics Systems, DOI 10.1007/978-3-319-07167-1_2,
© Springer International Publishing Switzerland 2015

disciplines. Researchers have focused both on the development of computational methods (Alexandrov and Lewis 2000; Cramer et al. 1994) and on the application of these methods to several types of multi-physics interaction, for example, fluid–structure (Belytschko 1980), thermal–structural (Thornton 1996), fluid–thermal–structural (Culler et al. 2009), etc. Studies have considered these methods and applications either for MDA or for multi-disciplinary optimization (MDO). The coupling between individual disciplinary analyses may be one-directional (feed-forward) or bi-directional (feedback). Feed-forward coupling is straightforward to deal with, since the output of one model simply becomes an output to another (Sankararaman 2012). On the other hand, the topic of bi-directional (feedback) coupling is challenging to deal with, because the output of the first model becomes an input to the second while the output of the second model becomes an input to the first; therefore, it is necessary to iterate until convergence between these models in order to analyze the whole system. Such analysis needs to account for the different sources of uncertainty in order to accurately estimate the reliability and ensure the safety of the multi-physics system.

Computational methods for MDA can be classified into three different groups of approaches (Felippa et al. 2001). The first approach, known as the field elimination method (Felippa et al. 2001), eliminates one or more coupling variables (referred to as "field" in the literature pertaining to fluid–structure interaction) using reduction/elimination techniques such as integral transforms and model reduction. This approach is restricted to linear problems that permit efficient and evident coupling. The second approach, known as the monolithic method (Felippa et al. 2001; Michler et al. 2004), solves the coupled analysis simultaneously using a single solver (e.g. Newton–Raphson). The third approach, known as the partitioned method, solves the individual analyses separately with different solvers. The well-known fixed point iteration (FPI) approach (repeated analysis until convergence of coupling variables) and the staggered solution approach (Felippa et al. 2001; Park et al. 1977) are examples of partitioned methods. While the field elimination and monolithic methods tightly couple the multi-disciplinary analyses together, the partitioned method does not.

Two major types of methods have been pursued for MDO—single level approaches and multi-level approaches. Single level approaches (Cramer et al. 1994) include the multi-disciplinary feasible (MDF) approach (also called fully integrated optimization or the all-in-one approach), the all-at-once (AAO) approach (also called simultaneous analysis and design (SAND)), and the individual disciplinary feasible (IDF) approach. Multi-level approaches for MDO include collaborative optimization (Braun 1996; Braun et al. 1997), concurrent subspace optimization (Sobieszczanski-Sobieski 1988; Wujek et al. 1997), bi-level integrated system synthesis (Sobieszczanski-Sobieski et al. 2003), analytical target cascading (Kokkolaras et al. 2006; Liu et al. 2006), etc.

An important factor in the analysis and design of multi-disciplinary systems is the presence of uncertainty in the system inputs. It is necessary to account for the various sources of uncertainty in both MDA and MDO problems. The MDA problem focuses on uncertainty propagation to calculate the uncertainty in the outputs. In the MDO problem, the objective function and/or constraints may become stochastic if the inputs are random. The focus of the present chapter is only on uncertainty propagation in MDA and not on optimization.

While most of the aforementioned methods for deterministic MDA can easily be extended to non-deterministic MDA using Monte Carlo sampling (MCS), this may be computationally expensive due to repeated evaluations of disciplinary analyses. Hence, researchers have focused on developing more efficient alternatives. Gu et al. (2000) proposed worst case uncertainty propagation using derivative-based sensitivities. Kokkolaras et al. (2006) used the advanced mean value method for uncertainty propagation and reliability analysis, and this was extended by Liu et al. (2006) by using moment-matching and considering the first two moments. Several studies have focused on uncertainty propagation in the context of reliability analysis. Du and Chen (2005) included the disciplinary constraints in the most probable point (MPP) estimation for reliability analysis. Mahadevan and Smith (2006) developed a multi-constraint first-order reliability method (FORM) for MPP estimation. While all the aforementioned techniques are probabilistic, non-probabilistic techniques based on fuzzy methods (Zhang and Huang 2010), evidence theory (Agarwal et al. 2004), interval analysis (Li and Azarm 2008), etc. have also been studied for MDA under uncertainty.

Similar to MDA, methods for MDO under uncertainty have also been investigated by several researchers. Kokkolaras et al. (2006) extended the analytical target cascading approach to include uncertainty. A sequential optimization and reliability analysis (SORA) framework was developed by Du et al. (2008) by decoupling the optimization and reliability analyses. Chiralaksanakul and Mahadevan (2007) integrated solution methods for reliability-based design optimization with solution methods for deterministic MDO problems to address MDO under uncertainty. Smith (2007) combined the techniques in Mahadevan and Smith (2006) and Chiralaksanakul and Mahadevan (2007) for the design of aerospace structures. As mentioned earlier, the focus of this chapter is only on MDA under uncertainty, and therefore, aspects of MDO will not be discussed hereafter.

Review of the above studies reveals that the existing methods for MDA under uncertainty are either computationally expensive or based on several approximations. Computationally expense is incurred in the following ways:

1. Using deterministic MDA methods with MCS (Haldar and Mahadevan 2000) require several thousands of evaluations of the individual disciplinary analyses.
2. Non-probabilistic techniques (Agarwal et al. 2004; Li and Azarm 2008; Zhang and Huang 2010) use interval-analysis-based approaches and also require substantial computational effort. Further they are also difficult to interpret in the context of reliability analysis; this is an important consideration for MDO which may involve reliability constraints.

Approximations are introduced in the following manner:

1. Probability distributions are approximated with the first two moments (Du and Chen 2005; Kokkolaras et al. 2006; Liu et al. 2006; Mahadevan and Smith 2006).

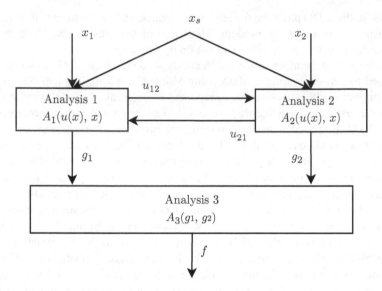

Fig. 1 A multi-disciplinary system

2. Approximations of individual disciplinary analyses may be considered using
   derivative-based sensitivities (Gu et al. 2000) or linearization at MPP for
   reliability calculation (Du and Chen 2005; Mahadevan and Smith 2006).

Some of these problems can be overcome by the use of a decoupled approach that
has been advocated by Du and Chen (2005) and Mahadevan and Smith (2006). In
this decoupled approach, Taylor's series approximation and the first-order second
moment (FOSM) method have been proposed to calculate the probability density
function (PDF) of the coupling variables.

For example, consider the multi-disciplinary system shown in Fig. 1. Here $x = \{x_1, x_2, x_s\}$ are the inputs, and $u(x) = \{u_{12}, u_{21}\}$ are the coupling variables. Note
that this is a not only a multi-disciplinary system, but also a multi-level system
where the outputs of the coupled analysis (g_1 and g_2) are used to compute a higher
level system output (f).

Once the PDFs of the coupling variables u_{12} and u_{21} are estimated using
the decoupled approach, the coupling between "Analysis 1" and "Analysis 2"
is removed. In other words, the variable u_{21} becomes an input to "Analysis 1"
and the variable u_{12} becomes an input to "Analysis 2," and the dependence
between the quantities u_{12}, u_{21}, and x is not considered any further. This "fully
decoupled" approach reduces the computational effort considerably by avoiding
repeated evaluations of the fully coupled system; however, this is still based on
approximations and more importantly, suitable only when the aim is to estimate the
statistics of g_1 or g_2.

In the case of a multi-level system, where the multi-disciplinary outputs (g_1 and
g_2 in this case) could be inputs to another model (Analysis 3 in Fig. 1), the fully

decoupled approach will not be applicable for the following reason. In Fig. 1, for a given x, there is a unique g_1, and a unique g_2; in addition, for a given u_{12}, there is a unique u_{21}, and hence for a given g_1, there is a unique g_2. This functional dependence between u_{12} and u_{21}, and hence between g_1 and g_2, cannot be ignored when estimating the probability distribution of f. In the fully decoupled approach, the functional dependence between u_{12} and u_{21} is not preserved in subsequent analysis; once the PDFs of u_{12} and u_{21} are estimated, independent samples of u_{12} and u_{21} are used to generate samples of g_1 (using only Analysis 1) and g_2 (using only Analysis 2) which in turn are used to compute the statistics of f. This will lead to an erroneous estimate of f, since g_1 and g_2 values are not related to each other as they should be in the original system. This "subsequent analysis" need not necessarily refer to a higher level output; this could even refer to an optimization objective which is computed based on the values of g_1 and g_2 (or even u_{12} and u_{21}). Thus, if the objective is only to get the statistics of g_1 and g_2 as considered in Du and Chen (2005) and Mahadevan and Smith (2006), then the fully decoupled approach is adequate. But if g_1 and g_2 are to be used in further analysis, then the one-to-one correspondence between u_{12} and u_{21} (and hence between g_1 and g_2) cannot be maintained in the fully decoupled approach. Hence, one would have to revert to the expensive Monte Carlo simulation (MCS) outside a deterministic MDA procedure to compute the statistics of the output f. Thus, it becomes essential to look for alternatives to the fully decoupled approach, especially when the complexity of the system increases.

In order to address the above challenges, Sankararaman and Mahadevan (2012) proposed a new likelihood-based approach for uncertainty propagation analysis in multi-level, multi-disciplinary systems. In this method, the probability of satisfying the inter-disciplinary compatibility is calculated using the principle of likelihood, which is then used to quantify the PDF of the coupling variables. This approach for MDA offers several advantages:

1. This method for the calculation of the PDF of the coupling variable is theoretically exact; the uncertainty in the inputs is accurately propagated through the disciplinary analyses in order to calculate the PDF of the coupling variable. No approximations of the individual disciplinary analyses or the moments of the coupling variable are necessary.
2. This approach requires no coupled system analysis, i.e. repeated iteration between individual disciplinary analyses until convergence (FPI), thereby improving the computational cost.
3. For multi-level systems, the difficulty in propagating the uncertainty in the feedback variables to the system output is overcome by replacing the feedback coupling with unidirectional coupling, thereby preserving the functional dependence between the individual disciplinary models. The direction of coupling can be chosen either way, without loss of generality. This semi-coupled approach is also useful in an optimization problem where the objective function is a function of the disciplinary outputs.

The goal of this chapter is to explain this likelihood-based methodology for uncertainty quantification in multi-physics systems in detail and illustrate its application through numerical examples. The rest of this chapter is organized as follows. Section 2 discusses a "sampling with optimization-based deterministic MDA" (SOMDA) approach, which is an example of using the partitioned method along with MCS. Certain ideas explained in this section are used to motivate the likelihood-based method. Then, the likelihood-based approach for multi-disciplinary analysis (LAMDA) is explained in Sect. 3 and its numerical implementation is discussed in Sect. 4. Section 5 illustrates the LAMDA methodology using a mathematical example and Sect. 6 uses the LAMDA methodology for a three-discipline analysis of a fire detection satellite (Zaman 2010).

2 Sampling with Optimization-Based Deterministic MDA

Consider the multi-disciplinary system shown earlier in Fig. 1. The overall goal is to estimate the probability distribution of the outputs g_1, g_2, and f, given the probability distributions of the inputs x. As explained in Sect. 1, an intermediate step is to calculate the PDFs of the coupling variables u_{12} and u_{21} and then use these PDFs for uncertainty propagation.

First consider the deterministic problem of estimating the converged u_{12} and u_{21} values corresponding to given values of x. The conventional FPI approach starts with an arbitrary value of u_{12} as input to "Analysis 2" and the resultant value of u_{21} serves as input to "Analysis 1." If the next output from "Analysis 1" is the same as the original u_{12}, then the analysis is said to have reached convergence and the inter-disciplinary compatibility is satisfied. However, if it is not, the conventional FPI approach treats the output of "Analysis 1" as input to "Analysis 2" and the procedure is repeated until convergence.

This search for the convergent values of u_{12} and u_{21} can be performed in an intelligent manner by formulating it as an optimization problem. For this purpose, define a new function G whose input is the coupling variable u_{12}, in addition to x. The output of "G" is denoted by U_{12}, which is obtained by propagating the input through "Analysis 2" followed by "Analysis 1," as shown in Fig. 2.

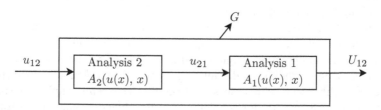

Fig. 2 Definition of G

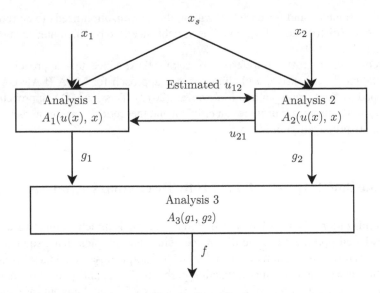

Fig. 3 A multi-disciplinary system: unidirectional coupling

The multi-disciplinary constraint is said to be satisfied if and only if $u_{12} = U_{12}$. For a given x, the convergent value of the coupling variable u_{12} can be obtained by minimizing the squared error $E = (u_{12} - G(u_{12}, x))^2$ for a given set of inputs x, where G is given by:

$$U_{12} = G(u_{12}, x) = A_1(u_{21}, x) \text{ where } u_{21} = A_2(u_{12}, x) \qquad (1)$$

Note that this is an unconstrained optimization problem. If the multi-disciplinary compatibility is satisfied, then $u_{12} = U_{12}$, and the optimum value of E will be equal to zero. In the rest of the chapter, it is assumed that it is possible to satisfy inter-disciplinary compatibility for each realization of the input x; in other words, the MDA has a feasible solution for each input realization. Once the converged value of u_{12} is estimated, then the bi-directional coupling can be removed and replaced with a uni-directional coupling from "Analysis 2" to "Analysis 1" as shown in Fig. 3.

If there are multiple coupling variables in one direction, i.e. if u_{12} is a vector instead of a scalar, then E is also a vector, i.e. $E = [E_1, E_2, E_3, \ldots E_n]$. If the MDA has a solution, then the optimal value of the vector u_{12} will lead to $E_i = 0$ for all i's. Since each $E_i = 0$ by definition, the optimal value of u_{12} can be estimated by minimizing the sum of all E_i's (instead of minimizing each E_i separately), and the minimum value of this sum will also be equal to zero.

This is a minor modification to the FPI approach; here the convergent value of the coupling variable is calculated based on an optimization which may choose iterations judiciously in comparison with the FPI approach. Hence, in terms of uncertainty propagation, the computational cost is still very high. The input values need to be sampled and for each realization, this optimization needs to be repeated and the entire distribution of the coupling variable needs to be calculated using many such samples.

Hereon, this approach is referred to as SOMDA. Since this approach is still computationally expensive, a likelihood-based approach for MDA (LAMDA) was developed by Sankararaman and Mahadevan (2012). This LAMDA approach does not require sampling and provides an efficient and theoretically accurate method for uncertainty propagation in MDA.

3   Likelihood Approach for Multi-Disciplinary Analysis

The optimization discussed in the previous section is much similar to a least-squares-based optimization; the difference being that a typical least squares optimization is posed as a summation problem with multiple observed data whereas this is not the case in the current optimization problem. The quantity to be estimated is the convergent value of u_{12} for a given set of inputs x. When the inputs are random, then the coupling variable u_{12} is also random and its probability distribution needs to be calculated. This can be viewed similar to a statistical parameter estimation problem.

Consider a typical parameter estimation problem where a generic parameter θ needs to be estimated based on some available data. According to Fisher (1912), one can "solve the real problem directly" by computing the "probability of observing the given data" conditioned on the parameter θ (Aldrich 1997; Fisher 1912). This quantity is referred to as the likelihood function of θ (Edwards 1984; Pawitan 2001). Singpurwalla (2006, 2007) explains that the likelihood function can be viewed as a collection of weights or masses and is meaningful up to a proportionality constant (Edwards 1984). In other words, if $L(\theta^{(1)}) = 10$, and $L(\theta^{(2)}) = 100$, then it is 10 ten times more likely for $\theta^{(2)}$ than $\theta^{(1)}$ to correspond to the observed data. While this likelihood function is commonly maximized to obtain the maximum likelihood estimate (MLE) of the parameter θ, the entire likelihood function can also be used to obtain the entire PDF of θ.

Now, consider the problem of estimating the PDF of the coupling variable u_{12} in MDA. This is purely an uncertainty propagation problem and there is no "data" to calculate the likelihood function of u_{12} which is defined as the "probability of observing the data." Hence, the definition of the likelihood function cannot be used directly.

However, the focus of the MDA problem is to satisfy the inter-disciplinary compatibility condition. Consider "the probability of satisfying the inter-disciplinary compatibility" conditioned on u_{12} which can be written as $P(U_{12} = u_{12}|u_{12})$. This definition is similar to the original definition of the likelihood function. It is a weight that is associated with a particular value of u_{12} to satisfy the multi-disciplinary constraint. In other words, if the ratio of $P(U_{12} = u_{12}^{(1)}|u_{12}^{(1)})$ to $P(U_{12} = u_{12}^{(2)}|u_{12}^{(2)})$

is equal to 0.1, then it is 10 ten times more likely for $u_{12}^{(2)}$ than $u_{12}^{(1)}$ to satisfy the inter-disciplinary compatibility condition. Thus, the properties of this expression are similar to the properties of the original likelihood function. Hence, this expression is defined to be the likelihood of u_{12} in this chapter, as shown in Eq. (2). Since the likelihood function is meaningful only up to a proportionality constant, Eq. (2) also uses only a proportionality sign.

$$L(u_{12}) \propto P(U_{12} = u_{12}|u_{12}) \tag{2}$$

Note that this definition is in terms of probability and hence the tool of likelihood gives a systematic procedure for including the uncertainty in the inputs during the construction of likelihood and estimating the probability distribution of the coupling variables, as explained below.

Note that there is a convergent value of u_{12} for every realization of x. If x is represented using a probability distribution, then one sample of x has a relative likelihood of occurrence with respect to another sample of x. Correspondingly, a given sample of u_{12} has a relative likelihood of being a convergent solution with respect to another sample of u_{12}, and hence u_{12} can be represented using a probability distribution. It is this likelihood function and the corresponding probability distribution that will be calculated using the LAMDA method.

For a given value of u_{12}, consider the operation $U_{12} = G(u_{12}, x)$ defined earlier in Eq. (1). When x is random, an uncertainty propagation method can be used to calculate the distribution of U_{12}. Let the PDF of U_{12} be denoted by $f_{U_{12}}(U_{12}|u_{12})$.

The aim is to calculate the likelihood of u_{12}, i.e. $L(u_{12})$ as the probability of satisfying the multi-disciplinary constraint, i.e. $U_{12} = u_{12}$. Since $f_{U_{12}}(U_{12}|u_{12})$ is a continuous PDF, the probability that U_{12} is equal to any particular value, u_{12} in this case, is equal to zero. Pawitan (2001) explained that this problem can be overcome by considering an infinitesimally small window $\left[u_{12} - \frac{\epsilon}{2}, u_{12} + \frac{\epsilon}{2}\right]$ around u_{12} by acknowledging that there is only limited precision in the real world.

$$L(u_{12}) \propto P(U_{12} = u_{12}|u_{12}) = \int_{u_{12}-\frac{\epsilon}{2}}^{u_{12}+\frac{\epsilon}{2}} f_{U_{12}}(U_{12}|u_{12})dU_{12} \propto f_{U_{12}}(U_{12} = u_{12}|u_{12})$$
$$\tag{3}$$

Note that this equation is very similar to the common practice of estimating the parameters of a probability distribution given observed data for the random variable. In other words, if X is a random variable whose PDF is given by $f_X(x|P)$ where P refers to the parameters to be estimated, and if the data available is denoted by x_i ($i = 1$ to n), then the likelihood of the parameters can be calculated as $L(P) \propto f_X(x_i|P)$ where i varies from 1 to n. The maximizer of this expression is referred to as the MLE of P. Details can be found in statistics textbooks (Haldar and Mahadevan 2000; Pawitan 2001).

Note that the likelihood function $L(u_{12})$ is conditioned on u_{12} and hence the PDF of U_{12} is always conditioned on u_{12}. Once the likelihood function of u_{12}, i.e the probability of satisfying the multi-disciplinary compatibility for a given value of u_{12}.

is calculated, the PDF of the converged value of the coupling variable u_{12} can be calculated as:

$$f(u_{12}) = \frac{L(u_{12})}{\int L(u_{12}) d u_{12}} \tag{4}$$

In the above equation, the domain of integration for the variable u_{12} is such that $L(u_{12}) \neq 0$. Note that Eq. (4) is a form of Bayes theorem with a non-informative uniform prior density for u_{12}. Once the PDF of u_{12} is calculated, the MDA with uni-directional coupling in Fig. 3 can be used in lieu of the MDA with bi-directional coupling in Fig. 1. The system output f can then be calculated using well-known methods of uncertainty propagation such as MCS, FORM, and second-order reliability method (SORM).

During the aforementioned uncertainty propagation, the converged u_{12} and x are considered as independent inputs in order to compute the uncertainty in u_{21}, g_1, g_2, and f. However, for every given value of x, there is only one value of u_{12}; this is not a statistical dependence but a functional dependence. The functional dependence between the converged u_{12} and x is not known and not considered in the decoupled approach. If the functional dependence needs to be explicitly considered, one would have to revert to the computationally expensive FPI approach for every sample of x. (An alternative would be to choose a few samples of x, run FPI analysis on each of them and construct a surrogate/approximation of the functional dependence between x and u_{12}, and explicitly use this surrogate in uncertainty propagation. Obviously, the surrogate could also be directly constructed for any of the responses—g_1, g_2, or f—instead of considering the coupling variable u_{12}. However, replacing the entire MDA by a surrogate model is a different approach and does not fall within the scope of the decoupled approach, which is the focus of this chapter.)

The above discussion calculated the PDF of u_{12} and cut the coupling from "Analysis 1" to "Analysis 2." Without loss of generality, the same approach can be used to calculate the PDF of u_{21} and cut the coupling from "Analysis 2" to "Analysis 1." This method has several advantages:

1. This method is free from first-order or second-order approximations of the coupling variables.
2. The equations of the individual disciplinary analyses are not approximated during the derivation of Eq. (3) and the calculation of the PDF of the coupling variables in Eq. (4) is exact from a theoretical perspective.
3. The method does not require any coupled system analysis, i.e. repeated iteration between "Analysis 1" and "Analysis 2" until convergence.

Though the computation of the PDF of u_{12} is theoretically exact, two issues need to be addressed in computational implementation. (1) The calculation of $L(u_{12})$ requires the estimation of $f_{U_{12}}(U_{12}|u_{12})$ which needs to be calculated by propagating the inputs x through G for a given value of u_{12}. (2) This likelihood function needs to be calculated for several values of u_{12} to perform the integration in

Eq. (4). These two steps, i.e. uncertainty propagation and integration, could make the methodology computationally expensive if a Monte Carlo-type approach is pursued for uncertainty propagation.

Therefore, the following section presents a methodology that makes the numerical implementation inexpensive for the above two steps. From here on, there are approximations made; note that these approximations are only for the purpose of numerical implementation and not a part of the mathematical theory. Here, "theory" refers to the derivation and use of Eqs. (3) and (4) for uncertainty quantification in MDA, and "implementation" refers to the numerical computation of $f_{U_{12}}(U_{12} = u_{12}|u_{12})$ in Eq. (3).

4 Numerical Implementation

This section addresses the two issues mentioned above in the numerical implementation of the LAMDA method.

4.1 Evaluation of the Likelihood Function $L(u_{12})$

The first task is to calculate the likelihood function $L(u_{12})$ for a given value of u_{12}. This requires the calculation of the PDF $f_{U_{12}}(U_{12}|u_{12})$. However it is not necessary to calculate the entire PDF. Based on Eq. (3), the calculation of likelihood $L(u_{12})$ only requires the evaluation of the PDF at u_{12}, i.e. $f_{U_{12}}(U_{12} = u_{12}|u_{12})$. Hence, instead of entirely evaluating the PDF $f_{U_{12}}(U_{12}|u_{12})$, only local analysis at $U_{12} = u_{12}$ needs to be performed. One method is to make use of FORM to evaluate this PDF value. This is the first approximation.

The FORM estimates the probability that a performance function $H = h(x)$ is less than or equal to zero, given uncertain input variables x. This probability is equal to the cumulative probability density (CDF) of the variable H evaluated at zero (Haldar and Mahadevan 2000). In this approach, the so-called MPP is calculated by transforming the variables x into uncorrelated standard normal space u and by determining the point in the transformed space that is closest to the origin. An optimization problem can be formulated as shown in Fig. 4.

Given PDFs of x

Minimize $\beta = u^T u$

such that $H \equiv h(x) = 0$

where standard normal $u = T(x)$

$$P(H < 0) = \Phi(-\beta)$$

Fig. 4 Use of FORM to estimate the CDF value

The details of the transformation $u = T(x)$ in Fig. 4 can be found in Haldar and Mahadevan (2000). This optimization can be solved by using the well-known Rackwitz–Fiessler algorithm (Rackwitz and Flessler 1978), which is based on a repeated linear approximation of the constraint $H = 0$. Once the shortest distance to the origin is estimated to be equal to β, then the CDF value is calculated in FORM as:

$$P(H \leq 0) = \Phi(-\beta) \tag{5}$$

FORM can also be used to calculate the CDF value at any generic value h_c, i.e. $P(h(x) \leq h_c)$ and the probability that $h(x)$ is less than or equal to h_c can be evaluated by executing the FORM analysis for the performance function $H = h(x) - h_c$. For the problem at hand, it is necessary to calculate the PDF value at u_{12} and not the CDF value. This can be accomplished by finite differencing, i.e. by performing two FORM analyses at $h_c = u_{12}$ and $h_c = u_{12} + \delta$, where δ is a small difference that can be chosen, for example, $0.001 \times u_{12}$. The resultant CDF values from the two FORM analyses are differenced and divided by δ to provide an approximate value of the PDF value at u_{12}. This is the second approximation.

Hence, the evaluation of the likelihood function $L(u_{12})$ is based on two approximations: (1) the PDF value is calculated based on finite differencing two CDF values; and (2) each CDF value is in turn calculated using FORM which is a first-order approximation (Eq. (5)).

4.2   Construction of PDF of u_{12}

Recall that Eq. (4) is used to calculate the PDF of u_{12} based on the likelihood function $L(u_{12})$. In theory, for any chosen value of u_{12}, the corresponding likelihood $L(u_{12})$ can be evaluated, and hence the integral in Eq. (4) can be computed. For the purpose of numerical implementation, the limits of integration need to be chosen. The first-order estimates of the mean and variance of u_{12} can be estimated by calculating the converged value of u_{12} at the mean of the uncertain input values using FPI. The derivatives of the coupling variables with respect to the inputs can be calculated using Sobieski's system sensitivity equations (Hajela et al. 1990), as demonstrated later in Sect. 4.1. These first order estimates can be then used to select the limits (for example, six sigma limits) for integration.

For the purpose of implementation, the likelihood function is evaluated only at a few points; a recursive adaptive version of Simpson's quadrature (McKeeman 1962) is used to evaluate this integral and the points at which the likelihood function needs to be evaluated are adaptively chosen until the quadrature algorithm converges.

This quadrature algorithm is usually applicable only in the case of one-dimensional integrals whereas in a typical multi-disciplinary problem, u_{12} may be a vector, where there are several coupling variables in each direction. Hence, the multi-dimensional integral can be decomposed into multiple one-dimensional integrals so that the quadrature algorithm may be applied.

$$\int L(\alpha, \beta)d\alpha d\beta = \int \left(\int L(\alpha, \beta)d\alpha \right) d\beta \tag{6}$$

Each one-dimensional integral is evaluated using recursive adaptive Simpson's quadrature algorithm (McKeeman 1962). Consider any general one-dimensional integral and its approximation using Simpson's rule as:

$$\int_a^b f(x)dx \approx \frac{b-a}{6} \left(f(a) + 4f\left(\frac{a+b}{2}\right) + f(b) \right) = S(a,b) \tag{7}$$

The adaptive recursive quadrature algorithm calls for subdividing the interval of integration (a,b) into two sub-intervals $((a,c)$ and (c,b), $a \leq c \leq b)$ and then, Simpson's rule is applied to each sub-interval. The error in the estimate of the integral is calculated by comparing the integral values before and after splitting. The criterion for determining when to stop dividing a particular interval depends on the tolerance level ϵ. The tolerance level for stopping may be chosen, for example as (McKeeman, 1962):

$$|S(a,c) + S(c,b) - S(a,b)| \leq 15\epsilon \tag{8}$$

Once the integral is evaluated, the entire PDF is approximated by interpolating the points at which the likelihood has already been evaluated.

This technique ensures that the number of evaluations of the individual disciplinary analyses is minimal. Would it be possible to approximately estimate the number of disciplinary analyses needed for uncertainty propagation? Suppose that the likelihood function is evaluated at ten points to solve the integration in Eq. (4). Each likelihood evaluation requires a PDF calculation, and hence two FORM analyses. Assume that the optimization for FORM converges in five iterations on average; each iteration would require $n + 1$ (where n is the number of input variables) evaluations of the individual disciplinary analysis (one evaluation for the function value and n evaluations for derivatives). Thus, the number of individual disciplinary analyses required will approximately be equal to $100(n + 1)$. This is computationally efficient when compared to existing approaches. For example, Mahadevan and Smith (2006) report that for a MDA with 5 input variables, the multi-constraint FORM approach required 69 evaluations for the evaluation of a single CDF value, which on average may lead to 690 evaluations for 10 CDF values. While the LAMDA method directly calculates the entire PDF, it also retains the functional dependence between the disciplinary analyses, thereby enabling uncertainty propagation to the next analysis level.

As the number of coupling variables increases, the integration procedure causes the computational cost to increase exponentially. For example, if there are ten coupling variables, each with five discretization points (for the sake of integration), then the number of individual disciplinary analyses required will approximately be equal to $105 \times 10 \times (n + 1)$. Alternatively, a sampling technique such as Markov Chain Monte Carlo (MCMC) sampling can be used to draw samples of the coupling variables; this method can draw samples of the coupling variable without evaluating the integration constant in Eq. (4). Further, since this is sampling approach, the computational cost does not increase exponentially with the number of coupling variables. In each iteration of the MCMC chain, two FORM analyses need to be conducted to evaluate the likelihood for a given value of u_{12} (which is now vector), and several thousands (say, Q) of evaluations of this likelihood function may be necessary for generating the entire PDFs of the coupling variables. Thus, the number of individual disciplinary analyses will be approximately equal to $10 \times (n + 1) \times Q$. Currently, the LAMDA method is demonstrated only for a small number of coupling variables. Future work needs to extend the methodology to field-type quantities (temperatures, pressures, etc. in finite element analysis) where the number of coupling variables is large.

5   Numerical Example: Mathematical MDA Problem

5.1   Description of the Problem

This problem consists of three analyses, two of which are coupled with one another. This is an extension of the problem discussed by Du and Chen (2005), and later by Mahadevan and Smith (2006) where only two analyses were considered. The functional relationships are shown in Fig. 5. In addition to the two analyses given in Mahadevan and Smith (2006), the current example considers a third analysis where a system output is calculated based on g_1 and g_2 as $f = g_2 - g_1$. All the five input quantities $x = (x_1, x_2, x_3, x_4, x_5)$ are assumed to be normally distributed (only for the sake of illustration) with unit mean and standard deviation equal to 0.1; there is no correlation between them. The goal in Du and Chen (2005) and Mahadevan and Smith (2006) was to calculate the probability $P(g_1 \leq 0)$, and now, the goal is to calculate the entire probability distributions of the coupling variables u_{12} and u_{21}, the outputs of the individual analyses g_1 and g_2, and the overall system output f.

A coarse approximation of the uncertainty in the output variables and coupling variables can be obtained in terms of first-order mean and variance using Taylor's series expansion (Haldar and Mahadevan 2000). For example, consider the coupling variable u_{12}; the procedure described for can be extended to u_{21}, g_1, g_2, and f.

The first-order mean of u_{12} can be estimated by calculating the converged value of u_{12} at the mean of the input values, i.e. $x = (1, 1, 1, 1, 1)$. The first-order mean

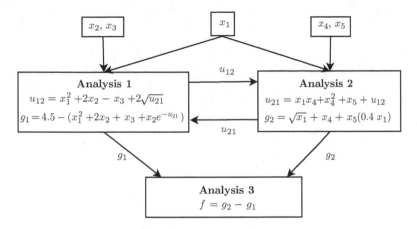

Fig. 5 Functional relationships

values of u_{12}, u_{21}, g_1, g_2, and f are calculated to be equal to 8.9, 11.9, 0.5, 2.4, and 1.9, respectively. The first-order variance of u_{12} can be estimated as:

$$\text{Var}(u_{12}) = \sum_{i=1}^{n} \left(\frac{d u_{12}}{d x_i}\right)^2 \text{Var}(x_i) \qquad (9)$$

where the first-order derivatives are calculated using Sobieski's system (or global) sensitivity equations (Hajela et al. 1990), by satisfying the multi-disciplinary compatibility as:

$$\frac{d u_{12}}{d x_i} = \frac{\partial u_{12}}{\partial x_i} + \frac{\partial u_{12}}{\partial u_{21}} \frac{\partial u_{21}}{\partial x_i} \qquad (10)$$

All the derivatives are calculated at the mean of the input values, i.e. $x = (1, 1, 1, 1, 1)$. The values of $\frac{\partial u_{12}}{\partial x_i}$ are 2, 2, -1, 0, and 0 ($i = 1$ to 5), respectively. The values of $\frac{\partial u_{21}}{\partial x_i}$ are 1, 0, 0, 3, and 1 ($i = 1$ to 5), respectively. The value of $\frac{\partial u_{12}}{\partial u_{21}}$ is $\frac{1}{\sqrt{u_{21}}}$, evaluated at the mean and therefore, is equal to 0.29. Hence, using Eqs. (9) and (10), the standard deviation of u_{12} is calculated to be 0.333.

The system sensitivity equation-based approach only provides approximations of the mean and variance, and it cannot calculate the entire PDF of u_{12}. The remainder of this section illustrates the LAMDA approach, which can accurately calculate the entire PDF of u_{12}. Though the system of equations in Fig. 5 may be solved algebraically by eliminating one variable, the forthcoming solution does not take advantage of this closed form solution and assumes each analysis to be a black-box. This is done to simulate the behavior of realistic multi-disciplinary analyses that may not have closed form solutions. For the same reason, finite differencing is used to approximate the gradients even though analytical derivatives can be calculated easily for this problem.

Fig. 6 PDF of u_{12}

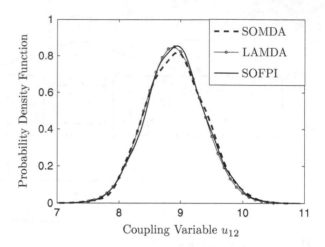

5.2   Calculation of the PDF of the Coupling Variable

In this numerical example, the coupling variable u_{12} is estimated for the sake of illustration, and the arrow from "Analysis 1" to "Analysis 2" is severed. The PDF of u_{12} is estimated using (1) SOMDA; and (2) LAMDA. In Fig. 6, the PDF using the LAMDA method uses ten integration points for the evaluation of Eq. (4). The resulting PDFs from the SOMDA method and the LAMDA method are compared with the benchmark solution which is estimated using 10,000 Monte Carlo samples of x and FPI (until convergence of Analysis 1 and Analysis 2) for each sample of x. The probability bounds on MCS results for the benchmark solution are also calculated using the formula $\mathrm{CoV}(F) = \sqrt{\frac{(1-F)}{F \cdot N}}$ where F is the CDF value (Haldar and Mahadevan 2000), and found to be narrow and almost indistinguishable from the solution reported in Fig. 6. Since the benchmark solution uses FPI for each input sample, it is indicated as SOFPI (sampling outside fixed point iteration) in Fig. 6.

In addition to the PDF in Fig. 6, the CDF of u_{12} is shown in Fig. 7. The CDF is plotted in linear and log-scale. Further, the tail probabilities are important in the context of reliability analysis; hence, the two tails of the CDF curves are also shown separately.

It is seen that the solutions (PDF values and CDF values) from the LAMDA method match very well with the benchmark (SOFPI) solution and the SOMDA approach. Note that the mean and standard deviation of the PDF in Fig. 6 agree well with the first-order approximations previously calculated (8.9 and 0.333). Obviously, the above PDF provides more information than the first-order mean and standard deviation and is more suitable for calculation of tail probabilities in reliability analysis.

The differences (maximum error is less than 1 %) seen in the PDFs and the CDFs from the three methods, though small, are accountable. The PDF obtained using SOMDA differs from the benchmark solution because it uses only 1,000 Latin

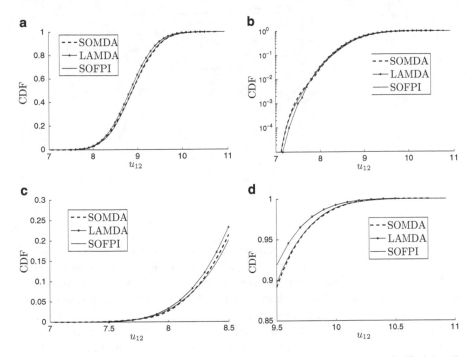

Fig. 7 Cumulative distribution function of u_{12}. (**a**) Linear; (**b**) log-scale; (**c**) left tail; (**d**) right tail

hypercube samples (realizations of inputs) whereas the benchmark solution used 10,000 samples. The PDF obtained using LAMDA differs from the benchmark solution because of two approximations—(1) finite differencing two CDF values to calculate the PDF value, and (2) calculating each CDF value using FORM.

The benchmark solution is based on FPI and required about 105 evaluations each of Analysis 1 and Analysis 2. The SOMDA method required 8,000–9,000 executions of each individual disciplinary analysis. (This number depends on the random samples of the input, since for each sample, the number of optimization iterations required for convergence is different.) Note that theoretically, the SOMDA method would produce a PDF that is identical to the benchmark solution if the same set of input samples were used in both the cases. This is because the SOMDA approach simply solves the deterministic MDA problem and then considers sampling in an outside loop. The solution approach in SOMDA is different from that in the benchmark solution approach; however, the treatment of uncertainty is the same. As discussed in Sect. 2, the SOMDA method is still expensive; replacing the brute force FPI in the benchmark solution by an optimization did not significantly improve the computational efficiency in this problem.

The LAMDA method treats the uncertainty directly in the definition of likelihood, and was found to be the least expensive, as it required only about 450–500 evaluations of each disciplinary analysis for the estimation of the entire PDF of

Fig. 8 PDF of f

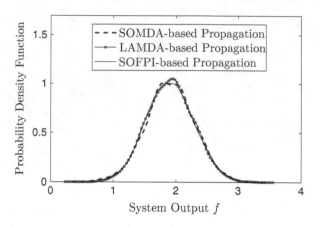

u_{12} in Fig. 6. The number of evaluations is given as a range because of three sources of variation: (1) different initial guesses for FORM analyses may require different numbers of function evaluations for convergence to MPP; (2) the number of integration points used for evaluation of Eq. (4); and (3) the actual values of the integration points used for evaluation of Eq. (4). In contrast, the multi-constraint FORM approach developed by Mahadevan and Smith (2006) required about 69 evaluations for the calculation of the CDF at one particular value. If the entire PDF as in Fig. 6 is desired, the multi-constraint FORM would take approximately $69 \times 2n$ function evaluations, where n is the number of points on the PDF and each PDF evaluation would require two CDF evaluations.

5.3   Calculation of PDF of the System Output

Once the PDF of u_{12} is calculated, the scheme in Fig. 3 can be used for uncertainty propagation and the PDF of the system output f is calculated. Note that this does not require any MDA (iterative analysis between the two subsystems) and it is now a simple uncertainty propagation problem. Well-known methods for uncertainty propagation such as MCS, FORM, and SORM (Haldar and Mahadevan 2000) can be used for this purpose. For the sake of illustration, MCS is used. The PDF of the system output f is shown in Fig. 8.

As the coupling variable u_{12} has been estimated here, the "arrow" from Analysis 1 to Analysis 2 alone is severed, whereas the arrow from Analysis 2 to Analysis 1 is retained. Hence, to solve for the system output f, the probability distributions of the inputs x and the probability distribution of the coupling variable u_{12} are used first in Analysis 2 (to calculate u_{21}), and then in Analysis 1 to calculate the individual disciplinary system outputs g_1 and g_2, followed by the overall system output f. As seen from Fig. 8, the solutions from the three different methods— SOMDA, LAMDA, and the benchmark solution (SOFPI)—compare well against each other.

6   Three-Discipline Fire Detection Satellite Model

This section illustrates the LAMDA methodology for system analysis of a satellite that is used to detect forest fires. First, the various components of the satellite system model are described, and then, numerical results are presented.

6.1   Description of the Problem

This problem was originally described by Wertz and Larson (1999). This is a hypothetical but realistic spacecraft consisting of a large number of subsystems with both feedback and feed-forward couplings. The primary objective of this satellite is to detect, identify, and monitor forest fires in near real time. This satellite is intended to carry a large and accurate optical sensor of length 3.2 m, weight 720 kg and has an angular resolution of 8.8×10^{-7} rad. This example considers a modified version of this problem considered earlier by Ferson et al. (2009) and Zaman (2010).

Zaman (2010) considered a subset of three subsystems of the fire detection satellite, consisting of (1) Orbit Analysis, (2) Attitude Control, and (3) Power, based on Ferson et al. (2009). This three-subsystem problem is shown in Fig. 9. There are nine random variables in this problem, as indicated in Fig. 9.

As seen in Fig. 9, the Orbit subsystem has feed-forward coupling with both Attitude Control and Power subsystems, whereas the Attitude Control and Power subsystems have feedback or bi-directional coupling through three variables P_{ACS}, $I_{\min}$, and $I_{\max}$. A satellite configuration is assumed in which two solar panels extend out from the spacecraft body. Each solar panel has dimensions L by W and the inner edge of the solar panel is at a distance D from the centerline of the satellite's body as shown in Fig. 10.

The functional relationships between the three subsystems are developed in detail by Wertz and Larson (1999) and summarized by Ferson et al. (2009) and Sankararaman and Mahadevan (2012). These functional relationships are briefly described in this section.

6.1.1   The Orbit Subsystem

The inputs to this subsystem are: radius of the earth (R_E); orbit altitude (H); earth's standard gravitational parameter (μ); and target diameter (ϕ_{target}).

The outputs of this subsystem are: satellite velocity (v); orbit period ($\Delta t_{\mathrm{orbit}}$); eclipse period ($\Delta t_{\mathrm{eclipse}}$); and maximum slewing angle (θ_{slew}). The relationships between these variables are summarized in the following equations:

$$v = \sqrt{\frac{\mu}{R_E + H}} \qquad (11)$$

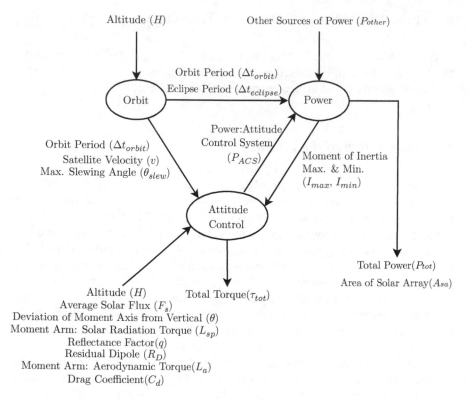

Fig. 9 A three-subsystem fire detection satellite

Fig. 10 Schematic diagram
for the satellite solar array

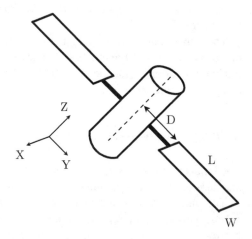

$$\Delta t_{\text{orbit}} = 2\pi \sqrt{\frac{(R_E + H)^3}{\mu}} = \frac{2\pi(R_E + H)}{v} \tag{12}$$

$$\Delta t_{\text{eclipse}} = \frac{\Delta t_{\text{orbit}}}{\pi} \arcsin\left(\frac{R_E}{R_E + H}\right) \tag{13}$$

$$\theta_{\text{slew}} = \arctan\left(\frac{\sin\left(\frac{\phi_{\text{target}}}{R_E}\right)}{1 - \cos\left(\frac{\phi_{\text{target}}}{R_E}\right) + \frac{H}{R_E}}\right) \tag{14}$$

6.1.2 The Attitude Control Subsystem

The 23 inputs to this subsystem are: earth's standard gravitational parameter (μ); radius of the earth (R_E); Altitude (H); maximum and minimum moment of inertia of the spacecraft ($I_{\max}$ and $I_{\min}$); deviation of major moment axis from local vertical (θ); moment arm for the solar radiation torque (L_{sp}); average solar flux (F_s); speed of light (c); reflectance factor (q); surface area off which solar radiation is reflected (A_s); Slewing time period (Δt_{slew}); magnetic moment of the Earth (M); residual dipole of the spacecraft (R_D); moment arm for aerodynamic torque (L_a); atmospheric density (ρ); maximum slewing angle (θ_{slew}); sun incidence angle (i); drag coefficient (C_d); cross-sectional surface area in the direction of flight (A); satellite velocity (v); rotation velocity of reaction wheel ($\omega_{\max}$); number of reaction wheels (n); and holding power (P_{hold}), i.e. the power required to maintain the constant velocity ($\omega_{\max}$).

The overall output of this subsystem is the total torque (τ_{tot}). The value of the total torque is computed based on slewing torque (τ_{slew}), disturbance torque (τ_{dist}), gravity gradient torque (τ_g), solar radiation torque (τ_{sp}), magnetic field interaction torque (τ_m), and aerodynamic torque (τ_a), as shown in the following equations.

$$\tau_{\text{tot}} = \max(\tau_{\text{slew}}, \tau_{\text{dist}}) \tag{15}$$

$$\tau_{\text{slew}} = \frac{4\theta_{\text{slew}}}{(\Delta t_{\text{slew}})^2} I_{\max} \tag{16}$$

$$\tau_{\text{dist}} = \sqrt{\tau_g^2 + \tau_{sp}^2 + \tau_m^2 + \tau_a^2} \tag{17}$$

$$\tau_g = \frac{3\mu}{2(R_E + H)^3} |I_{\max} - I_{\min}| \sin(2\theta) \tag{18}$$

$$\tau_{sp} = L_{sp} \frac{F_s}{C} A_s (1 + q) \cos(i) \tag{19}$$

$$\tau_m = \frac{2 M R_D}{(R_E + H)^3} \tag{20}$$

$$\tau_a = \frac{1}{2} L_a \rho C_d A v^2 \tag{21}$$

Note that this subsystem takes two coupling variables (I_{max} and I_{min}) as input and produces another coupling variable (Attitude control power: P_{ACS}) as output, as given in the following equation.

$$P_{ACS} = \tau_{tot} \omega_{max} + n P_{hold} \tag{22}$$

This coupling variable is an input to the power subsystem, as described in the following subsection.

6.1.3 The Power Subsystem

The 16 inputs to the power subsystem are: attitude control power (P_{ACS}); other sources of power (P_{other}); orbit period (Δt_{orbit}); eclipse period ($\Delta t_{eclipse}$); sun incidence angle (i); inherent degradation of the array (I_d); average solar flux (F_s); power efficiency (η); lifetime of the spacecraft (L_T); degradation in power production capability in % per year (ϵ_{deg}); length to width ratio of solar array (r_{lw}); number of solar arrays (n_{sa}); average mass density of solar arrays (ρ_{sa}); thickness of solar panels (t); distance between the panels (D); and moments of inertia of the main body of the spacecraft (I_{bodyX}, I_{bodyY}, I_{bodyZ}).

The overall outputs of this subsystem are the total power (P_{tot}), and the total size of the solar array (A_{sa}), as calculated below.

$$P_{tot} = P_{ACS} + P_{other} \tag{23}$$

Let P_e and P_d denote the spacecraft's power requirements during eclipse and daylight, respectively. For the sake of illustration, it is assumed that $P_e = P_d = P_{tot}$. Let T_e and T_d denote the time per orbit spent in eclipse and in sunlight, respectively. It is assumed that $T_e = \Delta t_{eclipse}$ and $T_d = \Delta t_{orbit} - T_e$. Then the required power output (P_{sa}) is calculated as:

$$P_{sa} = \frac{\left(\frac{P_e T_e}{0.6} + \frac{P_d T_d}{0.8} \right)}{T_d} \tag{24}$$

The power production capabilities at the beginning of life (P_{BOL}) and at the end of the life (P_{EOL}) are calculated as:

$$P_{BOL} = \eta F_s I_d \cos(i)$$
$$P_{EOL} = P_{BOL}(1 - \epsilon_{deg})^{LT} \tag{25}$$

The total solar array size, i.e. the second output of this subsystem, is calculated as:

$$A_{sa} = \frac{P_{sa}}{P_{EOL}} \tag{26}$$

Note that this subsystem takes a coupling variable (P_{ACS}) as input and produces the other two coupling variables (I_{max} and I_{min}) as output, to be fed into the attitude control subsystem described earlier.

The length (L), width (W), mass (m_{sa}), moments of inertia (I_{saX}, I_{saY}, I_{saZ}) of the solar array are calculated as follows:

$$L = \sqrt{\frac{A_{sa} r_{lw}}{m_{sa}}}$$

$$W = \sqrt{\frac{A_{sa}}{r_{lw} m_{sa}}} \tag{27}$$

$$m_{sa} = 2\rho_{sa} L W t$$

$$I_{saX} = m_{sa}\left[\frac{1}{12}(L^2 + t^2) + \left(D + \frac{L}{2}\right)^2\right] \tag{28}$$

$$I_{saY} = \frac{m_{sa}}{12}(t^2 + W^2) \tag{29}$$

$$I_{saZ} = m_{sa}\left[\frac{1}{12}(L^2 + W^2) + \left(D + \frac{L}{2}\right)^2\right] \tag{30}$$

The total moment of inertia (I_{tot}) can be computed in all three directions (X, Y, and Z), from which the maximum and the minimum moments of inertia (I_{max} and I_{min}) can be computed.

$$I_{tot} = I_{sa} + I_{body} \tag{31}$$

$$I_{max} = \max(I_{totX}, I_{totY}, I_{totZ}) \tag{32}$$

$$I_{min} = \min(I_{totX}, I_{totY}, I_{totZ}) \tag{33}$$

Table 1 List of deterministic quantities

Variable	Symbol	Unit	Numerical value
Earth's radius	R_E	m	6,378,140
Gravitational parameter	μ	$\mathrm{m^3\,s^{-2}}$	3.986×10^{14}
Target diameter	ϕ_{target}	m	235,000
Light speed	c	$\mathrm{m\,s^{-1}}$	2.9979×10^8
Area reflecting radiation	A_s	$\mathrm{m^2}$	13.85
Sun incidence angle	i	$^\circ$	0
Slewing time period	Δt_{slew}	s	760
Magnetic moment of earth	M	$\mathrm{A\,m^2}$	7.96×10^{15}
Atmospheric density	ρ	$\mathrm{kg\,m^{-3}}$	5.1480×10^{-11}
Cross-sectional in flight direction	A	$\mathrm{m^2}$	13.85
No. of reaction wheels	n	–	3
Maximum velocity of a Wheel	ω_{max}	rpm	6,000
Holding power	P_{hold}	W	20
Inherent degradation of array	I_d	–	0.77
Power efficiency	η	–	0.22
Lifetime of spacecraft	LT	Years	15
Degradation in power production capability	ϵ_{deg}	% per year	0.0375
Length to width ratio of solar array	r_{lw}	–	3
Number of solar arrays	n_{sa}	–	3
Average mass density to arrays	ρ_{sa}	$\mathrm{kg\,m^3}$	700
Thickness of solar panels	t	m	0.005
Distance between panels	D	m	2
Moments of inertia of spacecraft body	I_{body}	$\mathrm{kg\,m^2}$	$I_{\mathrm{body},X} = 4,700$
			$I_{\mathrm{body},Y} = 6,200$
			$I_{\mathrm{body},Z} = 4,700$

6.2 Numerical Details

Some of the input quantities are chosen to be stochastic while others are chosen to be deterministic. Table 1 provides the numerical details for the deterministic quantities and Table 2 provides the numerical details for the stochastic quantities. All the stochastic quantities are treated to be normally distributed, for the sake of illustration.

6.3 Uncertainty Propagation Problem

As seen in Fig. 9, this is a three-disciplinary analysis problem, with feedback coupling between two disciplines "power" and "attitude control." It is required to compute the uncertainty in three system output variables—total power P_{tot}, required solar array area A_{sa}, and total torque τ_{tot}.

Table 2 List of stochastic quantities

Variable	Symbol	Unit	Mean	Standard deviation
Altitude	H	m	18,000,000	1,000,000
Power other than ACS	P_{other}	W	1,000	50
Average solar flux	F_s	W/m^2	1,400	20
Deviation of moment axis	θ	$^\circ$	15	1
Moment arm for radiation torque	L_{sp}	m	2	0.4
Reflectance factor	q	–	0.5	1
Residual dipole of spacecraft	R_D	A m^2	5	1
Moment arm for aerodynamic torque	L_a	m	2	0.4
Drag coefficient	C_d	–	1	0.3

Prior to the quantification of the outputs, the first step is the calculation of the probability distribution of the coupling variables. The functional dependency can be severed in either direction, either from "power" to "attitude control" or from "attitude control" to "power," and this choice can be made without loss of generality. The probability distribution of P_{ACS}, i.e. the power of the attitude control system is chosen for calculation, and then, P_{ACS} becomes an independent input to the "power subsystem"; the functional dependency between "power" to "attitude control" is retained through the two coupling variables in the opposite direction. The following subsections present these results; Sect. 6.4 calculates the PDF of the feedback variable P_{ACS} and Sect. 6.5 calculates the PDFs of the system outputs.

6.4   Calculation of PDF of the Coupling Variable

Similar to the mathematical example presented in Sect. 5, this section calculates the PDF of the coupling variable P_{ACS} using sampling with SOMDA and the LAMDA. These results are compared with the benchmark solution in Fig. 11. In Fig. 11, the PDF using the LAMDA method uses ten integration points for the evaluation of Eq. (4).

Similar to the mathematical example in Sect. 5, it is seen from Fig. 11 that the results from SOMDA and LAMDA compare well with the benchmark solution (SOFPI). In addition to the PDFs, the CDFs and the tail probabilities are also in reasonable agreement. The benchmark solution is based on FPI and required about 200,000 evaluations each of the power subsystem and the attitude control subsystem. The SOMDA method required about 20,000 evaluations whereas the LAMDA method required about 900–1,000 evaluations. It is clear that the LAMDA approach provides an efficient and accurate alternative to sampling-based approaches.

Fig. 11 PDF of coupling
variable P_{ACS}

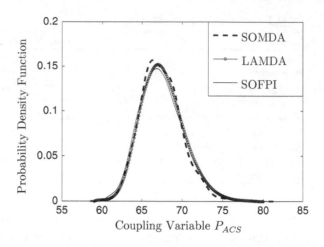

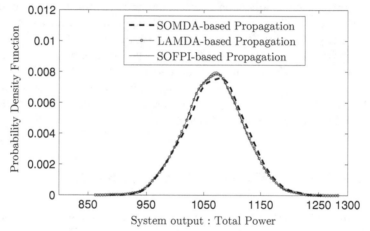

Fig. 12 PDF of total output power P_{tot}

6.5 Calculation of PDFs of the System Outputs

Once the probability distribution of the coupling variable P_{ACS} is calculated, the
system does not contain any feedback coupling and hence, methods for simple
forward uncertainty propagation can be used to estimate the PDFs of the three
system outputs total power (P_{tot}), required solar array area (A_{sa}), and total torque
(τ_{tot}). MCS is used for uncertainty propagation, and the resulting PDFs are plotted
in Figs. 12, 13, and 14.

As seen from Figs. 12, 13, and 14, the PDFs of the system outputs obtained
using both SOMDA and LAMDA compare very well with the benchmark solution
(SOFPI).

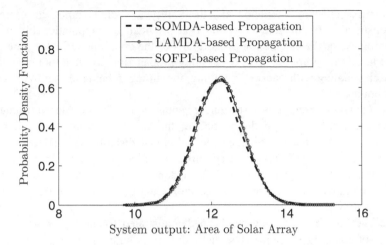

Fig. 13 PDF of area of solar array

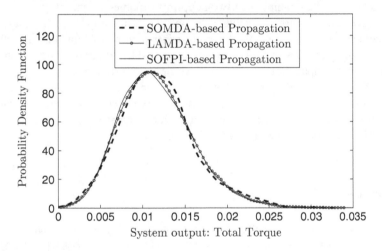

Fig. 14 PDF of total torque τ_{tot}

7   Conclusion

Existing methods for uncertainty propagation in multi-disciplinary system models are based on (1) MCS around FPI, which is computationally expensive; and/or (2), approximating the system equations; and/or (3) approximating the probability distributions of the coupling variables and then decoupling the disciplinary analyses. The fully decoupled approach does not preserve one-to-one correspondence between the individual disciplinary analyses and is not suitable for further downstream analysis using the converged MDA output.

The perspective of likelihood and the ability to include input uncertainty in the construction of the likelihood function provided a computationally efficient methodology for the calculation of the PDFs of the coupling variables. The MDA was reduced to a simple forward uncertainty propagation problem by replacing the feedback coupling with one-way coupling, the direction being chosen without loss of generality.

The LAMDA method has several advantages. (1) It provides a framework for the exact calculation of distribution of the coupling variables. (2) It retains the functional dependence between the individual disciplinary analyses, thereby utilizing the estimated PDFs of the coupling variables for uncertainty propagation, especially for downstream analyses. (3) It does not require any coupled system analysis (iterative analyses between the individual disciplines until convergence) for uncertainty propagation.

The LAMDA methodology has been demonstrated for problems with a small number of coupling variables. The methodology is straightforward to implement when there is a vector of coupling variables as explained earlier in Sect. 4.2. (Recall that the fire satellite example had two coupling variables in one of the directions.) However, if the coupling variable is a field-type quantity (e.g., pressures and displacements exchanged in a fluid–structure interaction problem at the interface of two disciplinary meshes), further research is needed to extend the LAMDA method for uncertainty propagation in such multi-disciplinary problems.

The likelihood-based approach can be extended to address MDO under uncertainty. Further, this chapter considered only aleatory uncertainty (natural variability) in the inputs. Future research may include different types of epistemic uncertainty such as data and model uncertainty in MDA and optimization.

References

Agarwal H, Renaud JE, Preston EL, Padmanabhan D (2004) Uncertainty quantification using evidence theory in multidisciplinary design optimization. Reliab Eng Syst Saf 85(1):281–294

Aldrich J (1997) R.A. Fisher and the making of maximum likelihood 1912–1922. Stat Sci 12(3):162–176

Alexandrov NM (2000) Lewis RM algorithmic perspectives on problem formulations in mdo. In: Proceedings of the 8th AIAA/USAF/NASA/ISSMO symposium on MA & O, Long Beach. AIAA, Reston

Belytschko T (1980) Fluid-structure interaction. Comput Struct 12(4):459–469

Braun RD (1996) Collaborative optimization: an architecture for large-scale distributed design. Ph.D. thesis, Stanford University

Braun RD, Moore AA, Kroo IM (1997) Collaborative approach to launch vehicle design. J Spacecr Rockets 34(4):478–486

Chiralaksanakul A, Mahadevan S (2007) Decoupled approach to multidisciplinary design optimization under uncertainty. Optim Eng 8(1):21–42

Cramer EJ, Dennis JE Jr, Frank PD, Lewis RM, Shubin GR (1994) Problem formulation for multidisciplinary optimization. SIAM J Optim 4:754–776

Culler AJ, Crowell AR, McNamara JJ (2009) Studies on fluid-structural coupling for aerothermoelasticity in hypersonic flow. AIAA J 48:1721–1738

Du X, Chen W (2005) Collaborative reliability analysis under the framework of multidisciplinary systems design. Optim Eng 6(1):63–84

Du X, Guo J, Beeram H (2008) Sequential optimization and reliability assessment for multidisciplinary systems design. Struct Multidiscip Optim 35(2):117–130

Edwards AWF (1984) Likelihood. Cambridge University Press, Cambridge

Felippa CA, Park KC, Farhat C (2001) Partitioned analysis of coupled mechanical systems. Comput Methods Appl Mech Eng 190(24–25):3247–3270

Ferson S, Tucker W, Paredis C, Bulgak Y, Kreinovich V (2009) Accounting for epistemic and aleatory uncertainty in early system design. Technical report, NASA SBIR

Fisher RA (1912) On an absolute criterion for fitting frequency curves. Messenger Math 41(1):155–160

Gu X, Renaud JE, Batill SM, Brach RM, Budhiraja AS (2000) Worst case propagated uncertainty of multidisciplinary systems in robust design optimization. Struct Multidiscip Optim 20(3):190–213

Hajela P, Bloebaum CL, Sobieszczanski-Sobieski J (1990) Application of global sensitivity equations in multidisciplinary aircraft synthesis. J Aircr (USA) 27:1002–1010

Haldar A, Mahadevan S (2000) Probability, reliability, and statistical methods in engineering design. Wiley, New York

Kokkolaras M, Mourelatos ZP Papalambros PY (2006) Design optimization of hierarchically decomposed multilevel systems under uncertainty. J Mech Des 128:503

Li M, Azarm S (2008) Multiobjective collaborative robust optimization with interval uncertainty and interdisciplinary uncertainty propagation. J Mech Des 130:081402

Liu H, Chen W, Kokkolaras M, Papalambros PY, Kim HM (2006) Probabilistic analytical target cascading: a moment matching formulation for multilevel optimization under uncertainty. J Mech Des 128:991

Mahadevan S, Smith N (2006) Efficient first-order reliability analysis of multidisciplinary systems. Int J Reliab Saf 1(1):137–154

McKeeman WM (1962) Algorithm 145: adaptive numerical integration by simpson's rule. Commun ACM 5(12):604

Michler C, Hulshoff SJ, Van Brummelen EH, De Borst R (2004) A monolithic approach to fluid–structure interaction. Comput Fluids 33(5):839–848

Park KC, Felippa CA, DeRuntz JA (1977) Stabilization of staggered solution procedures for fluid-structure interaction analysis. Comput Methods Fluid Struct Interact Problem 26:95–124

Pawitan Y (2001) In all likelihood: statistical modelling and inference using likelihood. Oxford University Press, Oxford

Rackwitz R, Flessler B (1978) Structural reliability under combined random load sequences. Comput Struct 9(5):489–494

Sankararaman S (2012) Uncertainty quantification and integration in engineering systems. Ph.D. thesis, Vanderbilt University

Sankararaman S, Mahadevan S (2012) Likelihood-based approach to multidisciplinary analysis under uncertainty. J Mech Des 134:031008

Singpurwalla ND (2006) Reliability and risk: a Bayesian perspective, vol 637. Wiley, Hoboken

Singpurwalla ND (2007) Betting on residual life: the caveats of conditioning. Stat Probab Lett 77(12):1354–1361

Smith N (2007) Probabilistic design of multidisciplinary systems. Ph.D. thesis, Vanderbilt University, Nashville

Sobieszczanski-Sobieski J (1988) Optimization by decomposition: a step from hierarchic to non-hierarchic systems. Recent Adv Multidiscip Anal Optim 1:51–78

Sobieszczanski-Sobieski J, Altus TD, Phillips M, Sandusky R (2003) Bilevel integrated system synthesis for concurrent and distributed processing. AIAA J 41(10):1996–2003

Thornton EA (1996) Thermal structures for aerospace applications. AIAA, Reston

Wertz JR, Larson WJ (1999) Space mission analysis and design. Microcosm Inc., Torrance

Wujek BA, Renaud JE, Batill SM (1997) A concurrent engineering approach for multidisciplinary design in a distributed computing environment. In: Multidisciplinary design optimization: state-of-the-art. Proceedings in applied mathematics, vol 80, pp 189–208

Zaman KB (2010) Modeling and management of epistemic uncertainty for multidisciplinary system analysis and design. Ph.D. thesis, Vanderbilt University, Nashville

Zhang X, Huang HZ (2010) Sequential optimization and reliability assessment for multidisciplinary design optimization under aleatory and epistemic uncertainties. Struct Multidiscip Optim 40(1):165–175

Bayesian Methodology for Uncertainty Quantification in Complex Engineering Systems

Shankar Sankararaman and Sankaran Mahadevan

Abstract This chapter presents a Bayesian methodology for system-level uncertainty quantification and test resource allocation in complex engineering systems. The various component, subsystem, and system-level models, the corresponding parameters, and the model errors are connected efficiently using a Bayesian network. This provides a unified framework for uncertainty analysis where test data can be integrated along with computational models across the entire hierarchy of the overall engineering system. The Bayesian network is useful in two ways: (1) in a forward problem where the various sources of uncertainty are propagated through multiple levels of modeling to predict the overall uncertainty in the system-level response; and (2) in an inverse problem where the model parameters of multiple subsystems are calibrated simultaneously using test data. Test data available at multiple data are first used to infer model parameters, and then, this information is propagated through the Bayesian network to compute the overall uncertainty in the system-level prediction. Then, the Bayesian network is used for test resource allocation where an optimization-based procedure is used to identify tests that can effectively reduce the overall uncertainty in the system-level prediction. Finally, the overall Bayesian methodology for uncertainty quantification and test resource allocation is illustrated using three different numerical examples. While the first example is mathematical, the second and third examples deal with practical applications in the domain of structural mechanics.

S. Sankararaman (✉)
SGT Inc., NASA Ames Research Center, Moffett Field, CA 94035, USA
e-mail: shankar.sankararaman@gmail.com

S. Mahadevan
Department of Civil and Environmental Engineering, Vanderbilt University, Nashville, TN 37235, USA

S. Kadry and A. El Hami (eds.), *Numerical Methods for Reliability and Safety Assessment: Multiscale and Multiphysics Systems*, DOI 10.1007/978-3-319-07167-1_3, © Springer International Publishing Switzerland 2015

1 Introduction

During the past 20 years, there has been an increasing need and desire to design and build engineering systems with increasingly complex architectures and new materials. These systems can be multi-level, multi-scale, and multi-disciplinary in nature, and may need to be decomposed into simpler components and subsystems to facilitate efficient model development, analysis, and design. The development and implementation of computational models is not only sophisticated and expensive, but also based on physics which is often not well understood. Therefore, when such models are used to design and analyze complex engineering systems, it is necessary to ensure their reliability and safety.

In order to facilitate efficient analysis and design, computational models are developed at the component-level, subsystem-level, and system-level. Each individual model may correspond to isolated features, or isolated physics, or simplified geometry of the original system. Typically, along the hierarchy of a multi-level system, the complexity of the governing physics increases, and hence, the complexity of the model increases, the cost of testing increases, and hence, the amount of available experimental data decreases. At the system-level, full-scale testing may not even be possible to predict the system performance under actual operating conditions. It is essential to quantify the overall uncertainty in the system-level prediction using the models and data available at all levels. The field of "quantification of margins and uncertainties" (QMU) has the goal of enabling this overall capability (Helton and Pilch 2011). This analysis is helpful to estimate the reliability and adjudge the safety of the overall engineering system.

An important challenge in this regard is to efficiently connect all of the models and experimental data available across the hierarchy of the entire system. This is not straightforward because there are several sources of uncertainty—physical variability, data uncertainty, and model uncertainty—at each level of the overall system. Further, the issue is complicated by the presence of individual model inputs, parameters, outputs, and model errors, all of which may be uncertain. It is important to use a computational approach that can not only facilitate integration across multiple levels but also provide a fundamental framework for the treatment of uncertainty.

This can be accomplished through the use of a Bayesian network that serves as an efficient and powerful tool to integrate multiple levels of models (including inputs, parameters, outputs, and errors of each and every model), the associated sources of uncertainty, and the experimental data at all different levels of hierarchy. The Bayesian network is based on Bayes' theorem, and can efficiently integrate all of the aforementioned information using the principles of conditional probability and total probability. It can be used for uncertainty propagation (forward problem), where the system-level prediction uncertainty is quantified by propagating all the sources of uncertainty at lower levels through the Bayesian network. The Bayesian network is also useful for model parameter calibration (inverse problem), where the data available at all levels can be simultaneously used to calibrate the underlying parameters at different levels of models. Different types of sampling techniques can be used in conjunction with a Bayesian network; while Monte Carlo-sampling-based approaches are often used for uncertainty propagation (forward problem), Markov Chain Monte Carlo-based approaches are often used for model parameter estimation (inverse problem). Since sampling methods require several hundreds of thousands of evaluations of the system models that may be complicated to evaluate, the Bayesian approach can also include cost-efficient surrogate models like Gaussian process, as demonstrated by Kennedy and O'Hagan (2001). This chapter explains the fundamentals of Bayesian methodology and illustrates the use of Bayesian networks for uncertainty quantification in complex multi-level, multi-physics engineering systems. Methods for uncertainty quantification, model parameter estimation, and system-level uncertainty quantification are presented in detail.

Finally, it is also illustrated as to how these capabilities of Bayesian networks can be exploited to guide test resource allocation in hierarchical systems (Sankararaman et al. 2013). Test data available at multiple levels of system hierarchy are used for model parameter calibration, which in turn leads to a reduction of uncertainty in the model parameters; this reduced uncertainty is represented through the posterior distributions of the model parameters. When these posterior distributions are propagated through the Bayesian network, the uncertainty in the system-level response also decreases. Thus, testing can be used to evolve the system performance prediction and a common concern is to select that test design which leads to maximum reduction in the uncertainty (usually expressed through variance) of the system performance prediction. The tests need to be selected and designed with adequate precision (measurement error and resolution), and the simulations need to be developed with adequate resolution (model fidelity) to achieve the project requirements. This can be performed by embedding the Bayesian network within an optimization algorithm where the decision variables correspond to the test data. This formulation is very interesting because, in the past, model parameters have been calibrated with available test data; the difference now is that it is required to perform Bayesian calibration and assess the reduction in uncertainty in the system-level response even before actual testing is performed.

Two types of questions are of interest: (1) what tests to do? and (2) how many tests to do? Tests at different levels of the system hierarchy have different costs and variance reduction effects. Hence, the test selection is not trivial and it is necessary to identify an analytical procedure that helps in the optimum test resource allocation. However, current practice for this is, at best, based on simplified analysis and relevant experience, and at worst based on ad hoc rules, any of which may or may not result in truly conservative estimates of the margin and uncertainty. For multi-level systems, a rational test selection procedure should also incorporate the information from component-level and subsystem-level tests towards overall system level performance prediction. Recently, Sankararaman et al. (2013) developed an optimization-based methodology to identify the tests that will lead to maximum reduction in the system-level uncertainty, while simultaneously minimizing the cost of testing. This methodology is presented in detail, towards the end of this chapter, and illustrated using numerical examples.

2   Fundamentals of Bayesian Methodology

The Bayesian methodology is based on subjective probabilities, which are simply considered to be degrees of belief and quantify the extent to which the "statement" is supported by existing knowledge and available evidence. This may be contrasted with the frequentist approach to probability (classical approach to statistics), according to which probabilities can be assigned only in the context random physical systems and experiments. Uncertainty arising out of the model parameter estimation procedure is expressed in terms of confidence intervals, and it is not statistically meaningful to assign probability distributions to estimation parameters, since they are assumed to be "deterministic but unknown" in the frequentist methodology. This is a serious limitation, since it is not possible to propagate uncertainty after parameter estimation, which is often necessary in the case of model-based quantification of uncertainty in the system-level response. For example, if the uncertainty in the elastic modulus had been estimated using a simple axial test, this uncertainty cannot be used for quantifying the response in a plate made of the same material. Another disadvantage of this approach is that, when a quantity is not random, but unknown, then the tools of probability cannot be used to represent this type of uncertainty (epistemic). The subjective interpretation of probability, on which the Bayesian methodology relies upon, overcomes both of these limitations.

In the Bayesian approach, even deterministic quantities can be represented using probability distributions which reflect the subjective degree of the analyst's belief regarding such quantities. As a result, probability distributions can even be assigned to parameters that need to be estimated, and therefore, this interpretation facilitates uncertainty propagation after parameter estimation; this aspect of the Bayesian methodology is helpful for uncertainty integration across multiple models after inferring the underlying model parameters.

For example, consider the case where a variable is assumed to be normally distributed and it is desired to estimate the mean and the standard deviation based on available point data. If sufficient data were available, then it is possible to uniquely estimate these distribution parameters. However, in some cases, data may be sparse and therefore, it may be necessary to quantify the uncertainty in these distribution parameters. Note that this uncertainty is an example of epistemic uncertainty; the quantities may be estimated deterministically with enough data. Though these parameters are actually deterministic, the Bayesian methodology can calculate probability distributions for the distribution parameters, which can be easily used in uncertainty propagation. The fundamentals of Bayesian philosophy are well established in several textbooks (Calvetti and Somersalo 2007; Lee 2004; Leonard and Hsu 2001; Somersalo and Kaipio 2004), and the Bayesian approach is being increasingly applied to engineering problems in recent times, especially to solve statistical inverse problems. This section provides an overview of the fundamentals of the Bayesian approach, and later sections illustrate the application of Bayesian methods to uncertainty quantification in complex engineering systems.

2.1 Bayes Theorem

Though named after the eighteenth century mathematician and theologian Thomas Bayes (Bayes and Price 1763), it was the French mathematician Pierre-Simon Laplace who pioneered and popularized what is now called Bayesian probability (Stigler 1986, 2002). For a brief history of Bayesian methods, refer to Fienberg (2006). The law of conditional probability is fundamental to the development of Bayesian philosophy:

$$P(AB) = P(A|B)P(B) = P(B|A)P(A) \tag{1}$$

Consider a list of mutually exclusive and exhaustive events A_i ($i = 1$ to n) that together form the sample space. Let B denote any other event from the sample space such that $P(B) > 0$. Based on Eq. (1), it follows that:

$$P(A_i|B) = \frac{P(B|A_i)P(A_i)}{\sum_j P(B|A_j)P(A_j)} \tag{2}$$

What does Eq. (2) mean? Suppose that the probabilities of events A_i ($i = 1$ to n) are known to be equal to $P(A_i)$ ($i = 1$ to n) before conducting any random experiments. These probabilities are referred to as prior probabilities in the Bayesian context. Suppose that a random experiment has been conducted and event B has been observed. In the light of this data, the so-called posterior probabilities $P(A_i|B)$ ($i = 1$ to n) can be calculated using Eq. (2).

The quantity $P(B|A_i)$ is the probability of observing the data conditioned on A_i. It can be argued that event B has "actually been observed," and there is no uncertainty regarding its occurrence, which renders the probability $P(B|A_i)$ meaningless. Hence, researchers "invented" new terminology in order to denote this quantity. In earlier days, this quantity was referred to as "inverse probability," and since the advent of Fisher (Aldrich 2008; Jeffreys 1998) and Edwards (1984), this terminology has become obsolete, and has been replaced by the term "likelihood." In fact, it is also common to write $P(B|A_i)$ as $L(A_i)$.

2.2 Bayesian Inference

The concept of Bayes theorem can be extended from the discrete case to the continuous case. Consider the context of statistical inference where a set of parameters θ needs to be inferred. All the current knowledge regarding this parameter is represented in the form of a prior distribution denoted by $f'(\theta)$. The choice of the prior distribution reflects the subjective knowledge of uncertainty regarding the variable before any observation. It is assumed that the prior distribution is able to explain the data with some degree of uncertainty; in other words, there exists a nonempty set E such that $\forall\ \theta \in E$, the prior probability density function (PDF) and likelihood values evaluated $\forall\ \theta \in E$ are both nonzero.

Measurement data (D) is collected on a quantity which depends on the parameter (θ). This information is then used to update the distribution of θ to arrive at the posterior distribution $(f''(\theta))$, as:

$$f''(\theta) = \frac{L(\theta)f'(\theta)}{\int L(\theta)f'(\theta)d\theta} \tag{3}$$

In Eq. (3), $L(\theta)$ is the likelihood function of θ and is proportional to $P(D|\theta)$, i.e. probability of observing the data D conditioned on the parameter θ. Typically, data D is available in the form of independent, paired input–output combinations (x_i versus x_i, where i varies from 1 to n), where the input X and output Y are related to one another as:

$$y = G(x, \theta) \tag{4}$$

Considering an observation error ϵ that is assumed to follow a Gaussian distribution with zero mean and standard deviation σ, the likelihood can be written as:

$$L(\theta) = \prod_{i=1}^{n} \frac{1}{\sigma\sqrt{(2\pi)}} \exp-\left[\frac{(y_i - G(x_i, \theta))^2}{2\sigma^2}\right] \tag{5}$$

Note that the above equation assumes that the numerical value of σ is known. If this quantity is unknown, σ may be considered to be an argument to the likelihood function and updated along with θ. Equation (5) is substituted in Eq. (3), and the posterior distribution $(f''(\theta))$ can be computed.

Note that the denominator on the RHS of Eq. (3) is simply a normalizing constant, which ensures that $f''(\theta)$ is a valid PDF, i.e., the integral of the PDF is equal to unity. So, Eq. (3) is sometimes written as:

$$f''(\theta) \propto L(\theta)f'(\theta) \tag{6}$$

The posterior in Bayesian inference is always known only up to a proportionality constant and it is necessary generate samples from this posterior for uncertainty analysis. When there is only one parameter, the proportionality constant can be calculated through one-dimensional integration. Often, multiple parameters may be present, and hence, multi-dimensional integration may not be affordable to calculate the proportionality constant. Therefore, a class of methods popularly referred to as Markov Chain Monte Carlo (MCMC) sampling is used to generate samples from the Bayesian posterior. In general, these methods can be used when it is desired to generate samples from a PDF which is known only up to a proportionality constant. The topic of MCMC will be discussed in detail later in this chapter, in Sect. 3.

2.3 Notes on the Likelihood Function

The likelihood function is defined as the probability of observing data conditioned on the parameters, i.e. $L(\theta) = P(D|\theta)$; note that, since the *data* (D) *has actually been observed*, the terminology "probability of observing the data" is physically meaningless. Therefore, as explained earlier in Sect. 2.1, this quantity was renamed as "the likelihood." The likelihood function does not follow the laws of probability, and must not be confounded with probability distributions or distribution functions. In fact, Edwards (1984) explains that the likelihood function is meaningful only up to a proportionality constant; the relative values of the likelihood function are alone significant and the absolute values are not of interest.

The concept of likelihood is used in the context of both physical probabilities (frequentist) and subjective probabilities, especially in the context of parameter estimation. In fact, Edwards (1984) refers to the likelihood method as the third or middle way.

From a frequentist point of view (the underlying parameters are deterministic), the likelihood function can be maximized in order to obtain the maximum likelihood estimate of the parameters. According to Fisher (1912), the popular least-squares-based optimization methodology is an indirect approach to parameter estimation and one can "solve the real problem directly" by maximizing the "probability of observing the given data" conditioned on the parameter θ (Aldrich 1997; Fisher 1912). Further, it is also possible to construct likelihood-based confidence intervals for the inferred parameters (Pawitan 2001).

On the other hand, the likelihood function can also be interpreted using subjective probabilities. Singpurwalla (2006, 2007) explains that the likelihood function can be viewed as a collection of "weights" or "masses" and therefore is meaningful only up to a proportionality constant (Edwards 1984). In other words, if $L(\theta^{(1)}) = 10$, and $L(\theta^{(2)}) = 100$, then it is 10 ten times more likely for $\theta^{(2)}$ than $\theta^{(1)}$ to correspond to the observed data. The entire likelihood function can be used in Bayesian inference, as in Eq. (3), in order to obtain the entire PDF of the parameters.

3 MCMC Sampling

The class of MCMC methods can be used to generate samples from an arbitrary probability distribution, especially when the CDF is not invertible or when the PDF is known only up to a proportionality constant. In Sect. 2.2, it was explained that the latter is the case in Bayesian inference, where the objective is to compute the posterior distribution. Therefore, MCMC sampling can be used to draw samples from the posterior distribution, and these samples can be used in conjunction with kernel density estimation (Rosenblatt 1956) procedure to construct the posterior distribution.

There are several algorithms which belong to the class of MCMC sampling methods. Two such algorithms, the Metropolis algorithm (Metropolis et al. 1953) and the slice sampling (Neal 2003) algorithm are discussed below.

3.1 The Metropolis Algorithm

Assume that a function that is proportional to the PDF is readily available, as $f(x)$; this means that $f(x)$ is not a valid PDF because $\int f(x)dx \neq 1$. For the purpose of illustration, consider the one-dimensional case, i.e. $x \in \mathbb{R}$. The following steps constitute the algorithm in order to generate samples from the underling PDF. Note that the function $f(x)$ is always evaluated at two points and the ratio is only considered; the effect of the unknown proportionality constant is therefore nullified.

1. Set $i = 0$ and select a starting value x_0 such that $f(x_0) \neq 0$.
2. Initialize the list of samples $X = \{x_0\}$.
3. Repeat the following steps; each repetition yields a sample from the underlying PDF.

   (a) Select a prospective candidate from the proposal density $q(x^*|x_i)$. The probability of accepting this sample is equal to $\frac{f(x^*)}{f(x_i)}$.
   (b) Calculate acceptance ratio $\alpha = \min\left(1, \frac{f(x^*)}{f(x_i)}\right)$.
   (c) Select a random number u, uniformly distributed on $[0, 1]$.
   (d) If $u < \alpha$, then set $x_{i+1} = x^*$, otherwise set $x_{i+1} = x_i$.
   (e) Augment the list of samples in X by x_{i+1}.
   (f) Increment i, i.e. $i = i + 1$.

4. After the Markov chain converges, the samples in X can be used to construct the PDF of X using kernel density estimation.

The common practice is to generate a few hundreds of thousands of samples and discard the first few thousand samples to ensure the convergence of the Markov Chain.

The Metropolis algorithm (Metropolis et al. 1953) assumes that the proposal density is symmetric, i.e. $q(x^*|x_i) = q(x_i|x^*)$. A generalization of this algorithm assumes asymmetric proposal density functions $q_1(x^*|x_i)$ and $q_2(x_i|x^*)$; this algorithm is referred to as Metropolis–Hastings algorithm (Hastings 1970). The only difference is that the probability of accepting the prospective candidate is calculated as $\frac{f(x^*)q_2(x_i|x^*)}{f(x_i)q_1(x^*|x_i)}$.

3.2 Slice Sampling

Consider the same function $f(x)$, i.e. the PDF of X, known up to a proportionality constant. The steps of the slice sampling algorithm are as follows:

1. Set $i = 0$ and select a starting value x_0 such that $f(x_0) \neq 0$.
2. Draw a random number y from the uniform distribution $[0, f(x)]$.
3. Consider the set $f^{-1}[y, \infty)$; note that this set may not be convex, especially when the target distribution is multi-modal. Select a sample which is uniformly distributed on this set. Assign $i = i + 1$, and call this sample x_i.
4. Repeat Steps 1–3 to generate multiple samples of X and construct the PDF of X using kernel density estimation.

In contrast with the previously discussed Metropolis algorithm, the slice sampling algorithm is not a acceptance–rejection algorithm.

3.3 MCMC Sampling: Summary

In addition to the above algorithms, other MCMC sampling algorithms such as Gibbs sampling (Geman and Geman 1984), multiple-try Metropolis (Liu et al. 2000), and Metropolis-within-Gibbs (Roberts and Rosenthal 2006) are also discussed in the literature. One critical disadvantage of MCMC sampling approaches is that they may require several hundreds of thousands of samples, and in turn, several hundreds of thousands of evaluations of G in Eq. (4), which may be computationally prohibitive. Therefore, it is common in engineering to replace G (which may be a complicated physics-based model) with an inexpensive surrogate, such as the Gaussian process surrogate model.

4 Gaussian Process Surrogate Modeling

The use of sampling techniques involves repeated evaluations of mathematical models, which may be computationally intensive. One approach to overcome this computational difficulty is to make use of surrogate models to replace the original physics-based model. A few evaluations of the original model are used to train this inexpensive, efficient surrogate model. Different types of surrogate modeling techniques such as polynomial response surface (Rajashekhar and Ellingwood 1993), polynomial chaos expansion (Ghanem and Spanos 1990), support vector regression (Boser et al. 1992), relevance vector regression (Tipping 2001), and Gaussian process interpolation (Bichon et al. 2008; Rasmussen 2004; Santner et al. 2003) have been investigated in the literature.

The Gaussian process interpolation is a powerful technique based on spatial statistics and is increasingly being used to build surrogates to expensive computer simulations for the purposes of optimization and uncertainty quantification (Bichon et al. 2008; Rasmussen 2004; Santner et al. 2003). The GP model is preferred in this research because (1) it is not constrained by functional forms; (2) it is capable of representing highly nonlinear relationships in multiple dimensions; and (3) can estimate the prediction uncertainty which depends on the number and location of training data points.

The basic idea of the GP model is that the response values Y evaluated at different values of the input variables X are modeled as a Gaussian random field, with a mean and covariance function. Suppose that there are m training points, $x_1, x_2, x_3 \ldots x_m$ of a d-dimensional input variable vector, yielding the output values $y(x_1)$, $y(x_2)$, $y(x_3) \ldots y(x_m)$. The training points can be compactly written as x_T vs. y_T where the former is a $m \times d$ matrix and the latter is a $m \times 1$ vector. Suppose that it is desired to predict the response (output values y_P) corresponding to the input x_P, where x_P is $p \times d$ matrix; in other words, it is desired to predict the output at n input combinations simultaneously. Then, the joint density of the output values y_P can be calculated as:

$$p(y_P|x_P, x_T, y_T; \Theta) \sim N(m, S) \tag{7}$$

where Θ refers to the hyperparameters of the Gaussian process, which need to be estimated based on the training data. The prediction mean and covariance matrix (m and S, respectively) can be calculated as:

$$m = K_{PT}(K_{TT} + \sigma_n^2 I)^{-1} y_T$$
$$S = K_{PP} - K_{PT}(K_{TT} + \sigma_n^2 I)^{-1} K_{TP} \tag{8}$$

In Eq. (8), K_{TT} is the covariance function matrix (size $m \times m$) amongst the input training points (x_T), and K_{PT} is the covariance function matrix (size $p \times m$) between the input prediction point (x_P) and the input training points (x_T). These covariance matrices are composed of squared exponential terms, where each element of the matrix is computed as:

$$K_{ij} = K(x_i, x_j; \Theta) = -\frac{\theta}{2} \left[\sum_{q=1}^{d} \frac{(x_{i,q} - x_{j,q})^2}{l_q} \right] \tag{9}$$

Note that all of the above computations require the estimate of the hyperparameters Θ; the multiplicative term (θ), the length scale in all dimensions (l_q, $q = 1$ to d), and the noise standard deviation (σ_n) constitute these hyperparameters ($\Theta = \{\theta, l_1, l_2 \ldots l_d, \sigma_n\}$). As stated earlier, these hyperparameters are estimated based on the training data by maximizing the following log-likelihood function:

$$\log p(y_T|x_T; \Theta) = -\frac{y_T^T}{2}(K_{TT} + \sigma_n^2 I)^{-1} y_T - \frac{1}{2} \log |(K_{TT} + \sigma_n^2 I)| + \frac{d}{2} \log(2\pi)$$
(10)

Once the hyperparameters are estimated, then the Gaussian process model can be used for predictions using Eq. (8). Note that the "hyperparameters" of the Gaussian process are different from the "parameters" of a generic parametric model (e.g., linear regression model). This is because, in a generic parametric model, it is possible to make predictions using only the parameters. On the contrary, in the case of the Gaussian process model, all the training points and the hyperparameters are both necessary to make predictions, even though the hyperparameters may have estimated previously. For details of this method, refer to Bichon et al. (2008); Chiles and Delfiner (1999); Cressie (1991); McFarland (2008); Rasmussen (1996, 2004); Santner et al. (2003), and Wackernagel (2003).

An important issue in the construction of the Gaussian process model is the selection of training points. In general, the training points may arise out of field experiments or may be generated using a computer code. Model parameter estimation considers the latter case and hence, there is no noise in the data, thereby eliminating σ_n from the above equations. Adaptive techniques can be used to select training points for the GP model, in order to construct the response surface to a desired level of accuracy or precision. Since the GP model is capable of estimating the variance in model output, a variance minimization algorithm proposed by McFarland (2008) identifies the next training point at the input variable value which corresponds to the largest variance. This selection algorithm is repeated and training points are adaptively identified until the estimated variance is below a desired threshold. Alternatively, another training point selection algorithm has been developed by Hombal and Mahadevan (2011), where the focus is to select successive training points so that the bias error in the surrogate model is minimized.

Once the training points are selected and the surrogate model is constructed, it can be used for several uncertainty quantification activities such as uncertainty propagation [through Monte Carlo simulation (MCS)], inverse analysis and parameter estimation (through MCMC simulation), and sensitivity analysis. It must be noted that the replacement of a complex computer simulation with an inexpensive surrogate leads to approximations; therefore, it is important to include the effect of this approximation in the procedure for overall uncertainty quantification (Sankararaman 2012).

5   Bayesian Networks

The previous sections of this chapter discussed certain fundamental concepts of the Bayesian methodology, in general. The Bayesian inference approach for model parameter estimation was presented, and the use of MCMC sampling and the importance of using Gaussian process surrogate models were explained. Most of this discussion dealt with single-level models that may represent a component

Fig. 1 Bayesian network
illustration

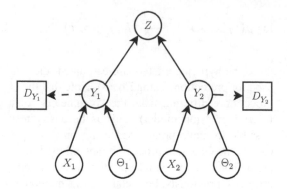

or a subsystem. Since complex engineering systems consist of many components
and subsystems, it is important to integrate information across all components
and subsystems, in order to compute the overall uncertainty in the system-level
prediction. This goal can be accomplished through the use of a Bayesian network.

A Bayesian network (Heckerman 2008; Jensen 1996) is an acyclic, graphical
representation of conditional probability relationships between uncertain quantities.
Each uncertain quantity is represented as a node and successive links are connected
to each other using unidirectional arrows that express dependence in terms of
conditional probabilities. Disconnected nodes imply independence between the
corresponding random variables. Figure 1 shows a conceptual Bayesian network
that aids in uncertainty quantification across multiple levels of models and observed
data. Circles correspond to uncertain variables and squares represent observed data.
A solid line arrow represents a conditional probability link, and a dashed line arrow
represents the link of a variable to its observed data if available.

In Fig. 1, a system level output Z is a function of two subsystem level quantities
Y_1 and Y_2; in turn, Y_1 is a function of subsystem-level input X_1 and model parameter
Θ_1, and similarly, Y_2 is a function of subsystem-level input X_2 and model parameter
Θ_2. For example, in a beam deflection study, the applied force is an input, the elastic
modulus is a model parameter, while the deflection is measured and a model is built
to predict the same. Experimental data D_{Y_1} and D_{Y_2} are available for comparison
with the respective model predictions Y_1 and Y_2.

5.1 Uncertainty Propagation: Forward Problem

In the forward problem, the probability distributions of the inputs (X_1 and X_2) and
model parameters (Θ_1 and Θ_2) are known or assumed, and these distributions are
used to calculate the PDF of Y_1 and Y_2, which in turn are used to calculate the PDF
of the system-level output Z, as:

$$f_Z(z) = \int f_Z(z|y_1, y_2) f_{Y_1}(y_1) f_{Y_2}(y_2) f_{Y_1}(y_1) f_{Y_2}(y_2) dy_1 dy_2$$

$$f_{Y_1}(y_1) = \int f_{Y_1}(y_1|x_1, \theta_1) f_{X_1}(x_1) f_{\Theta_1}(\theta_1) dx_1 d\theta_1$$

$$f_{Y_2}(y_2) = \int f_{Y_2}(y_2|x_2, \theta_2) f_{X_2}(x_2) f_{\Theta_2}(\theta_2) dx_2 d\theta_2 \tag{11}$$

Note that uppercase letters are used to denote random variables and the corresponding lowercase letters are used to denote realizations of those random variables. Equation (11) can be solved using methods of uncertainty propagation such as MCS, first-order reliability method (FORM), and second-order reliability method (SORM) (Haldar and Mahadevan 2000).

5.2 Inference: Inverse Problem

In the inverse problem, the probability densities of the model parameters (Θ_1 and Θ_2 in Fig. 1) can be updated based on the observed data (D_{Y_1} and D_{Y_2}) using Bayes' theorem as:

$$f_{\Theta_1,\Theta_2}(\theta_1, \theta_2|D_{Y_1}, D_{Y_2}) = C L(\theta_1, \theta_2) f_{\Theta_1}(\theta_1) f_{\Theta_2}(\theta_2) \tag{12}$$

In Eq. (12), the prior distributions of the model parameters Θ_1 and Θ_2 are given by $f_{\Theta_1}(\theta_1)$ and $f_{\Theta_2}(\theta_2)$, respectively. The choice of the prior distribution reflects the subjective knowledge of uncertainty regarding the variable before any testing. It is assumed that the prior distribution is able to explain the data with some degree of uncertainty; in other words, there exists a nonempty set E such that $\forall \{\Theta_1, \Theta_2\} \in E$, the prior PDF ($f_{\Theta_1}(\theta_1) f_{\Theta_2}(\theta_2)$) and likelihood ($L(\theta_1, \theta_2)$) values evaluated at $\{\Theta_1, \Theta_2\}$ are both non-zero.

The joint posterior density of the parameters is given by $f_{\Theta_1,\Theta_2}(\theta_1, \theta_2|D_{Y_1}, D_{Y_2})$. The likelihood function $L(\theta_1, \theta_2)$ is calculated as the probability of observing the given data (D_{Y_1}, D_{Y_2}), conditioned on the parameters being updated, i.e. Θ_1 and Θ_2. The likelihood function accounts for the uncertainty in the inputs X_1 and X_2. For details of the likelihood function, refer to Edwards (1984), Pawitan (2001), and Sankararaman (2012). As explained earlier in Sect. 3, Eq. (12) can be evaluated by generating samples of model parameters (Θ_1 and Θ_2) through MCMC sampling.

6 Test Resource Allocation

Typically, a multi-level, multi-physics system has several parameters that influence the overall system-level output, and the uncertainty in these parameters can be updated by tests at multiple levels of the system and multiple types of physics coupling. When the posterior distributions of the parameters are propagated through the system model to calculate the overall system-level output, the posterior variance of the overall system-level prediction can be computed. With more acquisition of data, a decreasing trend can be observed in the variance of the system-level output.

Two types of questions need to be answered: (1) What type of tests to do (which component, isolated physics, etc.)? and (2) How many repetitions of each type? Each type of test has a different testing cost and an associated reduction in the variance of system-level prediction. Further, the same type of test may need to be repeated on nominally identical specimens of the same component or subsystem. Such repetition is performed in order to account for the effect of natural variability across nominally identical specimens; while each repetition may have the same monetary cost, the associated reduction in the variance of system-level prediction may be different.

The test conducted on one subsystem is assumed to be statistically independent of another test on another subsystem; in other words, one type of test is independent of any other type. Further, for a given type of test, the repetitions across multiple replicas are also considered to be independent. It is assumed that a model is available to predict the quantity being measured in each type of test; the model may have several outputs but only that output which is measured is of concern. The overall objective is to identify how many tests of each type must be performed so as to achieve the required reduction in the variance of the system-level output. If there are several system-level outputs, either an aggregate measure or the most critical output can be considered. However, multi-objective optimization formulations to simultaneously reduce the variance of more than one system-level output have not yet been addressed in the literature.

6.1 Sensitivity Analysis

The method of sensitivity analysis has been used to quantify the sensitivity of model output to parameters. While derivative-based methods only compute local sensitivities, the method of global sensitivity analysis (Saltelli et al. 2008) to apportion the variance in the system-level output to the various sources of uncertainty, and thereby guide in the reduction of system-level prediction uncertainty.

The first step of the resource allocation methodology is to use sensitivity analysis and identify those parameters that have a significant influence on the variance of the overall system-level prediction. Once the "important" parameters are identified, only those tests that aid in reducing the uncertainty in these important parameters can be performed. For example, consider a system-level output that is very sensitive to the uncertainty in the parameters of sub-system-I but not sensitive to the parameters of sub-system-II, then it is logical to perform more sub-system-I tests than sub-system-II tests. Note that this procedure for test identification is only a preliminary approach. This approach can answer the question—"which tests to do?" In order to answer the question, "how many tests to do?", it is necessary to quantify the decrease in variance that may be caused due to a particular test. The effect of a particular test on variance reduction can be quantified by using Bayesian updating. Therefore, the resource allocation methodology first uses sensitivity analysis for selection of calibration parameters and then uses Bayesian updating to quantify the effect of a test on the variance of system-level prediction.

6.2 Optimization Formulation

In order to solve the resource allocation problem and identify the number of tests to be performed for each type, the optimization problem can be formulated in two ways, as explained below.

In the first formulation shown in Eq. (13), the goal is to minimize the variance of the system-level output subject to satisfying a budget constraint.

$$\underset{N_{\text{test}}}{\text{Minimize }} E(\text{Var}(R))$$

$$\text{s.t. } \sum_{i=1}^{q}(C_i N_i) \leq \text{Total Budget} \tag{13}$$

$$N_{\text{test}} = [N_1, N_2 \ldots N_q]$$

In Eq. (13), q refers to the number of different types of possible tests. The cost of the ith ($i = 1$ to q) type of test is equal to C_i, and N_i (decision variable) denotes the number of repetitions of the ith type of test. Let D_i denote all the data collected through the ith type of test. Let N_{test} denote the vector of all N_i's and let D denote the entire set of data collected from all q types of tests.

Alternatively, the resource allocation problem can be formulated by minimizing the cost required to decrease the variance of the system-level output below a threshold level, as:

$$\underset{N_{\text{test}}}{\text{Minimize }} \sum_{i=1}^{q}(C_i N_i)$$

$$\text{s.t. } E(\text{Var}(R)) \leq \text{Threshold Variance} \tag{14}$$

$$N_{\text{test}} = [N_1, N_2 \ldots N_q]$$

Sankararaman et al. (2013) pursued the first formulation (Eq. (13)) because the threshold level for the variance is assumed to be unknown. Using D, the model parameters are calibrated and the system-level response ($R(D)$) is computed. The optimization in Eq. (13) calculates the optimal values of N_i, given the cost values C_i, such that the expected value of variance of the system-level prediction ($E(\text{Var}(R))$) is minimized, while the budget constraint is satisfied.

This optimization formulation uses $E(\text{Var}(R))$ as the objective function because R is a function of $\boldsymbol{D}$, which is not available before testing. Hence, random realizations of the test data set ($\boldsymbol{D}$) are generated; each random realization is used to compute $\text{Var}(R|\boldsymbol{D})$, and the expectation over such random realizations is calculated to be the objective function, as:

$$E(\text{Var}(R)) = \int \text{Var}(R|\boldsymbol{D}) f(\boldsymbol{D}) d\boldsymbol{D} \tag{15}$$

where $f(\boldsymbol{D})$ is the density considered for the test data. Assuming that one type of test is performed independent of the another (i.e., a subsystem-level test is independent of a material-level test), Eq. (15) can be written as:

$$E(\text{Var}(R)) = \int \text{Var}(R|D_1, D_2 \ldots D_q) f(D_1) f(D_2) \ldots f(D_q) dD_1 dD_2 \ldots dD_q \tag{16}$$

where $f(D_i)$ is the density considered for the data obtained through the ith test. Before any testing is done, all prior knowledge regarding the model parameters and the mathematical models constitute the only information available for the calculation of $f(D_i)$. Therefore, $f(D_i)$ is calculated as:

$$f(D_i) = \int f(y_i|\boldsymbol{\theta}_i) f'(\boldsymbol{\theta}_i) d\boldsymbol{\theta}_i \tag{17}$$

where y_i represents the output of the mathematical model corresponding to the ith type of test, $\boldsymbol{\theta}_i$ represents the underlying parameters, and $f'(\boldsymbol{\theta}_i)$ represents the prior knowledge regarding those parameters. Note that Eq. (17) is simply an uncertainty propagation problem, where the other sources of uncertainty (such as physical variability in inputs, solution approximation errors, data uncertainty) can also be included in the computation of $f(y_i|\boldsymbol{\theta}_i)$.

Equations (15)–(17) are implemented using a numerical algorithm, where a finite number of realizations of $\boldsymbol{D}$ are generated and $E(\text{Var}(R))$ is computed over these realizations. Then, $E(\text{Var}(R))$ can be minimized using the optimization in Eq. (13), and the ideal combination of tests can be identified.

Note that an inequality constraint (for the budget), and not an equality constraint, is considered in Eq. (13). This means that the optimal solution which minimizes $E(\text{Var}(R))$ need not necessarily exhaust the budget. Consider the simple case where there are two possible test types ($C_1 = 2$ and $C_2 = 3$), and the budget is equal to six cost units. There are two test combinations which exhaust the budget: (1) [$N_1 = 3$, $N_2 = 0$], and (2) [$N_1 = 0$, $N_2 = 2$]. Suppose that these two combinations lead to a value of $E(\text{Var}(R))$ which is *greater* than that achieved through the test combination [$N_1 = 1$, $N_2 = 1$]. Then, obviously the combination [$N_1 = 1$, $N_2 = 1$] must be selected because it achieves the goal of reducing $E(\text{Var}(R))$ even though it may not exhaust the budget.

6.3   Solution of the Optimization Problem

Equation (13) is a complicated integer optimization problem, where Bayesian updating and forward propagation need to be repeated for each random realization of the test data in order to evaluate the objective function, thus increasing the computational cost several fold. In spite of the use of Gaussian process surrogate models to replace the expensive system model, high computing power is still needed to solve the optimization problem.

Integer optimization is sometimes solved using an approximation method, where the integer constraint is first relaxed, and the integers nearest to the resulting optimal solution are used in further solution of the original (un-relaxed) problem. Unfortunately, this approach is not applicable to the solution of Eq. (13), since the objective function (system-level prediction variance) is defined and computed only for integer-valued decision variables (number of tests). It is meaningless to have a non-integer number of tests.

A multi-step procedure for solving the optimization problem was proposed by Sankararaman et al. (2013). Within each step, the global optimal solution is computed using an exhaustive search process, whereas across steps, a greedy algorithm is pursued. The step size is chosen in cost units, and additional steps are added until the budget constraint is satisfied.

Let the size of the first step be equal to ϕ^1 cost units; the globally optimal testing combination for this cost ($= \phi^1$) is denoted by N_{test}^1, and is calculated using exhaustive search, as:

$$\underset{N_{\text{test}}^1}{\text{Minimize}} \ E(\text{Var}(R))$$

$$\text{s.t.} \ \sum_{i=1}^{q}(C_i N_i^1) \leq \phi^1 \tag{18}$$

$$N_{\text{test}}^1 = [N_1^1, N_2^1 \ldots N_q^1]$$

The optimization procedure in the second stage is dependent on the optimal solution from the first stage, i.e. N_{test}^1. In general, the optimization for the jth stage, given the solution in the previous stage (i.e., N_{test}^{j-1}), is performed for cost $= \phi^j$. Note that $\sum_{j} \phi^j = $ Total budget. The jth optimization is formulated as:

$$\underset{N_{\text{test}}^{j,\text{new}}}{\text{Minimize}} \ E(\text{Var}(R))$$

$$\text{s.t.} \ \sum(C_i N_i^{j,\text{new}}) \leq \phi^j \ (i = 1 \text{to} q) \tag{19}$$

$$N_{\text{test}}^j = N_{\text{test}}^{j-1} + N_{\text{test}}^{j,\text{new}}$$

$$N_{\text{test}}^{j,\text{new}} = [N_1^{j,\text{new}}, N_2^{j,\text{new}} \ldots N_q^{j,\text{new}}]$$

As seen in Eq. (19), the decision variables for the jth stage are $N_{\text{test}}^{j,\text{new}}$, i.e. those tests which need to be performed in the jth stage; therefore the total number of tests is equal to the sum of $N_{\text{test}}^{j,\text{new}}$ and N_{test}^{j-1}, i.e. the optimal number of tests in the previous stage. The same procedure is repeated until no additional test can be performed with the budget constraint satisfied.

The selection of step size for a given budget is an important issue. The true global optimal solution can be calculated by considering *one step* whose size is equal to the entire budget. However, due to the large number of possible testing combinations, this approach may be computationally infeasible. In a practical problem, several steps are considered, and the step sizes must be chosen judiciously based on (1) the costs of each type of test; (2) time required for each Bayesian update; (3) number of random realizations of data needed to compute $E(\text{Var}(R))$; and (4) the test combinations that are suitable for the chosen step size; a very small step size may not even include an expensive type of test.

6.4   Summary of the Optimization Methodology

The various steps of the optimization-based methodology for test resource allocation are summarized below:

1. **Construction of the Bayesian network:** The first step is to identify the various component-level, subsystem-level, and system-level models. Each model has an output quantity and correspondingly, a test can be performed to measure this quantity. All the models are connected through the Bayesian network, and the data available across the nodes is also indicated. The model errors, if available, can also be included in the Bayesian network. Though solution approximation errors can be calculated prior to testing and included in the Bayes network, model form error cannot be calculated before testing. It must be noted that the Bayesian network, due to its acyclic nature, cannot account for feedback coupling between models. When the system-level response is a coupled physics-based solution, the overall coupled solution is directly included in the Bayesian network instead of considering the individual physics separately.
2. **Sensitivity analysis:** The next step is to perform global sensitivity analysis and identify the "important" parameters that significantly contribute to the uncertainty in the system-level response. Then, those tests which can aid in the reduction of uncertainty in these "important" parameters are selected for consideration in the optimization for test resource allocation.
3. **Bayesian updating:** The next step is to perform Bayesian updating and calibrate parameters for a particular realization of measurement data. Then, this needs to be repeated by generating multiple realizations of measurement data in order to compute the expected value of variance, as in Eq. (15). Due to the required computational expense, the original physics models can be replaced with Gaussian process surrogates. Though this does not lead to analytical calculation of the posterior, it increases the computational efficiency several fold.

4. **Resource allocation optimization:** The final step is to perform the resource allocation optimization using the multi-step procedure developed in Sects. 6.2 and 6.3. It may be useful to verify that the resultant solution is actually optimal by computing $E(\text{Var}(R))$ for few other N_{test} values.

The rest of this chapter illustrates the optimization-based test resource allocation methodology using three different numerical examples. While Sect. 7 deals with a simple mathematical example, Sects. 8 and 9 consider multi-physics and multi-level engineering systems, respectively.

7   Illustration Using Mathematical Example

This subsection presents a simple illustrative example to illustrate the optimization-based methodology for test resource allocation. In order to focus on this objective, simple mathematical relationships are chosen (even the system-level response has no coupling), and measurement errors are assumed to be negligible. Other features such as coupled system response, measurements errors, and solution approximation errors (while replacing the underlying physics-based model with a Gaussian process approximation) are considered later in Sects. 9 and 8.

The Bayesian network for this problem is exactly the same as that in Fig. 1. There are four independent quantities and three dependent quantities; the numerical details of this problem are specified in Table 1. The notation $N(\mu, \sigma)$ is used to represent a normally distributed quantity with mean μ and standard deviation σ. Two types of tests (on two different lower levels) can be done and this information is used to update the uncertainty in the system-level response based on the tests.

Probability distributions are assumed to be available for the inputs X_1 and X_2; if this information was not available, and only sparse and/or interval data was available for the inputs, then the likelihood-based method developed in Sankararaman and Mahadevan (2011) can be used to construct a probability distributions for them.

Table 1 Numerical details

Quantity	Type	Description
X_1 (input)	Independent	N(100,5)
Θ_1 (parameter)	Independent	N(50, 10)
X_2 (input)	Independent	N(10,1)
Θ_2 (parameter)	Independent	N(15, 4)
Y_1	Dependent	Model : $y_1 = x_1 + x_2$
Y_2	Dependent	Model : $y_2 = x_3 + x_4$
Z	System-level response	Model : $z = y_1 - y_2$
Quantity to measure	Cost	No. of tests
Y_1	10	N_1
Y_2	5	N_2

Table 2 Resource
allocation: results

Cumulative cost	N_1	N_2	$E(\mathrm{Var}(z))$
$10	**1**	**0**	**62.0**
	0	2	127.0
$20	2	0	53.0
	1	**2**	**46.6**
$30	**2**	**2**	**37.6**
	1	4	46.1
$40	**3**	**2**	**34.0**
	2	4	37.6
$50	**4**	**2**	**32.5**
	3	4	33.8

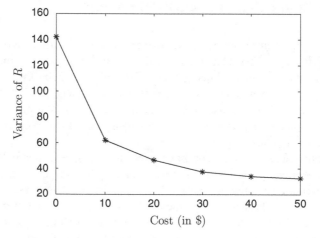

Fig. 2 Variance vs. cost

The variance of the system-level response quantity Z before conducting any test
(i.e., by propagating the above distributions of X_1, X_2, Θ_1, and Θ_2 through the
models) is 142 units. The objective is calculate the number of tests on Y_1 and Y_2
(N_1 and N_2), that will lead to a minimum variance in Z, subject to a total budget of
$50 cost units. Since there are only two parameters, global sensitivity analysis is not
necessary, and hence, both Θ_1 and Θ_2 are chosen for calibration. The optimization-
based methodology discussed in Sect. 6 is used for this purpose; five different stages
are considered and the available budget in each stage is considered to be $10. The
results of test prioritization are given in Table 2 (the optimal value in each stage is
indicated in bold) and Fig. 2.

At the end of the optimization procedure, the optimal combination is found
to be four tests on Y_1 and 2 tests on Y_2. Further, this solution was verified by
considering all other combinations (exhaustive search) of N_1 and N_2 and computing
the corresponding $E(\mathrm{Var}(R))$; for this illustrative example, this verification is
numerically affordable. However, for practical examples, a few random values of
$N_{\text{test}} = [N_1, N_2]$ (if not all) can be considered and it can be verified if the estimated
solution is really optimal.

8   Numerical Example: Multi-Physics System

8.1   Description of the Problem

This coupled-physics thermal vibration example illustrates a laboratory experiment which can be used to study and simulate the behavior in solar arrays of telescopes and spacecraft booms (Thornton 1996). In this experiment, a thin walled circular tube is rigidly attached at its top and supports a concentrated mass at its bottom. The tube and the mass are initially at rest and a constant heat flux is applied on one side along the length of the tube. The application of the heat flux causes an increase in the temperature on the incident surface while the unheated side remains at the initial temperature. The temperature gradient causes the beam to bend away from the lamp, due to the thermal moment. The displacement of the beam, in turn, changes the distribution of temperature along the length of the beam, leading to a change in the temperature gradient and the thermal moment, which in turn affects the flexural behavior. Thus the combination of heat transfer and flexural mechanics leads to oscillations of the beam. The setup of this experiment is shown in Fig. 3.

The temperature at the tip mass (T_m) is given by the following differential equation:

$$\frac{\partial T_m}{\partial t} + \frac{T_m}{\tau} = \frac{T^*}{\tau}\left(1 - \frac{v(x,t)}{\beta^*}\right) \qquad (20)$$

In Eq. (20), $v(x,t)$ represents the displacement of the beam as a function of length and time. Thornton (1996) explains how to calculate the parameters T^*, τ, β^* as a function of the incident solar flux (S).

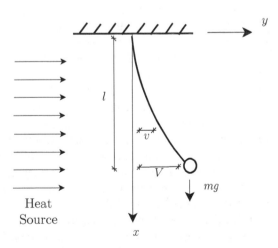

Fig. 3 Thermally induced vibration

The displacement $v(x,t)$ can be related to the displacement of the tip mass $V(t)$ as:

$$v(x,t) = \left(\frac{3x^2}{2l^2} - \frac{x^3}{2l^3}\right)V(t) \tag{21}$$

The tip mass displacement $V(t)$, in turn, depends on the forcing function as follows:

$$\ddot{V} + 2\xi\omega_0\dot{V} + \left(\omega_0^2 + \frac{6g}{5l}\right)V = \frac{F(t)}{m} \tag{22}$$

In Eq. (5), ξ is the damping ratio, and ω_0 is the angular frequency. The forcing function $F(t)$ depends on the thermal moment which in turn depends on the temperature, thereby casing coupling between the thermal equation and the structural equation. These relations are shown in the following equations:

$$F(t) = -\frac{3}{l^3}\int_0^l \int_0^x M(u,t)dudx \tag{23}$$

$$M(x,t) = \int E\alpha T_m(x,t)cos(\Phi)ydA \tag{24}$$

In Eq. (24), E is the elastic modulus, α is the coefficient of thermal expansion, Φ is the angle of incident flux on the cross section, y is the distance from the center of the cross section and the integral is over the area of the cross section A. Refer Thornton (1996) for a detailed description of this problem.

The overall objective of test resource allocation is to minimize the variance of the system-level output (R), which is defined to be the ratio of displacement amplitudes at two different time instants for the coupled system when the incident solar flux (S) is 2,000 W/m^2. If $R < 1$, the system is stable with oscillations diminishing as a function of time. If $R > 1$, the system is unstable, commonly referred to as flutter, an undesirable scenario. While a Gaussian process model is constructed to calculate the multi-physics response R, individual physics predictions are performed using the above described physics-based models.

There are several parameters (both thermal and structural) in the above equations that can be calibrated using test data. The method of sensitivity analysis is used to identify five parameters, which significantly contribute to the uncertainty in the system-level prediction. The prior means are based on Thornton (1996), and the assumed coefficients of variation (CoV) are tabulated in Table 3; note that the radius being a geometric property has a lower CoV. The calibrated parameters are then used to quantify the uncertainty in R.

The calibration parameters need to be estimated during test data; four different types of tests are considered, as shown in Table 4. The total budget available for testing is assumed to be $2,000. It is assumed that the entire multi-disciplinary system cannot be tested.

Table 3 Calibration quantities: thermal vibration problem

Symbol	Quantity	Property	Prior CoV
E	Elastic modulus	Structural	0.1
c	Independent	Thermal	0.1
ξ	Independent	Structural	0.1
r	Independent	Geometric	0.03
e	Dependent	Thermal	0.1

Table 4 Types of tests: thermal vibration problem

Test type	Physics	Calibrate	Input–Output	Cost	No. of tests
Material-level	Thermal	c	Heat-Temperature rise	$100	N_{m_1}
Material-level	Structural	ξ	Amplitude decay	$100	N_{m_2}
Subsystem-level	Thermal	c, e, r	Heat-Temperature rise	$500	N_T
Subsystem-level	Structural	ξ, E	Acceleration	$500	N_F

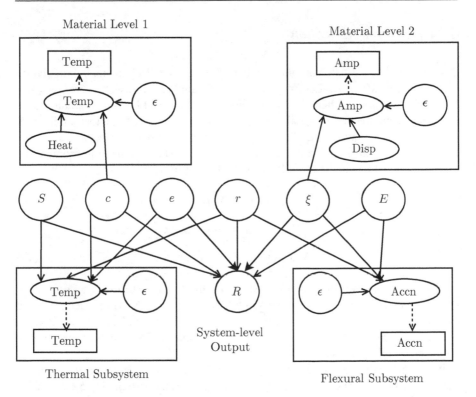

Fig. 4 Thermal vibration: Bayesian network

The calibration quantities, the model predictions, and the test data are connected through a Bayesian network, as shown in Fig. 4.

In the Bayesian network in Fig. 4, "Temp" refers to temperature, "Accn" refers to the acceleration, "Disp" refers to the displacement, and "Amp" refers to the

Table 5 Resource allocation results: thermal vibration problem

Stage no.	N_{m_1}	N_{m_2}	N_F	N_T	$E(\text{Var}(R))$ (in %)
No tests	0	0	0	0	100.0
Stage 1 : $500	1	4	0	0	74.6
Stage 2 : $1,000	1	4	1	0	51.4
Stage 3 : $1,500	1	4	1	1	44.8
Stage 4 : $2,000	1	9	1	1	44.2

amplitude of vibration. Measurement errors (ϵ) are assumed to have a standard deviation that is equal to 10 % of the model prediction. This Bayesian network is used for uncertainty quantification, Bayesian updating and resource allocation.

8.2   Resource Allocation

The objective is to calculate the number of tests that lead to maximum reduction in variance in R. Let N_{test} denote the number of tests, where $N_{\text{test}} = [N_{m_1}, N_{m_2}, N_F, N_T]$; where N_{m_1} is the number of material level temperature tests, N_{m_2} is the number of material level pluck tests, N_F is the number of flexural subsystem tests, and N_T is the number of thermal subsystem tests. Let $D = [D_{m_1}, D_{m_2}, D_F, D_T]$ denote the test measurements. The optimization problem for resource allocation can be formulated as shown in Eq. (25)

$$\underset{N_{\text{test}}}{\text{Minimize}} \ E(\text{Var}(R))$$

$$\text{s.t. } 100(N_{m_1} + N_{m_2}) + 500(N_F + N_T) \le 2000$$

$$N_{\text{test}} = [N_{m_1}, N_{m_2}, N_F, N_T] \tag{25}$$

The above optimization is solved using the multi-stage optimization procedure discussed in Sects. 6.2 and 6.3. Four stages and a budget of $500 for each stage are considered, thereby accounting for the total budget of $2,000. Each stage has eight options (as against two in the mathematical example in Sect. 7); only the optimal solution in each stage is shown.

Note that Table 5 expresses the expectation of variance of R in terms of percentage of the variance before any testing; this variance is equal to 5.69×10^{-7}; since R is a ratio, this variance is dimensionless.

For a $2,000 budget, it is seen that one temperature test, nine pluck tests, one thermal subsystem test and one flexural subsystem test are required to achieve the maximum reduction in the variance of R. The results show that while it is useful to do all the tests, repeating the pluck test which calibrates structural damping, is not only cheap but also leads to effective decrease in the variance of R. The decrease of variance with cost is shown in Fig. 5.

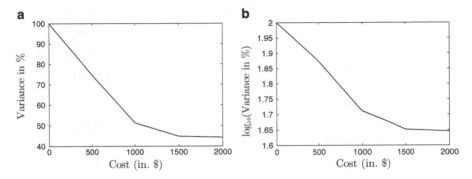

Fig. 5 Decrease of variance with cost. (**a**) Variance with cost; (**b**) log(Variance) with cost

It is seen that the reduction in variance using the last $1,000 (i.e., from $1,000 to $2,000) was much smaller when compared to the reduction in variance using the initial $1,000. Such information is very useful for budgeting purposes, since all the above computation (and practical resource allocation) is done before any test is actually conducted.

9 Numerical Example: Multi-Level System

9.1 Description of the Problem

A three-level structural dynamics developed at Sandia National Laboratories (Red-Horse and Paez 2008) is considered as shown in Fig. 6.

The first level (component) consists of a single spring-mass-damper. Three such spring-mass dampers are integrated to form a spring-mass-damper subsystem in the second-level. In the third level, the integrated spring-mass-damper subsystem is mounted on a beam to form the overall system.

The models to represent the first two levels are straightforward (Chopra 1995). Red-Horse and Paez (2008) describe in detail the modeling and simulation of the overall system (third-level). The overall objective is to test resource allocation to minimize the variance of the system level output (R) which is defined to be the maximum acceleration of mass m_3, when a random force is applied as specified in Red-Horse and Paez (2008). The first-level and second-level responses are computed using physics-based models while the third-level and system-level responses are computed by constructing two Gaussian process surrogate models.

In this numerical example, the stiffness values of the three masses, i.e. k_1, k_2, and k_3 are all the parameters that need to be calibrated with test data; since all parameters are calibrated, sensitivity analysis is not used in this example. The numerical values (in SI units) of three calibration parameters are summarized in Table 6.

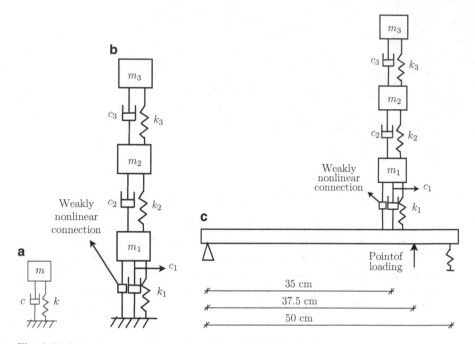

Fig. 6 Multi-level structural dynamics problem. (**a**) Level 1; (**b**) level 2; (**c**) level 3

Table 6 Model parameters: structural dynamics problem

Number	Mass (m) (in kg)	Damping (c) (in Ns/m)	Prior mean of stiffness (μ_k) (in N/m)	Prior std. dev. of mean (σ_k) (in N/m)
1	0.012529	0.023466	5,600	560
2	0.019304	0.021204	11,000	1,100
3	0.035176	0.031216	93,000	9,300

The mass of the beam is taken to be 0.1295. Further numerical details of the beam are given in Red-Horse and Paez (2008).

Data for calibration is assumed to be available through five different types of tests. The details of these different types of tests are provided in Table 7. For each test, a sinusoidal load (amplitude = 10,000 and angular velocity = 10 rad s^{-1}) is used. For the first and second level tests, the sinusoidal load is applied at the base; for the third level test, the sinusoidal load is applied as specified in Red-Horse and Paez (2008).

The model predictions, experimental data, and the calibration quantities are connected using the Bayesian network, shown in Fig. 7. The corresponding experimental errors are denoted by ϵ_{11}, ϵ_{12}, ϵ_{13}, ϵ_2, and ϵ_3, respectively, and assumed to be equal to 10 % of the prediction.

This Bayesian network can be used for uncertainty quantification, Bayesian updating, and resource allocation, as explained below.

Table 7 Model parameters: structural dynamics problem

Test type	Description	Model prediction	Data	Cost	No. tests
Level-1	Only mass m_1	Acceleration (x_{11})	D_{11}	\$100	N_{m_1}
Level-1	Only mass m_1	Acceleration (x_{12})	D_{12}	\$100	N_{m_2}
Level-1	Only mass m_1	Acceleration (x_{13})	D_{13}	\$100	N_{m_3}
Level-2	3-mass assembly	Acceleration of m_3 (x_2)	D_2	\$500	N_2
Level-3	3-mass assembly on beam	Acceleration of m_3 (x_3)	D_3	\$1,000	N_3

Fig. 7 Bayesian network

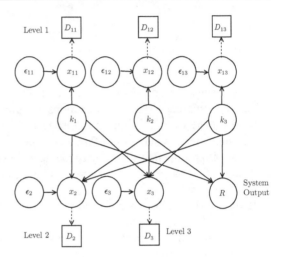

9.2   Resource Allocation

In the resource allocation problem, testing is yet to be done and hence realizations of future experimental data are generated randomly. Then, $E(\text{Var}(R))$ is computed so as to identify which set of tests will lead to the maximum reduction in variance. Let $N_{\text{test}} = [N_{m_1}, N_{m_2}, N_{m_3}, N_2, N_3]$. The optimization problem for resource allocation can be formulated as shown in Eq. (26).

$$\underset{N_{\text{test}}}{\text{Minimize}} \ E(\text{Var}(R))$$

$$\text{s.t. } 100(N_{m_1} + N_{m_2} + N_{m_3}) + 500N_2 + 1000N_3 \leq 1000$$

$$N_{\text{test}} = [N_{m_1}, N_{m_2}, N_{m_3}, N_2, N_3] \tag{26}$$

First, the resource allocation is solved for a budget of \$1000. There are 54 possible testing combinations and out of these 54, ten testing combinations lead to the same minimum variance of system-level output R, approximately 0.8 % of the variance before testing. These combinations are given in Table 8. The value of $E(\text{Var}(R))$ for these ten cases are close enough that it is not possible to determine whether the difference is due to reality or due to sampling/numerical errors.

Table 8 Resource allocation
results: structural dynamics
problem

N_{m_1}	N_{m_2}	N_{m_3}	N_2	N_3
0	1	4	1	0
0	4	1	1	0
0	3	2	1	0
0	2	3	1	0
1	1	3	1	0
3	1	1	1	0
1	3	1	1	0
1	2	2	1	0
2	1	2	1	0
2	2	1	1	0

It is a subjective decision as to which one of these ten test combinations is selected. However, all ten combinations unanimously suggest that no tests are needed for the overall system and one test is needed for the second level three spring-mass-damper subsystem. The first four rows in Table 8 suggest that testing is not needed for the first spring-mass-damper. However, it may be desirable to have at least one test for each component, and hence one amongst the latter six options may be preferred.

It was also found that an extra budget of $1,000 caused no further reduction in the variance of R. If the available budget is $2,000, a subjective decision may be made to conduct the full system test (which costs $1,000) in order to further improve the confidence in uncertainty quantification.

10   Conclusion

This chapter presented a Bayesian methodology for uncertainty quantification in complex engineering systems consisting of multiple physics behavior and multiple levels of integration. The various component, subsystem, and system models, and their inputs, parameters, and outputs, and experimental data were efficiently connected through a Bayesian network. Further, the various sources of uncertainty—physical variability, data uncertainty, and model uncertainty were also included in the Bayesian network. The Bayesian network was used for three different tasks: (1) calibrate the parameters of models at multiple levels using all available test data from multiple levels; (2) propagate the various sources of uncertainty (including the previously estimated model parameters) through the Bayes network to predict the overall uncertainty in the system-level response; and (3) aid in resource allocation for test selection, in order to identify the most effective tests to reduce the overall uncertainty in the system-level prediction. The procedure for test resource allocation required Bayesian calibration and assessment of system-level prediction uncertainty even before actual testing was performed. This was achieved by generating multiple samples of test data and estimating the expected reduction in variance of the system-level prediction.

The algorithm for test resource allocation leads to several insights. A lower level test can easily isolate individual components and hence, the model parameters can be effectively updated, leading to a significant reduction in the variance of the system-level prediction. However, such a test would not account for the interaction between the higher level models and the corresponding parameters. In contrast, a higher level test would readily include the effects of interaction between multiple subsystem-level and component-level models. However, the calibration of parameters across multiple models may be difficult and may not lead to a significant reduction in the variance of the system-level prediction. The optimization-based test resource allocation procedure trades off between lower level tests and higher level tests by accounting not only for the resultant reduction in variance of the system-level prediction but also for the costs involved in testing.

References

Aldrich J (1997) R.A. Fisher and the making of maximum likelihood 1912–1922. Stat Sci 12(3):162–176

Aldrich J (2008) R.A. Fisher on Bayes and Bayes' theorem. Bayesian Anal 3(1):161–170

Bayes T, Price M (1763) An essay towards solving a problem in the doctrine of chances, by the late Rev. Mr. Bayes, FRS communicated by Mr. Price, in a letter to John Canton, AMFRS. Philos Trans (1683–1775) 53:370–418

Bichon BJ, Eldred MS, Swile LP, Mahadevan S, McFarland JM (2008) Efficient global reliability analysis for nonlinear implicit performance functions. AIAA J 46(10):2459–2468

Boser BE, Guyon IM, Vapnik VN (1992) A training algorithm for optimal margin classifiers. In: Proceedings of the fifth annual workshop on computational learning theory. ACM, New York, pp 144–152

Calvetti D, Somersalo E (2007) Introduction to Bayesian scientific computing: ten lectures on subjective computing, vol 2. Springer, New York

Chiles JP, Delfiner P (1999) Geostatistics: modeling spatial uncertainty, vol 344. Wiley-Interscience, New York

Chopra AK (1995) Dynamics of structures. Prentice Hall, Englewood

Cressie N (1991) Spatial statistics. Wiley, New York

Edwards AWF (1984) Likelihood. Cambridge University Press, New York

Fienberg SE (2006) When did Bayesian inference become Bayesian? Bayesian Anal 1(1):1–40

Fisher RA (1912) On an absolute criterion for fitting frequency curves. Messenger Math 41(1):155–160

Geman S, Geman D (1984) Stochastic relaxation, gibbs distributions and the Bayesian restoration of images. IEEE Trans Pattern Anal Mach Intell 6:721–741

Ghanem R, Spanos PD (1990) Polynomial chaos in stochastic finite elements. J Appl Mech 57(1):197–202

Haldar A, Mahadevan S (2000) Probability, reliability, and statistical methods in engineering design. Wiley, New York

Hastings WK (1970) Monte carlo sampling methods using markov chains and their applications. Biometrika 57(1):97

Heckerman D (2008) A tutorial on learning with Bayesian networks. In: Innovations in Bayesian networks. Springer, Berlin, pp 33–82

Helton JC, Pilch M (2011) Quantification of margins and uncertainties. Reliab Eng Syst Saf 96(9):959–964

Hombal V, Mahadevan S (2011) Bias minimization in Gaussian process surrogate modeling for uncertainty quantification. Int J Uncertain Quantif 1(4):321–349

Jeffreys H (1998) Theory of probability. Oxford University Press, Oxford

Jensen FV (1996) An introduction to Bayesian networks, vol 210. UCL Press, London

Kennedy MC, O'Hagan A (2011) Bayesian calibration of computer models. J R Stat Soc Ser B Stat Methodol 63(3):425–464

Lee PM (2004) Bayesian statistics. Arnold, London

Leonard T, Hsu JSJ (2001) Bayesian methods. Cambridge Books, New York

Liu JS, Liang F, Wong WH (2000) The multiple-try method and local optimization in metropolis sampling. J Am Stat Assoc 95:121–134

McFarland JM (2008) Uncertainty analysis for computer simulations through validation and calibration. Ph.D. thesis, Vanderbilt University

Metropolis N, Rosenbluth AW, Rosenbluth MN, Teller AH, Teller E (1953) Equation of state calculations by fast computing machines. J Chem Phys 21(6):1087

Neal RM (2003) Slice sampling. Ann Stat 31(3):705–741

Pawitan Y (2001) In all likelihood: statistical modelling and inference using likelihood. Oxford University Press, New York

Rajashekhar MR, Ellingwood BR (1993) A new look at the response surface approach for reliability analysis. Struct Saf 12(3):205–220

Rasmussen CE (1996) Evaluation of Gaussian processes and other methods for non-linear regression. Ph.D. thesis, University of Toronto

Rasmussen CE (2004) Gaussian processes in machine learning. In: Advanced lectures on machine learning. Springer, Berlin, pp 63–71

Red-Horse JR, Paez TL (2008) Sandia national laboratories validation workshop: structural dynamics application. Comput Methods Appl Mech Eng 197(29–32):2578–2584

Roberts GO, Rosenthal JS (2006) Harris recurrence of metropolis-within-gibbs and trans-dimensional markov chains. Ann Appl Probab 16(4):2123–2139

Rosenblatt M (1956) Remarks on some nonparametric estimates of a density function. Ann Math Stat 27:832–837

Saltelli A, Ratto M, Andres T, Campolongo F, Cariboni J, Gatelli D, Saisana M, Tarantola S (2008) Global sensitivity analysis: the primer. Wiley, Hoboken

Sankararaman S (2012) Uncertainty quantification and integration in engineering systems. Ph.D. thesis, Vanderbilt University

Sankararaman S, Mahadevan S (2011) Likelihood-based representation of epistemic uncertainty due to sparse point data and/or interval data. Reliab Eng Syst Saf 96(7):814–824

Sankararaman S, McLemore K, Mahadevan S, Bradford SC, Peterson LD (2013) Test resource allocation in hierarchical systems using bayesian networks. AIAA J 51(3):537–550

Santner TJ, Williams BJ, Notz W (2003) The design and analysis of computer experiments. Springer, New York

Singpurwalla ND (2006) Reliability and risk: a Bayesian perspective, vol 637. Wiley, Hoboken

Singpurwalla ND (2007) Betting on residual life: the caveats of conditioning. Stat Probab Lett 77(12):1354–1361

Somersalo E, Kaipio J (2004) Statistical and computational inverse problems. Springer, New York

Stigler SM (1986) Laplace's 1774 memoir on inverse probability. Stat Sci 1(3):359–363

Stigler SM (2002) Statistics on the table: the history of statistical concepts and methods. Harvard University Press, Cambridge

Thornton EA (1996) Thermal structures for aerospace applications. AIAA J, Reston, VA

Tipping ME (2001) Sparse Bayesian learning and the relevance vector machine. J Mach Learn Res 1:211–244

Wackernagel H (2003) Multivariate geostatistics: an introduction with applications. Springer, New York

The Stimulus-Driven Theory
of Probabilistic Dynamics

Agnès Peeters

Abstract Probabilistic Safety Assessments (PSA) are widely used to evaluate the safety of installations or systems. Nevertheless, classical PSA methods—as fault trees or event trees—have difficulties dealing with time dependencies, competition between events and uncertainties. The Stimulus-Driven Theory of Probabilistic Dynamics (or SDTPD) copes with these dynamics aspects. It is a general theory based on dynamic reliability, supplemented by the notion of stimulus. Hence, each event is divided into two phases: the stimulus activation (as soon as all the conditions for the event occurrence are met) and a delay (before the actual occurrence of the event), with a possible stimulus deactivation if the conditions are no more met. It allows modeling in an accurate way the competitions between events. This chapter presents the theory of the SDTPD as well as the solving of a simple example, using the MoSt computer code.

1 Introduction

How to deal with dynamic systems, with strong interactions between their differen components or sub-systems, with physical processes, with continuous evolutions and discrete events, with uncertainties and with important timing aspects? The theory presented in this chapter takes up the challenge!

This theory is quite general and can be applied to a large range of situations. Nevertheless, the Stimulus-Driven Theory of Probabilistic Dynamics (or SDTPD) is particularly required for large systems whose the evolution is governed by dynamics phenomena.

A. Peeters (✉)
Institut Supérieur Industriel de Bruxelles (ISIB), Haute Ecole Paul-Henri Spaak,
150 rue Royale, 1000 Bruxelles, Belgium
e-mail: peeters@isib.be

S. Kadry and A. El Hami (eds.), *Numerical Methods for Reliability and Safety Assessment: Multiscale and Multiphysics Systems*, DOI 10.1007/978-3-319-07167-1_4,
© Springer International Publishing Switzerland 2015

Fig. 1 Interactions involved
in a system evolution

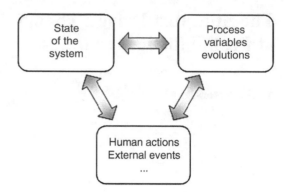

Let us consider a large system S composed of several sub-systems, s_i, considered as entities. This system is described by a set of process variables (physical or not) and the evolution of these variables is modeled by a set of equations.

The values of these variables influence the evolution of the system and the state of the system determines the way these variables evolve. Moreover some punctual actions or events may change the system state and/or the variables evolutions. These interactions are depicted in Fig. 1.

It is a big challenge to model all these interactions and to correctly combine their effects, especially when timing aspects play a significant role.

The following of this section presents these difficulties and highlights the reasons why the SDTPD was created.

Then, the second section of this chapter presents this theory and illustrates it via a very simple test-case.

The third section describes how to apply the SDTPD to a tank issue and presents some results obtained with the Monte Carlo computer code dedicated to the stimuli approach, MoSt.

And, finally, the last section presents some conclusions.

1.1 Background

Before building a structure or starting a system, we have to be sure that the structure will not collapse or that the system will work in a safe way. Consequently, the conception or design phase was historically based on studies using deterministic approaches, conservative parameters, safety margins and even enveloping scenarios (i.e. for large systems). Nowadays this approach is completed by probabilistic studies.

Probabilistic Safety Assessment (PSA) starts in 1975 with the WASH-1400 report (Rasmussen 1975) and has frequently demonstrated its usefulness and its effectiveness. One of the major proofs of the value and benefits of the probabilistic approach is the Three Mile Island accident (Rogovin and Frampton 1980), in 1979.

During this accidental transient, the nuclear power plant does not evolve as predicted during the design phase and led to a worse situation than the conservative scenarios considered in the deterministic studies.

PSA methodologies are based on a best estimate approach (rather than enveloping scenarios), they try to be as close as possible to the actual evolution of the system. They allow exploring in a systematic way the possible evolutions of a system, starting from a specific configuration and evaluating the likelihood of each scenario, while paying a particular attention to scenarios leading to a frightened event. Moreover these studies highlight the main contributors to the final risk.

In order to identify the possible sequences and to evaluate the probability of a specific event, several techniques or methodologies have been developed. The two main approaches are the event trees and the fault trees.

An event tree (Meneley 2001; Rausand 2004; Zio 2002) is an inductive and logical way to represent the different evolutions of a system by a succession of successes or failures (binary events) and to quantify the frequency (or probability) of each evolution.

The event tree starts with an initiating event that can be seen as the first significant deviation from a normal situation (system failure, start of fire, etc.). Then, the analyst has to list all the systems required to manage the consequences of the initiating event and to rank them according to the time they are solicited. Each system solicitation is represented by a node. For each node, there is an incoming branch, which represents the plant evolution before the considered event, and two outcoming branches, which represent the occurrence or success of the event, or its non-occurrence or failure. Of course, the global system evolution on a given branch is conditional to the success or failure of the previous systems.

Let us consider the example of a fire start in a room equipped with a fire detection system, sprinklers and a fire alarm, presented in Rausand (2004).

The initiating event is the fire start. The first solicited system is the fire detection system. If the fire is not detected, no other action is possible; if the fire is detected, the sprinklers should work. If they start, the fire is extinguished; if they do not start, a fire alarm has to sound in order to evacuate the room or building. If the alarm is not activated, no other action is possible. If this alarm is activated but does not work, the result will be the same. If this alarm is activated and works, a safe situation is possible.

After the identification of all these possible scenarios, probabilities are associated to each branch of the even tree. The results are presented in Fig. 2.

The frequency of the feared event is the sum of the frequencies of all the sequences leading to this feared event:

$$\text{Frequency} = F_3 + F_4 + F_5 \tag{1}$$

$$= 0.9 \times 0.15 \times 0.99 \times 0.999 + 0.9 \times 0.15 \times 0.01 + 0.1 \tag{2}$$

$$= 0.00013 + 0.00135 + 0.1 \tag{3}$$

$$= 0.10148 \tag{4}$$

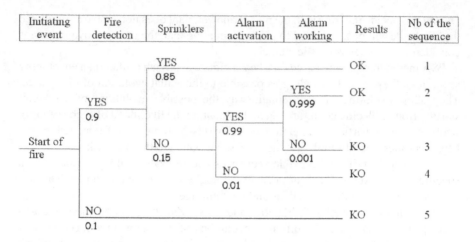

Initiating event	Fire detection	Sprinklers	Alarm activation	Alarm working	Results	Nb of the sequence
		YES 0.85			OK	1
	YES 0.9			YES 0.999	OK	2
Start of fire		NO 0.15	YES 0.99	NO 0.001	KO	3
			NO 0.01		KO	4
	NO 0.1				KO	5

Fig. 2 Event tree example, with frequencies (adapted from Rausand 2004)

On the basis of all the values introduced or calculated in the event tree, it is possible to determine the sequences which are the most important contributors to this event and the sequences may be ranked from the most dangerous to the less dangerous. The main contributor to the risk in this fire example is the fire detection system.

These data could then be used by analysts in order to improve the global safety.

Of course, an event tree has to be built for each considered initiating event. For large systems, it leads to many trees and an explosion of the total number of branches.

Unlike event trees, fault trees are a deductive and systematic way to represent the different causes of an undesired event, thanks to logical gates.

The start point is the considered event, called *top event*, as a system failure. Then, the next stage contains all the direct contributors to this top event. These contributors could be sufficient to lead to the top event or have to be combined with other contributors. In the latter case, all the involved contributors are linked to the top event via an AND gate. For the second stage, each contributor identified in the first stage has to be separately studied. It becomes the top event of a sub-fault tree, which traces backward the causes of this contributor (of course, if the contributor is a basic event, which cannot be further decomposed, the branch is stopped and the basic event is represented by a circle). The construction of the tree goes on stage by stage with further and deeper investigations and causal relations until each path is ended by a basic event.

Let us consider the very simple hydraulic system presented in Avontuur (2003) and illustrated in Fig. 3. A pump takes off water from barrel D and feeds barrel E via two pipes.

The studied event is the lack of water in barrel E. It must be due to the fact that no water goes out pipes B and C. If no water goes out pipe B, it could be due to the

Fig. 3 Hydraulic system, adapted from Avontuur (2003)

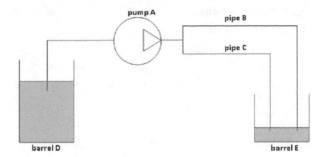

Fig. 4 Fault tree associated to the hydraulic system

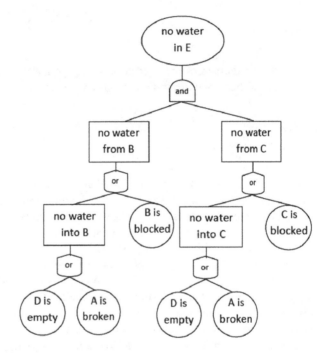

fact either that no water enters pipe B or that pipe B is broken. If no water enters pipe B, it is either because there is no water in barrel D or because pump A does not work. The same analysis is performed for pipe C.

The resulting fault tree is presented in Fig. 4 and a simplified fault tree, obtained thanks to Boolean algebra, is presented in Fig. 5.

Then, frequencies have to be assigned to each basic event and the frequency of each (non-basic) event is calculated by combining the frequencies of its sub-events according to the logical gates used in the tree (as long as there is no repeated event in the tree and no dependencies between the considered events). In this way, frequencies are brought up to the top event.

Like in event trees, it is possible to rank the contributors according to their importance.

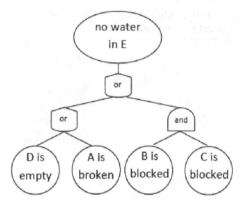

Fig. 5 Simplified fault tree associated to the hydraulic system

Generally, fault trees are used to study and quantify the events involved in the event trees. These two different methods are hence used in a complementary way.

1.2 Motivations

These methods are well known and used all over the world in different sectors (for nuclear power plants of course but also for chemical industries or aerospace systems, etc.). Nevertheless they are not adapted to dynamics systems and do not allow evaluating the impact of assumptions or approximations on the accident progression and on the final likelihood of the considered event.

Let us consider a very simple example, presented in Labeau and Izquierdo (2005a) and Peeters and Labeau (2004). A combustion inside a containment leads to a pressure increase. The undesired event is the containment failure if the pressure is higher than a specific value, p_c. In order to avoid such a situation, a pressure valve is assumed to open if the pressure is higher than a specific value, p_v.

The combustion duration belongs to the interval $[0, t_H]$ with a uniform distribution. The thresholds p_v and p_c, respectively, belong to intervals $[p_v^{min}, p_v^{max}]$ and $[p_c^{min}, p_c^{max}]$ with uniform distributions (with $p_v^{min} < p_c^{min}$).

The probability density functions associated to the combustion time, the valve opening setpoint and the containment rupture pressure are, respectively, noted f_H, f_v and f_c. The associated cumulative density functions are F_H, F_v and F_c.

The pressure evolution due to the combustion is:

$$\frac{dp}{dt} = c \tag{5}$$

The pressure decrease when the valve is opened is governed by the following equation:

$$\frac{dp}{dt} = -\kappa_v p \tag{6}$$

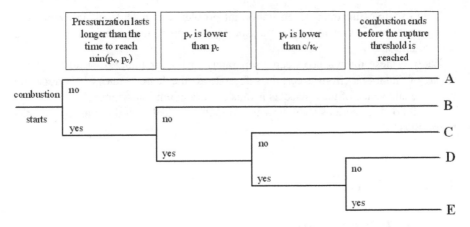

Fig. 6 Event tree of the containment pressurization problem, based on Labeau and Izquierdo (2005a)

Consequently, if the valve opens before the end of the combustion, the pressure evolution is given by the sum of the contributions of both phenomena:

$$\frac{dp}{dt} = c - \kappa_v p \qquad (7)$$

On the basis of this description, how can the system evolve?

The combustion may end before pressure reaches p_v or p_c. The valve does not open and no containment rupture is possible (scenario A).

The combustion may end before the pressure reaches the value p_v but after the pressure reached the value p_c (assuming that p_c is lower than p_v). In this case, the valve does not open but the containment fails (scenario B).

If the combustion ends after the pressure reached the value p_v, the valve opens and several scenarios are possible. If the pressure decreases after the valve opening, no rupture is possible (scenario C). In other cases, the pressure continues to increase after the valve opening and a rupture is possible. If the combustion ends before the pressure reaches the threshold p_c, there is no combustion (scenario D). If the combustion does not, the containment fails (scenario E).

All these scenarios are presented in the event tree of Fig. 6.

On this event tree, we can see that headers depend on the evolution of the process variables, on uncertainties on the combustion duration and thresholds, and on parameters (the velocity of the pressurization due to the combustion and the velocity of the depressurization due to the valve opening).

The competition between the times the values p_v and p_c are, respectively, reached is very important because it could lead to opposite situations (rupture of the containment or safe situation).

This example illustrates three important problems met while using event trees to model dynamic systems: the time aspects, the competition between events, and uncertainties.

Each of these three aspects can—separately or in combination—lead to a large range of situations (from a certain safe state to the feared state). A correct and accurate allowance of these aspects is thus of paramount importance.

The following sections show how it is possible to deal with these aspects thanks to the SDTPD.

2   The SDTPD

2.1   SDTPD: Principles

The SDTPD (Labeau and Izquierdo 2005a,b; Peeters 2012, 2013a,b) is based on the dynamic reliability concepts (Devooght and Smidts 1996; Marseguerra et al. 1998) and consequently describes the system evolutions as a succession of entries in different dynamics (as shown in Fig. 7). Each dynamics corresponds to a specific global state of the system and the evolution of this system is governed, in each dynamics, by a specific set of equations. Dynamic transitions are due to a change of the global state of the system as, for example, the failure of a component, the start of a safety system, a human action, etc. Each transition between two dynamics represents the occurrence of an event.

In the framework of the SDTPD, each event occurrence is divided into two different phases. The first one corresponds to the time needed to fulfil all the conditions required for the event occurrence. The second one corresponds to the time elapsed between the time the conditions are met and the time of the actual event occurrence (see Fig. 8).

But, during this time interval, the system continues to evolve and to change. These changes could lead the system in a state where the event occurrence is no

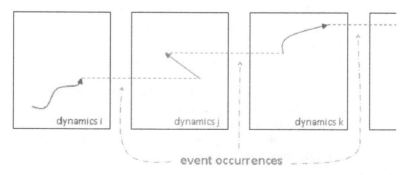

Fig. 7 Representation of the system evolution within the dynamic reliability approach

Fig. 8 The two phases of an event occurrence

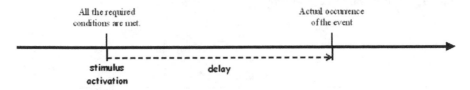

Fig. 9 The two phases of an event occurrence, described by a stimulus

Table 1 Examples of events and their possible modeling by stimulus

Event	Stimulus triggering	Delay
The start of a system, on the basis of a threshold	The crossing of the threshold	The mechanical inertia of the system
A human action if a threshold is crossed	The crossing of the threshold	Diagnosis and reaction
Combustion	Flammable mixture	Time before a spark

more possible. For example, in the case of an hydrogen combustion (Peeters and Labeau 2007), even if the gas mixture becomes flammable, we have to wait the ignition of the gases before the actual combustion. And before this ignition, a system which reduces the hydrogen concentration (as recombiners) may start. Hence, the gas mixture is no more flammable and no combustion is possible.

In the SDTPD formalism, an activation stimulus is associated to each event. The first phase corresponds to the triggering of the activation stimulus and, then an activation delay (long, short or even equal to zero; probabilistic or deterministic; etc.), associated to this stimulus, starts to elapse. Once the activation delay has elapsed, the event occurs (see Fig. 9). Hence, before the event occurrence, the activation stimulus has to be triggered; no event occurrence is possible before this triggering.

This delay can represent the inertia of a mechanical system, the uncertainty associated to a threshold, the time needed to perform a diagnosis in case of a human action, etc. Some examples are presented in Table 1.

If, during the elapsing of this activation delay, the system changes and prevents the event from happening, the activation stimulus has to be deactivated. Indeed, the activation stimulus can be seen as a flag, informing if an event is possible or not.

This deactivation process is also composed of two phases: the triggering of the deactivation stimulus (when the conditions required for the deactivation are met) and the elapsing of the deactivation delay (before the real deactivation).

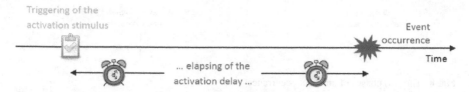

Fig. 10 Activation without deactivation

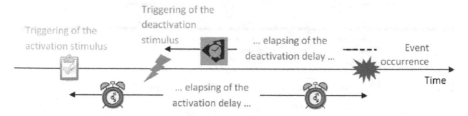

Fig. 11 Activation with triggering of the deactivation stimulus but occurrence of the event

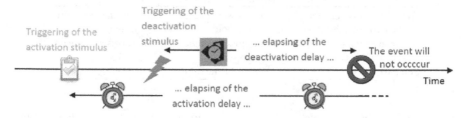

Fig. 12 Activation with triggering of the deactivation stimulus and without occurrence of the event

After the triggering of an activation or deactivation stimulus, an activation or a deactivation delay starts to elapse.

Of course, no triggering of a deactivation stimulus is possible before the triggering of the activation stimulus associated to the same event. The elapsing of an activation or deactivation delay may not start before the triggering of the corresponding activation or deactivation stimulus.

On the basis of these descriptions, three situations are possible.

In the first situation, there is no deactivation (see Fig. 10). No deactivation is possible or the deactivation conditions are not met. The event occurs.

In the second situation, the deactivation stimulus is triggered but the activation delay is elapsed before the elapsing of a deactivation delay (see Fig. 11). The event also occurs.

In the third situation, the deactivation stimulus is triggered and the deactivation delay is elapsed before the elapsing of the activation delay (see Fig. 12). The event is inhibited and does not occur.

Summarizing, a twofold stimulus has to be associated to each event: the first part is dedicated to the activation (called *activation stimulus*) and the second one is dedicated to the deactivation (called *deactivation stimulus*). Then a delay is

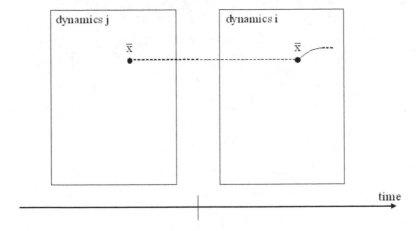

Fig. 13 Transition between dynamics j and i at time t with process variables $\overline{x}$

associated to each of these parts (called *activation delay* or *deactivation delay*); it starts to elapse as soon as the corresponding stimulus is triggered and the associated event (the actual event of the deactivation of the activation stimulus) occurs as soon as the delay is elapsed. Depending on the modeled event, delays could be equal to zero and a deactivation could be impossible. But the modeling of an event must at least contains an activation stimulus.

2.2 SDTPD: Mathematical Formulation

This section presents the main equations related to SDTPD and shows how the principles described in the previous section can be expressed with a mathematical formalism (using the following semi-markovian assumption: each entry of the system in a new dynamics is a regeneration point).

More details can be found in Labeau and Izquierdo (2005a).

The evolution of the process variable x_j in dynamics i is described by:

$$x_j(t) = g_{ji}(t, \overline{x}) \tag{8}$$

where $\overline{x}$ is the vector of process variables (and x_j is the jth component of this vector).

The system evolution depends on the succession of dynamics entered by the system and we would like to calculate the ingoing density of the system into dynamics i at time t with process variables $\overline{x}$, $\varphi(\overline{x}, i, t)$.

To enter dynamics i at time t with process variables $\overline{x}$, the system has to come from another dynamics, j, and to undergo a transition from dynamics i to dynamics j at time t with process variables $\overline{x}$ (as illustrated in Fig. 13).

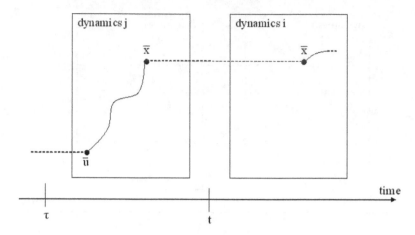

Fig. 14 Transitions at time τ with process variables $\bar{u}$ and at time t with process variables $\bar{x}$

In order to have this transition at time t with process variables $\bar{x}$, the system has had to evolve in dynamics j from its entry point $\bar{u}$ (at time τ) to point $\bar{x}$ during the time interval $t - \tau$ (see Fig. 14).

Dynamics j either has been entered at time τ with process variables $\bar{u}$ and with the ingoing density $\varphi(\bar{u}, j, \tau)$, or is the initial dynamics: the system starts in dynamics j with process variables $\bar{u}$ and with the probability density function $\pi(\bar{u}, j, 0)$—assuming that $\pi(\bar{x}, i, t)$ is the pdf of finding the system in dynamics i at time t with process variables $\bar{x}$.

In the first case, the ingoing density $\varphi(\bar{x}, i, t)$ is expressed by the following equation:

$$\varphi(\bar{x}, i, t) = \varphi(\bar{u}, j, \tau) \times \delta(\bar{x} - \bar{g}_j(t - \tau, \bar{u})) \times q_{ji}(t - \tau, \bar{u}) \tag{9}$$

where the first factor expresses that the system enters dynamics j at τ with $\bar{u}$, the second factor expresses that the system evolves in dynamics j from $\bar{u}$ to $\bar{x}$ in a time $t - \tau$ and the third one is the probability (per time unit) of a transition from dynamics j to dynamics i, a time $t - \tau$ after the entry of the system in dynamics j with process variables $\bar{u}$.

In the second case, we have:

$$\varphi(\bar{x}, i, t) = \pi(\bar{u}, j, 0) \times \delta(\bar{x} - \bar{g}_j(t, \bar{u})) \times q_{ji}(t, \bar{u}) \tag{10}$$

Taking into account that the system may come from any dynamics j, entered at any time τ (between 0 and t) and grouping both cases, we obtain:

$$\varphi(\bar{x}, i, t) = \sum_{j \neq i} \int_0^t d\tau \int \left[\pi(\bar{u}, j, \tau) \times \delta(\tau) + \varphi(\bar{u}, j, \tau)\right] \times \delta(\bar{x} - \bar{g}_j(t - \tau, \bar{u})) \times q_{ji}(t - \tau, \bar{u}) \, d\bar{u}$$

$$\tag{11}$$

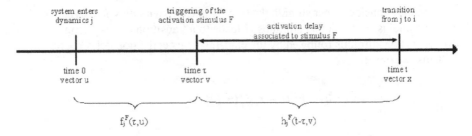

Fig. 15 Chronology of a transition from j to i

Now we have to determine the probability per time unit $q_{ji}(t,\overline{u})$ of having a transition from dynamics j to dynamics i at time t with process variables $\overline{x}$.

In the SDTPD framework, transitions between dynamics are governed by stimuli. No transition can occur before an activation delay has elapsed and no activation delay can elapse without the preliminary triggering of an activation stimulus.

As several stimuli F (associated to mutually exclusive events) can lead to the same dynamics transition from j to i, we have:

$$q_{ji}(t,\overline{u}) = \sum_F q_{ji}^F(t,\overline{u}) \tag{12}$$

where $q_{ji}^F(t,\overline{u})$ is the probability per time unit of a transition between dynamics j and i due to stimulus F a time t after entering dynamics j at point $\overline{u}$.

First, we assume there is no deactivation and no stimulus other than F comes into play.

If the system enters j at time 0 with process variables $\overline{u}$ (see Fig. 15), a transition to i at time t with process variables $\overline{x}$ can occur only if

1. the activation stimulus F is triggered at time τ ($< t$) with process variables $\overline{v}$ ($\equiv \overline{g}_j(\tau,\overline{u})$)
2. the activation delay has elapsed at time t with process variables $\overline{x}$ ($= \overline{g}_j(t - \tau,\overline{v}) = \overline{g}_j(t,\overline{u})$)

If $f_j^F(t,\overline{x})$ is the probability density function of triggering activation stimulus F (associated to an event which can occur in dynamics j and lead to the system entry into dynamics i) a time t after entering dynamics j at point $\overline{x}$ and $h_j^F(t,\overline{x})$ is the probability density function of elapsing of the activation delay associated to stimulus F in dynamics j a time t after the triggering of the activation stimulus F at point $\overline{x}$, we can write:

$$q_{ji}^F(t,\overline{u}) = \int_0^t f_j^F(\tau,\overline{u}) \times h_j^F(t - \tau,\overline{v}) \, d\tau \tag{13}$$

$$= \int_0^t f_j^F(\tau,\overline{u}) \times h_j^F(t - \tau,\overline{g}_j(\tau,\overline{u})) \, d\tau \tag{14}$$

If we also consider other stimuli, the transition from dynamics j to dynamics i is due to stimulus F if no other stimulus G leads to a transition out of j (triggering of this activation stimulus followed by the elapsing of its activation delay) before t.

Thus, we have:

$$q_{ji}^F(t,\overline{u}) = \int_0^t f_j^F(\tau,\overline{u}) \times h_j^F(t-\tau,\overline{g}_j(\tau,\overline{u}))\,d\tau$$

$$\times \prod_{G \neq F} \left[1 - \int_0^t dt' \int_0^{t'} f_j^G(\tau',\overline{u}) \times h_j^G(t'-\tau',\overline{g}_j(\tau',\overline{u}))\,d\tau' \right] \quad (15)$$

If we add a possible deactivation of stimulus F, we have to take into account that the transition between j and i occurs only if F is not deactivated.

However, F is deactivated if its deactivation stimulus is triggered at time τ^* (between τ and t) and if the deactivation delay is elapsed before t.

If $k_j^F(t,\overline{x})$ is the probability density function of triggering deactivation stimulus F a time t after triggering the associated activation stimulus at point $\overline{x}$ in dynamics j and $l_j^F(t,\overline{x})$ is the probability density function of elapsing of the deactivation delay associated to stimulus F in dynamics j a time t after triggering the deactivation stimulus F at point $\overline{x}$, we can write:

$$q_{ji}^F(t,\overline{u}) = \left[\int_0^t f_j^F(\tau,\overline{u}) \times h_j^F(t-\tau,\overline{g}_j(\tau,\overline{u})) \right.$$

$$\times \left[1 - \int_\tau^t d\tau^* \int_{\tau^*}^t k_j^F(\tau^*-\tau,\overline{g}_j(\tau,\overline{u})) \times l_j^F(t^*-\tau^*,\overline{g}_j(\tau^*,\overline{u}))\,dt^* \right] d\tau \right]$$

$$\times \prod_{G \neq F} \left[1 - \int_0^t dt' \int_0^{t'} f_j^G(\tau',\overline{u}) \times h_j^G(t'-\tau',\overline{g}_j(\tau',\overline{u}))\,d\tau' \right] \quad (16)$$

if a deactivated stimulus is not allowed to be reactivated.

If we also consider that the other stimuli can be deactivated, they will not lead to a transition if the activation delay has not elapsed at time t or if a deactivation occurs before time t. So, we have:

$$q_{ji}^F(t,\overline{u}) = \left[\int_0^t f_j^F(\tau,\overline{u}) \times h_j^F(t-\tau,\overline{g}_j(\tau,\overline{u})) \right.$$

$$\times \left[1 - \int_\tau^t d\tau^* \int_{\tau^*}^t k_j^F(\tau^*-\tau,\overline{g}_j(\tau,\overline{u})) \times l_j^F(t^*-\tau^*,\overline{g}_j(\tau^*,\overline{u}))\,dt^* \right] d\tau \right]$$

$$\times \prod_{G \neq F} \left[1 - \int_0^t dt' \left[\int_0^{t'} f_j^G(\tau',\overline{u}) \times h_j^G(t'-\tau',\overline{g}_j(\tau',\overline{u})) \right. \right.$$

$$\left. \left. \times \left[1 - \int_{\tau'}^{t'} d\tau'' \int_{\tau''}^{t'} k_j^G(\tau''-\tau',\overline{g}_j(\tau',\overline{u})) \times l_j^G(t''-\tau'',\overline{g}_j(\tau'',\overline{u}))\,dt'' \right] d\tau' \right] \right] \quad (17)$$

Finally, the ingoing density of the system into dynamics i at time t with process variables $\bar{x}$ has the following expression:

$$\varphi(\bar{x}, i, t) = \sum_{j \neq i} \sum_{F} \int_0^t d\tau \int d\bar{u} \times \left[\pi(\bar{u}, j, \tau) \times \delta(\tau) + \varphi(\bar{u}, j, \tau) \right]$$

$$\times \delta(\bar{x} - \bar{g}_j(t - \tau, \bar{u})) \times \left[\int_0^t f_j^F(\tau, \bar{u}) \times h_j^F(t - \tau, \bar{g}_j(\tau, \bar{u})) \right.$$

$$\times \left[1 - \int_\tau^t d\tau^* \int_{\tau^*}^t k_j^F(\tau^* - \tau, \bar{g}_j(\tau, \bar{u})) \right.$$

$$\left. \times l_j^F(t^* - \tau^*, \bar{g}_j(\tau^*, \bar{u})) \, dt^* \right] d\tau$$

$$\times \prod_{G \neq F} \left[1 - \int_0^t dt' \left[\int_0^{t'} f_j^G(\tau', \bar{u}) \times h_j^G(t' - \tau', \bar{g}_j(\tau', \bar{u})) \right. \right.$$

$$\times \left[1 - \int_{\tau'}^{t'} d\tau'' \int_{\tau''}^{t'} k_j^G(\tau'' - \tau', \bar{g}_j(\tau', \bar{u})) \right.$$

$$\left. \left. \times l_j^G(t'' - \tau'', \bar{g}_j(\tau'', \bar{u})) \, dt'' \right] d\tau' \right] \right]$$

(18)

It is possible to release the non-reactivation assumption if any deactivation point is considered as a regeneration point.

2.3 Shocks

In probabilistic safety studies, the time durations of the considered phenomena can be very different and vary from a few seconds to several days. In such situations, short events could be considered as instantaneous in comparison with the transient duration.

Consequently, the notion of shock has been introduced in the SDTPD.

The effect of a shock is to instantaneously change the value of the process variables (without change the time value).

A shock acts in the following way, illustrated in Fig. 16. The system leaves dynamics j at time t with process variables $\bar{x}$ and enters dynamics s (corresponding to a shock) at the same time with the same values. Then, the system is instantaneously brought to point $\bar{x}^*$. Finally, the system enters dynamics i at time t with process variables $\bar{x}^*$.

The way the ingoing density function has to be modified is explained in Labeau and Izquierdo (2005a).

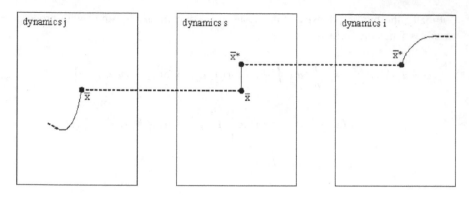

Fig. 16 The notion of shock

2.4 A Pressurization Test-Case

In this section, the SDTPD will be applied to the containment pressurization issue described in Sect. 1.2.

2.4.1 Problem Analysis

According to the description of this problem, there are four possible situations and each of them corresponds to a specific pressure evolution law:

1. Pressurization
   The combustion inside the containment leads to a linear pressure increase:

$$\frac{dp}{dt} = c \tag{19}$$

2. Equilibrium
   The combustion inside the containment has stopped; the pressure value does not change:

$$\frac{dp}{dt} = 0 \tag{20}$$

3. Pressurization and depressurization
   The combustion goes on (and leads to a pressure increase) and the valve is opened (and leads to a pressure decrease):

$$\frac{dp}{dt} = c - \kappa_v p \tag{21}$$

4. Depressurization
   The valve is opened and the combustion has stopped:

$$\frac{dp}{dt} = -\kappa_v p \tag{22}$$

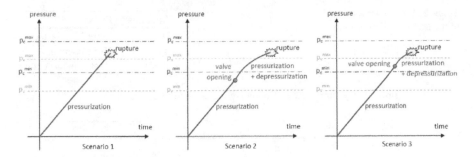

Fig. 17 Scenarios leading to a containment rupture

During a transient, three events can occur:

1. end of combustion
2. valve opening
3. containment rupture

In this example, we are only interested in situations where $\frac{c}{\kappa_v}$ is greater than $p_c^{\min}$ (in order to keep a pressure increase after the valve opening) and where $p_c^{\min}$ is lower than $p_v^{\max}$ (otherwise no rupture is possible).

2.4.2 Analytical Solution (Without Stimulus)

There are three scenarios likely to lead to a containment rupture (see Fig. 17):

1. The combustion is going on and the valve remains closed.
2. The combustion is going on and the valve has been opened before reaching $p_c^{\min}$.
3. The combustion is going on and the valve has been opened after reaching $p_c^{\min}$.

Hence, the rupture probability is the sum of three terms:

$$P_{\text{rupt}}(t) = P_1(t) + P_2(t) + P_3(t) \tag{23}$$

$P_1(t)$ is the rupture probability associated to the first scenario and is given by:

$$P_1(t) = \int_{p_c^{\min}}^{p_c^{\max}} f_c(p_c) \times \left[1 - F_H\left(\frac{p_c}{c}\right)\right] \times \left[1 - F_v\left(p_c\right)\right] \times \theta\left(t - \frac{p_c}{c}\right) dp_c \tag{24}$$

where p_c is the rupture pressure.

$P_2(t)$ is the rupture probability associated to the second scenario and is given by:

$$P_2(t) = \int_{p_v^{\min}}^{p_c^{\min}} dp_v \, f_v\left(p_v\right) \times \theta\left(p\left(t \mid p_v\right) - p_c^{\min}\right)$$

$$\int_{p_c^{\min}}^{p(t \mid p_v)} dp_c \, f_c\left(p_c\right) \times \left[1 - F_H\left(t_c\left(p_c \mid p_v\right)\right)\right] \tag{25}$$

Table 2 Stimuli associated to the pressurization test-case

Stimulus	Event
0	End of combustion
1	Valve opening
2	Containment rupture

Table 3 Dynamics and evolution laws associated to the pressurization test-case

Dynamics	Description	Pressure evolution
0	Pressurization	$\frac{dp}{dt} = c$
1	Equilibrium	$\frac{dp}{dt} = 0$
2	Pressurization and depressurization	$\frac{dp}{dt} = c - \kappa_v p$
3	Depressurization	$\frac{dp}{dt} = -\kappa_v p$
4	Rupture	–

where p_v is the valve opening pressure, $p(t|p_v)$ is the pressure value at time t given that the valve opened at p_v $\left(\text{and at time } t_v = \frac{p_v}{c}\right)$ and is given by solving Eq. (21) from p_v:

$$t - t_v = \frac{1}{\kappa_v} \ln \left(\frac{c - \kappa_v p_v}{c - \kappa_v p(t|p_v)} \right) \tag{26}$$

and $t_c(p_c|p_v)$ is the time pressure rupture is reached given that the valve opened at time t_v and pressure p_v:

$$t_c(p_c|p_v) = t_v + \frac{1}{\kappa_v} \ln \left(\frac{c - \kappa_v p_v}{c - \kappa_v p_c} \right) = \frac{p_v}{c} + \frac{1}{\kappa_v} \ln \left(\frac{c - \kappa_v p_v}{c - \kappa_v p_c} \right) \tag{27}$$

$P_3(t)$ is the rupture probability associated to the third scenario and is given by:

$$P_3(t) = \theta \left(t - \frac{p_c^{\min}}{c} \right) \int_{p_c^{\min}}^{\min\left(\frac{c}{\kappa_v}, ct\right)} dp_v \, f_v(p_v) \int_{p_v}^{p(t|p_v)} dp_c \, f_c(p_c) \times \left[1 - F_H\left(t_c(p_c|p_v)\right) \right] \tag{28}$$

2.4.3 Solution Using SDTPD

This problem can also be described and solved through the SDTPD formalism (Labeau and Izquierdo 2005a).

Firstly, we have to associate a stimulus to each event; they are described in Table 2.

Then, a dynamics has to be associated to each different pressure evolution (see Table 3), with an evolution equation for each dynamics.

The event of which we want to evaluate the probability is the containment pressure. So, we add a last dynamics corresponding to this rupture.

Finally, we have to define the probability density functions of triggering of the activation stimuli (for each dynamics) and of elapsing of the activation delays (for each dynamics). No deactivation is possible.

$$f_0^0(t, p') = \frac{f_H\left(t + \frac{p'}{c}\right)}{1 - F_H\left(\frac{p'}{c}\right)} \quad \text{and} \quad h_0^0(t, p'') = \delta(t) \tag{29}$$

$$f_1^0(t, p') = \delta(t - t_{AD}) \tag{30}$$

$$f_2^0(t, p') = \frac{f_H\left(t + \frac{p'}{c}\right)}{1 - F_H\left(\frac{p'}{c}\right)} \quad \text{and} \quad h_2^0(t, p'') = \delta(t) \tag{31}$$

$$f_3^0(t) = \delta(t - t_{AD}) \tag{32}$$

$$f_0^1(t, p') = \delta\left(t - \frac{p_v^{\min} - p'}{c}\right).\theta\left(p_v^{\min} - p'\right) \quad \text{and} \quad h_0^1(t, p'') = c.f_v(p'' + ct) \tag{33}$$

$$f_1^1(t, p') = \delta(t - t_{AD}) \tag{34}$$

$f_2^1(t, p')$: F_2 has already been activated (triggering and delay)

as the system stays in dynamics 3

$$f_3^1(t, p') = \delta(t - t_{AD}) \tag{35}$$

$$f_0^2(t, p') = \delta\left(t - \frac{p_c^{\min} - p'}{c}\right) \times \theta\left(p_c^{\min} - p'\right) \quad \text{and} \quad h_0^2(t, p'') = c\, f_c(p'' + ct) \tag{36}$$

$$f_1^2(t, p') = \delta(t - t_{AD}) \tag{37}$$

$$f_2^2(t, p') = \theta\left(p_c^{\min} - p'\right) \times \delta\left(t - \frac{1}{\kappa_v} \times \ln\left(\frac{c - \kappa_v p'}{c - \kappa_v p_c^{\min}}\right)\right) \quad \text{and} \tag{38}$$

$$h_2^2(t, p'') = \left(c - \kappa_v p''\right) \times e^{-\kappa_v t} \times f_c\left(\frac{c}{\kappa_v} - \left(\frac{c}{\kappa_v} - p''\right)e^{-\kappa_v t}\right) \tag{39}$$

$$f_3^2(t, p') = \delta(t - t_{AD}) \tag{40}$$

Fig. 18 Transitions involved
in the pressurization test-case

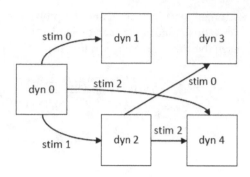

where

t_{AD} is an upper bound of the maximum duration of the transient
$\quad p'$ is the pressure value while entering the considered dynamics
$\quad p''$ is the pressure value when the concerned stimulus has been triggered

On the basis of this description, we can conclude it is not possible to reach the rupture pressure from dynamics 1 and 3. Indeed, these two dynamics correspond to a pressure decrease (see Fig. 18).

As dynamics 0 is the initial dynamics and is entered at time $t = 0$ with the initial pressure value $p = 0$, we have:

$$\varphi(p_c, 4, t_c) = \pi(0, 0, 0) \times \delta(p_c - c.t_c) \times q_{04}(t_c, 0)$$

$$+ \int_0^{t_c} dt_v \int dp_v \varphi(p_v, 2, t_v) \times q_{24}(t_c - t_v, p_v)$$

$$\times \delta\left(p_c - p_v e^{-\kappa_v\left(t_c - \frac{p_v}{c}\right)}\right) \times \left(\left(1 - \frac{c}{p_v \kappa_v}\right) + \frac{c}{\kappa_v}\right) \tag{41}$$

where $\pi(0, 0, 0) = 1$.

A transition from dynamics 0 to dynamics 4 is possible only thanks to stimulus 2. So, we have:

$$q_{04}(t, 0) = q_{04}^2(t, 0) \tag{42}$$

$$= \int_0^t f_0^2(\tau, 0) \times h_0^2(t - \tau, c\tau) \, d\tau \tag{43}$$

$$\times \left[1 - \int_0^t dt' \int_0^{t'} f_0^0(\tau', 0) \times h_0^0\left(t' - \tau', c\tau'\right) d\tau'\right]$$

$$\times \left[1 - \int_0^t dt' \int_0^{t'} f_0^1(\tau', 0) \times h_0^1\left(t' - \tau', c\tau'\right) d\tau'\right]$$

$$= \int_0^t \delta\left(\tau - \frac{p_c^{min}}{c}\right) \times \theta\left(p_c^{min}\right) \times c \, f_c(ct) \, d\tau \tag{44}$$

$$\times \left[1 - \int_0^t dt' \int_0^{t'} f_H(\tau') \delta(t' - \tau') \, d\tau' \right]$$

$$\times \left[1 - \int_0^t dt' \int_0^{t'} \delta\left(\tau' - \frac{p_v^{\min}}{c} \right) \cdot \theta(p_v^{\min}) \times c \, f_v(ct) \, d\tau' \right]$$

$$= f_c(ct_c) \times [1 - F_H(t_c)] \times [1 - F_v(ct_c)] \tag{45}$$

Dynamics 2 can only be entered from dynamics 0 (entered at time $t = 0$, with pressure $p = 0$). Hence, we have:

$$\varphi(p_v, 2, t_v) = \pi(0, 0, 0) \times \delta(p_v - c \cdot t_v) \times q_{02}(t_v, 0) \tag{46}$$

with $\pi(0, 0, 0) = 1$.

A transition from 0 to 2 is possible only thanks to stimulus 1. So, we have:

$$q_{02}(t_v, 0) = q_{02}^1(t_v, 0) \tag{47}$$

$$= \int_0^{t_v} f_0^1(\tau, 0) \times h_0^1(t_v - \tau, c\tau) \cdot d\tau \tag{48}$$

$$\times \left[1 - \int_0^{t_v} dt' \int_0^{t'} f_0^0(\tau', 0) \times h_0^0(t - -\tau', c\tau') d\tau' \right]$$

$$\times \left[1 - \int_0^{t_v} dt' \int_0^{t'} f_0^2(\tau', 0) \times h_0^2(t - -\tau', c\tau') d\tau' \right]$$

$$= \int_0^{t_v} \delta\left(\tau - \frac{p_v^{\min}}{c} \right) \times \theta(p_c^{\min}) \times c \, f_v(ct_v) \, d\tau \tag{49}$$

$$\times \left[1 - \int_0^{t_v} dt' \int_0^{t'} f_H(\tau') \times \delta(t' - \tau') \, d\tau' \right]$$

$$\times \left[1 - \int_0^{t_v} dt' \int_0^{t'} \delta\left(\tau' - \frac{p_c^{\min}}{c} \right) \times \theta(p_c^{\min}) \times c \, f_c(ct') \, d\tau' \right]$$

$$= c \, f_v(ct_v) \times [1 - F_H(t_v)] \times [1 - F_c(ct_v)] \tag{50}$$

A transition from 2 to 4 is possible only thanks to stimulus 2 and stimulus 1 has no more to be considered (because it has already been activated). So, we have:

$$q_{24}(t_c - t_v, p_v) = q_{24}^2(t_c - t_v, p_v) \tag{51}$$

While combining these different results and changing the integration variable, we have the same expression as $P_{\text{rupt}}(t)$, obtained by adding $P_1(t)$, $P_2(t)$ and $P_3(t)$.

3 Application: The Tank Issue

In the previous section, a very simple pressurization test-case has been presented. This example concerns only one process variable (the pressure inside the containment) and no deactivation is possible. Moreover, the number of dynamics and events are small (5 dynamics, including the containment rupture, and 3 stimuli). Hence, it is quite easy to consider all the possible evolution of the system and to calculate the associated probabilities.

Of course, for bigger or more complex systems (and real systems are much more complex), it is not feasible. Consequently, a computer code, called MoSt, has been developed (see Peeters 2012). It is dedicated to the probability evaluation of a specific event, with, on the one hand, a probabilistic aspect based on a combination of the SDTPD and the Monte Carlo simulations and, on the other hand, a deterministic aspect associated to the process and physical variables evolutions. It allows simulating the evolution of any system described with the SDTPD formalism.

The following subsections present a more complex (but still limited) issue (Peeters 2013a) and the way this issue has been described using the SDTPD formalism in order to be solved with MoSt. Some results are then presented.

3.1 General Description

Let us consider a tank with a volume V, used to supply water to a system S requiring water according to three different flows. Each flow is required at a specific time of the day:

- From 0:00 am to 6:00 am
      No water is used (flow 1)
- From 6:00 am to 2:00 pm
      The water flow is constant (flow 2)
- From 2:00 pm to 6:00 pm
      The water flow is exponential (flow 3)
- From 6:00 pm to 0:00 am
      The water flow is constant (flow 2)

This tank is filled in a continuous way thanks to the continuous feeding system (CF) and extra water is provided by the auxiliary feeding system (AF) if a low level in the tank is detected. A contrario, if a high level in the tank is detected, the feeding systems are stopped. Moreover, if the auxiliary tank is used during the day, it will be fulfil during the night (at midnight). The failure of the continuous feeding system is taken into account.

The different parts of this issue are illustrated in Fig. 19.

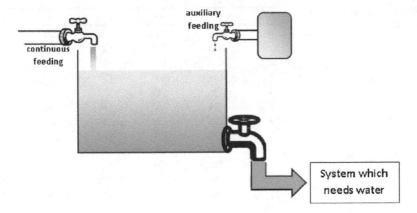

Fig. 19 Configuration of the tank issue

Table 4 Uncertainties on the timing of the water needs

From	To	Pdfs values
0:00 am	5:00 am	0
5:00 am	5:55 am	0.0027
5:55 am	6:05 am	0.07
6:05 am	7:00 am	0.0027
7:00 am	1:30 pm	0
1:30 pm	2:30 pm	0.0167
2:30 pm	5:00 pm	0
5:00 pm	5:30 pm	0.0067
5:30 pm	6:30 pm	0.0083
6:30 pm	7:00 pm	0.01
7:00 pm	0:00 am	0

There are two situations to avoid: to be unable to supply water to the system, and an overflowing tank (with possible damages to other installations). The likelihood of these two situations will be studied as well as their evolution over time.

3.2 Uncertainties

Of course, uncertainties are linked to these situations.

The timing of the water needs may change from 1 day to another; the probability density functions are constant on each time interval (see Table 4).

The level sensors are not perfect and the associated uncertainties are represented by gaussian functions.

Table 5 Description of the 14 dynamics

Nb	CF	AF	F1	F2	F3
0	Off	Off	On	Off	Off
1	Off	Off	Off	On	Off
2	Off	Off	Off	Off	On
3	On	Off	On	Off	Off
4	On	Off	Off	On	Off
5	On	Off	Off	Off	On
6	Off	On	On	Off	Off
7	Off	On	Off	On	Off
8	Off	On	Off	Off	On
9	On	On	On	Off	Off
10	On	On	Off	On	Off
11	On	On	Off	Off	On
12	Overflow				
13	Empty tank				

Table 6 Evolution equations of the levels according to each dynamic

Dyn nb	Tank level	Aux. level
0	$L(t) = L_0$	$l(t) = l_0$
1	$L(t) = L_0 - d \cdot t$	$l(t) = l_0$
2	$L(t) = L_0 - e^{b \cdot t}$	$l(t) = l_0$
3	$L(t) = L_0 + c \cdot t$	$l(t) = l_0$
4	$L(t) = L_0 + (c - d) \cdot t$	$l(t) = l_0$
5	$L(t) = L_0 - e^{b \cdot t} + c \cdot t$	$l(t) = l_0$
6	$L(t) = L_0 + a \cdot t$	$l(t) = l_0 - a \cdot t$
7	$L(t) = L_0 + (a - d) \cdot t$	$l(t) = l_0 - a \cdot t$
8	$L(t) = L_0 - e^{b \cdot t} + a \cdot t$	$l(t) = l_0 - a \cdot t$
9	$L(t) = L_0 + (a + c) \cdot t$	$l(t) = l_0 - a \cdot t$
10	$L(t) = L_0 + (a + c - d) \cdot t$	$l(t) = l_0 - a \cdot t$
11	$L(t) = L_0 - e^{b \cdot t} + (a + c) \cdot t$	$l(t) = l_0 - a \cdot t$

3.3   SDTPD Formalism

3.3.1   Dynamics

As explained in Sect. 2.1, the ways the system evolves are described through dynamics. Hence, a dynamics has to be associated to each kind of evolution of the system. In the tank issue, fourteen dynamics (numbered from 0 to 13) have been identified; they are described in Table 5. The two studied events (overflow and empty tank) are, respectively, represented by dynamics 12 and 13.

Three variables are needed to describe the evolution of the system; they are: the water level in the tank (L), the water level in the auxiliary tank (l) and the hour. The hour evolves of course in the same way independently of the dynamics. The evolution of the two other variables depends on the dynamics and is described in Table 6.

The entry of the system in dynamics 12 or 13 stops the simulation. Consequently, there is no evolution equation associated to these dynamics.

Table 7 Description of the 11 stimuli

Stimulus	Description	Allowed transitions
Stimulus 0	Transition from flow 1 to flow 2	From dynamics 0 to dynamics 1
		From dyn. 3 to dyn. 4
		From dyn. 6 to dyn. 7
		From dyn. 9 to dyn. 10
Stimulus 1	Transition from flow 2 to flow 3	From dyn. 1 to dyn. 2
		From dyn. 4 to dyn. 5
		From dyn. 7 to dyn. 8
		From dyn. 10 to dyn. 11
Stimulus 2	Transition from flow 2 to flow 1	Opposite transitions to stim. 0
Stimulus 3	Transition from flow 3 to flow 2	From dyn. 2 to dyn. 1
		From dyn. 5 to dyn. 4
		From dyn. 8 to dyn. 7
		From dyn. 11 to dyn. 10
Stimulus 4	Start of the continuous feeding system	From dyn. 0 to dyn. 3
		From dyn. 1 to dyn. 4
		From dyn. 2 to dyn. 5
		From dyn. 6 to dyn. 9
		From dyn. 7 to dyn. 10
		From dyn. 8 to dyn. 11
Stimulus 5	Normal stop of the continuous feeding system	Opposite transitions to stim. 4
Stimulus 6	Start of the auxiliary feeding system	From dyn. 0 to dyn. 6
		From dyn. 1 to dyn. 7
		From dyn. 2 to dyn. 8
		From dyn. 3 to dyn. 9
		From dyn. 4 to dyn. a 10
		From dyn. 5 to dyn. 11
Stimulus 7	Stop of the auxiliary feeding system	Opposite transitions to stim. 6
Stimulus 8	Tank overflow	From any dyn. to dyn. 12
Stimulus 9	Empty tank	From any dyn. to dyn. 13
Stimulus 10	Stop of the continuous feeding system, due to a failure	Same transitions as stim. 5

3.3.2 Stimuli

The occurrence of an event is represented by an activation stimulus and the elapsing of the delay associated to such a stimulus leads to a change of dynamics.

In this issue, the events are the start or stop (normal or accidental) of the continuous feeding system, the start or stop of the auxiliary feeding system, a change of flow, an overflow or an empty tank (combined with water needs). The 11 stimuli associated to these events are presented in Table 7, with the transitions allowed after the elapsing of their activation delays.

In order to complete the definition of each stimulus, we have to describe the probability density functions associated to the triggering of the activation stimulus, the delay of the activation stimulus, the triggering of the deactivation stimulus and the delay of the deactivation stimulus, respectively labeled f, h, k and l.

Stimulus 0 represents the transition from flow 1 to flow 2 and should occur at six o'clock, with an uncertainty of $\pm 1u$. Thus, the transition is possible since five o'clock. Consequently, the activation stimulus is triggered at this time; it corresponds to a threshold. The associated delay is given by the following function:

$$h_i^0 = \begin{cases} 0.15/55 & \text{if } 5:00 < \text{clock} < 5:55 \\ 0.07 & \text{if } 5:55 \le \text{clock} < 6:05 \\ 0.15/55 & \text{if } 6:05 \le \text{clock} < 7:00 \end{cases}$$

where i corresponds to any dynamics (the dynamics the system evolves does not influence this probability density function).

Stimuli 1, 2 and 3 are described in a similar way.

Stimulus 4 represents the normal stop of the continuous feeding system and is consecutive to a high level signal. This signal occurs when the water volume in the tank is equal to 50 % of the tank volume. But the sensibility of the sensor leads to an uncertainty modeled by the following gaussian function:

$$h_i^4 = \frac{1}{0.02V\sqrt{2\pi}} \cdot e^{-\frac{(L-0.5V)^2}{2(0.02V)^2}} \tag{52}$$

if the water level in the tank is between $0.45V$ and $0.55V$ (and $h_i^4 = 0$ otherwise).

This level could decrease due to the water flow to the system S or increase thanks to the auxiliary feeding system. If the water level becomes too high before the end of the activation delay, the associated deactivation stimulus has to be triggered and its delay is equal to zero.

Other stimuli linked to the level sensors are described in a similar way.

3.4 Results

The MoSt program (presented in Peeters 2012) has been used to perform simulations. This program is based on the SDTPD and on Monte Carlo simulations. Hence, the accuracy of the final result depends on the number of histories used for the simulations. Figures 20 and 21 show the fluctuations of these results. We can see that simulations with a low number of histories lead to visible fluctuations. For simulations with at least 100,000 histories, the fluctuations are negligible.

Figure 22 shows the succession of the dynamics entered by the system and the timing of these dynamics changes. This is related to Fig. 23 which shows the evolution of the water levels in the tank and in the auxiliary feeding system. The first decrease of the tank level corresponds to the change from flow 1 to flow 2 (change from dynamics 0 to dynamics 1). Then, the auxiliary feeding system starts,

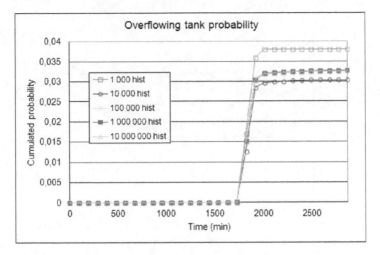

Fig. 20 Evolution of the overflowing tank probability, for different numbers of histories in the Monte Carlo simulations

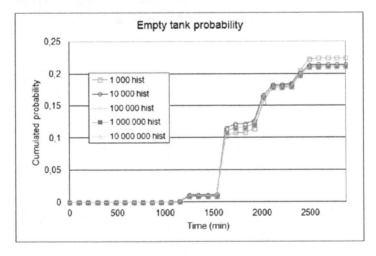

Fig. 21 Evolution of the empty tank probability, for different numbers of histories in the Monte Carlo simulations

the dynamics changes from 1 to 7 and the water level decrease is slowed. After a specific time, the auxiliary tank is empty and the level in the tank decreases in the same way as during the first phase. Then, the auxiliary tank is instantaneously filled and its level reaches the maximal value. These figures are only one example of the results obtained after a simulation, only one possible scenario. Indeed, MoSt prints all the data (process variables evolutions, activation, deactivation, delays of each stimulus, etc.) for a specific number of histories, randomly chosen before the beginning of the simulation.

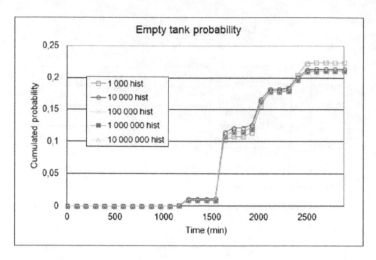

Fig. 22 Dynamics entered by the system

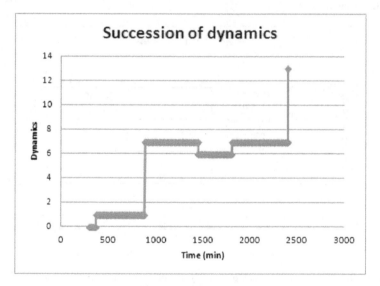

Fig. 23 Evolution of the water levels in the tank ($L(t)$) and in the auxiliary feeding system ($l(t)$)

Moreover, MoSt uses a memorization technique whose purpose is to save computation time. Indeed, each history starts with the same initial conditions and keeps the same evolution until the first event. So, it is useless to repeat (and compute again) the same system evolution at the beginning of each history of a simulation. The first event that will occur during a history must be the triggering of an activation stimulus. No triggering of a deactivation stimulus, no end of (de)activation delay (and, consequently, no change of dynamics) is possible before the triggering of an activation stimulus, by definition of these events. Consequently, for each history, the system evolution will be exactly identical until the first triggering of an activation stimulus.

4 Conclusions

As shown in Sect. 1, current probabilistic safety assessment methodologies are very useful to study systematic processes or system but are not adapted to correctly deal with dynamics systems or phenomenological aspects. Indeed, in these last cases, the time is of paramount importance and the chronology of the events is not so easy to predict. Hence, new methodologies or approaches are needed.

This observation is the starting point of the Stimulus-Driven Theory of Probabilistic Dynamics (SDTPD). This theory is a recent development of dynamic reliability where transitions between dynamics (due to the occurrence of probabilistic or deterministic events) are modeled by stimuli. This transition process is divided into two phases (the triggering of an activation stimulus and the elapsing of the corresponding activation delay) in order to be as close as possible to the actual process progress. Moreover, deactivation of an activation stimulus is also possible in case of change of the system state. Other techniques such as cell-to-cell mapping can also be used to model the system evolution; this approach is presented in Labeau and Izquierdo (2005b) and Peeters and Labeau (2004).

The SDTPD and the associated computer code MoSt are consequently really powerful tools to evaluate the likelihood of feared events for complex systems and to assess the safety of dynamic systems.

5 Variables and Functions Used in This Chapter

a Slope of the linear increase of the water level in the tank due to the auxiliary feeding system (tank issue)

b Parameter of the exponential decrease of the water level in the tank in case of flow 3 (tank issue)

c Slope of the linear pressure increase (pressurization test-case)

c Slope of the linear increase of the water level in the tank due to the continuous feeding system (tank issue)

d Slope of the linear decrease of the water level in the tank in case of flow 2 (tank issue)

f_c Probability density function associated to the containment rupture pressure (pressurization test-case)

F_c Cumulative density function associated to the containment rupture pressure (pressurization test-case)

f_H	Probability density function associated to the combustion time (pressurization test-case)
F_H	Cumulative density function associated to the combustion time (pressurization test-case)
$f_j^F(t, \overline{x})$	Probability density function of triggering activation stimulus F a time t after entering dynamics j with process variables $\overline{x}$
f_v	Probability density function associated to the valve opening setpoint (pressurization test-case)
F_v	Cumulative density function associated to the valve opening setpoint (pressurization test-case)
g_{ji}	Evolution law of the jth process variable in dynamics i
$h_j^F(t, \overline{x})$	Probability density function of elapsing the activation delay associated to stimulus F a time t after entering dynamics j with process variables $\overline{x}$
$k_j^F(t, \overline{x})$	Probability density function of triggering deactivation stimulus F a time t after triggering the associated activation stimulus in dynamics j with process variables $\overline{x}$
l	Water level in the auxiliary tank (tank issue)
L	Water level in the tank (tank issue)
L_0	Initial water level in the tank (tank issue)
$l_j^F(t, \overline{x})$	Probability density function of elapsing the deactivation delay associated to stimulus F a time t after triggering the deactivation stimulus F in dynamics j with process variables $\overline{x}$
p	Pressure
p_c	Containment rupture pressure (pressurization test-case)
p_c^{max}	Upper bound of the containment rupture pressure interval (pressurization test-case)
p_c^{min}	Lower bound of the containment rupture pressure interval (pressurization test-case)
p_v	Valve opening pressure (pressurization test-case)
p_v^{max}	Upper bound of the valve opening pressure interval (pressurization test-case)
p_v^{min}	Lower bound of the valve opening pressure interval (pressurization test-case)
q_{ji}	Probability per time unit of having a transition from dynamics j to dynamics i
q_{ji}^F	Probability per time unit of having a transition from dynamics j to dynamics i due to stimulus F
t	Time
t_{AD}	Upper bound of the maximum duration of the transient (pressurization test-case)
t_c	Time of the containment rupture pressure (pressurization test-case)
t_H	Maximal duration of the combustion (pressurization test-case)
t_v	Time if the valve opening ($= \frac{p_v}{c}$) (pressurization test-case)
V	Volume of the tank (tank issue)

$\overline{x}$ Vector of the process variables

x_j jth component of $\overline{x}$

κ_v Parameter of the pressure decrease (pressurization test-case)

$\varphi(\overline{x}, i, t)$ Ingoing density function of the system into dynamics i at time t with process variables $\overline{x}$

$\pi(\overline{x}, j, 0)$ Probability density function of finding the system in dynamics i at time t with process variables $\overline{x}$

$\theta(x)$ Heaviside function

Acknowledgements A part of the work presented in this chapter has been developed in the framework of a collaboration between ULB (Université Libre de Bruxelles, Ecole Polytechnique) and IRSN (Institut de Radioprotection et de Sûreté Nucléaire, France).

References

Avontuur GC (2003) Reliability analysis in mechanical engineering design. Thesis, Delft University of Technology

Devooght J, Smidts C (1996) Probabilistic dynamics as a tool for dynamic PSA. Reliab Eng Syst Saf 52:185–196

Labeau PE, Izquierdo JM (2005a) Modeling PSA problems: I. The stimulus-driven theory of probabilistic dynamics (SDTPD). Nucl Sci Eng 150(2):115–139

Labeau PE, Izquierdo JM (2005b) Modeling PSA problems. II. A cell-to-cell transport theory approach. Nucl Sci Eng 150(2):140–154

Marseguerra M, Zio E, Devooght J, Labeau PE (1998) A concept paper on dynamic reliability via Monte Carlo simulation. Math Comput Simul 47:371–382

Meneley D (2001) Nuclear safety and reliability. CANTEACH Project

Peeters A (2012) MoSt: a new tool for dynamic reliability. In: Proceedings of European nuclear conference 2012 (ENC 2012), Manchester

Peeters A (2013a) Modeling and propagation of uncertainties in the framework of the stimulus-driven theory of probabilistic dynamics. In: Proceedings of the 11th international conference on structural safety & reliability (ICOSSAR 2013), New York

Peeters A (2013b) The stimulus-driven theory of probabilistic dynamics: future developments. In: Proceedings of the 9th workshop on cooperation for higher education on radiological and nuclear techniques (CHERNE), Salamanca

Peeters A, Labeau P-E (2004) Cell-to-cell mapping applied to the stimulus-driven theory of probabilistic dynamics. A pressurization test-case. In: Proceedings of Colloque de Maîtrise des Risque et de Sûreté de Fonctionnement, $\lambda\mu 14$, Bourges

Peeters A, Labeau P-E (2007) The stimulus-driven theory of probabilistic dynamics applied to the level-2 psa hydrogen combustion issue. In: Proceedings of Colloque de Maîtrise des Risque et de Sûreté de Fonctionnement, $\lambda\mu 15$, Lille

Rasmussen (1975) An assessment of accident risk in U.S. commercial nuclear power plants. USNRC document, NUREG-75/014

Rausand M (2004) System analysis. Event tree analysis. In: System reliability theory. Wiley, Hoboken

Rogovin M, Frampton G (1980) Three mile island: a report to the commissioners and to the public. NUREG/CR-1250, vols I-II

Zio E (2002) Event tree analysis. Dipartimento di Ingeneria Nucleare Politecnico di Milano (Technical University-Italy). http://www.cesnef.polimi.it/corsi/sicura/Event.pdf

The Pavement Performance Modeling: Deterministic vs. Stochastic Approaches

Md. Shohel Reza Amin

Abstract The pavement performance modeling is an essential part of pavement management system (PMS). It estimates the long-range investment requirement and the consequences of budget allocation for maintenance treatments of a particular road segment on the future pavement condition. The performance models are also applied for life-cycle economic evaluation and for the prioritization of pavement maintenance treatments. This chapter discusses various deterministic and stochastic approaches for calculating the pavement performance curves. The deterministic models include primary response, structural performance, functional performance, and damage models. The deterministic models may predict inappropriate pavement deterioration curves because of uncertain pavement behavior under fluctuating traffic loads and measurement errors. The stochastic performance models assume the steady-state probabilities and cannot consider the condition and budget constraints simultaneously for the PMS. This study discusses the Backpropagation Artificial Neural Network (BPN) method with generalized delta rule (GDR) learning algorithm to offset the statistical error of the pavement performance modeling. This study also argues for the application of reliability analyses dealing with the randomness of pavement condition and traffic data.

1 Introduction

The appropriate and effective pavement performance curves are the fundamental components of pavement management system (PMS), and ensure the accuracy of pavement maintenance and rehabilitation (M&R) operations (Jansen and Schmidt 1994; Johnson and Cation 1992; Attoh-Okine 1999). The performance models

M.S.R. Amin (✉)
Department of Building, Civil and Environmental Engineering,
Concordia University, Montreal, QC H3G 2W1, Canada
e-mail: shohelamin@gmail.com; md_amin@encs.concordia.ca

S. Kadry and A. El Hami (eds.), *Numerical Methods for Reliability and Safety Assessment: Multiscale and Multiphysics Systems*, DOI 10.1007/978-3-319-07167-1_5,
© Springer International Publishing Switzerland 2015

calculate the future conditions of pavement based on which PMS optimizes several maintenance treatments, and estimates the consequences of maintenance operations on the future pavement condition during the life-span of the pavement (George et al. 1989; Li et al. 1997). Early PMSs did not have performance curves rather they evaluated only the current pavement condition. Later, the simplified performance curves were introduced based on the engineering opinions on the expected design life of different M&R actions (Kulkarni and Miller 2002). The only predictive variable of these performance curves was the pavement age. The development of performance curve is explicitly complicated as the pavement performance is subjected to a large number of parameters of pavement performance.

There are two streams of pavement performance modeling—deterministic and stochastic approaches. The major differences between deterministic and stochastic performance prediction models are model development concepts, modeling process or formulation, and output format of the models (Li et al. 1996). This study discusses various deterministic and stochastic approaches of pavement performance modeling, and elucidates the advantages and disadvantages of these methods.

2   Deterministic Pavement Performance Modeling

The deterministic models include primary response, structural performance, function performance, and damage models (George et al. 1989). Three are different types of deterministic models, such as mechanistic models, mechanistic-empirical models, and regression models. The mechanistic models draw the relationship between response parameters such as stress, strain, and deflection (Li et al. 1996). The mechanistic-empirical models draw the relationship between roughness, cracking, and traffic loading. On the other hand, the regression models draw the relationship between a performance parameter (e.g., riding comfort index, RCI) and the predictive parameters (e.g., pavement thickness, pavement material properties, traffic loading, and age) (Li et al. 1996). A large number of deterministic models have been developed for regional or local PMSs such as traffic-related models, time-related models, interactive-time-related models, and generalized models (Attoh-Okine 1999).

The general function of a deterministic pavement performance model can be expressed by Eq. (1) (Li et al. 1996).

$$\text{PCS}_t = f\left(P_0, \text{ESALs}_t, H_e \text{ or SN}, M_R, C, W, I\right) \tag{1}$$

Where PCS_t is the generalized pavement condition state (PCS) at year t, P_0 is the initial pavement condition state, ESALs_t is the accumulated equivalent single axle loads (ESALs) applications at age t, H_e is the total equivalent granular thickness of the pavement structure, SN is the structural number index of total pavement thickness, M_R is the subgrade soil resilient modulus, W is the set of climatic or environmental effects, I is the interaction effects of the preceding effects, and C

is the set of construction effects. The PCS represents RCI, present serviceability index (PSI), pavement quality index (PQI), roughness, and cracking. The SN is also known as pavement strength that can be calculated by $\sum a_i h_i + SN_g$, where a_i is the material layer coefficients and h_i is the layer thicknesses. The SN_g is a subgrade contribution that can be calculated by $3.51 \log CBR - 0.85 (\log CBR)^2 - 1.43$; CBR is the in situ California bearing ratio of subgrade (Li et al. 1996).

The American Association State Highway and Transportation Officials (1985) developed the PSI for the flexible pavement. The PSI and 18 kip ESALs are the main factors of pavement performance along with other factors such as materials properties, drainage and environmental conditions, and performance reliability (Eq. (2)) (Abaza et al. 2001).

$$\log_{10}(ESAL_t) = Z_R \times S_0 + 9.36 \times \log_{10}(SN + 1) - 0.2 + \frac{\log_{10}\left[\frac{\Delta PSI}{4.2 - 1.5}\right]}{0.40 + \frac{1094}{(SN+1)^{5.19}}}$$

$$+ 2.32 \times \log_{10}(M_R) - 8.07$$

(2)

Where ΔPSI is the difference between the initial design serviceability index (PSI_0) and the serviceability index at year t (PSI_t), and Z_R and S_0 are the standard normal deviate and combined standard error of the traffic prediction and performance prediction, respectively.

Lee et al. (1993) also developed the PSI for the flexible pavements shown in Eq. (3) (Lee et al. 1993).

$$\log_{10}(4.5 - PSI) = 1.1550 - 1.8720 \times \log_{10}SN + 0.3499 \times \log_{10}t$$

$$+ 0.3385 \times \log_{10}ESAL$$

(3)

The Ontario Pavement Analysis of Costs (OPAC) developed the deterministic flexible pavement deterioration model of pavement condition index (PCI), which is expressed by Eq. (4) (Jung et al. 1975; Li et al. 1997).

$$\Delta PCI = PCI_0 - (P_T + P_E) = PCI_0 - \left[(2.4455\Psi + 8.805\Psi^3)\right.$$

$$\left. + \left(PCI_0 - \frac{PCI_0}{1 + \beta_w}\right)(1 - e^{-\alpha t})\right]$$

(4)

$$w = \frac{9000 \times 25.4}{2M_s\left(0.9H_e \sqrt[3]{\frac{M_2}{M_R}}\right)\sqrt{1 + \frac{6.4}{0.9H_e\sqrt[3]{\frac{M_2}{M_R}}}}} \quad \text{and} \quad \Psi = 3.7238 \times 10^{-6}w^6 ESAL$$

Where w is subgrade deflection, PCI_0 is as-built PCI, P_T and P_E are the traffic- and environment-induced deteriorations of pavement condition, M_2 is the modulus of granular base layer, β is the regional factor 1 ($\beta = 60$ in southern Ontario), and α is the regional factor 2 ($\alpha = 0.006$ in southern Ontario).

The Nevada Department of Transportation (NDOT) developed 16 deterministic performance models for different pavement rehabilitation and maintenance treatments in 1992 (Sebaaly et al. 1996). The factors of the performance models are traffic, environmental, materials, and mixtures data in conjunction with actual performance data (PSI). The performance model for the asphalt concrete (AC) overlays is given by Eq. (5) (Sebaaly et al. 1996).

$$PSI = -0.83 + 0.23\,DPT + 0.19\,PMF + 0.27\,SN + 0.078\,TMIN$$

$$+ 0.0037\,FT - (7.1e - 7\,ESAL) - 0.14\,t \tag{5}$$

Where DPT is the depth of overlay, PMF is the percent mineral filler, TMIN is the average minimum annual air temperature (°F), and FT is the number of freeze-thaw cycles per year. The PSI was calculated by using a modified version (Eq. (6)) of the AASHTO performance method (Eq. (1)) (Sebaaly et al. 1996).

$$PSI = 5 \times e^{-0.0041*IRI} - 1.38RD^2 - 0.03(C + P)^{0.5} \tag{6}$$

Where IRI is the international roughness index, RD is the rut depth, C is the cracking and P is the patching. The IRI is a key property of road condition considered in any economic evaluation of design and maintenance standards for pavements, and also in any functional evaluation of the standards desire of road users (Paterson 1987; Ockwell 1990). Haugodegard et al. (1994) derived that the IRI function followed the parabolic distribution (Eq. (7)).

$$IRI_t = IRI_i + (IRI_1 - IRI_0) * \left(\frac{A_d}{A_1}\right)^{1.5} \tag{7}$$

Where IRI_t is the predicted roughness at year t, IRI_0 is the roughness just after the latest rehabilitation, IRI_1 is the latest recorded roughness, A_t is the age of the pavement surface at year t, A_1 is the age of the pavement surface when the latest roughness recording made.

Saleh et al. (2000) developed a mechanistic roughness model relating the roughness with the number of load repetitions, axle load, and asphalt layer thickness (Eq. (8)). The model applied vehicle dynamic analysis to estimate the dynamic force profile. The model also used the finite element structural analysis to estimate the change of pavement surface roughness for each load repetition. The statistical relationships in Eq. (8) show that initial roughness (IRI_0) is the most significant factor that affects roughness at later ages. The other important factors are axle load (P), asphalt thickness (T), and the number of load repetitions (ESALs) (Saleh et al. 2000).

$$IRI = -1.415 + 2.923\sqrt{IRI_0} + 0.00129\sqrt{ESALs} + 0.000113T$$

$$- 5.485 \times 10^{-10}P^4 - 10^{-3}T\sqrt{ESALs} + 5.777 \times 10^{-12}P^4\sqrt{ESALs} \tag{8}$$

George et al. (1989) carried out various regression analyses to develop empirical-mechanistic performance models for the highways in Mississippi based on the pavement condition data during the period of 1986–1988. The constructed performance models were evaluated based on the rational formulation, behavior of the models, and statistical parameters. The exponential and power functions of both concave and convex shapes were identified as the statistically significant functions. The best-fit models for the performance prediction (PCI_t) of the flexible pavement with no overlay (Eq. (9)), with overlay (Eq. (10)), and composite pavement (Eq. (11)) are given below (George et al. 1989):

$$PCI_t = 90 - a \left[\exp\left(t^b\right) - 1\right] \log \left[\frac{ESAL}{SN^c}\right] \forall a = 0.6349; \ b = 0.4203; \ C = 2.7062$$

$$(9)$$

$$PCI_t = 90 - a \left[\exp\left(t^b\right) - 1\right] \log \left[\frac{ESAL}{SN^c \times T}\right] \forall a = 0.8122; \ b = 0.3390; \ C = 0.8082$$

$$(10)$$

$$PCI_t = 90 - a \left[\exp\left(\left(\frac{t}{T}\right)^b\right) - 1\right] \log [ESAL] \ \forall a = 1.7661; \ b = 0.2826$$

$$(11)$$

The prediction models identified t, SN, and T as the most significant factors of pavement performance. The computational accuracy along with the direct influence of SN and T on the mechanistic parameters (e.g., stress, strain, and deflection) were the reasons for their significance in the performance model (George et al. 1989). The attribute ESALs was identified as the less important factor of pavement performance. George et al. (1989) argued that ESALs would be the weakest link in the cumulative traffic computation because several questionable input parameters (e.g., traffic count, traffic growth factor, and truck factor) are associated with the ESALs estimation. George et al. (1989) applied the same argument for the exclusion of the environmental loads which include thermal effects, subgrade movements in the expansive clays, freeze-thaw effects, and bitumen aging.

Smadi and Maze (1994) determined the PCI for the Iowa Interstate 80 based on the 10 years traffic data. The performance curve of PCI was a function of only the total number of 18 kip ESALs that the pavement had experienced (Eq. (12)) (Smadi and Maze 1994):

$$PCI = 100 - \alpha \, (ESALs) \, , \quad \alpha \text{ is constant depends on surface type} \qquad (12)$$

de Melo e Siva et al. (2000) proposed the logistic growth pavement performance curve for local government agencies in Michigan. These agencies commonly use a PMS called RoadSoft (de Melo e Siva et al. 2000). The model of de Melo e Siva et al. (2000) was based on the Kuo's pavement model considering the ascending distress index with different design service life values (Kuo 1995). In this formulation, the starting distress index of a reconstructed or resurfaced pavement was established as 0. The boundary condition of Kuo's logistic growth model (Kuo 1995) is expressed by Eq. (13) (de Melo e Siva et al. 2000).

$$DI = \alpha \left[\frac{(\alpha + \beta)}{(\alpha + \beta e^{-\gamma t})} - 1 \right] \tag{13}$$

Where DI is distress index, α is the potential initial of DI, β is the limiting of DI, t is the age (years), $\gamma = -\frac{1}{DSL} \ln \left(\left\{ \left[\frac{(\alpha+\beta)}{(\alpha+cDP)} \right] - 1 \right\} \frac{\alpha}{\beta} \right)$ is the deterioration pattern index, DSL is the design service life, and cDP is the predetermined DI (de Melo e Siva et al. 2000).

de Melo e Siva et al. (2000) argued that the parameter values, in the logistic growth model, had to be inverted to meet the constraints of the PASER and RoadSoft data. In the PASER and RoadSoft data, the values range from 1 to 10, and the starting DI of a distress-free pavement (reconstructed or resurfaced) is 10. To reflect this, the boundary condition was reconstructed as Eq. (14) (de Melo e Siva et al. 2000).

$$\text{PASER Rating} = \alpha - \beta \left[\frac{(\alpha + \beta)}{(\beta + \alpha e^{-\gamma t})} - 1 \right]$$

$$\forall \gamma = -\frac{1}{DSL} \ln \left(\left\{ \left[\frac{(\alpha + \beta)}{(\alpha + \beta - cDP)} \right] - 1 \right\} \frac{\beta}{\alpha} \right) \tag{14}$$

Sadek et al. (1996) developed a distress index (DI), which was a composite index reflecting severity and frequency of the observed distresses in the pavement surface. This index is a function of average yearly ESALs, age (t) of the pavement and thickness of the overlay (T) (Eq. (15)).

$$DI = 100 - 5.06 \times t^{0.48} ESALs^{1.29} T^{-0.20} \tag{15}$$

Robinson et al. (1996) developed a sigmoidal form of the distress model for the Texas Pavement Management Information System, where D_1 predicted the punch-outs per mile and D_2 predicted the Portland cement concrete patches per mile (Eq. (16)) (Pilson et al. 1999).

$$D_1 = 101.517 \exp - \left[\frac{538.126}{t} \right]^{0.438} \text{ and } D_2 = 1293.840 \exp - \left[\frac{399.932}{t} \right]^{0.536} \tag{16}$$

Pilson et al. (1999) developed the pavement deterioration model. The fundamental concept of this model is that the rate of the deterioration of one component of a system is a function of the level of deterioration of itself and other components in the system. The coefficients describing these functions can be summarized as an interactivity matrix (**C**) (Eq. (17)). The additional deterioration of surface (S) during the current year (dS) is proportional to its own current level of deterioration and the deterioration levels of the base (B) and subbase (Sb) (Eq. (18)) (Pilson et al. 1999).

$$\mathbf{C} = \begin{bmatrix} C_{s0} & C_{s1} & C_{s2} & C_{s3} \\ C_{B0} & C_{B1} & C_{B2} & C_{B3} \\ C_{Sb0} & C_{Sb1} & C_{Sb2} & C_{Sb3} \end{bmatrix} \tag{17}$$

$$dS = C_{s0} + C_{s1}S + C_{s2}B + C_{s3}Sb \tag{18}$$

Where C_{s0}, C_{s1}, C_{s2}, C_{s3} are the linear proportional constants for surface. The effect of different maintenance actions on each component can be measured by the same interactivity matrix. The assumption is that the maintenance actions will reduce the deterioration level to a specific fraction of the current value (Pilson et al. 1999).

However, the deterministic approaches of performance model cannot explain some issues such as (a) randomness of traffic loads and environmental conditions, (b) the difficulties in quantifying the factors or parameters that substantially affect pavement deterioration, and (c) the measurement errors associated with pavement condition, and the bias from subjective evaluations of pavement condition (Li et al. 1997). For example, in Eq. (1), each of the factors of pavement performance index can further be subdivided into a set of individual factors. Total equivalent granular thickness of the pavement structure (H_e) is determined by the properties of pavement materials, equivalent layer factors defined for the pavement materials, and construction quality. The effect of ESALs applied on the pavement for t years is not the same because of the traffic growth rate, percentage of trucks, and traffic distribution on the pavement (Li et al. 1997). These constraints of deterministic approaches broker for the application of stochastic pavement performance modeling.

3 Stochastic Pavement Performance Modeling

The stochastic models recently have received considerable attentions from pavement engineers and researchers (Wang et al. 1994; Karan 1977). Typically, a stochastic model of pavement performance curve is represented by the Markov transition process (Li et al. 1997). Knowing the "before" condition or state of pavement, the Markov process predict the "after" state (George et al. 1989). The main challenge for these stochastic models is to develop the transition probability matrices (TPMs).

Wang et al. (1994) developed the Markov TPMs for the Arizona Department of Transportation by using a large number of observed pavement performance historical data for categorized highways with several initial pavement condition states. The pavement probabilistic behavior is expressed by Eq. (19) for all $i, j, l,$ n, and $0 \leq v \leq N$, $0 \leq n \leq N$ (Wang et al. 1994):

$$P_{ij}^{(n)} = \sum_{k=0}^{M} P_{ik}^{(1)} P_{kj}^{(n-1)} \forall n \leq v \quad \text{and} \quad P_{ij}^{(n)} = \sum_{i=0}^{M} \sum_{k=0}^{M} \left(P_{ik}^{(v)} \cdot P_{kl}^{(1)a} \right) P_{lj}^{(n-v-1)} \forall n > v$$

(19)

Where $P_{ij}^{(n)}$ is the n-step transition probability from condition state i to j for the entire design period (N), $M + 1$ is the total number of pavement condition states, v is the period when the rehabilitation is applied; $P_{ik}^{(v)}$ is the v-step transition probability from condition state i to k under the routine maintenance; $P_{kl}^{(1)a}$ is the one-step transition probability from condition k to l at period v; and $P_{lj}^{(n-v-1)}$ is the $(n - v - 1)$ step transition probability from condition l to j under the routine maintenance (Wang et al. 1994).

The n-step transition probability matrix $(P^{(n)})$ is given by Eq. (20) (Wang et al. 1994).

$$P^{(n)} = \begin{bmatrix} P_{00}^{(n)} & \cdots & P_{0M}^{(n)} \\ \vdots & & \vdots \\ P_{M0}^{(n)} & \cdots & P_{MM}^{(n)} \end{bmatrix} = \begin{cases} P_{\text{routine}}^{(n)} & n \leq v \\ P_{ik}^{(v)} \times P_{kl}^{(1)a} \times P_{lj}^{(n-v-1)} & n > v \end{cases}$$

(20)

Where $P_{\text{routine}}^{(n)}$ is the n-step transition probability matrix before the rehabilitation when $n \leq v$ (Wang et al. 1994). Equations (19) and (20) can easily be expanded to analyze pavement probabilistic behavior where more than one rehabilitation actions are applied.

Karan (1977) developed pavement deterioration functions by means of Markov process modeling for the PMS of the Waterloo (Ontario) regional road network. In this study, the pavement performance deterioration versus age was modeled as a time-independent Markov process (Eq. (21)).

$$V(n) = V(0) \times M^n$$

(21)

Where $V(n)$ is the predicted condition state matrix at year n, $V(0)$ is the initial condition state matrix at year 0, and M is the one-step transition probability matrix (Wang et al. 1994).

For the stochastic performance modeling for different pavement categories of roads, a large amount of measured performance data for all pavement categories in a road network have to be obtained and processed, which are time-consuming and costly (Li et al. 1997).

4 Transition from Deterministic to Stochastic Performance Modeling

Since the deterministic methods are widely applied by different studies and organization for the pavement performance modeling, the provision of transitional process from deterministic to stochastic modeling can be useful. Li et al. (1997) discussed the principles of system conversion from a deterministic to a stochastic model. Li et al. (1997) considered the AASHTO (Eq. (2)) and OPAC (Eq. (4)) deterministic performance models to convert into stochastic models. Li et al. (1997) assumed that the predicted actual traffic (ESALs) is normally distributed with $p^s_{(ESALs)_t}$ (ESALs) probability density function for a pavement section (s) in t years. The $\overline{ESALs}$ is the mean value of the traffic (ESALs) that drives the pavement condition state to deteriorate from the initial state i to state j. The $(ESALs)^s_{ij}$, a random variable, can be defined as the maximum numbers of ESALs that a pavement section s can carry before it drops from condition state i to state j. The transition of $(ESALs)^s_{ij}$ from deterministic to stochastic numbers can be expressed by Eq. (22) (Li et al. 1997).

$$P^s_{ij}(t) = P\left((ESALs)^s_{i,j+1} < \overline{ESALs} < (ESALs)^s_{i,j}\right) = P\left(\overline{ESALs} < (ESALs)^s_{i,j}\right)$$

$$- P\left(\overline{ESALs} < (ESALs)^s_{i,j+1}\right)$$

$$= \int_0^\infty \left[\int_0^y p^s_{(ESALs)_t}(x)dx\right] P_{(ESALs)^s_{ij}}(y)\, dy - \sum_{k=i}^{j+1} p^s_{ik}(t),\, j < i \tag{22}$$

By applying Eq. (22) to each specific section of pavement in a road network, the non-homogeneous Markov TPM for pavement section s at stage t can be calculated by Eq. (23) (Li et al. 1997).

$$P^s(t) = \begin{bmatrix} P^s_{10,10}(t) & P^s_{10,9}(t) \ldots & P^s_{10,0}(t) \\ P^s_{9,10}(t) & P^s_{9,9}(t) \ldots & P^s_{9,0}(t) \\ \vdots & \vdots & \vdots \\ P^s_{0,10}(t) & P^s_{0,9}(t) \cdots & P^s_{0,0}(t) \end{bmatrix} \tag{23}$$

Traditionally, the TPMs have been assumed to be time independent through the analysis period. Li et al. (1997) developed time-dependent nonhomogeneous Markov transition process. The modeling process was governed by three components: state, stage, and transition probability. First, stages were considered a series of consecutive equal periods of time (e.g., each year). Second, states were used to measure pavement functional and structural deterioration in terms of PCS. Finally, a set of TPMs was calculated to predict the pavement condition state (assuming 10 condition states) at year t (Eq. (24)) (Li et al. 1997).

$$P(t) = P^{(1)} P^{(2)} \ldots P^{(t-1)} P^{(t)}$$

$$p(t) = \begin{bmatrix} P^{(1)}_{10,10} & P^{(1)}_{10,9} \cdots P^{(1)}_{10,0} \\ P^{(1)}_{9,10} & P^{(1)}_{9,9} \cdots P^{(1)}_{9,0} \\ \vdots & \vdots & \vdots \\ P^{(1)}_{0,10} & P^{(1)}_{0,9} \cdots P^{(1)}_{0,0} \end{bmatrix} \begin{bmatrix} P^{(2)}_{10,10} & P^{(2)}_{10,9} \cdots P^{(2)}_{10,0} \\ P^{(2)}_{9,10} & P^{(2)}_{9,9} \cdots P^{(2)}_{9,0} \\ \vdots & \vdots & \vdots \\ P^{(2)}_{0,10} & P^{(2)}_{0,9} \cdots P^{(2)}_{0,0} \end{bmatrix} \cdots \begin{bmatrix} P^{(t)}_{10,10} & P^{(t)}_{10,9} \cdots P^{(t)}_{10,0} \\ P^{(t)}_{9,10} & P^{(t)}_{9,9} \cdots P^{(t)}_{9,0} \\ \vdots & \vdots & \vdots \\ P^{(t)}_{0,10} & P^{(t)}_{0,9} \cdots P^{(t)}_{0,0} \end{bmatrix}$$

(24)

This nonhomogeneous Markov transition process can be applied to simulate the probabilistic behavior of pavement deterioration in predicting pavement serviceability level (Li et al. 1997).

5 Drawbacks of Markov Decision Process

The main drawback of Markov Decision Process (MDP) approach is that it does not accommodate budget constraints (Liebman 1985). Another important drawback of this approach is that pavement sections have to be grouped into a large number of roughly homogeneous families based on pavement characteristics (Li et al. 2006). A large number of families mean fewer sample of pavement sections in each family, which compromises the reliability and validity of the TPMs generated for each family (Li et al. 2006). There are equally large numbers of M&R treatments for each family of pavement sections. It is suggested that all pavement sections should be categorized into small numbers of families. As the MDP addresses the performance evaluation of the pavement section as a group, it is not possible to address the performance condition of individual pavement section. Similarly, the optimization programming of M&R strategies are determined for a group of pavement sections rather than an individual section under a given budget. Moreover, the optimization programming of M&R strategies are calculated from the steady-state probabilities. However, in reality, the pavements under a given maintenance policy usually takes many years to reach the steady state and the proportion of the pavements are changing year by year. Therefore, the use of steady-state probabilities in the optimization objective function does not fully reflect reality, especially when this transition period is very long (Li et al. 2006).

6 Backpropagation Neural Network for Dealing with Uncertainties

In reality, many uncertain factors are involved in pavement performance curves. Ben-Akiva et al. (1993) developed the latent performance approach to report the problem of forecasting condition when multiple technologies are used to collect condition data. In that approach, a facility's condition is represented by a latent/unobservable variable which captures the ambiguity that exists in defining (and consequently in measuring) infrastructure condition (Durango-Cohen 2007). Unfortunately, this proposed model suffers from computational limitations. The process of finding an optimal action for a given period involves estimating and assigning a probability to every possible outcome of the data-collection process. The number of outcomes, the number of probabilities, and the computational effort to obtain M&R policies increases exponentially with the number of distresses being measured (Durango-Cohen 2007).

Durango-Cohen (2007) applied the polynomial linear regression model to define the dynamic system of infrastructure deterioration process. At the start of every period, the agency collects sets of condition data (X_t), and decides to take an action (A_t). The structure of deterioration process is determined by the material and construction quality, environmental conditions, and so on. These factors are represented by the vector $\overrightarrow{\beta}_t$. The deterioration model is given in Eq. (25) (Durango-Cohen 2007).

$$D_t\left(X_t, A_t, \overrightarrow{\beta}_t\right) = g_t X_t + h_t A_t + \overrightarrow{\beta}_t + \varepsilon_t \qquad (25)$$

This model assumed that ε_t accounted for the systematic and random errors in the data-collection process. The relationship between the latent condition and the distress measurements can be expressed by Eq. (26) (Durango-Cohen 2007). The measurement error model not only included the condition data (X_t), but also includes a set of exogenous (deterministic or stochastic) inputs captured in the matrices Γ_t (one vector associated with each distress measurement). The vector $\overrightarrow{\xi}_t$ is assumed to follow a Gaussian distribution with finite covariance matrix (Durango-Cohen 2007).

$$M(X_t, \Gamma_t) = H_t X_t + I_t \Gamma_t + \overrightarrow{\xi}_t \qquad (26)$$

The Durango-Cohen's model cannot define the proportion of errors contributed by each of the factors to the distress outcome. This study proposes the Back-propagation Artificial Neural Network (BPN) method to estimate the pavement performance for each year during the life-span of the pavement. The estimated pavement performance for each year can, later on, be plotted with respect to the pavement-age to determine pavement-deterioration during the life-span of the pavement.

The fundamental concept of BPN networks for a two-phase propagate-adapt cycle is that predictive variables (e.g., traffic loads, structural number) are applied as a stimulus to the input layer of network units that is propagated through each upper layer until an output (e.g., PCI, IRI) is generated. This estimated output can then be compared with the desired output to estimate the error for each output unit. These errors are then transferred backward from the output layer to each unit in the intermediate layer that contributes directly to the output. Each unit in the intermediate layer will receive only a portion of the total error signal, based roughly on the relative contribution the unit made to the original output. This process will repeat layer-by-layer until each node in the network will receive an error representing its relative contribution to the total error. Based on the error received, connection weights will then be updated by each unit to cause the network to converge toward a state allowing all the training patterns to be encoded (Freeman and Skapura 1991).

Attoh-Okine (1994) proposed the use of Artificial Neural Network (ANN) for predicting the roughness progression in the flexible pavements. However, some built-in functions, including learning rate and momentum term of the neural network algorithm, were not investigated properly. The inaccurate application of these built-in functions may affect the accuracy capability of the prediction models (Attoh-Okine 1999). Attoh-Okine (1999) analyzed the contribution of the learning rate and the momentum term in BPN algorithm for the pavement performance prediction using the pavement condition data from Kansas department of transportation network condition survey 1993 (Kansas Department of Transportation 1993). In that model, IRI was the function of rutting, faulting distress, transverse cracking distress, block cracking, and ESALs (Attoh-Okine 1999).

Shekharan (1999) applied the partitioning of connection weights for ANN in order to determine the relative contribution of structural number, age of pavement, and cumulative 80-KN ESALs to the prediction of pavement's present serviceability rating (PSR) (Shekharan 1999). The output layer connection weights are partitioned into input node shares. The weights, along the paths from the input to the output node, indicate the relative predictive importance of input variables. These weights are used to partition the sum of effects on the output layer (Shekharan 1999).

The built-in functions of ANN proposed by Attoh-Okine (1994) and the partitioning of connecting weights of ANN applied by Shekharan (1999) may affect the accurate capability of the prediction models. As we know, a neural network is a mapping network to compute the functional relationship between its input and output; and these functional relationships are defined as the appropriate set of weights (Freeman and Skapura 1991). The generalized delta rule (GDR) algorithm of BPN can deal with these problems. It is a generalization of the least-square-mean (LMS) rule. This chapter discusses the GDR to learn the algorithm for the neural network because the relationship is likely to be nonlinear and multidimensional. Suppose we have a set of P vector-pairs in the training set, (x_1, y_1), (x_2, y_2), $\ldots (x_p, y_p)$, which are examples of a functional mapping $y = \phi(x) : x \in R^N, y \in R^M$. We also assume that (x_1, d_1), (x_2, d_2), $\ldots (x_p, d_p)$ is some processing function that associates input vectors, $\mathbf{x}_k$ (rutting, faulting distress, transverse cracking distress, block cracking, ESALs, and environmental conditions) with the desired output value, d_k (e.g., IRI). The mean square error, or expectation value of error, is defined by Eq. (27) (Freeman and Skapura 1991).

$$\varepsilon_k^2 = \theta_k = (d_k - y_k)^2 = \left(d_k - \mathbf{w}^t \mathbf{X}_N\right)^2 \text{ where } y = \mathbf{w}^t \mathbf{X} \tag{27}$$

The weight vector at time-step t is $\mathbf{w}^t$. As the weight vector is an explicit function of iteration, R, the initial weight vector is denoted $\mathbf{w}(0)$, and the weight vector at iteration R is $\mathbf{w}(R)$. At each step, the next weight vector is calculated according to Eq. (28) (Freeman and Skapura 1991).

$$\mathbf{w}(R+1) = \mathbf{w}(R) + \Delta \mathbf{w}(R) = \mathbf{w}(R) - \mu \nabla \theta_k (\mathbf{w}(R))$$

$$= \mathbf{w}(R) + 2\mu \varepsilon_N \mathbf{X}_N \forall \nabla \theta (\mathbf{w}(R)) \approx \nabla \theta (\mathbf{w}) \tag{28}$$

Equation (28) is the LMS algorithm, where $\Delta \mathbf{w}(R)$ is the change in $\mathbf{w}$ at the Rth iteration, and μ is the constant of negative gradient of the error surface. The error surface is assumed as a paraboloid. The cross section of the paraboloidal weight surface is usually elliptical, so the negative gradient may not point directly at the minimum point, at least initially. The constant variable (μ) determines the stability and speed of convergence of the weight vector toward the minimum error value (Freeman and Skapura 1991).

The input layer of input variables distributes the values to the hidden layer units. Assuming that the activation of input node is equal to the net input, the output of this input node (I_{pj}) is given by Eq. (29). Similarly, the output of output node (O_{pk}) is given by Eq. (30), where the net output from the jth hidden unit to kth output units is net$_{pk}$ (Freeman and Skapura 1991).

$$I_{pj} = f_j \left(\text{net}_{pj}\right) \qquad \text{net}_{pj} = \sum_{i=1}^{N} w_{ji} x_{pi} + \theta_j \tag{29}$$

$$O_{pk} = f_k \left(\text{net}_{pk}\right) \qquad \forall \text{net}_{pk} = \sum_{j=1}^{L} w_{kj} I_{pj} + \theta_k \tag{30}$$

Where net$_{pj}$ is the net input to the jth hidden unit, net$_{pk}$ is the net input to kth output unit, w_{ji} is the weight on the connection from the ith input unit to jth hidden unit, w_{kj} is the weight on the connection from the jth hidden unit to pth output unit, and θ_j is the bias term derived from Eq. (27). The weight is determined by taking an initial set of weight values representing a first guess as the proper weight for the problem. The output values are calculated applying the input vector and initial weights. The output is compared with the correct output and a measure of the error is determined. The amount to change each weight is determined and the iterations with all the training vectors are repeated until the error for all vectors in the training set is reduced to an acceptable value (Freeman and Skapura 1991).

Equations (29) and (30) are the expressions of output of input and output nodes, respectively. In reality, there are multiple units in a layer. A single error value (θ_k) is not suffice for BPN. The sum of the squares of the errors for all output units can be calculated by Eq. (31) (Freeman and Skapura 1991).

$$\theta_{pk} = \frac{1}{2}\sum_{k=1}^{M} \varepsilon_{pk}^2 = \frac{1}{2}\sum_{k=1}^{M} \left(y_{pk} - O_{pk}\right)^2$$

$$\Delta_p \theta_p \left(\mathbf{w}\right) = \frac{\partial \left(\theta_p\right)}{\partial w_{kj}} = -\left(y_{pk} - O_{pk}\right)\frac{\partial}{\partial w_{kj}} \left(O_{pk}\right)$$

$$= -\left(y_{pk} - O_{pk}\right)\frac{\partial f_k}{\partial \left(\text{net}_{pk}\right)}\frac{\partial \left(\text{net}_{pk}\right)}{\partial w_{kj}} \tag{31}$$

Combining Eqs. (29), (30), and (31), the change in weight of output layer can be determined by Eq. (32) (Freeman and Skapura 1991).

$$\frac{\partial \left(\theta_p\right)}{\partial w_{kj}} = -\left(y_{pk} - O_{pk}\right)\frac{\partial f_k}{\partial \left(\text{net}_{pk}\right)}\frac{\partial}{\partial w_{kj}}\left(\sum_{j=1}^{L} w_{kj} I_{pj} + \theta_k\right)$$

$$= -\left(y_{pk} - O_{pk}\right) f_k'\left(\text{net}_{pk}\right) I_{pj} \tag{32}$$

In Eq. (32), $f_k'(\text{net}_{pk})$ is the differentiation of Eq. (30); this differentiation eliminates the possibility of using a linear threshold unit, since the output function for such a unit is not differentiable at the threshold value. Following Eq. (32), the weights on the output layer can be written as Eq. (33) (Freeman and Skapura 1991).

$$w_{kj}\left(R + 1\right) = w_{kj}\left(R\right) + \tau\left(y_{pk} - O_{pk}\right) f_k'\left(\text{net}_{pk}\right) I_{pj} \tag{33}$$

Where τ is a constant, and is also known as learning-rate parameter. However, $f_k(\text{net}_{pk})$ needs to be differentiated to derive f_k'. There are two forms of output functions for paraboloid [$f_k(\text{net}_{jk}) = \text{net}_{jk}$] and sigmoid or logistic function $\left[f_k\left(\text{net}_{jk}\right) = (1 + e^{-\text{net}_{jk}})^{-1}\right]$. The sigmoid or logistic function is for binary output units and the paraboloid function is for continuous output units. As the output of this model (pavement condition index) is continuous, paraboloid function can be applied for output function and can be expressed by Eq. (34) (Freeman and Skapura 1991).

$$w_{kj}\left(t + 1\right) = w_{kj}\left(t\right) + \tau\left(y_{pk} - O_{pk}\right) O_{pk}\left(1 - O_{pk}\right) I_{pj} = w_{kj}\left(t\right) + \tau\delta_{pk} I_{pj} \tag{34}$$

The estimated output, from connection weight, is compared to the desired output, and an error is computed for each output unit. These errors are then transferred backward from the output layer to each unit in the intermediate layer that contributes directly to the output. Each unit in the intermediate layer receives only a portion of the total error signal, based roughly on the relative contribution the unit made to the original output. This process repeats layer-by-layer until each node in the network has received an error that represents its relative contribution to the total error. Based on the error received, connection weights are then updated by each unit to cause the network to converge toward a state that allows all the training patterns to be encoded. Reconsidering Eqs. (30), (31), and (34) for Backpropagation algorithm, change of weights on hidden layer is expressed by Eq. (35) (Freeman and Skapura 1991).

$$\theta_p = \frac{1}{2}\sum_{k=1}^{M} \left(y_{pk} - O_{pk}\right)^2 = \frac{1}{2}\sum_{k=1}^{M} \left(y_{pk} - f_k\left(\text{net}_{pk}\right)\right)^2$$

$$= \frac{1}{2}\sum_{k=1}^{M} \left(y_{pk} - f_k\left(\sum_{j=1}^{L} w_{kj} I_{pj} + \theta_k\right)\right)^2$$

$$\frac{\partial\theta_p}{\partial w_{ji}} = -\sum_{k=1}^{M} \left(y_{pk} - O_{pk}\right)\frac{\partial O_{pk}}{\partial w_{ji}}$$

$$= -\sum_{k=1}^{M} \left(y_{pk} - O_{pk}\right)\frac{\partial O_{pk}}{\partial\left(\text{net}_{pk}\right)}\frac{\partial\left(\text{net}_{pk}\right)}{\partial I_{pj}}\frac{\partial I_{pj}}{\partial\left(\text{net}_{pj}\right)}\frac{\partial\left(\text{net}_{pj}\right)}{\partial w_{ji}}$$

$$\frac{\partial\theta_p}{\partial w_{ji}} = -\sum_{k=1}^{M} \left(y_{pk} - O_{pk}\right) f_k'\left(\text{net}_{pk}\right) w_{kj}\, f_j'\left(\text{net}_{pj}\right) x_{pi}$$

$$\Delta_p w_{ji} = \frac{\partial\theta_p}{\partial w_{ji}} = \tau f_j'\left(\text{net}_{pj}\right) x_{pi}\sum_{k=1}^{M} \left(y_{pk} - O_{pk}\right) f_k'\left(\text{net}_{pk}\right) w_{kj}$$

$$= \tau f_j'\left(\text{net}_{pj}\right) x_{pi}\sum_{k=1}^{M} \partial_{pk} w_{kj} \tag{35}$$

Equation (35) explains that every weight-update on hidden layer depends on all the error terms (∂_{pk}) on the output layer, which is the fundamental essence of the backpropagation algorithm. By defining the hidden layer error term as $\delta_{pj} = f_j'(\text{net}_{pj})\sum_{k=1}^{M}\partial_{pk} w_{kj}$, we can update the weight equations to become analogous to those for the output layer (Eq. (36)). Equations (34) and (36) have the same form of delta rule (Freeman and Skapura 1991).

$$w_{ji}(t+1) = w_{ji}(t) + \tau \delta x_{pi} \tag{36}$$

7 Reliability Analysis of the Traffic Data and Estimated Pavement Deterioration

The BPN can properly deal with the statistical randomness. The uncertainty is not only associated with the statistical analysis but also with the pavement condition and traffic data. How can we confirm that the traffic data for each year are reliable? To overcome these uncertainties, the reliability analysis (R_i) of the traffic data (ESALs) can be performed. The reliability analysis of ESALs is expressed at Eq. (37) by comparing the potential ESALs that the pavement structure can withstand before its condition state drops to a defined level (ESAL$_{pcs(i)}$) and the actual predicted annual ESALs (Li et al. 1996).

$$R_{i(\text{ESAL})} = P\left[\left(\log \text{ESAL}_{\text{pcs}(i)} - \log \text{ESAL}_t\right) > 0\right]$$

$$= \Phi\left[\frac{\log \overline{\text{ESAL}_{\text{pcs}(i)}} - \overline{\log \text{ESAL}_t}}{\sqrt{S^2_{\log \text{ESAL}_{\text{pcs}(i)}} + S^2_{\log \text{ESAL}_t}}}\right] = \Phi(z) \tag{37}$$

Where $\Phi(z)$ is the probability distribution function for standard normal random variable, $\log \overline{\text{ESAL}_{\text{pcs}(i)}}$ is the mean value of $\log \text{ESAL}_{\text{pcs}(i)}$, $\overline{\log \text{ESAL}_t}$ is the mean value of $\log \text{ESAL}_t$, and $S^2_{\log \text{ESAL}_{\text{pcs}(i)}}$ and $S^2_{\log \text{ESAL}_t}$ are the standard deviations of $\log \overline{\text{ESAL}_{\text{pcs}(i)}}$ and $\overline{\log \text{ESAL}_t}$, respectively.

8 Conclusion

The pavement performance modeling is an essential part of pavement management system (PMS) because it estimates the long-range investment requirement and the consequences of budget allocation for maintenance treatments of a particular road segment on the future pavement condition. This chapter discusses various deterministic and stochastic approaches for calculating pavement performance curves. The deterministic models include primary response, structural performance, functional performance, and damage models. The deterministic models may predict inappropriate pavement deterioration curves because they cannot explain some issues, such as: randomness of traffic loads and environmental conditions, difficulties in quantifying the factors or parameters that substantially affect pavement deterioration, and the measurement errors. The stochastic performance models, usually apply Markov transition process (MTP), predict the "after" state condition knowing the "before" state condition of pavement. This chapter also shows the transition methods from deterministic to stochastic pavement performance modeling. These stochastic methods cannot address the performance condition of individual pavement section. Another major drawback of stochastic models is that the optimization programming of M&R strategies are calculated from the steady-state probabilities. However, in reality, the pavements under a given maintenance policy usually takes many years to reach the steady state, and the proportion of the pavements are changing year by year. Therefore, the use of steady-state probabilities in the optimization objective function does not fully reflect reality, especially when this transition period is very long. This chapter proposes Backpropagation Artificial Neural Network (BPN) method with generalized delta rule (GDR) learning algorithm to offset the statistical error of pavement performance modeling. This chapter also suggests the application of reliability analyses to deal with the uncertainty of pavement condition and traffic data.

References

AASHTO (1985) AASHTO guide for the design of pavement structures (appendix EE). AASHTO, Washington, DC

Abaza KA, Ashur SA, Abu-Eisheh SA et al (2001) Macroscopic optimum system for management of pavement rehabilitation. J Transp Eng 127:493–500

Attoh-Okine NO (1994) Predicting roughness progression in flexible pavements using artificial neural networks. In: Third international conference on Managing Pavements, Transportation Research Board, National Research Council, San Antonio, TX, p 55–62

Attoh-Okine NO (1999) Analysis of learning rate and momentum term in backpropagation neural network algorithm trained to predict pavement performance. Adv Eng Softw 30:291–302

Ben-Akiva M, Humplick F, Madanat S et al (1993) Infrastructure management under uncertainty: the latent performance approach. J Transp Eng 119(1):43–58

de Melo e Siva F, Van Dam TJ, Bulleit WM et al (2000) Proposed pavement performance models for local government agencies in Michigan. Transp Res Rec 1699:81–86

Durango-Cohen PL (2007) An optimal estimation and control framework for the management of infrastructure facilities. In: Karlsson C, Anderson WP, Johansson B, Kobayashi K (eds) The management and measurement of infrastructure: performance, efficiency and innovation. Edward Elgar, Cheltenham, pp 177–197

Freeman JA, Skapura DM (1991) Neural networks. Algorithms, applications, and programming techniques. Addison-Wesley, Boston, MA, pp 56–125

George KP, Rajagopal AS, Lim LK (1989) Models for predicting pavement deterioration. Transp Res Rec 1215:1–7

Haugodegard T, Johansen JM, Bertelsen D et al (1994) Norwegian Public Roads Administration: a complete pavement management system in operation. In: Third international conference on Managing Pavements, Transportation Research Board, National Research Council, San Antonio, TX

Jansen JM, Schmidt B (1994) Performance models and prediction of increase overlay need in Danish state highway pavement management system, BELMA. In: Third international conference on Managing Pavements, San Antonio, TX, p 74–84

Johnson KD, Cation KA (1992) Performance prediction development using three indexes for North Dakota pavement management system. Transp Res Rec 1344:22–30

Jung FW, Kher R, Phang WA (1975) A performance prediction subsystem—flexible pavements, research report 200. Ontario Ministry of Transportation and Communications, Downs View

Kansas Department of Transportation (1993) Pavement management system NOS condition survey. Kansas Department of Transportation, Kansas, MO

Karan M (1977) Municipal pavement management system. Dissertation, Department of Civil Engineering, University of Waterloo

Kulkarni RB, Miller RW (2002) Pavement management systems past, present, and future. Transp Res Rec 1853:65–71

Kuo WH (1995) Pavement performance models for pavement management system of Michigan Department of Transportation. Michigan Department of Transportation, Lansing, MI

Lee Y-H, Mohseni A, Darter MI (1993) Simplified pavement performance models. Transp Res Rec 1397:7–14

Li N, Xie W-C, Haas R (1996) Reliability-based processing of Markov chains for modeling pavement network deterioration. Transp Res Rec 1524:203–213

Li N, Haas R, Xie W-C (1997) Investigation of relationship between deterministic and probabilistic prediction models in pavement management. Transp Res Rec 1592:70–79

Li Y, Cheetham A, Zaghloul S et al (2006) Enhancement of Arizona pavement management system for construction and maintenance activities. Transp Res Rec 1974:26–36

Liebman J (1985) Optimization tools for pavement management. In: North American Pavement Management conference. Ontario Ministry of Transportation and Communication, U.S. Federal Highway Administration, Washington, DC, p 6.6–6.15

Ockwell A (1990) Pavement management: development of a life cycle costing technique. Occasional Paper 100, Bureau of Transport and Communications Economics, Department of Transport and Communications. Australian Government Publishing Service, Canberra

Paterson DE (1987) Pavement management practices. Transportation Research Board, National Research Council, Washington, DC

Pilson C, Hudson R, Anderson V (1999) Multiobjective optimization in pavement management by using genetic algorithms and efficient surfaces. Transp Res Rec 1655:42–48

Robinson C, Beg MA, Dosseu T, Hudson WR (1996) Distress prediction models for rigid pavements for Texas pavement management information system. Transp Res Rec 1524:145–151

Sadek AW, Freeman TE, Demetsky MJ (1996) Deterioration prediction modeling of Virginia's interstate highway system. Transp Res Rec 1524:118–129

Saleh MF, Mamlouk MS, Owusu-Antwi EB (2000) Mechanistic roughness model based on vehicle-pavement interaction. Transp Res Rec 1699:114–120

Sebaaly PE, Hand A, Epps J et al (1996) Nevada's approach to pavement management. J Transp Res Rec 1524:109–117

Shekharan AR (1999) Assessment of relative contribution of input variables to pavement performance prediction by artificial neural networks. Transp Res Rec 1655:35–41

Smadi OG, Maze TH (1994) Network pavement management system using dynamic programming: application to Iowa State Interstate Network. Paper presented at third international conference on Managing Pavements, Transportation Research Board, National Research Council, San Antonio, TX

Wang KCP, Zaniewski J, Way G (1994) Probabilistic behavior of pavements. J Transp Eng 120(3):358–375

Probabilistic Considerations in the Damage Analysis of Ship Collisions

Abayomi Obisesan, Srinivas Sriramula, and John Harrigan

Abstract Ship collision events are often analyzed by following the approach of internal mechanics and external dynamics. The uncertainties in collision scenario parameters, which are used in the calculation of external dynamics, are usually quantified during ship collision analysis. However, uncertainties in the material and geometric properties are often overlooked during the analysis of internal mechanics. Consequently, it may lead to overestimation or underestimation of ship structural design capacity, which could impact on system performance.

This study aims to show a framework for assessing the reliability of ship hull structures during collision events. Finite element analysis using ABAQUS software and simplified analytical methods have been utilized to model the resistance of ship hull plates against the impact from the bulbous bow of a striking ship. The particulars of a general cargo vessel involved in a real-life collision have been used in the study to determine the external force required for the considered hull plate to resist. Based on Monte Carlo simulation, reliability analysis has been carried out to model the uncertainties of hull plate displacement. Two thousand design sets of the geometric and material properties were propagated through the simplified analytical model to obtain the resulting displacement data. These data were subsequently analyzed to obtain a probabilistic model for hull plate displacement, along with the variable sensitivity.

A. Obisesan • S. Sriramula (✉) • J. Harrigan
Lloyd's Register Foundation (LRF) Centre for Safety and Reliability Engineering,
School of Engineering, University of Aberdeen, Fraser Noble Building,
Aberdeen AB24 3UE, UK
e-mail: s.sriramula@abdn.ac.uk

S. Kadry and A. El Hami (eds.), *Numerical Methods for Reliability and Safety Assessment: Multiscale and Multiphysics Systems*, DOI 10.1007/978-3-319-07167-1_6, © Springer International Publishing Switzerland 2015

197

1 Introduction

Due to the increase in energy and commodity demand worldwide, there have been an increase in sea transportation by Cargo and tanker vessels. The greatest risk associated with sea transportation is loss of containment, which then poses a threat to marine biodiversity, human lives, structural assets, and the reputation of the companies involved. According to IHS Fairplay world casualty statistics for 2011 (IHS Fairplay 2012), 126 ships with a total of approximately 770,000 gross tonnage were lost at sea in 2011. Approximately 19 % of the total ship losses were due to ship collisions and contact. Despite the advancement in proactive measures such as the ship navigation systems, ship collisions are still being recorded every year. This reiterates the importance of analyzing the structural performance of ships during collisions.

The two most common methods for analyzing ship collisions are external dynamics and internal mechanics. External dynamics is concerned with the rigid body motion of the striking ship and the struck ship as well as the energy dissipated by the ship structures due to collision. External dynamics are calculated either by numerical or analytical methods (Zhang 1999). Using an analytical approach, closed-form expressions can be derived for the impact impulses and the kinetic energy loss during ship collisions. With the application of the principles of conservation of momentum and energy, the analysis is achieved by calculating the deformation due to the body motion of the struck ship and the striking ship, based on the assumption that the structural response comes from the local contact only. Detailed discussion on the internal mechanics is provided in Sect. 2 of this study.

There are existing models which incorporate the processes of internal mechanics and external dynamics for ship collision analysis. The most common approaches in the literature are ALPS/SCOL (Paik and Pedersen 1996), DAMAGE (Simonsen 1999), SIMCOL (Chen 2000), and DTU (Lutzen 2001) models. The common denominator for these models is the assigned stopping criteria which compare the energy dissipation calculated from the external dynamics and the energy absorption calculated from the internal mechanics. Once an equilibrium condition is met, these models evaluate output parameters such as the maximum penetration and damage length. External dynamics and internal mechanics are estimated either separately or simultaneously in the models.

The outcome of the ship collision analysis provides input to decisions on design specifications, for example, on the quality of steel or the appropriate plate thickness to be used in shipbuilding. Improved understanding that leads to design improvements can then reduce the risk to ship structures and the marine environment. To improve our understanding, uncertainties that are inherent in the analysis variables need to be considered. Models such as SIMCOL (Chen 2000) and DTU (Lutzen 2001) quantified the uncertainties in collision scenarios during the estimation of output characteristics of the external dynamics. The variables of the collision scenario were identified as collision angle, speed, length, type, draught of the striking ship and struck ship.

However, the uncertainties associated with the internal mechanics of ship collisions also need to be analyzed. The design strength of ship structures, which is a major factor in the analysis of internal mechanics, possesses random behavior. Such behavior could be attributed to the basic strength and geometry variables of structural members. Unfortunately, these variables tend to be represented by their deterministic or design values only, with possible variances often ignored. However, these values are applied with the inclusion of a generalized safety margin between the applied load and the strength of structural members, called safety factor. Due to the nonflexible nature of the safety factor, safety margin of structural members may not be able to accommodate possible changes to certain factors, such as strength and load variability, possibility of correlation among load and strength parameters, and uncertainties in structural analysis. It is important to understand how such changes may affect the characteristics and reliability of structural members in order to achieve optimal design.

Uncertainty quantification of basic random variables for reliability assessment can contribute to the advancement and improvement of decisions made in ship structural design procedures by providing new design measures and a reliability framework that is applicable to both conventional and advanced ship structural members. It is worth noting that the application of reliability assessment to ship structures is not a new phenomenon in ship structural design. The study by Nikolaidis et al. (1993) proposed a method for assessing the reliability of ship deck panels to identify and rank the most important uncertainties and the most effective design improvement. However, the assessment is limited to the type of ship assessed as a result of the deterministic values assigned to the main geometric variables of the deck panel. The study by Fang and Das (2005) assessed and compared the reliability of damaged ships during collision and grounding scenarios by fitting suitable distributions to model the uncertainties from load and strength variables. It was concluded that the risk caused by collision is far greater than that of grounding when a damaged ship continued en route.

Chowdhury (2007) assessed the failure probability of the mid-ship section due to yielding by quantifying uncertainties from the load and strength bending moments. The reliability assessment study by Khan and Das (2008) followed a similar approach and identified that particular types of double hull tanker and bulk carriers in two different service conditions (intact and damage) would have a higher failure probability in sagging than in hogging during side collision and grounding scenarios. They quantified the model uncertainties that are attributed to the variables of the ultimate/residual strength of the mid-ship section and also to the load variables such as still water, wave-induced vertical and horizontal bending moments.

The literature discussed earlier justified the necessity to consider various uncertainties of ship collision analysis in a reliability framework. Failure of the ship structure, such as buckling of hull girder members, can be defined and modeled using the corresponding limit state functions. The functional relations can be solved using structural reliability techniques such as First and Second Order Methods (FORM/SORM) and Monte Carlo Simulation (MCS).

In this study, it is intended to apply reliability assessment for the purpose of improving existing knowledge on the performance of ship hull structures during collisions. The uncertainties from the strength and geometry variables of ship hull plating are quantified using data from existing literature. The probabilistic characteristics of the strength and geometry variables are then propagated through both the finite element and simplified analytical models using a suitable sampling technique. Reliability and sensitivity assessment are then performed to model the outcome: MCSs to analyze the structural performance and Sobol's first-order indices to determine the contribution of input variables on the model output variation.

2   Internal Mechanics

The internal mechanics approach is concerned with the local deformation of the structural members at the bow of the striking ship and at the side of the struck ship and how they would respond during loading, in terms of energy absorption, displacement, and failure load. The bow of the striking ship is usually modeled to be perfectly rigid, hence internal mechanics mostly consider the deformation of the structural members of the stuck ship.

The first step in the internal mechanics analysis is the identification of ship structural members affected by the impact and the classification of their energy absorbing deformation mechanisms. This is followed by the derivation of a closed-form expression of their failure load and energy absorbed. Ship structural members may be classified into basic elements as follows:

- The side shell panels, which are the outer and the inner shell panels.
- The web girders, such as longitudinal stringers, transverse frames, transverse bulkheads, decks, and floors.
- The intersection elements, created at the junction between transverse and longitudinal members.

Failure of these members is generally defined beyond material yield points, hence failure and energy absorbing mechanisms are governed by a complex mixture of buckling, folding, tearing, rupture, and crushing of the members.

In recent years, the finite element method and simplified method have been the preferred technique employed to predict ship damage due to a collision. Due to its costly nature, the experimental method is mostly used to have a better understanding of the deformation process and to validate results from other methods. Computer advancement has made finite element method a useful tool for ship collision analysis. For example, Amdahl and Kavlie (1992), Kitamura (1997), and Paik and Thayamballi (2003) have investigated the behavior of ship structures using nonlinear finite element methods. However, due to the huge computational effort required for FE analysis of ship collision problems, simplified methods are commonly employed for the verification of ship impact resistance.

Minorsky (1959) made the first attempt to derive a simplified expression for collision resistance. An empirical equation was derived which relates the collision energy absorbed with the material volume of the damaged ship structures based on past history of collision cases. Good agreement was achieved for high energy collision cases. Following this pioneering work, several authors have developed improved analytical solutions. These solutions have produced closed-form expressions that give a good prediction of the basic features of structural deformations.

Simplified analytical models assess the resistance of ship hull structures when subjected to collision. The structural members that make up the hull structure are outer shell plating, transverse frames, longitudinal stringers, transverse bulkhead, and the inner shell plating. Several authors have developed simplified analytical formulae for the resistance force during deformation of these members. The resistance force is calculated from the consideration of the internal energy dissipated due to a kinematically admissible deformation of a structural member. Common assumptions in this method include:

- Structural members are independent of each other, but they can be combined to derive the total collision resistance of the ship structure.
- Loading by the bow of the striking ship is normal to the hull structure; the material of the structural member is rigid-perfectly plastic and rate independent.
- Most of the simplified analytical formulae from the literature consider the loading regime for which ship structures deform in a quasi-static mode.

3 Simplified Analytical Formulae for Impact Resistance

The evaluation of the impact resistance of ship structures using simplified analytical formulae is usually computed using energy conservation techniques. This involves the application of idealized assumptions and empirical data.

The shape of the striking ship bow, which collides with the hull structure of the struck ship is considered first in the evaluation of impact resistance. This is particularly important when establishing the resistance and displacement relationship for the hull shell plating. In the literature, the contact between the bow and side panel has been idealized as a point load as well as a spherical or elliptical paraboloid. Zhang (1999) and Buldgen et al. (2012) established a resistance formula for a rectangular plate subjected to transverse and oblique impact, respectively, by a point load. Simonsen and Lauridsen (2000) and Wang (2002) developed a resistance formula for a circular plate impacted by a spherical-shaped indenter. Zhang (1999) and Haris and Amdahl (2012) idealized the striking ship bow as elliptical paraboloid and evaluated the resistance of a rectangular plate to the bow impact.

Zhang (1999) further simplified the bow shape into a circular paraboloid using the assumption that vertical and horizontal radii are equal, but this idealization may not be appropriate for all ship bulbous bows. Haris and Amdahl (2012) introduced two parameters to their resistance formulae in order to consider the possibility of

Fig. 1 An idealized model of bulbous bow impact on shell plate

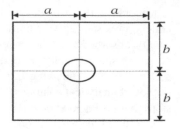

unequal curvatures of the bow in the x- and y-directions. The resistance of a square shell plate (see Fig. 1) to impact by an elliptical paraboloid is given as follows (Zhang 1999):

$$F_p = 1.5396\sigma_0 t_p \delta \left[1 + 3\frac{R^2\delta}{R_L a^2} + 6\left(\frac{R^2\delta}{R_L a^2}\right)^2 \right],$$ (1)

where F_p is the external force, σ_0 is the plate material flow stress usually defined as the average of the material yield stress and ultimate stress, t_p is the plate thickness, δ is the displacement, a represents the half-length and half-breadth of the square plate, R_L is the bulb length, and R is the bulb radius.

4   Numerical Analysis of Damage to Hull Structure

4.1   Finite Element Models

In this study, a square shell plate and the bulb section of a ship bow were modeled. The dimensions applied to the two models are given in Table 1. The bulb was idealized as a circular paraboloid (see Fig. 2) and modeled using the following parametric equation:

$$\frac{x}{R_L} = \frac{y^2 + z^2}{R^2},$$ (2)

where x, y, and z are the coordinates of the bulb. The shell plate and the bulb were modeled using Abaqus/CAE (ABAQUS 2011). Numerical analysis was carried out using the nonlinear FE software, Abaqus/Standard. The shell plate was modeled by a general-purpose shell element in order to incorporate changes in plate thickness as the plate elements deform. Therefore, the shell plate was modeled using an S4R element, which is a quadrilateral shell element with linear interpolation and reduced integration. Frictionless contact was assumed between the bulb and plate.

Table 1 Geometrical variables for square plate impact

R_L	R	a	t_p
8 m	2.83 m	10 m	0.04 m

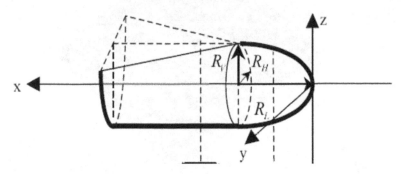

Fig. 2 Geometry of the idealized bulb (Zhang 1999)

Table 2 Material properties of the considered square plate

E	v	σ_y	σ_u	k	ε_{σ_u}
210 GPa	0.3	400 MPa	600 MPa	400 MPa	0.5

4.2 Material Properties

Shipbuilding steel, suchas ordinary strength steel (NVA, ABS A, ASTM A36), was considered as the material of construction for the shell plate.

The shell plate material was assumed to have linear isotropic hardening, which is a simple and common form (Yaw 2012). The plastic stress was modeled using the expression:

$$\sigma\left(\varepsilon_p\right) = \sigma_y + k\varepsilon_p, \tag{3}$$

where σ_y is the yield stress, ε_p is the equivalent plastic strain, and k is the strength index, which for linear hardening, can be defined as plastic modulus. Other key materials are the ultimate stress (σ_u) and its equivalent plastic strain (ε_{σ_u}). The elastic properties are the Young modulus (E) and Poisson ratio (v). The values of the material properties used in the analysis are provided in Table 2.

4.3 Numerical Simulation

The plate deformation due to transverse loading by the bulb section of a bow is largely governed by membrane action. This was observed on the plating when the equivalent plastic strain distribution was analyzed, as shown in Fig. 3. Itwas also observed that the largest strain occurs in a localized area of the plate, close to the contact area. Figure 4 shows the simulation result of the finite element

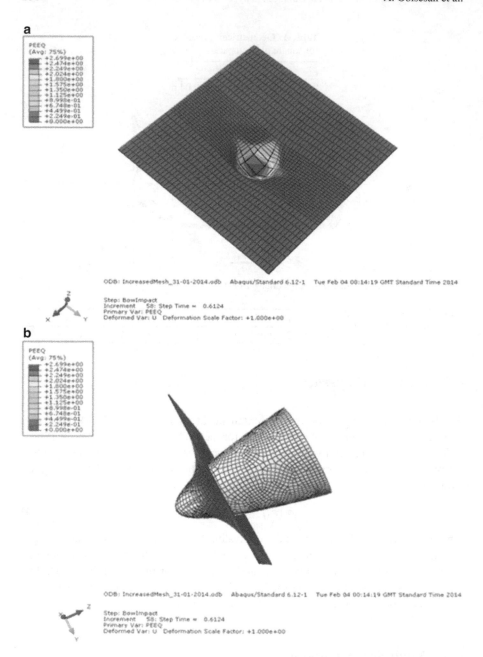

Fig. 3 Indentation of the square plate showing (**a**) plastic strain distribution and (**b**) the deformation of the plate

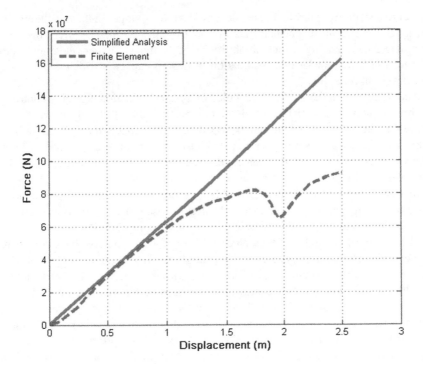

Fig. 4 Force–displacement plot of square plate impact

method plotted alongside the result of the force–displacement relationship of the simplified analytical method (Eq. (1)). The two methods show large differences after approximately 1.0 m displacement.

The linear trend of the simplified method predicts the early load–displacement behavior well. The simplified method diverges from the FE prediction at a displacement of approximately 1.0 m. This is because, simplified models assume that plate thickness is constant as the plate deforms, whereas the FE model captures the large local strains in the contact area of the plate. This is illustrated in a deformed mesh plot shown in Fig. 3a. This is accompanied by reductions in thickness in the contact region. At this stage, damage (ductile failure) is not incorporated into either of these models. Future investigations will incorporate a damage model in the FE analysis. At present, the change in slope at 1.0 m displacement is assumed to correspond to failure of the square plate.

5 Uncertainty Quantification

The outcome of most structural designs and their performances are largely governed by expert judgments. These judgments are made with the qualitative and/or quantitative consideration of uncertainties that could be inherent in the design from the conceptual stage right through to the decommissioning stage. In the end, judgments are made on the most advantageous option for the design. However, without a complete understanding of the variability of performance, expert judgment could lead to overestimation or underestimation of design capacity, which could impact on the system reliability.

According to Lindley (1971), the choice of a course of action is largely controlled by human activities which depend on the personality of the decision makers. However, the decision taken can be positively influenced through quantitative approach. Uncertainty is mostly introduced into a system through parameters which take values within a number of ranges. Parameters with varying values, also called variables, could be either from the load or strength factors and could also be from both factors. The variables would therefore require single representations when analyzing system performance and safety. These representations are currently achievable by quantifying the variable probabilistic characteristics.

The structural strength and geometry variables of a ship shell plate are considered in the study. The strength variables are Young's modulus, yield strength, ultimate strength, and the strength index. The geometric variables are the plate length, breadth, and thickness. The quantification process of the variables from the two classes usually involves data collection and analysis for the determination of probabilistic characteristics such as probability distribution function (PDF), mean, standard deviation, and coefficient of variation (C.O.V). Fortunately, this process has already been performed for different kinds of shipbuilding steel and is available in the open literature.

Hess et al. (2002) and Guo et al. (2012) compiled statistical information and data for several shipbuilding steels. Collectively, the literature provided a data bank with more than 10,500 tests carried out on 11 different steel grades used for marine applications. However, the common steel types considered are ordinary, high tensile, high-strength low-alloy (c), and high yield steels. Hess et al. (2002) achieved the generalization of the compiled probabilistic characteristics using a normalization technique called bias. The bias was calculated as the ratio of the variable mean value to its nominal (or design) value. The recommended probabilistic characteristics were derived by taking an average of the weighted values. The goodness-of-fit tests were applied to the output of the 10,500 tests to identify suitable PDFs for the strength and geometric variables. It was observed that normal and lognormal distributions are valid choices for describing the considered variables.

The study by Guo et al. (2012) used the statistical information to develop probabilistic models of geometric and material properties for the purpose of determining the failure probability level for the corroding deck plate of aging tankers. Their study provided the most up-to-date list of statistical information for the following strength variables: Young's modulus, yield strength, and plate thickness. For the purpose of this study, the statistical information for mild steel variables are used. The bias factors and ranges for probabilistic characteristics of basic strength variables recommended by Hess et al. (2002) as well as the nominal values specified in Tables 1 and 2 are applied to the finite element and simplified models to carry out a reliability analysis on a square plate impacted by the bulb

Table 3 Probabilistic characteristics of strength random variables (Hess et al. 2002)

Variable	Distribution	Mean	C.O.V
E (GPa)	Normal	$0.987E$	0.076
σ_y (MPa)	Normal	$1.3\sigma_y$	0.124
σ_u (MPa)	Normal	$1.05\sigma_u$	0.075
k (MPa)	Deterministic	400	–
t_p (m)	Lognormal	$1.05t_p$	0.044
a (m)	Lognormal	$0.988a$	0.046

section of a striking ship bow. These characteristics are presented in Table 3 and their application is discussed in the following sections. It is worth noting that the mean of the strength random variables is from quoted values of the design steel material, but multiplied by a bias factor, so that the probabilistic characteristics would be a representative of a wide spectrum of steel used for shipbuilding.

6 External Force

The striking ship dissipates a considerable amount of kinetic energy during a collision. The energy dissipated is calculated by taking the time integral of the force exerted by the striking ship. However, the force value varies in terms of the striking ship characteristics and sea condition at the moment of impact. The combination of physical laws of motion and the energy absorbed by the deformation modes in the structural components are used to predict the extent of the structural damage.

The external force is usually calculated for the xyz-coordinates of the striking ship and in terms of the ship motion for each coordinate which are surge, sway, and yaw motions. For simplicity, the surge motion is only considered in this study, and the expression for the external force is given as:

$$F = M\left(1 + m_x\right)a_x, \tag{4}$$

where M is the mass of the striking ship, m_x is the added mass coefficient, and a_x is the acceleration of the striking ship. The added mass coefficient represents the effects of ship interaction with its surrounding waters. For the current study, a real-life collision case is used to estimate the contact force between the two ships.

The collision event selected for this study is a 2013 collision event where the general cargo vessel, Sulpicio Express Siete, impacted on a ferry with about 870 passengers on board. The accident, which claimed the lives of 55 people with 65 more passengers reported missing, happened outside a major port in Talisay. The general cargo vessel which had approximately 160,000 L of hydrocarbon spilled a considerable amount of oil into the sea (The Philippine Star 2013).

The main particulars of the striking vessel required for the study are provided in Table 4. It should be noted that due to the unavailability of sufficient data for the studied ship in the literature, assumptions were made for two ship parameters.

Table 4 Main particulars for Sulpicio Express Siete (Marine Connector 2013; Marine Traffic 2013)

Ship particular	Value
Deadweight tonnage (DWT)	11,464 t
Length (L)	146 m
Breadth (B)	21 m
Draught (D)	6.8 m
Block coefficient (C_B)	0.7
Lightship displacement (LD)	14,714.06 t
Ship mass (M)	26,178.06 t
Maximum speed recorded (v)	7.3 knots

These parameters are the ship acceleration and its block coefficient. However, the assumptions were made from both ship guidelines and data from a similar vessel. A block coefficient value of 0.7 was assigned to the vessel and this was used to calculate the lightship displacement using the following expression:

$$LD = 1.008 \times L \times B \times D \times C_B. \tag{5}$$

The definition of the parameters in Eq. (5) is provided in Table 4. The total mass of the ship is calculated by doing the sum of the DWT and LD. The added mass coefficient usually takes a value ranging from 0.02 to 0.07. A value of 0.05 was assigned to the striking ship, based on the study by Zhang (1999). The ship acceleration was calculated by taking an average of the acceleration at different units of a cargo vessel, following the guideline for cargo stowage and securing (see DNV 2004). For ships with lengths greater than 100 m, a correction factor should be applied to the derived acceleration value. The correction factor (i) is calculated from the expression (DNV 2004):

$$i = \left(\frac{0.345v}{\sqrt{L}} \right) + \left(\frac{58.62L - 1034.5}{L^2} \right). \tag{6}$$

With the 146 m length of Sulpicio Express Siete general cargo vessel, a correction factor of 0.56 was applied to the ship acceleration to give a value of 3.2 m/s². The external force was then calculated to be 87.96 MN.

7 Random Variable Generation and Propagation

Abaqus/CAE used in this study can be interfaced with the programming language tool—Python (ABAQUS 2011). Python can be applied to an existing model through one of two different methods, the Python scripting interface method and the scripting parametric method. The use of scripting parametric method for the current study would require the use of the minimum and maximum values of the random variables to generate, execute, and derive results of multiple analyses. The method requires the development of a python script file (*.psf) and the creation of an input file (*.inp) from Abaqus/CAE model. The method first generates multiple design points for the variables by using one of the available sampling techniques. These techniques are interval, number, reference, and values. The random values generated for each variable are combined to generate design points that can be propagated one by one through the model input file.

The second method is through the use of the python development environment (PDE) which is accessible from the menu bar of Abaqus/CAE. Python scripting command provides basic functions—*random( )*—for generating pseudo-random numbers for various PDFs. The parameters required by the function are the distribution type, mean, and standard deviation of the basic random variable. The developed basic function for the random variables can be applied to the journal file (*.jnl) of an existing model by using them to replace their corresponding deterministic values to generate outputs of multiple analyses. This method was adopted for propagating the strength and geometry variables through the finite element model in this study.

The FE model was run with 2,000 sampled strength and geometry variable design sets by using the Python script developed for the model. The reactive force across the plate and the displacement of the reference nodal point were derived for each design set. MCSs were performed by propagating the generated design sets for the strength and geometry variables through the simplified model using MATLAB. First, the model (Eq. (1)) was rewritten to make displacement the subject of the cubic polynomial:

$$\left(6\frac{R^2}{R_L}\right)\delta^3 + \left(3a^2\frac{R^2}{R_L}\right)\delta^2 + a^4\delta - \frac{0.406a^4F_p}{\sigma_0 t_p} = 0. \tag{7}$$

This means that the resulting displacement value would be a vector with three roots, which contains both real and imaginary values. Second, a MATLAB code was written to propagate the design sets through the cubic polynomial. Complex roots of Eq. (7) were rejected, leaving only the real numbers as possible solutions. The results of the analysis are discussed in the following section. It is worth noting that the displacement term used in this section and subsequent sections relate to the local hull plate indentation and not the global ship displacement.

8 Stochastic Analysis

8.1 Determination of Displacement Probability Model

In order to understand the variation in the displacement values for the sampled strength and geometry variable design sets, the values were represented using a histogram plot, as shown in Fig. 5. It was observed that the displacement data from

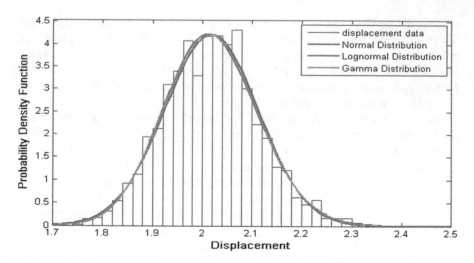

Fig. 5 Probability distributions fitted to the displacement frequency distribution

the MCS varied within the range of 1.7234 and 2.3386 m. Approximately 66 % of the displacement data lied within the range of 1.9 and 2.1 m. The mean and standard deviation of the displacement data were also calculated to be 2.018 and 0.0946 m, respectively.

A statistical approach was implemented by investigating the distribution of the displacement data. Several PDFs were fitted to the frequency distribution of the displacement data. By visual observation, three distributions—Normal, Lognormal, and Gamma distributions—provided good fits for the displacement data with minimum parameter standard error, as shown in Fig. 5. The estimated mean and standard deviation of the three distributions was similar to that derived from the displacement data. Further verification was required to discern the probability distribution model that would be plausible for the displacement data.

To this end, goodness-of-fit tests were used to statistically identify the validity and superiority among Normal, Lognormal, and Gamma distributions for the displacement data. There are several types of goodness-of-fit tests available in the literature. Prominent among them are the Kolmogorov–Smirnov (KS) and the Anderson–Darling (AD) tests, due to their applicability to both small and large data sets. KS and AD tests both use the cumulative distribution function approach to assess the suitable distribution of a data set (Romeu 2003). The AD test is particularly useful because it is more sensitive to the tails of a distribution. However, both KS and AD tests were applied to the displacement data using MATLAB. A significance level of 5 % was considered for both tests. The results of KS and AD tests showed that the null hypothesis that the displacement data could be represented by either Gamma or Lognormal distributions should be rejected, while that of Normal distribution should not be rejected at a significance level of 0.05.

8.2   Sensitivity Analysis

Due to the presence of various uncertainties, uncertainty exists in the calculated ship hull plate displacement. To characterize this, sensitivity analysis was carried out to investigate the contribution or the importance of the input variables on the variance in displacement values. The importance measure, Sobol's first-order indices were used to quantify the expected amount of variance reduction in the output data if a deterministic value of an input factor was known (Wagner 2007). The approach was implemented by defining a simplified model of displacement (δ) as a function of the random variables:

$$\delta = f\left(a, t_p, \sigma_y, \sigma_u\right). \tag{8}$$

To measure how much the variance of δ would change by fixing a variable, the variance of δ given that a variable is fixed was calculated. For example, the contribution of a fixed value of t_p on output variance would be quantified using the sensitivity index:

$$S_{t_p} = \frac{V\left(\delta | t_p\right)}{V\left(\delta\right)}, \tag{9}$$

where $V(\delta | t_p)$ is the variance of δ for a known value of t_p and $V(\delta)$ is the variance of δ. However, assigning a fixed value for a random variable in the model can have a strong influence on the model outcome. In the context of the present example, the study by Saltelli et al. (2006) avoided the dependency of fixing a variable on the model outcome by replacing the numerator of Eq. (9) with the variance of the expected values of δ, obtained for each of the randomly generated values of thickness, t_{p_i}. The new expression for the sensitivity index of t_p was defined as:

$$S_{t_p} = \frac{V\left[E\left(\delta | t_{p_i}\right)\right]}{V\left(\delta\right)}. \tag{10}$$

The numerator of Eq. (10) was calculated by applying the law of total variance:

$$V\left(\delta\right) = V\left[E\left(\delta | t_{p_i}\right)\right] + E\left[V\left(\delta | t_{p_i}\right)\right], \tag{11}$$

where $E\left[V\left(\delta | t_{p_i}\right)\right]$ is the average of the variance of δ which is calculated by fixing t_p with all its possible values. A graphical representation of the percentage influence of all the variables on the displacement data is shown in Fig. 6. It was observed that t_p is the most sensitive of the variables. The variable **a** is the least sensitive to the

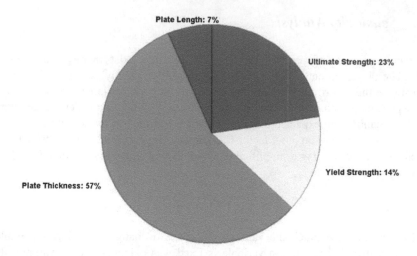

Fig. 6 Sensitivity indices of random variables on displacement data

displacement data with a sensitivity index of approximately 7 %. This result means that the variation in *a* is not significant and could be represented by a deterministic value in the simplified model.

The results of the stochastic analysis can be used to develop a probabilistic framework to predict the reliability index and failure probability of a ship hull plate under different collision scenarios. For this purpose, a limit state function can be developed to estimate the likelihood of a ship hull plate displacement exceeding a chosen design displacement capacity.

9 Discussion and Conclusion

In this study, finite element analysis of a ship hull plate impacted by a bulbous bow was combined with statistical and reliability techniques, in order to assess the effect of random strength and geometry variables on the extent of the plate deformation. The availability of Python scripting interface and sampling from pseudo-random numbers in Abaqus/CAE makes it valuable to generate multiple simulations. MATLAB was used to perform reliability assessments using MCS.

The analysis results provided knowledge and quantification of the variation in the displacement of shell plates during collisions. Normal distribution was found to be the best fit distribution to the displacement data. Parameters such as the mean value and standard deviation were identified for the displacement variable. Normal distribution and its identified parameters can then be used to represent the displacement of shell plates in any probabilistic framework.

The following are the conclusions from the reliability analysis of ship shell plate:

- The uncertainties of strength and geometry variables need to be considered during ship collision analyses and structural reliability analyses.
- The most probable displacement value of a ship shell plate would lie within the range of 1.9 and 2.1 m for the considered case study.
- Normal distribution is the best fit distribution to model the variation in displacement of a shell plate.
- Variations in the thickness of the shell plate are observed to significantly effect the hull plate displacement.

The presented study is being improved further by validating the probabilistic characteristics of displacement through the application of other sampling and reliability techniques.

Acknowledgements The authors are grateful for the support provided through the Lloyd's Register Foundation Centre. The Foundation helps to protect life and property by supporting engineering-related education, public engagement, and the application of research.

References

ABAQUS (2011) ABAQUS 6.11 analysis user's manual. *Online documentation help.* Dassault Systèmes

Amdahl J, Kavlie D (1992) Experimental and numerical simulation of double hull stranding. DNV-MIT workshop on mechanics of ship collision and grounding

Buldgen L, Le Sourne H, Besnard N, Rigo P (2012) Extension of the super-elements method to the analysis of oblique collision between two ships. Mar Struct 29(1):22–57

Chen D (2000) Simplified collision model (SIMCOL). Master's thesis, Department of Ocean Engineering, Virginia Tech

Chowdhury M (2007) On the probability of failure by yielding of hull girder midship section. Ships Offshore Struct 2(3):241–260

DNV (2004) Cargo securing model manual (no. 97-0161)

Fang C, Das PK (2005) Survivability and reliability of damaged ships after collision and grounding. Ocean Eng 32(3–4):293–307

Guo JJ, Wang GG, Perakis AN, Ivanov L (2012) A study on reliability-based inspection planning— application to deck plate thickness measurement of aging tankers. Mar Struct 25(1):85–106

Haris S, Amdahl J (2012) An analytical model to assess a ship side during a collision. Ships Offshore Struct 7(4):431–448

Hess PE, Bruchman D, Assakkaf IA, Ayyub BM (2002) Uncertainties in material and geometric strength and load variables. Nav Eng J 114(2):139–165, +209

IHS Fairplay (2012) World casualty statistics 2011

Khan IA, Das PK (2008) Reliability analysis of intact and damaged ships considering combined vertical and horizontal bending moments. Ships Offshore Struct 3(4):371–384

Kitamura O (1997) Comparative study on collision resistance of side structure. Mar Technol SNAME News 34(4):293

Lindley DV (1971) Making decisions. Wiley, London

Lutzen M (2001) Ship collision damage. Technical university of Denmark

Marine Connector (2013) Sulpicio Express Siete—7724344—general cargo. http://maritime-connector.com/ship/sulpicio-express-siete-7724344/. Accessed 3 Jan 2014

Marine Traffic (2013) Sulpicio Express Siete. https://www.marinetraffic.com/en/ais/details/ships/7724344/vessel:SPAN_ASIA_17. Accessed 3 Jan 2014

Minorsky VU (1959) An analysis of ship collision with reference to protection of nuclear power ships. J Ship Res 3(2):1–4

Nikolaidis E, Hughes O, Ayyub BM, White GJ (1993) A methodology for reliability assessment of ship structures. In: Ship structure symposium, SSC/SNAME, Arlington, VA, 17–17 Nov 1993, pp H1–H10

Paik JK, Pedersen PT (1996) Modelling of the internal mechanics in ship collisions. Ocean Eng 23(2):107–142

Paik JK, Thayamballi AK (2003) A concise introduction to the idealized structural unit method for nonlinear analysis of large plated structures and its application. Thin Wall Struct 41(4):329–355

Romeu JL (2003) Anderson-Darling: a goodness of fit test for small samples assumptions. RAC START 10(5). http://src.alionscience.com/pdf/A_DTest.pdf

Saltelli A, Ratto M, Tarantola S, Campolongo F (2006) Sensitivity analysis practices: strategies for model-based inference. Reliab Eng Syst Saf 91(10–11):1109–1125

Simonsen BC (1999) Theory and validation for the DAMAGE collision module. Technical report 67, Joint MIT-Industry Program on Tanker Safety

Simonsen BC, Lauridsen LP (2000) Energy absorption and ductile failure in metal sheets under lateral indentation by a sphere. Int J Impact Eng 24(10):1017–1039

The Philippine Star (2013) 55 confirmed dead in Cebu ship collision. http://www.philstar.com/nation/2013/08/19/1108751/updated-55-confirmed-dead-cebu-ship-collision. Accessed 3 Jan 2014

Wagner S (2007) Global sensitivity analysis of predictor models in software engineering. Paper presented at the proceedings—ICSE 2007 workshops: third international workshop on predictor models in software engineering, PROMISE'07

Wang G (2002) Some recent studies on plastic behaviour of plates subjected to large impact loads. J Offshore Mech Arct Eng 124(3):125–131

Yaw LL (2012) Nonlinear static—1D plasticity—various forms of isotropic hardening. Walla Walla University, College Place, WA

Zhang S (1999) The mechanics of ship collisions. Technical University of Denmark, Kongens Lyngby

An Advanced Point Estimate Method for Uncertainty and Sensitivity Analysis Using Nataf Transformation and Dimension-Reduction Integration

Xiaohui H. Yu and Dagang G. Lu

Abstract This article presents an advanced point estimate method (APEM) with the purposes of estimating the moments of model outputs and identifying the sensitivities of inputs variables. The APEM consists of four ingredients, i.e. (1) Nataf transformation, (2) a generalized dimension-reduction method, (3) the Gauss–Hermite integration (GHI), and (4) a global sensitivity index. Nataf transformation enables the APEM to deal with the practical engineering problems generally involving the random variables often given the marginal distributions and correlations. The generalized dimension-reduction method with the univariate and the bivariate formulas is adopted to decompose a high-dimension function into several one- and two-dimension functions, respectively. Consequently, the moments of the high-dimension function are approximated by that of the decomposed low-dimension functions, which are further calculated by the GHI. As a complement to the APEM, a global sensitivity index is additionally proposed. Through three numerical examples and two applications, the APEM shows good performance at estimating the moments of random functions, and the global sensitivity index is validated for defining the influence of a variable's variation in standard normal space on model responses.

1 Introduction

The probability-based design and performance evaluation of civil facilities has been widely used. The use of probabilistic concepts in this context stems from the recognitions that the applied loads from occupant usage and man-made and natural hazards and the construction materials are uncertain in nature. Uncertainty

X.H. Yu (✉) • D.G. Lu
School of Civil Engineering, Harbin Institute of Technology, 73 Huanghe Road,
Nangang District, Harbin, Heilongjiang 150090, China
e-mail: yxhhit@126.com

S. Kadry and A. El Hami (eds.), *Numerical Methods for Reliability and Safety
Assessment: Multiscale and Multiphysics Systems*, DOI 10.1007/978-3-319-07167-1_7,
© Springer International Publishing Switzerland 2015

propagation and analysis, therefore, provide necessary information and useful insights to civil engineers. In mathematics, computation of probability moments is a common problem in uncertainty analysis, which entails calculating a multidimension integral to determine the probabilistic outputs of a random model. However, the large uncertainties and nonlinearities involved in the engineering problems prevent such an integral to be evaluated analytically. As a consequence, the approximate methods instead of the numerical or analytical integration methods are often used. Current existing methods to approximate the probability moments of random models can be roughly classified into three major categories: (1) samplingbased methods, (2) Taylor series expansion methods (also referred as perturbation methods), and (3) point estimate methods (PEMs).

The sampling-based methods, e.g., Monte Carlo simulation (Robinstein 1981) and Latin Hypercube sampling (Olsson et al. 2003), are simple but useful tools to derive the exact statistical results with high confidence by performing sufficient simulations. However, the sampling-based methods are not attractive to civil engineers since even performing one time of analysis is often time consuming for a complex engineering system. In contrast, the perturbation methods, e.g., the first and second order reliability methods (Fiessler et al. 1979; Madsen et al. 1996), are highly efficient since they approximate a complex model by its first and second order Taylor series expansions, respectively. Nevertheless, such methods cannot derive moment estimation with adequate accuracy; moreover, they require the computation of gradients of random functions. In reality, the gradient computation generally imposes an additional calculation process known as the finite element sensitivity analysis that is actually an intensive computational task (Haukaas 2003).

Firstly proposed by Rosenblueth (1975, 1981), the PEMs are a kind of moment matching methods through replacing a continuous variable with a discrete one having the same moments. Compared with the aforementioned two categories of methods, the PEMs cost lower computational efforts than the sampling-based method and avoid conducting finite element sensitivity analysis that is the indispensable component of the Taylor series expansion methods. Over the past years, a number of PEMs have been published, which will be reviewed in the next section.

For an optimum PEM, there are two main challenges: (1) how to efficiently generate estimate points for the models involving a large number of uncertain parameters; and (2) how to effectively deal with problems always involving the variables submitting to nonnormal and dependent probability distributions. To face the above two challenges, this study develops an advanced point estimate point method (APEM) with the purposes of estimating the probability moments of model responses and identifying the sensitivities of input variables to model outputs. The APEM consists of four ingredients, i.e. (1) Nataf transformation (Liu and Der Kiureghian 1986), (2) a generalized dimension-reduction method (Xu and Rahman 2004), (3) the Gauss–Hermite integration (GHI), and (4) a global sensitivity index. With Nataf transformation, the APEM is feasible to solve the problems with known the variables' marginal distributions and the correlations. The generalized dimension-reduction method is adopted for coping with the complex models usually in forms of the function involving a large number of variables.

Considering computational efforts, only the univariate and bivariate dimension-reduction formulas are employed, and the moments of a high-dimensional function can be approximated by that of the decomposed one- and two-dimension functions, which are further estimated by GHI in order to avoid solving nonlinear equations to generate estimate points. To the best knowledge of the authors, nearly all the existing PEMs only focus on estimating probability moments of random functions, while they don't attach sufficient importance to determining the sensitivities of random variables to model responses. In this context, a global sensitivity index is presented only using the estimated variances by the APEM without requiring additional computations. Through three numerical examples and two engineering applications, the performance of the APEM is investigated by comparing with seven existing PEMs, and the validity of the presented sensitivity index is examined by comparing with the widely used sensitivity index in FORM.

2   Reviews of Existing PEMs

2.1   PEMs for Univariate Functions

Given a real random and continuous variable X, a total of $2m$ nonlinear equations derived by matching the first $2m - 1$ moments of X are required to be solved for generating the equivalent discrete variable with the corresponding probabilistic mass function (PMF) of m concentrations (Hong 1998). Rosenblueth (1975) and Gorman (1980) have provided the concentrations for the cases of only knowing the first two and three moments of X, respectively. In general cases, the accuracy of the methods with only two and three concentrations is quite low especially for the performance functions with strong nonlinearities. To remove this weakness, the following researchers naturally prefer to obtain more concentrations through matching the higher order moments of X. However, the moments with the order higher than two or three are seldom provided in practical engineering problems. Even if they are given for X, it is not an easy task to solve the corresponding nonlinear equations for defining the concentrations. In light of this, Zhou and Nowak (1988) and Zhao and Ono (2000) recommended to use the GHI points for the normal variable and then these points can be extended for the nonnormal variable through probability transformation. Since GHI has been well studied, the corresponding abscissas and weights can be found in many works [e.g., Davis and Rabinowitz 1983]. Consequently, the tedious process of solving nonlinear equations is avoided.

It has been proven that the concentrations derived by the formulas in Rosenblueth (1975) and Gorman (1980) are identical with that generated by GHI for a standard normal variable (Christian and Baecher 1999). According to above description, this study adopts GHI in the method to generate approximate points.

2.2 PEMs for Multiple Normal-Variate Functions

Consider a general multivariate performance function $g(X)$, where $X = X_1$, $X_2, \ldots, X_n$ is the collection of random variables. To approximate the moments of $g(X)$, Rosenblueth (1975, 1981) developed a PEM by assuming the joint probability distribution function (JPDF) of X to be concentrated at each combination of one standard deviation above and below the mean value of X_i. There are two major limitations for this method. Primarily, only the first two moments of X_i are considered, thus it has low accuracy to solve the complex problems involving a large number of random variables. Secondly, 2^n estimate points which also mean 2^n function evaluations are required. For engineering problems, one time of function evaluation is one time of system analysis that is often time consuming, which will lead to the unaffordable computational efforts.

To overcome the above two limitations, several alternative PEMs have been presented. The proposed improvements aim at either incorporating higher order moments of variables or reducing the needed 2^n estimating points. Panchalingam and Harr (1994) developed a PEM accounting for the skewness coefficients of X_i; however, this method still requires 2^n estimate points. Li (1992) presented an explicit expression to assess the expected value of $g(X)$. This expression considers the kurtosis coefficients of X_i, while it only requires $(n^2 + 3n + 2)/2$ evaluations of $g(X)$. Analogous to Li's method, Tsai and Franceschini (2005) further derived the formula to estimate the expectation of $g(\mathbf{X})$ with the consideration of the skewness coefficients of random variables. According to the generalized principle detailed in Hong (1998), $m \times n$ concentrations can be determined by solving nonlinear equations if the first $2m - 1$ moments are given for X_i.

In fact, the concentrations of Rosenblueth method are located at the corners of an n-dimensional hypercube in the space defined by $\mathbf{X}$. To reduce computation efforts, Lind (1983) suggested selecting points near the centers of the $2n$ faces of the hypercube. While Harr (1989) defined a hypersphere that touches the corners of the hypercube and then generated estimating points by passing the orthogonal vectors of the correlation matrix of X with this hypersphere. To simplify computation process, Chang et al. (1995) modified Harr's method by selecting estimating points in the standardized space of X that is determined through the rotational transformation. According to the opinion in He and Sallfors (1994), it is an optimum option to select points from the hypersphere that is located in the standardized space with a radius equal to $\sqrt{n}$.

2.3 PEMs for Multiple Non-normal-variate Function

Strictly speaking, the PEMs mentioned in Sect. 2.2 are totally limited to the problems involving normal variables, because they don't incorporate the variables' probability distributions. In engineering applications, however, random variables are commonly given with different nonnormal probability distributions. To handle with this case, probability transformations are often adopted in the PEMs [e.g., Zhou and Nowak 1988; Zhao and Ono 2000; Chang et al. 1997; Wang and Tung 2009]. The widely used probability transformations consist of Rosenblatt transformation (Rosenblatt 1952) and Nataf transformation (Liu and Der Kiureghian 1986), in which the former is useful to deal with the random variables with the known JPDF, while the other one is specially used for the case of incomplete probabilistic information, only knowing the marginal probability density functions (MPDFs) and the correlations. Due to the absence of statistical data and the nature of engineering, JPDF is rarely given while the MPDFs and the correlations are often provided. Therefore, Nataf transformation is more appropriate for practical engineering applications than Rosenblatt transformation. Based on this recognition, this study employs Nataf transformation in the presented method.

2.4 PEMs Using Dimension-Reduction Methods

Reducing estimating points while keeping accuracy is the permanent target pursued by all the PEMs. As an effective tool, dimension-reduction methods have been used to promote the computational efficiency. Rosenblueth (1975) presented a product formula to approximate the first two moments of $g(\mathbf{X})$, which decomposed the n-variable function into n functions of a single variable. However, this product formula is special for the uncorrelated normal variables. To circumvent this computational disadvantage, Zhou and Nowak (1988) modified the Rosenblueth's method by use of the product formula in the standard normal space, and further extended it to be suitable to the problems involving nonnormal variables through probability transformations. Different from Zhou and Nowak (1988), Zhao and Ono (2000) presented a new PEM by adopting a summation formula. Actually, if $\ln[g(\mathbf{X})]$ is the function in question, the corresponding summation-expressed approximation can be derived through making logarithmic transformation for the product formula of $g(\mathbf{X})$. From this viewpoint, the product formula can be viewed as a special case of the summation formula. In fact, either the product formula or the summation one is the special result of the univariate dimension-reduction methods that are not sufficiently accurate to estimate the high-order moments of complex models. That is the main reason why this study uses a generalized dimension-reduction formula in the presented method.

3 Advanced Point Estimate Method

3.1 Raw Moments and Central Moments of Performance Functions

Let μ_k and M_{kg} denote the kth raw and central moments of $g(\mathbf{X})$, which can be calculated by

$$\mu_k = E\left\{[g(\mathbf{x})]^k\right\} = \int_{-\infty}^{\infty} \cdots \int_{-\infty}^{\infty} [g(\mathbf{x})]^k f_\mathbf{X}(\mathbf{x})\, d\mathbf{x}, \quad k \geq 0 \tag{1}$$

and

$$M_{kg} = E\left\{[g(\mathbf{x}) - \mu_g]^k\right\} = \int [g(\mathbf{x}) - \mu_g]^k f_\mathbf{X}(\mathbf{x})\, d\mathbf{x}, \quad k \geq 2 \tag{2}$$

respectively, where $E[\cdot]$ is the expectation operator, $f_\mathbf{X}(\mathbf{x})$ is the JPDF of $\mathbf{X}$, and μ_g is the mean value of $g(\mathbf{X})$, which is evaluated by

$$\mu_g = E[g(\mathbf{x})] = \int g(\mathbf{x})\, f_\mathbf{X}(\mathbf{x})\, d\mathbf{x} \tag{3}$$

The central moment of $g(\mathbf{X})$ can be determined by the corresponding raw moments, yielding

$$M_{kg} = \left(\sum_{j=0}^{k} (-1)^j \binom{j}{k} \mu_{k-j} \mu_j^{k-j}\right), \quad k \geq 2 \tag{4}$$

In the following descriptions, for simplicity, the presented method is focused on estimating the raw moments of random functions due to the relationships shown in Eq. (4). In particular, the first four center moments of $g(\mathbf{X})$ are determined by

$$\mu_g = \mu_1 \tag{5a}$$

$$\sigma_g = \sqrt{\mu_2 - \mu_1^2} \tag{5b}$$

$$\alpha_{3g} = \left(\mu_3 - 3\mu_2\mu_1 + 2\mu_1^3\right)/\sigma_g^3 \tag{5c}$$

$$\alpha_{4g} = \left(\mu_4 - 4\mu_3\mu_1 + 6\mu_2\mu_1^2 - 3\mu_1^4\right)/\sigma_g^4 \tag{5d}$$

where μ_g, σ_g, α_{3g}, and α_{4g} are the mean value, the standard deviation (std), the coefficient of skewness, and the coefficient of kurtosis of $g(X)$, respectively.

Estimating μ_k raises the problem of how to solve the multiple integral in Eq. (1). That is actually not an easy task due to two major limitations: (1) the JPDF $f_{\mathbf{X}}(\mathbf{x})$ is rarely known; and (2) $g(\mathbf{X})$ often has strong nonlinearities and large uncertainties. To avoid these limitations, two corresponding measures are taken, i.e. (1) employing Nataf transformation instead of Rosenblatt transformation to transform the integral into the corresponding standard normal space (U space); and (2) utilizing dimension-reduction method to decompose $g(\mathbf{X})$ into a simple form only involving lower dimension functions. The following two sections will describe the above two measures in detail.

3.2 Calculation of Probability Moments of High-Dimension Functions Using Nataf Transformation

Let $T_N(\cdot)$ and $T_N^{-1}(\cdot)$ denote the operators of the forward and the inverse Nataf transformations that are performed through

$$\mathbf{u} = T_N(\mathbf{x}) = \mathbf{L}_0^{-1} \Phi^{-1} [\mathbf{F}_{\mathbf{X}}(\mathbf{x})] \tag{6}$$

and

$$\mathbf{x} = T_N^{-1}(\mathbf{u}) = \mathbf{F}_{\mathbf{X}}^{-1} [\Phi(\mathbf{L}_0 \mathbf{u})] \tag{7}$$

respectively, where $F_{\mathbf{X}}(\cdot)$ is the cumulative distribution function (CDF) of $\mathbf{X}$ with the elements $F_{X_1}(\cdot), F_{X_2}(\cdot), \ldots, F_{X_n}(\cdot)$ representing the CDFs of $X_1, X_2, \ldots, X_n$, respectively, $\Phi^{-1}(\cdot)$ is the inverse normal CDF, and $\mathbf{L}_0$ is the lower triangular matrix obtained from Cholesky decomposition of the U space correlation matrix, $\mathbf{R}_0 = [\rho_{0,ij}]$. As noted by Liu and Der Kiureghian (1986), the correlation coefficient ρ_{ij} of X_i and X_j is not the same with $\rho_{0,ij}$, which can be solved iteratively for given $F_{X_1}(X_i)$ and ρ_{ij}. To avoid such a tedious process, a set of semiempirical formulas for the ratio $F = \rho_{0,ij}/\rho_{ij}$ have been developed, which can be found in Melchers (1999).

Substituting Eq. (7) into $g(\mathbf{X})$, a new performance function is defined in U space by

$$L(\mathbf{u}) = \left\{ g\left[T_N^{-1}(\mathbf{u}) \right] \right\}^k \tag{8}$$

and then the multiple integral in Eq. (1) is rewritten as

$$\mu_k = \int L(\mathbf{u}) \varphi_{\mathbf{U}}(\mathbf{u}) d\mathbf{u} = \int L(\mathbf{u}) \varphi_{U_1}(u_1) \cdots \varphi_{U_n}(u_n) du_1 \cdots du_n \tag{9}$$

where $\varphi_U(\cdot)$ is the JPDF of vector U. Since the variables in the U space are totally uncorrelated, thus $\varphi_U(\cdot)$ is equal to the product of the standard normal PDFs $\varphi_{U_1}(\cdot), \varphi_{U_2}(\cdot), \ldots, \varphi_{U_n}(\cdot)$ for $U_1, U_2, \ldots, U_n$, respectively. Compared with Eq. (1), Eq. (9) avoids the process of determining $f_X(x)$ since Nataf transformation is used to map the variables into U space.

3.3 Decomposition of High-Dimension Function by Dimension Reduction Method

Xu and Rahman (2004) have derived a generalized dimension-reduction method for high-dimension function. Through this method, an n-dimension function $L(U)$ can be expressed in a summation form involving a series of at most s-dimensional functions, yielding

$$L(U) \cong L_s(U) = \sum_{i=0}^{s} (-1)^i \binom{n-s+i-1}{i} \sum_{k_1 < \cdots < k_{s-i}} y_{s-i} \tag{10}$$

where s is less than n, $L_s(U)$ is the s-dimensional approximation of $L(U)$, and

$$y_{s-i} = L\left(c_1', \ldots, c_{k_1-1}', U_{k_1}, c_{k_1+1}', \ldots, c_{k_{s-i}-1}', U_{k_{s-i}}, c_{k_{s-i}+1}', \ldots, c_n'\right) \tag{11}$$

As seen from Eq. (11), y_{s-i} is a general $(s-i)$-dimension function, in which the variables without being treated as random ones are all assumed as the specified values in terms of "reference point," which is determined in U space by c',

$$c' = \left[c_1', c_2', \ldots, c_n'\right] = T_N(c) \tag{12}$$

where the vector c in the X space can be conveniently adopted as the means of X_i for $i = 1, 2, \ldots, n$.

Replacing $L(U)$ with $L_s(U)$, the n-dimension integral in Eq. (9) is approximated by several no more than s-dimension integrals:

$$\mu_k \cong \int L_s(u) \varphi_U(u) \, du$$

$$= \sum_{i=0}^{s} (-1)^i \binom{n-s+i-1}{i} \sum_{k_1 < \cdots < k_{s-i}} E(y_{s-i})$$

$$= \sum_{i=0}^{s} (-1)^i \binom{n-s+i-1}{i} \sum_{k_1 < \cdots < k_{s-i}} \int_{-\infty}^{+\infty} \cdots \int_{-\infty}^{+\infty} y_{s-i}' \varphi_{U_{k_1}}(u_{k_1})$$

$$\cdots \varphi_{U_{k_{s-i}}}(u_{k_{s-i}}) \, du_{k_1} \cdots du_{k_{s-i}} \tag{13}$$

In particular, the univariate and bivariate dimension-reduction formulas are defined by

$$L\left(\mathbf{U}\right) \cong L_1\left(\mathbf{U}\right) = \sum_{i=1}^{n} L_i\left(\mathbf{U}_i\right) - (n-1) L\left(\mathbf{c}'\right) \tag{14}$$

and

$$L\left(\mathbf{U}\right) \cong L_2\left(\mathbf{U}\right) = \sum_{i<j}^{n} L_{ij}\left(\mathbf{U}_{ij}\right) - (n-2) \sum_{i}^{n} L_i\left(\mathbf{U}_i\right) + \frac{(n-1)(n-2)}{2} L\left(\mathbf{c}'\right) \tag{15}$$

respectively, in which the decomposed univariate and bivariate functions, $L_i(\mathbf{U}_i)$ and $L_{ij}(\mathbf{U}_{ij})$, are expressed by

$$L_i\left(\mathbf{U}_i\right) = L\left(c_1', \ldots, c_{i-1}', U_i, c_{i+1}', \ldots, c_n'\right) \tag{16}$$

and

$$L_{ij}\left(\mathbf{U}_{ij}\right) = L\left(c_1', \ldots, c_{i-1}', U_i, c_{i+1}', \ldots, c_{j-1}', U_j, c_{j+1}', \ldots, c_n'\right) \tag{17}$$

respectively.

It is valuable to note that Eq. (14) has a similar expression with the summation approximation used in Zhao and Ono (2000) except for Rosenblatt transformation instead of Nataf transformation was used. According to the above descriptions, it is not difficult to find that the summation-formed approximation used in Zhao and Ono (2000) is actually the univariate case of the generalized dimension-reduction method.

Substituting $L_1(\mathbf{U})$ and $L_2(\mathbf{U})$ into Eq. (13) gives

$$\mu_k \cong E\left[L_1\left(\mathbf{u}\right)\right] = \sum_{i=1}^{n} \mu_{L1i} - (n-1) L\left(\mathbf{c}'\right) \tag{18}$$

$$\mu_k \cong E\left[L_2\left(\mathbf{u}\right)\right] = \sum_{i<j}^{n} \mu_{L2ij} - (n-2) \sum_{i}^{n} \mu_{L1i} + \frac{(n-1)(n-2)}{2} L\left(\mathbf{c}'\right) \tag{19}$$

where μ_{L1i} and μ_{L2ij} are the raw moments of $L_i(\mathbf{U}_i)$ and $L_{ij}(\mathbf{U}_{ij})$, respectively.

Table 1 Gauss points and
the corresponding weights

m	z_j	w_j
3	0	1.1816
	±1.2247	0.2954
5	0	0.9453
	±2.0202	0.0200
	±0.9586	0.3936
7	0	0.8103
	±2.6520	0.0010
	±1.6736	0.0545
	±0.8163	0.4256

3.4 Calculation of Probability Moments of Lower Dimension Functions via Gauss–Hermite Integration

Consider an integral $\int_{-\infty}^{+\infty} y(x)e^{-x^2}dx$, where $y(x)$ is a general and real univariate function. Using GHI, the integral with weight function e^{-x^2} can be approximated by (Davis and Rabinowitz 1983)

$$\int_{-\infty}^{+\infty} y(x)e^{-x^2}dx = \sum_{j=1}^{m} w_j\, y\left(z_j\right) \tag{20}$$

where z_j and w_j are the abscissas (also referred as Gauss points) and weights, respectively, and m is the quadrature order (also referred as the number of Gauss points). Table 1 lists the Gauss points and weights which will be used in the following sections, where $m =3, 5$, and 7.

If y is a function of a single standard normal variable, then the corresponding expectation can be approximated by

$$E(y) = \int y(u)\varphi_U(u)du = \int y(u)\frac{1}{\sqrt{\pi}}e^{-u^2/2}du = \sum_{j=1}^{m} \frac{w_j}{\sqrt{\pi}} y\left(\sqrt{2}z_j\right) \tag{21}$$

Since **U** space has a characteristics of rotational symmetry, Eq. (21) can be extended to solve the expectation of a function involving n standard normal variables by

$$E(y) = \int_{-\infty}^{+\infty} \cdots \int_{-\infty}^{+\infty} y\left(u_1, \ldots, u_n\right) \varphi_{U_1}\left(u_1\right) \cdots \varphi_{U_n}\left(u_n\right) du_1 \cdots du_n$$

$$\cong \sum_{j_1=1}^{m_1} \cdots \sum_{j_n=1}^{m_n} \frac{w_{j_1} \cdots w_{j_n}}{\sqrt{\pi}} y\left(\sqrt{2}z_{j_1}, \ldots, \sqrt{2}z_{j_n}\right) \tag{22}$$

where $y(z_1, \ldots, z_n)$ is the n-dimension function of standard normal variables, $z_1, \ldots, z_n$, $m_1, \ldots, m_n$ are the numbers of Gauss points used in $z_1, \ldots, z_n$ directions, $z_{j1}, \ldots, z_{jn}$ are Gauss points, and $w_{j1}, \ldots, w_{jn}$ are weights corresponding to $z_{j1}, \ldots, z_{jn}$.

Use Eq. (22) to approximate $E(y_{s-i})$ and then we have

$$
\mu_k \cong \sum_{i=0}^{s} (-1)^i \binom{n-s+i-1}{i} \sum_{k_1 < \cdots < k_{s-i}} E\left(y_{s-i}\right)
$$

$$
\cong \sum_{i=0}^{s} (-1)^i \binom{n-s+i-1}{i} \times \cdots \times \cdots \sum_{k_1 < \cdots < k_{s-i}} \sum_{j_{k_1}=1}^{m_{k_1}} \cdots \sum_{j_{k_{s-i}}=1}^{m_{k_{s-i}}}
$$

$$
\frac{w_{j_{k_1}} \cdots w_{j_{k_{s-i}}}}{\sqrt{\pi}} L\left(c_1', \ldots, c_{k_1-1}', z_{k_1}, c_{k_1+1}', \ldots, c_{k_{s-i}-1}', z_{k_{s-i}}, c_{k_{s-i}+1}', \ldots, c_n'\right)
$$

$$
(23)
$$

3.5 Computational Efforts

According to Eq. (23), $\prod_{i=1}^{s-i} m_{k_i}$ function evaluations are required to estimate $E(y_{s-i})$ through GHI. Hence the total computational cost for the approximation of μ_k can be determined by

$$
N_s = \sum_{i=0}^{s} \binom{n-s+i-1}{i} \prod_{i=1}^{s-i} m_{k_i} \tag{24}
$$

where N_s is the number of function evaluations. If $m_{k1} = m_{k2} =, \ldots, = m_{kn} = m$, then Eq. (24) can be rewritten as

$$
N_s = \sum_{i=0}^{s} \binom{n-s+i-1}{i} m^{s-i} \tag{25}
$$

It is evident for the dimension-reduction formula in Eq. (10) that $L_s(\mathbf{U})$ has increasingly higher accuracy to approximate $L(\mathbf{U})$ with s getting closer to n. As a result, the required number of function evaluations N_s will increase rapidly. Consider a general function involving ten variables. The corresponding values of N_s are calculated in Table 2, where $m = 3, 5$, and 7, respectively.

As seen from this table, unaffordable numbers of function evaluations are always required when s is more than 2. For example, a maximum of 3,675 function evaluations are required for the case of $s = 3$ and $m = 3$, while this number is reduced to 435 when $s = 2$ and $m = 3$. In engineering problems, finite element

Table 2 Values of N_s for the functions involving ten variables

	N_s		
s	$m=3$	$m=5$	$m=7$
1	30	50	70
2	435	1,175	2,275
3	3,675	16,175	43,435
4	22,143	158,675	590,863
5	77,547	871,175	4,422,859
6	188,355	3,246,175	22,305,507
7	330,822	8,335,460	75,953,451
8	437,673	14,697,068	169,837,350
9	473,289	18,231,294	242,858,170

analyses are often needed to evaluate function values. If quite a large number of function evaluations are required, then the presented method in this study will be restricted to engineering applications and less attractive for engineers. On the basis of this consideration, only the univariate and bivariate dimension reduction formulas (Eqs. (14) and (15)) are used, and the corresponding computational costs are calculated by

$$N_1 = \binom{n}{1} \times m + \binom{n}{0} \times m^0 = n \times m + 1 \tag{26}$$

and

$$N_2 = \binom{n}{2} \times m^2 + \binom{n}{1} \times m + \binom{n}{0} \times m^0 = \frac{n \times (n-1)}{2} \times m^2 + n \times m + 1 \tag{27}$$

respectively.

3.6 Selection of Reference Points

Two algorithms are used to select reference points, namely, the mean-reference algorithm and median-reference algorithm. Let $\boldsymbol{\mu}_X$ and $\mathbf{m}_X$ denote the mean vector and the median vector of $\mathbf{X}$, respectively. According to the mean-reference algorithm, the reference points in $\mathbf{U}$ space are mapped from $\boldsymbol{\mu}_X$ by

$$\mathbf{c}'_{Mn} = \left[c'_{1,Mn}, \ldots, c'_{n,Mn} \right] = T_N \left(\boldsymbol{\mu}_X \right) \tag{28}$$

where the subscript "Mn" is the abbreviation for "mean."

Different from the mean-reference algorithm, $\mathbf{m}_X$ is adopted in the median-reference algorithm as the reference point in the $\mathbf{X}$ space. Through Nataf transformation, an n-dimension zero vector is derived to be the corresponding reference point in the $\mathbf{U}$ space, yielding

$$\mathbf{c}'_{\mathrm{Md}} = \left[c'_{1,\mathrm{Md}}, \dots, c'_{n,\mathrm{Md}} \right] = T_N\left(\mathbf{m_X}\right) = \left[\underbrace{0, \dots, 0}_{n} \right] \tag{29}$$

where the subscript "Md" is the abbreviation for "median."

In fact, the n-dimension zero vector $\left[\underbrace{0, \dots, 0}_{n} \right]$ is not only the mapped vector of $\mathbf{m_X}$ in the $\mathbf{U}$ space but also the mean collection of the n-dimension standard normal vector. From this aspect, the median-reference algorithm in the $\mathbf{X}$ space can also be viewed as the mean-reference algorithm in the $\mathbf{U}$ space.

3.7 A Global Sensitivity Index

For complex engineering problems, it is necessary to investigate the sensitivities of the input variables to the model outputs in order to simplify the process of uncertainty propagation and analysis. For the variable with a quite small value of sensitivity, it may well be treated as deterministic rather than random. In the recent past, the literature has assisted to the fast growth of sensitivity analysis methods. A variety of sensitivity analysis methods for solving different problems have been developed, from local methods to nonparametric methods (Helton 1993), to screening methods (Morris 1991), to variance-based methods (Saltelli and Tarantola 2002), to distribution-based (Borgonovo 2006). For the variation-based methods, the Sobol' sensitivity indices (Sobol' 1990), which are in fact the global indices (Saltelli et al. 2008), maybe are the mostly adopted measures (Dimov and Georgieva 2010; Satelli et al. 2010; Nossent et al. 2011; Xu et al. 2012; Glen and Isaacs 2012).

Similar to the first-order Sobol' sensitivity index, an measure denoted by η_i is used to define the sensitivities of X_i for $i = 1, 2, \dots, n$ by

$$\eta_i = \frac{\sigma_i^2}{\sigma_g^2} \quad i = 1, 2, \dots, n \tag{30}$$

where σ_i^2 and σ_g^2 are the estimated variances of $L_i(\mathbf{U}_i)$ and $L(\mathbf{U})$, respectively.

According to Eq. (18), it is not difficult to derive

$$\sigma_g^2 = \sum_{i=1}^{n} \sigma_i^2 \tag{31}$$

Based on the above equation, η_i satisfies

$$\sum_{i=1}^{n} \eta_i = 1 \tag{32}$$

Besides Eq. (32), there are three other useful properties for η_i: (1) it can identify the effect of the variation of a random variable in the $\mathbf{U}$ space on model outputs; (2) it can consider the correlations between input random variables; and (3) it requires no additional computational efforts, and thus it is easier to compute than the first-order Sobol sensitivity index.

4 Numerical Examples

4.1 Performance Evaluation

Three numerical examples (Chang et al. 1995; Wang and Tung 2009) are adopted herein to demonstrate the performance of the APEM. The tested models include: (1) $W_1 = \sum i X_i$ (linear), (2) $W_2 = \sum X_i^2$ (quadratic), and (3) $W_3 = \prod X_i$ (multiplicative). These numerical examples mainly aim at investigating the effects of the increasing model complexities on the performance of the APEM. For simplicity, four variables are involved in the tested models, and they are assumed to follow the identical lognormal distribution with the mean value and the standard deviation that are equal to 1.0 and 0.3, respectively. In addition, the correlation coefficient between X_i and X_j is set as $\rho_{ij} = 0.9^{|i-j|}$.

A total of 100,000 samples are generated according to the Nataf-transformation-based multivariate MCS (Chang et al. 1994). The first four central moments from the MCS are adopted as the "true values," based on which seven different PEMs are used as the comparisons for better examining the performance of the APEM. In the following descriptions, the central moments are called moments for short. These PEMs for comparison include Rosenblueth method (Rosenblueth 1975, 1981), Li method (Li 1992), Harr method (Harr 1989), Modify–Harr method (Chang et al. 1995), Chang method (employing the median-expansion algorithm in Chang et al. (1997)), He–Sallfors method (adopting $2n + 2^n$ points in He and Sallfors (1994)), and Zhou–Nowak method (adopting $n + 1$ points in Zhou and Nowak (1988)).

Two criteria are adopted to evaluate the accuracy of the PEMs. Let e_M denote the error of the estimated moments, which is described by

$$e_M = \frac{\theta_{PEM} - \theta_{MCS}}{\theta_{MCS}} \tag{33}$$

where θ_{PEM} and θ_{MCS} are the estimated moments by PEMs and MCS, respectively.

The goodness of the fitted distributions is also compared to represent the collective behaviors of different methods. Let ω_{PEM} and ω_{MCS} represent the fitted distributions by the estimated moments θ_{PEM} and θ_{MCS}, respectively, and the error e_D of the fitted distribution is defined as

$$e_D = \int_{-\infty}^{\infty} |\omega_{\text{PEM}}(x) - \omega_{\text{MCS}}(x)| \, dx \qquad (34)$$

where ω_{PEM} and ω_{MCS} are derived through the maximum entropy method by use of the estimated first four moments. The mathematical details of the maximum entropy method can be found in Zellner and Highfield (1988).

Also it is necessary to examine the validity of the presented sensitivity index (η_i in Eq. (30)). To match this demand, the widely used sensitivity index α_i in FORM is used to compare with η_i. This comparison is based on two similarities between α_i and η_i. Firstly, α_i also demonstrates the influence of a variable's variation in the $\mathbf{U}$ space on model response since it is calculated by Melchers (1999)

$$\alpha_i = -\frac{\partial_{u_i} G\left(\mathbf{u}^*\right)}{\left\| \partial_{u_i} G\left(\mathbf{u}^*\right) \right\|} \qquad i = 1, 2, \dots, n \qquad (35)$$

where u^* is the "design point" or the "check point" of the standardized limit state function $G(\mathbf{U}) = 0$.

Secondly, the summation of α_i^2 is also equal to one since α_i represents the direction cosine of $G(\mathbf{U}) = 0$ at $\mathbf{u}^*$. During calculating α_i, the limit state functions G_j corresponding to the numerical models W_j for $j = 1, 2, 3$ are defined by

$$G_j = \mu_{\text{MCS},j} - W_j \qquad j = 1, 2, 3 \qquad (36)$$

where $\mu_{\text{MCS},j}$ is the estimated mean value of W_j by MCS.

4.2 Moment Results

Tables 3, 4, and 5 compare the performances of various PEMs based on the results of e_M (see Eq. (33)) and e_D (see Eq. (34)) for model W_1, W_2, and W_3, respectively. In the APEM, five Gauss points are employed; both the mean-reference algorithm (see Eq. (28)) and the median-reference algorithm (see Eq. (29)) are used and are shortly named "Mn" and "Md," respectively; and both the univariate and the bivariate dimension-reduction formulas are adopted and are abbreviated as "Uni" and "Bi," respectively.

For the linear model W_1, Table 3 shows that different PEMs yield very close estimations to the exact means. Rosenblueth method, Harr method, Modify–Harr method, and Li method all can provide the stds with high accuracy, while these methods cannot give good estimations of the coefficients of skewness and kurtosis.

Table 3 Comparison of estimated moments and fit goodness of distributions for W_1

Methods	Function evaluations	e_M (%)				e_D (%)
		Mean	Std	Skewness	Kurtosis	
Rosenblueth	16	0.15	0.37	100.00	76.88	150.93
Harr	8	0.15	0.37	100.00	76.88	150.93
Modify–Harr	8	0.15	0.37	100.00	56.10	38.11
Li	12	0.15	0.37	80.99	15.50	140.74
Chang	8	0.19	3.21	57.49	55.30	36.66
He–Sallfors	24	0.15	0.37	100.00	66.69	67.32
Zhou–Nowak	5	0.13	0.26	61.87	66.69	77.56
APEM (Mn + Uni)	21	0.15	0.96	16.01	8.55	3.76
APEM (Md + Uni)		0.17	1.53	15.34	8.25	3.59
APEM (Mn + Bi)	171	0.15	0.46	0.31	0.55	0.00
APEM (Md + Bi)		0.15	0.46	0.31	0.55	0.00

Table 4 Comparison of estimated moments and fit goodness of distributions for W_2

Methods	Function evaluations	e_M (%)				e_D (%)
		Mean	Std	Skewness	Kurtosis	
Rosenblueth	16	0.36	13.16	100.00	91.21	153.51
Harr	8	0.36	13.16	100.00	90.17	124.15
Modify–Harr	8	0.36	10.82	44.35	61.16	14.58
Li	12	0.36	8.92	51.09	90.46	146.95
Chang	8	0.40	2.70	13.13	54.78	60.07
He–Sallfors	24	0.36	12.37	80.45	80.54	37.27
Zhou–Nowak	5	0.75	7.67	60.88	77.69	42.31
APEM (Mn + Uni)	21	0.20	4.98	27.87	36.84	6.10
APEM (Md + Uni)		0.14	6.82	26.08	35.43	5.57
APEM (Mn + Bi)	171	0.03	0.68	2.18	6.03	0.02
APEM (Md + Bi)		0.03	0.88	3.34	7.17	0.01

Table 5 Comparison of estimated moments and fit goodness of distributions for W_3

Methods	Function evaluations	e_M (%)				e_D (%)
		Mean	Std	Skewness	Kurtosis	
Rosenblueth	16	5.88	49.39	99.05	99.72	195.54
Harr	8	5.38	47.59	98.22	99.74	210.64
Modify–Harr	8	4.24	27.66	79.44	98.61	107.72
Li	12	5.88	36.16	96.96	99.90	206.23
Chang	8	3.33	3.80	78.47	98.56	135.01
He–Sallfors	24	5.14	41.32	87.09	98.83	109.39
Zhou–Nowak	5	1.59	32.30	87.34	99.28	138.93
APEM (Mn + Uni)	21	2.25	18.27	57.90	92.69	78.32
APEM (Md + Uni)		3.87	20.79	57.03	92.51	77.14
APEM (Mn + Bi)	171	0.05	2.59	44.18	86.94	0.81
APEM (Md + Bi)		0.11	3.14	44.72	87.20	0.83

In contrast, the APEM can provide not only the sufficiently accurate means and stds but also the reasonably well estimated higher order moments. Consequently, the APEM provides the better fitted distributions than the other PEMs. The good performance of the APEM to estimate models' high-order moments, especially the coefficients of skewness and kurtosis, can also be observed in the comparison results for the models of W_2 (see Table 4) and W_3 (see Table 5) although they have increasingly larger nonlinearity than W_1. In particular, for the most complex model W_3, only the APEM using the bivariate dimension-reduction method can provide the relatively exact estimations of moments and corresponding fitted distributions.

It can also been seen from the above tables that reference points haven't shown significant influence on the estimated moments and the corresponding fitted distributions. Actually, it is not difficult to predict that the APEM using more GHI points can give more accurate estimation results. However, adopting more GHI points means more extensive computational efforts. Actually five GHI points are adequate for the APEM to provide sufficiently accurate estimation. Besides that it is easy to understand that using the bivariate dimension-reduction formula will obtain better estimated moments than using the univariate dimension-reduction formula. But the univariate dimension-reduction formula can estimate the means and stds of model outputs with acceptable accuracy as illustrated in the above tables. In practical engineering applications, the first two moments of model responses are often paid much more attention to than the other higher order moments. From this aspect, the APEM using the univariate dimension-reduction formula is especially useful for engineers since it requires favorable less computational efforts than the bivariate dimension-reduction formula.

4.3   Sensitivity Results

Figure 1 shows the comparisons of the sensitivities of various random variables defined by the proposed sensitivity index (η_i in Eq. (30)) and that by the FORM index (α_i^2 in Eq. (35)). To calculate η_i, five Gauss points and both the mean- and median-reference algorithms are used in the APEM. Figure 1 reveals that the mean- and median-reference algorithms generate almost the same results of sensitivity. The APEM generates almost the same sensitivity data with the FORM through comparing η_i and α_i^2. Such a consistence validates the presented sensitivity index at measuring the influence of a variable's variation in the U space on model responses. Based on the results of η_i and α_i^2, we can give the same quantitative judgments of variables' sensitivities. However, the calculation of gradient that is necessary for α_i^2 is avoided during determining η_i.

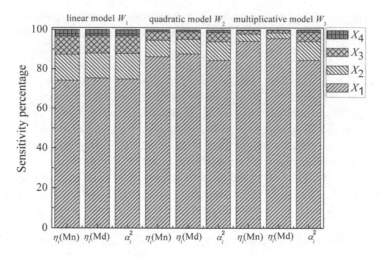

Fig. 1 Comparisons of the sensitivity results from the APEM and FORM for different models

5   Applications

In this section, two simple models to respectively predict the terminal velocity and the potential scour depth around bridge piers are used for uncertainty and sensitivity analysis with the consideration of the randomness of model input variables. The first four moments of model outputs and the sensitivities of input random variables are evaluated by the APEM. For further examination of the performance of the APEM, the seven existing PEMs used in the numerical examples (Sect. 4) are also used as comparisons through the two performance criteria in Eqs. (33) and (34), where the exact results are calculated by the MCS involving 100,000 samples. For further validation of the presented sensitivity index η_i, it is also compared with α_i^2, where five Gauss points and both the mean- and median-reference algorithm are adopted in the APEM, and the limit state functions needed by FORM are defined according to Eq. (36).

5.1   Terminal Velocity Model

The concept of terminal velocity is an important aspect in several fields of environmental engineering. Tsai and Franceschini (2005) used a probabilistic model to perform uncertainty analysis for the terminal velocity v_t. According to this model, v_t is predicted by

$$v_t = \left[\frac{4g\left(\rho_p - \rho_f\right)d}{3C_D\rho_f} \right]^{0.5} \tag{37}$$

Table 6 Random variables of the terminal velocity model v_t

Variables	Distribution	Probability moments			
		Mean	Std	Skewness	Kurtosis
d (cm)	Lognormal	2.49	1.19	1.22	4.91
ρ_p (g/cm^3)	Weibull	2.02	0.69	1.30	6.16
C_D	Lognormal	0.97	0.06	−0.13	2.77

Table 7 Comparisons of estimated moments and the corresponding distributions of v_t by various PEMs

PEMs	e_M (%)				e_D (%)
	Mean	Std	Skewness	Kurtosis	
Rosenblueth	2.85	32.88	78.69	13.35	13.18
Harr	3.01	30.69	17.17	17.93	13.62
Modify–Harr	3.56	23.73	55.46	57.17	73.84
Li	1.47	55.31	7,311.65	8,772.78	199.98
Chang	2.99	9.17	100.00	70.35	151.29
He–Sallfors	3.33	26.34	76.90	47.25	40.78
Zhou–Nowak	0.65	11.70	22.71	56.53	81.92
APEM (Mn + Uni)	0.27	0.45	96.00	12.06	10.59
APEM (Md + Uni)	0.02	0.03	95.86	12.04	10.58
APEM (Mn + Bi)	0.05	1.42	51.73	0.57	0.02
APEM (Md + Bi)	0.05	1.43	83.71	0.70	0.04
Rosenblueth	2.85	32.88	78.69	13.35	13.18
Harr	3.01	30.69	17.17	17.93	13.62
Modify–Harr	3.56	23.73	55.46	57.17	73.84
Li	1.47	55.31	7,311.65	8,772.78	199.98
Chang	2.99	9.17	100.00	70.35	151.29
He–Sallfors	3.33	26.34	76.90	47.25	40.78

where the gravity acceleration g, the particle density ρ_p, the particle density ρ_f, the particle diameter d, and the drag coefficient C_D are considered as the main influence factors. According to Tsai and Franceschini (2005), g and ρ_p are taken as constant values: $g = 981$ cm/s^2 and $\rho_p = 1$ g/mm^3, while ρ_p, d, and C_D are considered as random with the corresponding distributions listed in Table 6. The correlation coefficients for ρ_p and d, ρ_p and C_D, and d and C_D are assumed as −0.5, −0.31, and 0.68, respectively.

Table 7 presents the comparisons of the estimated moments and the corresponding distributions for v_t by the APEM and the other PEMs. It is clear that the APEM yields the best estimated mean, std, and the coefficient of kurtosis, while Harr method gives the best estimation of the coefficient of skewness with less than 20 error percentages. Nevertheless, the APEM yields the better fitted distribution compared with the other PEMs. This observation demonstrates the better performance of the APEM to estimate moments from a collective viewpoint.

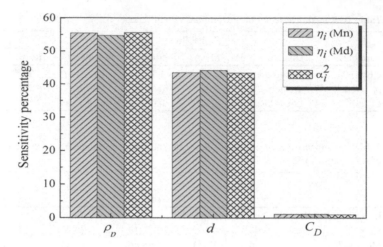

Fig. 2 Sensitivities defined by η_i and α_i^2 for the variables, i.e., the particle density ρ_p, the particle diameter d, and the drag coefficient C_D, which are involved in the probabilistic model of the terminal velocity v_t

The APEM using the bivariate dimension-reduction formula is more accurate than the univarite dimension-reduction formula to estimate the coefficients of skewness and kurtosis. That is the reason why the fitted distribution by the bivariate dimension-reduction formula is more close to the exact distribution than that by the univarite dimension-reduction formula. However, the estimated means and stds by the bivariate dimension-reduction formula are almost identical with that by the univariate dimension-reduction formula. In addition, the APEM adopting either the mean-reference algorithm or the median-reference algorithm can give almost the identical and accurate moments and the corresponding distributions.

Figure 2 illustrates the comparisons of the sensitivities of ρ_p, d, and C_D defined by both η_i and α_i^2. It is evident that the η_i-defined sensitivities either from the mean-algorithm or from the median-algorithm are very close to that defined by α_i^2. The particle density ρ_p has the most prominent influence on v_t, while the variation of drag coefficient C_D illustrates negligent influence on v_t. Therefore, we can view the drag coefficient as a deterministic value rather than a random variable for further performing uncertainty analysis by use of the probabilistic model in Eq. (37).

5.2 Pier Scouring Model

Bed scouring is a frequent phenomenon in a river caused by the interaction of flow and the river bed. Hydraulic structures such as bridge piers are susceptible to failure under long-term and continuous bed scouring. Therefore, it's important to predict

Table 8 Random variables for the pier scouring model D_s

Variables	Distribution types	Mean	COV
λ	Lognormal	1.00	0.18
y	Gamma	4.25	0.20
F	Weibull	0.54	0.38
σ	Lognormal	4.00	0.20

Table 9 Correlation coefficients among random variables for the pier scouring model D_s

Variables	λ	y	F	σ
λ	1.00	0.00	0.00	0.00
y	0.00	1.00	−0.33	−0.79
F	0.00	−0.33	1.00	0.29
σ	0.00	−0.79	0.29	1.00

Table 10 Comparisons of the estimated moments and the fitted distributions of D_s by various PEMs

PEMs	e_M (%) Mean	Std	Skewness	Kurtosis	e_D (%)
Rosenblueth	0.04	3.91	64.80	52.58	49.46
Harr	0.20	1.73	96.99	22.89	14.17
Modify–Harr	0.15	2.40	91.39	22.88	13.32
Chang	0.38	2.72	18.20	27.73	14.72
He–Sallfors	0.13	2.72	73.66	42.91	27.55
Zhou–Nowak	0.29	7.31	98.86	24.83	80.94
APEM (Mn + Uni)	0.06	0.14	42.85	6.34	6.09
APEM (Md + Uni)	0.05	0.07	41.61	6.21	5.92
APEM (Mn + Bi)	0.06	0.11	0.49	1.34	0.06
APEM (Md + Bi)	0.06	0.09	0.42	1.12	0.03

the potential scour depth around bridge piers. A simple probabilistic model has been used to perform uncertainty analysis for the predicted scour depth D_s by (Chang et al. 1994, 1997)

$$D_s = 2.02\lambda y \left(\frac{b}{y}\right)^{0.98} F^{0.21} \sigma^{-0.34} \tag{38}$$

where the model correction factor λ, the flow depth y, the Froude number F, and the sediment gradation σ are all considered as random variables, while the pier width b is determined as a constant value of $2m$. Table 8 presents the set of distributions assumed for λ, y, F, and σ. All these random parameters except λ are correlated and the corresponding correlation coefficients are presented in Table 9.

Table 10 presents the error percentages of the estimated moments and the corresponding distributions. It is found that the APEM yields better moments and distributions than the other PEMs. In particular, the APEM with the bivariate

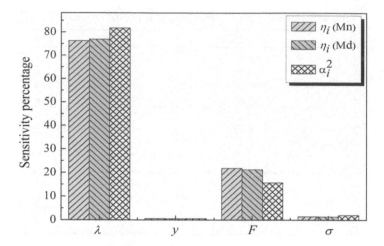

Fig. 3 Sensitivities defined by η_i and α_i^2 for the variables, i.e., the correction factor λ, the flow depth y, the Froude number F, and the sediment gradation σ, which are involved in the probabilistic model of the predicted scour depth D_s

dimension-reduction formula can greatly improve the accuracy of the estimated coefficients of skewness and kurtosis and the distribution; however it requires large computation efforts. Compared to the APEM using the bivariate dimension-reduction formula, the method adopting the univariate dimension-reduction formula needs less computation resource and can provide the mean and std estimations with adequate accuracy. The estimated results by the mean-reference algorithm are nearly identical with that by the median-reference algorithm.

Figure 3 shows the comparisons of sensitivities of λ, y, F, and σ by η_i and α_i^2. The calculated values of η_i are very close to the corresponding values of α_i^2. According to the defined sensitivities, the model correction factor λ and the Froude number F show the dominant influences on the predicted D_s, while the flow depth y and the sediment gradation σ can be treated as deterministic due to the corresponding quite low sensitivities.

In fact, most conclusions from the observations of the above two applications are also obtained in the numerical examples (see Sect. 4). However, the probabilistic models in Eqs. (37) and (38) involve the random variables submitting to a collection of different nonnormal distributions instead of the identical lognormal distribution used in the numerical examples. From this point, these two applications again verify the good performance of the APEM.

6 Conclusions

An advanced point estimate method (APEM) is presented for uncertainty and sensitivity analysis. Three numerical examples and two practical applications are provided to demonstrate the accuracy of the APEM from the estimated moments and the defined sensitivities. According to the results, several main conclusions are drawn as follows:

- Compared with other PEMs, the APEM provides better estimated moments, especially the higher order ones, i.e., the coefficients of skewness and kurtosis. Correspondingly, the APEM gives the best fitted distributions from the estimated moments. These results indicate the good performance of the APEM for estimating moments of random functions.
- The sensitivities defined by the presented index η_i are very close to that defined by the FORM index α_i^2. Such a consistence verifies the validity of η_i at measuring the significances of a variable's variation in the U space on model responses. However, the calculation of η_i only requires the estimated variance while without needing to calculate the gradient always required by FORM to compute α_i^2. As a beneficial complement, the presented index η_i enables the APEM to not only estimate the model moments but also identify the sensitivities.
- In the APEM, selecting different reference points doesn't illustrate significant influence on the estimated results. Compared with the univariate dimension-reduction formula, the bivariate dimension-reduction formula can greatly promote the accuracy of the estimated high-order moments. However, the univariate dimension-reduction formula is sufficiently accurate to estimate the means and the stds of random functions. This observation enhances the feasibility of the APEM to employ the univariate dimension-reduction formula for practical engineering problems, since the first two moments of model responses are often paid much attention to, and the univariate dimension-reduction formula saves reasonable computational efforts than the bivariate dimension-reduction formula.

Acknowledgments The financial support received from the China Postdoctoral Science Foundation (2014M551251) and the Specialized Research fund for the Fundamental Research funds for the Central Universities (HIT.NSRIF.2015.099) is gratefully appreciated.

References

Borgonovo E (2006) Measuring uncertainty importance: investigation and comparison of alternative approaches. Risk Anal 26:1349–1362

Chang CH, Tung YK, Yang JC (1994) Monte Carlo simulation for correlated variables with marginal distributions. ASCE J Hydraul Eng 120:313–331

Chang CH, Tung YK, Yang JC (1995) Evaluation of probability point estimate methods. Appl Math Model 19:95–105

Chang CH, Yang JC, Tung YK (1997) Uncertainty analysis by point estimate methods incorporating marginal distributions. ASCE J Hydraul Eng 123:244–250

Christian JT, Baecher GB (1999) Point-estimate method as numerical quadrature. ASCE J Geotech Geoenviron Eng 125:779–786

Davis PJ, Rabinowitz P (1983) Methods of numerical integration, 2nd edn. Academic, London

Dimov I, Georgieva R (2010) Monte Carlo algorithms for evaluating Sobol' sensitivity indices. Math Comput Simulat 81:506–514

Fiessler B, Neumann HJ, Rackwitz R (1979) Quadratic limit states in structural reliability. ASCE J Eng Mech 105:661–676

Glen G, Isaacs K (2012) Estimating Sobol sensitivity indices using correlation. Environ Model Software 37:157–166

Gorman MR (1980) Reliability of structural systems. Dissertation, Case Western Reserve University

Harr ME (1989) Probability estimates for multivariate analyses. Appl Math Model 13:313–318

Haukaas T (2003) Finite element reliability and sensitivity methods for performance-based engineering. Dissertation, University of California, Berkeley

He J, Sallfors G (1994) An optimal point estimate method for uncertainty studies. Appl Math Model 18:494–499

Helton JC (1993) Uncertainty and sensitivity analyses techniques for use in performance assessment for radioactive waste disposal. Reliab Eng Syst Safety 42:327–367

Hong HP (1998) An efficient point estimate method for probabilistic analysis. Reliab Eng Syst Safety 59:261–267

Li KS (1992) Point-estimate method for calculating statistical moments. ASCE J Eng Mech 118:1506–1511

Lind NC (1983) Modeling uncertainty in discrete dynamic systems. Appl Math Model 7:146–152

Liu PL, Der Kiureghian A (1986) Multivariate distribution models with prescribed marginals and covariances. Probab Eng Mech 1:105–112

Madsen HO, Krenk S, Lind NC (1996) Methods of structural safety. Prentice-Hall, Englewood Cliffs, NJ

Melchers RE (1999) Structural reliability: analysis and prediction, 2nd edn. Wiley, New York

Morris MD (1991) Factorial sampling plans for preliminary computational experiments. Technometrics 33:161–174

Nossent J, Elsen P, Bauwens W (2011) Sobol' sensitivity analysis of a complex environment model. Environ Model Software 26:1515–1525

Olsson A, Sandberg G, Dahlblom O (2003) On Latin hypercube sampling for structural reliability analysis. Struct Safety 25:47–68

Panchalingam G, Harr ME (1994) Modeling of many correlated and skewed random variables. Appl Math Model 18:635–640

Robinstein RY (1981) Simulation and the Monte Carlo method. Wiley, New York

Rosenblatt M (1952) Remarks on a multivariate transformation. Ann Math Stat 23:470–472

Rosenblueth E (1975) Point estimates for probability moments. Proc Natl Acad Sci 72:3812–3814

Rosenblueth E (1981) Two-point estimates in probability. Appl Math Model 5:329–335

Saltelli A, Tarantola S (2002) On the relative importance of input factors in mathematical models: safety assessment for nuclear waste disposal. J Am Stat Assoc 97:702–709

Saltelli A, Ratto M, Andres T, Campolongo T et al (2008) Global sensitivity analysis: the primer. Wiley, New York

Satelli A, Annoni P, Azzini I et al (2010) Variance based sensitivity analysis of model output. Design and estimator for the total sensitivity index. Comput Phys Commun 181:259–270

Sobol' IM (1990) On sensitivity estimation for nonlinear mathematical models. Matem Mod 2:112–118

Tsai CW, Franceschini S (2005) Evaluation of probabilistic point estimate methods in uncertainty analysis for environmental engineering applications. ASCE J Environ Eng 131:387–395

Wang Y, Tung YK (2009) Improved probabilistic point estimation schemes for uncertainty analysis. Appl Math Model 33:1042–1057

Xu H, Rahman S (2004) A generalized dimension-reduction method for multidimensional integration in stochastic mechanics. Int J Numer Methods Eng 61:1992–2019

Xu JH, Shu H, Jiang HH et al (2012) Sobol's sensitivity analysis of parameters in the common land model for simulation of water and energy fluxes. Earth Sci Inform 5:167–179

Zellner A, Highfield RA (1988) Calculation of maximum entropy distributions and approximation of marginal posterior distributions. J Economet 37:195–209

Zhao YG, Ono T (2000) New point estimates for probability moments. ASCE J Eng Mech 126:433–436

Zhou JH, Nowak AS (1988) Integration formulas to evaluate functions of random variables. Struct Safety 5:267–284

Part III
Reliability and Risk Analysis

Risk Assessment of Slope Instability Related Geohazards

Mihail E. Popescu, Aurelian C. Trandafir, and Antonio Federico

Abstract The paramount importance of slope instability hazards assessment and management is by and large recognized. The general mechanisms of slope instability processes are now fairly well understood but there remains the problem of establishing the risks to lives and property. This is being tackled by relating the local ground conditions to the regional geological surveys and integrating this with site-specific information to produce a hazard potential estimate.

Herein lies the guiding principle of the current chapter, i.e., to describe slope instability related geohazards and methods to estimate the associated risks in an appropriate and effective way. A case study is presented to illustrate the need and tools of a probabilistic framework for slope instability analysis and emphasize that deterministic and probabilistic approaches can often be regarded as complementary.

1 Foreword

At present society is greatly concerned with both natural and man-made hazards and the environmental impact of any engineering activity is considered to be extremely important all over the world.

To cope with hazards, whether natural or man-made, it is necessary to understand risk and try to quantify it. One site may be exposed to a spectrum of different

M.E. Popescu (✉)
Illinois Institute of Technology, 3300 South Federal Street, Chicago, IL 60616, USA
e-mail: mihail.e.popescu@gmail.com

A.C. Trandafir
Fugro GeoConsulting, Inc., Houston, TX, USA

A. Federico
Faculty of Engineering, Politecnico di Bari, Taranto, Italy

S. Kadry and A. El Hami (eds.), *Numerical Methods for Reliability and Safety
Assessment: Multiscale and Multiphysics Systems*, DOI 10.1007/978-3-319-07167-1_8,
© Springer International Publishing Switzerland 2015

hazardous events, such as storms, floods, earthquakes, rock avalanches, and debris flows, which must be assessed separately, both in respect with their magnitude and probability of occurrence.

Frequently occurring cyclic events, such as storms, floods, and earthquakes, can be observed over a period of time to obtain statistical data. Hazard probability can then be derived from an expected frequency. Some events, such as many landslides, are noncyclic. Their probability of occurrence is not a frequency, but a measure of uncertainty whether they will ever take place or not (Popescu and Zoghi 2005).

Risk engineering process includes (Boyd 1994):

1. Identification of hazards
2. Understanding the causes and sources of hazard
3. Assessing the consequences which might arise as a result of the hazard occurring
4. Assessing the probability of the hazard occurring
5. Developing precautions to minimize the risk or mitigate the consequences
6. Assessing residual risk and its tolerability

Landslides and slope instability phenomena are frequently responsible for considerable losses of both money and lives (Petley 2012). The severity of the landslide problem intensifies with increased urban development and change in land use.

In view of this consideration, it is not surprising that landslides are rapidly becoming the focus of major scientific research, engineering study and practice, and land-use policy throughout the world. International cooperation among various individuals concerned with the fields of geology, geomorphology, and soil and rock mechanics has recently contributed to improvement of our understanding of landslides in recent years.

Landslides and related slope instability phenomena plague many parts of the world. Japan leads other nations in landslide severity with projected combined direct and indirect losses of $4 billion annually (Schuster 1996). The USA, Italy, and India follow Japan, with an estimated annual cost ranging between $1 billion to $2 billion. Landslide disasters are also common in developing countries and economical losses sometimes equal or exceed their gross national products.

2 Risk Management Process

The risk management process comprises two components: risk assessment and risk treatment.

Landslide and slope engineering has always involved some form of risk management, although it was seldom formally recognized as such. This informal type of risk management was essentially the exercise of engineering judgment by experienced engineers and geologists.

Figure 1 shows the process of landslide risk management in a flow chart form (Australian Geomechanics Society 2007). In simple form, the process involves answering the following questions:

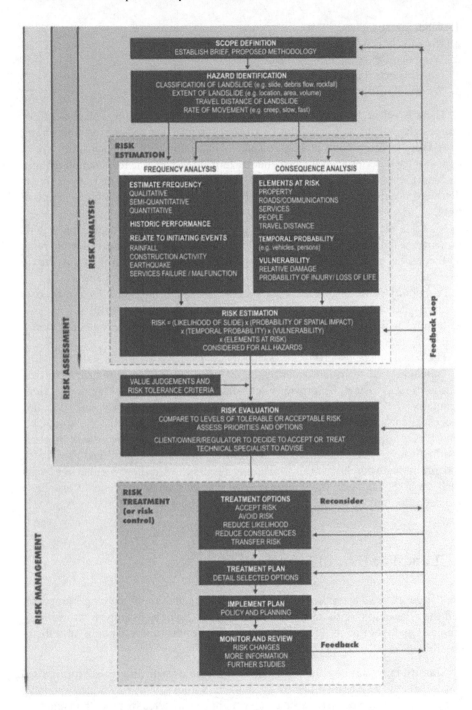

Fig. 1 Process of landslide risk management (from Australian Geomechanics Society, 2007, reprinted with permission)

1. What might happen?
2. How likely is it?
3. What damage or injury may result?
4. How important is it?
5. What can be done about it?

There is a clear distinction between hazard, risk, and probability. Fell et al. (2008) defines the hazard as a condition with the potential for causing an undesirable economic, environmental or human safety consequence. When landslides are of concern, the description of the landslide hazard should include the location, volume (or area), classification and velocity of the potential landslides and any resultant detached material, and the probability of their occurrence within a given period of time.

Risk is a measure of the probability and severity of an adverse effect to health, property or the environment. Risk is often estimated by the product of probability and consequences. However, a more general interpretation of risk involves a comparison of the probability and consequences in a non-product form.

Probability is the likelihood of a specific outcome, measured by the ratio of specific outcomes to the total number of possible outcomes. Probability is expressed as a number between 0 and 1, with 0 indicating an impossible outcome, and 1 indicating that an outcome is certain.

The intent of a landslide hazard assessment is to identify a region's susceptibility to landslides and their consequences based on a few key or significant physical attributes comprising the previous landslide activities, bedrock features, slope geometry, and hydrologic characteristics. In a development program (planning process) concerning a landslide-prone area, one needs to determine the acceptable risk. It is indispensable to recognize the vulnerability and degrees of risk involved and instigate a systematic approach in avoiding, controlling, or mitigating existing and future landslide hazards in a planning process. Accordingly, either a planner shall avoid the landslide-susceptible areas if it is deemed appropriate, or else he or she needs to implement strategies to reduce risk (Fell et al. 2000).

3 Landslide Hazard Identification

Landslide hazard identification requires an understanding of the slope processes and the relationship of those processes to geomorphology, geology, hydrogeology, climate, and vegetation (Courture 2011). From this understanding it will be possible to:

- Classify the types of potential landsliding—the classification system proposed by Varnes (1978) as modified by Cruden and Varnes (1996) constitutes a suitable system. It should be recognized that a site may be affected by more than one type of landslide hazard. For example, existence of deep-seated landslides occurs on the site, whereas, rockfall and debris flow will initiate from above the site.

- Assess the physical extent of each potential landslide being considered, including the location, areal extent, and volume involved.
- Assess the likely causal factor(s), the physical characteristics of the materials involved, and the slide mechanics.
- Estimate the resulting anticipated travel distance and velocity of movement.
- Address the possibility of fast-acting processes, such as flows and falls, from which it is more difficult to escape (Hungr et al. 2007).

Methods which may be used to identify hazards include geomorphological mapping, gathering of historic information on slides in similar topography, geology, and climate (e.g., from maintenance records, air photographs, newspapers, review of analysis of stability). Some forms of geological and geomorphological mapping are considered to be an integrated component of the fieldwork stage when assessing natural landslides, which requires understanding the site whilst inspecting it.

As stated by Varnes (1978): "The processes involved in slope movements comprise a continuous series of events from cause to effect." When assessing landslide hazard for a particular site, of primary importance is the recognition of the conditions which caused the slope to become unstable and the processes which triggered that movement. Only an accurate diagnosis makes it possible to properly understand the landslide mechanisms and thence to propose effective treatment measures.

In every slope there are forces which tend to promote downslope movement and opposing forces which tend to resist movement. A general definition of the factor of safety of a slope results from comparing the downslope shear stress with the available shear strength of the soil, along an assumed or known rupture surface. Starting from this general definition, Terzaghi (1950) divided landslide causes into external causes which result in an increase of the shearing stress (e.g., geometrical changes, unloading the slope toe, loading the slope crest, shocks and vibrations, drawdown, changes in water regime) and internal causes which result in a decrease of the shearing resistance (e.g., progressive failure, weathering, seepage erosion). However, Varnes (1978) points out that there are a number of external or internal causes which may be operating either to reduce the shearing resistance or to increase the shearing stress. There are also causes affecting simultaneously both terms of the factor of safety ratio.

The great variety of slope movements reflects the diversity of conditions that cause the slope to become unstable and the processes that trigger the movement. It is more appropriate to discuss causal factors (including both "conditions" and "processes") than "causes" per se alone. Thus ground conditions (weak strength, sensitive fabric, degree of weathering, and fracturing) are influential criteria but are not causes. They are part of the conditions necessary for an unstable slope to develop, to which must be added the environmental criteria of stress, pore water pressure and temperature. It does not matter if the ground is weak as such—failure will only occur as a result if there is an effective causal process which acts as well. Such causal processes may be natural or anthropogenic, but effectively change the static ground conditions sufficiently to cause the slope system to fail, i.e., to adversely change the state of stability (Popescu 1996).

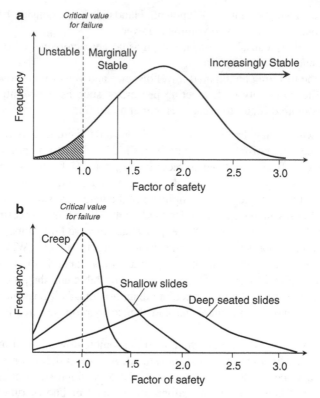

Fig. 2 Probability distribution curves of the factor of safety (**a**) Probabilistic representation of various slope stability stages and (**b**) Probability distribution curves of the factor of safety for various slope instability processes

Seldom, if ever, can a landslide be attributed to a single causal factor. The process leading to the development of the slide has its beginning with the formation of the rock itself, when its basic properties are determined and includes all the subsequent events of crustal movement, erosion, and weathering. The computed value of the factor of safety is a clear and simple distinction between stable and unstable slopes. However, from the physical point of view, it is better to visualize slopes existing in one of the following three stages: stable, marginally stable, and actively unstable (Crozier 1986). Stable slopes are those where the margin of stability is sufficiently high to withstand all destabilizing forces. Marginally stable slopes are those which will fail at some time in response to the destabilizing forces attaining a certain level of activity. Finally, actively unstable slopes are those in which destabilizing forces produce continuous or intermittent movement.

The three stability stages must be seen to be part of a continuum, with the probability of failure being minute at the stable end of the spectrum, but increasing through the marginally stable range to reach certainty in the actively unstable stage. This is qualitatively illustrated in Fig. 2a by the probability distribution curve of the factor of safety for any set of slopes in a specified environment. Figure 2b shows that in any one area, it is likely that more slopes will be subjected to minor forms of mass movement, such as creep, than to large-scale displacements such as deep-seated failures (Crozier and Glade 2005).

The three stability stages provide a useful framework for understanding the causal factors of landslides and classifying them into two groups on the basis of their function (WP/WLI 1994):

- Preparatory causal factors which make the slope susceptible to movement without actually initiating it and thereby tending to place the slope in a marginally stable state.
- Triggering causal factors which initiate movement. The causal factors shift the slope from a marginally stable to an actively unstable state.

A particular causal factor may inflict either or both functions, depending on its degree of activity and the margin of stability. Although it may be possible to identify a single triggering process, an explanation of ultimate causes of a landslide invariably involves a number of preparatory conditions and processes. Based on their temporal variability, the destabilizing processes may be grouped into slow changing (e.g., weathering, erosion) and fast changing processes (e.g., earthquake, drawdown). In the search for landslide causes, attention is often focused on those processes within the slope system, which provoke the greatest rate of change. Although slow changes act over a long period of time to reduce the resistance/shear stress ratio, often a fast change can be identified as having triggered movement.

4 Regional Landslide Hazard Assessment

A landslide risk assessment, whether regional or local, has to be preceded by a corresponding landslide hazard assessment, from which it is then derived. The ordinary landslide hazard assessment techniques include deterministic, statistical, and heuristic approaches (Barredo et al. 2000). Deterministic approaches are based on stability models and are utilized to map landslide hazards at large scales, typically for construction purposes. These models, however, necessitate detailed geotechnical and groundwater field data, which may not be always available.

The statistical approach, bivariate or multivariate, requires a large database, obtained from combination of factors that have initiated landsliding in the past. This model is generally useful for prediction of future landslides at medium scale and it is less suitable for reactivated slides.

The heuristic technique, also known as the knowledge-driven approach, is based on the analyst's expertise in identifying the type and degree of hazard for a designated area based on direct or indirect mapping. The direct method is accomplished either by directly mapping the degree of hazard in the field or by recording a geomorphic map. The indirect method utilizes an integration technique by combining several parameter maps based on qualitative weighing values assigned to each class of parameter map. This procedure is facilitated by compilation of various parameter layers via GIS for establishing the hazard assessment.

Before embarking on a regional landslide hazard assessment, the following preparatory steps are to be taken (Hutchinson 2001):

1. Identify the user and purpose of the proposed assessment. Involve the user in all phases of the program.
2. Define the area to be mapped and decide the appropriate scale of mapping. This may range from 1: 100,000 or smaller to 1:5,000 or larger.
3. Obtain, or prepare, a good topographical base map of the area, preferably contoured.
4. Construct a detailed database of the geology (solid and superficial), geomorphology, hydrogeology, pedology, meteorology, mining and other human interference, history and all other relevant factors within the area, and of all known mass movements including all published work, newspaper articles and the results of interviewing the local population.
5. Obtain all available air photo cover, satellite imagery and ground photography of the area. Photography of various dates can be particularly valuable, both because of what can be revealed by differing lighting and vegetation conditions and to delineate changes in the man-made and natural conditions, including slide development.

Barredo et al. (2000) employed the preceding GIS-assisted direct and indirect heuristic multicriteria evaluation procedure to evaluate the hazards from the Barranco de Tirajana basin, composed of several large landslides. The aforementioned investigators revealed that the above techniques are relatively simple and cost-effective for landslide hazard at medium scales, in particular when costly geotechnical and groundwater data are not readily available.

Landslide hazard has been defined (Varnes and IAEG 1984) as "the probability of occurrence within a specified period of time and within a given area of a potentially damaging phenomenon." Furthermore, Varnes and IAEG (1984) describe the landslide risk assessment as "the expected degree of loss due to a landslide (specific risk) and the expected number of live lost, people injured, damage to property and disruption of economic activity (total risk)." As shown in Fig. 3, the integrated assessment of landslide hazard and risk requires a broad-based knowledge from a wide spectrum of disciplines including geosciences, geomorphology, meteorology, hydrogeology, and geotechnical engineering (Chowdhury et al. 2001).

Landslide hazards are commonly delineated on inventory maps, which display distributions of hazard classes and identify areas where potential landslides may be generated. Inventory maps, exhibit the location and, where applicable, the date of occurrence and historical records of landslides in a region. They are prepared by different techniques and, ideally, provide information concerning the spatial and temporal probability, type, magnitude, velocity, runout distance, and retrogression limit of the mass movements predicted in a designated area (Nadim and Lacasse 2003). Details of inventory maps depend on available resources and are based on the scope, extent of study area, scales of base maps, aerial photographs, and future land use (Guzzetti 2003).

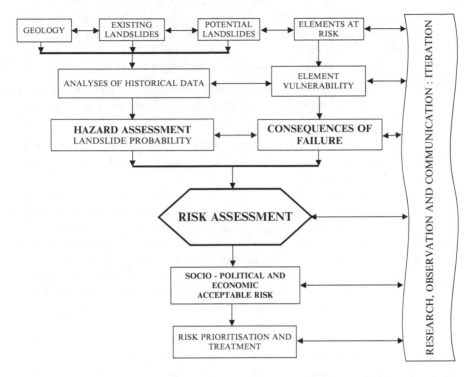

Fig. 3 Methodology for landslide risk assessment (from Chowdhury et al. 2001, reprinted with authors' permission)

Evidently, the extent of information required concerning the landslide hazard analysis would depend on the level and nature of proposed development for a region. The negligence of incorporating the impact of potential landsliding on a project or the prospects of new development on landslide potential may lead to increased risk. The assessment of future landslide susceptibility in a region will require the evaluation of relevant conditions and processes controlling landslides, thus causative factors. These include historical events, slope geometry, bedrock characteristics, and hydrologic features of the designated area.

Einstein (1997) has presented a comprehensive mapping procedure for landslide management. Following are key features of mapping procedures proposed by Einstein:

- *State-of-nature maps*—these maps, used to characterize site, present data without interpretation, such as geologic and topographic maps, precipitation data, and results of site investigation.
- *Danger maps*—are utilized to identify the failure modes involving debris flows, rock falls, etc.

- *Hazard maps*—exhibit the probability of failure related to the possible modes of failure on danger maps. Alternatively, the results can be expressed qualitatively as high, medium, or low.
- *Management map*—entails a summary of management decisions.

Furthermore, the International Association of Engineering Geology has outlined the following scales of analysis for landslide hazard zonation (Soeters and van Westen 1996):

- *National Scale* (<1:1 Million)—this is a low-level detail map intended to provide a general inventory of nationwide hazard. It is used to notify national policy makers and general public.
- *Regional Scale* (1:100,000–1:500,000)—since landslide hazards are considered to be undesirable factors as far as the planners are concerned, the regional mapping scale is employed in evaluating possible constraints due to instability related to the development of large engineering projects and regional development plans. In general, these types of maps are constructed in early phases of regional development projects with low level details and cover large study areas, on the order of 1,000 km^2 or more. They are used to identify areas where landsliding could be a constraint concerning the development of rural or urban transportation projects.
- *Medium Scale* (1:25,000–1:50,000)—this range is considered to be a suitable scale range for the landslide hazard maps. As such, they are utilized to identify the hazard zones in developed areas with large structures, roads, and urbanization. Considerably greater level of details is required to prepare the maps at this scale and the details should encompass slopes in adjacent sites in the same lithology with possibility of having different hazard scores depending on their characteristics. Furthermore, distinction should be made between various slope segments, located within the same terrain unit, such as rating of a concave slope as opposed to a convex slope.
- *Large Scale* (1:5,000–1:15,000)—these types of maps are generally prepared for limited areas based on both interpretation of aerial photographs and extensive field investigations utilizing various techniques applied in routine geotechnical engineering, engineering geology and geomorphology.

Guzzetti (2003) makes the following recommendations concerning the preparation and use of landslide inventory maps. He states that the landslide cartography should be increasingly utilized and landslide inventory maps should be created for entire regions based on consistent and reproducible methods. On a regional scale, Carrara and Guzzetti (1995) indicate, "the temporal dimension of landsliding is essentially a function of the triggering mechanisms which are climatic (due to extreme rainfall) or geodynamic (earthquakes) in nature." Thus, it would be advantageous to prepare landslide inventory maps following each landslide-triggering event such as a rainstorm, a snowmelt event, or an earthquake for the entire affected region. They will provide valuable information regarding type(s), extent and severity of damage caused by the event and will help assessing the impact of the landslide events on infrastructures in that region.

It is important that the quality, reliability and sensitivity of landslide hazard models and maps be carefully examined. Guzzetti (2003) suggests that the created models need to be checked against high-quality inventory maps and reliable historical catalogues of landslide events. Guzzetti (2003) also recommended that both quantitative and qualitative methods should be increasingly performed regarding total risk assessments at local and regional scales. There is a lack of sufficient data and case histories for critical evaluation of techniques and models concerning regional and local landslide risk assessments. Thus, there is a need for compilation of relevant data that will help in comparison of the qualitative and quantitative risk assessment procedures and outcomes. It is anticipated that the recent development in statistical analyses utilizing GIS techniques will enhance analyses of spatial data sets and, thus, quantitative representation of landslide potential along with graphical depictions (Carrara and Guzzetti 1995).

5 Landslide Risk Assessment

By and large, the elements at risk involve property, people, services, such as water supply or drainage or electricity supply, roads and communication facilities, and vehicles on roads. The consequences may not, however, be limited to property damage and injury/loss of life. Other factors include public outrage, political effects, loss of business confidence, effect on reputation, social upheaval, consequential costs, such as litigation (Australian Geomechanics Society 2007).

Many of these may not be readily quantifiable and will require considerable judgment if they are to be included in the assessment. Consideration of such consequences may constitute part of the risk evaluation process by the client/owner/regulator.

Risk estimation may be carried out quantitatively, semi quantitatively or qualitatively (Leroi et al. 2005). Wherever possible, the risk estimate should be based on a quantitative analysis, even though the results may be summarized in a qualitative terminology. Quantitative risk estimation involves integration of the frequency analysis and the consequences.

A full risk analysis involves consideration of all landslide hazards at a designated site (e.g., large, deep-seated landsliding, smaller slides, boulder falls, debris flows) and relevant elements at risk. For total risk (in relation to property and/or life), the risk for each hazard of each element is summed. As estimates made for an analysis will be imprecise, sensitivity analyses are useful to evaluate the effect of changing assumptions or estimates. Variation in the estimate of risk by one or two orders of magnitude, or perhaps three orders of magnitude at low risks, will not be uncommon (Einstein and Karam 2001). The resulting sensitivity may aid judgment as to the critical aspects requiring further investigation or evaluation.

6 Landslide Risk Treatment

Engineering community recognizes that the degree of inherent risk is significant in any landslide project. By its nature, risk sometimes defies definition. Often the most onerous risks are those that were not initially anticipated by anyone, including the owner, the designer, and the contractor. Therefore our ability to identify and mitigate risks ultimately defines a successful project.

The key to establishing a robust yet useful risk based approach for landslide treatment is to find a proper balance between simplicity and methodological rigor. Risk is the combination of the probability of an adverse performance outcome and the consequences that ensue if that outcome is realized. It is becoming common in industrial and regulatory applications of risk screening to portray information in a "risk matrix" of the form shown in Fig. 4. The cells are ranked vertically according to how likely it is that some adverse event would occur, and horizontally by how severe the consequences would be if the event occurred. Events to the lower left are low in probability and low in consequence, and thus of low risk. Those to the upper right are high in probability and high in consequence, and thus of high risk. This discrimination allows priority decisions to be made in a simple way.

Risk treatment is the final stage of the risk management process and provides the methodology of controlling the risk. At the end of the evaluation procedure, it is up to the client or policy makers to decide whether to accept the risk or not, or to decide that more detailed study is required. The landslide risk analyst can provide background data or normally acceptable limits as guidance to the decision maker but should not be making the decision. Part of the specialist's advice may be to

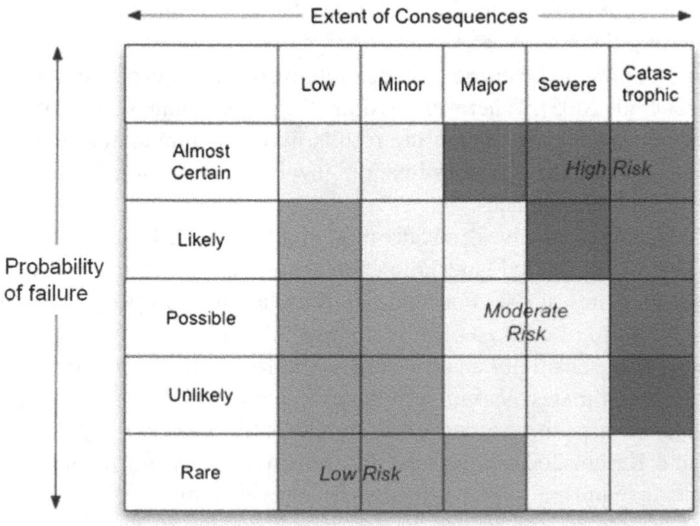

Fig. 4 Conventional risk matrix

identify the options and methods for treating the risk. Typical options would include Australian Geomechanics Society (2007):

- *Accept the risk*—this would usually require the risk to be considered to be within the acceptable or tolerable range.
- *Avoid the risk*—this would require abandonment of the project, seeking an alternative site or form of development such that the revised risk would be acceptable or tolerable.
- *Reduce the likelihood*—this would require stabilization measures to control the initiating circumstances, such as re-profiling the surface geometry, groundwater drainage, anchors, stabilizing structures, or protective structures.
- *Reduce the consequences*—this would require provision of defensive stabilization measures, amelioration of the behavior of the hazard or relocation of the development to a more favorable location to achieve an acceptable or tolerable risk.
- *Monitoring and warning systems*—in some situations monitoring (such as by regular site visits, or by survey), and the establishment of warning systems may be used to manage the risk on an interim or permanent basis. Monitoring and warning systems may be regarded as another most efficient means of reducing the consequences.
- *Transfer the risk*—by requiring another authority to accept the risk or to compensate for the risk such as by insurance.
- *Postpone the decision*—if there is sufficient uncertainty, it may not be appropriate to make a decision on the data available. Further investigation or monitoring would be required to provide data for better evaluation of the risk.

The relative costs and benefits of various options need to be considered so that the most cost effective solutions, consistent with the overall needs of the client, owner and regulator, can be identified. Combinations of options or alternatives may be appropriate, particularly where relatively large reductions in risk can be achieved for relatively small expenditure. Prioritization of alternative options is likely to assist with selection.

7   Levels of Effectiveness and Acceptability That May Be Applied in the Use of Landslide Remedial Measures

Terzaghi (1950) stated that "if a slope has started to move, the means for stopping movement must be adapted to the processes which started the slide." For example, if erosion is a causal process of the slide, an efficient remediation technique would involve armoring the slope against erosion, or removing the source of erosion. An erosive spring can be made non-erosive by either blanketing with filter materials or drying up the spring with horizontal drains, etc.

The greatest benefit in understanding landslide-producing processes and mechanisms lies in the use of the above understanding to anticipate and devise measures to minimize and prevent major landslides (Popescu and Schaefer 2008). The term major should be underscored here because it is neither possible nor feasible, nor even desirable, to prevent all landslides. There are many examples of landslides that can be handled more effectively and at less cost after they occur. Landslide avoidance through selective locationing is obviously desired—even required—in many cases, but the dwindling number of safe and desirable construction sites may force more and more the use of landslide—susceptible terrain.

Selection of an appropriate remedial measure depends on: (a) engineering feasibility, (b) economic feasibility, (c) legal/regulatory conformity, (d) social acceptability, and (e) environmental acceptability. A brief description of each method is presented herein:

(a) Engineering feasibility involves analysis of geologic and hydrologic conditions at the site to ensure the physical effectiveness of the remedial measure. An often-overlooked aspect is making sure the design will not merely divert the problem elsewhere.
(b) Economic feasibility takes into account the cost of the remedial action to the benefits it provides. These benefits include deferred maintenance, avoidance of damage including loss of life, and other tangible and intangible benefits.
(c) Legal-regulatory conformity provides for the measure meeting local building codes, avoiding liability to other property owners, and related factors.
(d) Social acceptability is the degree to which the remedial measure is acceptable to the community and neighbors. Some measures for a property owner may prevent further damage but could be an unattractive eyesore to neighbors.
(e) Environmental acceptability addresses the need for the remedial measure to not adversely affect the environment. De-watering a slope to the extent it no longer supports a plant community may not be environmentally acceptable solution.

Just as there are a number of available remedial measures, so are there a number of levels of effectiveness and levels of acceptability that may be applied in the use of these measures. We may have a landslide, for example, that we simply choose to live with; one that poses no significant hazard to the public, whereas it will require periodic maintenance for example, through removal, due to occasional encroachment onto the shoulder of a roadway. The permanent closure of the Manchester—Sheffield road at Mam Tor in 1979 (Skempton et al. 1989) is well known example of abandonment due to the effects of landslides where repair was considered uneconomical.

Most landslides, however, must usually be dealt with sooner or later. How they are handled depends on the processes that prepared and precipitated the movement, the landslide type, the kinds of materials involved, the size and location of the landslide, the place or components affected by or the situation created as a result of the landslide, available resources, etc. The technical solution must be in harmony with the natural system, otherwise the remedial work will be either short lived or excessively expensive. In fact, landslides are so varied in type and size, and in most instances, so dependent upon special local circumstances, that for a given landslide problem there is more than one method of prevention or correction that can be successfully applied. The success of each measure depends, to a large extent, on the degree to which the specific soil and groundwater conditions are prudently recognized in an investigation and incorporated in design.

8 Monitoring and Warning as Tools of Landslide Hazard Management

Monitoring of landslides plays an increasingly important role in the context of living and coping with these natural hazards. The classical methods of land surveys, inclinometers, extensometers, and piezometers are still the most appropriate ones. In future, the emerging techniques based on remote sensing and remote access techniques should be of primary interest (Popescu 2001).

DOE: Department of the Environment (1994) identifies the following categories of monitoring, designed for slightly differing purposes but generally involving similar techniques:

1. Preliminary monitoring involves provision of data on pre-existing landslides so that the dangers can be assessed and remedial measures properly designed or the site abandoned.
2. Precautionary monitoring is carried out during construction in order to ensure safety and to facilitate redesign if necessary.
3. Post-construction monitoring in order to check on the performance of stabilization measures and to focus attention on problems that require remedial measures.

Observational methods based on careful monitoring coupled with back analysis methods (Popescu and Habimana 2010) are essential in achieving reliable and cost–effective remedial measures.

Leroi (1996) defined the following four possible different stages of landslide activity:

1. Pre-failure stage, when the soil mass is still continuous. This stage is mostly controlled by progressive failure and creep
2. Onset of failure characterized by the formation of a continuous shear surface through the entire soil or rock mass
3. Post-failure stage which includes movement of the soil or rock mass involved in the landslide, from just after failure until it essentially stops
4. Reactivation stage when the soil or rock mass slides along one or several pre-existing shear surfaces. This reactivation can be occasional or continuous with seasonal variations of the rate of movement

Several methods have been proposed for the prediction concerning the time of occurrence of landslides (Federico et al. 2012, 2014). In engineering practice, phenomenological methods, that infer the time to failure by means of the monitored surface displacements, are preferred for a prediction, given that they remove all the uncertainties involved in these problems.

When dealing with a slope of precarious stability and/or presenting a risk which is considered too high, a possible option is to do nothing but install a warning system in order to insure or improve the safety of people (Baum and Godt 2010). It is worth noting that warning systems do not modify the hazard but contribute to reducing the consequences of the landslide and thus the risk, in particular the risk associated to the loss of life.

Various types of warning systems have been proposed and the selection of an appropriate one should take into account the stage of landslide activity:

1. At pre-failure stage, the warning system can be applied either to revealing factors or to aggravating factors. The revealing factors can be for example the opening of fissures or the movement of given points on the slope; in such cases, the warning criterion will be the magnitude or rate of movement. When the warning system is associated to triggering or aggravating factors, there is a need to firstly define the relation between the magnitude of the controlling factors and the stability condition or the rate of movement of the slope. The warning criterion can be a given hourly rainfall or the cumulative rainfall during a certain period of time, pore water pressure, a given stage of erosion, a minimum negative pore pressure in a loess deposit, etc.

   At failure stage, the warning system can only be linked to revealing factors, generally a sudden acceleration of movements or the disappearance of a target.
2. At post-failure stage, the warning system has to be associated to the expected consequences of the movement. It is generally associated with the rate of movement and run out distance.

   Majority of the techniques, outlined above, could be cost-prohibitive and may be socially and politically unpopular. As a result, there may be a temptation to adopt and rely instead upon the installation of apparently cheaper and much less disruptive monitoring and warning systems to "save" the population from future catastrophes. However, for such an approach to be successful it is necessary to fulfill satisfactorily each of the following steps (Hutchinson 2001):

   (a) The monitoring system shall be designed to record the relevant parameters, to be in the right places, to be sound in principle and effective in operation.
   (b) The monitoring results need to be assessed continuously by suitable experts.
   (c) A viable decision shall be made, with a minimum of delay, that the danger point has been reached.
   (d) The decision should be passed promptly to the relevant authorities, with a sufficient degree of confidence and accuracy regarding the forecast place and time of failure for those authorities to be able to act without the fear of raising a false alarm.
   (e) Once the authorities decide to accept the technical advice, they must pass the warning on to the public in a way that will not cause panic and possibly exacerbate the situation.

(f) The public, who need to be very well informed and prepared in advance, to respond in an orderly and pre-arranged manner.

In view of the preceding discussion, it is not surprising that, though there have been few successes with monitoring and warning systems, particularly in relatively simple, site-specific situations, there have been many cases where these have failed, on one or more of the grounds (a) through (f) above, often with tragic and extensive loss of life. It is concluded, therefore, that sustained good management of an area, as outlined above, should be our primary response to the threat of landslide hazards and risks, with monitoring and warning systems in a secondary, supporting role.

9   Probabilistic Risk Assessment of Slope Instability

The keyword in hazard assessment is uncertainty and hence probabilistic methods are the most appropriate in both defining the areas at risk and analyzing the mechanisms. The basis of a probabilistic framework is the recognition that the estimated factor of safety reflects imperfect knowledge and is, therefore, a variable.

The most important probabilistic method used in geotechnical risks assessment is the first order second moment method. One of the advantages of this method is that one can derive moments of the dependent variable from moments of the independent variables, without knowledge of probability density functions. The method is based on the mean value and coefficient of variation only.

The statistics characterizing the soil variability and the variability of the spatial average values are necessary inputs into probabilistic methods for quantifying risk and reliability of soil structures. Soil properties should be modeled as spatially correlated variables or "random variables." The use of perfectly correlated soil properties gives rise to unrealistically large values of failure probabilities for geotechnical structures.

The stability of slopes is undoubtedly the most popular application of probabilistic and reliability methods judging from the number of publications on this subject. Slope stability mechanics and analysis procedures are felt to be so well understood that variability of input parameters is considered to be the only significant unknown. Statistical distribution of input data such as soil strength and pore-pressure is analyzed to estimate failure probability. However, "ignorance factors" are more important than natural variability, yet they are almost never accounted for in probabilistic stability studies.

D'Appolonia (1977) proposed that an important consideration in a probabilistic analysis was what he referred to as « unknown unknowns ». These might include unanticipated geological conditions or some influence of the construction process. It is likely that provision for this component should be greater for new types of structures, new geological conditions and where either the structure or the geology is highly complex.

With the availability of friendly software, geotechnical reliability evaluations can be greatly facilitated. Recent changes in both computer hardware and software allow the responsible engineer to be now able to be more involved in personally running analyses and at the same time to be more readily able to construct a reasonably detailed model of the problem.

Current procedures for evaluating the safety of slopes consist in determining a factor of safety which is compared with allowable values found to be satisfactory on the basis of previous experience. The factor of safety suffers from the following:

1. Elements of uncertainty in analyses are not quantified when the factor of safety is used.
2. The scale of the factor of safety is not known. For example, a slope with a factor of safety of 3.0 is not necessarily twice as safe as another with a factor of safety of 1.5.
3. Allowable values to be selected for the factor of safety are the result of experience. In dealing with new or different problems for which there is no previous experience, there is no allowable factor of safety.

There is much to be gained from applying concepts of risk analysis to supplement conventional procedures for determining the factor of safety against shear failure of soil slopes.

The reliability index of a slope can be defined in several ways on the basis of the probability distribution of the random variable(s) governing the slope failure mechanism(s). Thus considering the factor of safety, one of the simplest definitions of the reliability index is:

$$\beta = (E\,[F] - 1.0)\,/\sigma\,[F] \tag{1}$$

in which $E[F]$ is the best estimate of the factor of safety and $\sigma[F]$ is the standard deviation of the factor of safety.

The reliability index provides a better indication of how close the slope is to failure than does the factor of safety alone. This is because it incorporates contributions due to randomness of soil parameters, geometry, environmental loading, and other physical effects as well as due to uncertainties in the computational method.

Slopes with large values of the reliability index are farther from failure than slopes with small values of the reliability index, regardless of the value of the best estimate of the factor of safety. This is illustrated in Fig. 5 by the example of two slopes with different probability distribution functions for the factors of safety. The slope with the higher $E[F]$ has a larger probability of failure than the slope with the lower $E[F]$ though the conventional approach, as well as many regulations, would regard the former slope as significantly safer than the latter.

A variety of techniques for evaluating the reliability index and the probability of failure is available. Popescu et al. (1998) have used both Monte Carlo method and the Point Estimate method to determine the probability of failure of simple slopes and found a good agreement between the results obtained with the two methods.

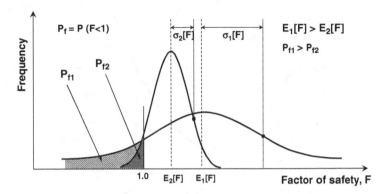

Fig. 5 Example of two slopes with different probability distribution functions of the factor of safety

The Monte Carlo simulation method determines the mean and standard deviation of a performance function of random variables by performing repeated computations with the randomly generated values for the component variables. There are four stages in a Monte Carlo simulation:

1. Generate random numbers (i.e., independent random variables uniformly distributed over the unit interval between 0 and 1), transform the random numbers from a uniform distribution to the distribution applicable to the component variable and calculate values of all component variables based on the appropriate random numbers.
2. Using the randomly generated values of the component variables, compute the system performance function (i.e., factor of safety).
3. Repeat steps (1) and (2) a large numbers of times. The number of times depends on the variability of the input and output parameters and the desired accuracy of the output.
4. Create a cumulative distribution of the system performance function (i.e., factor of safety) using the data obtained from the above simulations. Interpretation of the distribution of the system performance function provides the mean and standard deviation of the factor of safety.

The Monte Carlo simulation requires a high-speed computer so that a large number of trials can be conducted. An alternative approach for calculating the mean and standard deviation of the factor of safety is the Point Estimate method (Rosenblueth 1975).

The basic principle of the Point Estimate method is illustrated in Fig. 6 where the probability distribution function (pdf) of a random variable X is approximated by a two-point probability mass function. The mass function consists of concentrations P_+ and P_- at X_+ and X_-, respectively. If $F(X)$ is a function of X, a two-point

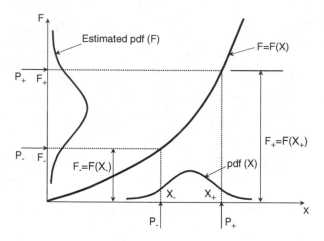

Fig. 6 Basic principle of the point estimate method

approximation of the pdf of F is obtained by evaluating the function $F(X)$ at X_+ and X_- ($F_+ = F(X_+)$ and $F_- = F(X_-)$):

$$E[F] = P_+ F_+ + P_- F_- \tag{2a}$$

$$E[F]^2 = P_+ F_+^2 + P_- F_-^2 \tag{2b}$$

$$\sigma[F] = \sqrt{E[F^2] - (E[F])^2} \tag{2c}$$

In general if F is a function of N random independent variables, then 2^N points are needed to approximate the multivariate mass function and the entire procedure can be summarized as follows:

$$E[F] = \sum (P_{1\pm} \times P_{2\pm} \times \dots P_{N\pm}) \ F(X_{1\pm}, X_{2\pm}, \dots X_{N\pm}) \tag{3a}$$

$$E[F^2] = \sum (P_{1\pm} \times P_{2\pm} \times \dots P_{N\pm}) \ F^2(X_{1\pm}, X_{2\pm}, \dots X_{N\pm}) \tag{3b}$$

For correlated random variables, additional adjustment must be made to the probability concentrations, P_i. Where symmetrically distributed variables are assumed, the point estimates X_{i+} and X_{i-} are taken at one standard deviation above and below the expected value, respectively. The probability concentration or weighting factor P for the case of uncorrelated random variables with symmetrical distribution is equal to $(1/2)^N$.

A major benefit of the probabilistic approach is that the sensitivity of the solution to various parameters can be determined. The parameters of most relevance for slope stability studies are the standard deviations of the cohesion, angle of internal friction and pore-water pressure ratio.

A guide of decision making and interpreting open-pit slope performance in the context of a probabilistic framework is given in Table 1 according with Priest and Brown (1983).

In the following, an example of probabilistic slope stability analysis using the Point Estimate method is presented. The analysis addresses the stability of spoil pile slopes of a strip coal mine located in southwestern Turkey that has been subject to detailed geotechnical investigations reported by Ulusay et al. (1996). This case history was selected for analysis because a large amount of geotechnical data was available that allowed a statistical analysis of the shear strength parameters, required in any probabilistic slope stability study. As seen in Fig. 7, the spoil pile geometry is characterized by a height, H, of 20–50 m, and a slope angle, β, of 30–50°. The slope failure mechanism consists of a biplanar wedge instability mode, with the basal portion of the sliding surface located in a weak clay layer representing the foundation material of the spoil pile, while the upper portion of the sliding surface was located in the spoil material (Ulusay et al. 1996).

The input random variables considered for the probabilistic slope stability assessment included the shear strength parameters in terms of effective stresses (i.e., c'—effective cohesion intercept and ϕ'—effective friction angle) for the weak basal clay layer (i.e., c_1' and ϕ_1') and for the spoil material (i.e., c_2' and ϕ_2'), and the angle relative to the horizontal of the basal (α) and upper (δ) portions of the biplanar sliding surface. The estimated correlation coefficient between parameters c' and ϕ' is -0.779, and the unit weight of the spoil material utilized in slope stability calculations is 16.5 kN/m^3.

The analysis was performed for a groundwater table coincident with the basal sliding surface. Table 2 provides the best estimate and standard deviation values of the input random variables considered in the present probabilistic study. A flowchart of the probabilistic slope stability analysis is presented in Fig. 7. The analysis was performed assuming that the input random variables are normally distributed. Point estimates of the safety factor required in evaluations of the probability distribution function pdf(F) were determined using a two-wedge limit equilibrium slope stability model illustrated in Fig. 8.

Figure 9 displays the probabilistic analysis results, including best estimate values of the safety factor ($E[F]$), probability of safety factor being less 1.0 ($P(F < 1.0)$), and probability of safety factor being less 1.5 ($P(F < 1.5)$). These quantities are plotted on three dimensional charts as functions of slope angle (β) and slope height (H) in order to assess the slope performance according to the probabilistic slope design criteria falling under the slope category 1 in Table 1. As seen on the probability chart in Fig. 9, all of the analyzed slope configurations violate the probabilistic criterion $P(F < 1.5) \leq 20\ \%$, and only a few combinations of H and β parameters, with H not greater than 23 m and β not greater than 34°, satisfy the probabilistic criterion $P(F < 1.5) \leq 10\ \%$. The shaded area located in the horizontal (H, β) plane on the $E[F]$ chart comprises the slope configurations characterized by $E[F] \geq 1.3$ thus satisfying the minimum F criterion from Table 1. Based on the probabilistic analysis results from Fig. 9, slope configurations characterized by combinations of H and β falling outside the shaded area corresponding to

Table 1 Probabilistic slope design criteria

Category of slope	Consequences of failure	Examples	Acceptable values Minimum mean **F**	Maxima $P(F < 1.0)$	$P(F < 1.5)$
1	Not serious	Individual benches; small[a] temporary slopes not adjacent to haulage roads	1.3	0.1	0.2
2	Moderately serious	Any slope of a permanent or semipermanent nature	1.6	0.01	0.1
3	Very serious	Medium-size and high slopes carrying major haulage roads or underlying permanent mine installations	2.0	0.003	0.05

[a]Small, height <50 m; medium, height 50–150 m; high, height >150 m

Guide to interpretation of slope performance

Performance of slope	Interpretation
Satisfies all three criteria	Stable slope
Exceeds minimum mean *F*, but violates one of both probabilistic criteria	Operation of slope presents risk that may or may not be acceptable; level of risk can be reduced by comprehensive monitoring program
Falls below minimum mean *F*, but satisfies both probabilistic criteria	Marginal slope: minor modifications of slope geometry required to raise mean *F* to satisfactory level
Falls below minimum mean *F* and violates one of both probabilistic criteria	Unstable slope: major modifications of slope geometry required; rock improvement and slope monitoring may be necessary

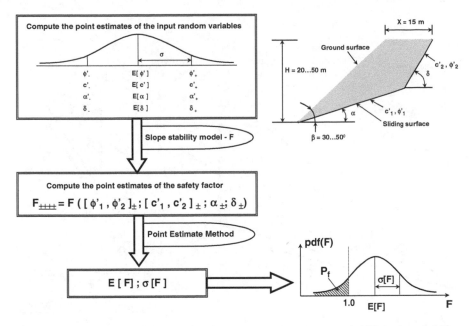

Fig. 7 Geometry of slope failure mechanism and flow chart for the probabilistic slope stability analysis

Table 2 Best estimate and standard deviation values of the input random variables considered for the example probabilistic slope stability analysis

Random variable	Best estimate, $E[]$	Standard deviation, $\sigma[]$
ϕ'_1 (o)	15.4	5.1
c'_1 (kPa)	5.9	5.6
ϕ'_2 (o)	33.0	5.0
c'_2 (kPa)	8.9	3.6
α (o)	7.4	5.2
δ (o)	66.8	2.5

$E[F] \geq 1.3$ on the $E[F]$ plot can be interpreted as potentially unstable, whereas slope configurations falling inside the $E[F] \geq 1.3$ shaded area may still carry a certain level of operational risk that could be mitigated by extensive monitoring activities.

10 Concluding Remarks

Assessing landslide hazard is a most important step in landslide risk management. Once that has been done, it is feasible to assess the number, size and vulnerability of the fixed elements at risk (structures, roads, railways, pipelines, etc.) and thence the damage they will suffer. The various risks have to be combined to arrive at a total risk in financial terms. Comparison of this with, for instance, cost-benefit studies of relocation of populations and facilities, or mitigation of the hazard by engineering countermeasures, provides a very useful tool for management and decision-making.

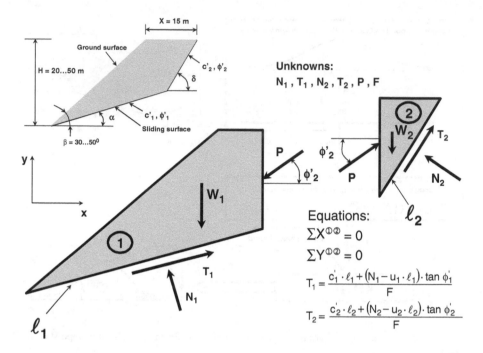

Fig. 8 Two-wedge limit equilibrium slope stability model for safety factor evaluation

Slope stability engineering is concerned with decision-making based on information and analysis combined with observation. Concepts of statistics and probability have been used from time to time but, as a formal basis for analysis, the use of probabilistic framework has been advocated only in the last few decades.

Probabilistic slope stability is a developing field with various models including different effects. A complete model for probabilistic analysis of a soil slope is still some way off. Nevertheless even a simplified probabilistic analysis can add valuable information to guide the slope design in decision making.

Much progress has been made in developing techniques to minimize the impact of landslides, although new, more efficient, quicker and cheaper methods could well emerge in the future. There are a number of levels of effectiveness and levels of acceptability that may be applied in the use of these measures, for while one slide may require an immediate and absolute long-term correction, another may only require minimal control for a short period.

Whatever the measure chosen, and whatever the level of effectiveness required, the geotechnical engineer and engineering geologist have to combine their talents and energies to solve the problem. Solving landslide related problems is changing from what has been predominantly an art to what may be termed an art-science. The continual collaboration and sharing of experience by engineers and geologists will no doubt move the field as a whole closer toward the science end of the art-science spectrum than it is at present.

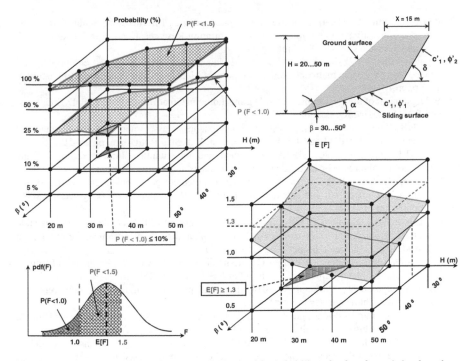

Fig. 9 Computed mean safety factor ($E[F]$) along with probability of safety factor being less than 1.0 ($P(F < 1.0)$) and probability of safety factor being less than 1.5 ($P(F < 1.5)$)

References

Australian Geomechanics Society (2007) A national landslide risk management framework for Australia, AGS Landslide Taskforce, Australian Geomechanics, V42N1, March 2007, Introduction to the suite of papers contained within the issue. http://australiangeomechanics. org/resources/downloads/#dlLRM2007

Barredo JI, Benavides A, Hervás J, Van Westen CJ (2000) Comparing heuristic landslide hazard assessment techniques using GIS in the Tirajana Basin, Gran Canaria Island, Spain. Int J Appl Earth Obs Geoinf 2(1):9–23

Baum RL, Godt JW (2010) Early warning of rainfall-induced shallow landslides and debris flows in the USA. Landslides 7(3):259–272

Boyd RD (1994) Managing risk. Ground Eng 27(5):30–33

Carrara A, Guzzetti F (eds) (1995) Geographical information systems in assessing natural hazards. Kluwer Academic Publisher, Dordrecht, p 353

Chowdhury R, Flentje P, Ko Ko C (2001) A focus on hilly areas subject to the occurrence and effects of landslides. Global blueprint for change, 1st edn—prepared in conjunction with the International Workshop on Disaster Reduction convened on August 19–22, 2001

Courture R (2011) Landslide terminology—national technical guidelines and best practices on landslides. Geological Survey of Canada, Ottawa, p 12

Crozier MJ (1986) Landslides—causes, consequences and environment. Croom Helm, London

Crozier MJ, Glade T (2005) Landslide hazard and risk: issues, concepts and approach. In: Landslide Hazard and Risk, Glade T, Anderson M, Crozier MJ (eds), Wiley, pp 2–40

Cruden DM, Varnes DJ (1996) Landslide types and processes. In: Turner K, Schuster RL (eds) Landslides investigation and mitigation. Transportation Research Board Special Report 247, National Research Council, Washington, DC

D'Appolonia ED (1977) Relationship between design and construction in soil engineering. In: Proceedings of the International Conference on Soil Mechanics and Foundation Engineering, Tokyo, pp 479–485

DOE: Department of the Environment (1994) Landsliding in Great Britain (edited by D.K.C. Jones and E.M. Lee). HMSO, London

Einstein HH (1997) Landslide risk—systematic approaches to assessment and management. In: Cruden DM, Fell R (eds) Landslide risk assessment. Balkema, Rotterdam, pp 25–50

Einstein HH, Karam KS (2001) Risk assessment and uncertainties. In: Proceedings of the international conference on landslides—causes, impacts and countermeasures, Davos

Fell R, Hungr O, Leroueil S, Riemer W (2000) Keynote lecture-geotechnical engineering of the stability of natural slopes, and cuts and fills in soil. In: Proceedings of geoengineering 2000, Melbourne, pp 21–120

Fell R, Corominas J, Bonnard C, Cascini L, Leroi E, Savage W (2008) Guidelines for landslide susceptibility, hazard and risk zoning for land use planning—Joint Technical Committee on Landslides and Engineered Slopes. Eng Geol 102:85–98

Federico A, Popescu M, Elia G, Fidelibus C, Interno G, Murianni A (2012) Prediction of time to slope failure: a general framework. Environ Earth Sci J 66:245–256

Federico A, Popescu M, Murianni A (2014) Temporal prediction of landslide occurrence: A possibility or a challenge? (In preparation)

Guzzetti F (2003) Landslide cartography, hazard assessment and risk evaluation: overview, limits and perspective. http://www.mitch-ec.net/workshop3/Papers/paper_guzzetti.pdf

Hungr O, McDougall S, Wise M, Cullen M (2007) Magnitude–frequency relationships of debris flows and debris avalanches in relation to slope relief. Geomorphology 96:355–365

Hutchinson JN (2001) Landslide risk—to know, to foresee, to prevent. Geologia Tecnica e Ambientale n 9:3–24

Leroi E (1996) Landslide hazard—risk maps at different scales: objectives, tools and developments. In: Senneset K (ed) Landslides, Proceedings of the international symposium on landslides, Trondheim, 17–21 June, pp 35–52

Leroi E, Bonnard C, Fell R, McInnes R (2005) Risk assessment and management. In: Hungr O, Fell R, Couture R, Eberhardt E (eds) Proceedings of the international conference on landslide risk management, Vancouver

Nadim F, Lacasse S (2003) Review of probabilistic methods for quantification and mapping of geohazards. Geohazards, Edmonton, pp 279–286

Petley DN (2012) Global patterns of loss of life from landslides. Geology 40:927–930

Popescu ME (1996) From landslide causes to landslide remediation, special lecture, vol 1. In: Senneset K (ed) Landslides, proceedings of the international symposium on landslides, Trondheim, 17–21 June, pp 75–96.

Popescu ME (2001) A suggested method for reporting landslide remedial measures. IAEG Bull 60(1):69–74

Popescu ME, Trandafir A, Federico A, Simeone V (1998) Probabilistic risk assessment of landslide related geohazards. In: Geotechnical hazards, Proceedings of the 11th Danube European conference soil mechanics and geotechnical engineering. A.A.BalkemaPublishers, Porec, pp 863–870

Popescu ME, Zoghi M (2005) Landslide risk assessment and remediation. In: Taylor C, VanMarke E (eds) Monograph on natural, accidental, and deliberate hazards. American Society of Civil Engineering, Council on Disaster Risk Management, Monograph No. 1, Reston, VA, pp 161–193

Popescu M, Schaefer VR (2008) Landslide stabilizing piles: a design based on the results of slope failure back analysis. In: Proceedings of the 10th international symposium on landslides and engineered slopes, Xi'an, pp 1787–1793

Popescu M, Habimana J (2010) Limit equilibrium and stress-deformation back analysis of some geotechnical problems. In: Proceedings of the international conference on forensic approach to analysis of geohazard problems. ISFGE, Mumbai, pp 300–313

Priest SD, Brown ET (1983) Probabilistic stability analysis of variable rock slopes. Trans Instn Min Metall A92:1–12

Rosenblueth E (1975) Point estimates for probability moments. Proc Natl Acad Sci USA 72(10):3812–3814

Schuster RL (1996) Socioeconomic significance of landslides. In: Turner AK, Schuster RL (eds) Chapter 2, Landslides—Investigation and mitigation, Special Report 247, Transportation Research Board, National Research Council. National Academy Press, Washington, DC

Skempton AW, Leadbeater AD, Chandler RJ (1989) The Mam Tor Landslide, North Derbyshire. Phil Trans R Soc Lond A329:503–547

Soeters R, van Westen CJ (1996) Slope instability recognition, analysis, and zonation. in landslides: investigations and mitigation. Special Report 247, Transportation Research Board, National Research Council, Washington, DC, pp 129–177

Terzaghi K (1950) Mechanisms of landslides. Geological Society of America, Berkley, pp 83–123

Ulusay R, Caglan D, Arlkan F, Yoleri MF (1996) Characteristics of biplanar wedge spoil pile instabilities and methods to improve stability. Can Geotech J 33:58–79

Varnes DJ (1978) Slope movements types and processes. In: Landslides analysis and control, Transportation Research Board Special Report, vol 176, pp 11–33

Varnes DJ, IAEG (The International Association of Engineering Geology) Commission on Landslides and Other Mass Movements (1984) Landslide hazard zonation: a review of principles and practice. Natural Hazards, vol 3. UNESCO, Paris, p 63

WP/WLI: International Geotechnical Societies' UNESCO Working Party on World Landslide Inventory. Working Group on Landslide Causes—Popescu ME, Chairman (1994) A suggested method for reporting landslide causes. Bull IAEG 50:71–74

Advances in System Reliability Analysis Under Uncertainty

Chao Hu, Pingfeng Wang, and Byeng D. Youn

Abstract In order to ensure high reliability of complex engineered systems against deterioration or natural and man-made hazards, it is essential to have an efficient and accurate method for estimating the probability of system failure regardless of different system configurations (series, parallel, and mixed systems). Since system reliability prediction is of great importance in civil, aerospace, mechanical, and electrical engineering fields, its technical development will have an immediate and major impact on engineered system designs. To this end, this chapter presents a comprehensive review of advanced numerical methods for system reliability analysis under uncertainty. Offering excellent in-depth knowledge for readers, the chapter provides insights on the application of system reliability analysis methods to engineered systems and gives guidance on how we can predict system reliability for series, parallel, and mixed systems. Written for the professionals and researchers, the chapter is designed to awaken readers to the need and usefulness of advanced numerical methods for system reliability analysis.

1 Introduction

Failures of engineered systems (e.g., vehicle, aircraft, and material) lead to significant maintenance/quality-care costs and human fatalities. Examples of such system

C. Hu (✉)
Department of Mechanical Engineering, University of Maryland College Park, MD, USA
e-mail: huchaostu@gmail.com

P. Wang
Department of Industrial and Manufacturing Engineering,
Wichita State University, Wichita, KS, USA

B.D. Youn
Seoul National University, Seoul, South Korea

S. Kadry and A. El Hami (eds.), *Numerical Methods for Reliability and Safety Assessment: Multiscale and Multiphysics Systems*, DOI 10.1007/978-3-319-07167-1_9,
© Springer International Publishing Switzerland 2015

failures have been found in many engineering fields: DC-10-10 aircraft engine loss (1979), the explosion of the Challenger space shuttle (1986), Ford Explorer rollover (1998–2000), the interstate 35W bridge failure in Minneapolis, MN (2007), etc. Today, US industry spends $200 billion each year on reliability and maintenance. Many system failures can be traced back to various difficulties in evaluating and designing complex systems under highly uncertain manufacturing and operational conditions and our limited understanding of physics of failures. Thus, reliability analysis under uncertainty, which assesses the probability that a system performance (e.g., fatigue, corrosion, fracture) meets its marginal value while taking into account various uncertainty sources (e.g., material properties, loads, geometries), has been recognized as of significant importance in product design and development (Haldar and Mahadevan 2000).

Reliability analysis involving a single performance function is referred to as component reliability analysis. In engineering practice, it is also very likely to encounter reliability analysis problems involving multiple performance functions and these performance functions often describe different physical phenomena (associated with system performances) that are coupled together via the common random variables shared by the functions. We refer to this type of reliability analysis as system reliability analysis. System reliability analysis aims at analyzing the probability of system success while considering multiple system performances (e.g., fatigue, corrosion, and fracture). For example, the design of a truss structure requires both the displacement at a critical node and the stress of a critical truss element satisfy the reliability requirements. Here we have two performance functions in reliability analysis, i.e., the nodal displacement and the elemental stress. This example has two failure criteria, namely, displacement and stress. Another example is the design of a lower control A-arm in a vehicle. In this example, even if we only consider a single failure criterion (i.e., stress), we still need to deal with multiple performance functions which are the stresses at multiple hotspots of the control arm.

The task may become more challenging if we have different system configurations (e.g., series, parallel, and mixed). In order to ensure high reliability of complex engineered systems against deterioration or natural and man-made hazards, it is essential to have an efficient and accurate method for estimating the probability of system failure regardless of different system configurations (series, parallel, and mixed systems). Although tremendous advances have been made in component reliability analysis and design optimization, the research in system reliability analysis has been stagnant due to the complicated nature of the multiple system failure modes and their interactions, as well as the costly computational expense of system reliability evaluation (Youn and Wang 2009; Wang et al. 2011). Since system reliability prediction is of great importance in civil, aerospace, mechanical, and electrical engineering fields, its technical development will have an immediate and major impact on engineered system designs. This chapter is devoted to providing an in-depth discussion of the recently developed numerical methods for system reliability analyses of series, parallel, and mixed systems with an aim to give insights into the merits and limitations of these methods.

2 Overview of Reliability Analysis Under Uncertainty

The formal definition of reliability is the probability that an engineered system will perform its required function under prescribed conditions (for a specified period of time). We intentionally use a bracket in this definition to indicate the existence of two different types of reliabilities: *time-independent reliability* and *time-dependent reliability*. The former type is often used in designing an engineered system to provide a high built-in reliability at the very beginning of operation (i.e., it generally does not consider the health degradation during the life cycle); the later type is often employed in supporting an engineered system to ensure a high operational reliability (i.e., the health degradation during the life cycle is taken into account to estimate the reliability). The discussion in this chapter only focuses on the time-independent reliability which can be defined as the probability that the actual performance of an engineered system meets the required or specified design performance under various uncertainty sources (e.g., material properties, loads, geometric tolerances). This section discusses the types of uncertainty and provides an overview of component and system reliability analyses under uncertainty.

2.1 Types of Uncertainty

Uncertainty present in engineering applications can be formally classified into two categories: aleatory uncertainty and epistemic uncertainty (Swiler and Giunta 2007). Aleatory uncertainty characterizes the inherent uncertainty in a random input of the performance function under study. Aleatory uncertainty is objective and irreducible and is used when sufficient data on the random input are available. Aleatory uncertainties can be characterized by using appropriate probability distributions. Epistemic uncertainty, on the other hand, characterizes the lack of knowledge on the appropriate value to use for an input that has a fixed value. Epistemic uncertainty is subjective and can be reduced by gathering more data for the input. Epistemic uncertainty reflects the degree of "belief" and can be represented by fuzzy sets (Möller and Beer 2004), possibility theory (Youn et al. 2007), or imprecise probability (Ferson et al. 2003).

This chapter assumes that sufficient random input data are available and, thereafter, only considers aleatory uncertainty. In engineering practice, the aleatory uncertainty of a random input X can be characterized with three sequentially executed steps (Xi et al. 2010):

- Step 1: Obtain optimal distribution parameters for candidate probability distributions using the maximum likelihood method. It can be formulated as

$$\text{maximize} \quad L\left(X|\delta\right) = \sum_{l=1}^{K} \log_{10}\left[f\left(x_l|\delta\right)\right] \tag{1}$$

where δ is the unknown distribution parameter vector; x_l is the lth random data point (or realization) of X; $L(\cdot)$ is the likelihood function; K is the number of random data points (or realizations); and f is the probability density function (PDF) of X for the given δ.
- Step 2: Perform the Chi-Square goodness-of-fit tests on the candidate distribution types with the optimum distribution parameters obtained in Step 1. It is noted that, depending on the specific engineering application, the Kolmogorov–Smirnov (K–S) test or the Anderson–Darling (AD) test may be more appropriate than the Chi-Square goodness-of-fit test.
- Step 3: Select the distribution type with the maximum p-value as the optimal distribution type for X.

2.2 Overview of Component Reliability Analysis

The (time-independent) component reliability can be defined as the probability that the actual performance of an engineered system meets the required or specified design performance under various uncertainty sources (e.g., material properties, loads, geometric tolerances). This definition is often used in reliability-based design of civil structural systems, mechanical systems, and aerospace systems. In order to formulate the component reliability in a mathematical framework, random variables are often used to model uncertainty sources in engineered systems. The time-independent reliability can then be formulated as

$$R(\mathbf{X}) = P(G(\mathbf{X}) < 0) = 1 - P(G(\mathbf{X}) \geq 0) \tag{2}$$

where the random vector $\mathbf{X} = (X_1, X_2, \ldots, X_N)^{\mathrm{T}}$ models uncertainty sources such as material properties, loads, geometric tolerances; $G(\mathbf{X})$ is a system performance function and the system success event $E_{\text{sys}} = \{G(\mathbf{X}) < 0\}$. The uncertainty of the vector $\mathbf{X}$ further propagates and leads to the uncertainty in the system performance function G. In reliability analysis, equating the system performance function G to zero, i.e., $G = 0$, gives us the so-called limit-state function which separates

Fig. 1 Concept of time-independent reliability analysis

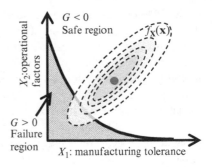

the safe region $G(\mathbf{X}) < 0$ from the failure region $G(\mathbf{X}) > 0$. Depending on the specific problems, a wide variety of system performance functions can be defined to formulate component reliabilities. The most well-known example is the safety margin between the strength and load of an engineered system.

The concept of component reliability analysis in a two-dimensional case is illustrated in Fig. 1. The dashed lines represent the contours of the joint PDF of the two random variables X_1 (operational factors) and X_2 (manufacturing tolerance). The basic idea of component reliability analysis is to compute the probability that $\mathbf{X}$ is located in the safety region $\{G < 0\}$. Mathematically, this probability can be expressed as a multidimensional integration of the performance function over the safety region

$$R(\mathbf{X}) = P(G(\mathbf{X}) < 0) = \int \cdots \int_{\Omega^S} f_{\mathbf{X}}(\mathbf{x}) d\mathbf{x} \qquad (3)$$

where $\mathbf{X} = (X_1, X_2, \ldots, X_N)^{\mathrm{T}}$ models uncertainty sources such as material properties, loads, geometric tolerances; $f_{\mathbf{X}}(\mathbf{x})$ denotes the joint PDF of this random vector; the safety domain Ω^S is defined by the limit-state function as $\Omega^S = \{\mathbf{X}: G(\mathbf{X}) < 0\}$.

Neither analytical multidimensional integration nor direct numerical integration is computationally affordable for large-scale engineering problems where the numbers of random variables are relatively large. The search for efficient computational procedures to estimate the component reliability has resulted in a variety of numerical and simulation methods. In general, these methods can be categorized into four groups: (1) expansion methods; (2) most probable point (MPP)-based methods; (3) sampling methods; and (4) stochastic response surface methods (SRSMs). In what follows we intend to give an overview of these methods.

Expansion methods obtain the second-moment statistics of the performance function based on the first- or second-order Taylor series expansion of this function at the mean values of the input random variables (Haldar and Mahadevan 2000). Reliability can be computed by assuming that the performance function follows a normal distribution. It can be seen, therefore, that expansion methods involve two approximations, i.e., the first-order (linear) or second-order (quadratic) approximation of the performance function at the mean values and the normal approximation to the PDF of the performance function. The approximations lead to the fact that these methods are only applicable for engineering problems with relatively small input uncertainties and weak output nonlinearities.

Among many reliability analysis methods, the first- or second-order reliability method, FORM (Hasofer and Lind 1974) or SORM (Breitung 1984; Tvedt 1984), is most commonly used. The FORM/SORM uses the first- or second-order Taylor expansion to approximate a limit-state function at the most probable failure point (MPP) where the limit-state function separates failure and safety regions of a product (or process) response. Some major challenges of the FORM/SORM include (1) it is very expensive to build the probability density function (PDF) of the response and (2) structural design can be expensive when employing a large number of the responses.

The sampling methods include the direct or smart Monte Carlo simulation (MCS) (Rubinstein 1981; Fu and Moses 1988; Au and Beck 1999; Hurtado 2007; Naess et al. 2009). Assuming that we know the statistical information (PDFs) of the input random variables, the direct MCS generally involves the following three steps:

- Step 1: The MCS starts by randomly generating a large number of samples based on the PDFs of the random inputs.
- Step 2: In this step, the performance function is evaluated at each of the random samples. Simulations or experiments need to be conducted for this purpose. Upon the completion of this step, we obtain a large number of random values or realizations of the performance function.
- Step 3: We extract from these random realizations the probabilistic characteristics of the performance function, including statistical moments, reliability, and PDF.

Although the direct MCS (Rubinstein 1981) produces accurate results for reliability analysis and allows for relative ease in the implementation, it demands a prohibitively large number of simulation runs. Thus, it is often used for the purpose of a benchmarking in reliability analysis. To alleviate the computational burden of the direct MCS, researchers have developed various smart MCS methods, such as the (adaptive) importance sampling methods (Fu and Moses 1988; Au and Beck 1999; Hurtado 2007) and the enhanced MCS method with an optimized extrapolation (Naess et al. 2009). Despite the improved efficiency than the direct MCS, these methods are still computationally expensive.

The SRSM is an emerging technique for reliability analysis under uncertainty. As opposed to the deterministic response surface method whose input variables are deterministic, the SRSM employs random variables as its inputs. The aim of the SRSM is to alleviate the computational burden required for accurate uncertainty quantification (i.e., quantifying the uncertainty in the performance function) and reliability analysis. This is achieved by constructing an explicit multidimensional response surface approximation based on function values given at a set of sample points. Generally speaking, uncertainty quantification and reliability analysis propagation of input uncertainty through a model using the SRSM consists of the following steps: (1) determining an approximate functional form for the performance function (possible based on the statistical information of the input

Fig. 2 Concept of system reliability analysis (two performance functions)

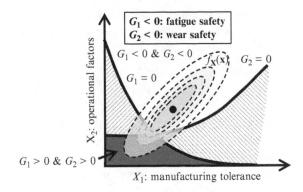

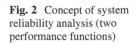

random variables); (2) evaluating the parameters of the functional approximation based on function values at a set of sample points; (3) conducting MCS or numerical integration based on the functional approximation to obtain the probabilistic characteristics (statistical moments, reliability, and PDF) of the performance function. The current state-of-the art SRSMs for uncertainty quantification include the dimension reduction (DR) methods (Rahman and Xu 2004; Xu and Rahman 2004; Youn et al. 2008, Youn and Xi 2009), stochastic spectral methods (Ghanem and Spanos 1991; Wiener 1938; Xiu and Karniadakis 2002; Foo et al. 2008; Foo and Karniadakis 2010), and stochastic collocation methods (Smolyak 1963; Grestner and Griebel 2003; Klimke 2006; Ganapathysubramanian and Zabaras 2007; Xiong et al. 2010; Hu and Youn 2011).

2.3 Overview of System Reliability Analysis

System reliability analysis aims at analyzing the probability of system success while considering multiple system performances (e.g., fatigue, corrosion, and fracture). Figure 2 illustrates the concept of system reliability analysis with a simple series system involving two performance functions (i.e., fatigue safety G_1 and wear safety G_1) and two random variables (i.e., operational factors X_1 and manufacturing tolerance X_2). We have two limit state functions $G_1 = 0$ and $G_2 = 0$ which divides the input random space into four subspaces $\{G_1 < 0$ and $G_2 < 0\}$, $\{G_1 < 0$ and $G_2 > 0\}$, $\{G_1 > 0$ and $G_2 < 0\}$, $\{G_1 > 0$ and $G_2 > 0\}$. Component reliability analysis aims at quantifying the probability that a random sample $\mathbf{x}$ falls into the component safety region (i.e., $\{G_1 < 0\}$ or $\{G_2 < 0\}$) while system reliability analysis (assuming a series system) aims at quantifying the probability that a random sample $\mathbf{x}$ falls into the system safety region (i.e., $\{G_1 < 0$ and $G_2 < 0\}$). Clearly, the component reliability (for $\{G_1 < 0\}$ or $\{G_2 < 0\}$) is larger than the system reliability since the component safety region has a larger area than the system safety region by the area of an intersection region $\{G_1 < 0$ and $G_2 > 0\}$ or $\{G_1 > 0$ and $G_2 < 0\}$.

The aforementioned discussion leads to a mathematical definition of system reliability as a multidimensional integration of a joint probability density function over a system safety region, expressed as

$$R_{sys} = \int \cdots \int_{\Omega^S} f_{\mathbf{X}}(\mathbf{x})\, d\mathbf{x} \tag{4}$$

where $\mathbf{X} = (X_1, X_2, \ldots, X_N)^T$ models uncertainty sources such as material properties, loads, geometric tolerances; $f_{\mathbf{X}}(\mathbf{x})$ denotes the joint PDF of this random vector; Ω^S denotes the system safety domain. We can see that this formula bears a striking resemblance to that of component reliability analysis. The only difference between these two formulae lies in the definition of the safety domain. For component reliability analysis, the safety domain can be defined in terms of a single limit-state function as $\Omega^S = \{\mathbf{x}: G(\mathbf{x}) < 0\}$. For system reliability analysis involving nc performance functions, the safety domains can be expressed as

$$\Omega^S = \{x : \cap_{i=1}^{nc} G_i(x) < 0\} \text{ series system}$$

$$\Omega^S = \{x : \cup_{i=1}^{nc} G_i(x) < 0\} \text{ parallel system}$$

$$\Omega^S = \{x : \cup_{k=1}^{np} \cap_{i \in P_k} G_i(x) < 0\} \text{ mixed system} \tag{5}$$

where P_k is the index set in the kth path set and np is the number of mutually exclusive path sets.

It can be observed that a series system requires all the performance functions satisfy the reliability requirements, resulting in the system safety events being an intersection of component safety events, expressed as

$$E_{series} = \cap_{i=1}^{nc} G_i(\mathbf{x}) < 0 \tag{6}$$

In this case, the system survives if and only if all of its constraints satisfy the reliability requirements.

In contrast to a series system, a parallel system has multiple path sets with each being its component safety event, expressed as

$$E_{parallel} = \cup_{i=1}^{nc} G_i(\mathbf{x}) < 0 \tag{7}$$

In this case, the component survives if any of its constraints satisfy the reliability requirement. A comparison between a series system and a parallel system is graphically shown in Fig. 3, where we observe that the safety domain of a parallel system contains two more regions $\{G_1 > 0 \text{ and } G_2 < 0\}$ and $\{G_1 < 0 \text{ and } G_2 > 0\}$, resulting in a higher system reliability value.

The logic becomes more complicated for a mixed system. We often need to describe the system success event of a mixed system in terms of the mutually

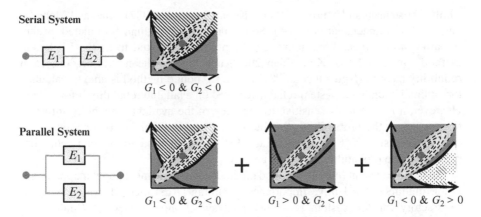

Fig. 3 Comparison between a series system and a parallel system (two performance functions)

exclusive path sets of which each path set Path$_k$ is a series system with multiple component safety events specified in Path$_k$. Thus, we have the following expression

$$E_{\text{mixed}} = \cup_{k=1}^{np} \cap_{i \in \text{Path}_k} G_i\,(\mathbf{x}) < 0 \tag{8}$$

In probability theory, two events are said to be mutually exclusive if they cannot occur at the same time or, in other words, the occurrence of any one of them automatically implies the nonoccurrence of the other. Here, system path sets are said to be mutually exclusive if any two of them are mutually exclusive.

By employing the system safety event E_{sys}, we can derive another important formula for system reliability analysis as

$$R_{\text{sys}} = \Pr\left(E_{\text{sys}}\right) \tag{9}$$

where E_{sys} represents E_{series}, E_{parallel}, and E_{mixed}, for a series system, a parallel system, and a mixed system, respectively.

We note that, in practice, it is extremely difficult to perform the multidimensional numerical integration for system reliability analysis in Eq. (4) due to the high nonlinearity and complexity of the system safety domain. In contrast to the tremendous advances in component reliability analysis as discussed in Sect. 2.2, the research in system reliability analysis has been stagnant, mainly due to the complicated nature of the multiple system failure modes and their interactions, as well as the costly computational expense of system reliability evaluation (Wang et al. 2011).

Due to the aforementioned difficulties, most system reliability analysis methods provide the bounds of system reliability. Ditlevsen proposed the most widely used second-order system reliability bound method (Ditlevsen 1979), which gives much tighter bounds compared with the first-order bounds for both series and parallel systems. Other equivalent forms of Ditlevsen's bounds were given by

Thoft-Christensen and Murotsu (1986), Karamchandani (1987), Xiao and Mahade-van (1998), Ramachandran (2004). Song and Der Kiureghian formulated system reliability to a Linear Programming (LP) problem, referred to as the LP bound method (Song and Der Kiureghian 2003) and latterly the matrix-based system reliability method (Nguyen et al. 2010). The LP bound method is able to calculate the optimal bounds for system reliability based on available reliability information. However, it is extremely sensitive to accuracy of the available reliability information, which is the probabilities for the first-, second-, or higher-order joint safety events. To assure high accuracy of the LP bound method for system reliability prediction, the probabilities must be given very accurately.

Besides the system reliability bound methods, one of the most popular approaches is the multimodal Adaptive Importance Sampling (AIS) method, which is found satisfactory for the system reliability analysis of large structures (Mahadevan and Raghothamachar 2000). The integration of surrogate model techniques with Monte Carlo Simulation (MCS) can be an alternative approach to system reliability prediction as well (Zhou et al. 2000). This approach, which can construct the surrogate models for multiple limit-state functions to represent a joint failure region, is quite practical but accuracy of the approach depends on fidelity of the surrogate models. It is normally expensive to build accurate surrogate models.

Most recently, Youn and Wang (2009) introduced a new concept of the complementary intersection event and proposed the Complementary Intersection Method (CIM) for series system reliability analysis. The CIM provides not only a unique formula for system reliability but also an effective numerical method to evaluate the system reliability with high efficiency and accuracy. The CIM decomposes the probabilities of high-order joint failure events into probabilities of complementary intersection events. For large-scale systems, a CI matrix was proposed to store the probabilities of component safety and complementary intersection events. Then, series system reliability can be efficiently evaluated by advanced reliability methods, such as dimension reduction method and stochastic collocation method. Later, the GCIM framework was proposed to generalize the original CIM so that it can be used for system reliability analysis regardless of system structures (series, parallel, and mixed systems) (Wang et al. 2011). In the subsequent sections, we will review in details the most widely used system reliability bound methods (Ditlevsen 1979) as well as the recently developed point estimation method (Youn and Wang 2009; Wang et al. 2011).

3 System Reliability Analysis for Serial System

A series system succeeds only if all of its components succeed and, in other words, the system fails if any of its components fails. Let us start with a simple series system, namely, a steel portal frame structure shown in Fig. 4 (Ditlevsen 1979). The structure is subjected to a vertical load V at the center of the top beam and a horizontal load F at the hinge joint between the top and bottom-left beams.

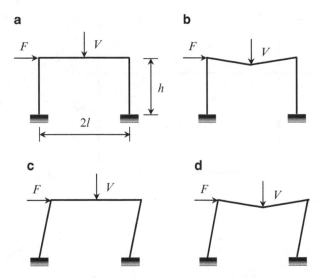

Fig. 4 Loading condition and three failure modes of the portal frame structure (Ditlevsen 1979). (**a**) Loading condition, (**b**) beam failure, (**c**) sway failure, and (**d**) combined failure

These two loads are assumed to be Gaussian random variables. According to the yield hinge mechanism theory, the frame structure has three distinct failure modes: beam failure, sway failure, and combined failure, as shown in Fig. 4.

If we assume an identical yield moment M_y for all three beams, we then have the following performance functions for the three failure modes

$$
\begin{aligned}
&\text{Beam}: G_1 = Vl - 4M_y \\
&\text{Sway}: G_2 = Fh - 4M_y \\
&\text{Combined}: G_3 = Vl + Fh - 6M_y
\end{aligned}
\tag{10}
$$

The limit-state functions for the three failure modes are graphically shown in Fig. 5. The failure domain for each failure mode, or the component failure domain, can be expressed as $\{(F, V)|\ G_i > 0\}$, for $i = 1, 2, 3$, as shown in Fig. 5. Since the occurrence of any of the three failure modes causes system failure, the system failure domain is a union of the component failure domains and has a larger area than any of the component failure domain (see Fig. 5). If we define the failure event of the ith failure mode as $\bar{E}_i$, the probability of system failure for the portal frame structure can be expressed as

$$
p_{fs} = P\left(\bar{E}_1 \cup \bar{E}_2 \cup \bar{E}_3\right)
\tag{11}
$$

where p_{fs} represents the probability of system failure. The above equation can be further derived in terms of the probabilities of component and joint failure events, expressed as

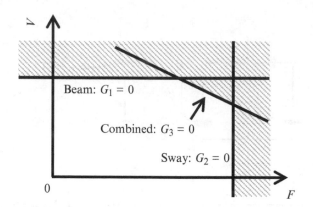

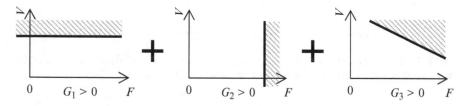

Fig. 5 Limit-state functions and system failure domain of the portal frame structure

$$p_{fs} = P\left(\overline{E}_1\right) + P\left(\overline{E}_2\right) + P\left(\overline{E}_3\right)$$
$$- P\left(\overline{E}_1 \cap \overline{E}_2\right) - P\left(\overline{E}_2 \cap \overline{E}_3\right) - P\left(\overline{E}_1 \cap \overline{E}_3\right)$$
$$+ P\left(\overline{E}_1 \cap \overline{E}_2 \cap \overline{E}_3\right) \tag{12}$$

where $\overline{E}_i \cap \overline{E}_j$ is the second-order joint failure event composed of the component failure events $\overline{E}_i$ and $\overline{E}_j$, for $1 \leq i < j \leq 3$, and $\overline{E}_1 \cap \overline{E}_2 \cap \overline{E}_3$ is the third-order joint failure event composed of the component failure events $\overline{E}_1$, $\overline{E}_2$, and $\overline{E}_3$.

As the number of component events increases, we may have fourth- and higher-order joint events. Considering a series system with m components, the probability of system failure can be expressed as

$$p_{fs} = P\left(\bigcup_{i=1}^{m} \overline{E}_i\right) = \sum_{j=1}^{n}(-1)^{j+1}P_j \quad \text{with} \quad P_j = \sum_{1 \leq i_1 < \cdots < i_j \leq n} P\left(\bigcap_{k \in \{i_1, \ldots, i_j\}} \overline{E}_k\right) \tag{13}$$

where p_{fs} represents the probability of system failure and $\overline{E}_i$ denotes the failure event of the ith component. It can be observed that, to exactly analyze system reliability with m series connected components, we generally need to compute the probabilities of joint events up to the mth order. Unlike the computation of the probability of a component failure or safety event discussed earlier, the computation of a joint failure or safety event is very difficult and even practically impossible unless we employ the very expensive MCS or direct numerical integration. The research efforts to alleviate the computational burden have resulted in a set of system reliability bound methods, such as the first- and second-order bound methods, and the point estimation method or the complementary intersection method (CIM) (Youn and Wang 2009).

3.1 First- and Second-Order Bound Methods

The simplest system reliability bounds are the so-called first-order bounds. Based on the well-known Boolean bounds in Eq. (14), the first-order bounds of probability of system failure are given in Eq. (15).

$$\max_i \left(P\left(\overline{E}_i\right) \right) \leq P\left(\bigcup_{i=1}^m \overline{E}_i \right) \leq \sum_{i=1}^m P\left(\overline{E}_i\right) \tag{14}$$

$$\max\left[P\left(\overline{E}_i\right) \right] \leq p_{fs} \leq \min\left[\sum_{i=1}^m P\left(\overline{E}_i\right), 1 \right] \tag{15}$$

The lower bound in Eq. (15) is obtained by assuming the component events are perfectly independent and the upper bound is derived by assuming the component events are mutually exclusive. Despite the simplicity (only component reliability analysis required), the first-order bound method provides very wide bounds of system reliability that are not practically useful. Thus, the second-order bound method was proposed by Ditlevsen (1979) in Eq. (16) to give much narrower bounds of probability of system failure.

$$P\left(\overline{E}_1\right) + \sum_{i=2}^m \max\left\{ \left[P\left(\overline{E}_i\right) - \sum_{j=1}^{i-1} P\left(\overline{E}_i\overline{E}_j\right) \right], 0 \right\} \leq p_{fs}$$

$$\leq \min\left\{ \left[\sum_{i=1}^m P\left(\overline{E}_i\right) - \sum_{i=2}^m \max_{j<i} P\left(\overline{E}_i\overline{E}_j\right) \right], 1 \right\} \tag{16}$$

where E_1 is the event having the largest probability of failure.

Fig. 6 Statically determinate truss structure system (Ditlevsen 1979; Song and Der Kiureghian 2003)

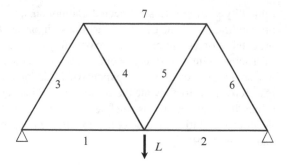

Case Study 1

Consider a statically determinate truss structure (Ditlevsen 1979; Song and Der Kiureghian 2003) in Fig. 6. In this structure, the failure of any truss member leads to the failure of the truss. Thus, the truss structure can be treated as a series system with the seven truss members as its components. For this illustration, we assume the following probabilities of the component and joint failure events (with $P_i = P(\bar{E}_i)$, and $P_{ij} = P(\bar{E}_i \cap \bar{E}_j)$): $P_1 = 1.88\text{E}{-}4$, $P_2 = 1.88\text{E}{-}4$, $P_3 = 1.88\text{E}{-}4$, $P_4 = 1.88\text{E}{-}4$, $P_5 = 1.88\text{E}{-}4$, $P_6 = 1.88\text{E}{-}4$, and $P_7 = 1.88\text{E}{-}4$, $P_{12} = 5.73\text{E}{-}5$, $P_{13} = 4.35\text{E}{-}5$, $P_{14} = 5.42\text{E}{-}5$, $P_{15} = 4.59\text{E}{-}5$, $P_{16} = 5.13\text{E}{-}5$, $P_{17} = 4.85\text{E}{-}5$, $P_{23} = 6.08\text{E}{-}5$, $P_{24} = 7.79\text{E}{-}5$, $P_{25} = 6.47\text{E}{-}5$, $P_{26} = 7.42\text{E}{-}5$, $P_{27} = 6.87\text{E}{-}5$, $P_{34} = 5.75\text{E}{-}5$, $P_{35} = 4.86\text{E}{-}5$, $P_{36} = 5.43\text{E}{-}5$, $P_{37} = 5.14\text{E}{-}5$, $P_{45} = 6.10\text{E}{-}5$, $P_{46} = 6.88\text{E}{-}5$, $P_{47} = 6.48\text{E}{-}5$, $P_{56} = 5.76\text{E}{-}5$, $P_{57} = 5.44\text{E}{-}5$, and $P_{67} = 6.11\text{E}{-}5$. Compute the first- and second-order bounds for the probability of failure of the truss.

Solution Let us first compute the first-order bounds with Eq. (15) as:

$$\max\left[P\left(\bar{E}_i\right)\right] \le p_{fs} \le \min\left[\sum_{i=1}^{m} P\left(\bar{E}_i\right), 1\right]$$

The lower and upper bounds can be computed as

$$\max\left[P\left(\bar{E}_i\right)\right] = 1.88\text{E} - 4$$
$$\min\left[\sum_{i=1}^{m} P\left(\bar{E}_i\right), 1\right] = \min\left[1.32\text{E} - 3, 1\right] = 1.32\text{E} - 3$$

Thus, the first-order bounds are [1.88E–4, 1.32E–3]. Then we compute the second-order bounds with Eq. (16). The lower and upper bounds are computed as 4.02E–4 and 9.12E–4, respectively. Thus, the second-order bounds are [4.02E–4, 9.12E–4]. We can clearly see that, compared to the first-order bound method, the second-order bound method gives much narrower bounds of probability of system failure (and thus system reliability).

3.2 Point Estimation Method

In reliability-based design, it is more desirable to have a unique point estimate of system reliability than an interval estimate. In what follows, we introduce a recently developed point estimate method (Wang et al. 2011), namely, the generalized complementary intersection method (GCIM).

Since the probabilities of all events are nonnegative, the following inequalities must be satisfied as

$$\max_i \left(P \left(\overline{E}_i \right) \right) \leq \sqrt{\sum_{i=1}^{m} \left[P \left(\overline{E}_i \right) \right]^2} \leq \sum_{i=1}^{m} P \left(\overline{E}_i \right) \tag{17}$$

Based on Eqs. (16) and (17), the probability of system failure (p_{fs}) of a series system failure can be simplified to a unique explicit formula as

$$p_{fs} \cong P \left(\overline{E}_1 \right) + \sum_{i=2}^{m} \left\langle P \left(\overline{E}_i \right) - \sqrt{\sum_{j=1}^{i-1} \left[P \left(\overline{E}_i \overline{E}_j \right) \right]^2} \right\rangle \tag{18}$$

It can be proved that this approximate probability lies in the second-order bounds in Eq. (16). Based on Eq. (18), serial system reliability can be assessed as (1 − the probability of system failure) and formulated as

$$R_{sys} = P \left(E_1 E_2 \cdots E_{m-1} E_m \right) \cong P \left(E_1 \right) - \sum_{i=2}^{m} \left\langle P \left(\overline{E}_i \right) - \sqrt{\sum_{j=1}^{i-1} \left[P \left(\overline{E}_i \overline{E}_j \right) \right]^2} \right\rangle$$

$$\text{where} \quad \langle A \rangle \equiv \begin{cases} A, & \text{if } A > 0 \\ 0, & \text{if } A \leq 0 \end{cases} \tag{19}$$

Note that the terms inside the bracket, $\langle \cdot \rangle$, should be ignored if it is less than zero and R_{sys} should be set to zero if the approximated one given by Eq. (19) is less than zero. It is noted that Eq. (19) provides an explicit and unique formula for system reliability assessment based on the second-order reliability bounds shown in Eq. (16) and an inequality Eq. (17).

Case Study 2

Consider an internal combustion engine for series system reliability analysis (Liang et al. 2007). Five random variables are considered in this example: the cylinder bore b, compression ratio c_r, exhaust valve diameter d_E, intake valve diameter d_I,

Table 1 Statistical information of input random variables for internal combustion engine

Random variable	Mean	Standard deviation	Distribution type
b (mm)	–	0.40	Normal
c_r (mm)	–	0.15	Normal
d_E (mm)	–	0.15	Normal
d_I	–	0.05	Normal
ω ($\times 10^{-3}$)	–	0.25	Normal

Table 2 Eight different design points for system reliability analysis

| Design points | Mean values for random variables | | | | |
	B	c_r	d_E	d_I	ω
1	82.1025	35.8039	30.3274	9.3397	5.2827
2	82.3987	36.1754	30.4835	9.3684	5.5983
3	82.5511	36.3630	30.5676	9.3811	5.7550
4	82.6770	36.5187	30.6334	9.3920	5.8901
5	82.8234	36.7006	30.7121	9.4049	5.9498
6	82.8750	36.7655	30.7407	9.4096	5.9754
7	82.9204	36.8222	30.7657	9.4137	5.9772
8	82.9977	36.9197	30.8084	9.4204	5.9795

and the revolutions per minute (rpm) at peak power, ω. From a thermodynamic viewpoint, nine component safety events are defined as follows:

$E_1 = \{1.2N_c b - 400 \leq 0\}$ (min .bore wall thickness)

$E_2 = \left\{ [8V/(200\pi N_c)]^{0.5} - b \leq 0 \right\}$ (max .engine height)

$E_3 = \{d_I + d_E - 0.82b \leq 0\}$ (valve geometry and structure)

$E_4 = \{0.83d_I - d_E \leq 0\}$ (min .value diameter ratio)

$E_5 = \{d_E - 0.89d_I \leq 0\}$ (max .value diameter ratio)

$E_6 = \left\{ 9.428 \times 10^{-5} [4V/(\pi N_c)] \left(\omega/d_I^2 \right) - 0.6C_s \leq 0 \right\}$ (max .Mech/Index)

$E_7 = \{0.045b + c_r - 13.2 \leq 0\}$ (knock-limit compression ratio)

$E_8 = \{\omega - 6.5 \leq 0\}$ (max .torque converter rpm)

$E_9 = \{230.5Q\eta_{tw} - 3.6 \times 10^6 \leq 0\}$ (max .fuel economy)

where

$$\eta_{tw} = 0.85951 \left(1 - c_r^{-0.33} \right) - S_v, \quad V = 1.859 \times 10^6 \text{ mm}^3$$
$$Q = 43{,}958 \text{ kJ/kg}, C_s = 0.44, \quad \text{and} \quad N_c = 4$$

All the random variables are assumed to follow normal distribution with statistical information presented in Table 1. Perform system reliability analyses at the eight reliability-based optimum design points as listed in Table 2 using the first-order bounds (FOB), second-order bounds (SOB), and GCIM methods (Wang et al. 2011).

Table 3 Results of system reliability analysis with MCS, FOB using MCS, SOB using MCS, and GCIM using MCS ($M = 1,000,000$)

Analysis method		System reliability level at each design							
		1	2	3	4	5	6	7	8
FOB	Upper	0.9989	0.9899	0.9745	0.9495	0.8984	0.8742	0.8490	0.7988
	Lower	0.9949	0.9506	0.8744	0.7432	0.5367	0.4318	0.3410	0.1513
SOB	Upper	0.9949	0.9520	0.8822	0.7741	0.6224	0.5554	0.4987	0.3967
	Lower	0.9949	0.9517	0.8798	0.7653	0.5929	0.5190	0.4418	0.3049
GCIM		0.9949	0.9518	0.8805	0.7674	0.6026	0.5312	0.4612	0.3371
MCS		0.9949	0.9520	0.8820	0.7731	0.6179	0.5476	0.4871	0.3748
GCIM error		0.0000	0.0002	0.0015	0.0057	0.0153	0.0164	0.0259	0.0377

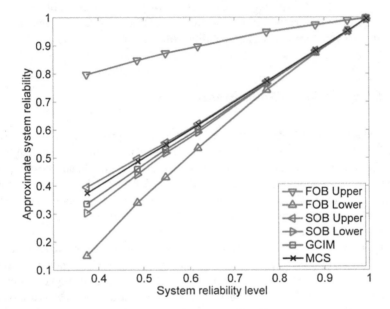

Fig. 7 Results of system reliability analysis at eight different reliability levels (Wang et al. 2011)

Solution Equations (15), (16), and (19) are used to compute the first-order system reliability bounds in the FOB, the second-order system reliability bounds in the SOB, and the point system reliability estimate in the GCIM, respectively. The probabilities of component and second-order joint failure events are computed with the direct MCS. The results of system reliability analysis at the eight design points are summarized in Table 3 and also graphically shown in Fig. 7 (Wang et al. 2011). From the results, it is found that the first-order bound method gives too wide bounds to be of practical use. On the contrary, the second-order bound method gives tighter bounds. It is expected based on the results that the GCIM can predict system reliabilities accurately at various reliability levels and the estimation errors tend to be lower at high system reliability levels (e.g., greater than 0.95), which are often encountered in engineering practices, than those at low system reliability levels.

This case study considers the first- and second-order joint failure events. The GCIM produces numerical error because of the ignorance of the probabilities of the third- or higher-order joint failure events. The effects of the third- or higher-order joint failure events tend to increase as the system reliability decreases, simply because the probabilities of joint failure events are usually bigger at low reliability level than at high reliability level. This is also true for the reliability bound methods. As can be observed from Table 3 and Fig. 7, the lower the system reliability, the larger the GCIM estimation error and the wider the bounds produced by the FOB and SOB. Thus for series systems, the GCIM produces smaller numerical error at a high system reliability level than that at a lower level. This is valid only for series systems. When only probabilities of the first- and the second-order joint events are used for system reliability analysis, the GCIM will provide comparable results with the average of SOBs. However, compared with SOBs, the GCIM provides system reliability analysis formula with probabilities of any-order joint events.

3.3   Computation of Joint Events

In Case Study 2, we use the direct MCS to compute the probabilities of joint events $(\bar{E}_i \cap \bar{E}_j = \{G_1 > 0 \text{ and } G_2 > 0\}$. In practice, however, the direct MCS requires an intolerably large number of function evaluations. On the other hand, the component reliability analysis methods (e.g., FORM/SORM, DR methods, stochastic spectral methods, and stochastic collocation methods) discussed earlier cannot be directly used to compute these probabilities since there are neither explicit nor implicit performance functions associated with the joint events. Therefore, the primary challenge in system reliability analysis lies in efficient and accurate determination of the probabilities of joint safety events. In what follows, we review a newly developed method to efficiently evaluate the probabilities of the second- or higher-order joint safety events. This method is embedded in the aforementioned GCIM as a solver for the probabilities of joint events (Youn and Wang 2009; Wang et al. 2011).

The second-order CI event can be denoted as $E_{ij} \equiv \{X | G_i \cdot G_j \leq 0\}$. The CI event can be further expressed as $E_{ij} = \bar{E}_i E_j \cup E_i \bar{E}_j$ where the component failure events are defined as $\bar{E}_i = \{X | G_i > 0\}$, $\bar{E}_j = \{X | G_j > 0\}$. The CI event E_{ij} is thus composed of two events: $E_i \bar{E}_j = \{X | G_i \leq 0 \cap G_j > 0\}$ and $\bar{E}_i E_j = \{X | G_i > 0 \cap G_j \leq 0\}$. Since the events, $\bar{E}_i E_j$ and $E_i \bar{E}_j$, are disjoint, the probability of the CI event E_{ij} can be expressed as (Youn and Wang 2009)

$$
P\left(E_{ij}\right) \equiv P\left(X | G_i \cdot G_j \leq 0\right)
$$

$$
= P\left(X \Big| G_i > 0 \cap G_j \leq 0\right) + P\left(X | G_i \leq 0 \cap G_j > 0\right)
$$

$$
= P\left(\bar{E}_i E_j\right) + P\left(E_i \bar{E}_j\right) \tag{20}
$$

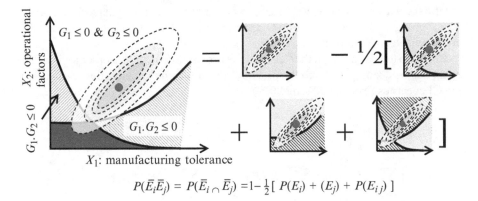

$$P(\bar{E}_i\bar{E}_j) = P(\bar{E}_i \cap \bar{E}_j) = 1 - \tfrac{1}{2}[\; P(E_i) + (E_j) + P(E_{ij}) \;]$$

Fig. 8 Decomposition of a joint failure event into component safety and CI events

Based on the probability theory, the probability of the second-order joint safety event $E_i \cap E_j$ can be expressed as

$$P\left(E_i E_j\right) = P\left(E_i\right) - P\left(E_i \overline{E}_j\right)$$
$$= P\left(E_j\right) - P\left(\overline{E}_i E_j\right) \tag{21}$$

From Eqs. (20) and (21), the probabilities of the second-order joint safety and failure events can be decomposed as

$$P\left(E_i E_j\right) = \frac{1}{2}\left[P\left(E_i\right) + P\left(E_j\right) - P\left(E_{ij}\right)\right] \tag{22}$$

$$P\left(\overline{E}_i \overline{E}_j\right) = 1 - \frac{1}{2}\left[P\left(E_i\right) + P\left(E_j\right) + P\left(E_{ij}\right)\right] \tag{23}$$

It is noted that each CI event has its own limit state function, which enables the use of any component reliability analysis method. The decomposition of joint failure events into component safety and CI events is graphically shown in Fig. 8. We observe that a joint failure event without any limit-state function is decomposed into two component safety events and one CI events, all of which have their own limit-state function and thus allows for the use of any component reliability analysis method.

We have discussed the definition of the second-order CI event. In general, this definition can be generalized to any higher-order event. Let an Kth-order CI event denote $E_{12\ldots K} \equiv \{X|G_1 \cdot G_2 \cdot \ldots G_K \leq 0\}$, where the component safety (or first-order CI) event is defined as $E_i = \{X|G_i \leq 0, i = 1, 2, \ldots, K\}$. The defined Kth-order CI event is actually composed of K distinct intersections of component events E_i and their complements $\bar{E}_j$ in total where $i, j = 1, \ldots, K$ and $i \neq j$. For example, for the second-order CI event E_{ij}, it is composed of two distinct intersection events, $E_1\bar{E}_2$ and $\bar{E}_1E_2$. These two events are the intersections of E_1 (or E_2) and the complementary event of E_2 (or E_1).

Based on the definition of the CI event, the probability of an Nth-order joint safety event can be decomposed into the probabilities of the component safety events and the CI events as (Youn and Wang 2009)

$$
P\left(\bigcap_{i=1}^{N} E_i\right)
$$

$$
= \frac{1}{2^{N-1}}
\left[
\begin{array}{l}
\displaystyle\sum_{i=1}^{N} P\left(E_i\right) - \sum_{\substack{i=1;\\ j=2;\\ i<j}}^{N} P\left(E_{ij}\right) + \sum_{\substack{i=1;\\ j=2;\\ k=3\\ i<j<k}}^{N} P\left(E_{ijk}\right) + \cdots + \\[4em]
(-1)^{m-1} \sum_{\substack{i=1;\\ j=2;\\ \vdots\\ l=m\\ i<j<\cdots<l}}^{N} P\left(E_{\underbrace{ij\,\ldots\,l}_{m}}\right) + \cdots + (-1)^{N-1} P\left(E_{12\cdots N}\right)
\end{array}
\right]
\tag{24}
$$

It is again noted that each CI event has its own limit state function, which enables the use of any component reliability analysis methods. In general, higher-order CI events are expected to be highly nonlinear. As a good trade-off between computational efficiency and accuracy, the use of the first- and second-order CI events in Eq. (24) is suggested for system reliability analysis of most engineered systems. However, we still note that more terms in Eq. (24) can be obtained within the same computational budget as advanced component reliability analysis methods are developed in future.

Case Study 3

Consider the steel portal frame structure shown in Fig. 4. In the three performance functions for the three failure modes in Eq. (10), the vertical load G and the horizontal load F are assumed to be Gaussian random variables with means both means being 35,000 N and standard deviations being 7,000 N. The other variables in the performance functions are assumed to be deterministic and take the following values: $l = h = 5$ m, and $M_y = 60,000$ Nm.

1. Compute the probabilities of component safety events and second-order CI events with the direct MCS.
2. Based on the results from (1), compute the probabilities of second-order safety events.
3. Determine the first- and second-order bounds.

Solution (1) If we conduct a direct MCS with M random samples $x_1, x_2, \ldots, x_M$, the probabilities of component safety events can be computed as

$$P(E_i) = E(I(G_i(\mathbf{x}) < 0))$$

$$\approx \frac{1}{M} \sum_{k=1}^{M} I(G_i(\mathbf{x}_j) < 0), \quad \text{for} \quad i = 1, 2, 3$$

The probabilities of second-order CI events can be computed as

$$P(E_{ij}) = E\left(I(G_i(\mathbf{x}) G_j(\mathbf{x}) < 0)\right)$$

$$\approx \frac{1}{M} \sum_{k=1}^{M} I(G_i(\mathbf{x}_k) G_j(\mathbf{x}_k) < 0), \quad \text{for} \quad (i, j) = (1, 2), (1, 3), (2, 3)$$

We can conveniently write these probabilities in a CI matrix. For this example with three components in total, the CI matrix can be defined as

$$\mathrm{CI} = \begin{bmatrix} P(E_1) & P(E_{12}) & P(E_{13}) \\ - & P(E_2) & P(E_{23}) \\ - & - & P(E_3) \end{bmatrix}$$

In the upper triangular CI matrix, the diagonal elements correspond to the component reliabilities (or probabilities of the first-order CI events) and the element on ith row and jth column corresponds to the probability of the second-order CI event E_{ij} if $j < i$. The CI matrix computed with a direct MCS with 1,000,000 random samples reads

$$\mathrm{CI} = \begin{bmatrix} 0.9686 & 0.0610 & 0.3891 \\ - & 0.9682 & 0.3891 \\ - & - & 0.5811 \end{bmatrix}$$

It is noted that all the probabilities in the CI matrix (the probabilities of component events and second-order CI events) can be computed using any component reliability analysis method (e.g., FORM/SORM, DR, PCE) instead of the direct MCS.

(2) Based on the CI matrix obtained from (1) and according to Eq. (23), we can obtain the probabilities of second-order failure events as $P(\bar{E}_1 \bar{E}_2) = 0.0011$,

$P(\bar{E}_1\bar{E}_3) = 0.0306$, and $P(\bar{E}_2\bar{E}_3) = 0.0308$. For example, the probability of the second-order failure event $P(\bar{E}_2\bar{E}_3)$ can be computed as

$$P\left(\overline{E}_2\overline{E}_3\right) = 1 - \frac{1}{2}\left[P\left(E_2\right) + P\left(E_3\right) + P\left(E_{23}\right)\right]$$

$$= 1 - \frac{1}{2}\left[0.9682 + 0.5811 + 0.3891\right]$$

$$= 0.0308$$

(3) The first-order bounds can be computed with Eq. (15) as:

$$1 - \min\left[\sum_{i=1}^{m} P\left(\overline{E}_i\right), 1\right] \le R_{sys} \le 1 - \max\left[P\left(\overline{E}_i\right)\right]$$

The lower and upper bounds can be computed as

$$1 - \min\left[\sum_{i=1}^{m} P\left(\overline{E}_i\right), 1\right] = 1 - \min\left[0.4812, 1\right] = 0.5188$$

$$1 - \max\left[P\left(\overline{E}_i\right)\right] = 1 - 0.4189 = 0.5811$$

Thus, the first-order bounds read [0.5188, 0.5811]. Note that the above bounds are the first-order bounds of system reliability. The corresponding bounds of system probability failure can be easily obtained by exchanging the lower and upper bounds and subtracting both from 1. Then we compute the second-order bounds with Eq. (16) as [0.5793, 0.5801]. We can again clearly see that, compared to the first-order bound method, the second-order bound method gives much narrower bounds of system reliability.

4   System Reliability Analysis for Parallel System

Unlike a series system whose success requires the success of all its components, a parallel system succeeds as long as one of its components succeeds. In other words, a parallel system fails only if all its components fail, and the probability of system failure is the probability of the intersection of all component failure events, expressed as

$$p_{fs} = P\left(\bigcap_{i=1}^{m} \overline{E}_i\right) \tag{25}$$

Consider a 10-bar parallel system in Fig. 9 (Wang et al. 2011), where 10 brittle bars are connected in parallel to sustain a vertical load applied at one end. Ten bars

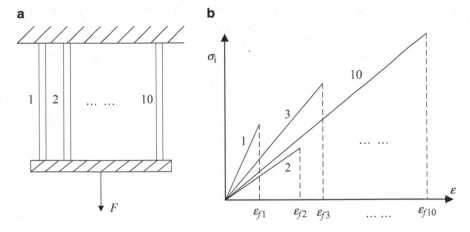

Fig. 9 Ten brittle bar parallel system: (**a**) system structure model; (**b**) brittle material behavior in a parallel system (Wang et al. 2011)

are all brittle with different fracture strain limits ε_{fi}, $1 \le i \le 10$, which are sorted in an ascending order. If the exerted strain ε is between the $(i-1)$th and ith fracture strain limits, i.e., $n_{i-1)} \le \varepsilon < \varepsilon_{fi}$, bar components with fracture strains below ε_{fi} will fail, and the allowable load is then the sum of the strength of components with fracture strains equal to or above ε_{fi}. Therefore, the strain level corresponding to the overall maximum allowable load is among the 10 fracture strain limits.

Ten success scenarios where the tem-bar system can withstand the vertical load are listed as (Wang et al. 2011):

- First success scenario ($\varepsilon = \varepsilon_{f1}$): No fracture occurs, and the system strength R_T, as the sum of strength of all the 10 brittle bars, is larger than the load F. The performance function can be expressed as

$$G_1 = F - \sum_{j=1}^{10} R_j \left(\varepsilon_{f1} \right) = F - \sum_{j=1}^{10} \left(E_j A_j \right) \cdot \varepsilon_{f1} \qquad (26)$$

where R_j represents the allowable load that can be sustained by the jth brittle bar, A_j the cross section area of the jth brittle bar, and E_j the Young's modulus of the jth brittle bar.

- Second success scenario ($\varepsilon = \varepsilon_{f2}$): The first brittle bar fails due to the fracture and no longer contributes to the overall system strength. The system strength R_T, as the sum of strength of the other nine brittle bars, is larger than the load F. The performance function can be expressed as

$$G_1 = F - \sum_{j=2}^{10} R_j \left(\varepsilon_{f2} \right) = F - \sum_{j=1}^{10} \left(E_j A_j \right) \cdot \varepsilon_{f2} \qquad (27)$$

- Tenth success scenario ($\varepsilon = \varepsilon_{f10}$): The first nine brittle bars fail due to the fracture and no longer contribute to the overall system strength. The system strength R_T, as the sum of strength of the remaining one bar, is larger than the load F. The performance function can be expressed as

$$G_1 = F - R_{10}\left(\varepsilon_{f10}\right) = F - (E_{10}A_{10}) \cdot \varepsilon_{f10} \qquad (28)$$

The brittle bar system fails to sustain the load F only if we have the nonoccurrence of all the 10 success scenario or, in other words, the system strength at any of the 10 fracture strains is smaller than the load F. Therefore, this is a parallel system with 10 components, corresponding to the 10 fracture strains.

4.1 First- and Second-Order Bound Methods

A parallel system reliability formula can be obtained based on the formula of series system reliability by using the De Morgan's law (Wang et al. 2011). According to the De Morgan's law, the probability of parallel system failure can be expressed as

$$P\left(\bigcap_{i=1}^{m}\overline{E}_i\right) = 1 - P\left(\overline{\bigcap_{i=1}^{m}\overline{E}_i}\right) = 1 - P\left(\bigcup_{i=1}^{m}E_i\right) \qquad (29)$$

where $\bar{E}_i$ is the ith component failure event. Equation (29) relates the probability of parallel system failure with the probability of series system safety (reliability). If we treat E_i as the ith component failure event in a series system, the right side of Eq. (29) is then the series system reliability.

Based on this relationship and the first-order bounds for a series system in Eq. (15), the first-order bounds for a parallel system can be derived as

$$\max\left\{\left[1 - \sum_{i=1}^{m}P\left(E_i\right)\right], 0\right\} \le p_{fs} \le \min\left[P\left(\overline{E}_i\right)\right] \qquad (30)$$

The lower bound is obtained by assuming the component events are mutually exclusive and the upper bound is derived by assuming the component events are perfectly independent.

Similarly, based on the second-order bounds for a series system in Eq. (16), the second-order bounds for a parallel system can be derived as

$$\max\left\{\left[\sum_{i=1}^{m} P\left(\overline{E}_i\right) - \sum_{i=2}^{m} \min_{j<i} P\left(\overline{E}_i \cup \overline{E}_j\right)\right], 0\right\} \leq p_{fs}$$

$$\leq P\left(\overline{E}_1\right) - \sum_{i=2}^{m} \max\left\{\left[\begin{array}{c} 2-i-P\left(\overline{E}_i\right) \\ +\sum_{j=1}^{i-1} P\left(\overline{E}_i \cup \overline{E}_j\right) \end{array}\right], 0\right\} \tag{31}$$

where E_1 is the event having the largest probability of failure.

4.2 Point Estimation Method

Based on the aforementioned relationship between a series system and a parallel system, the probability of parallel system failure can be obtained from Eq. (19) by treating the safe events in the series system as the failure events in the parallel system as (Wang et al. 2011)

$$p_{fs} \cong P\left(\overline{E}_1\right) - \sum_{i=2}^{m} \left\langle P\left(E_i\right) - \sqrt{\sum_{j=1}^{i-1} \left[P\left(E_i E_j\right)\right]^2} \right\rangle, \quad \langle A \rangle \equiv \begin{cases} A, & \text{if } A > 0 \\ 0, & \text{if } A \leq 0 \end{cases} \tag{32}$$

Finally, parallel system reliability can be obtained from Eq. (32) by one minus the probability of system failure as

$$R_{sys} \cong P\left(E_1\right) + \sum_{i=2}^{m} \left\langle P\left(E_i\right) - \sqrt{\sum_{j=1}^{i-1} \left[P\left(E_i E_j\right)\right]^2} \right\rangle, \quad \langle A \rangle \equiv \begin{cases} A, & \text{if } A > 0 \\ 0, & \text{if } A \leq 0 \end{cases} \tag{33}$$

5 System Reliability Analysis for Mixed Systems

A mixed system may have various system structures as mixtures of series and parallel systems. The success and failure logics of such systems are more complicated than those of series and parallel systems. Consider a cantilever beam-bar system (Song and Der Kiureghian 2003; Wang et al. 2011) which is an ideally elastic–plastic cantilever beam supported by an ideally rigid–brittle bar, with a load applied at the midpoint of the beam, as shown in Fig. 10. There are three failure modes and five independent failure events $\overline{E}_1$–$\overline{E}_5$. These three failure modes are formed by different combinations of failure events as:

Fig. 10 A cantilever
beam-bar system (Song and
Der Kiureghian 2003)

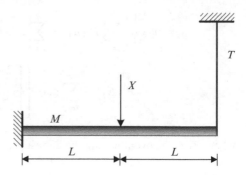

- First failure mode: The fracture of the brittle bar (event $\bar{E}_1$) occurs, and subsequently the formation of a hinge at the fixed point of the beam (event $\bar{E}_2$).
- Second failure mode: The formation of a hinge at the fixed point of the beam (event $\bar{E}_3$) followed by the formation of another hinge at the midpoint of the beam (event $\bar{E}_4$).
- Third failure mode: The formation of a hinge at the fixed point of the beam (event $\bar{E}_3$) followed by the fracture of the brittle bar (event $\bar{E}_5$).

The five safety events can be expressed as:

$$
\begin{aligned}
E_1 &= \left\{ X, T \,\middle|\, 5X/16 - T \le 0 \right\} \\
E_2 &= \left\{ X, L, M \,\middle|\, LX - M \le 0 \right\} \\
E_3 &= \left\{ X, L, M \,\middle|\, 3LX/8 - M \le 0 \right\} \\
E_4 &= \left\{ X, L, M \,\middle|\, LX/3 - M \le 0 \right\} \\
E_5 &= \left\{ X, L, M, T \,\middle|\, LX - M - 2LT \le 0 \right\}
\end{aligned}
\tag{34}
$$

Considering these three failure modes, the system success event can be obtained as (Wang et al. 2011):

$$
E_S = (E_1 \cup E_2) \cap \{E_3 \cup (E_4 \cap E_5)\}
\tag{35}
$$

It is not possible to derive any bounds or point estimates of system reliability based on the system success event in Eq. (35) which contains a mixture of intersection and union logics.

One way to tackle this difficulty is to decompose the mixed system success event into mutually exclusive success events or path sets (see an example in Fig. 11), of which each is a series system. As a result, system reliability of this mixed system can be expressed as a sum of the probabilities of these mutually exclusive series events. This method is embedded in the GCIM (Wang et al. 2011).

Considering a mixed system with N components, the following procedure can be proceeded to derive the mutually exclusive path sets and conduct system reliability analysis in search for a point system reliability estimate (Wang et al. 2011).

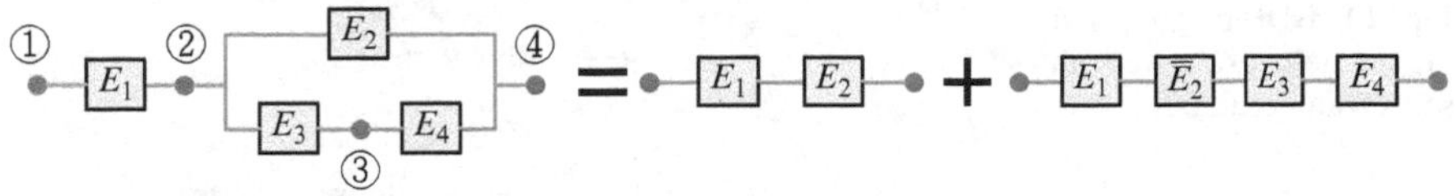

Fig. 11 Decomposition of a mixed system success event into mutually exclusive series system success events

Fig. 12 Conversion of the system block diagram to SS matrix (Wang et al. 2011)

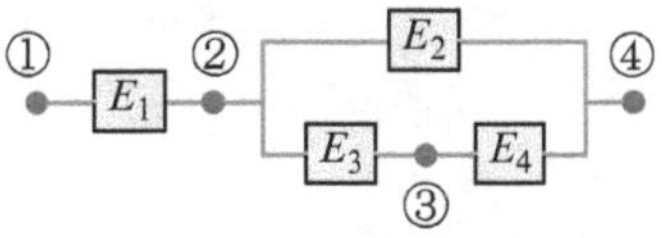

$$\rightarrow \text{SS-matrix} = \begin{bmatrix} 1 & 2 & 3 & 4 \\ 1 & 2 & 2 & 3 \\ 2 & 4 & 3 & 4 \end{bmatrix} \begin{matrix} 1^{\text{st}} \text{ row: component no.} \\ 2^{\text{nd}} \text{ row: starting node} \\ 3^{\text{rd}} \text{ row: end node} \end{matrix}$$

Step I: Constructing a System Structure Matrix

An SS matrix, a 3-by-M, can be used to characterize any system structural configuration (components and their connections) in a matrix form. The SS matrix contains the information about the constituting components and their connection. The first row of the matrix contains component numbers, while the second and third rows correspond to the starting and end nodes of the components. Generally, the total number of columns of a SS matrix, M, is equal to the total number of system components, N. In the case of complicated system structures, one component may repeatedly appear in between different sets of nodes and, consequently, M could be larger than N, for example a 2-out-of-3 system.

Let us consider the mixed system example shown in Fig. 11. The SS matrix for the system can be constructed as a 3×4 matrix, as shown in Fig. 12. The first column of the system structure matrix $[1, 1, 2]^{\text{T}}$ indicates that the first component connects nodes 1 and 2.

Step II: Finding Mutually Exclusive System Path Sets

Based on the SS matrix, the Binary Decision Diagram (BDD) technique (Lee 1959) can be employed to find the mutually exclusive system path sets, of which each path set is a series system. As discussed in Chap. 2, two events are said to be mutually exclusive if they cannot occur at the same time or, in other words, the occurrence of any one of them automatically implies the nonoccurrence of the other. Here, system path sets are said to be mutually exclusive if any two of them are mutually exclusive. We note that, without the SS matrix, it is not easy for the BDD technique to automate the process to identify the mutually exclusive path sets. The mixed system shown

Fig. 13 BDD diagram and
the mutually exclusive path
sets (Wang et al. 2011)

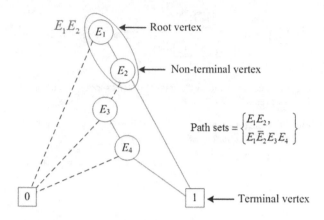

in Fig. 12 can be decomposed into the two mutually exclusive path sets using the
BDD, which is shown in Fig. 13. Although the path sets $E_1\overline{E}_2E_3E_4$ and $E_1E_3E_4$
represent the same path that go through from the left terminal 1 to the right terminal
0 in Fig. 13, the former belongs to the mutually exclusive path sets while the latter
does not. This is due to the fact that the path sets $E_1E_3E_4$ and E_1E_2 are not mutually
exclusive. We also note, however, that we could still construct another group of
mutually exclusively path sets, $\{E_1E_3E_4,\ E_1E_2\overline{E}_3\}$, which contains the path set
$E_1E_3E_4$ as a member. This is due to the fact that a mixed system may have multiple
BDDs with different configurations depending on the ordering of nodes in BDDs
and these BDDs result in several groups of mutually exclusive path sets, among
which the one with the smallest number of path sets is desirable. Another point
deserving of notice is that the mixed system considered here consists of only two
mutually exclusive path sets. In cases of more than two mutually exclusive path sets,
any two path sets are mutually exclusive. This suggests that the system path sets can
equivalently be said to be pairwise mutually exclusive.

Step III: Evaluating All Mutually Exclusive Path Sets and System Reliability

Due to the property of the mutual exclusiveness, the mixed system reliability is the
sum of the probabilities of all paths as

$$R_{sys} = P\left(\bigcup_{i=1}^{np} \text{Path}_i\right) = \sum_{i=1}^{np} P\left(\text{Path}_i\right) \tag{36}$$

Table 4 Statistical information of input random variables for the cantilever beam-bar system

Random variable	Mean	Standard deviation	Distribution type
L	5.0	0.05	Normal
T	1,000	300	Normal
M	150	30	Normal
X	u_X	20	Normal

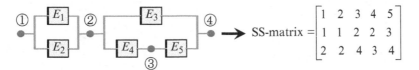

Fig. 14 System block diagram and SS matrix for the cantilever beam-bar example (Wang et al. 2011)

where Path_i is the ith mutually exclusive path set obtained by the BDD and np is the total number of mutually exclusive path sets. For the mixed system shown in Fig. 12, the system reliability can be calculated as

$$R_{sys} = P\left(\bigcup_{i=1}^{2} \text{Path}_i\right) = \sum_{i=1}^{2} P\left(\text{Path}_i\right)$$

$$= P\left(E_1 E_2\right) + P\left(E_1 \overline{E}_2 E_3 E_4\right) \qquad (37)$$

where the probability of each individual path set can be calculated using the point estimate formula for the series system reliability given by Eq. (19).

Case Study 4

Consider the cantilever beam-bar system (Song and Der Kiureghian 2003; Wang et al. 2011)) shown in Fig. 10. In the performance functions for the five component safety events, random variables and their random properties are summarized in Table 4. Compute the system reliability with the GCIM method at 10 different reliability levels that correspond to 10 different loading conditions (X), 100, 90, 85, 80, 70, 60, 50, 40, 20, and 10.

Solution The reliability block diagram along with the SS matrix is shown in Fig. 14 (Wang et al. 2011). Based on this SS matrix, the BDD diagram can be constructed as shown in Fig. 15. The BDD indicates the following mutually exclusive system path sets as (Wang et al. 2011)

$$\text{Path sets} = \left\{E_1 E_3, \ \overline{E}_1 E_2 E_3, \ E_1 \overline{E}_3 E_4 E_5, \ \overline{E}_1 E_2 \overline{E}_3 E_4 E_5\right\}$$

Fig. 15 BDD diagram for
the cantilever beam-bar
example (Wang et al. 2011)

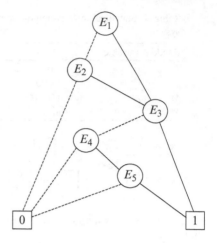

Table 5 Results of different system reliability analysis methods: GCIM and MCS
($ns = 1,000,000$)

Analysis method	System reliability level at each design									
	1	2	3	4	5	6	7	8	9	10
u_x	100	90	85	80	70	60	50	40	20	10
GCIM	0.3546	0.4981	0.5724	0.6444	0.7708	0.8666	0.9308	0.9681	0.9954	0.9995
MCS	0.3548	0.4982	0.5725	0.6445	0.7708	0.8667	0.9309	0.9681	0.9954	0.9995

The system reliability can be calculated as

$$R_{sys} = \sum_{i=1}^{4} P\left(\text{Path}_i\right)$$

$$= P\left(E_1 E_3\right) + P\left(\overline{E}_1 E_2 E_3\right) + P\left(E_1 \overline{E}_3 E_4 E_5\right) + P\left(\overline{E}_1 E_2 \overline{E}_3 E_4 E_5\right)$$

We can then use Eq. (33) to compute the reliability of each path set to derive a point system reliability estimate of this mixed system.

The system reliability analysis is carried out with 10 different loading conditions (10 different u_x values for the X) as presented in Table 5 (Wang et al. 2011). The probabilities of component and second-order joint failure events are computed with the direct MCS. The MCS is used for a benchmark solution and the results are also summarized in Table 5. We expect based on the results that the GCIM can give accurate system reliability estimates for mixed systems at various reliability levels.

6 Conclusion

The chapter reviews advanced numerical methods for system reliability analysis under uncertainty, with an emphasis on the system reliability bound methods and the GCIM (point estimation method). The system reliability bound methods provide system reliability estimates in the form of two-sided bounds for a series or parallel system, while the GCIM offers system reliability estimates in the form of single points for any system structure (series, parallel, and mixed systems). The GCIM generalizes the original CIM so that it can be used for system reliability analysis regardless of system structures. Four case studies are employed to demonstrate the effectiveness of the system reliability bound methods and the GCIM in assessing system reliability. As observed from the case studies, the GCIM offers a generalized framework for system reliability analysis and thus shows a great potential to enhance our capability and understanding of system reliability analysis.

References

Au SK, Beck JL (1999) A new adaptive importance sampling scheme for reliability calculations. Struct Saf 21(2):135–158

Breitung K (1984) Asymptotic approximations for multinormal integrals. ASCE J Eng Mech 110(3):357–366

Ditlevsen O (1979) Narrow reliability bounds for structural systems. J Struct Mech 7(4):453–472

Ferson S, Kreinovich LRGV, Myers DS, Sentz K (2003) Constructing probability boxes and Dempster–Shafer structures. Technical report SAND2002-4015, Sandia National Laboratories, Albuquerque, NM

Foo J, Wan X, Karniadakis GE (2008) The multi-element probabilistic collocation method (ME-PCM): error analysis and applications. J Comput Phys 227:9572–9595

Foo J, Karniadakis GE (2010) Multi-element probabilistic collocation method in high dimensions. J Comput Phys 229:1536–1557

Fu G, Moses F (1988) Importance sampling in structural system reliability. In: Proceedings of ASCE joint specialty conference on probabilistic methods, Blacksburg, VA, pp 340–343

Ganapathysubramanian B, Zabaras N (2007) Sparse grid collocation schemes for stochastic natural convection problems. J Comput Phys 225(1):652–685

Ghanem RG, Spanos PD (1991) Stochastic finite elements: a spectral approach. Springer, New York

Grestner T, Griebel M (2003) Dimension-adaptive tensor-product quadrature. Computing 71(1):65–87

Haldar A, Mahadevan S (2000) Probability, reliability, and statistical methods in engineering design. Wiley, New York

Hasofer AM, Lind NC (1974) Exact and invariant second-moment code format. ASCE J Eng Mech Div 100(1):111–121

Hu C, Youn BD (2011) An asymmetric dimension-adaptive tensor-product method for reliability analysis. Struct Saf 33(3):218–231

Hurtado JE (2007) Filtered importance sampling with support vector margin: a powerful method for structural reliability analysis. Struct Saf 29(1):2–15

Karamchandani A (1987) Structural system reliability analysis methods. Report no. 83, John A. Blume Earthquake Engineering Center, Stanford University, Stanford, CA

Klimke A (2006) Uncertainty modeling using fuzzy arithmetic and sparse grids. PhD thesis, Universität Stuttgart, Shaker Verlag, Aachen

Lee CY (1959) Representation of switching circuits by binary-decision programs. Bell Syst Tech J 38:985–999

Liang JH, Mourelatos ZP, Nikolaidis E (2007) A single-loop approach for system reliability-based design optimization. ASME J Mech Des 126(2):1215–1224

Mahadevan S, Raghothamachar P (2000) Adaptive simulation for system reliability analysis of large structures. Comput Struct 77(6):725–734

Möller B, Beer M (2004) Fuzzy randomness: uncertainty in civil engineering and computational mechanics. Springer, Berlin

Naess A, Leira BJ, Batsevych O (2009) System reliability analysis by enhanced Monte Carlo simulation. Struct Saf 31(5):349–355

Nguyen TH, Song J, Paulino GH (2010) Single-loop system reliability-based design optimization using matrix-based system reliability method: theory and applications. ASME J Mech Des 132(1):011005(11)

Rahman S, Xu H (2004) A univariate dimension-reduction method for multi-dimensional integration in stochastic mechanics. Prob Eng Mech 19:393–408

Ramachandran K (2004) System reliability bounds: a new look with improvements. Civil Eng Environ Syst 21(4):265–278

Rubinstein RY (1981) Simulation and the Monte Carlo method. Wiley, New York

Smolyak S (1963) Quadrature and interpolation formulas for tensor product of certain classes of functions. Sov Math Doklady 4:240–243

Song J, Der Kiureghian A (2003) Bounds on systems reliability by linear programming. J Eng Mech 129(6):627–636

Swiler LP, Giunta AA (2007) Aleatory and epistemic uncertainty quantification for engineering applications. Sandia technical report SAND2007-2670C, Joint Statistical Meetings, Salt Lake City, UT

Thoft-Christensen P, Murotsu Y (1986) Application of structural reliability theory. Springer, Berlin

Tvedt L (1984) Two second-order approximations to the failure probability. Section on structural reliability. Hovik: A/S Vertas Research

Wang P, Hu C, Youn BD (2011) A generalized complementary intersection method for system reliability analysis and design. J Mech Des 133(7):071003(13)

Wiener N (1938) The homogeneous chaos. Am J Math 60(4):897–936

Xi Z, Youn BD, Hu C (2010) Effective random field characterization considering statistical dependence for probability analysis and design. ASME J Mech Des 132(10):101008(12)

Xiao Q, Mahadevan S (1998) Second-order upper bounds on probability of intersection of failure events. ASCE J Eng Mech 120(3):49–57

Xiong F, Greene S, Chen W, Xiong Y, Yang S (2010) A new sparse grid based method for uncertainty propagation. Struct Multidisc Optim 41(3):335–349

Xiu D, Karniadakis GE (2002) The Wiener–Askey polynomial chaos for stochastic differential equations. SIAM J Sci Comput 24(2):619–644

Xu H, Rahman S (2004) A generalized dimension-reduction method for multi-dimensional integration in stochastic mechanics. Int J Numer Methods Eng 61:1992–2019

Youn BD, Choi KK, Du L (2007a) Integration of possibility-based optimization and robust design for epistemic uncertainty. ASME J Mech Des 129(8):876–882. doi:10.1115/1.2717232

Youn BD, Xi Z, Wang P (2008) Eigenvector dimension reduction (EDR) method for sensitivity-free probability analysis. Struct Multidisc Optim 37(1):13–28

Youn BD, Wang P (2009) Complementary intersection method for system reliability analysis. ASME J Mech Des 131(4):041004(15)

Youn BD, Xi Z (2009) Reliability-based robust design optimization using the eigenvector dimension reduction (EDR) method. Struct Multidisc Optimiz 37(5):474–492

Youn BD, Wang P (2009) Complementary intersection method for system reliability analysis. ASME Journal of Mechanical Design 131(4):041004(15)

Zhou L, Penmetsa RC, Grandhi RV (2000) Structural system reliability prediction using multi-point approximations for design. In: Proceedings of 8th ASCE specialty conference on probabilistic mechanics and structural Reliability, PMC2000-082

Reliability of Base-Isolated Liquid Storage Tanks under Horizontal Base Excitation

S.K. Saha and V.A. Matsagar

Abstract Reliability of base-isolated liquid storage tanks is evaluated under random base excitation in horizontal direction considering uncertainty in the isolator parameters. Generalized polynomial chaos (gPC) expansion technique is used to determine the response statistics, and reliability index is evaluated using first order second moment (FOSM) theory. The probability of failure (p_f) computed from the reliability index, using the FOSM theory, is then compared with the probability of failure (p_f) obtained using Monte Carlo (MC) simulation. It is concluded that the reliability of broad tank, in terms of failure probability, is more than the slender tank. It is observed that base shear predominantly governs the failure of liquid storage tanks; however, failure due to overturning moment is also observed in the slender tank. The effect of uncertainties in the isolator parameters and the base excitation on the failure probability of base-isolated liquid storage tanks is studied. It is observed that the uncertainties in the isolation parameters and the base excitation significantly affect the failure probability of base-isolated liquid storage tank.

1 Introduction

Liquid storage tanks are one of the most important structures in several industries, such as oil refinery, aviation, chemical industries, power generation, etc. Failure of such tanks may lead to enormous losses directly or indirectly. Several researchers reported the catastrophic failure of liquid storage tanks during past earthquakes, leading to loss of human lives as well as massive economic loss (Haroun 1983a; Rammerstorfer et al. 1990). To safeguard such important structures against devastating earthquake, base isolation is considered as an efficient technique

S.K. Saha (✉) • V.A. Matsagar
Department of Civil Engineering, Indian Institute of Technology (IIT) Delhi,
Hauz Khas, New Delhi - 110 016, India
e-mail: sandipksh@civil.iitd.ac.in

S. Kadry and A. El Hami (eds.), *Numerical Methods for Reliability and Safety
Assessment: Multiscale and Multiphysics Systems*, DOI 10.1007/978-3-319-07167-1_10,
© Springer International Publishing Switzerland 2015

(Kelly and Mayes 1989; Jangid and Datta 1995a; Malhotra 1997; Deb 2004; Shrimali and Jangid 2004; Matsagar and Jangid 2008). Several international design guidelines (AWWA D-100-96 1996; EN 1998-4 2006; API 650 2007; AIJ 2010) are available which take into account the seismic action for analysis and design of liquid storage tanks deterministically. However, the reliability evaluation of structures under dynamic loading has drawn significant attention over the recent years (Chaudhuri and Chakraborty 2006; Gupta and Manohar 2006; Padgett and DesRoches 2007; Rao et al. 2009). Few studies were carried out on seismic fragility analysis of ground-supported fixed-base and base-isolated liquid storage tanks under base excitation due to earthquake (O'Rourke and So 2000; Iervolino et al. 2004; Saha et al. 2013a). However, only limited studies were reported on the reliability assessment of the base-isolated liquid storage tanks, under base excitation (Mishra and Chakraborty 2010). On the other hand, the current design codes are gradually shifting toward the reliability-based design philosophy, which requires probabilistic analysis of structures. It therefore mandates systematic reliability analysis of liquid storage tanks.

Herein, a detailed methodology for seismic reliability analysis of base-isolated liquid storage tanks, under base excitation in horizontal direction, is proposed duly accounting for uncertainties. Failure modes for the liquid storage tanks are considered from earlier research works in accordance with international guidelines. The uncertainties of the base isolator characteristics parameters are also considered in the evaluation of the seismic reliability. Generalized polynomial chaos (gPC) expansion technique is used to determine the response statistics under random horizontal base excitation considering the uncertainties in the isolation parameters. First order second moment (FOSM) theory was used to carry out probabilistic analyses of structures in several research works (Shinozuka 1983; Ayyub and Haldar 1984; Bjerager 1990). The FOSM theory is used here for reliability assessment of base-isolated liquid storage tanks. Monte Carlo (MC) simulation is also carried out to compare applicability of the FOSM theory to estimate the probability of failure of base-isolated liquid storage tanks.

The major objectives of this book chapter are: (1) to formulate the reliability of base-isolated liquid storage tanks under base excitation in horizontal direction; (2) to compare the effectiveness of the FOSM theory, with the MC simulation, to determine the reliability of base-isolated liquid storage tanks; and (3) to investigate the effect of the uncertainties in the isolator and the excitation parameters on the reliability of base-isolated liquid storage tanks under base excitation in horizontal direction.

2 Reliability Analysis of Structures

In structural design and analysis, reliability (R_0) is conveniently described as the complement of the probability of failure (p_f). Reliability is defined as the probability that a structure will not exceed a specified limiting criterion during considered reference period or life of the structure (Ranganathan 1999). In mathematical form it is expressed as,

$$R_0 = 1 - p_f. \tag{1}$$

Let the resistance (capacity or strength) is represented by *Cap* and the demand (action of the load, i.e., shear force, moment, etc.) is represented by *Dem*. The objective for the design is to achieve an acceptable condition, i.e., $Cap \geq Dem$. Hence, the probability of failure is written as the probability of the case when $Dem > Cap$, in mathematical form,

$$p_f = P\,(Dem > Cap)\,. \tag{2}$$

Considering *Cap* and *Dem* both as random variables, the probability of failure can be computed as (Ranganathan 1999),

$$p_f = 1 - \int_{-\infty}^{\infty} f_1(Cap)\,F_2(Cap)\,d(Cap) = \int_{-\infty}^{\infty} f_2(Dem)\,F_1(Dem)\,d(Dem) \tag{3}$$

where, f_1 and f_2 denote the probability density functions (PDF), and F_1 and F_2 denote the cumulative distribution functions (CFD) of the capacity and demand, respectively. However, in real life situations, obtaining solution of such integrals may be intractable. Moreover, many times proper identification of the PDF of the demand or capacity may not even be possible. Therefore, several numerical techniques are developed over the years to estimate the reliability of structures when a closed form analytical solution is unavailable.

2.1   First Order Second Moment (FOSM) Theory

The reliability evaluation procedures are divided into three levels (Ranganathan 1999), namely (1) 1st level procedure, where the reliability is defined simply in terms of safety factors; (2) 2nd level procedure, where safety checks are carried out at the selected points on the failure boundary or the failure surface to estimate the reliability; and (3) 3rd level procedure, also known as higher order reliability analysis, where all the points on the failure surface or failure boundary are considered. The higher order reliability analyses are capable of estimating the reliability of structures most accurately. Nevertheless, the FOSM theory, which is categorized as 2nd level procedure, is widely used for its simplicity.

In the FOSM theory, the failure function $[F(Cap, Dem)]$ is defined in terms of the safety margin (S_m) as,

$$S_m = F\,(Cap, Dem) = Cap - Dem. \tag{4}$$

The reliability is estimated in terms of first and second moments of the failure function (i.e., mean and variance of S_m). It is also to be noted that when $F(Cap, Dem)$ is a nonlinear function, made up of several basic input variables, then the first order approximation is used to evaluate the mean and variance of S_m. Because of this reason, the method is known as the first order second moment (FOSM) theory. In this theory, the reliability is commonly expressed in the form of reliability index (β). Cornell (1969) defined the reliability in terms of the reliability index (β) as the ratio of the mean (μ_{S_m}) to the standard deviation (σ_{S_m}) of the safety margin as,

$$\beta = \frac{\mu_{S_m}}{\sigma_{S_m}}. \tag{5}$$

The probability of failure can be computed as,

$$p_f = \Phi^{-1}(-\beta) \tag{6}$$

where, Φ^{-1} is the inverse of the standard normal density function. For normally distributed random variables and linear failure function, this relation (Eq. (6)) calculates accurate probability of failure. Moreover, this relation gives a preliminary estimation of probability of failure for other types of distributions as well.

2.2 Monte Carlo (MC) Simulation

In many cases, the probability of failure determined from the reliability index, using the FOSM theory, provides a reasonable estimate. However, the actual distribution of the demand may not be sufficiently represented by the first and second moment, i.e., the mean and standard deviation. In those cases, in the absence of the higher order reliability theory, the MC simulation is invariably used for variety of reliability analysis problems. Although this technique is not computationally efficient, with the help of the modern computing facilities the MC simulation is a widely used technique in risk and reliability engineering. In the MC simulation, a large number of sample points are generated from the predefined probability distributions of input random variables. Failure function is formulated in terms of the demand and capacity, which consists of several input random variables. Response of the structure is obtained deterministically for each set of the input variables to check the safety. The reliability analysis procedure using the MC simulation is summarized in the following steps.

1. The failure function (F) is written in terms of n input random variables (X_i) as,

$$F = f(X_1, X_2, X_3, \ldots, X_n) \tag{7}$$

where, the probability distribution of each random variable is known.

2. Realization (x_{ik}) of each input variable is generated from its distribution, and by substituting it in Eq. (7), N_{sim} number of realizations of F is obtained. The kth realization, F_k is obtained as,

$$F_k = f(x_{1k}, x_{2k}, x_{3k}, \ldots, x_{nk}) \tag{8}$$

where, k takes the values from 1 to N_{sim}.

3. The failure criterion is checked for each set of input random variables, and the cases for which the failure occurs are counted, say N_{fail}. Then the probability of failure is computed as,

$$p_f = \frac{N_{fail}}{N_{sim}}. \tag{9}$$

Total number of the simulations (i.e., N_{sim}) depends on the required accuracy and the size of the problem. If sufficient computational facilities are available, and the probability distributions of the input random variables are known, then practically any reliability problem can be solved by the MC simulation.

Herein, the probability of failure (p_f) of base-isolated liquid storage tank is estimated from the reliability index (β) using the FOSM theory and compared with the probability of failure obtained using the MC simulation. Based on the distribution and statistics of the input random variables, the mean and standard deviation of the demand are computed using the gPC expansion technique. Subsequently, the reliability index for the base-isolated liquid storage tank is computed and the probability of failure is estimated. However, when the failure function is not linearly related to the input parameters, and not normally distributed, the FOSM theory may not provide accurate estimate of the probability of failure. In such cases, it is necessary to compare the probability of failure estimated using the FOSM theory with the probability of failure evaluated using other higher order reliability theory or the MC simulation. Here, the MC simulation is carried out using the same distributions and statistics of the input random variables, as considered in the gPC expansion technique, to evaluate and compare the probability of failure.

3   Reliability Analysis of Base-Isolated Structures

Stochastic response and the reliability analysis of base-isolated structures have received considerable attention among the research community (Jangid and Datta 1995b; Pagnini and Solari 1999; Jangid 2000; Jacob et al. 2013). To study the reliability problem of a hysteretic system, Spencer and Bergman (1985) developed a procedure using Petrov-Galerkin finite element method for the determination of statistical moments. They compared the statistical moments obtained using the proposed method with direct MC simulation, however the reliability evaluation was not carried out. Pradlwarter and Schuëller (1998) carried out reliability analysis of multi-degree-of-freedom (MDOF) system equipped with hysteresis devices such as isolators, dampers, etc. They used direct MC simulation to compute the reliability and proposed a controlled MC simulation to reduce the sample sizes with better accuracy of the reliability estimate. Scruggs et al. (2006) proposed an optimization procedure for base isolation system with active controller, considering the system reliability under stochastic earthquake. They optimized the probability of failure considering the uncertain earthquake model parameters. Mishra et al. (2013) presented a reliability-based design optimization (RBDO) procedure considering the uncertainty in the earthquake parameters as well as in the isolation system. They observed significantly higher probability of failure in case of the RBDO approach, as compared to the deterministic approach, due to uncertainty involved in the system parameters. They concluded that the optimum design parameters obtained using deterministic approach overestimate the structural reliability.

Buildings remained the major concern while analyzing the reliability of base-isolated structures in most of the previous research works. However, only limited studies reported the reliability analysis of liquid storage tanks. Mishra and Chakraborty (2010) investigated the effect of uncertainties in the isolator parameters on the seismic reliability of tower mounted base-isolated liquid storage tanks. They concluded that the uncertainty in the earthquake motion dominates the variability in the reliability; however, the uncertainties in the isolator parameters also play a crucial role in the reliability estimation. Saha et al. (2013c) presented stochastic analysis of ground-supported base-isolated liquid storage tanks considering the uncertain isolator parameters under random base excitation. They used generalized polynomial chaos (gPC) expansion technique to consider the uncertainty in isolation parameters and base excitation, and compared the probability distributions of the peak response quantities. They demonstrated the necessity of considering the uncertainty in the dynamic analysis of fixed-base and base-isolated liquid storage tanks.

4 Failure of Steel Liquid Storage Tank

Selection of failure mechanism of the structure and defining the limiting criteria are important steps in any reliability analysis. Typical earthquake induced failures observed in cylindrical steel liquid storage tanks are: (1) buckling of the tank wall, (2) rupture of tank wall in hoop tension, (3) tank roof failure, (4) sliding and uplifting of tank base, (5) failure of base plate, (6) anchorage failure, and (7) failure of connecting accessories.

Out of all the above-mentioned failure modes the buckling of tank wall received most attention in the research works, and the international guidelines differ in many ways to address this issue. Buckling of liquid filled thin walled steel tanks, under horizontal component of earthquake, is categorized into two types (EN 1998-4 2006), namely (1) elastic buckling and (2) elasto-plastic buckling. The elastic

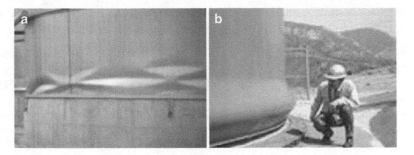

Fig. 1 (**a**) Diamond shape and (**b**) elephant foot buckling of tanks (source: NISEE e-Library)

buckling is also referred to as diamond buckling which mainly occurs under severe vertical stress induced by the overturning moment and other vertical loads. When diamond shape buckling occurs in a tank wall, the buckled region generally bends inward, forming several wrinkles in circumferential direction on the wall surface (Fig. 1a). This kind of buckling mainly occurs when the hoop tension in the tank wall is less. Such type of tank wall buckling is more common in case of slender tank, i.e., height to radius ratio is high (Niwa and Clough 1982).

Outward bulging of the tank wall under the horizontal earthquake excitation is known as elephant foot buckling (Fig. 1b). Niwa and Clough (1982) concluded from their experimental studies that elephant foot buckling occurs due to the combined action of hoop stress and axial compressive stress. When the axial compressive stress exceeded the axial buckling stress, at the same time the hoop stress was close to the material yield strength, elephant foot buckling was observed (Niwa and Clough 1982). Later, Akiyama (1992) also validated this observation by conducting a set of experiments on steel tanks. Elephant foot buckling was predominantly observed in broad tanks, i.e., for low height to radius ratio (Hamdan 2000). Such kind of tank wall buckling is classified as elasto-plastic buckling in EN 1998-4 (2006).

Some design guidelines relate the buckling of the tank wall to the axial compression developed due to the overturning moment (AWWA D-100-96 1996; API 650 2007). However, studies are reported which relate the buckling of the tank wall to the base shear in horizontal direction. Okada et al. (1995) presented a method to evaluate the effect of the shear force on the elasto-plastic buckling of cylindrical tank. Tsukimori (1996) examined the effect of interaction between the shear and bending loads on the buckling of thin cylindrical shell. Some of the present tank design guidelines provide checks for the shear buckling, along with the axial buckling (AIJ 2010).

The elastic buckling stress of a thin cylindrical tank is given by Timoshenko and Gere (1961),

$$\sigma_{cr} = \frac{1}{[3\,(1 - \nu^2)]^{0.5}} \frac{E_s t_s}{R} \qquad (10)$$

where, the Poisson's ratio and modulus of elasticity of the tank wall material are denoted by E_s and v, respectively; t_s is the tank wall thickness; and R is the radius of the tank wall. For steel tank wall ($v = 0.3$) the expression becomes,

$$\sigma_{cr} = 0.605 \frac{E_s t_s}{R}. \tag{11}$$

The elastic shear buckling stress is given by,

$$\tau_{cr} = 0.07708 \frac{\pi^2 E_s}{(1 - v^2)^{5/8} \sqrt{\frac{H}{R}}} \left(\frac{R}{t_s}\right)^{-5/4} \tag{12}$$

where, H is height of the liquid column. The limiting overturning moment ($M_{b,cr}$) and base shear ($V_{b,cr}$), based on the elastic buckling stresses, are respectively written as (Okada et al. 1995),

$$M_{b,cr} = \sigma_{cr} \pi R^2 t_s \tag{13}$$

and

$$V_{b,cr} = \tau_{cr} \pi R t_s. \tag{14}$$

5 Modeling of Base-Isolated Liquid Storage Tank

Appropriate modeling of base-isolated liquid storage tank is essential for dynamic analysis and response evaluation. Several international codes and design guidelines (AWWA D-100-96 1996; EN 1998-4 2006; API 650 2007; AIJ 2010) recommend the lumped mass mechanical analog to model cylindrical liquid storage tank. The lumped mass mechanical analog is also recommended for seismic analysis of liquid storage tanks using response spectrum approach. Simplified representation of the liquid storage tank is always required for using it routinely in the design offices. Haroun and Housner (1981) proposed a mechanical analog, with three degrees-of-freedom (DOF), for the dynamic analysis of liquid storage tanks. As per the analog, the liquid column is discretized into three lumped masses, namely (1) convective mass (m_c), lumped at height H_c above the base; (2) impulsive mass (m_i), lumped at height H_i above the base; and (3) rigid mass (m_r), lumped at height H_r above the base. The lumped mass model of a base-isolated liquid storage tank is shown in Fig. 2a. The lumped masses (m_c, m_i, and m_r) are computed from the total mass of the liquid column ($= \pi R^2 H$), neglecting the mass of the tank wall. The deterministic dynamic behavior of liquid storage tanks, using this model, was validated with experimental results by Haroun (1983b). The model was widely used in earlier research works (Shrimali and Jangid 2002, 2004; Saha et al. 2013b) for the dynamic analyses of base-isolated liquid storage tanks. Here, laminated rubber bearing (LRB)

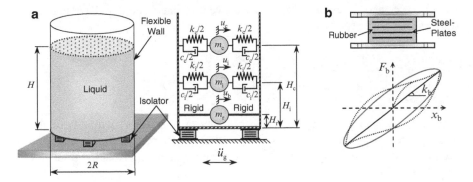

Fig. 2 (**a**) Model of base-isolated liquid storage tank and (**b**) laminated rubber bearing (LRB) with its force-deformation behavior

is considered as the isolator in the base-isolated liquid storage tank system. The linear force-deformation behavior of the LRB is shown in Fig. 2b, where F_b is the restoring force and x_b is the isolator level displacement, relative to the ground.

The matrix form of the equations of motion for the base-isolated liquid storage tank is written as,

$$\overline{M}\{\ddot{X}\} + \overline{C}\{\dot{X}\} + \overline{K}\{X\} = -\overline{M}\{r\}\ddot{u}_g \tag{15}$$

where, $\{X\} = \{x_c \ x_i \ x_b\}^T$ is the displacement vector; $x_c = (u_c - u_b)$, $x_i = (u_i - u_b)$ and $x_b = (u_b - u_g)$ are the relative displacements of the convective, impulsive, and rigid masses, respectively; and $\{r\} = \{0 \ 0 \ 1\}^T$ is the influence coefficient vector. Here, u_c, u_i, and u_b represent the absolute displacements of the convective mass, the impulsive mass, and the isolator level, respectively. The uni-directional horizontal base acceleration is denoted by $\ddot{u}_g$. The mass matrix ($\overline{M}$), the damping matrix ($\overline{C}$), and the stiffness matrix ($\overline{K}$) are expressed as follows.

$$\overline{M} = \begin{bmatrix} m_c & 0 & m_c \\ 0 & m_i & m_i \\ m_c & m_i & M \end{bmatrix} \tag{16}$$

where, $M = m_c + m_i + m_r$.

$$\overline{C} = \begin{bmatrix} c_c & 0 & 0 \\ 0 & c_i & 0 \\ 0 & 0 & c_b \end{bmatrix} \tag{17}$$

where, c_c, c_i, and c_b are damping of the convective mass, the impulsive mass, and the base isolator, respectively.

$$\overline{K} = \begin{bmatrix} k_c & 0 & 0 \\ 0 & k_i & 0 \\ 0 & 0 & k_b \end{bmatrix} \tag{18}$$

where, k_c, k_i, and k_b are stiffness of the convective mass, the impulsive mass, and the base isolator, respectively. The LRB is characterized by its viscous damping ($c_b = 4\pi M \xi_b / T_b$), where ξ_b is the damping ratio and isolation time period $\left(T_b = 2\pi \sqrt{M/k_b} \right)$.

Saha et al. (2013c) presented stochastic modeling of the base-isolated liquid storage tank using the gPC expansion technique. For simplicity, the base excitation is represented by a uni-directional sinusoidal acceleration input with random amplitude and frequency. Apart from the base excitation, the randomness in the characteristic parameters of the isolator is also considered in the stochastic modeling of the base-isolated liquid storage tank. Considering the uncertain parameters, the dynamic equations of motion (Eq. (15)) are rewritten in the matrix form as,

$$\overline{M} \left\{ \ddot{X}\left(t, \overline{\zeta}\right) \right\} + \overline{C}\left(\zeta_c\right) \left\{ \dot{X}\left(t, \overline{\zeta}\right) \right\} + \overline{K}\left(\zeta_k\right) \left\{ X\left(t, \overline{\zeta}\right) \right\} = -\overline{M}\{r\}\ddot{u}_g\left(t, \overline{\zeta}_g\right) \tag{19}$$

where, $\left\{ X\left(t, \overline{\zeta}\right) \right\}$ is the unknown displacement vector which is random in nature; ζ_c and ζ_k represent the randomness in the damping and stiffness of the base-isolation system, respectively; and the vector $\overline{\zeta}_g$ represents the randomness in the base excitation. The vector $\overline{\zeta}$ represents all the random variables involved in the system response.

6   Solution of the Stochastic Equations of Motion

Truncated gPC expansions are used to represent the uncertain damping and stiffness matrices as given by Saha et al. (2013c).

$$\overline{C}\left(\overline{\zeta}_c\right) = \sum_{i_1=0}^{N_1} \overline{c}_{i_1} \psi_{i_1}\left(\zeta_c\right) \tag{20}$$

$$\overline{K}\left(\overline{\zeta}_k\right) = \sum_{i_2=0}^{N_2} \overline{k}_{i_2} \psi_{i_2}\left(\zeta_k\right) \tag{21}$$

where, $\bar{c}_{i_1}$ and $\bar{k}_{i_2}$ are deterministic unknown coefficient matrices; $\psi_{i_1}(\zeta_c)$ and $\psi_{i_2}(\zeta_k)$ are the stochastic basis functions for damping and stiffness, respectively. Similarly, the random base acceleration and the unknown displacement vector are modeled as random fields and represented by the truncated gPC expansions as,

$$\ddot{u}_g\left(t, \bar{\zeta}_g\right) = \sum_{i_3=0}^{N_3} u_{g_{i_3}}(t)\psi_{i_3}\left(\bar{\zeta}_g\right) \tag{22}$$

$$\left\{x\left(t, \bar{\zeta}\right)\right\} = \sum_{i_4=0}^{N_4} \bar{x}_{i_4}(t)\psi_{i_4}\left(\bar{\zeta}\right) \tag{23}$$

where, at a particular time instant, $u_{g_{i_3}}(t)$ represents the deterministic unknown base excitation coefficient, and $\bar{x}_{i_4}(t)$ represents the deterministic unknown response coefficient vector. The stochastic basis functions for the excitation and the response are defined by $\psi_{i_3}\left(\bar{\zeta}_g\right)$ and $\psi_{i_4}\left(\bar{\zeta}\right)$, respectively. Here, the uncertain stiffness (k_b) of the LRB is represented in terms of the isolation time period (T_b). The base excitation in horizontal direction is assumed as sinusoidal acceleration as,

$$\ddot{u}_g\left(t, \bar{\zeta}_g\right) = A_m\left(\zeta_a\right)\sin\left[\omega\left(\zeta_\omega\right)t\right] \tag{24}$$

where, random amplitude and frequency of the base excitation is denoted by $A_m(\zeta_a)$ and $\omega(\zeta_\omega)$, respectively.

Substitution of these expansions (Eqs. 20–23) in Eq. (19) yields an approximated stochastic form of the system equations. The stochastic approximation error, denoted by $\varepsilon\left(t, \bar{\zeta}\right)$, is defined as,

$$\varepsilon\left(t, \bar{\zeta}\right) = M\sum_{i_4=0}^{N_4} \ddot{\bar{x}}_{i_4}(t)\psi_{i_4}\left(\bar{\zeta}\right) + \sum_{i_1=0}^{N_1}\bar{c}_{i_1}\psi_{i_1}\left(\zeta_c\right)\sum_{i_4=0}^{N_4}\dot{\bar{x}}_{i_4}(t)\psi_{i_4}\left(\bar{\zeta}\right)$$
$$+ \sum_{i_2=0}^{N_2}\bar{k}_{i_2}\psi_{i_2}\left(\zeta_k\right)\sum_{i_4=0}^{N_4}\bar{x}_{i_4}(t)\psi_{i_4}\left(\bar{\zeta}\right) + M\{r\}\sum_{i_3=0}^{N_3}u_{g_{i_3}}(t)\psi_{i_3}\left(\bar{\zeta}_g\right). \tag{25}$$

To solve the stochastic equation of motion, the stochastic basis function $\psi\left(\bar{\zeta}\right)$ of each input random variable must be known or defined.

Once the $\psi\left(\overline{\zeta}\right)$ s are defined, the solution of the equations is reduced to the determination of the unknown displacement vector $\overline{x}_{i_4}(t)$ by minimizing the error $\varepsilon\left(t,\overline{\zeta}\right)$. The error is deterministically equated to zero at specific points using a nonintrusive method. The nonintrusive method is same as the method of collocation points. The collocation points are generally chosen from the roots of the similar polynomials as used for the basis function. When more numbers of collocation points are required, roots of the higher order polynomials are chosen.

The response quantities of the base-isolated liquid storage tank, under base excitation in horizontal direction, are considered as the base shear (V_b) and the overturning moment (M_b). Deterministically, the base shear and the overturning moment are computed as,

$$V_b = m_c\ddot{u}_c + m_i\ddot{u}_i + m_r\ddot{u}_b \tag{26}$$

and

$$M_b = (m_c\ddot{u}_c)\,H_c + (m_i\ddot{u}_i)\,H_i + (m_r\ddot{u}_b)\,H_r. \tag{27}$$

Here, all the input random variables are assumed to be normally distributed and uncorrelated. Using 3rd order Hermite polynomial, the uncertain response quantities (V_b and M_b) are written in the following forms (Saha et al. 2013c).

$$V_b(\zeta, t) = y_0^v(t) + y_1^v(t)\zeta + y_2^v(t)\left(\zeta^2 - 1\right) + y_3^v(t)\left(\zeta^3 - 3\zeta\right) \tag{28}$$

and

$$M_b(\zeta, t) = y_0^m(t) + y_1^m(t)\zeta + y_2^m(t)\left(\zeta^2 - 1\right) + y_3^m(t)\left(\zeta^3 - 3\zeta\right) \tag{29}$$

where, $y_i^v(t)$ and $y_i^m(t)$ are the unknown deterministic coefficients at each time step corresponding to the base shear and the overturning moment, respectively.

Once the polynomial coefficients are determined, they are substituted back into Eqs. (28) and (29). Now, the response of the base-isolated liquid storage tank is expressed in terms of the uncertain input random variables, at each time step, and the response statistics are obtained. The mean and standard deviation of the base shear (μ_{V_b} and σ_{V_b}) are calculated using the following equations (Sepahvand et al. 2010).

$$\mu_{V_b} = y_0^v \tag{30}$$

and

$$\sigma_{V_b} = \sqrt{\sum_{i=1}^{3}\left(y_i^v\right)^2 h_i^2} \tag{31}$$

where, h_i^2 is the norm of the polynomial. For one-dimensional Hermite polynomial, with normally distributed uncertain parameters, $h_1^2 = 1$, $h_2^2 = 2$ and $h_3^2 = 6$.

Similarly, equations to calculate the mean and standard deviation of the overturning moment (μ_{M_b} and σ_{M_b}) are given as,

$$\mu_{M_b} = y_0^m \qquad (32)$$

and

$$\sigma_{M_b} = \sqrt{\sum_{i=1}^{3} (y_i^m)^2 h_i^2}. \qquad (33)$$

7 Numerical Studies

The reliability of base-isolated liquid storage tank under base excitation in horizontal direction is assessed through analysis of ground-supported cylindrical steel tanks with different slenderness ratio ($S = H/R$). The geometrical and material properties of the broad and slender tanks are summarized in Table 1, where ρ_s and ρ_w denote the mass density of the tank wall material and the liquid, respectively. The damping, corresponding to the convective mass and the impulsive mass, is assumed as 0.5 % and 2 %, respectively (Haroun 1983b). The type of distribution and statistics of the input parameters, considered in the present study, are presented in Table 2. The duration of the base excitation is considered as 15 s, with time increment 0.02 s, whereas the amplitude and the frequency are considered uncertain (Table 2). The uncertain parameters, which are assumed to be independent and normally distributed, are represented by the Hermite polynomial. Galerkin projection technique is used to represent the input parameters in terms

Table 1 Geometrical and material properties of tanks

Configuration	S	t_s/R	H (m)	Tank wall material	Contained liquid
Broad	0.6	0.001	14.5	Steel: $\rho_s = 7{,}800 \text{ kg/m}^3$; $E_s = 2 \times 10^5 \text{ MPa}$	Water: $\rho_w = 1{,}000 \text{ kg/m}^3$
Slender	1.85	0.001	11.3		

Table 2 Considered distribution and statistics of input parameters

Uncertain parameter	Distribution	Mean (μ)	Standard deviation (σ)
Isolation damping (ξ_b in %)	Normal	0.1	0.02
Isolation time period (T_b in sec)	Normal	2.5	0.50
Excitation amplitude (A_m in m/sec^2)	Normal	3.6	0.72
Excitation frequency (ω in rad/sec)	Normal	10	2

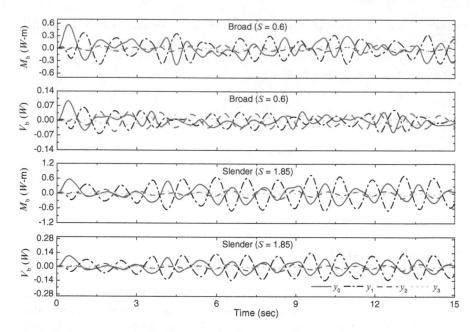

Fig. 3 Time history of the gPC expansion coefficients

of their mean and standard deviation. Nine collocation points (0, ± 0.742, ± 2.334, ± 1.3556, and ± 2.875) are chosen from the roots of the 4th and 5th order Hermite polynomials. Nine sets of the uncertain input parameters are generated from these collocation points using the Galerkin projection technique. Deterministic analyses are carried out, by numerically solving Eq. (15) using Newmark's-β method, to obtain the response of the base-isolated liquid storage tanks for the nine sets of the uncertain input parameters. Regression analysis is performed to determine the unknown coefficients of the base shear and the overturning moment (Eqs. (28) and (29)).

Figure 3 shows the time histories of the polynomial coefficients for the peak base shear and peak overturning moment, both for the broad and slender tank configurations. The coefficient y_0 represents the mean, whereas the coefficient y_1 largely contributes to the deviation of the response from the mean response. Convergence in the response calculation using the gPC expansion technique is achieved when the higher order coefficients (i.e., y_2 and y_3) are smaller in amplitude as compared to y_0 and y_1. If desired convergence in the response calculation is not achieved, then higher order approximating polynomial is to be considered. Moreover, large amplitudes of the higher order coefficients also signify the nonlinear relation between the response and the input parameters. It is observed that the coefficients y_0 and y_1 are comparable in both the response quantities for the broad and slender tank configurations. This shows the significance of the considering uncertainty in the input parameters for the dynamic analysis of base-isolated liquid

Table 3 Statistics of tank response quantities using gPC expansion

	Base shear (W)			Overturning moment (W-m)		
Configuration	μ_{V_b}	σ_{V_b}	Capacity	μ_{M_b}	σ_{M_b}	Capacity
Broad ($S = 0.6$)	0.09	0.03	0.19	0.58	0.18	14.89
Slender ($S = 1.85$)	0.11	0.04	0.11	0.58	0.20	3.84

storage tanks. Moreover, the lower contributions from the higher order coefficients, y_2 and y_3, indicate that even with 3rd order Hermite polynomial, convergence of the response calculation can be achieved in the gPC expansion technique. Nevertheless, nonzero y_2 and y_3 indicate the nonlinear relation between the peak response quantities of the base-isolated liquid storage tanks and the considered input parameters.

7.1 Computation of Reliability Index (β)

The peak of the mean base shear (μ_{V_b}) and mean overturning moment (μ_{M_b}) are determined from the time history of the response quantities, and the respective standard deviations (σ_{V_b} and σ_{M_b}) are computed at the corresponding time instant. The mean and standard deviation of the response quantities, computed using the gPC expansion technique, and the corresponding limiting values (capacity) are presented in Table 3. The limiting values are computed from Eqs. (13) and (14). The response quantities and the capacities are presented in normalized form with respect to the total weight ($W = Mg$), where g is the gravitational acceleration. The safety margin is expressed in terms of two criteria based on the limiting base shear and limiting overturning moment as,

$$S_m = V_{b,cr} - V_b \tag{34a}$$

or

$$S_m = M_{b,cr} - M_b. \tag{34b}$$

The limiting base shear ($V_{b,cr}$) and limiting overturning moment ($M_{b,cr}$) are considered as deterministic, therefore the mean of the safety margin (μ_{S_m}) is computed as,

$$\mu_{S_m} = V_{b,cr} - \mu_{V_b} \tag{35a}$$

or

$$\mu_{S_m} = M_{b,cr} - \mu_{M_b}. \tag{35b}$$

Table 4 Reliability index
(β) and probability of failure
(p_f) using FOSM theory

Configuration	Based on V_b		Based on M_b	
	β	p_f	β	p_f
Broad ($S = 0.6$)	3.157	0.0008	79.748	No failure
Slender ($S = 1.85$)	−0.015	0.5080	16.462	No failure

The standard deviation of the safety margin (σ_{S_m}) is given by,

$$\sigma_{S_m} = \sigma_{V_b} \tag{36a}$$

or

$$\sigma_{S_m} = \sigma_{M_b}. \tag{36b}$$

Once the first and second moment (i.e., the mean and standard deviation) of the safety margin are known, the reliability index (β) is computed using Eq. (5).

The reliability indices for the broad and slender tanks are computed and presented in Table 4. The reliability index corresponding to the exceedance of the base shear is considerably low which results in a high probability of failure in both the broad and slender tank configurations. However, the reliability index corresponding to the exceedance of the overturning moment is observed significantly high. Nevertheless, the probability of failure, under base excitation in horizontal direction, is more in slender tank as compared to the broad tank.

7.2 Computation of Probability of Failure (p_f) using MC Simulation

A set of realizations is generated from the considered distribution of the input parameters, as given in Table 2. For each set of the input parameters, the tank model is analyzed deterministically to obtain the peak response quantities. The peak response quantities (V_b and M_b) are then compared with the limiting base shear ($V_{b,cr}$) and limiting overturning moment ($M_{b,cr}$). The total number of the simulations, when the demand exceeds the capacity is counted as N_{fail}. The probability of failure is computed as the ratio between the numbers of failures to the total number of simulations (Eq. (9)). The number of simulations plays a crucial role in accurate estimation of the probability of failure; hence, a convergence study is carried out to find out the sufficient number of simulations. The probabilities of failure (p_f) with respect to the number of simulations are presented in Table 5. To investigate the critical failure mode, number of failures due to exceedance of the limiting base share (n_v) and limiting overturning moment (n_m) are also counted and presented in Table 5. It is observed from Table 5 that the base shear criterion governs the failure in both the broad and slender tank configurations. Figure 4 also shows the

Table 5 Convergence of probability of failure (p_f) using MC simulation

No. of simulations	Broad ($S = 0.6$)			Slender ($S = 1.85$)		
	p_f	n_v	n_m	p_f	n_v	n_m
10	0.1	1	0	0.6	5	0
100	0.06	6	0	0.46	46	0
1,000	0.079	79	0	0.517	517	9
2,000	0.0807	164	0	0.5065	1,013	11
5,000	0.0832	426	0	0.5058	2,529	33
10,000	0.0827	827	0	0.5173	5,173	86
50,000	0.0826	4,130	0	0.517	25,850	569

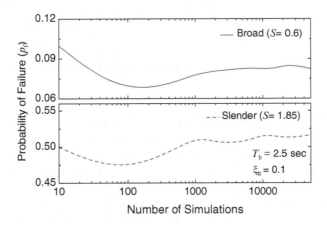

Fig. 4 Convergence of probability of failure using (p_f) MC simulation

convergence of the probability of failure with number of simulations. It is observed that with 10,000 simulations, estimation of the probability of failure converges for both broad and slender tank configurations. Further, it is observed that the probability of failure (0.0008), estimated using the FOSM theory for broad tank, is much lesser as compared to the probability of failure (0.0826) obtained from the MC simulations. However for slender tank, probability of failure estimated using the FOSM theory (0.508) is similar to the probability of failure obtained from the MC simulations.

To explain this observation, the probability distribution of the peak base shear is plotted from the peak response computed using 50,000 MC simulations and compared with the distribution obtained through the gPC expansion technique. To obtain the probability distribution of the base shear using the gPC expansion, 50,000 standard normal variates are generated, and the response time history of the base shear is generated using Eq. (28). Figure 5 shows the comparison of the probability distributions of the base shear for both the broad and slender tank configurations. The limiting base shears (capacities) of the broad and slender tanks are also plotted for comparison purpose. It is observed that the overall distribution of the peak

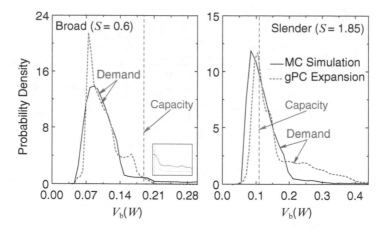

Fig. 5 Comparison of capacity and peak base shear distribution using gPC expansion and MC simulation

base shear, obtained using the gPC expansion technique closely matches to that predicted by the MC simulation. The computed limiting base shear (capacity) of the broad tank is significantly higher than the peak base shear demand, providing a higher safety margin. Furthermore, ordinates of the probability density for the broad tank, obtained from the MC simulation, differ from that obtained using the gPC expansion technique, specifically near the peak region and beyond the base shear capacity. Beyond the base shear capacity, the MC simulation distribution curve has considerably lower ordinates, as compared to the distribution curve using the gPC expansion technique. As the peak responses, greater than the capacity, are only considered in the computation, the deviation in the distribution in this region significantly influences the failure probability estimation. Hence, significant difference in the probability of failure estimation, using the FOSM theory and the MC simulation, for the broad tank, is observed.

On the other hand, in case of the slender tank the probability distributions of the peak base shear, obtained using the FOSM theory and the MC simulation, are matching closely near the peak region. Moreover, the base shear capacity of the slender tank is close to the demand with lesser safety margin. Hence, the marginal deviation in the peak response distribution, beyond the capacity, does not significantly increase the numbers of failures due to base shear exceedance. Owing to this fact, similar failure probability estimations are obtained, by the FOSM theory using the gPC expansion technique and the MC simulation, for the slender tank. It is concluded that the accuracy in estimating the probability of failure, for a base-isolated liquid storage tank, largely depends on the actual distribution of the peak response quantities and the available safety margin. Probability of failure estimated from the mean and standard deviation, using the FOSM theory, may not provide reasonable accuracy for base-isolated liquid storage tanks. Therefore, only the MC simulation is used to estimate the probability of failure in the following studies.

Table 6 Influence of individual uncertain parameter on probability
of failure (p_f) using MC simulation

	Probability of failure (p_f)	
Uncertain parameter	Broad ($S = 0.6$)	Slender ($S = 1.85$)
Isolation damping (ξ_b)	0	0.809
Isolation time period (T_b)	0.0200	0.5207
Excitation amplitude (A_m)	0.0668	0.5323
Excitation frequency (ω)	0.0174	0.519

7.3   Effect of Individual Parameter Uncertainty on Probability of Failure (p_f)

The effect of uncertainty in each parameter on the probability of failure of the base-isolated liquid storage tanks is investigated. The peak response quantities are obtained considering uncertainty in one parameter only at a time, while the other parameters are considered deterministic. The values of each uncertain parameter, taken in the analysis, are presented in Table 2. The mean values are taken as the deterministic inputs, while the standard deviations of the parameters are taken as zero, except for the parameter under consideration. The MC simulation is used to obtain the probability of failure (p_f) with 10,000 realizations of the input variables. The number of simulations is considered 10,000 to avoid unnecessary computational effort since reasonable convergence of the probability of failure (p_f) is observed in Fig. 4. Table 6 presents the variation of the probability of failure with respect to the individual uncertain parameters. It is observed that the effect of the individual uncertain parameters is significant on the probability of failure (p_f) of the tanks. It is also observed that for the broad tank, probability of failure, estimated considering uncertainty only in the isolation damping, is lower as compared to the probability of failure when uncertainty is considered in the other input parameters. However for the slender tank, probability of failure, estimated considering uncertainty only in the isolation damping, is higher as compared to the probability of failure when uncertainty is considered in the other input parameters.

To explain this disparity, the distribution of the peak base shear is plotted in Fig. 6 for broad and slender tank configurations. It is observed that the simulated values of the peak base shear are distributed around the deterministic peak value (0.098 W). However, the effect of the uncertain damping is insignificant on the distribution of the peak base shear, and the simulated peak responses are distributed within a narrow band. For the broad tank, base shear capacity (0.19 W) is considerably higher than the mean peak base shear (0.099 W) when uncertainty is considered only in the isolation damping. Therefore, no failure is observed in case of the broad tank. However for the slender tank, the base shear capacity is same as the deterministic peak base shear (0.11 W). The maximum probability density of the peak base shear (with the mean value as 0.112 W) is also observed around the deterministic value for the slender tank when the uncertainty is considered only in the isolation damping.

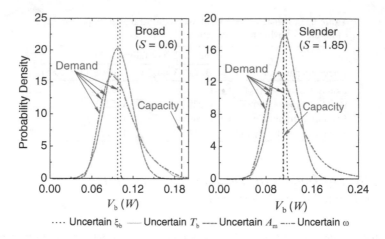

Fig. 6 Comparison of capacity and peak base shear distribution for uncertainty in individual input parameter using MC simulation

With marginal variation in the isolation damping, most of the peak responses exceed the capacity. Consequently, the probability of failure in case of the uncertain damping is evaluated to be significantly higher in case of the slender tank. Therefore, it is concluded that when the capacity is significantly more than the demand, the effect of uncertainty in the isolation damping is negligible. However, when the demand is marginally more than the capacity, uncertainty in the isolation damping significantly affects probability of failure.

7.4   Effect of Level of Uncertainty on Probability of Failure (p_f)

The effect of the level of uncertainty on the probability of failure of the base-isolated liquid storage tanks is also investigated. The standard deviation, in terms of the % mean, is used to quantify the levels of uncertainty in each parameter. The range of the standard deviation is taken as 5–20 %, with an increment of 5 %, simultaneously for all the input parameters. In Fig. 4, it is shown that 10,000 MC simulations are sufficient to obtain convergence in the evaluation of probability of failure (p_f). Therefore, the MC simulation is used to obtain the probability of failure (p_f) with 10,000 realizations of the input variables at each level of the uncertainty for broad ($S = 0.6$) and slender ($S = 1.85$) tank configurations. Table 7 presents the variation in the probability of failure of the base-isolated broad and slender tanks with increasing level of uncertainty. It is observed that the probability of failure in the broad tank increases with the increase in the uncertainty, whereas the probability of failure in the slender tank decreases with increase in the uncertainty. The distributions of the peak base shear for the broad and slender tank configurations,

Table 7 Effect of uncertainty level on probability of failure (p_f) using MC simulation

Uncertainty level (all parameters, in % mean)	Probability of failure (p_f)	
	Broad ($S = 0.6$)	Slender ($S = 1.85$)
5	0	0.5643
10	0.0032	0.5283
15	0.0367	0.5177
20	0.0827	0.5173

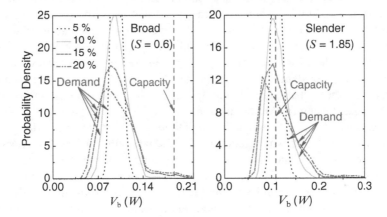

Fig. 7 Comparison of capacity and peak base shear distribution for different levels of uncertainty in input parameters using MC simulation

at different levels of uncertainty in the input parameters, are shown in Fig. 7 to explain the observation. In case of the broad tank, with increase in the uncertainty level, more number of times the peak base shear exceeds the capacity which leads to increase in the probability of failure. It is also observed that with increasing uncertainty level in the isolator and the excitation parameters, the peak region of the peak base shear distribution shifts toward lower value, disturbing the symmetry of the distribution. Moreover, for the slender tank base shear capacity is same as the deterministic peak base shear (0.11 W). Owing to this fact, higher uncertainty in the input parameters leads to lesser numbers of cases, when the base shear demand exceeds the capacity. As a result, lower probability of failure is observed with increasing uncertainty in the input parameters for slender tank. Therefore, it is concluded that the higher level of uncertainties in the isolator and the excitation parameters disturb the symmetry of the peak base shear distribution of base-isolated liquid storage tanks. It is also concluded that the effects of the level of uncertainty on the probability of failure of the base-isolated liquid storage tanks depend on the difference between the designed capacity and demand (or safety margin) for the tanks.

8   Summary and Conclusions

The first order second moment (FOSM) theory, in combination with the generalized polynomial chaos (gPC) expansion technique, is used to determine the reliability of base-isolated liquid storage tanks under base excitation in horizontal direction. The failure of the ground-supported cylindrical steel tanks is defined in terms of the limiting base shear and overturning moment in the elastic range. The effectiveness of the FOSM theory to estimate the seismic reliability is compared with the MC simulation. It is concluded that the accuracy in estimating the probability of failure, for base-isolated liquid storage tank, largely depends on the actual distribution of the peak response quantities and the available safety margin. Probability of failure estimated using the FOSM theory may not provide reasonable accuracy for base-isolated liquid storage tanks. The effect of uncertainties in the isolator and excitation parameters on the probability of failure (p_f) of base-isolated liquid storage tanks is also investigated using the MC simulation.

It is concluded that the probability of failure (p_f) is more in slender tank as compared to broad tank which indicates that the reliability of broad tank is more than the slender tank. The base shear predominantly governs the failure in both broad and slender tank configurations. The uncertainty of the isolation time period and the base excitation significantly influence the failure probability of the base-isolated broad and slender tanks. Further, it is concluded that when the demand is marginally more than the capacity, failure probability increases with the increase in isolation damping uncertainty. However, when the capacity is significantly more than the demand, the effect of uncertainty in the isolation damping is negligible. It is observed that the higher level of uncertainties in the isolator and excitation parameters disturb the symmetry of the peak base shear distribution of base-isolated liquid storage tanks. It is concluded that the effects of the level of uncertainty on the probability of failure of the base-isolated liquid storage tanks depend on the safety margin available for the tanks.

References

AIJ (2010) Design recommendation for storage tanks and their supports with emphasis on seismic design. Architectural Institute of Japan, Tokyo

Akiyama H (1992) Earthquake resistant limit state design for cylindrical liquid storage tanks. 10th World conference on earthquake engineering, Madrid, Spain

API 650 (2007) Welded storage tanks for oil storage. American Petroleum Institute (API) Standard, Washington, DC

AWWA D-100-96 (1996) Welded steel tanks for water storage. American Water Works Association (AWWA), Denver, CO

Ayyub B, Haldar A (1984) Practical structural reliability techniques. J Struct Eng ASCE 110(8):1707–1724

Bjerager P (1990) On computation methods for structural reliability analysis. Struct Saf 9(2):79–96

Chaudhuri A, Chakraborty S (2006) Reliability of linear structures with parameter uncertainty under non-stationary earthquake. Struct Saf 28(3):231–246

Cornell CA (1969) A probability-based structural code. J Am Concr Inst 66(12):974–985

Deb SK (2004) Seismic base isolation-an overview. Curr Sci 87(10):1426–1430

EN 1998-4 (2006) Eurocode 8: design of structures for earthquake resistance-part 4: silos, tanks and pipelines, Brussels, Belgium

Gupta S, Manohar CS (2006) Reliability analysis of randomly vibrating structures with parameter uncertainties. J Sound Vib 297(3–5):1000–1024

Hamdan FH (2000) Seismic behaviour of cylindrical steel liquid storage tanks. J Construct Steel Res 53(3):307–333

Haroun MA (1983a) Behavior of unanchored oil storage tanks: Imperial Valley earthquake. J Tech Top Civil Eng 109(1):23–40

Haroun MA (1983b) Vibration studies and tests of liquid storage tanks. Earthq Eng Struct Dyn 11(2):179–206

Haroun MA, Housner GW (1981) Earthquake response of deformable liquid storage tanks. J Appl Mech 48(2):411–418

Iervolino I, Fabbrocino G, Manfredi G (2004) Fragility of standard industrial structures by a response surface based method. J Earthq Eng 8(6):927–945

Jacob CM, Sepahvand K, Matsagar VA, Marburg S (2013) Stochastic seismic response of an isolated building. Int J Appl Mech 5(1):1350006-1–1350006-21. doi:10.1142/S17588 25113500063

Jangid RS (2000) Stochastic seismic response of structures isolated by rolling rods. Eng Struct 22(8):937–946

Jangid RS, Datta TK (1995a) Seismic behavior of base-isolated buildings: a state-of-the-art-review. Struct Build 110(2):186–203

Jangid RS, Datta TK (1995b) Performance of base isolation systems for asymmetric building subject to random excitation. Eng Struct 17(6):443–454

Kelly TE, Mayes RL (1989) Seismic isolation of storage tanks. In: Seismic engineering: research and practice (structural congress 89), CA, USA, pp 408–417

Malhotra PK (1997) Method for seismic base isolation of liquid-storage tanks. J Struct Eng ASCE 123(1):113–116

Matsagar VA, Jangid RS (2008) Base isolation for seismic retrofitting of structures. Pract Period Struct Des Construct ASCE 13(4):175–185

Mishra SK, Chakraborty S (2010) Reliability of base isolated liquid storage tank under parametric uncertainty subjected to random earthquake. In: 14th Symposium on earthquake engineering, Roorkee, India, pp 869–883

Mishra SK, Roy BK, Chakraborty S (2013) Reliability-based-design-optimization of base isolated buildings considering stochastic system parameters subjected to random earthquakes. Int J Mech Sci 75:123–133

NISEE e-Library The earthquake engineering online archive: Karl V. Steinbrugge collection. University of California, Berkeley, USA. http://nisee.berkeley.edu/elibrary

Niwa A, Clough RW (1982) Buckling of cylindrical liquid-storage tanks under earthquake loading. Earthq Eng Struct Dyn 10(1):107–122

O'Rourke MJ, So P (2000) Seismic fragility curves for ongrade steel tanks. Earthq Spectra 16(4):801–815

Okada J, Iwata K, Tsukimori K, Nagata T (1995) An evaluation method for elastic-plastic buckling of cylindrical shells under shear forces. Nucl Eng Des 157(1–2):65–79

Padgett JE, DesRoches R (2007) Sensitivity of response and fragility to parameter uncertainty. J Struct Eng ASCE 133(12):1710–1718

Pagnini LC, Solari G (1999) Stochastic analysis of the linear equivalent response of bridge piers with aseismic devices. Earthq Eng Struct Dyn 28(5):543–560

Pradlwarter HJ, Schuëller GI (1998) Reliability of MDOF-systems with hysteretic devices. Eng Struct 20(8):685–691

Rammerstorfer FG, Fisher FD, Scharf K (1990) Storage tanks under earthquake loading. Appl Mech Rev 43(11):261–282

Ranganathan R (1999) Structural reliability analysis and design. Jaico Publishing House, Mumbai

Rao BN, Chowdhury R, Prasad AM, Singh RK, Kushwaha HS (2009) Probabilistic characterization of AHWR inner containment using high dimensional model representation. Nucl Eng Des 239(6):1030–1041

Saha SK, Matsagar VA, Jain AK (2013a) Seismic fragility of base-isolated industrial tanks. 11th International conference on structural safety and reliability, New York, USA

Saha SK, Matsagar VA, Jain AK (2013b) Comparison of base-isolated liquid storage tank models under bi-directional earthquakes. Nat Sci Special Issue Earthq 5(8A1):27–37

Saha SK, Sepahvand K, Matsagar VA, Jain AK, Marburg S (2013c) Stochastic analysis of base-isolated liquid storage tanks with uncertain isolator parameters under random excitation. Eng Struct 57:465–474

Scruggs JT, Taflanidis AA, Beck JL (2006) Reliability-based control optimization for active base isolation systems. Struct Control Health Monitor 13(2–3):705–723

Sepahvand K, Marburg S, Hardtke H-J (2010) Uncertainty quantification in stochastic systems using polynomial chaos expansion. Int J Appl Mech 2(2):305–353

Shinozuka M (1983) Basic analysis of structural safety. J Struct Eng ASCE 109(3):721–740

Shrimali MK, Jangid RS (2002) Non-linear seismic response of base-isolated liquid storage tanks to bi-directional excitation. Nucl Eng Des 217(1–2):1–20

Shrimali MK, Jangid RS (2004) Seismic analysis of base-isolated liquid storage tanks. J Sound Vib 275(1–2):59–75

Spencer BF Jr, Bergman LA (1985) On the reliability of a simple hysteretic system. J Eng Mech 111(12):1502–1514

Timoshenko SP, Gere JM (1961) Theory of elastic stability. McGraw-Hill Book Company, Singapore

Tsukimori K (1996) Analysis of the effect of interaction between shear and bending loads on the buckling strength of cylindrical shells. Nucl Eng Des 167(1):23–53

Robust Design of Accelerated Life Testing and Reliability Optimization: Response Surface Methodology Approach

Taha-Hossein Hejazi, Mirmehdi Seyyed-Esfahani, and Iman Soleiman-Meigooni

Abstract Due to cost and time savings and improving reliability, accelerated life tests are commonly used; in which some external stresses are conducted on items at higher levels than normal. Estimation and optimization of the reliability measure in the presence of several controllable and uncontrollable factors becomes more difficult especially when the stresses interact. The main idea of this chapter is employing different phases of response surface methodology to obtain a robust design of accelerated life testing. Since uncontrollable variables are an important part of accelerated life tests, stochastic covariates are involved in the model. By doing so, a precise estimation of reliability measure can be obtained. Considering the covariates as well as response surface methodology simultaneously are not addressed in the literature of accelerated life test. This methodology can be used on the conditions that a broad spectrum of variables is involved in the accelerated life test and the failed units have a massive cost for producers. Though considering covariates in the experiments, the optimization of reliability can generate more realistic results in comparison with noncovariates model. For the first step of this study, experimental points using D-optimal approach are designed to decrease the number of experiments as well as the prediction variance. The reliability measure is estimated under right censoring scheme by Maximum likelihood estimator (MLE) assuming that lifetime data have an exponential distribution with parameter, λ, depending on the design and stress variables as well as covariates. In order to find the best factor setting that leads to the most reliable design, response surface methodology is applied to construct the final mathematical program. Finally, a numerical example is analyzed by the proposed approach and sensitivity analyses are performed on variables.

T.-H. Hejazi (✉) • M. Seyyed-Esfahani • I. Soleiman-Meigooni
Department of Industrial Engineering and Management Systems, Amirkabir University
of Technology (Tehran Polytechnic), Tehran, Iran
e-mail: t.h.hejazi@aut.ac.ir

S. Kadry and A. El Hami (eds.), *Numerical Methods for Reliability and Safety Assessment: Multiscale and Multiphysics Systems*, DOI 10.1007/978-3-319-07167-1_11, © Springer International Publishing Switzerland 2015

1 Introduction

In the recent years, various industries are encountered with a tremendous growth in competition. In this regard, producers have paid a worthy attention to develop and design new products that satisfy the consumers' expectations. Producers should consider different characteristics in their products to increase their share in the market. Reliability is one of the most important quality dimensions and plays a pivotal role in increasing lifetime of products, reducing warranty costs, and achieving expected product's functions. Therefore, manufacturers have focused their attention to design products with high quality and reliability to decrease the number of failures in the warranty period. For this purpose, various tests should be applied and many constraints would be considered in this field such as the high cost of products that fail tests, test duration, and material limitations. Accelerated life testing (ALT) is a well-grounded method to save cost and achieve lifetime data in a shorter time period. Units in ALT are subjected to higher stress levels, and the output data of ALT are used for the estimation of product life at normal conditions. ALT can generate many opportunities for manufacturers, such as a proper maintenance scheduling for products to be kept in an acceptable level of reliability and determining a reasonable warranty period to reduce the warranty costs (Smith 1983).

Reliability can be defined as a probability of expected performance of a unit in a normal and operating condition as well as a predefined time interval. Reliability improvement is a vital part of product quality development. Although quality can be illustrated in quite a few ways, one statement which always is acceptable among the definitions is: a not reliable product is not a high quality one. Reliability function is defined as Eq. (1)

$$R(t) = 1 - F(t) \tag{1}$$

where $F(t)$ is the cumulative distribution function of lifetime. Several methods have been applied to predict and evaluate the various aspects of product's reliability. The mentioned methods collect different data and use them to estimate and analyze reliability measurements.

One of the most important issues in reliability analysis is the use of historical or experimental data to estimate product's lifetime. Due to the mentioned constraints such as long time needed for running the test and high cost of failed units, accelerated life test has been introduced to overcome the problems. Accelerated life testing is based on the principle that products have the same behavior in both conditions of high stress in short time and low stress in longer time. The aim of such testing is to quickly obtain failure data which, properly modeled and analyzed, yield desired information on product life or performance under normal use (Nelson 2009). It can be inferred from the definition of ALT that in the first phase examiner specifies the lifetime of units at the high stress condition and in the second phase estimates the lifetime of units at the normal condition. It is clear that the key point

Fig. 1 Constant stress
accelerated life test

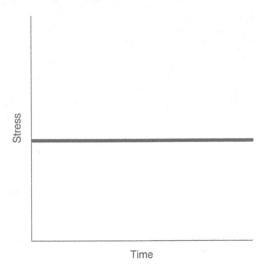

in designing an optimal ALT is to construct a robust and appropriate relationship model between estimated life time in tests and the normal conditions. With all these taken into account, in order to design an adequate ALT considering some prerequisites, namely types of stress loading, relationships between life and stress, types of variables, censored data, and types of estimation methods are absolutely essential. In this regard, in the next subsections a brief review on the mentioned topics is presented.

1.1 Stress Loading

To begin with, in order to construct an ALT a broad spectrum of loads can be considered. The different types of stress loading can be categorized into two main classes with respect to dependency of stresses to time factor. The first class includes those types of stresses loading which are independent from time, in more precisely definition, in Constant Stress ALT (CSALT), items are tested at a specified constant stress level which does not vary during the experiments. Since these tests are simple to perform and have many merits such as available models, CSALT is widely used in reliability tests. Tang et al. (2002) considered two alternative ways of planning CSALT with three stress levels which optimize both stress levels and sample allocations. In Fig. 1 a constant stress accelerated life test is shown.

In second class, the stresses on the units are dependent on time and increase in a discrete way. In other words, at the initial time of test, units are subjected to a low stress level and when a specific time period is passed, the stress level is increased to a higher value. This process can be continued based on practical constraints of experiments. This type of ALT is called Step Stress ALT (SSALT). Firstly, Geol in 1971 introduced the implication of step stress partially accelerated life test.

Fig. 2 Step stress
accelerated life test

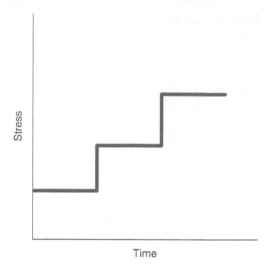

Miller and Nelson (1983) presented the optimal design of simple SSALT with considering some assumptions, namely, one stress variable, complete data, and exponential lifetime distribution. Khamis (1997) proposed optimal design of accelerated life test by considering K stress variables and M levels. It is assumed that lifetime has an exponential distribution and a complete information about life–stress relationship has been existed. Li and Fard (2007) proposed an optimum step-stress accelerated life test for two stress variables which include censored data. Fard and Li (2009) derived a simple Step-stress ALT (SSALT) model to determine the optimal hold time at which the stress levels changed; they assumed a Weibull distribution for failure time. In Fig. 2 a SSALT is demonstrated.

The last class of stress loading is Progressive Stress ALT (PSALT). In PSALT units are exposed under a stress which is a nondecreasing continuous function of time. In Fig. 3 a PSALT is shown.

1.2 Relationship Between Stress and Life

After determining the type of stress loading, experimenters should make a very important decision about the relationship between stress and life. In fact, in this phase of designing an accelerated life test and to achieve an accurate estimation of life time, practical conditions of problem should be considered in the model. Hence, selecting a proper relationship model between stress and life is the core of ALT design. Therefore, verification of other computations would be dependent on the validation of this model. Two main steps should be completed to construct such a relationship model. In the first step an appropriate lifetime distribution should

Fig. 3 Progressive accelerated life test

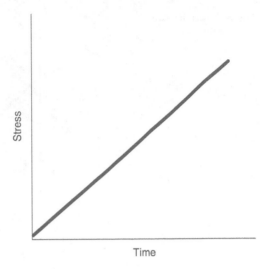

Fig. 4 Life–stress relationship

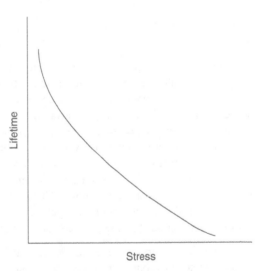

be determined for units based on the practical conditions and experts' opinions. At the second step the appropriate relationship between stress variables and the parameter(s) of lifetime distribution is derived. It should be considered that the parameter(s) of lifetime is a function of stress and not the own lifetime. In Fig. 4 a relationship between lifetime and stress is demonstrated. According to this figure, by increasing the stress the lifetime will decrease. Through a change to logarithmic scale, Fig. 5 is obtained.

Some of the most common models for the relationship between stress and life are summarized in Table 1.

Fig. 5 Linear life–stress
relationship

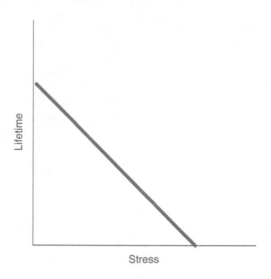

1.3   Types of Variables

In order to find an appropriate relationship model between stress and life, variables of model should be chosen with respect to experts' ideas and historical data. Classification of variables can be viewed from different aspects. In a point of view, some of the variables are input variables and some of them are output variables (responses) which are dependent on the inputs. However, this classification should be studied more precisely. In the one hand, some variables are known, controllable, and are of interest so should be set at ideal levels to optimize response variables. On the other hand, some variables are not of interest as a factor but exist in the model, which are called nuisance variables. These variables are very common in practical conditions and ignoring them may generate a noticeable inaccuracy in the model. Montgomery (2005) has categorized such variables into the following three groups.

1. In the first group, variables are known and controllable but are not of interest as a factor in model. Different techniques are available to handle these kinds of variables such as blocking method. These statistical techniques can eliminate effects of such variables on the results of experiments. The variation generated through changing the shifts of workers is an example of such variables.
2. In the second group, variables are known and uncontrollable. This type of variables is usually called covariates, which have a significant effect on the results of model. It should be noted that the value of such variables is measured during the experiments. Considering covariates and whose interactions with other variables have a profound impact on improving the modeling of responses. The chemical and physical properties of raw materials are examples for such variables.

Table 1 Life–stress models

Relationship	Use conditions	Model name
$\ln(L(V)) = \ln(C) + \frac{B}{V}$	When stress is thermal (for example, temperature)	Arrhenius
$L(V) = \frac{1}{V} \times e^{-\left(A - \frac{B}{V}\right)}$	When stress is thermal or humidity	Eyring
$\ln(L) = -\ln(K) - n\ln(V)$	When stress is nonthermal (for example, voltage)	Inverse power law
$\ln(L(V, U)) = \ln(A) + \frac{\varnothing}{V} + \frac{b}{U}$	When stresses are temperature and humidity	Temperature–humidity
$\ln(L(U, V)) = \ln(C) - n\ln(U) + \frac{B}{V}$	When a temperature stress and a nonthermal stress are conducted on units	Temperature–nonthermal
$\ln(L) = \alpha_0 + \alpha_1 X_1 + \alpha_2 X_2 + \cdots + \alpha_n X_n$	When experiment is involved of multiple stresses	General log-linear
$\lambda\left(t, X\right) = \lambda_0(t) \cdot e^{\alpha_1 X_1 + \alpha_2 X_2 + \cdots + \alpha_m X_m}$	Widely used in medical fields and has new applications in reliability engineering. Cox considers a great number of stresses in model	Proportional hazards (Cox)

3. In the last group, variables not only are unknown but also are uncontrollable. This
   type of variables is not detectable and their levels randomly change during the
   experiment. The randomization technique is the main approach used to decrease
   the effects of such variables on the outputs. Humidity and climate variables are
   examples for this group of nuisance variables.

1.4 Censored Data

Through considering three last implications, a parametric relationship model
between stress and life can be constructed. However, the parameters of this basic
model are to be estimated. In an ideal condition, models' parameters (coefficients)
are estimated using complete life data. The term of complete indicates that the
failure data for all of test units are available. However, due to quite a few practical
constraints such as time and cost limitation often a few number of failure data are
missed. These missed data are called censored data (Chenhua 2009). Censored
data play a crucial role in reliability testing, in general, and ALT, in particular.
Therefore, removing censored data from the analysis will decrease the validity of
estimations and accuracy of the results. Considering the mentioned reasons, it is
clear that censored data should be considered in a model to reach a better design
and optimized accelerated life test. Through considering the censored data a proper
modification can be performed on estimation methods, which are introduced in next
subsection, based on the types of censored data. These modifications significantly
affect the accuracy of estimation in a model. In this regard, some important types of
censored data will be presented as follows (Nelson 2009; Lewis 1987).

1. *Left censoring*: A failure time is lower than a point of inspection so it cannot be
   determined.
2. *Right censoring*: Some units are not failed in the experiment period. Right
   censoring is most common censoring scheme in reliability testing. The reasons
   for right censoring are as follows: (a) units exit from experiment for any reasons
   before that failure occurs. (b) In the end of test some units work correctly. (c)
   Units exit from experiment due to failure modes out of the purpose of the study.
3. *Interval censoring*: Time is missed because the units fail between two specific
   inspection times.

Ling et al. (2011) presented a SSLAT for two stress variables to find optimal
times for to change stress levels based on type-censoring plan. In the type-censoring
the censoring scheme has a predetermined termination time. They supposed that
lifetime has an exponential distribution and used MLE as the estimator method. Ng
et al. (2012) studied the estimation of three-parameter Weibull distribution under
progressively type II censored samples. In the type censoring the experiment will
be terminated after occurrence of a specific number of failures. To obtain reliable
initial estimates for iterative procedures they also proposed application of censored

estimation technique with one-step bias correction. Wang et al. (2012) presented a step-stress partially accelerated life tests to estimate the parameters of Weibull distribution which considered multiply censored data.

1.5 Methods of Estimation

In the last step, the approximate relationship model between responses and independent variables should be constructed. An adequate regression model such as first and second order based on the aforementioned criteria should be selected and the unknown parameters are estimated. The following three methods are prevalent for obtaining the point estimator: (a) method of moments, (b) Bayes method, and (c) method of maximum likelihood (Ramachandran and Tsokos 2009). The most important and common method of point estimation is maximum likelihood estimation (MLE). MLE has many suitable properties and can be applied for a vast area of models and data. MLE such as other estimation methods has a simple and effective theorem behind their often massive computations. In MLE method the likelihood function of the occurred events is maximized. Although MLE has a specific algorithm to obtain estimations, a great number of techniques and methods can be used in its different phases. The process of MLE can be divided into two main phases. In the first phase, the likelihood function is demonstrated by the occurrences' probability of observed events and in the second phase the obtained function is maximized with respect to unknown parameters of the distribution function. The maximization process can be done through a wide range of techniques. In the following section, the mentioned process will be illustrated more precisely.

2 Problem Definition

2.1 Purposes of Study

The main purpose of this study is to obtain a precise estimation of reliability measurements considering stochastic covariates. To explain in more details, covariates are mostly ignored in ALT models, and this problem can drastically decline the quality of models' estimations. With respect to the mentioned reason, it can be inferred that covariates should be considered in ALT models with an equal importance as other variables. In addition, quite a few number of issues are involved in constructing a robust accelerated life test such as censored data, interaction between variables, high cost of failed units, and too name but a few. To satisfy these objectives quite a few of methods and techniques are used in the presented methodology and most of practical conditions of ALT are considered in the model. In order to cope with the mentioned problems in this study, Response

Surface Methodology is applied to design ALT models. In normal situations units perform whose functions in presence of more than one stress variable. However, the mentioned operational condition is ignored due to difficulty of computations in estimation process usually in researches. As a result, considering more than one stress variable in the model can develop the design of ALT to practical conditions. In addition, in the proposed approach ALT is designed with considering the censored data. By accepting this as a true that due to the lengthy time of many experiments, censored data is not avoidable, paying attention to censored data has a significant effect on accuracy of predictions. Through performing different phases of RSM a significant improvement in results of estimation and a noticeable decrease in number of tests will be achieved. Moreover, an optimal region that minimizes the prediction variance of estimations will be obtained using optimal designs in the first phase of RSM. In Sect. 3, the steps of RSM are explained and some examples are shown for more illustration.

2.2 Model Assumptions

Quite a few of assumptions are considered in the proposed approach to design an optimal ALT. Some of which are principal assumptions that the validation of model is dependent on these. One of the most important problems in designing an ALT is extrapolation errors result from the difference between the operational stress and accelerated stress. In order to overcome this error, some notions should be considered before designing ALTs. To begin with, the relationship between stress and life should be modeled with respect to experts' opinions as well as physical properties of units. In this regard, various information about the failures mechanism of units should be collected and analyzed by the experts. Furthermore, some essential assumptions should be considered to design ALTs. First, the life time distribution is known and remains unchanged for any level of stresses. Second, the obtained model for failure data in accelerated levels has the same behavior in operational level. As a result, the extrapolation for data in operational levels has an acceptable validation for calculated parameters. The other aspect that should be noticed to design ALTs would be the dependency of lifetimes' parameter to time. To explain in more details, units or products have a life cycle based on their failure rate. The cycle is divided into three periods. In "burn-in" period, the failure rate will be decreased with passage of time. In the "useful life" period, units have a constant failure rate. In fact, the failure rate in the useful life period is independent of time. In the last period, which is called wear out, with passage of time the failure rate will be increased.

The duration of useful life period is much greater than burn-in and wear-out periods. In addition, the failures that occur in burn-in period are due to initial problems and mistakes in design of product. Usually by a proper redesign process the mentioned failures can be removed after passing a short time. The wear-out period includes units approaching to the end of life. In this regard, designing reliability tests in burn-in and wear-out periods has less importance than in useful-life period. Hence, mostly the designs of ALT models are performed in useful life period. Therefore, accepting the assumption that failure rate is independent of time is a reasonable hypothesis.

In addition to the mentioned principal assumptions, this study includes additional assumptions on the conditions of experiments. To begin with, it is assumed that stresses are constant during running of the experiment. In other words, the model is a constant stress accelerated life test, in which stresses do not vary with passage of time. A right censoring plan is assumed for data regarding the censoring scheme. In addition, it is assumed that the lifetime parameters are a function of design and stress variables, as well as covariates.

The proposed approach has the advantage that is not dependent on the lifetime distribution and according to condition of problem different distributions can be chosen for lifetime data.

3 Methodology

With respect to mentioned issues in the last section, an effective approach to design an accelerated life test is proposed in this section. In the presented approach to design an optimized ALT, three main steps should be performed. The steps can be explained as follows: firstly an appropriate approximate relationship among the response (parameters of lifetime distribution) and input variables (design and stress variables as well as covariates) should be modeled. Secondly, the unknown parameters in the approximate relationship should be estimated. In the last step, an optimization process should be conducted on the obtained relationship with respect to input variables. In this regard, the RSM is an absolutely well-designed method for covering and handling the mentioned steps. RSM is a collection of mathematical and statistical techniques which are applied to construct a proper functional relationship between response and input variables (Myers et al. 2004). Moreover, RSM is a powerful tool for improving and optimizing the response variables. Initial works in the field of RSM were proposed in 1950. At the mentioned years RSM was widely applied in the chemical industries. However, in recent years the applications of RSM have a tremendous growth in a broad spectrum of systems and processes (Myers et al. 2011). This methodology consists of three major phases that are consistent with ALT design phases. At first step, a design of experiment should be selected to run the tests. The aim of this phase is to select design points where response variable should be evaluated. Through applying the first phase two major advantages come up to design ALT. Firstly by choosing a statistical design to conduct the experiment, the numbers of test unit which might be failed during the test would be significantly decreased. Secondly, by considering the best set of design points to run the experiments, the prediction error of unknown parameters would be reduced. As a result, the accuracy of estimations would be meaningfully increased. At second step, the unknown parameters in the designed relationship between response and input variables would be estimated by using an appropriate method. By doing so, lifetime of the units can be predicted using the obtained relationship model. In the last step of the methodology, an optimization method would be applied with respect to different objectives such as improving the reliability of units and involved constraints in the practical condition of experiments. The main three steps are illustrated by simple examples in the following subsections.

3.1 Design of Experiments

The basic model to design a relationship between response(s) and input variables can be defined as Eq. (2).

$$y = f'(x)\beta + \epsilon \tag{2}$$

where $= (x_1, x_2, \ldots, x_k)'$, y is the response variable, and $f(x)$ is a P-elements vector function, which contains powers and Cartesian products of $x_1, x_2, \ldots, x_k$. In addition, β is a p-element vector of unknown coefficients related to input variables and ϵ is the experimental random error, which is assumed to have a zero mean. With respect to presented model, it is clear that the value of $f'(x)\beta$ is the expected value of response variable (y). In other words, $f'(x)\beta$ is the mean of response variable and can be shown as $\mu(x)$. Two useful and important equations that are commonly used in response surface methodology are first order and second order equations. These equations are shown in Eqs. (3) and (4).

$$y = \beta_0 + \sum_{i=1}^{k} \beta_i x_i + \epsilon \tag{3}$$

$$y = \beta_0 + \sum_{i=1}^{k} \beta_i x_i + \sum \sum_{i<j} \beta_{ij} x_i x_j + \sum_{i=1}^{k} \beta_{ii} x_i^2 + \epsilon \tag{4}$$

In order to achieve the mentioned goals in the last section for first phase of response surface methodology, initially a number of experiments should be conduct on units. By doing so, the values of input variables and its response variable can be determined. The input variables can be presented in a design matrix as follows.

$$\mathcal{D} = \begin{bmatrix} x_{11} & x_{12} & \cdots & x_{1k} \\ x_{22} & x_{22} & \cdots & x_{2k} \\ \cdot & \cdot & \cdots & \\ \cdot & \cdot & \cdots & \\ \cdot & \cdot & \cdots & \\ x_{n1} & x_{n2} & \cdots & x_{nk} \end{bmatrix}$$

where X_{ui} is the Uth design set of input variable X_i ($i = 1, 2, \ldots, k$; $u = 1, 2, \ldots, n$). Each row of this matrix presents a design point. If y be a response that is obtained by applying Uth design point, it can be calculated through Eq. (5)

$$y = f'(x_u)\beta + \epsilon_u, \quad u = 1, 2, \ldots, n \tag{5}$$

where ϵ_u presents the error of Uth design point. Equation (5) can be rewritten as Eq. (6).

$$y = X\beta + \epsilon \tag{6}$$

where $y_u = (y_1, y_2, \ldots, y_n)'$, X is a $n \times p$ matrix in which Uth row is $f'(x_u)$, and $\epsilon = (\epsilon_1, \epsilon_2, \ldots, \epsilon_n)'$. The elements of the first column are valued by 1 s to enable fitting with nonzero intercept. This is assumed that the mean of ϵ is zero and its variance is $\sigma^2 I_n$ (Rencher and Schaalje 2008). In this regard, OLS can provide a linear estimator of unknown parameters as shown in Eq. (7)

$$\widehat{\beta} = (X'X)^{-1}X'y \tag{7}$$

with the following variance–covariance matrix.

$$\mathrm{Var}(\widehat{\beta}) = \sigma^2(X'X)^{-1} \tag{8}$$

Accordingly, Eq. (9) estimates the response mean at the design point u.

$$\widehat{\mu}(x_u) = f'(x_u)\widehat{\beta}, \quad u = 1, 2, \ldots, n \tag{9}$$

In addition, by considering the mentioned point at the previous section, this is clear that $f'(x_u)\widehat{\beta}$ is the value of response prediction. On the whole, in each design point in region of experiment, the prediction of response can be achieved as Eq. (10).

$$\widehat{y}(x) = f'(x)\widehat{\beta}, \quad x \in R \tag{10}$$

where R is the feasible region of experimental factors.

Since $\widehat{\beta}$ is a unbiased estimator for B, $\widehat{y}(x)$ is a unbiased estimator for $f'(x)\beta$. As a result, the variance for $\widehat{y}(x)$ can be obtained by applying Eq. (11) (Díaz-García et al. 2005).

$$\mathrm{Var}[\widehat{y}(x)] = \sigma^2 f'(x)(X'X)^{-1} f(x) \tag{11}$$

As shown earlier, we can only decide on $X'X$ since other elements are supposed to be fixed, constant, or predetermined. Therefore, values in the design matrix have

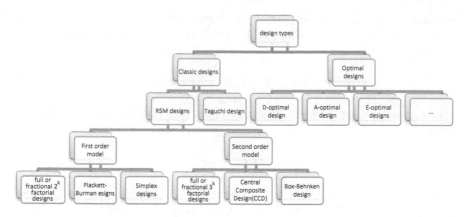

Fig. 6 Classification of design methods

a meaningful effect on the variance of prediction. The vital point of first phase of response surface methodology is to construct the proper design matrix. With respect to Eq. (11), it can be inferred that by choosing a proper design matrix that satisfy some criteria, a relevant decrease in prediction variance would be obtained, and as a result a significant improvement in accuracy of estimations would be provided.

In order to reach the mentioned purposes, the design matrix should have some features, among which orthogonality and rotatability have a more importance in comparison to others. If the matrix $X'X$ be a diagonal, then the design matrix is orthogonal. The great merit of orthogonal matrix is that the elements of $\widehat{\beta}$ will be uncorrelated. The ability of rotation can be defined in terms of the constant prediction variance for each design point. In simple words, each design point in a matrix with the ability of rotation has a same distance from the design center. The main merit for such a matrix is that the prediction variance remains unchanged under any rotation of axes. Since the prediction variance is constant and different responses can only be compared based on their mean values, the process of comparison among them can become easier than normal condition (Ai and Mukhopadhyay 2010).

Various types of design can be categorized into two main classes: classic and optimal designs. Figure 6 shows these two main groups and their subgroups.

To begin with, in classic RSM design various methods for first order designs are available, among which 2^k factorial, Plackett–Burman, and simplex design are the most prevalent ones. In a 2^k factorial design each input variable is set at two levels, coded by $+1$ and -1. By doing so, a design matrix will be constructed including all possible combinations of input variables. Thus, this design requires to run $n = 2^k$. For example, suppose the lifetime of a lamp depends on two factors, namely, voltage and temperature. The condition of experiments is as follows: The voltage is set at 120 and 220 V and the temperature falls into 30 or 40 °C. By scaling process the mentioned values can be converted to 1 and −1. The required number of runs for this example equals to $2^2 = 4$. The design matrix for this example has been shown below.

$$\begin{bmatrix} 1 & 1 \\ 1 & -1 \\ -1 & 1 \\ -1 & -1 \end{bmatrix}$$

The 2^k factorial designs result in an orthogonal matrix. However, by increasing the number of input variables the number of required design points grows drastically. Hence, a significant increase in cost will be generated. By considering this matter of fact that in a first order model only $k + 1$ unknown parameters should be estimated, to overcome the mentioned problem the 2^k factorial design can be conducted by a fraction of design such as half or one-fourth fraction design which is greater than $k + 1$. The last design is called fractional factorial design and is very common in design of experiments. The other method in this subgroup is Plackett–Burman designs which have a vast application in design of experiments. Plackett–Burman designs such as 2^k factorial designs assume two levels for each input variables but require fewer number of design points (precisely $n = k + 1$) compared to 2^k factorial design. Therefore, Plackett–Burman designs can reduce cost of experiments. The last design is simplex in which $k + 1$ design points are required. Although the simplex designs based on aforementioned note are economical, because of its difficult computations usually are not of interest. All the mentioned designs are orthogonal and it is crystal clear that orthogonal design matrix provides best results for estimation of unknown parameters.

The most prevalent designs for second order models are, namely, the 3^k factorial designs, the Central Composite Design (CCD), and Box–Behnken designs. The 3^k factorial designs have a same procedure like 2^k factorial designs but set variables in three levels, namely, 1, 0, −1. Because of the mentioned reasons in the last section fractional factorial designs are considered in these designs. The minimum number of design points which are required in a second order model like Eq. (4) is $h = 1 + 2k + \frac{1}{2}k(k - 1)$. Probably, the most applicable design for second order models is CCD. CCD has well-organized steps to construct design matrix. In the first part CCD, 2^k design points are selected based on first order model. Hence, the first part is complete by using first order model information. In the second part of CCD, more details about the second order model are added to design matrix. Finally, Box–Behnken designs need a small part of a 3^k factorial design. Therefore, this design has a great cost benefit and is adequate for real industrial problems (Ai and Mukhopadhyay 2010).

The other subgroup for classic designs is Taguchi design that mainly deals with nuisance factors or dynamic environments. Due to a simple procedure, this group of design is very applicable for practitioners. According to Taguchi's viewpoint of robustness and minimum variations in quality issues, these designs are constructed with a special focus on orthogonality measure (Taguchi et al. 2005).

Optimal designs are another important class of design matrix. This type of design is based on some optimality criteria. The various criteria have a same goal which is the closeness of the estimated response to the mean response (Atkinson et al. 2007). In other words, such criteria consider the confidence region of unknown coefficient, B, and try to minimize this region in different indices. The confidence region is a general form of a confidence interval in a multidimensional space. Commonly, this region has an ellipsoid shape near the estimated point. Each point in the confidence region has a probability to occur. Therefore, by minimizing the area of this region the accuracy of estimation would be improved. All the different optimality criteria using different methods try to make this region as small as possible. By doing the last step, the response prediction variance would be minimized and a proper design matrix will be obtained. To explain in more details, the variance–covariance matrix of predicted unknown parameters vector, $\widehat{\beta}$, is presented as follows.

$$\sum\widehat{\beta} = \begin{pmatrix} \text{Var}\left(\widehat{B}_0\right) & \cdots & \text{Cov}\left(\widehat{B}_0, \widehat{B}_k\right) \\ \vdots & \ddots & \vdots \\ \text{Cov}\left(\widehat{B}_0, \widehat{B}_k\right) & \cdots & \text{Var}\left(\widehat{B}_0\right) \end{pmatrix}$$

As can be seen from the variance–covariance matrix of $\widehat{\beta}$, the main diagonal elements are the variances of estimated coefficients. Hence, the minimizing process should be performed on these elements. With respect to Eq. (8) if it supposed that the value of σ^2 is one, then for minimizing the predicted variance of model coefficient only the $(X'X)^{-1}$ must be considered. The inverse of variance–covariance matrix is called information matrix (Goos and Jones 2011). Information matrix presents the available information of the model. It is crystal clear that by more information the accuracy of estimation will increase. If a criterion is defined to minimize the predicted variance of factors, it can have their application to maximize the information in the information matrix. Therefore, it can be claimed that all optimality criteria can be defined in two different ways that have same target.

Now that the aim of optimal designs is illuminated, different optimality criteria will be introduced. Probably, the most popular optimality criterion is D-optimality. The D is a clue of determinant. In a D-optimal design the determinant of $(X'X)^{-1}$ would be minimized. In other words, the determinant of information matrix should be maximized. The D-optimality criterion has a logical assumption by which, minimization of the determinant of $(X'X)^{-1}$ leads to smaller variances in the prediction step. It should be considered that if $(X'X)^{-1}$ be a diagonal matrix, then the determinant of $(X'X)^{-1}$ is equal to product of diagonal elements of this matrix. This case makes the last mentioned assumption easy to understand. In addition, when $(X'X)^{-1}$ is orthogonal the unknown parameters can be estimated independently. However, the orthogonally assumption for $(X'X)^{-1}$ might be violated in real–world situations especially where some nuisances are to be included in the statistical model. In this regard, one of the most prominent advantages of optimal designs is that such designs are robust against not satisfying of orthogonally assumption.

Fig. 7 Coordinate exchange
algorithm steps

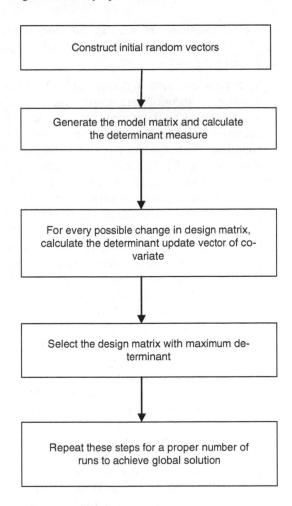

In fact, optimal designs will remain a proper method for nonorthogonal matrices. Although nonorthogonal matrix might generate inaccuracy in model by variance inflation as well as covariance among estimated parameters, optimal designs obtain best combinations of factors among available designs so that the violation would be small as possible. Compared to the factorial design, optimal design would consider the variance of prediction by assuming some relationships models between response and factors. Therefore, the design points are determined so that the variance of prediction is minimized.

A great number of algorithms are available to obtain D-optimal designs, one of which is coordinate exchange algorithm. Coordinate exchange algorithm is a useful algorithm with clear steps to minimize the determinant of the variance–covariance matrix and has a wide application in constructing of D-optimal designs. The steps of coordinate exchange algorithm are shown in Fig. 7.

For the mentioned reasons, in this study the coordinate exchange algorithm has been used for constructing a *D*-optimal design. The details of these steps are explained as follows.

1. To begin with a random design matrix should be constructed. Before starting to construct a random design matrix a recommended prerequisite which is factors scaling should be introduced. Although scaling of variables is not an absolute necessity for generating an optimal design, this process brings some merits to coordinate exchange algorithm. By doing so, the continuous variables are set between −1 and 1. In addition, categorical variables which have two levels can be set at −1 and +1. The other kinds of categorical variables can also be coded based on the condition of experiment. Scaling of variables has an advantage for comparing factors' effect. For instance, when two factors X_1 and X_2 have a same scale and the model coefficient of which are $B_1 = 3$ and $B_2 = 6$, it can be inferred that X_2 has a twice impact than X_1 on the response variable. By taking all these into account, to construct a random design matrix random values between −1 and +1 must be assigned to the controllable scaled variables. In addition to controllable variables, the design matrix is contained covariates which might have a stochastic behavior. Commonly in practice a specific distribution function such as normal distribution is defined for covariates. Hence, a random value with respect to the mentioned distribution function should be generated for covariates. By doing so, a random design matrix is constructed and the others steps would be performed on this matrix.

2. Since the optimality criterion is to minimize the determinant of variance–covariance matrix, firstly the information matrix should be calculated. In this regard, the model matrix X would be generated. The model matrix is obtained through using design matrix. In fact, each row of the model matrix represents the value of each term in the model for a specific design point. After obtaining the model matrix, the information matrix can be constructed through the mentioned method. This should be considered when the computation of determinant for a small information matrix is simple but for more complicated matrices these calculations become very time consuming. Therefore, this process must be performed by using computer programs. By obtaining the determinant of initial information matrix a start point for algorithm is created. The rest of the algorithm includes replicating steps designed to maximize the mentioned determinant.

3. In third step, every possible change in design matrix would be performed. For any unique design matrix the determinant of information matrix should be calculated and saved. To do so, an element-by-element process should be applied. In this regard, when a change occurs in an element the other elements should be remained constant to make possible the comparison and evaluation of each change in design matrix. This must be considered when covariates are not controllable variables. Thus, the effects of such variables should be removed from the determinant of information matrix. In this regard, for every change in design matrix a sufficient number of random values would be generated for each covariate. By considering the mentioned point, for each change in design

matrix a great number of design matrices with different covariate column would be obtained. Now the determinant measure for the obtained design matrix must be computed. The sample average of these determinants can be proposed as the determinant of the information matrix for every change. By doing so, the effect of covariates will be justified in the computations. Note that the updating process of covariates should be performed equal to a sufficient number for every change in design matrix. This process would be continued till all the possible changes in design matrix terminate. It is worthy to note that this procedure is a kind of Monte-Carlo approach. The design matrix associated to the maximum value of determinant of information matrix would be chosen as the best combination of factors for running the experiments.

4. The obtained solution from the last step is a local optimum solution. To explain in more details, this solution is the best one among the neighborhood of this point. However, it can be possible that by starting with a different initial random design matrix a better solution is obtained. In this regard, to achieve the global optimum solution the last three steps should be repeated for a sufficient number of runs. It is like a mountain climber who has climbed a hill, by looking around the hill he cannot come to the conclusion that it is the highest point in that mountain. However, to find the highest point in that mountain he should try climbing different hills. This action should be repeated until the highest hill in the mountain can be recognized. By doing so, the global optimum solution is likely to be a design matrix which will obtain the greatest determinant of information matrix among all the repeated loops in the algorithm. The final solution of algorithm is the best combination of factors for conducting the experiments.

3.2 Parameter Estimation

In order to achieve an appropriate relationship between response and factors in an accelerated life test model, an adequate method of estimation should be selected. More precisely, after obtaining the best design points for constructing the model, the unknown coefficients of model must be estimated through a point estimator. Since the process of estimation is performed in the obtained design from the last step, the predicted variance of response variable becomes as small as possible. With respect to aforementioned methods of point estimation, in this subsection four main methods of point estimation are introduced, namely, least square error (LSE), the method of moments, MLE, and Bayes method. To begin with LSE is a useful method to minimize the difference between the real value of response variable and the estimated value of it. This difference can be called as error, because the ideal estimated points are those of ones which are nearest to the real value of response variables.

The other main method of estimation is the method of moments. The method of moments is one of the oldest methods of estimation. This method contains some equivalency equations between the initial moments of sample and population.

The number of equations is equal to the number of unknown parameters that must be estimated in the model. For instance, the number of equations to estimate the mean and variance of a normal distribution is two. The method of moments has a broad application in statistical and other fields of science.

Bayesian estimators are applied when prior information about the unknown parameters is available. The risk of a bad estimation is defined by some loss function such as mean square/absolute errors. The estimation has a minimum expected loss function over the prior distribution of the parameters (El and Casella 1998).

The last but the most prominent method of estimation in reliability analysis is MLE. The MLE has a probabilistic basis for computations. In the MLE a likelihood function would be constructed. The likelihood function is a relationship that demonstrates the occurrence probability of real response variable values. An illustrative example is presented to explain the process of constructing the likelihood function. For example, in ALT parts which are conducted in stress levels have a Bernoulli distribution with P as the probability of failures. In this test five parts are conducted in an accelerated life test. In the inspection stations two first parts are failed. Hence, the probability of happening the mentioned situation (likelihood function) is as follows.

$$L = p\,(x_1 = 0, x_2 = 0, x_3 = 1, x_4 = 1, x_5 = 1)$$

$$= P\,(x_1 = 0)\,P\,(x_2 = 0)\,P\,(x_3 = 1)\,P\,(x_4 = 1)\,P\,(x_5 = 1) \qquad (12)$$

where if $x_i = 0$ the part i is failed and reversely. Since the parts have a Bernoulli distribution the likelihood function can be rewritten as follows

$$L = (1 - P)\,(1 - P)\,(1 - P)\,(P)(P) = (1 - P)^2 P^3 \qquad (13)$$

The obtained likelihood function demonstrates the occurrence probability of observed results in the sample. Therefore, the implication of likelihood function can be understood by the mentioned example.

The next step in the MLE method is to maximize the obtained likelihood function with respect to unknown parameters of distribution. By doing so, the estimated parameters achieve the best prediction for response variables. In other words, the value that maximizes the likelihood function estimates the unknown parameters of model based on observed results of experiments. Thus, the response variables estimated by these parameters are properly estimated. To explain in more details, the mentioned example is extended to the second step. In this step, the likelihood function should be maximized. Therefore, the obtained likelihood function should be derived with respect to P and the result would be equal to zero. In this regard the following equation is achieved.

$$\frac{\partial L}{\partial P} = P^2\left(5P^2 + 8P + 3\right) = 0 \qquad (14)$$

By solving the above-mentioned equation, the P is estimated by 0.6. It can be claimed that if $P = 0.6$ then the occurrence probability of observed events is maximum. Therefore, the best possible estimation would be achieved.

Although two main steps of MLE are constant and clear, some details can be varied in the method with respect to the problem conditions. In the case of constructing an accelerated life test, experimenters are confronting with two major issues. First, commonly ALT involves censored data which have a profound impact on estimation of unknown parameters. Second, due to a great number of unknown parameters in the relationship between response variable and input variables, which are contained stress and design variables as well as covariates, the system of equations for maximization step in MLE become very large. In these regards, to make an adequate estimation for an ALT some additional steps should be considered in the two main steps of MLE.

By considering the type of censored data which are discussed in the previous section, a proper likelihood function should be constructed for available data as well as censored data. To achieve this objective, depending on the type of censored data the probability of each event is evaluated in the likelihood function. Right censored data can be seen in a broad spectrum of accelerated life tests. In this regard, in this study assumed that data are right censored. As mentioned in the previous section, the failure time of a right censored observation is not available for estimation of parameters. If censored data are ignored or analyzed similar to complete data a significant error in estimations will be generated. In order to cope up with this problem, the censored data should be considered in a model with a proper likelihood function. When a failure time is censored from right side, it is clear that the lifetime of the part is equal or greater than the censoring time. By accepting this fact as true the mentioned part is not failed before censoring time. Hence, the cumulative distribution function is a proper representative term for right censored data in likelihood function. More precisely, cumulative distribution function demonstrates the summation of probabilities before a specific value, which in the case of right censored data in ALT is equal to censoring time. If $f(t)$ presents the lifetime distribution function of the parts, then the lifetime cumulative distribution function, $F(t)$, for censoring time can be obtained through Eq. (15).

$$F(t = T) = p(t \leq T) = \int_0^T f(t)dt \tag{15}$$

where T is the censoring time. In view of these considerations, the likelihood function can be constructed for an accelerated life test model with the right censored data as Eq. (16).

$$L(\theta_1, \ldots, \theta_p, \ldots, \theta_k)$$

$$= \prod_{i=1}^{n-m} f(t_i, \theta_1, \ldots, \theta_p, \ldots, \theta_k) \prod_{i=n-m+1}^{n} F(t_i = T, \theta_1, \ldots, \theta_p, \ldots, \theta_k) \tag{16}$$

Fig. 8 Lifetime cycle

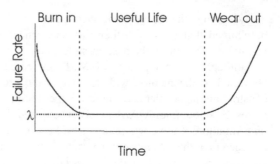

where $i = 1, \ldots, n$, n is the number of experiments, and m is the number of right censored observations. In addition, θ_p is the parameter of distribution functions and k is the number of these unknown parameters. It is clear that in this step the lifetime distribution function is required to be known. In the design of experiments step, the process can be performed by applying LSE which is independent of the lifetime distribution of parts. In this regard, a great number of choices are available, among which log-normal, Weibull, and exponential distribution have more applications in accelerated life testing and reliability analysis. Some notes should be considered before choosing the life time distribution. Firstly, the region of interest in accelerated life tests commonly located in useful time period. In the useful period that are shown in Fig. 8 the failure rate is approximately constant and this not dependent on time. Therefore, because of features of negative exponential distribution, it is a proper choice for lifetime of parts. In addition, this choice simplifies the future calculations for estimating process.

Although the proposed methodology has not any limitations about the type of lifetime distributions, based on the mentioned reasons considering exponential distribution for parts in the next section is a logical choice. If the lifetime distribution has only one unknown parameter, then the likelihood function can be rewritten as Eq. (17).

$$L(t, \Theta) = \prod_{i=1}^{n-m} f_t(t_i; \Theta) \prod_{j=n-m+1}^{n} F_t(t_j; \Theta) \tag{17}$$

where F_t is the cumulative distribution function for exponential distribution. As mentioned in the previous sections, in an accelerated life test the unknown parameters of lifetime distribution are a function of input variables. In fact, although the unknown parameter of lifetime distribution is not dependent on time, it is dependent on stress and design variables as well as covariates.

$$\theta_i = B_0 + \sum_{j=1}^{s} B_{1j} S_{ji} + \sum_{t=1}^{x} B_{2t} X_{ti} + \sum_{d=1}^{c} B_{3d} C_{di} + \sum_{j=1}^{s}\sum_{t=1}^{x} B_{4jt} S_{ji} X_{ti}$$

$$+ \sum_{j=1}^{s}\sum_{d=1}^{c} B_{5jd} S_{ji} C_{di} + \sum_{t=1}^{x}\sum_{d=1}^{c} B_{6td} X_{ti} C_{di} + E_i \tag{18}$$

where S, X, and C represent the stress variables, design variables, and covariates, respectively. In addition, S, X, and C are the number of stress variables, design variables, and covariates. The vector of coefficients $B = (B_0, \ldots, B_w)$, where w is equal to summation of main effects and first order interactions, is the unknown parameters of model. Therefore, using the MLE method these unknown parameters should be estimated. Before constructing the likelihood function one point should be considered that the type of relationship between the parameters and input variables should be selected based on features which are described in previous section. Thus, the first order interaction model chosen for this study is an arbitrary choice which can be varied depending on the purpose of the research. However, in the proposed first order interaction model the interaction between covariates and other controllable variables is considered which can improve the accuracy of estimation, because some of deviations from the mean of response variable are based on these terms in the model. All in all, to construct the likelihood function for proposed problem the relationship for distribution parameter should be located in Eq. (14). By doing so, the likelihood function would be depending on the coefficients stress variables, design variables, covariates, and their interactions. By doing so, the likelihood function would be constructed and the second step of MLE method will be started.

The aim of this step is to maximize the value of likelihood function. The usual method in this step is to analyze the derivations of likelihood function with respect to vector B and find its roots. Broad spectrums of methods are provided to solve this system of equations. In the one hand, some methods such as Gradient methods and Simlex (Nelder–Mead) are methods which will obtain local optimum solutions. On the other hand, Simulated Annealing (SA) and Genetic algorithm are methods which will obtain global optimum solutions. In addition, the first group is not effective for large-scale models with a great number of parameters, while the second group contains methods that are appropriate for the mentioned models. More precisely, by increasing the number of unknown parameters in the likelihood function, the number of equations in the system of equations will increase. Hence, solving the system of equation becomes a major problem. In this situation using the second group of methods is more appropriate.

Hence, to overcome the mentioned problem about difficulty in calculation for maximization of likelihood function, the second group and particularly SA algorithm can be used in this study and an optimization model can be constructed. It is clear that the objective function of this model is to maximize the likelihood function. However, due to the simplicity in calculations a natural logarithm of likelihood function is usually used as the objective function. This should be considered that the optimization process should be performed in the region obtained in the design of experiments phase. Hence, the design variables, stress variables, and covariates set in their design points value for each run and the lifetime of parts will be calculated for every design point. The estimation process is performed based on these design points value and the related responses.

In order to solve this optimization model, a meta-heuristic algorithm which is called Simulated Annealing is applied in this study. Simulated Annealing (SA) is a local optimization method for solving hard optimization problems. Firstly ideas about this method were proposed by the work of Kirkpatrick et al. (1983) who applied some similarities between simulating the annealing process of solids in optimization problems. Afterwards, SA has been used to optimize variety of problems in areas such as locational analysis, image processing, molecular physics and chemistry, and job shop scheduling (Eglese 1990).

The main steps of SA in analyzing the optimization problem can be expressed as follows.

- Define set, S, of feasible solutions
- Calculate objective function $f(s)$ for each solution
- Define a neighborhood structure which can be obtained by making a move in the current solution, a move being the change in value of one or more variables.
- Do the following iterative steps

  - Starting with an initial solution s generated by other means
  - Repeatedly move to a neighboring solution that meet one of the following conditions:

    - Having a better objective function
    - Having a worse objective function but the neighborhood solution meets certain probability. The probability of accepting an uphill move is normally set to exp $(-\Delta/T)$ where T is a control parameter which corresponds to temperature in the analogy with physical annealing, and Δ is the change in the objective function value.

  - Update the parameters such as T.

- Stop algorithm when one of the following criteria are met

  - Specified number of iterations
  - Specified change in two successive objective function
  - Specified process time

After solving the optimization model by SA algorithm the most accurate possible estimations of coefficient of models vector B would be obtained. Hence, locating this vector in the failure rate relationship, the estimation of failure rate will be created. By doing so, in different times the reliability of parts can be estimated. Therefore, the first main goal of this study is satisfied. In this study the MATLAB optimization toolbox is used for solving the optimization model by SA algorithm. Hence, more information about the steps of this algorithm can be found in software. More detailed for this section is presented in numerical example.

3.3 Multiobjective Optimization

In this phase, a multiobjective optimization model has been proposed to maximize the expected value of the predicted response variable and to maximize the probability of the covariates. After obtaining the proper life–stress relationship, by a few converts the reliability function can be constructed. It is clear that the improvement of reliability is the most important aim of this study. Hence, firstly the proper model for this optimization problem should be designed. Secondly, the adequate method to solve the obtained optimization model will be selected. By considering this matter of fact as a true that the reliability is a function of lifetime, the reliability function simply will be generated by using the obtained lifetime model from the last step. The other objective of the proposed optimization model is to maximize the probability of covariates. Since the covariates are stochastic and uncontrollable variables, the second objective improves the region of optimization process. In other words, by applying the second objective, results would be given with more likely values for the covariates. It is accepted as a true that considering the real experiment conditions in optimization process will develop the quality of results. All in all, the goal of the proposed optimization model is to maximize the reliability, while it is tried to consider the most probable region of the stochastic covariates.

The reliability $R(t, \Theta)$ is depended to the time in lifetime distribution, which t is the time and Θ is the parameter of lifetime distribution for parts. In addition, based on the aforementioned discussions in the previous sections the lifetime is a function of stress and design variables as well as covariates. Thus, the reliability function can be rewritten as $R(t; S, X, C)$, which S, X, and C are representatives of variables and defined in the previous section. Consider the reliability function given below (Smith 1983):

$$R(t, \Theta) = \int\limits_{t}^{\infty} f(t, \Theta) \, d(t) \tag{19}$$

where $f(t, \Theta)$ is the lifetime distribution function. One might be interested in analyzing the reliability by available information about the failure rate. In this case, failure rate is a function of stress variables, design variables, covariates, and time. By considering the implication of failure rate the reliability function can be rewritten as equation given below (Smith 1983).

$$R(t, s, x, c) = e^{-\int_{0}^{t} \lambda(t, s, x, c) \, dt} \tag{20}$$

which should be maximized as the first objective function of the optimization model.

A decision about the distribution of covariates should be made to construct the second objective. In this regard, a broad spectrum of distributions can be assigned to the covariates. Hence, to select best distribution for such variables different aspects of the covariates should be evaluated. With respect to the mentioned points about the covariates, the second objective can be constructed as maximization of probability distribution function of covariates (Hejazi et al. 2011; Salmasnia et al. 2013). In addition to objective functions, the model contains some constraints about the acceptable region and limitations for stress variables, design variables, and covariates. Therefore, the optimization model can be constructed as follows

$$
\text{maximize } \mu\left(\widehat{R}\right)(t,s,x,c) = \mathrm{e}^{-\int_0^t \mu\left(\widehat{\lambda}(t,s,x,c)\right)dt}
$$
$$
\text{maximize } f\left(C_t\right) \quad \forall t = 1,2,\ldots,l
$$
$$
\text{Subject to :}
$$
$$
L < X < U
$$
$$
S \text{ fixed in specified use stress}
$$
$$
C \in \Omega
$$

(21)

where $\mu\left(\widehat{R}\right)$ and $\mu\left(\widehat{\lambda}(t,s,x,c)\right)$ are the means of estimated reliability and failure rate. In addition, L is the lower bound and U is the upper bound for design variable X and Ω is the feasible region for covariates.

To explain the constraints of model, the first constraint defines acceptable limits for the design variables. The design variables should be fallen between their lower and upper bounds. This interval is specified with respect to the operational condition of problem and the experimental design. The second constraint is about the stress levels. In an ALT stress variables are set at some accelerated levels. Therefore, to optimize the reliability function these variables should change to the operational or design condition. The last constraint refers to acceptable region of covariates. The mentioned region is determined with respect to domain of the covariates' distribution function.

Stress variables are fixed at their use levels in the model. In addition the nature of covariates, which is stochastic, does not allow control of such variables. Therefore, optimization should be performed with respect to design variables. In the first objective the reliability function, which can be obtained in the last section by using the estimated lifetime, would be maximized with respect to design variables. In this regard, by using the obtained information from the accelerated life test and locating the stress variables in their use levels the reliability in operational condition can be optimized. The second objective causes that the stochastic condition consider in the problem. In simple words, by increasing the occurrence probability of covariates the problem will approach to their stochastic condition. To provide more analyses on the effects of the occurrence probability of covariates on the results, some constraints can be added so that the reliability optimization will be conducted as near as possible to the real condition of problem.

After constructing the optimization model a proper approach to solve the model should be selected. Since the optimization model contains two different objectives, the multiobjective optimization methods are adequate tools to solve the model.

The objectives in the presented model have conflict with each other. It is crystal clear that by increasing the occurrence probability of covariates the reliability will be decreased. Through increasing the occurrence probability of covariates the optimization region of the model becomes small and smaller. Therefore, the available points to optimize the reliability function would be limited and a decrease in reliability will be provided. Multiobjective methods are commonly used in such situations. These methods handle quite a few of conflicting objectives in an optimization model to achieve a satisfactory solution. The weighted p-norm, displaced ideal method, goal programming, global criterion, neutral compromise solution, weighted method, ϵ-constraint method, value function, loss function, and desirability are the methods that are frequently applied for multiobjective optimization (Ardakani and Wulff 2013).

Bounded objective method is one of the well-grounded approaches of multiobjective optimization which can be used to get more information and sensitivity analysis to the above-mentioned problem. In this method, the main objective function is maximized while the others considered in constraints with some satisfactory bounds. Since in the mentioned problem the main goal is maximization of reliability, the probability of covariate can be bounded by proper constraints. The key point in this approach is the evaluation of the bound associated to the covariates' occurrence probability. Thus, a sensitivity analysis should be performed on this factor. The optimization model for this approach is demonstrated in Eq. (22).

$$
\begin{aligned}
\text{Maximize} \quad & Z = \mu\left(\widehat{R}\right)(t, s, x, c) \\
& \text{Subject to :} \\
& f(c) \geq \omega \\
& L < X < U \\
& S \text{ fixed in specified use stress} \\
& C \in \Omega
\end{aligned}
\tag{22}
$$

where ω is the lower bound for constraint of covariate probability. This bound could be expressed by the decision maker as a minimum acceptable confidence of results (Hejazi et al. 2011).

4 Numerical Example

In this section a hypothetical example is studied to illustrate the applications of the proposed approach. An exponential distribution with mean 120 h is assumed for the lifetime probability. With respect to the aforementioned reasons the exponential distribution is adequate ones for lifetime data (Eq. (23)).

$$
f(t) = \lambda e^{-\lambda t}
\tag{23}
$$

where t is the lifetime and λ is the parameter of lifetime distribution which is called failure rate. In this case the response variable λ is considered to be a function of one design variable, two stress variables, and one covariate. The covariate is also assumed to follow a standard normal probability distribution. In addition, stress and design variables are located between -1 and 1. It has also been assumed that the relationship between inputs and outputs is expressed by linear and interacts effects as follows:

$$\lambda_i = B_0 + B_1 S_{1i} + B_2 S_{2i} + B_3 X_i + B_4 C_i + B_5 S_{1i} S_{2i} + B_6 S_{1i} X_i + B_7 S_{1i} C_i$$
$$+ B_8 S_{2i} X_i + B_9 S_{2i} C_i + B_{10} X_i C_i + E_i \qquad (24)$$

where i is the index of observations, S_1 and S_2 are stress variables, X is design variables, and C is the covariates. In addition, the vector $B = (B_0, B_1, \ldots, B_{10})$ is a vector of unknown coefficients. Finally, E is the modeling random error. According to the proposed approach, in the first step the optimum design points for conducting the experiments are obtained by modified exchange algorithm described in the previous sections. To do so, firstly an initial solution for the algorithm should be generated. Hence, all the stress and design variables are located in their lower bounds and covariates would be generated using the standard normal distribution. After obtaining the initial solution all the possible changes should be performed on the initial design matrix and the D-optimality criterion would be evaluated for each different design matrix. In this regard, element-by-element changes would be conducted on design matrix and best possible solution will be obtained. To reach the global solution this process should be repeated for a sufficient number of runs, for example, 1,000 runs. The design table and related response variable are presented in Table 2.

Assuming that the experiment was to be terminated after 120 h, observations that show lifetimes greater than 120 h have been censored.

In the next step likelihood function constructed as below:

$$L(\lambda) = \prod_{i=1}^{17} \lambda_i e^{-\lambda_i t_i} \prod_{i=18}^{20} e^{-\lambda_i t_i} \qquad (25)$$

Because of the mentioned reasons in the last sections, a natural logarithm is performed on aforementioned equation. SA algorithm is a powerful method for mathematical optimization problems. In order to find the constraints for this optimization model some notes should be considered. Since the failure rate cannot get negative value a limitation will be imposed in optimization model. In order to avoid assigning a negative value to failure rate λ, natural logarithm is performed on the relationship between failure rate and input variables and the objective function is maximized with respect to it. The developed likelihood function is maximized using this algorithm in Matlab mathematical software package (Global Optimization Toolbox) with following parameters setting (Table 3).

Table 2 Design table and related responses for numerical example

Number of observations	Design variable, X	Stress variable, $S1$	Stress variable, $S2$	Covariate, C	Response, t
1	1.0000	1.0000	1.0000	−0.5688	599.248
2	1.0000	1.0000	−1.0000	0.3793	315.6477
3	0.4921	−1.0000	−1.0000	2.2987	127.7357
4	0.5531	−1.0000	−1.0000	0.2241	99.2245
5	0.5472	0.0006	0.5108	0.9217	89.8678
6	0.0319	−1.0000	0.0524	1.9916	98.8301
7	0.1317	0.0660	−1.0000	−1.2066	98.9784
8	−1.0000	−1.0000	−1.0000	0.1919	70.1959
9	0.4374	−1.0000	0.3851	0.6947	47.5488
10	0.0705	−1.0000	0.9022	−0.8877	36.0969
11	−1.0000	1.0000	−1.0000	−0.4162	38.9341
12	0.3135	0.7381	0.0404	0.5055	44.2734
13	0.3871	0.5501	0.2439	1.6344	32.5114
14	−1.0000	0.3541	1.0000	−1.0601	16.0658
15	0.0750	0.5073	0.4156	1.8357	22.4051
16	1.0000	0.1897	0.0920	0.2532	17.8965
17	0.6292	0.0224	1.0000	0.0706	29.177
18	0.0407	0.1193	1.0000	−0.3146	4.2173
19	−1.0000	0.9231	0.1504	2.0359	1.5249
20	−1.0000	0.8537	−1.0000	0.8313	0.7068

Table 3 Parameter setting of SA algorithm performed in MATLAB

Parameter	Function tolerance	Annealing function	Temperature updating	Initial temp.
Setting	$1e^{-6}$	Boltzmann function	Exponentially	Default: 100

In order to avoid assigning a negative value to failure rate λ, natural logarithm function is performed on Eq. (24) and likelihood function is maximized.

$$\mathrm{Ln}\left(\widehat{\lambda}_1\right) = B_0' + B_1' S_{1i} + B_2' S_{2i} + B_3' X_i + B_4' C_i$$
$$+ B_5' S_{1i} S_{2i} + B_6' S_{1i} X_i + B_7' S_{1i} C_i$$
$$+ B_8' S_{2i} X_i + B_9' S_{2i} C_i + B_{10}' X_i C_i + E_i' \qquad (26)$$

The resulted MLE of B' is presented as Table 4.

After the statistical model was being estimated, values of stress variables related to the normal use condition should be utilized for freezing the variables S. Therefore, Eq. (26) would be only a function of the design variables and the covariate.

The goal of this problem is maximization of the reliability. Reliability can be obtained from formula that become in the next.

$$R(t) = \exp - \int_0^t \lambda(t)dt \qquad (27)$$

Table 4 Result of maximization on likelihood function

B_0'	B_1'	B_2'	B_3'	B_4'	B_5'	B_6'	B_7'	B_8'	B_9'	B_{10}'
-4.55428	0.712225	0.612446	1.118208	1.075125	0.438235	0.3018	0.657322	0.48	0.661788	-0.25823

Since in an exponential distribution, parameter lambda is constant, Eq. (27) could be rewritten as an exponential distribution like Eq. (29). Obviously, the minimum point of the lambda function corresponds to the maximum point of the reliability function.

$$P(t) = \lambda \exp(-\lambda t) \tag{28}$$

$$\exp\left(-\int_0^t \lambda dt\right) = \exp\left(-\lambda \int_0^t \lambda dt\right) = \exp(-\lambda t) \tag{29}$$

For the mentioned example, stress variables can be replaced by predetermined operational levels and consequently the relationship between natural logarithm of λ and design variable and covariate would be constructed. For this purpose, stress variables assigned to -2 and using Eq. (24) and Table 2, Eqs. (30) and (31) can be obtained

$$\text{Ln}(\lambda_i) = -5.45068 - 0.44539X_i - 1.5631C_i + 0.25823X_iC_i + E_i \tag{30}$$

and consequently,

$$E\left(\text{Ln}\left(\widehat{\lambda_i}\right)\right) = -5.45068 - 0.44539X_i - 1.5631C_i + 0.25823X_iC_i$$

$$E\left(\widehat{R}(t)\right) = \exp(-t(\exp(\text{Ln}(-5.45068 - .44539X$$

$$-1.5631C + 0.25823XC))) \tag{31}$$

Above equations represent the life parameter and the related reliability function at normal use condition. For mentioned situation, λ is a function of design variable and covariate. Figure 9 shows this relationship.

As shown in Fig. 9, by increase of design variable to their upper bound, λ has been decreased. Thus, estimation of reliability in region of upper bound is reasonable. If X is fixed in 1, reliability is a function of covariate and time so can be obtained from Eq. (32).

$$R(t) = \exp(-t(\exp(\text{Ln}(-5.89607 - 1.30487C)))) \tag{32}$$

In order to simplify the calculations, time unit is changed from hour to day. So Eq. (33) is obtained (Fig. 10).

$$R(t) = \exp(-t(24\exp(\text{Ln}(-5.89607 - 1.30487C))) \tag{33}$$

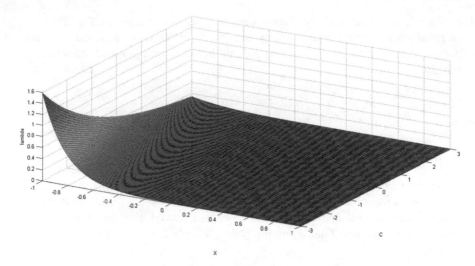

Fig. 9 Failure rate of exponential distribution as a function of the design factor and the covariate

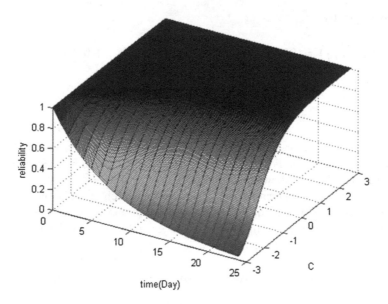

Fig. 10 A plot of reliability function with respect to time and covariate

Table 5 Probabilities of covariate and related mean of lifetimes

Minimum probability value of covariates	Maximum mean lifetime
0.26	585.58
0.3	577.15
0.35	537.55
0.39	454.34
0.4	Infeasible

In next step the mathematical model of the optimization problem can be formulated as follows.

$$\text{maximize} \quad \mu\left(\widehat{R}\right)(t,s,x,c) = e^{-\int_0^t \mu(\widehat{\lambda})t\,dt}$$

$$= e^{-\int_0^t \left(\frac{\sum_{i=1}^{20} \widehat{B_0}+\widehat{B_1}S_{1i}+\widehat{B_2}S_{2i}+\widehat{B_3}X_i+\widehat{B_4}C_i+\widehat{B_5}S_{1i}S_{2i}+\widehat{B_6}S_{1i}X_i+\widehat{B_7}S_{1i}C_i+\widehat{B_8}S_{2i}X_i+\widehat{B_9}S_{2i}C_i+\widehat{B_{10}}X_iC_i}{20}\right)t\,dt}$$

$$\text{maximize} \quad f(C)$$
$$\text{Subject to}:$$
$$-1 < X < 1$$
$$S_1 = S_2 = -2$$
$$C \in \Omega$$

$$(34)$$

where $\left(\widehat{B}_0, \ldots, \widehat{B}_{10}\right)$ are estimated unknown parameters. The optimum results of this model are obtained using generalized reduced gradient method of optimization built in Microsoft Excel 2010 software. By solving this model the desired probability value for covariate is obtained at 0.26. In addition, according to the results the maximum mean of expected value for lifetime is 585.57 h and the design variable is equal to one that is upper bound of X. By using the obtained value for lower bound of covariate probability the later sensitivity analysis will be performed on this value and its effect on the results. In this regard, the second approach to solve the problem should be performed.

As it can be seen in Table 5 with an increase in the probability values of covariate the mean of lifetime decreases. In addition, when this value exceeds from 0.4 the problem does not have a feasible solution. Therefore, the cost of increase in probability of covariates is to reduce the reliability of products so manufacturer should make an important decision about this matter to hold the reliability in the desired goal whereas the probability of covariates locate in an acceptable range. The probability values of covariates and related mean of lifetimes are presented in Table 5.

Table 6 A comparison between optimization model considering and not considering occurrence probability of covariate

	Maximum mean of lifetime	Occurrence probability of covariate	Maximum mean of expected value for lifetime
Proposed model	2,252	0.26	585.58
Non covariate constraints model	15,358	0.0044	68.06

Now, a model in which covariates are not considered is constructed to make a comparison between an ALT with covariate probability constraint and ones without any covariate probability constraint. This model is shown in Eq. (35).

$$\text{Maximize} \quad \mu\left(\widehat{R}\right)(t,s,x,c)$$
$$\text{Subject to}:$$
$$-1 < X < 1 \tag{35}$$
$$-3 \leq C \leq +3$$

As it can be seen from the recent model, the only objective function is maximizing the expected value of reliability. In addition, there is no constraint for covariates; since the covariate has a standard normal distribution 99 % of data are located between −3 and 3. Hence, the covariates can assign any value between the mentioned values. In the recent model covariates can get values with a very low probability of occurrence. Therefore, it can be predicted that a significant increase in the value of reliability will be provided. However, by solving the recent model the covariate is set at 3 that has a very low occurrence probability. As a result, if the constraint of covariate probability is not considered, then the solutions would be located in regions which have a few chances to occur. Hence, the proposed approach obtains more logical and realistic solutions than the recent model. Table 6 presents the differences between two mentioned approaches.

The cases, which are claimed in Table 6, are based on the fact that by limiting the region of optimization process to the most probable region of covariates, a significant decrease in the maximum mean of lifetime will be provided logically. On the other hand, a noticeable increase in the maximum mean of the expected value for lifetime will be provided. Since in the noncovariate constraints model, covariates can get value, which have a very low probability of occurrence, it is clear that the obtained results have a logical guarantee.

5 Conclusion

Reliability functions and related measures in complex systems should be derived with respect to operational/technical factors as well as environmental variables. For this purpose, several statistical and computational algorithms have been suggested.

Among them, response surface methodology as a mathematical-statistical approach enables finding an optimal operational condition after it was estimated. In this chapter a constant accelerated life test has been developed to find the settings of variables that optimize the reliability of products when the stochastic covariate is affecting the performance. In this model interactions between covariate and other variables are considered. This model consists of censored data and MLE method used for parameters estimation. Afterwards, a mathematical program was constructed to maximize the reliability of the estimated life performance and to maximize the probability of covariate's occurrences. This approach would take the best settings of design variables and consider the reliability measure as well as the stochastic covariate as an unavoidable part of the experiments. The results show the superiority of the proposed approach either in process identification by proposing new design construction method or in reliability optimization by considering the stochastic covariates that affect the performance.

References

Abdel-Ghaly AA, Attia AF, Abdel-Ghani MM (2002) The maximum likelihood estimates in step partially accelerated life tests for the weibull parameters in censored data. Commun Stat Theory Methods 31(4):551–573. doi:10.1081/Sta-120003134

Ai K, Mukhopadhyay S (2010) Response surface methodology. Wiley Interdisc Rev Comput Stat 2(2):128–149

Ardakani MK, Wulff SS (2013) An overview of optimization formulations for multiresponse surface problems. Qual Reliab Eng Int 29(1):3–16. doi:10.1002/Qre.1288

Atkinson A, Donev A, Tobias R (2007) Optimum experimental designs, with SAS. Oxford University Press, Oxford, New York

Chenhua L (2009) Optimal step-stress plans for accelerated life testing considering reliability/life prediction. Northeastern University, Boston, MA

Díaz-García JA, Ramos-Quiroga R, Cabrera-Vicencio E (2005) Stochastic programming methods in the response surface methodology. Comput Stat Data Anal 49(3):837–848

El L, Casella G (1998) Theory of point estimation, vol 31. Springer, New York

Eglese RW (1990) Simulated annealing: a tool for operational research. Eur J Oper Res 46(3):271–281. doi:10.1016/0377-2217(90)90001-R

Fard N, Li C (2009) Optimal simple step stress accelerated life test design for reliability prediction. J Stat Plan Inference 139(5):1799–1808

Goos P, Jones B (2011) Optimal design of experiments: a case study approach. Wiley, New Delhi

Hejazi TH, Bashiri M, Noghondarian K, Atkinson AC (2011) Multiresponse optimization with consideration of probabilistic covariates. Qual Reliab Eng Int 27(4):437–449. doi:10.1002/Qre.1133

Khamis IH (1997) Optimum M-Step, step-stress design with K stress variables. Commun Stat Simul Comput 26(4):1301–1313

Kirkpatrick S, Gelatt CD, Vecchi MP (1983) Optimization by simulated annealing. Science 220(4598):671–680. doi:10.1126/Science.220.4598.671

Lewis EE (1987) Introduction to reliability engineering. Wiley, New York

Li C, Fard N (2007) Optimum bivariate step-stress accelerated life test for censored data. IEEE Trans Reliab 56(1):77–84. doi:10.1109/Tr.2006.890897

Ling L, Xu W, Li M (2011) Optimal bivariate step-stress accelerated life test for type-I hybrid censored data. J Stat Comput Simul 81(9):1175–1186

Miller R, Nelson W (1983) Optimum simple step-stress plans for accelerated life testing. IEEE Trans Reliab 32(1):59–65

Montgomery DC (2005) Design and analysis of experiments, 6th edn. Wiley, Hoboken, NJ

Myers RH, Montgomery DC, Vining GG, Borror CM, Kowalski SM (2004) Response surface methodology: a retrospective and literature survey. J Qual Technol 36(1):53–78

Myers RH, Montgomery DC, Anderson-Cook CM (2011) Response surface methodology: process and product optimization using designed experiments. Wiley, New York

Nelson WB (2009) Accelerated testing: statistical models, test plans, and data analysis. Wiley, New York

Ng H, Luo L, Hu Y, Duan F (2012) Parameter estimation of three-parameter weibull distribution based on progressively type-II censored samples. J Stat Comput Simul 82(11):1661–1678

Ramachandran KM, Tsokos CP (2009) Mathematical statistics with applications. Elsevier (Online)

Rencher AC, Schaalje GB (2008) Linear models in statistics, 2nd edn. Wiley-Interscience, Hoboken

Salmasnia A, Baradaran Kazemzadeh R, Seyyed-Esfahani M, Hejazi TH (2013) Multiple response surface optimization with correlated data. Int J Adv Manuf Technol 64(5–8):841–855. doi:10.1007/S00170-012-4056-9

Smith CO (1983) Introduction to reliability in design. R.E. Krieger, Malabar, FL

Taguchi G, Chowdhury S, Wu Y (2005) Taguchi's quality engineering handbook. Wiley, New York

Tang L-C, Tan A-P, Ong S-H (2002) Planning accelerated life tests with three constant stress levels. Comput Ind Eng 42(2–4):439–446. doi:10.1016/S0360-8352(02)00040-2

Wang F-K, Cheng Y, Lu W (2012) Partially accelerated life tests for the weibull distribution under multiply censored data. Commun Stat Simul Comput 41(9):1667–1678

Reliability Measures Analysis of a Computer System Incorporating Two Types of Repair Under Copula Approach

Nupur Goyal, Mangey Ram, and Ankush Mittal

Abstract In this chapter, the authors have studied the reliability characteristics of a home or office based computer system constructed with hardware connectivity. The system contains multi possible stages that can be repaired. The designed system is studied by using the Markov process, supplementary variable technique, Laplace transformation, and Gumbel–Hougaard family of copula to obtain the various reliability measures such as transition state probabilities, availability, reliability, cost analysis, and sensitivity.

1 Introduction

Reliability theory has become a great anxiety in recent years, because high-tech industry processes, computer networking with increasing levels of sophistication comprise most engineering systems today (Verma et al. 2010; Ram 2013). Reliability can be defined as the probability that it will produce correct outputs up to given time period, according to McClusky and Mitra (2004). Reliability is enhanced by features that help to avoid, detect, and repair hardware faults. A reliable system does not mutely continue and deliver results that include corrupted data.

In the field of reliability theory, the remarkable work has been done by many researchers. Soi and Aggarwal (1980) discussed the future trends in the digital communication system and presented a system analysis model in the form of a state diagram to study the overall availability behavior of next generation

N. Goyal • M. Ram (✉)
Department of Mathematics, Graphic Era University, Dehradun, Uttarakhand, India
e-mail: drmrswami@yahoo.com

A. Mittal
Department of Computer Science and Engineering, Graphic Era University, Dehradun, Uttarakhand, India

S. Kadry and A. El Hami (eds.), *Numerical Methods for Reliability and Safety Assessment: Multiscale and Multiphysics Systems*, DOI 10.1007/978-3-319-07167-1_12, © Springer International Publishing Switzerland 2015

digital communication system. Goel et al. (1993) investigated a model for satellite based computer communication network system. In that work, a master station is connected to the remote micro earth stations in the country. A micro earth station fails due to a transient fault. Azaron et al. (2005) discussed reliability evaluation and optimization of dissimilar component cold standby redundant system which was the combination of series–parallel subsystems combination. They applied the shortest path technique for reliability evaluation. Elyasi-Komari et al. (2011) described the techniques and basic principles of dependable development and deployment of computer networks that are based on the results of FME(C)A (Failure Modes and Effects (Criticality) Analysis) analysis. Further, Nagiya and Ram (2013) investigated the various reliability characteristics of a satellite communication system which includes the earth station and terrestrial system and found the important reliability analysis.

In the context of computer systems, it is a universal purpose of device that can be planned to carry out every work in daily life. In today's fast life, everything depends on computer based systems. Now-a-days, it is quite impossible to overestimate the importance of computer systems in the environment around us. Embedded computer systems can be found in many devices around the home. Televisions, refrigerators, washing machines, telephones, just the few names. As a source of communication, computer plays a very crucial role. Information can be shared by anyone in the rest of the world and email has made written communication with anyone in the world potentially instantaneous. Usually, a computer system consists of at least one processing element, a central processing unit (CPU), and some form of memory. The processing element carries out arithmetic and logic operations, and a sequencing and control unit that can change the order of operations based on stored information. Peripheral devices allow information to be retrieved from an external source, and the result of operations saved and retrieved (Rajaraman 2010).

In computer hardware, availability refers to the overall uptime of the system. Reliability in general is likelihood of a failure occurring in a running system. A perfectly reliable system will also enjoy perfect availability within an intended period of time. The industry uses the concept of "high availability" to refer the systems and technologies specially engineered for reliability, availability, and sensitivity such systems include redundant hardware. By Lyu (1996), the demand for complex hardware systems has increased for more speedily than the ability to design implement, test, and maintain them. When the requirements for and dependencies on computer increases, the possibilities of calamities from computer failure also increase. The impact of these failures ranges from inconvenience, economic, damages, to loss of life. Hence the reliable performance of the computer systems has become a major concern. Hardware reliability can be described by exponential distribution. Also, hardware reliability decreases with time. The hardware reliability theory relies on the analysis of stationary processes because only physical faults are considered.

In the field of reliability-copula concept, Ram and Singh (2008, 2010a, b) studied the reliability indices of complex systems under two types of failure and repair using Gumbel–Hougaard family copula. Recently, the authors (Singh et al.

2013a, b) studied the complex systems under k-out-of-n types, which consist of two subsystems in series configuration using Gumbel–Hougaard family copula distribution in repair. Although they have done a good work by applying the copula approach, they did not think about the home or office based computer system performance under copula approach, which is a very important issue in today's work culture.

The present chapter reflects the performance of a home or office based computer system constructed with hardware connectivity under the concept of Gumbel–Hougaard family copula. The designed system is studied by using the Markov process, supplementary variable technique, and the Laplace transformation to obtain the various reliability measures.

2 Brief Introduction of Gumbel–Hougaard Family Copula

Several authors, including Nelsen (2006), have studied the family of copulas extensively. The Gumbel–Hougaard family copula is defined as:

$$C_\theta (u_1, u_2) = \exp \left(-\left((-\log u_1)^\theta + (-\log u_2)^\theta \right)^{1/\theta} \right), \quad 1 \leq \theta \leq \infty$$

For $\theta = 1$ the Gumbel–Hougaard copula models independence, for $\theta \to \infty$ it converges to comonotonicity.

Gumbel–Hougaard family copula gives the good results, when the system is in the complete failure mode. The best policy is to repair the failed system as soon as it is possible by Gumbel–Hougaard family copula when two distributions are coupled.

3 Mathematical Model Details

3.1 Nomenclature

Notations associated with work are shown in Table 1.

3.2 System Description

This chapter represents the reliability based mathematical modeling of a home or office based computer system under copula technique. Although the problem looks like general in daily routine, but here authors applied Gumbel–Hougaard family copula, makes the problem interesting. A general computer system has been converted into multi-states, which are good, degraded, and failed. The system has two types of failure, namely minor and major. From the good state after minor failure, the system goes to degraded state and after major failure, the system goes into a complete failure mode. Minor failure means any component of the system failed partially and the system could be worked with less efficiency while the major failure means the failure of any important component of the system, without which system could not be workable. After repairing, the system comes back in good state. A failed state could be repaired with the help of Gumbel–Hougaard family copula (Ram and Singh 2008, 2010; Ram 2010; Ram et al. 2013). The configuration and state transition diagram of the designed system have been shown in Fig. 1a, b.

3.3 State Description

All the states of the state transition diagram are described in Table 2.

3.4 Assumptions

The following assumptions are associated with the model

1. Initially, all the components are working that means system is in good state.
2. At any time, the system can cover from degraded or failed states.
3. All the components can be repaired.
4. Sufficient repair facilities are available.
5. After repair, the system works like a new one.

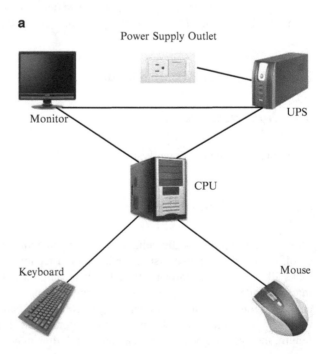

Fig. 1 (**a**) System configuration. (**b**) Transition state diagram

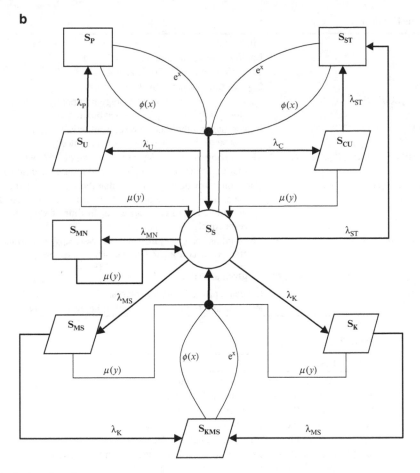

Fig. 1 (continued)

6. The failure and repair rate are constant. The value of different failure and repair rates are based on previous literature and work experience.
7. The expression for the joint probability distribution of repair of the complete failed states S_P, S_{ST} and degraded state S_{KMS} are computed with the help of Gumbel–Hougaard family copula.

3.5 Formulation and Solution of Model

On the basis of the transition state diagram by the consideration of possible transition state, we can obtain the following set of differential equations for the present model after applying Markov process:

Table 1 Notations

t	Time scale
s	Laplace transform variable
S_i	Transition state for $i = 0, 1, 2, 3, 4$
$\overline{P}(s)$	Laplace transformation of $P(t)$
$\lambda_C/\lambda_{ST}/\lambda_P/\lambda_U/\lambda_{MN}/\lambda_{MS}/\lambda_K$	Failure rates for control unit/storage unit/power supply/UPS/monitor/mouse/keyboard
$\mu(y)$	Repair rates for the state when control unit, UPS, monitor, mouse, and keyboard unit has been failed
$P_U(t)/P_{CU}(t)/P_{MS}(t)/P_K(t)/P_{KMS}(t)$	The probability of the stage at time t when UPS/control unit/mouse/keyboard/keyboard and mouse have failed
$P_P(x,t)/P_{ST}(x,t)/P_{MN}(y,t)$	The probability density function that the system is in the state, when power supply/storage unit/monitor is failed, at epoch t and has an elapsed repair time of x/y, respectively
$u_1 = e^x, u_2 = \varphi(x)$	The joint probability (failed state S_P, S_{ST}, S_{KMS} to normal state S_S) according to Gumbel–Hougaard family is given as $\exp\left[x^\theta + \{\log \varphi(x)\}^\theta\right]^{\frac{1}{\theta}}$
$E_p(t)$	Expected profit during the interval $[0, t)$
C_1, C_2	Revenue and service cost per unit time, respectively

Table 2 State description of the system

State	Description
S_S	All units are in good working condition
S_U	State of the system when UPS has failed
S_{CU}	State of the system when control unit has failed
S_{MS}	State of the system when mouse has failed
S_K	State of the system when the keyboard has failed
S_{KMS}	State of the system when keyboard and mouse both have failed
S_{ST}	State of the system when storage unit has failed
S_P	State of the system when the power supply has failed
S_{MN}	State of the system when the monitor has failed

$$\left[\frac{\partial}{\partial t} + \lambda_C + \lambda_{ST} + \lambda_U + \lambda_{MN} + \lambda_{MS} + \lambda_K\right] P_S(t) = \mu(y)\left[P_{CU}(t) + P_U(t)\right.$$

$$+ \left. P_{MS}(t) + P_K(t)\right] + \exp\left[x^\theta + \{\log \varphi(x)\}^\theta\right]^{\frac{1}{\theta}} P_{KMS}(t)$$

$$+ \int_0^\infty P_{MN}(y,t)\,\mu(y)dy + \int_0^\infty P_{ST}(x,t)\exp\left[x^\theta + \{\log \varphi(x)\}^\theta\right]^{\frac{1}{\theta}} dx$$

$$+ \int_0^\infty P_P(x,t)\exp\left[x^\theta + \{\log \varphi(x)\}^\theta\right]^{\frac{1}{\theta}} dx$$

$$(1)$$

$$\left[\frac{\partial}{\partial t} + \lambda_{\mathrm{ST}} + \mu(y)\right] P_{\mathrm{CU}}(t) = \lambda_{\mathrm{C}} P_{\mathrm{S}}(t) \tag{2}$$

$$\left[\frac{\partial}{\partial t} + \lambda_{\mathrm{P}} + \mu(y)\right] P_{\mathrm{U}}(t) = \lambda_{\mathrm{U}} P_{\mathrm{S}}(t) \tag{3}$$

$$\left[\frac{\partial}{\partial t} + \lambda_{\mathrm{K}} + \mu(y)\right] P_{\mathrm{MS}}(t) = \lambda_{\mathrm{MS}} P_{\mathrm{S}}(t) \tag{4}$$

$$\left[\frac{\partial}{\partial t} + \lambda_{\mathrm{MS}} + \mu(y)\right] P_{\mathrm{K}}(t) = \lambda_{\mathrm{K}} P_{\mathrm{S}}(t) \tag{5}$$

$$\left[\frac{\partial}{\partial t} + \exp\left[x^{\theta} + \{\log \varphi(x)\}^{\theta}\right]^{\frac{1}{\theta}}\right] P_{\mathrm{KMS}}(t) = \lambda_{\mathrm{K}} P_{\mathrm{MS}}(t) + \lambda_{\mathrm{MS}} P_{\mathrm{K}}(t) \tag{6}$$

$$\left[\frac{\partial}{\partial t} + \frac{\partial}{\partial x} + \exp\left[x^{\theta} + \{\log \varphi(x)\}^{\theta}\right]^{\frac{1}{\theta}}\right] P_{\mathrm{ST}}(x, t) = 0 \tag{7}$$

$$\left[\frac{\partial}{\partial t} + \frac{\partial}{\partial x} + \exp\left[x^{\theta} + \{\log \varphi(x)\}^{\theta}\right]^{\frac{1}{\theta}}\right] P_{\mathrm{P}}(x, t) = 0 \tag{8}$$

$$\left[\frac{\partial}{\partial t} + \frac{\partial}{\partial y} + \mu(y)\right] P_{\mathrm{MN}}(y, t) = 0 \tag{9}$$

Boundary conditions

$$P_{\mathrm{ST}}(0, t) = \lambda_{\mathrm{ST}}\left[P_{\mathrm{S}}(t) + P_{\mathrm{CU}}(t)\right] \tag{10}$$

$$P_{\mathrm{P}}(0, t) = \lambda_{\mathrm{P}} P_{\mathrm{U}}(t) \tag{11}$$

$$P_{\mathrm{MN}}(0, t) = \lambda_{\mathrm{MN}} P_{\mathrm{S}}(t) \tag{12}$$

Initial condition

$$P_{\mathrm{S}}(0) \quad \text{and other state probabilities are zero at } t = 0 \tag{13}$$

Solving Eqs. (1–12) with the help of Laplace transformation, and using Eq. (13), we obtain

$$[s + \lambda_\text{C} + \lambda_\text{ST} + \lambda_\text{U} + \lambda_\text{MN} + \lambda_\text{MS} + \lambda_\text{K}] \overline{P}_\text{S}(s) = 1 + \mu(y) \left[\overline{P}_\text{CU}(s) \right.$$

$$+ \overline{P}_\text{U}(s) + \overline{P}_\text{MS}(s) + \overline{P}_\text{K}(s)] + \exp[x^\theta + \{\log \varphi(x)\}^\theta]^{\frac{1}{\theta}} \overline{P}_\text{KMS}(s)$$

$$+ \int_0^\infty \overline{P}_\text{MN}(y,s) \mu(y) dy + \int_0^\infty \overline{P}_\text{ST}(x,s) \exp\left[x^\theta + \{\log \varphi(x)\}^\theta \right]^{\frac{1}{\theta}} dx$$

$$+ \int_0^\infty \overline{P}_\text{P}(x,s) \exp\left[x^\theta + \{\log \varphi(x)\}^\theta \right]^{\frac{1}{\theta}} dx \tag{14}$$

$$[s + \lambda_\text{ST} + \mu(y)] \overline{P}_\text{CU}(s) = \lambda_\text{C} \overline{P}_\text{S}(s) \tag{15}$$

$$[s + \lambda_\text{P} + \mu(y)] \overline{P}_\text{U}(s) = \lambda_\text{U} \overline{P}_\text{S}(s) \tag{16}$$

$$[s + \lambda_\text{K} + \mu(y)] \overline{P}_\text{MS}(s) = \lambda_\text{MS} \overline{P}_\text{S}(s) \tag{17}$$

$$[s + \lambda_\text{MS} + \mu(y)] \overline{P}_\text{K}(s) = \lambda_\text{K} \overline{P}_\text{S}(s) \tag{18}$$

$$\left[s + \exp\left[x^\theta + \{\log \varphi(x)\}^\theta \right]^{\frac{1}{\theta}} \right] \overline{P}_\text{KMS}(s) = \lambda_\text{K} \overline{P}_\text{MS}(s) + \lambda_\text{MS} \overline{P}_\text{K}(s) \tag{19}$$

$$\left[s + \frac{\partial}{\partial x} + \exp\left[x^\theta + \{\log \varphi(x)\}^\theta \right]^{\frac{1}{\theta}} \right] \overline{P}_\text{ST}(x,s) = 0 \tag{20}$$

$$\left[s + \frac{\partial}{\partial x} + \exp\left[x^\theta + \{\log \varphi(x)\}^\theta \right]^{\frac{1}{\theta}} \right] \overline{P}_\text{P}(x,s) = 0 \tag{21}$$

$$\left[s + \frac{\partial}{\partial y} + \mu(y) \right] \overline{P}_\text{MN}(y,s) = 0 \tag{22}$$

$$\overline{P}_\text{ST}(0,s) = \lambda_\text{ST} \left[\overline{P}_\text{S}(s) + \overline{P}_\text{CU}(s) \right] \tag{23}$$

$$\overline{P}_\text{P}(0,s) = \lambda_\text{P} \overline{P}_\text{U}(s) \tag{24}$$

$$\overline{P}_\text{MN}(0,s) = \lambda_\text{MN} \overline{P}_\text{S}(s) \tag{25}$$

Solving Eqs. (14–22) with the help of Eqs. (23–25), we get

$$\overline{P}_S(s) = \frac{1}{D(s)} \tag{26}$$

$$\overline{P}_{CU}(s) = \frac{\lambda_C}{s + \lambda_{ST} + \mu(y)} \overline{P}_S(s) \tag{27}$$

$$\overline{P}_U(s) = \frac{\lambda_U}{s + \lambda_P + \mu(y)} \overline{P}_S(s) \tag{28}$$

$$\overline{P}_{MS}(s) = \frac{\lambda_{MS}}{s + \lambda_K + \mu(y)} \overline{P}_S(s) \tag{29}$$

$$\overline{P}_K(s) = \frac{\lambda_K}{s + \lambda_{MS} + \mu(y)} \overline{P}_S(s) \tag{30}$$

$$\overline{P}_{KMS}(s) = \frac{\lambda_K \lambda_{MS}}{\left(s + \exp\left[x^\theta + \{\log \varphi(x)\}^\theta \right]^{\frac{1}{\theta}} \right)}$$
$$\times \left(\frac{1}{s + \lambda_K + \mu(y)} + \frac{1}{s + \lambda_{MS} + \mu(y)} \right) \overline{P}_S(s) \tag{31}$$

$$\overline{P}_{ST}(s) = \frac{1}{s} \left(\lambda_{ST} + \frac{\lambda_{ST}}{s + \lambda_{ST} + \mu(y)} \right)$$
$$\times \left(1 - \frac{\exp\left[x^\theta + \{\log \varphi(x)\}^\theta \right]^{\frac{1}{\theta}}}{\left(s + \exp\left[x^\theta + \{\log \varphi(x)\}^\theta \right]^{\frac{1}{\theta}} \right)} \right) \overline{P}_S(s) \tag{32}$$

$$\overline{P}_{MN}(s) = \frac{1}{s} \lambda_{MN} \left(1 - \frac{\mu(y)}{(s + \mu(y))} \right) \overline{P}_S(s) \tag{33}$$

$$\overline{P}_P(s) = \frac{\lambda_P \lambda_U}{s \{ s + \lambda_P + \mu(y) \}} \left(1 - \frac{\exp\left[x^\theta + \{\log \varphi(x)\}^\theta \right]^{\frac{1}{\theta}}}{\left(s + \exp\left[x^\theta + \{\log \varphi(x)\}^\theta \right]^{\frac{1}{\theta}} \right)} \right) \overline{P}_S(s) \tag{34}$$

where

$$D(s) = \left[(s + C_1) - \mu(y) \left[\frac{\lambda_C}{(s + C_2)} + \frac{\lambda_U}{(s + C_3)} + \frac{\lambda_{MS}}{(s + C_4)} + \frac{\lambda_K}{(s + C_5)} \right] \right]$$

$$- \frac{\exp\left[x^\theta + \{\log \varphi(x)\}^\theta \right]^{\frac{1}{\theta}}}{\left(s + \exp\left[x^\theta + \{\log \varphi(x)\}^\theta \right]^{\frac{1}{\theta}} \right)} \lambda_K \lambda_{MS} \left[\frac{1}{(s + C_4)} + \frac{1}{(s + C_5)} \right]$$

$$- \frac{\exp\left[x^\theta + \{\log \varphi(x)\}^\theta \right]^{\frac{1}{\theta}}}{\left(s + \exp\left[x^\theta + \{\log \varphi(x)\}^\theta \right]^{\frac{1}{\theta}} \right)} \left[\lambda_{ST} + \frac{\lambda_{ST} \lambda_C}{(s + C_2)} + \frac{\lambda_P \lambda_U}{(s + C_3)} \right]$$

$$- \frac{\mu(y)}{\{s + \mu(y)\}} \lambda_{MN}$$

$$C_1 = \lambda_C + \lambda_{ST} + \lambda_U + \lambda_{MN} + \lambda_{MS} + \lambda_K, \quad C_2 = \lambda_{ST} + \mu(y),$$

$$C_3 = \lambda_P + \mu(y), \quad C_4 = \lambda_K + \mu(y), \quad C_5 = \lambda_{MS} + \mu(y)$$

The Laplace transformation of the probabilities that the system is in upstate (i.e., either good or degraded):

$$\overline{P}_{up}(s) = \overline{P}_S(s) + \overline{P}_{CU}(s) + \overline{P}_U(s) + \overline{P}_{MS}(s) + \overline{P}_K(s) + \overline{P}_{KMS}(s)$$

$$= \left\{ 1 + \frac{\lambda_C}{(s + C_2)} + \frac{\lambda_U}{(s + C_3)} + \frac{\lambda_{MS}}{(s + C_4)} + \frac{\lambda_K}{(s + C_5)} \right.$$

$$\left. + \frac{\lambda_K \lambda_{MS}}{s + \exp\left[x^\theta + \{\log \varphi(x)\}^\theta \right]^{\frac{1}{\theta}}} \left(\frac{1}{(s + C_4)} + \frac{1}{(s + C_5)} \right) \right\} \overline{P}_S(s)$$

$$(35)$$

The Laplace transformation of the probabilities that the system is in downstate (i.e., failed):

$$\overline{P}_{\text{down}}(s) = \overline{P}_{\text{P}}(s) + \overline{P}_{\text{ST}}(s) + \overline{P}_{\text{MN}}(s)$$

$$= \left\{ \frac{1}{s} \left(1 - \frac{\exp\left[x^{\theta} + \{\log \varphi(x)\}^{\theta} \right]^{\frac{1}{\theta}}}{\left(s + \exp\left[x^{\theta} + \{\log \varphi(x)\}^{\theta} \right]^{\frac{1}{\theta}} \right)} \right) \right.$$

$$\left. \times \left(\lambda_{\text{ST}} + \frac{\lambda_{\text{ST}}\lambda_{\text{C}}}{(s + C_2)} + \frac{\lambda_{\text{P}}\lambda_{\text{U}}}{(s + C_3)} \right) + \lambda_{\text{MN}} \frac{1}{s} \left[1 - \frac{\mu(y)}{s + \mu(y)} \right] \right\} \overline{P}_{\text{S}}(s)$$

$$(36)$$

4 Particular Cases and Numerical Computations

4.1 Availability Analysis

4.1.1 When the System in Comprehensive State

Initially, the system works properly, for this, setting the value of different failure and repair rates as $\lambda_{\text{C}} = 0.2$, $\lambda_{\text{ST}} = 0.3$, $\lambda_{\text{U}} = 0.2$, $\lambda_{\text{MN}} = 0.1$, $\lambda_{\text{MS}} = 0.5$, $\lambda_{\text{K}} = 0.4$, $\lambda_{\text{P}} = 0.3$, $\varphi(x) = 1$, $\mu(y) = 1$ in Eq. (35), one can obtain the availability of the system

$$P_{\text{up}}(t) = \left\{ 0.08747435461 \ e^{(-2.903649803t)} \cos\left(0.3135682026t \right) \right.$$

$$- 0.3968940408 \ e^{(-2.903649803t)} \sin \ (0.3135682026t)$$

$$+ 0.0007440467225 \ e^{(-1.458316302t)} + 0.01570933388 \ e^{(-1.018061332t)}$$

$$\left. + 0.003352035390 \ e^{(-1.334602761t)} + 0.8927202294 \right\}$$

$$(37a)$$

4.1.2 When No Failure in Control and Storage Unit

The control and storage units are perfect, i.e., no failure occurrence in both the units, putting other failure and repair rates as $\lambda_{\text{U}} = 0.2$, $\lambda_{\text{MN}} = 0.1$, $\lambda_{\text{MS}} = 0.5$, $\lambda_{\text{K}} = 0.4$, $\lambda_{\text{P}} = 0.3$, $\varphi(x) = 1$, $\mu(y) = 1$ in Eq. (35), we have,

$$P_{\text{up}}(t) = \left\{ 0.03538863784 \ e^{(-2.660026210t)} \cos \ (0.6818499005t) \right.$$

$$+ 0.2778239624 \ e^{(-2.660026210t)} \sin \ (0.6818499005t)$$

$$+ 0.0007092452156 \ e^{(-1.455748534t)} + 0.01988553985 \ e^{(-1.021910991t)}$$

$$\left. + 0.002616287337 \ e^{(-1.320568054t)} + 0.9414002897 \right\}$$

$$(37b)$$

4.1.3 When No Failure in Power and Monitor

The power and monitor of the system are in perfect working condition, then setting different failure and repair rates as: $\lambda_C = 0.2$, $\lambda_{ST} = 0.3$, $\lambda_U = 0.2$, $\lambda_{MN} = 0$, $\lambda_{MS} = 0.5$, $\lambda_K = 0.4$, $\lambda_P = 0$, $\varphi(x) = 1$, $\mu(y) = 1$. Substituting all these values in Eq. (35), one can obtain

$$
\begin{aligned}
P_{up}(t) = \Big\{ & 0.5785670001 \; e^{(-2.850684518t)} \cos\left(0.2437004362t\right) \\
& - 0.5523608006 \; e^{(-2.850684518t)} \sin\ (0.0.2437004362t) \\
& - 0.002282954416 \; e^{(-1.040489494t)} + 0.002044908119 \; e^{(-1.320708716t)} \\
& + 0.00009129208258 \; e^{(-1.455712754t)} + 0.9422900542 \Big\}
\end{aligned}
$$

(37c)

4.1.4 When No Failure in Keyboard and Mouse

When keyboard and mouse have no failure, putting the value of different failure and repair rates as $\lambda_C = 0.2$, $\lambda_{ST} = 0.3$, $\lambda_U = 0.2$, $\lambda_{MN} = 0.1$, $\lambda_{MS} = 0$, $\lambda_K = 0$, $\lambda_P = 0.3$, $\varphi(x) = 1$, in Eq. (35), one can obtain

$$
\begin{aligned}
P_{up}(t) = \Big\{ & 0.1110157824 \; e^{(-3.038692672t)} + 0.0206201604 \; e^{(-1.745678579t)} \\
& + 0.02578659225 \; e^{(-1.033908748t)} + 0.8425774650 \Big\}
\end{aligned}
$$

(37d)

Varying the time unit t from 0 to 20 in each case of availability, the computed value in all four cases of availability is shown in Table 3 and demonstrated by the graphs in Fig. 2, respectively.

4.2 Reliability Analysis

The reliability of the design system can be found by fixing the repair rates equal to zero.

4.2.1 When the System in Comprehensive State

When the system is fully functioning, taking the value of different failure rates as $\lambda_C = 0.2$, $\lambda_{ST} = 0.3$, $\lambda_U = 0.2$, $\lambda_{MN} = 0.1$, $\lambda_{MS} = 0.5$, $\lambda_K = 0.4$, $\lambda_P = 0.3$. Substituting all these values in Eq. (35), one can obtain the reliability of the system

$$
R(t) = 0.1848739496 \; e^{(-1.7t)} + 0.5294117647 + 0.2857142857 \; e^{(-0.3t)} \quad \text{(38a)}
$$

Table 3 Availability as function of time

| Time (t) | Availability $P_{up}(t)$ | | | |
	37a	37b	37c	37d
0	1.00000	1.00000	1.00000	1.00000
1	0.89730	0.95257	0.93759	0.86066
2	0.89456	0.94437	0.94146	0.84672
3	0.89349	0.94239	0.94217	0.84386
4	0.89300	0.94175	0.94226	0.84301
5	0.89282	0.94152	0.94228	0.84273
6	0.89275	0.94144	0.94229	0.84263
7	0.89273	0.94142	0.94229	0.84260
8	0.89272	0.94141	0.94229	0.84258
9	0.89272	0.94140	0.94229	0.84258
10	0.89272	0.94140	0.94229	0.84258
11	0.89272	0.94140	0.94229	0.84258
12	0.89272	0.94140	0.94229	0.84258
13	0.89272	0.94140	0.94229	0.84258
14	0.89272	0.94140	0.94229	0.84258
15	0.89272	0.94140	0.94229	0.84258
16	0.89272	0.94140	0.94229	0.84258
17	0.89272	0.94140	0.94229	0.84258
18	0.89272	0.94140	0.94229	0.84258
19	0.89272	0.94140	0.94229	0.84258
20	0.89272	0.94140	0.94229	0.84258

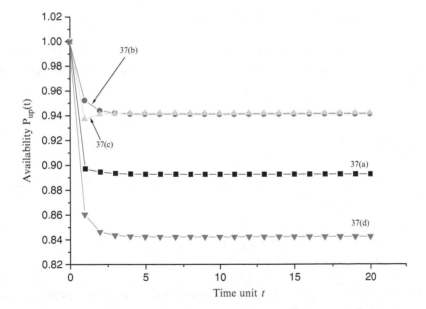

Fig. 2 Availability as function of time

4.2.2 When No Control Unit and Storage Unit Are failed

The control unit and storage unit are not failed, then their corresponding failure rates are zero, and other rates as $\lambda_C = 0$, $\lambda_{ST} = 0$, $\lambda_U = 0.2$, $\lambda_{MN} = 0.1$, $\lambda_{MS} = 0.5$, $\lambda_K = 0.4$, $\lambda_P = 0.3$. Putting all values in Eq. (35), we have

$$R(t) = 0.02777777778 \ e^{(-1.2t)} + 0.75 + 0.2222222222 \ e^{(-0.3t)} \qquad (38b)$$

4.2.3 When No Power and Monitor Are failed

Taking the value of different failure rates as $\lambda_C = 0.2$, $\lambda_{ST} = 0.3$, $\lambda_U = 0.2$, $\lambda_{MN} = 0$, $\lambda_{MS} = 0.5$, $\lambda_K = 0.4$, $\lambda_P = 0$. Substituting all values in Eq. (35), we can obtain the reliability of the system as

$$R(t) = 0.1586538462 \ e^{(-1.6t)} + 0.6875 + 0.1538461538 \ e^{(-0.3t)} \qquad (38c)$$

4.2.4 When No Keyboard and Mouse Are failed

The keyboard and mouse are not failed, setting the value of different failure rates as $\lambda_C = 0.2$, $\lambda_{ST} = 0.3$, $\lambda_U = 0.2$, $\lambda_{MN} = 0.1$, $\lambda_{MS} = 0$, $\lambda_K = 0$, $\lambda_P = 0.3$,. Putting all the values in Eq. (35), we can obtain the reliability of the system as:

$$R(t) = 0.2 \ e^{(-0.55t)} \ (5. \cosh(0.25) + 3. \sinh(0.25t)) \qquad (38d)$$

Varying the time unit t from 0 to 20 in each case of reliability, the computed numeric values are given in Table 4 and correspondingly shown the graph of reliability with respect to time in Fig. 3.

4.3 Expected Profit

For an organization and an official point of view, the expected profit during the interval $[0, t)$ is given as

$$E_P(t) = C_1 \int_0^t P_{up}(t)dt - tC_2 \qquad (39)$$

Using Eq. (37a) for a the comprehensive state only in Eq. (39), the cost for the same set of parameters is obtained as

Table 4 Reliability as function of time

| Time (t) | Reliability $R(t)$ | | | |
	38a	38b	38c	38d
0	1.00000	1.00000	1.00000	1.00000
1	0.77485	0.92299	0.83350	0.68252
2	0.69238	0.87448	0.77840	0.47943
3	0.64670	0.84111	0.75135	0.34340
4	0.61567	0.81716	0.73410	0.24910
5	0.59320	0.79965	0.72188	0.18217
6	0.57665	0.78675	0.71294	0.13389
7	0.56440	0.77722	0.70634	0.09870
8	0.55533	0.77016	0.70146	0.07291
9	0.54861	0.76494	0.69784	0.05391
10	0.54364	0.76106	0.69516	0.03990
11	0.53995	0.75820	0.69317	0.02954
12	0.53722	0.75607	0.69170	0.02187
13	0.53520	0.75450	0.69061	0.01620
14	0.53369	0.75333	0.68981	0.01200
15	0.53259	0.75247	0.68921	0.00889
16	0.53176	0.75183	0.68877	0.00658
17	0.53115	0.75135	0.68844	0.00488
18	0.53070	0.75100	0.68819	0.00361
19	0.53037	0.75074	0.68801	0.00268
20	0.53012	0.75055	0.68788	0.00198

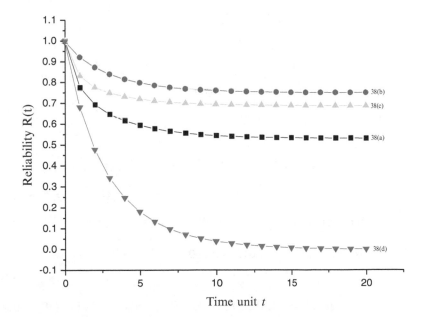

Fig. 3 Reliability as function of time

Table 5 Expected profit as function of time

Time (t)	$E_p(t)$				
	$C_2 = 0.1$	$C_2 = 0.2$	$C_2 = 0.3$	$C_2 = 0.4$	$C_2 = 0.5$
0	0.00000	0.00000	0.00000	0.00000	0.00000
1	0.82155	0.72155	0.62155	0.52155	0.42155
2	1.61707	1.41707	1.21707	1.01707	0.81707
3	2.41104	2.11104	1.81104	1.51104	1.21104
4	3.20425	2.80425	2.40425	2.00425	1.60425
5	3.99714	3.49714	2.99714	2.49714	1.99714
6	4.78993	4.18993	3.58993	2.98993	2.38993
7	5.58267	4.88267	4.18267	3.48267	2.78267
8	6.37540	5.57540	4.77549	3.97540	3.17540
9	7.16812	6.26812	5.36812	4.46812	3.56812

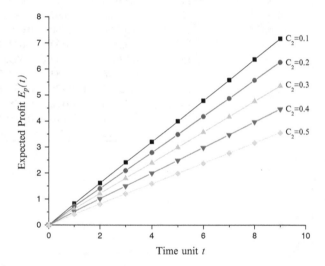

Fig. 4 Expected profit as function of time

$$E_P(t) = C_1 \left\{ -0.01518745908 \ e^{(-2.903649803t)} \cos\left(0.3135682026t \right) \right.$$
$$+ 0.1383280947 \ e^{(-2.903649803t)} \sin\left(0.3135682026t \right) + 0.3363993984$$
$$- 0.0005102094254 \ e^{(-1.458316302t)} - 0.002511635288 \ e^{(-1.334602761t)}$$
$$+ 0.8927202294t - 0.01543063604 \ e^{(-1.018061332t)} \left. \right\} - tC_2$$

$$(40)$$

Setting $C_1 = 1$ and $C_2 = 0.1, 0.2, 0.3, 0.4, 0.5$, respectively, in Eq. (40), we get the Table 5 and obtained results are demonstrated by Fig. 4.

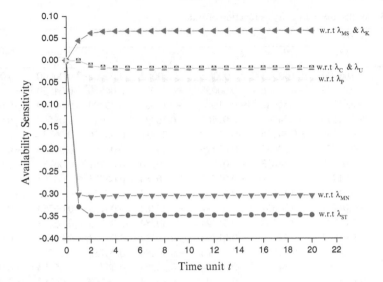

Fig. 5 Availability sensitivity as function of time

4.4 Sensitivities

Sensitivity of a function is explained as the partial derivative of the function with respect to their input factors. Sensitivity analysis, also called importance analysis (Henley and Kumamoto 1992; Andrews and Moss 1993), help detect which parameter contribute most to system performance and thus would be good ones for elevate. Sensitivity to a factor is defined as the partial derivative of the function with respect to input parameters. Here, these input parameters are the failure rates of the system.

4.4.1 Availability Sensitivity

Availability sensitivity can be obtained by partial differentiation of Eq. (35) with respect to the failure rates of control unit, storage unit, UPS, monitor, mouse, keyboard, power supply, respectively after taking unity as the repair rates. Using the values of the failure rates as $\lambda_C = 0.2$, $\lambda_{ST} = 0.3$, $\lambda_U = 0.2$, $\lambda_{MN} = 0.1$, $\lambda_{MS} = 0.5$, $\lambda_K = 0.4$, $\lambda_P = 0.3$, we have obtained the values of partial derivatives $\frac{\partial P_{up}(t)}{\partial \lambda_C}$, $\frac{\partial P_{up}(t)}{\partial \lambda_{ST}}$, $\frac{\partial P_{up}(t)}{\partial \lambda_U}$, $\frac{\partial P_{up}(t)}{\partial \lambda_{MN}}$, $\frac{\partial P_{up}(t)}{\partial \lambda_{MS}}$, $\frac{\partial P_{up}(t)}{\partial \lambda_K}$, $\frac{\partial P_{up}(t)}{\partial \lambda_P}$. Taking the time unit from 0 to 20, we obtain the Table 6 and correspondingly Fig. 5.

Table 6 Availability sensitivity as function of time

Time (t)	$E_p(t)$						
	$\frac{\partial P_{up}(t)}{\partial \lambda_C}$	$\frac{\partial P_{up}(t)}{\partial \lambda_{ST}}$	$\frac{\partial P_{up}(t)}{\partial \lambda_U}$	$\frac{\partial P_{up}(t)}{\partial \lambda_{MN}}$	$\frac{\partial P_{up}(t)}{\partial \lambda_{MS}}$	$\frac{\partial P_{up}(t)}{\partial \lambda_K}$	$\frac{\partial P_{up}(t)}{\partial \lambda_P}$
0	0.00000	0.00000	0.00000	0.00000	0.00000	0.00000	0.00000
1	−0.00038	−0.32766	−0.00038	−0.29998	0.04550	0.04550	−0.02768
2	−0.01070	−0.34680	−0.01070	−0.30581	0.06289	0.06289	−0.04100
3	−0.01570	−0.34746	−0.01569	−0.30379	0.06647	0.06647	−0.04366
4	−0.01730	−0.34706	−0.01730	−0.30310	0.06726	0.06726	−0.04396
5	−0.01776	−0.34683	−0.01776	−0.30291	0.06746	0.06746	−0.04392
6	−0.01789	−0.34673	−0.01789	−0.30286	0.06751	0.06751	−0.04387
7	−0.01793	−0.34669	−0.01793	−0.30284	0.06753	0.06753	−0.04384
8	−0.01793	−0.34668	−0.01793	−0.30284	0.06753	0.06753	−0.04384
9	−0.01794	−0.34667	−0.01794	−0.30284	0.06753	0.06753	−0.04383
10	−0.01794	−0.34667	−0.01794	−0.30284	0.06753	0.06753	−0.04383
11	−0.01794	−0.34667	−0.01794	−0.30284	0.06753	0.06753	−0.04383
12	−0.01794	−0.34667	−0.01794	−0.30284	0.06753	0.06753	−0.04383
13	−0.01794	−0.34667	−0.01794	−0.30284	0.06753	0.06753	−0.04383
14	−0.01794	−0.34667	−0.01794	−0.30284	0.06753	0.06753	−0.04383
15	−0.01794	−0.34667	−0.01794	−0.30284	0.06753	0.06753	−0.04383
16	−0.01794	−0.34667	−0.01794	−0.30284	0.06753	0.06753	−0.04383
17	−0.01794	−0.34667	−0.01794	−0.30284	0.06753	0.06753	−0.04383
18	−0.01794	−0.34667	−0.01794	−0.30284	0.06753	0.06753	−0.04383
19	−0.01794	−0.34667	−0.01794	−0.30284	0.06753	0.06753	−0.04383
20	−0.01794	−0.34667	−0.01794	−0.30284	0.06753	0.06753	−0.04383

4.4.2 Reliability Sensitivity

Sensitivity of reliability can be analyzed by partial differentiation of Eq. (38a) with respect to the failure rates of control unit, storage unit, UPS, monitor, mouse, keyboard, power supply, respectively. Using the values of the failure rates $\lambda_C = 0.2$, $\lambda_{ST} = 0.3$, $\lambda_U = 0.2$, $\lambda_{MN} = 0.1$, $\lambda_{MS} = 0.5$, $\lambda_K = 0.4$, $\lambda_P = 0.3$, we have obtained the values of $\frac{\partial R(t)}{\partial \lambda_C}$, $\frac{\partial R(t)}{\partial \lambda_{ST}}$, $\frac{\partial R(t)}{\partial \lambda_U}$, $\frac{\partial R(t)}{\partial \lambda_{MN}}$, $\frac{\partial R(t)}{\partial \lambda_{MS}}$, $\frac{\partial R(t)}{\partial \lambda_K}$, $\frac{\partial R(t)}{\partial \lambda_P}$. Taking the time unit from 0 to 20 units, one can obtain the Table 7 and corresponding Fig. 6.

5 Result Discussion

From Fig. 2, we have analyzed that when the system is in the comprehensive state, the availability of the system first decreases quickly and then becomes constant. When control and storage unit are not failed, then availability of the system decreases quickly and then becomes constant. But in this case, availability is high as compare to comprehensive state. In the same manner, when power supply and monitor are not failed, availability first decreases sharply and coincide with the availability of the system when control and storage unit are not failed. Further, when the keyboard and mouse are not failed, availability of the system is lowest. Firstly, it decreases quickly and then becomes constant.

Table 7 Reliability sensitivity as function of time

Time (t)	Reliability sensitivity						
	$\frac{\partial R(t)}{\partial \lambda_C}$	$\frac{\partial R(t)}{\partial \lambda_{ST}}$	$\frac{\partial R(t)}{\partial \lambda_U}$	$\frac{\partial R(t)}{\partial \lambda_{MN}}$	$\frac{\partial R(t)}{\partial \lambda_{MS}}$	$\frac{\partial R(t)}{\partial \lambda_K}$	$\frac{\partial R(t)}{\partial \lambda_P}$
0	0.00000	0.00000	0.00000	0.00000	0.00000	0.00000	0.00000
1	−0.00354	−0.45108	−0.00354	−0.40221	0.07857	0.07857	−0.04887
2	−0.05039	−0.52276	−0.05039	−0.41856	0.15005	0.15005	−0.10421
3	−0.10858	−0.52801	−0.10858	−0.39463	0.19002	0.19002	−0.13338
4	−0.15879	−0.51463	−0.15879	−0.37314	0.21444	0.21444	−0.14149
5	−0.19781	−0.49367	−0.19781	−0.35704	0.23108	0.23108	−0.13663
6	−0.22713	−0.47000	−0.22713	−0.34518	0.24304	0.24304	−0.12482
7	−0.24895	−0.44638	−0.24895	−0.33642	0.25182	0.25182	−0.10996
8	−0.26514	−0.42435	−0.26514	−0.32993	0.25830	0.25830	−0.09442
9	−0.27713	−0.40468	−0.27713	−0.32513	0.26310	0.26310	−0.07954
10	−0.28602	−0.38762	−0.28602	−0.32158	0.26666	0.26666	−0.06604
11	−0.29260	−0.37314	−0.29260	−0.31895	0.26929	0.26929	−0.05419
12	−0.29748	−0.36105	−0.29748	−0.31699	0.27124	0.27124	−0.04405
13	−0.30109	−0.35108	−0.30109	−0.31555	0.27269	0.27269	−0.03552
14	−0.30377	−0.34294	−0.30377	−0.31448	0.27376	0.27376	−0.02846
15	−0.30575	−0.33636	−0.30575	−0.31369	0.27455	0.27455	−0.02267
16	−0.30722	−0.33107	−0.30722	−0.31310	0.27514	0.27514	−0.01797
17	−0.30831	−0.32685	−0.30831	−0.31266	0.27557	0.27557	−0.01418
18	−0.30911	−0.32349	−0.30911	−0.31234	0.27589	0.27589	−0.01115
19	−0.30971	−0.32084	−0.30971	−0.31210	0.27613	0.27613	−0.00874
20	−0.31015	−0.31875	−0.31015	−0.31192	0.27631	0.27631	−0.00683

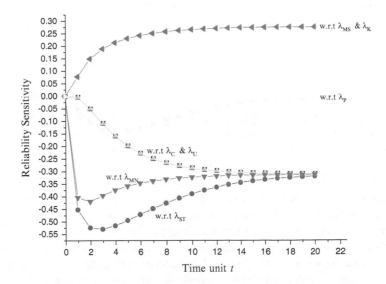

Fig. 6 Reliability sensitivity as function of time

Figure 3 shows the reliability of the system. When the system is in a comprehensive state, the reliability of the system first decreases smoothly and then becomes constant. Similarly, when storage and control unit are not failed, and power supply with the monitor are not failed, reliability of the system first decreases smoothly and then becomes constant, but the reliability of the system in case when storage and control unit are not failed, is highest. When keyboard and mouse are not failed reliability of the system is lowest. In this case also, reliability first decreases, quickly in the shape of a curve and then becomes constant.

Figure 4 shows the expected profit, when the revenue per unit time fixed at one and varying service cost from 0.1 to 0.5. It is clear from the graph that the profit decrease as the service cost increase.

From Fig. 5, availability sensitivity of the system decreases swiftly (as a straight line) and then becomes constant as time increases with respect to the failure rates of storage unit and monitor. Availability sensitivity with respect to the failure rate of keyboard and mouse increase, but after a short curve, it becomes constant as time increase. With respect to the failure rate of power supply, UPS and control unit, it decreases shortly and then becomes constant. From Fig. 6, reliability sensitivity of the system with respect to the failure rate of the monitor and storage unit, are first to decrease as a straight line but again after some increment, it becomes constant. With respect to the failure rate of power supply, reliability sensitivity decreases in the form of a smooth curve and then it comes back to near zero and then becomes constant as time increases. Reliability sensitivity with respect to the failure rate of keyboard and mouse increases and after some times, it also becomes constant. Reliability sensitivity with respect to the failure rate of UPS firstly becomes constant for a very short time after that, this decrease as time unit increases in the form of the hyperbola and becomes constant.

6 Conclusion

In this chapter, we have analyzed the availability, reliability, cost, and sensitivity of the home or office based computer system by introducing a mathematical model. The availability, reliability, and sensitivity become constant over a certain period of time. The system based profit decreases as service cost increases. It is also noticeable, the system could make less sensitive by controlling its failure rates. With the help of this developed model, one can conclude that the results achieved in this work are valuable in the study of improving the performance of the computer systems that contain multi-stages. Hence the present work evidently shows the importance of copula repair modeling, which seems very much to be possible at home or office based computer systems. The future work in this area can be keen to the elaboration of more complexities for specific computer networks.

References

Andrews JD, Moss TR (1993) Reliability and risk assessment. Longman Scientific and Technical, London

Azaron A, Katagiri H, Kato K, Sakawa M (2005) Reliability evaluation and optimization of dissimilar-component cold-standby redundant systems. J Oper Res Soc Jpn 48(1):71–88

Elyasi-Komari I, Gorbenko A, Kharchenko VS, Mamalis A (2011) Analysis of computer network reliability and criticality: technique and features. Int J Comm Netw Syst Sci 4(11):720–726

Goel LR, Gupta R, Rana VS (1993) Reliability analysis of a satellite-based computer communication network system. Microelectron Reliab 33(2):119–126

Henley EJ, Kumamoto H (1992) Probabilistic risk assessment. IEEE Press, Piscataway

Lyu MR (1996) Handbook of software reliability engineering, vol 3. IEEE Computer Society Press, Los Alamitos

McCluskey EJ, Mitra S (2004) Fault tolerance. In: Tucker AB (ed) Computer science handbook, 2nd edn. Chapman and Hall/CRC Press, London, Chapter 25

Nagiya K, Ram M (2013) Reliability characteristics of a satellite communication system including earth station and terrestrial system. Int J Perform Eng 9(6):667–676

Nelsen RB (2006) An introduction to copulas, 2nd edn. Springer, New York

Rajaraman V (2010) Fundamentals of computers. PHI Learning Pvt Ltd, New Delhi

Ram M (2010) Reliability measures of a three-state complex system: a copula approach. Appl Appl Math 5(10):1483–1492

Ram M (2013) On system reliability approaches: a brief survey. Int J Syst Assur Eng Manag 4(2):101–117

Ram M, Singh SB (2008) Availability and cost analysis of a parallel redundant complex system with two types of failure under preemptive-resume repair discipline using Gumbel-Hougaard family copula in repair. Int J Reliab Qual Saf Eng 15(04):341–365

Ram M, Singh SB (2010a) Analysis of a complex system with common cause failure and two types of repair facilities with different distributions in failure. Int J Reliab Saf 4(4):381–392

Ram M, Singh SB (2010b) Availability, MTTF and cost analysis of complex system under preemptive-repeat repair discipline using Gumbel-Hougaard family copula. Int J Qual Reliab Manag 27(5):576–595

Ram M, Singh SB, Singh VV (2013) Stochastic analysis of a standby system with waiting repair strategy. IEEE Trans Syst Man Cybern Syst 43(3):698–707

Singh VV, Ram M, Rawal DK (2013a) Cost analysis of an engineering system involving subsystems in series configuration. IEEE Trans Autom Sci Eng 10(4):1124–1130

Singh VV, Singh SB, Ram M, Goel CK (2013b) Availability, MTTF and cost analysis of a system having two units in series configuration with controller. Int J Syst Assur Eng Manag4(4):341–352

Soi IM, Aggarwal KK (1980) On human reliability trends in digital communication systems. Microelectron Reliab 20(6):823–830

Verma AK, Ajit S, Karanki DR (2010) Reliability and safety engineering, 1st edn. Springer, London. ISBN 978-1-84996-231-5

Reliability of Profiled Blast Wall Structures

Mohammad H. Hedayati, Srinivas Sriramula, and Richard D. Neilson

Abstract Stainless steel profiled walls have been used increasingly in the oil and gas industry to protect people and personnel against hydrocarbon explosions. Understanding the reliability of these blast walls greatly assists in improving the safety of offshore plant facilities. However, the presence of various uncertainties combined with a complex loading scenario makes the reliability assessment process very challenging. Therefore, a parametric model developed using ANSYS APDL is presented in this chapter. The significant uncertainties are combined with an advanced analysis model to investigate the influence of loading, material and geometric uncertainties on the response of these structures under realistic boundary conditions. To review and assess the effects of the dynamics and nonlinearities, four types of analyses including linear static, nonlinear static, linear transient dynamic, and nonlinear transient dynamic are carried out. The corresponding reliability of these structures is evaluated with a Monte Carlo simulation (MCS) method, implementing the Latin hypercube sampling (LHS) approach. The uncertainties related to dynamic blast loading, material properties, and geometry are represented in terms of probability distributions and the associated parameters. Dynamic, static, linear, and nonlinear responses of the structure are reviewed. Stochastic probabilistic analysis results are discussed in terms of the probability of occurrence, the cumulative distribution functions (CDFs), and the corresponding variable sensitivities. It is

M.H. Hedayati (✉) • R.D. Neilson
School of Engineering, University of Aberdeen, 3 Burnieboozle Crescent, Aberdeen, AB15 8NN, UK
e-mail: mohammad.hedayati@abdn.ac.uk; r.d.neilson@abdn.ac.uk

S. Sriramula
Lloyd's Register Foundation (LRF) Centre for Safety and Reliability Engineering, School of Engineering, University of Aberdeen, Aberdeen, UK
e-mail: s.sriramula@abdn.ac.uk

S. Kadry and A. El Hami (eds.), *Numerical Methods for Reliability and Safety Assessment: Multiscale and Multiphysics Systems*, DOI 10.1007/978-3-319-07167-1_13, © Springer International Publishing Switzerland 2015

observed that using the approach taken in this study can help identify the important variables and parameters to optimize the design of profiled blast walls, to perform risk assessments, or to carry out performance-based design for these structures.

1 Introduction

In modern structural engineering design, it is always recommended to assess the performance of complex structures, such as blast walls, under the effects of material, loading, and geometric uncertainties. The existence of these uncertainties cannot be avoided in many stages of structural integrity assessments or design and it may not be possible to justify some of the design decisions without considering them. If the randomness of a variable is relatively small, it can be considered as a deterministic variable. In the real world, most design variables have inherent uncertainties and it is required to consider them properly in assessing the structural performance, either in terms of random variables or random processes. The traditional safety factor-based design may not capture the structural behavior appropriately.

Stainless steel profiled walls are widely used in offshore facilities for protection against hydrocarbon explosions (Louca and Boh 2004). Understanding the reliability of these blast walls greatly assists in improving the safety of offshore personnel and facilities. However, the presence of various uncertainties combined with a complex dynamic loading scenario make the reliability assessment process very challenging. In recent years, probabilistic analysis methods have increasingly been studied (Haldar and Mahdavan 2000) and implemented to provide a new tool with which inherent uncertainties and variations in complex systems such as offshore structures can be considered. Nevertheless, there are various immature areas, methods and approaches that need to be developed and enhanced.

At the start of this research, a preliminary reliability approach was conducted implementing a static approach to perform finite element reliability analyses on profiled barrier blast walls (Hedyati and Sriramula 2012). It was observed that the response variables have been influenced strongly by pressure and depth of section to the same extend. Then a reliability study was developed considering the dynamic effects and nonlinearities in geometry and material properties (Hedayati et al. 2013). It was noticed that considering the dynamic and nonlinearity effects, the correlation sensitivity results are not similar at different time steps. Consequently, more investigations and studies were carried out on linear dynamic analysis, without implementing any nonlinearity effects, to review and understand linear dynamic behavior of profiled blast walls under explosion loading (Hedayati et al. 2014). In the previous studies associated with this research, it was found that performing linear, nonlinear, static, and dynamic analyses are crucial to assess dynamic and nonlinearity effects while implementing reliability approach. Accordingly, linear static (SLMG), nonlinear static (SNMG), linear dynamic (DLMG), and nonlinear dynamic (DNMG) analyses were carried out and the associated results are reviewed and discussed in this chapter. MG indicates variation of material and geometry during the analysis runs.

This chapter presents the reliability analysis of a blast wall profiled barrier, under explosion (pressure) loading, using a parametric model developed with the ANSYS Parametric Design Language (APDL), with and without consideration of the dynamics and any nonlinearity. By defining the uncertainties in different properties as random variables, it is possible to efficiently implement simulation strategies in assessing the structural performance. The response parameters can be obtained by linking the simulated values with the finite element models. These values can be used to perform reliability analysis directly or by considering an implicit performance function (e.g., using response surface methods). In the present study, a Latin hypercube sampling (LHS) approach has been used to study the performance of blast walls. This study will later be extended to consider implicit performance functions to implement in a stochastic finite element framework.

2 Design of Profiled Blast Wall Structures

Compared to other possible ways of protection against explosions, blast walls have lower cost/strength ratio and can be installed very quickly (Haifu and Xueguang 2009). Blast wall structures can be formed of stiffened or unstiffened panels; however, stainless steel profiled walls have increasingly been used in the offshore industry because of their excellent energy absorption and temperature-dependent properties (Brewerton 1999; Louca and Boh 2004).

In general, considering the deterministic response of profiled barrier structures, two approaches are usually recommended for the design of blast wall structures: the traditional single degree of freedom (SDOF) approach or the more sophisticated multi degree of freedom (MDOF) approach. The simplified SDOF approach is widely used in the offshore industry for predicting the dynamic structural response by implementing the Biggs method (Biggs 1964). This is a simple approach which idealizes the actual structure as a spring mass model and is thus very useful in routine design procedures to obtain accurate results for relatively simple structures with limited ductility (Louca and Boh 2004). The SDOF approach is a useful technique for conceptual or basic design of the profiled barrier structures under explosion loadings, whereas, the MDOF method, which is based on a finite element analysis (FEA) approach, provides a detailed analysis of the blast wall and is more accurate compared to the SDOF approach, but is computationally very intensive and, as a result, more expensive. However, with recent developments in computing technology, performing FEA is easier and faster than it was in the past. There have also been some preliminary studies to verify SDOF results against MDOF results (Liang et al. 2007).

The Design Guide for stainless steel blast walls, known as the Technical Note 5 (TN5) (Brewerton 1999), prepared by the Fire and Blast Information Group (FABIG) and API Recommended Practice 2FB (API 2FB 2006) are the two common technical industrial guidelines for the design of profiled blast walls based on the SDOF method.

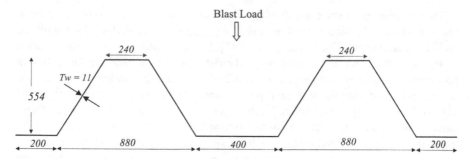

Fig. 1 Geometry of considered section (mm)

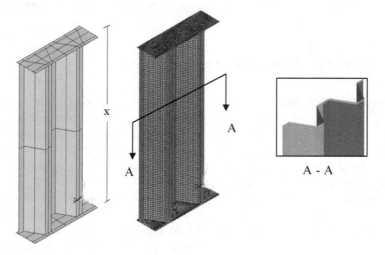

Fig. 2 Finite element representation

In the present study, in accordance with the design guidance from TN5, a profiled wall section that satisfies the geometric limits to be an appropriate structural element is considered. The geometry of the considered profiled barrier section is shown in Fig. 1. As mentioned earlier, the maximum capacities and deflections of these sections are of interest. The stainless steel section considered is assumed to have a Young's modulus of 200 GPa, Poisson's ratio of 0.3, and material density of 7,850 kg/m^3. The end plate thickness is 12.5 mm and the wall has a span, X, 6,000 mm as shown in Fig. 2 (Louca and Boh 2004).

3   Monte Carlo Simulation

Reliability analysis based on Monte Carlo simulation (MCS) is widely used because of the ease of implementation and the ability to handle complex engineering problems. It is particularly amenable to finite element-based studies when the performance functions are not readily available in closed form. However, when the associated probabilities of failure are very small, the computational effort increases significantly. In such cases, depending on the performance function, it is possible to reduce the required number of simulations by using an appropriate variation reduction scheme such as importance sampling, LHS, or directional simulation (Choi et al. 2006).

It is possible to implement different sampling methods within ANSYS (ANSYS 2012; Reh et al. 2006). In the present study, an LHS scheme is used. A major advantage of LHS is that it avoids repeated sample reliability evaluations (ANSYS 2012), thus drastically reducing the number of simulations. LHS also considers the tails of the distributions more accurately. This is very important for most structural engineering applications where extreme values are essential.

The LHS technique was first introduced by McKay et al. (1979). Later on, further developments were explained by other researchers, for example Iman et al. (1981). A typical LHS selects n different values from each of the k variables $X_1 \ldots X_k$ as per the following routine (Wyss and Jorgensen 1998):

- The range of each variable is divided into n non-overlapping intervals on the basis of equal probability.
- One value from each interval is selected at random with respect to the probability density in the interval.
- The n values thus obtained for X_1 are paired in a random manner (equally likely combinations) with the n values of X_2. These n pairs are combined in a random manner with the n values of X_3 to form n triplets, and so on, until n k-tuplets are formed; these n k-tuplets are the same as the n k-dimensional input vectors.

It is convenient to think of this Latin hypercube sample (or any random sample of size n) as forming an $(n \times k)$ matrix of inputs where the ith row contains specific values of each of the k input variables to be used on the ith run of the computer model. A more detailed description of LHS and the associated computer codes and manuals are given by Wyss and Jorgensen (1998).

This scheme has been further developed for different purposes by several researchers, e.g., Helton and Davis (2003) and Olsson et al. (2003).

4   Model Uncertainties and Probability Distributions

In assessing the reliability of profiled barrier blast walls, different types of uncertainties can be introduced. These uncertainties have various sources which may be grouped as follows:

- Uncertainties associated with the blast loadings including peak pressure, shape, and duration of the loading.
- Uncertainties due to imperfections and idealizations made in the physical model formulations.

- Uncertainties related to mechanical properties such as elastic modulus, minimum yield stress, ultimate tensile strength, and elongation.
- Uncertainties associated with the model geometry.
- Uncertainties in connectivity and boundary conditions, including support connections.

In addition to the above uncertainties, while implementing the probabilistic approaches, the following uncertainties can also be added:

- Uncertainties in the choices of probabilistic or reliability methods
- Uncertainties in the choices of probability distribution types.

In this research study, some of the above uncertainties including pressure blast loadings, duration of loading, geometric properties such as dimensional imperfections and thickness and material properties are considered in the reliability analyses by modeling the variables as random variables. These random variables have been represented by probability distributions. Reliability analysis results can be sensitive to the tail of the probability distribution and therefore, an adequate approach/method to select the proper distribution type is necessary (DNV 1992). However, in this study, except for profiled barrier thickness, Tw, for all other random variables, the normal or Gaussian distribution is assumed, for demonstrative purposes.

5 Finite Element Probabilistic Modeling

5.1 Geometry and Meshing

A parametric model was developed using the APDL available in ANSYS. A four-node quadrilateral shell element type, SHELL181 was used for modeling the profiled barrier. This element has six degrees of freedom at each node: translations in the x, y, and z directions, and rotations about the x-, y-, and z-axes. This type of element is well suited to linear, large rotation, and/or large strain nonlinear applications (Haifu and Xueguang 2009), when change in shell thickness is taken into account. In the element domain, both full and reduced integration schemes are supported and SHELL181 accounts for follower (load stiffness) effects of distributed pressures.

The corrugated profile shown in Fig. 1 and the connecting end plates were modeled by means of first-order shell elements. Figure 2 gives an overall view of the finite element model of the profiled barrier. It can be seen that two corrugation bays were modeled for the probabilistic analysis.

As the validity of the results obtained from a finite element model depends on the mesh capturing the real system characteristics accurately, a set of mesh sensitivity studies were carried out. This was to identify initially the minimum size of the mesh required before starting a probabilistic analysis. From the results of this analysis, the maximum mesh or element size is limited to 60 mm with a maximum aspect ratio

Fig. 3 Triangular blast
pressure load
pulse—($t_r = t_d/2$)

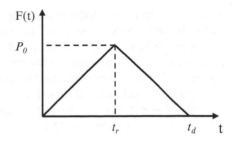

of 1.5. The size of the mesh is variable for each set of analysis run, as the dimension of the element is one of the considered random variables. On average, 12,500 shell elements were generated in each of the simulation cycles.

5.2   Explosion Loading

In the past decades, many theoretical developments, and research studies associated with the dynamic behavior of structures under load pulse or explosion loadings have been discussed, e.g., in Biggs 1964; Clough and Penzien 1995; Chopra 1995; Beards 1996; and Li and Chen 2009.

Dynamic pressure loading generated by explosions varies with time, and the resulting response of the structure is therefore time dependent (Louca and Boh 2004; Brewerton 1999). This loading causes the structure to vibrate at its natural period and large intensity loading can cause large deformation of the structure (Biggs 1964). Therefore, in this study, for the dynamic analyses, as a comprehensive approach for probabilistic analysis, full transient dynamic explosion loadings have been applied to the structure. Consequently, the dynamic responses including deflection, strain, and stresses are expected to be more accurate and reliable than those determined by applying the peak load statically.

For the dynamic analyses, a triangular load pulse with a peak dynamic pressure (P_0) of 2.0 bar is used. The total time duration (t_d) for this load pulse is 0.15 s. The analyses are continued up to 2 t_d, i.e., 0.3 s. In other words, the time for free vibration is between 0.15 and 0.3 s. The peak pressure load pulse and the time are both introduced as variable inputs for the probabilistic analysis. Figure 3 shows a typical explosion loading shape used in this study.

For the static analyses, a peak pressure load (P_0) of 2.0 bar is considered and applied to the structure; however, this pressure load is introduced as a variable input for the probabilistic analysis.

It should be noted that the maximum response of a structure under a load pulse is reached in a very short time. As such, damping does not absorb much energy from the structure and consequently can be ignored for load pulse or blast loadings (Clough and Penzien 1995) for dynamic analyses. Therefore, no damping effect is considered in this study.

Before performing reliability analyses, two sets of studies, including modal and linear dynamic analyses, were carried out to verify the baseline dynamic model. The mean values of input variables, shown in the Table 1, were considered to check model accuracy. Figure 4 presents the first dynamic mode shape of structure. The natural period of the structure for this mode is 0.03064 s (or a frequency of 32.6328 Hz), which is 0.2 t_d (or 0.4 t_r).

Figure 5 also shows the transient displacement and highlights that the maximum response is at 0.08 s, close to 0.5 t_d ($t_r = t_d/2$).

5.3   Material Nonlinearity

Stress–strain curves are an extremely important graphical measure of a material's mechanical properties (Roylance 2001), and all structural engineers will have to deal with them, especially for advanced structural design or assessments. In this study, for the nonlinear analyses, the material behavior of a profiled barrier is described by a bilinear stress–strain curve, starting at the origin with positive stress and strain values. The initial slope of the curve is taken as the elastic or Young's modulus of the material. At the minimum yield stress, F_y, the curve continues along the second slope defined by the tangent modulus E_t (having the same units as the elastic modulus). The tangent modulus can be neither less than zero nor greater than the elastic modulus (ANSYS 2012).

To calculate E_t, initially true stress and true strain should be determined implementing tensile strength and elongation and then, elastic strain needs to be specified based on F_y and elastic modulus. Finally, E_t can be developed and calculated using true stress, true strain, and F_y. Average yield stress can also be replaced for the minimum yield stress; however, in this study minimum yield stress has been used. Since ultimate tensile strength, minimum yield stress, elongation, and elastic modulus are defined as random variables, E_t is changed for each individual run or analysis. In fact, for each run, there is a material curve with random stress–strain inputs. This provides a more accurate and realistic condition for the probabilistic assessments. For the linear analyses, only a Young's modulus (E) of 200 GPa is considered as representative of material. However, this Young's modulus is introduced as a variable input for the probabilistic analysis. Figure 6 gives an overview on bilinear stress–strain curved used for the nonlinear analyses (SNMG and DNMG), considering the mean values of input variables, shown in the Table 1.

It should be noted that in all nonlinear analyses the effects of geometry nonlinearity are also included.

Table 1 Parametric variables for probabilistic analysis

	Geometry		Impulse loading		Material			
Random variable	Height, H (mm)	Thickness, Tw (mm)	Time duration, t_d (s)	Peak pressure load P_0 (bar)	Minimum Yield Stress F_y (MPa)	Ultimate Tensile Strength, F_u (MPa)	Young's modulus, E (GPa)	Elongation (%)
Mean	554	11	0.15	2.0 (0.2 MPa)	460	740	200	25
Coefficient of variation	0.1	0.1 (±1 mm)	0.1	0.15	0.1	0.1	0.05	0.05
Probability distribution	Gaussian	Uniform	Gaussian	Gaussian	Gaussian	Gaussian	Gaussian	Gaussian

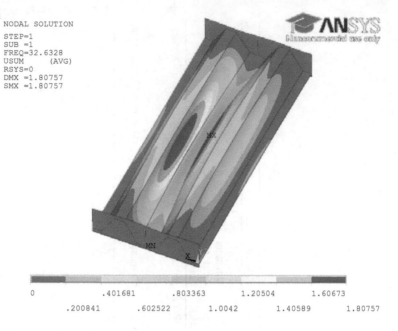

Fig. 4 First mode shape of structure

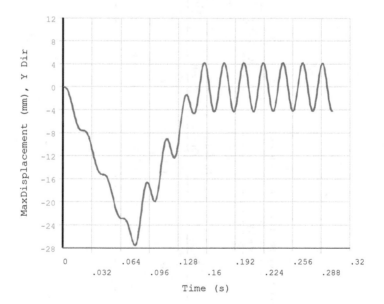

Fig. 5 Transient displacement, DLMG

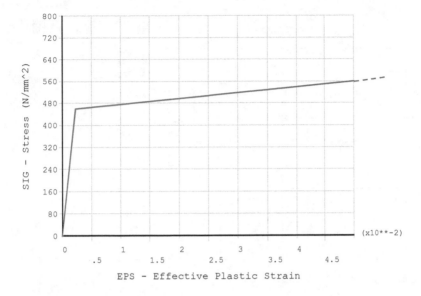

Fig. 6 Bilinear stress–strain curve–nonlinear analyses (SNMG and DNMG)

5.4 Probabilistic Models

The deterministic baseline model has eight parameters that are regarded as random input variables. However, for the linear and static analyses only some of these random variables are implemented. For instance, for linear assessments, only Young's modulus (E) is considered as a material characteristic.

The variables along with the assumed distribution models and parameters are given in Table 1. The random input variables are assumed to be statistically independent. Typical probability density function and cumulative distribution function (CDF) of peak pressure are shown in Fig. 7.

6 Reliability Analysis

In this study, before performing any probabilistic analyses, some sensitivity studies were carried out to make sure that the model inputs were correctly defined and to have a better understanding of the structural behavior under general loads and boundary conditions. Five hundred simulation loops were considered for the analyses. The maximum deflection of the profiled barrier is considered as the limiting property.

After performing probabilistic analyses, it is crucial to review the statistical results to check that the simulation loops are adequate. If the number of simulations

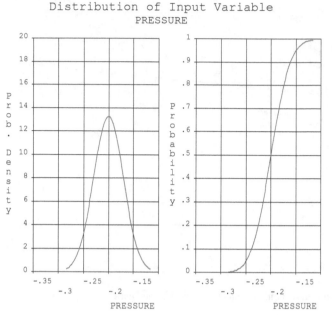

Fig. 7 Distribution of peak pressure, P_0

is sufficient, the mean value plots for random output variables converge (i.e., the curve flattens out). Figure 8 presents the level of satisfaction of the number of loops considered with regard to mean values of maximum deflections, for dynamic analyses. It can be seen that the values are converging.

For the assessment purposes including structural integrity and risk, it is useful to identify the probability that the maximum response (e.g., Maximum deflection, strain, or stress) remains below a specified limit or value. Furthermore, for design purposes, it is always useful to determine the probability corresponding to the occurrence of maximum response that satisfies the design requirements. This information can be readily obtained from the CDFs of the variables of interest.

Figure 9 shows the CDFs associated with maximum deflection at mid-span, respectively, for the dynamic analyses. The probability of having a specific response can be identified using CDFs plots. For instance, the probability of having a maximum displacement up to 16 mm (absolute) is 99 and 99.50 % for linear and nonlinear dynamic analyses, respectively. Similar figures can be produced for studying maximum strain and stress.

Table 2 presents a sample of the probability of having a maximum response, displacement greater than specified values including 20, 30, and 40 mm. As can be seen, the probability of having a response greater than 20 mm, for all types of analyses, including SLMG, SNMG, DLMG, and DNMG, are very close. In fact, it can be concluded that the effects of nonlinearity and dynamics is negligible,

Table 2 Probability of maximum response exceeding specified values

Maximum response (displacement) >	Analysis	Probability	lower bound	Upper bound
20 mm	SLMG	0.8847	0.8548	0.9107
	SNMG	0.8825	0.8523	0.9087
	DLMG	0.8727	0.8416	0.9000
	DNMG	0.8936	0.8646	0.9186
30 mm	SLMG	0.2822	0.2439	0.3227
	SNMG	0.3404	0.2998	0.3827
	DLMG	0.2764	0.2384	0.3167
	DNMG	0.3423	0.3016	0.3846
40 mm	SLMG	0.0471	0.0309	0.0680
	SNMG	0.0933	0.0699	0.1209
	DLMG	0.0475	0.0312	0.0685
	DNMG	0.0989	0.0748	0.1270

whereas, for having a response greater than 30 mm, the associated probabilities are not similar. Comparing the nonlinear and linear static results, SLMG (2.82E-01) and SNMG (3.40E-01), there is about 20 % difference between the probability results. Undertaking a similar comparison for the response greater than 40 mm, there is around 98 % difference between the probability results.

This gives a good understanding of dynamic and nonlinearity effects on responses including deflections, stresses, strains, and support reactions. Table 2 also presents lower and upper bounds of probabilities.

6.1 Sensitivity Analysis

The next step is to quantify the sensitivity of the output variables with respect to the variability of the input parameters. By generating scatter plots of the output variables as a function of the most important random input variables, it is possible to determine the correlation coefficients between the output and input variables. The evaluation of the probabilistic sensitivities is based on the correlation coefficients between all random input variables and a particular random output parameter. The correlation coefficients could be represented either in terms of the widely used Pearson linear correlation or in terms of the Spearman rank correlation coefficient. For this study, it is observed that the maximum response of the structure is within $(0.4$–$0.6)$ t_d; as such, the sensitivity results associated with this time domain are similar. Therefore, the correlation sensitivities related to 0.5 t_d are presented in this chapter. Figure 10 presents the correlation sensitivities associated with the input variables and output response (maximum deflections), for nonlinear static (SNMG) and nonlinear dynamic (DNMG) analyses. It can be seen that the response variable is dependent significantly on depth of the barrier section (H) and the pressure load applied (P_0). However, Fig. 10 also highlights that the maximum response is not

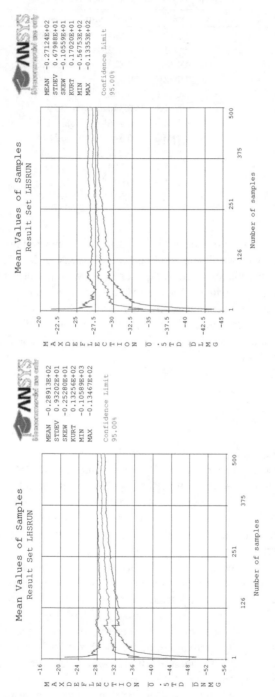

Fig. 8 Mean value of samples–maximum deflection–SLMG (*left*), DNMG (*right*)

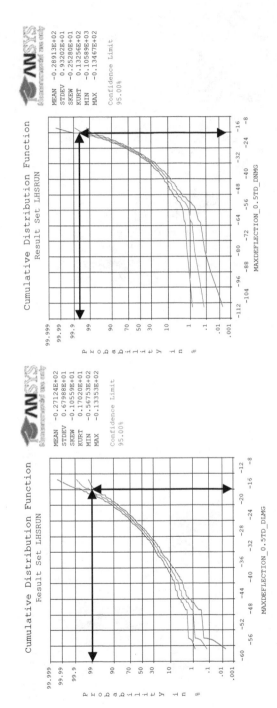

Fig. 9 CDF of maximum deflection–DLMG (*left*), DNMG (*right*)

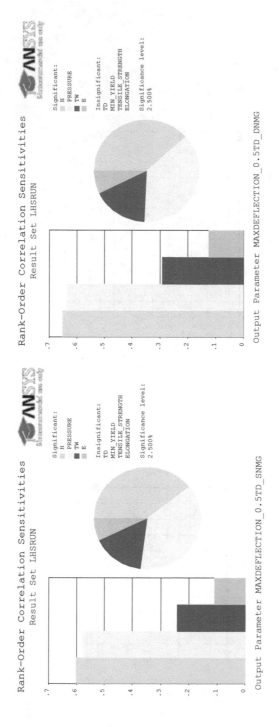

Fig. 10 Correlation sensitivities–maximum deflection–SNMG (*left*) and DNMG (*right*)

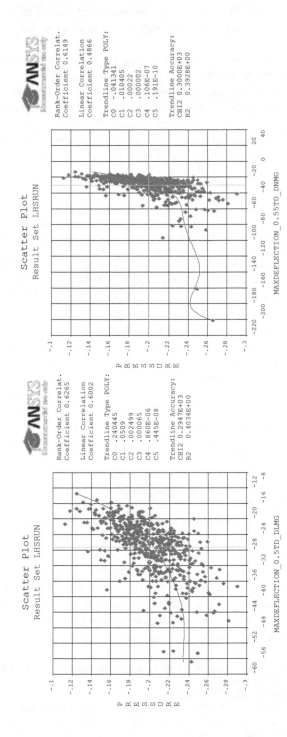

Fig. 11 Correlation scatter plot, DLMG (*left*) and DNMG (*right*)

sensitive to duration of loading. Moreover, from the correlation sensitivities, it can be seen that the maximum response is not very sensitive to thickness and even less sensitive to elastic modulus.

Correlation scatter plots can also be implemented to find out the relation between an input variable, such as blast loading, and the maximum response. Figure 11 shows two typical correlation scatters between the dynamic peak pressure load and maximum deflection. Nonlinear analysis shows wide range of parameters compared to linear analysis.

This correlation information can also be used to generate response surfaces (i.e., to develop implicit performance functions) to be implemented in structural reliability framework.

7 Conclusions

This chapter presents an approach for the reliability analysis of a profiled barrier under dynamic blast loading by implementing LHS in APDL. A profiled barrier with two corrugation bays is analyzed probabilistically by considering linear, nonlinear, static, and dynamic responses. For the reliability analyses, the maximum response in terms of deflection is used as a limiting factor and the results are discussed in terms of the corresponding variable sensitivities. These sensitivities are evaluated based on the correlation coefficients between considered random input variables and a particular random output variable.

For this case study, it is found that the maximum response is not very sensitive to thickness and even less sensitive to elastic modulus. To understand the influence of dynamics and nonlinearity, probability of occurrence of specified or target responses are reviewed and discussed. Based on the probability results and discussions, it can be concluded that the effects of nonlinearities are crucial to be considered in the assessments. It is also noticed that there is no profound difference between probabilistic responses for static and dynamic analyses. This indicates that the dynamic effects on the responses are less influential than the nonlinearity effects.

Based on the studies which have been carried out so far for this research and presented in this chapter, it has been noticed that further investigations are also necessary to consider different types of materials, section sizes, and panel heights. Moreover, strain-rate effects can also be considered in the assessments as these effects are disregarded in the current study due to complexity of combining dynamic and nonlinearity effects into a stochastic finite element reliability approach. As such, further investigations and assessments have been planned to be carried out implementing these considerations.

Acknowledgments This Ph.D. research is funded by Advanced Structural Analysis and Management Group (ASAMG) Ltd., Aberdeen.
Sriramula's work within the Lloyd's Register Foundation Centre for Safety and Reliability Engineering at the University of Aberdeen is supported by Lloyd's Register Foundation. The Foundation helps to protect life and property by supporting engineering-related education, public engagement, and the application of research.

References

ANSYS Inc. (2012) Documentation. Release 14.0

API 2FB (2006) Recommended practice for the design of offshore facilities against fire and blast loading

Beards C (1996) Structural vibration analysis and damping. Elsevier Science, Amsterdam

Biggs JM (1964) Introduction to structural dynamics. McGraw-Hill, New York

Brewerton R (1999) Design guide for stainless steel blast wall: technical note 5. Fire and Blast Information Group (FABIG), London

Choi S-K, Grandhi RV, Canfield RA (2006) Reliability-based structural design. Springer, London

Chopra AK (1995) Dynamics of structures. Theory and applications to earthquake engineering. Prentice Hall, Englewood Cliffs

Clough RW, Penzien J (1995) Dynamics of structures. McGraw-Hill Education, New York

DNV (1992) Structural reliability analysis of marine structures: classification note no. 30.6. Det Norske Veritas (DNV), Norway

Haifu Q, Xueguang L (2009) Approach to blast wall structure computing in ocean engineering. In: Proceedings of the international conference on industrial and information systems, IIS '09

Haldar A, Mahdavan S (2000) Reliability assessment using stochastic finite analysis. John Wiley & Sons Inc., New York

Hedayati MH, Sriramula S, Neilson RD (2013) Non-linear dynamic reliability analysis of profiled blast walls. 11th international conference on structural safety and reliability (ICOSSAR), New York, USA

Hedayati MH, Sriramula S, Neilson RD (2014) Linear dynamic reliability analysis of profiled blast walls, to be published, ASCE-ICVRAM-ISUMA conference, Liverpool, UK

Hedyati MH, Sriramula S (2012) Finite element reliability analysis of blast wall. 6th International, ASRANet (Advanced Structural Reliability Analysis Network) conference, Croydon, London, UK

Helton JC, Davis FJ (2003) Latin hypercube sampling and the propagation of uncertainty in analyses of complex systems. Reliab Eng Syst Saf 81(1):23–69

Iman RL, Helton JC, Campbell JE (1981) An approach to sensitivity analysis of computer models, part 1. Introduction, input variable selection and preliminary variable assessment. J Qual Tech 13(3):174–183

Li J, Chen J (2009) Stochastic dynamics of structures. Chichester, John Wiley & Sons (Asia) Pte Ltd

Liang YH, Louca LA, Hobbs RE (2007) Corrugated panels under dynamic loads. Int J Impact Eng 34:1185–1201

Louca LA, Boh JW (2004) Analysis and design of profiled blast walls. Research report 146, HSE, Prepared by Imperial College London for the Health and Safety Executive 2004

McKay MD, Beckman RJ, Conover WJ (1979) A comparison of three methods for selecting values of input variables in the analysis of output from a computer code. Technometrics 21(2):239–245

Olsson A, Sandberg G, Dahlblom O (2003) On Latin hypercube sampling for structural reliability analysis. Struct Saf 25(1):47–68

Reh S, Beley J-D, Mukherjee S, Khor EH (2006) Probabilistic finite element analysis using ANSYS. Struct Saf 28(1–2):17–43

Roylance D (2001) Stress-strain curves. Massachusetts Institute of Technology, Department of Materials Science and Engineering, Cambridge

Wyss GD, Jorgensen KH (1998) A user's guide to LHS: Sandia's Latin hypercube sampling software, SAND98-0210. Sandia National Lab, Albuquerque

Reliability Assessment of a Multi-Redundant Repairable Mechatronic System

Carmen Martin, Vicente Gonzalez-Prida, and François Pérès

Abstract The reliability modelling of redundant systems is an important step to estimate the ability of a system to meet the required specifications. Markov chains have characteristics making it very simple the graphic representation of this type of model. They however have the disadvantage of being quickly unworkable because of the size of the matrices to be manipulated when systems become complex in terms of number of components or states. This issue, known as of combinatorial explosion is discussed in this chapter. Two methods are proposed. The first one uses the concept of decoupling between phenomena driven by different dynamics. The second is based on a principle of iteration after cutting the model into classes of membership. Both are based on the principles of approximating the exact result by reducing the scale of the problem to be solved. A case study is eventually carried out, dealing with the reliability modelling and assessing of a mechatronic subsystem used for an Unmanned Aerial Vehicle flight control with a triple modular redundancy. Results are discussed.

C. Martin (✉)
Mechanics, Materials, Structure and Process Division, ENIT-INPT, Toulouse University,
47, avenue d'Azereix, 65016 Tarbes Cedex, France
e-mail: Carmen.Martin@enit.fr

V. Gonzalez-Prida
University of Seville, Seville, Spain
e-mail: vicente.gonzalezprida@gdels.com

F. Pérès
Decision Making and Cognitive System Division, ENIT-INPT, Toulouse University,
47, avenue d'Azereix, 65016 Tarbes Cedex, France
e-mail: Francois.Peres@enit.fr

S. Kadry and A. El Hami (eds.), *Numerical Methods for Reliability and Safety
Assessment: Multiscale and Multiphysics Systems*, DOI 10.1007/978-3-319-07167-1_14,
© Springer International Publishing Switzerland 2015

1 Introduction

The growing complexity of industrial systems is mainly due to the increasing variety of technologies involved in their implementation, e.g. mechanical, electronic and software components. Therefore, the reliability analysis, one of the most important problems for industrial systems, becomes extremely difficult.

Redundancy incorporated into systems generally leads to increased reliability. Hardware redundancy is widespread in areas where dependability is critical to people and environment safety as in aerospace, space, defence or nuclear industries. Generally, real systems consist of several components and have several failure modes. Such systems are called complex and their analysis can become difficult when the number of components increases.

The calculation of reliability based on Markov model has been largely used by scholars and research institutions. However the inclusion of multi-resource systems quickly leads to complex Markov models, difficult to be carried out by conventional means of calculation. When analysing the system for the study of its asymptotic behaviour, it can be made, through the application of certain assumptions, adjustments to the treatment model to reduce the size of calculation.

We present in this chapter methods to reduce the complexity of the reliability assessment issue by separating the global solving system in different independent subsystems of reduced size whose particular results are aggregated to give an approximate value of the original problem.

2 Problem Issue

The requirements for the reliability and the availability of technological systems may be very high. Within this context, redundancy is a technique widely used to improve the reliability and the availability of a system (Vujosevic and Meade 1985). The principle of redundancy leads to use several resources to perform a single function or a single task.

A redundant system improving the dependability of a given architecture is based on parallel structures: a system consisting of n redundant elements is reliable or available if at least one of its elements works properly (Nourelfath et al. 2012). Two types of redundancy can be implemented: active (warm) redundancy when the means are implemented simultaneously and, in case of failure of the main component, the redundancy is able to take over; and passive (cold) redundancy when the means are implemented on request and, in case of failure of the main component, the redundancy is committed to do the task (Amari and Dill 2010).

Different configurations can be established by combinations of serial, parallel or hybrid structures.

The study of the reliability and availability of complex systems, with reconfiguration capabilities and redundancy or other types of dependencies between components requires the use of behavioural models. The characterize the random process making the system evolve until it reaches an undesirable outcome (Godichaud et al. 2012).

Within this modelling context, Markov processes are an ideal support for mathematical developments. Their ease of use and processing and their computing power make it a very effective tool for dependability assessment. There is an extensive scientific literature on the subject (White 1993).

A finite state space Markov process can be entirely defined by a matrix of real numbers such as the product $a_{ij}\ dt$ of the term i,j of the matrix by dt represents the probability that the process goes from state i to state j within a time dt. These processes may be characterized through graphic representation called Markov graph or Markov chain. Most often the Markov chain is used to represent the states in which the system will be during the observation period and to assess the corresponding occupation probability at a given time (transient conditions) or when reaching an equilibrium (steady-state conditions).

The use of Markov chain is convenient as long as the number of states to be considered does not exceed a certain value. Unfortunately, as many other tools, Markov chains suffer from the problem of combinatorial explosion which corresponds to the situation where the dimension of Markov transition matrix increases exponentially along with the growing number of redundant components.

To deal with this problem, the Monte Carlo simulation which is a method very general and insensitive to the number of states can be carried out. However, the results can be imprecise and ask for prohibitive calculation time for highly reliable systems (Diaconis 2008; Andrieu et al. 2010; Godichaud et al. 2010).

This limitation of the simulation explains why analytical calculations on Markov graphs, which are all the more effective as systems are more reliable, still have a great interest. Consequently, to keep using Markov chains to assess complex system reliability, some modelling techniques allowing the reducing of the matrix sizes must be performed. We present here below two methods for that.

3 "Product-Form" Calculation

When a weak interaction between concurrent processes acting on the system is established, the model can be divided into subsystems treated separately for the calculation of internal states by solving the equilibrium distribution of the reduced Markov chain. Regardless of these calculations, the conditions of the overall balance between subsystems are considered and the asymptotic probabilities associated with the state occupation are assessed by applying product-form calculation (Buchholz 2008; Gupta and Albright 1992; Papazoglou and Gyftopoulos 1977).

These results have been established for models with a certain regularity of structure that allows easy expression of the equilibrium equations between subsystems (transitions of all states of one subsystem to another subsystem being identical). Such models accept submerged chains. We extend the method to systems that do not have this property by providing an appropriate expression of the transitions between subsystems. The approach is as follows:

- Isolate the different subsystems by removing transitions between states on which applies the hypothesis of decoupling: $E = \{E_1, E_2 \ldots E_n\}$, where E is the state space of the complete model and E_i the state space of subsystem i
- Process every subsystem separately by establishing the state occupation probabilities of each subsystem in steady-state conditions by resolution of:

$$P_{E_i}.\Lambda_{ii} = 0 \quad \forall i \in \{1, 2, \ldots, n\} \tag{1}$$

- Determine the transition rates between subsystems by aggregating in a same expression, the sum of each rate connecting two subsystems, weighted by the occupation probability of their state of origin:

$$T_{IJ} = \sum_{e_i \in E_I} \left[P_{e_i} . \sum_{e_j \in E_J} t_{ij} \right]$$

with $t_{ij} = $ transition from state e_i to state e_j and $I, J \in 1, 2, \ldots, n$

- Calculate the occupation probability of each subsystem:

$$P_K.\Lambda_{KK} = 0 \quad \text{with } P_K = [P_1, P_2, \ldots, P_n] \quad \text{and} \quad \Lambda_{KK} = \text{rate matrix } t_{IJ} \tag{2}$$

- Infer the final state occupation probabilities by multiplying the matching state occupation probability inside the subsystem to which it belongs (resulting of Eq. (1)) by the occupation probability of the corresponding subsystems (resulting of Eq. (2))

Applied to non-standard models, this calculation gives approximate results. The accuracy of the approximation is strongly dependent on the ratio between the internal dynamics of each subsystem and the dynamic evolution between subsystems. When the internal balance (subsystems) is reached before the external balance (between subsystems), the method gives excellent results. When the dynamics are similar, the method still works, but the accuracy is difficult to quantify. In the case of very mixed internal and external dynamics, the method can lead to large deviations.

An illustration of this is given on a three-state model represented in (Fig. 1).

On (Fig. 2) a comparison between exact and approximate methods is made by estimating the state occupation probability for several ratios between λ and β. The application corresponds to the following data: $\alpha = 20$, $\beta = 4$, $\mu = 0.2$. The dynamic is dependent on λ. The calculation using the approximate methods is based on a separation between, on the one hand, states E_1 and E_2, and, on the other hand, state E_3. The results are convincing, our approach (referred to as CPM) being consistently better than that of the conventional product form (referred to as CP).

Fig. 1 Three state model

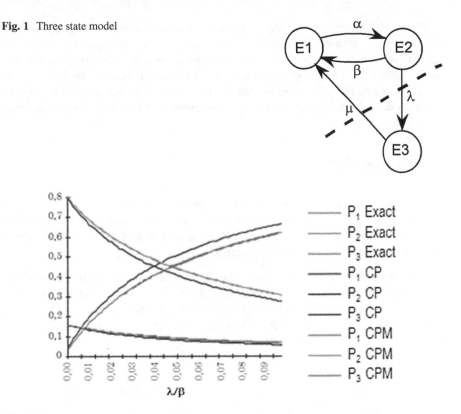

Fig. 2 Exact versus approximate methods

Kemeny and Snell (1967) a long time ago suggested some tests to assess the ability of a chain to accept partitions but the specific characteristics of each situation make this test not sufficient to validate the choice of a decoupling mode. Gupta and Albright (1992) propose from the spectral analysis of the full model transition matrix and a pre-established partition, a value criterion t (unitless), describing the relationship between internal and external dynamics. Three cases were considered. If the ratio does not exceed a certain threshold s_1, the method is considered to be directly applicable. If the ratio exceeds a threshold s_2, it is necessary to reverse the dynamics (the internal dynamics becoming external and vice versa). If the ratio is between s_1 and s_2, the two previous solutions are combined by introducing a weight depending on the degree of remoteness of t from s_1 and s_2.

Finally, Balakrishnan and Reibman (1993) denounced the lack of common criterion of convergence accuracy f related to this kind of methods. They proposed a criterion for evaluation of the error based on the comparison between a state occupation probability inside a subsystem (in internal equilibrium) and a ratio between the number of visits of the same state and the number of visits of the subsystem. If the difference is 0, the Markov model is said to be decomposable otherwise it gives a measure characterizing the inaccuracy of the approximation.

The model specificity and the complexity of combinations prevent however to define a generic framework ensuring, in compliance with certain conditions that the results obtained have a minimum level of precision. In some (few) situations where the dynamic partition is made difficult by the mix of heterogeneous form processes, we emphasized that this form of calculation approach could lead to significant accuracy deviations. It is then necessary to complete the method by other treatments.

4   Iterative Method

This method developed in (Nahman 1984) is all the more rapid if there is an a priori knowledge of the behaviour of the modelled system. Nevertheless we will note that ignorance of the system is not an obstacle to the accuracy of the results but affects the convergence speed. The principles are as follows:

- Decompose the system into classes corresponding to the group of states whose occupation probabilities are, if possible, of similar order (hence the importance of this method to be associated with the product-form calculation). $E = \{E_1, E_2 \ldots E_n\}$, with $E =$ state space of the full model and E_i class i of dimension N_i
- Order the matrix A of the model with $A = [\alpha_{ik}]$ where:

$$\alpha_{ik} = 1 \quad \text{for } i = 1$$
$$\alpha_{ik} = \lambda_{ik} \text{ for } i = 2, \ldots, N; k = 1, 2, \ldots, N; \text{with } i \neq k \text{ and } N$$
number of states of the full model

$$\alpha_{ik} = - \sum_{\substack{j=1 \\ j \neq i}}^{N} \lambda_{ik} \text{ for } i = 2, \ldots, N \text{ and } i = k$$

- Calculate the initial occupation probabilities of the states of the first class from:

$$p_{E_1} = \theta_{11}^{-1}.\beta_1 - \sum_{i=2}^{n} \theta_{11}^{-1}.\theta_{1i}.\tilde{p}_{E_i} \tag{3}$$

with:

$$[\theta_{uv}] = [\alpha_{ik}] \quad \text{for } i = a, \ldots, a + N_{u-1}; \quad k = b, \ldots, b + N_{v-1};$$
$$\text{and } a = 1 + \sum_{h=0}^{u-1} N_h; \quad b = 1 + \sum_{h=0}^{v-1} N_h \quad \text{and} \quad N_0 = 1$$

β_1, vector of equal size number of state N_1 of class i, with $\beta_1 = (1, 0, \ldots, 0)$
$\tilde{p}$, vector p calculated at the previous iteration ($\tilde{p}_{E_i} =$ vector null in the first iteration)

- Calculate the occupation probabilities of the states of the other classes from

$$p_{E_k} = -\sum_{i=1}^{k-1} \theta_{kk}^{-1}.\theta_{ki}.p_{E_i} - \sum_{i=k+1\leq n}^{n} \theta_{kk}^{-1}.\theta_{ki}.\tilde{p}_{E_i} \tag{4}$$

- Repeat calculations from (Eqs. (3) and (4)) until a convergence criterion is reached

Equations (3) and (4) greatly reduce the dimension of the system of equations to be solved. When an initial estimate allows the identification of classes by grouping states of equal probability, the method converges very quickly. If this is not the case, there are two options. Either the classes are modified taking into account the values of the first iteration and the method is resumed at the beginning or the calculations corresponding to the following iterations are performed until convergence which, nevertheless, appears relatively quickly.

5   Case Study

The case study will deal with a component of a drone long-range, long-dwell dedicated to direct operational control by army field commanders (Fig. 3). Its mission set includes wide-area Intelligence Surveillance, Reconnaissance (ISR), convoy protection, Improvised Explosive Device (IED) detection and defeat, close air support, communications relay and weapons delivery missions. For such a system, reliability is a key factor. To reach this objective, the drone features a

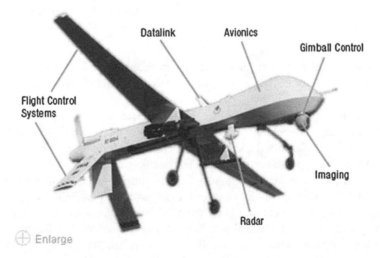

Fig. 3   UAV: unmanned aerial vehicle (drone)

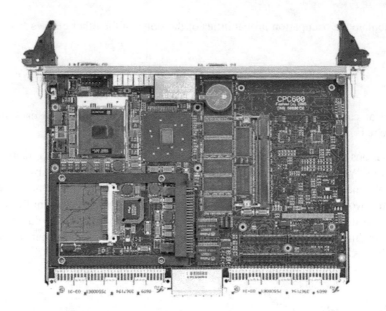

Fig. 4 VME card 6U

Fig. 5 VME card reliability
block diagram

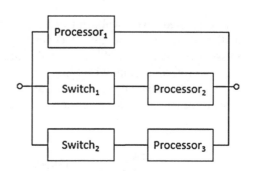

fault-tolerant control system and a triple-redundant avionic system architecture
which meets and exceeds manned aircraft reliability standards (Andrews et al.
2013).

The system under study is a mechatronic component controlling through the on-
board computer architecture the flight control unit. Three VME (Verso Module
Europa) cards, specifically dedicated to critical applications in harsh environment
are used in active redundancy (Fig. 4). Each one is itself based on three processors
offering a triple-redundant device securing the implementation of deterministic
applications (running in parallel on three CPUS) with the possibility of an immediate resynchronization of the data in the event of failure of one of the processors.

Taking into account the possible failure of a processor to start (represented by
a switch component which would collapse), the reliability of one card and the
corresponding Markov chain are given in (Fig. 5) and (Fig. 6).

Fig. 6 VME card Markov chain

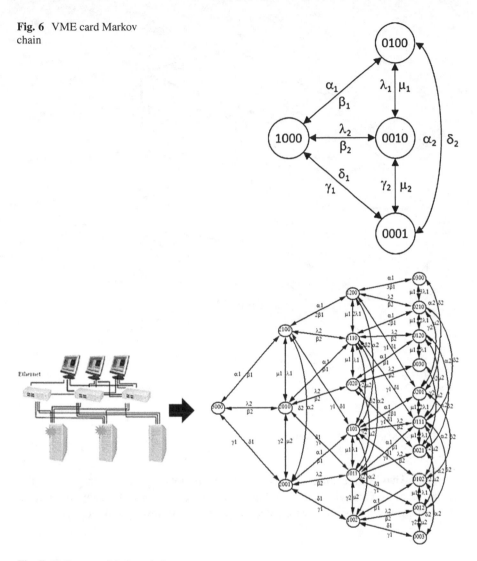

Fig. 7 Full system Markov chain

Related to the system made of three VME cards the corresponding Markov chain is given in (Fig. 7).

Each state is given a number whose digits represent, respectively, the type and number of operative processors. The nominal situation corresponds to state "3000" (the three main processors of the corresponding VME cards are operative). State "1101" would imply for instance that among the three VME cards, one card is running on the main processor, the second card has switched on the first redundancy and the third card is disabled.

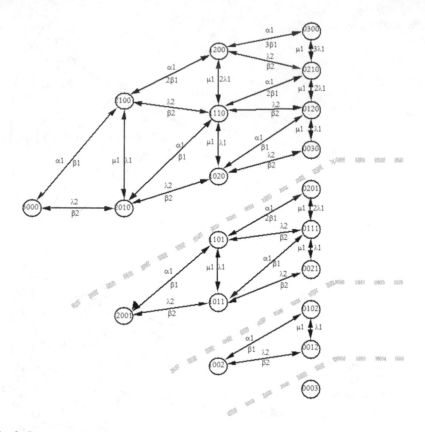

Fig. 8 Super-state inner Markov chain

Because it is not the object of the chapter we will not comment the meaning of the transition rates. One will understand however that a rate may aggregate the failure of a processor with the impossibility to switch on the following redundancy.

5.1 Product-Form Calculation

The principle of the approximate method based on "Product-form" calculation requires to separate processes according to their relative dynamics (Taylor 1992; Strelen 1997). Applied to the system, the high reliability of the second switch and the third processor (to prevent the complete system breakdown) leads to consider the value of rate δ_1 as lower than the other rates. Based on this assumption, we will then separately consider four Markov chains representing the evolution of the system when, respectively, 0, 1, 2 or 3 VME cards are out of order. The corresponding model is given in (Fig. 8).

Fig. 9 Super-state outer
Markov chain

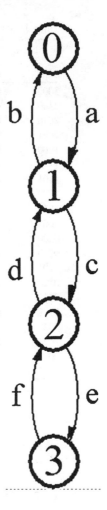

Each subsystem can be associated to a super-state. The model combining the
different subsystems is given in (Fig. 9). We can then calculate the occupation
probability of each super-state and eventually determine, by simple product, the
state occupation probability vector corresponding to the whole chain. Rates between
subsystems are as follows:

$$a = P_{0003} \cdot (\gamma_1 + \alpha_2 + \gamma_2)$$

$$b = P_{0012} \cdot \mu_2 + P_{0102} \cdot \delta_2 + P_{1002} \cdot \delta_1$$

$$c = (P_{0012} + P_{0102} + P_{1002}) \cdot (\gamma_1 + \alpha_2 + \gamma_2)$$

$$d = \mu_2 (P_{0021} + P_{0111} + P_{1011}) + \delta_2 (P_{0111} + P_{0201} + P_{1101})$$
$$+ \delta_1 (P_{1011} + P_{1101} + P_{2001})$$

Table 1 State occupation probability

State	P_{e_i} Exact result	P_{e_i} "Product-form" calculation (1 decomposition)	P_{e_i} "Product-form" calculation (2 decompositions)
3000	0.075230	0.075422	0.074818
2100	0.224645	0.224941	0.224455
2010	0.002795	0.002801	0.002794
2001	0.000810	0.000777	0.000366
1200	0.335458	0.335435	0.336883
1110	0.008345	0.008354	0.008383
1101	0.002323	0.002318	0.001099
1020	0.0000104	0.000104	0.000104
1011	0.000030	0.000029	0.000001
1002	0.000009	0.000009	0.000004
0300	0.333947	0.333470	0.336683
0210	0.012461	0.012457	0.012575
0201	0.003413	0.003457	0.001649
0120	0.000310	0.000310	0.000313
0111	0.000088	0.000086	0.000041
0102	0.000026	0.000026	0.000001
0030	0.000004	0.000004	0.000004
0021	0.000001	0.000001	0.000000
0012	0.000000	0.000000	0.000000
0003	0.000000	0.000000	0.000000

$$e = (P_{0021} + P_{0111} + P_{1011} + P_{0201} + P_{1101} + P_{2001}) \cdot (\gamma_1 + \alpha_2 + \gamma_2)$$

$$f = \mu_2 (P_{0030} + P_{0120} + P_{1020} + P_{0210} + P_{1110} + P_{2010})$$

$$+ \delta_2 (P_{0120} + P_{0210} + P_{1110} + P_{0300} + P_{1200} + P_{2100})$$

$$+ \delta_1 (P_{1020} + P_{1110} + P_{2010} + P_{1200} + P_{2100} + P_{3000})$$

The proposed approach can be applied recursively. If there is a weak interaction between some other internal processes, each super-state can itself be decomposed and processed on the same way. To illustrate this, we propose a second internal decomposition. The results are presented for both decompositions in (Table 1).

The application corresponds to the following data: $\alpha_1 = 3$, $\alpha_2 = 0.5$, $\beta_1 = 1$, $\beta_2 = 0.5$ $\delta_1 = 0.005$, $\delta_2 = 0.001$, $\gamma_1 = 0.125$, $\gamma_2 = 0.001$, $\lambda_1 = 0.01$, $\lambda_2 = 0.001$, $\mu_1 = 0.33$, $\mu_2 = 0.001$.

In this table, we associate with each state of the full model, the exact state occupation probability as well as the corresponding results when proceeding to one or two partitions. The full model calculation requires to deal with a (20×20) matrix. This complexity is halved when considering one partition (10×10) and reduced by 4 when taking into account a second form of decomposition (4×4).

Dispersion between approximate and exact results is low. For a single decomposition, the mean relative error is 5.7 %. It is less than 1 % if the last four values corresponding to the states with the lowest occupancy probability are not taken into account. In the case of two successive decompositions, differences are overall more important but the results remain, for the most significant states, in the same order of accuracy of 1 % up to 2 %.

It is worth noting however, that the model characteristics were not very favourable to the application of the decomposition principles since the full model makes appear a maximum connection and the dynamics between the values of the different rates remain limited.

Sensitivity studies show that the precision of the product-form calculations is, of course, all the greater as the evolution dynamics within the subsystems is high compared to the one between subsystems.

5.2   Iterative Method

For the calculation by iteration, we have grouped the states into classes according to the number of out of order VME cards. Consequently $E = \{E_1, E_2, E_3, E_4\}$ with: $E_1 = \{3000, 2100, 1200, 0300\}$, $E_2 = \{2010, 2001, 1110, 1101, 0210, 0201\}$, $E_3 = \{1020, 1011, 1002, 0120, 0111, 0102\}$, $E_4 = \{0030, 0021, 0012, 0003\}$ and:

$$\alpha = [a_{jk}] = \begin{bmatrix}
1 & 1 & 1 & 1 & 1 & 1 & 1 & 1 & 1 & 1 & 1 & 1 & 1 & 1 & 1 & 1 & 1 & 1 & 1 & 1 \\
\alpha_1 & a_{22} & 2\beta_1 & 0 & \mu_1 & \alpha_2 & \beta_2 & \gamma_1 & 0 & 0 & 0 & 0 & 0 & 0 & 0 & 0 & 0 & 0 & 0 & 0 \\
0 & \alpha_1 & a_{33} & 3\beta_1 & 0 & 0 & \mu_1 & \alpha_2 & \beta_2 & \gamma_1 & 0 & 0 & 0 & 0 & 0 & 0 & 0 & 0 & 0 & 0 \\
0 & 0 & \alpha_1 & a_{44} & 0 & 0 & 0 & \mu_1 & \alpha_2 & 0 & 0 & 0 & 0 & 0 & 0 & 0 & 0 & 0 & 0 & 0 \\
\lambda_2 & \lambda_1 & 0 & 0 & a_{55} & \gamma_2 & \beta_1 & 0 & 0 & 0 & \beta_2 & \gamma_1 & 0 & 0 & 0 & 0 & 0 & 0 & 0 & 0 \\
\delta_1 & \delta_2 & 0 & 0 & \mu_2 & a_{66} & 0 & \beta_1 & 0 & 0 & 0 & \beta_2 & \gamma_1 & 0 & 0 & 0 & 0 & 0 & 0 & 0 \\
0 & \lambda_2 & 2\lambda_1 & 0 & \alpha_1 & 0 & a_{77} & \gamma_2 & 2\beta_1 & 0 & \mu_1 & \alpha_2 & 0 & \beta_2 & \gamma_1 & 0 & 0 & 0 & 0 & 0 \\
0 & \delta_1 & \delta_2 & 0 & 0 & \alpha_1 & \mu_2 & a_{88} & 0 & 2\beta_1 & 0 & \mu_1 & \alpha_2 & 0 & \beta_2 & \gamma_1 & 0 & 0 & 0 & 0 \\
0 & 0 & \lambda_2 & 3\lambda_1 & 0 & 0 & \alpha_1 & 0 & a_{99} & \lambda_2 & 0 & 0 & 0 & \mu_1 & \alpha_2 & 0 & 0 & 0 & 0 & 0 \\
0 & 0 & \delta_1 & \delta_2 & 0 & 0 & 0 & \alpha_1 & \mu_2 & a_{1010} & 0 & 0 & 0 & 0 & \mu_1 & \alpha_2 & 0 & 0 & 0 & 0 \\
0 & 0 & 0 & 0 & \lambda_2 & 0 & \lambda_1 & 0 & 0 & 0 & a_{1111} & \gamma_2 & 0 & \beta_1 & 0 & 0 & \beta_2 & \gamma_1 & 0 & 0 \\
0 & 0 & 0 & 0 & \delta_1 & \lambda_2 & \delta_2 & \lambda_1 & 0 & 0 & \mu_2 & a_{1212} & \gamma_2 & 0 & \beta_1 & 0 & 0 & \beta_2 & \gamma_1 & 0 \\
0 & 0 & 0 & 0 & 0 & \delta_1 & 0 & \delta_2 & 0 & 0 & 0 & \mu_2 & a_{1313} & 0 & 0 & \beta_1 & 0 & 0 & \beta_2 & \gamma_1 \\
0 & 0 & 0 & 0 & 0 & 0 & \lambda_2 & 0 & 2\lambda_1 & 0 & \alpha_1 & 0 & 0 & a_{1414} & \gamma_2 & 0 & \mu_1 & \alpha_2 & 0 & 0 \\
0 & 0 & 0 & 0 & 0 & 0 & \delta_1 & \lambda_2 & \delta_2 & 2\lambda_1 & 0 & \alpha_1 & 0 & \mu_2 & a_{1515} & \gamma_2 & 0 & \mu_1 & \alpha_2 & 0 \\
0 & 0 & 0 & 0 & 0 & 0 & 0 & \delta_1 & 0 & \delta_2 & 0 & 0 & \alpha_1 & 0 & \mu_2 & a_{1616} & 0 & 0 & \mu_1 & \alpha_2 \\
0 & 0 & 0 & 0 & 0 & 0 & 0 & 0 & 0 & 0 & \lambda_2 & 0 & 0 & \lambda_1 & 0 & 0 & a_{1717} & \gamma_2 & 0 & 0 \\
0 & 0 & 0 & 0 & 0 & 0 & 0 & 0 & 0 & 0 & \delta_1 & \lambda_2 & 0 & \delta_2 & \lambda_1 & 0 & \mu_2 & a_{1818} & \gamma_2 & 0 \\
0 & 0 & 0 & 0 & 0 & 0 & 0 & 0 & 0 & 0 & 0 & \delta_1 & \lambda_2 & 0 & \delta_2 & \lambda_1 & 0 & \mu_2 & a_{1919} & \gamma_2 \\
0 & 0 & 0 & 0 & 0 & 0 & 0 & 0 & 0 & 0 & 0 & 0 & \delta_1 & 0 & 0 & \delta_2 & 0 & 0 & \mu_2 & a_{2020}
\end{bmatrix}$$

with:

$$a_{kk} = -\sum_{j \neq k} \lambda_{kj}$$

The results for the first two iterations of the method are presented in (Table 2).

Approximate values in (Table 2) are close to the exact values from the first iteration. A priori knowledge of the behaviour of the system proves to be very influential in the speed of convergence of the method (Peres and Martin 1999).

Table 2 State occupation probability

State	P_{e_i} Exact result	P_{e_i} Iteration 1	P_{e_i} Iteration 2
3000	0.075230	0.0760979	0.0752189
2100	0.224645	0.2287500	0.2245980
2010	0.002795	0.0028195	0.0027940
2001	0.000810	0.0008045	0.0008098
1200	0.335458	0.3452980	0.3353520
1110	0.008345	0.0084543	0.0083428
1101	0.002323	0.0023102	0.0023217
1020	0.000010	0.0001040	0.0001038
1011	0.000030	0.0000300	0.0000302
1002	0.000009	0.0000088	0.0000086
0300	0.333947	0.3498540	0.3337670
0210	0.012461	0.0126954	0.0124569
0201	0.003413	0.0034075	0.0034112
0120	0.000310	0.0003120	0.0003099
0111	0.000088	0.0000875	0.0000876
0102	0.000026	0.0000250	0.0000255
0030	0.000004	0.0000039	0.0000038
0021	0.000001	0.0000012	0.0000012
0012	0.000000	0.0000003	0.0000003
0003	0.000000	0.0000001	0.0000001

6 Discussion

Back to the method of product calculations. As we said previously, a certain dispersion can be observed when the dynamics of the model do not lend themselves to the decoupling principles. From the rates of the previous model, we progressively change the dynamic between subsystems. Each step corresponds to a new series of rates obtained by multiplying the external rate dynamics (between subsystems) by a constant. Gradually, the relationship between the internal and external dynamics is reversed. For each series of rates, the probability of occupation of each state of the model is calculated.

We present on (Fig. 10) the difference between exact and approached values (curves of same grey level) for several ratios between dynamics. Only the most significant states (in the sense of the greatest occupation probability) are here considered.

The last step of calculation corresponds to the following set of values: $\alpha_1 = 3$, $\alpha_2 = 256$, $\beta_1 = 1$, $\beta_2 = 0.5$ $\delta_1 = 2.56$, $\delta_2 = 0.512$, $\delta_1 = 64$, $\gamma_2 = 0.512$, $\lambda_1 = 0.01$, $\lambda_2 = 0.001$, $\mu_1 = 0.33$, $\mu_2 = 0.512$.

This set of values corresponds to a multiplication by 500 of the rates between subsystems. For these values, the application of the product form method provides

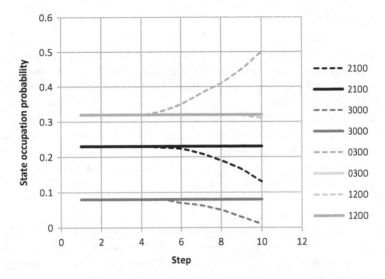

Fig. 10 Dispersions noted on the results achieved by product-form calculations

Fig. 11 Exact versus
approximate methods on a
particular case

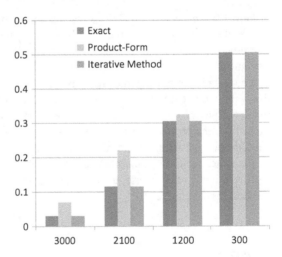

inaccurate or even erroneous results.[1] The use of the iterative method allows then the
compensation for the shortcomings of the first approach. (Eqs. (3) and (4)) stopped
after only six iterations give quite acceptable results. The comparison between exact
and approximate values by both methods is presented in (Fig. 11) for the most
significant states. Note that the distribution into classes was not influenced by the
results of the first method. If that were the case, the convergence would have been
even faster.

---

[1] It would be logical in such a case, to reverse the process decoupling but our intention here was to
observe the behaviour of the iterative method when product calculations do not give satisfactory
results. Let us note that the rates considered here are meaningless.

7 Conclusion

For many systems, reliability is of great importance to ensure the performance of duties or tasks for which they were designed. In practice, the easiest way to increase the dependability of a system is to add redundancy that is to say to make several identical systems in parallel.

To assess the level of reliability achieved and therefore the risk of not being able to meet these objectives, various tools exist. Among them, Markov processes and their graphical representation under the form of chains do not have to prove themselves. However, when the system is complex due to the number of components to consider or because of the many states that may describe the component condition, Markov chains face the classic problem of combinatorial explosion.

We have presented in this chapter two methods to tackle this problem by decreasing the matrix size corresponding to the mathematical issue to be dealt with. The first method of product-form calculation is based on the demonstration of a decoupling from the observation of a difference in dynamics between the events making evolve the system over time. The second relies on an iteration-based approach.

When the respective conditions for the deployment of these approaches are met, the results are very similar to those obtained by the complete calculation. When the dynamic gap is not sufficient, the product calculation method can lead to significant deviations.

The iterative method comes then as reinforcements of the product-form calculation method, but it can also be an excellent test of applicability of the latter. The results of product-form calculation method can indeed be taken over by the iterative method. If the product-form calculation method is applicable, the iterative method should converge to the same results from the first iteration (the classes composed of the states of same probability levels being deduced from the results of the first method). If it is not the case, the process decoupling in the product-form calculation method is not suitable.

References

Amari SV, Dill G (2010) Redundancy optimization problem with warm-standby redundancy. Reliab Maintainab Symp (RAMS), 2010 Proc—Annu. doi: 10.1109/RAMS.2010.5448068

Andrews JD, Poole J, Chen WH (2013) Fast mission reliability prediction for unmanned aerial vehicles. Reliab Eng Syst Saf 120:3–9. doi: http://dx.doi.org/10.1016/j.ress.2013.03.002

Andrieu C, Doucet A, Holenstein R (2010) Particle Markov chain Monte Carlo methods. J R Stat Soc Ser B Stat Methodol 72:269–342. doi:10.1111/j.1467-9868.2009.00736.x

Balakrishnan M, Reibman A (1993) Characterizing a lumping heuristic for a Markov network reliability model. FTCS-23 twenty-third international symposium on fault-tolerant computing. doi: 10.1109/FTCS.1993.627308

Buchholz P (2008) Product form approximations for communicating Markov processes. 2008 Fifth international conference on the quantitative evaluation of systems. doi: 10.1109/QEST.2008.23

Diaconis P (2008) The Markov chain Monte Carlo revolution. Bull Am Math Soc 46:179–205. doi:10.1090/S0273-0979-08-01238-X

Godichaud M, Pérès F, Tchangani A (2010) Disassembly process planning using Bayesian network. Engineering Asset Lifecycle Management. Springer, London, pp 280–287

Godichaud M, Tchangani A, Pérès F, Iung B (2012) Sustainable management of end-of-life systems. Prod Plann Contr 23:216–236. doi:10.1080/09537287.2011.591656

Gupta A, Albright SC (1992) Steady-state approximations for a multi-echelon multi-indentured repairable-item inventory system. Eur J Oper Res 62:340–353, doi: http://dx.doi.org/10.1016/0377-2217(92)90123-Q

Kemeny JG, Snell JL (1967) Excessive functions of continuous time Markov chains. J Comb Theory 3:256–278, doi: http://dx.doi.org/10.1016/S0021-9800(67)80074-7

Nahman JM (1984) Iterative method for steady state reliability analysis of complex Markov systems. IEEE Trans Reliab. doi:10.1109/TR.1984.5221880

Nourelfath M, Châtelet E, Nahas N (2012) Joint redundancy and imperfect preventive maintenance optimization for series–parallel multi-state degraded systems. Reliab Eng Syst Saf 103:51–60. doi:10.1016/j.ress.2012.03.004

Papazoglou IA, Gyftopoulos EP (1977) Markov processes for reliability analyses of large systems. IEEE Trans Reliab. doi:10.1109/TR.1977.5220125

Peres F, Martin C (1999) Design methods applied to the selection of a rapid prototyping resource. 1999 7th IEEE international conference on emerging technologies and factory automation proceedings, ETFA'99 (Cat No. 99TH8467). doi: 10.1109/ETFA.1999.815386

Strelen JC (1997) Approximate product form solutions for Markov chains. Perform Eval 30:87–110, doi: http://dx.doi.org/10.1016/S0166-5316(96)00053-3

Taylor PG (1992) Algebraic criteria for extended product form in generalised semi-Markov processes. Stoch Process Their Appl 42:269–282, doi: http://dx.doi.org/10.1016/0304-4149(92)90039-S

Vujosevic M, Meade D (1985) Reliability evaluation and optimization of redundant dynamic systems. IEEE Trans Reliab. doi:10.1109/TR.1985.5221983

White DJ (1993) A survey of applications of Markov decision processes. J Oper Res Soc 44:1073–1096. doi:10.1057/jors.1993.181

Infrastructure Vulnerability Assessment Toward Extreme Meteorological Events Using Satellite Data

Yuriy V. Kostyuchenko

Abstract In the chapter is discussed some aspects of the theoretical and methodological basis for using of remote sensing data of snow cover for hazards assessment related to meteorological, climatic, hydrological, and hydrogeological risks over urban areas. Main focus is on the urban infrastructure risk assessment toward extreme snowstorms using satellite data. A method for determining the snow cover parameters on remote sensing data base (in particular *MOD10A1* and *SWE* products, normalized index of snow depth (NDSI), and local meteorological observations) has been proposed. A method for data integrating from various sources on the basis of the modified Ensemble Transform Kalman Filter (ETKF) and Kernel Principal Component Analysis (KPCA) of data distributions has been proposed. It is shown that the proposed approach has quite high relative accuracy in comparison with existing methods, algorithms, and products (if *MOD10A1* and *SWE* products have been used separately) with application for urban agglomeration. As an example, the approach have been used for analysis of extreme snowfall in Kiev (March 21–23, 2013). Quantitative assessments of risk indicators and vulnerability parameters of municipal infrastructure under emergency effects have been proposed.

Y.V. Kostyuchenko (✉)
Scientific Centre for Aerospace Research of the Earth, National Academy of Sciences of Ukraine, 55-b, O. Honchar Street, Kiev 01601, Ukraine

Department of Earth Sciences and Geomorphology, Faculty of Geography, Taras Shevchenko National University of Kiev, 2A Glushkov Avenue, 03075 Kiev, Ukraine
e-mail: yvk@casre.kiev.ua; yuriy_kostyuchenko@ukr.net

S. Kadry and A. El Hami (eds.), *Numerical Methods for Reliability and Safety Assessment: Multiscale and Multiphysics Systems*, DOI 10.1007/978-3-319-07167-1__15, © Springer International Publishing Switzerland 2015

1 Introduction: General Methodological Remarks

Satellite observation for quantitative risk assessment is essentially new research direction. Existing wide range of sensors with different spectral, spatial, temporal, and radiometrical resolution provide a near real-time information on whole Earth system and infrastructure. Validity and reliability of data is quite high with using of ground calibration data. Considering methodological variety of data sources the approach to multi-source data coupling is required.

Satellite observation, ground calibration, and modeling data may be represented in framework of formalization "information—response" in security management systems. This formalization includes variable I—information obtained from direct ground measurements and modeling, and $H_I(i|\theta)$—probability distribution function of I, where θ—state of observed system or object. In general case this state cannot be determined with full reliability, so we should consider probability distribution $p(\theta)$ and function $H_I(i|\theta)$ to describe a priory incompleteness of existing information. Decision making could be formalized as the reaction to input information by decision function $d(I)$.

With indeterminate state θ or under observed changes of this state the losses could be formalized as $l(d(I), \theta)$. For decision function d the expected losses or risks could be defined through minimization of optimal decision function, for example, Bayes' function, $d*(I)$. Risks in this case are determined by implementation of decisions d from strategy A, based on information received.

Therefore we analyze set of data I (ground and modeling data) and $i*$—information, which optimize decision function d to $d*$, and so minimize corresponding risk. So this additional (in our case—information from satellite observations) made information I nominally full ($I*$).

General definition of risk in these terms might be presented as:

$$R\left(I^*, d^*\right) = \int \min_{a \in A} l\left(a, \theta\right) p\left(\theta\right) d\theta \tag{1}$$

where $l\left(b, \phi\left(i^*\right)\right) = \min_a l\left(a, \phi\left(i^*\right)\right)$, and $d*\left(i*\right) = b$, $\phi\left(i*\right)$—state of observed system, when $H_I(i|\theta) \neq 0$.

So for optimal security management we need correct models of surface objects and processes, ground point measurements data and satellite observations.

Using this general formalization we able to construct applied tool for risk analysis for different cases.

Description here aimed to demonstration of possibility to developing of suitable theoretic and methodological framework for quantitative estimation of parameters of meteorological, hydrological, hydrogeological, and climatic related risks in urban agglomerations using satellite observations of snow cover.

Urban areas are high vulnerable toward varied impacts through uncertain heterogeneities of infrastructure, population, buildings distribution, and its dynamic. So the highest risks measures are characterizing the urbanized agglomerations.

Natural disasters in such territories influencing to the interlinked spatially distributed objects and subsystems. Therefore the optimal disaster control system requires the quick access and updating of spatial data.

Spatially integrated, independent, and external data sources such as the remote sensing data allows to analyze risk parameters more effectively and so to optimize disaster mitigation strategies.

The approach presented is aimed to propose the methodologically correct algorithm for coupling of satellite observation data, snow cover modeling, ground calibration measurement, and meteorological data to snowstorm risks parameters estimation in urban agglomerations.

2 Case Study

Studying of the risks associated to emergencies with hydrological, meteorological, and climatic origin is quite well-known problem. But in the cases of the vulnerability analysis over the large urban areas, in particular over urban agglomerations, there are certain methodological features that must be considered for successful decision of tasks of safety.

It is obvious that emergencies, influence areas of which cover the areas of mass dwelling of people, infrastructure conglomerations and industrial objects, in particular large urban agglomerations, have the highest level of risks.

During 22–24 of March 2013 in Kyiv fell a record amount of snow (over 600 mm). The average depth of snow cover in the city during the second half of the March 22 to end of March 23 was about 50 cm. In parks it was about 35–45 cm, within built-up area 35–55 cm. There were numerous 65–85 cm height snowdrifts. After snow removal, 1.2–1.4 m height snow heaps had been accumulated. This was cause of significant obstacles for the vehicles and pedestrians. In the evening on March 22 this led to public transport stoppage, local energy interrupting, appreciable disruptions in provision of local food shops, in particular by bakeries products. And 23rd March this led to the full traffic collapse in the city.

Such course of events led to deterioration in the quality and conditions of citizens life, namely to deterioration in the quality of nutrition, increasing the number of hypothermia and injury cases, referrals to hospitals with cold symptoms and exacerbations of chronic diseases caused by synoptic stress, dangerous accumulation of domestic waste, uncontrolled use of reagents for snow melting. Integral impact of these processes on urban ecosystem is uncertain.

There are possibility to assess as level of risks from the current emergency and the effectiveness of risk management at the regional level using a set of objective quantitative indicators. Scilicet we can assess the level of efficiency for regional strategies of emergency risk management, preparedness of regional systems of emergency risk management, provision of timely assistance to victims and so on.

3 Problem-Oriented Model of Snow Cover

Risks are defined by burden indicators of excessive precipitation for engineer networks, life support systems, and transport infrastructure of the city. In the simplest case, it is possible to calculate the weight of snow and thus determine the resources to clean it up, estimate the mass of water that can be formed after melting, and so on. The total weight of the snow is a dynamic value which depends on the density of the snow cover. Dynamic density of the snow cover during the observation interval can be defined as:

$$\rho_{\text{snow}} = \frac{\sum_i (\rho_i \Delta z_i + \rho_0 \Delta z_0)}{\Delta z_{\text{snow}}} \tag{2}$$

where ρ_i, Δz_i—density and depth of snow cover recorded during the observation period i; ρ_0, Δz_0—density and depth of snow cover at the beginning of the observation period. In this case, dynamic of density defined by set of meteorological parameters recorded during the observation period i:

$$\rho_i = \rho_i^{\text{min}} + \kappa \frac{1 - Q_{\text{p}}}{f_l^{\text{max}}} \tag{3}$$

Here ρ_i^{min}—initial level of precipitation; κ—empirical coefficient (usually is assumed to be 181); Q_{p}—thermal functional of precipitation depend on temperature, atmospheric pressure, and humidity; f_l^{max}—maximum precipitation water content (for predictive calculations is assumed 0.5).

A set of empirical coefficients are determined by the complex of field measurements. Changes of local meteorological parameters are calculated using meteorological measurements, and spatial distributions of parameters can be calculated by remote sensing.

In this case, the weight of snow M_{snow}^{ij} for period i and definite area S_j can be calculated as:

$$M_{\text{snow}}^{ij} = \sum_{ij} z_{ij} \rho_{ij} S_j \tag{4}$$

Water equivalent of accumulated snow W_{ij} also should be calculated based on current and forecasted weather conditions, as well as the spatial heterogeneity of snow cover distribution. In the simplest case it can be represented as:

$$W_{ij} = \frac{\rho_{\text{w}} H_{\text{s}}}{c_{\text{s}}} \sum_{ij} \frac{d_{ij}^{\text{cc}}}{\left(0^{\circ}\text{C} - T_{ij}^{\text{s}}\right)} \tag{5}$$

where ρ_w—water density; T_{ij}^s—snow cover temperature; H_s—latent heat of snow cover (assumed 80 cal/g); c_s—snow diathermancy (generally is assumed 0.5 cal/g °C); d_{ij}^{cc}—depth of solid cover.

For the calculation of the whole catchment area, it can be used the integrated form of the water equivalent equation. According to the above equations, and by analogy with (Anderson 1976), the general equation for catchment areas is:

$$W_{\text{tot}} = \left[\frac{1 - A_0}{A_i - A_0} (W_i - W_0) \right] + W_0 \tag{6}$$

In this equation W_{tot}—total moisture content in the snow cover (water equivalent of snow cover); W_0—the water equivalent of snow cover up to melting; W_i—water equivalent of snow cover during the observation period i; A_0—snow cover area up to snow melting; A_i—current snow cover area (at the observation period i). So the key variables are the snow cover area and parameters that determine the water equivalent of snow cover during the observation period.

In above described approach is used four-component model of snow cover. This model describes snow cover as a set of ice, water, air, and water vapor fractions. This model provides more accurate forecasts for dynamic spatial–temporal scenarios, in contrast to two-component (ice and air) and three-component (ice, water, air) models. This model can be used for remote sensing, but it should be calibrated by ground-based measurements.

4 Remote Sensing for Snow Cover Detection and Analysis

The problem of snow cover analysis within urban areas is more complicated than estimation of snow cover parameters within natural and agricultural landscapes. Usually there are two key issues: efficiency and spatial resolution. Requirements to efficiency of information depend on the problem formulation. For standard cases set of daily parameters of snow cover distribution is enough. But for analysis of the situation in the city, information should be provided with 5–8 h intervals. Meteorological satellites provide such efficiency, but they don't provide a set of snow cover parameters. In addition, such survey rate supposes measurement of snow cover parameters during snow accumulation. But it is impossible with using of optical sensors because dense clouds.

Another problem is the low resolutions (spatial and spectral) of sensors used to analyze the snow cover parameters. Spatial resolution in several hundred meters is quite suitable for nature and agriculture landscapes. But urban areas require higher resolutions (50–150 m). Spectral resolution should provide robustness of recognition algorithm, i.e., the ability to detecting the snow-covered areas with significant changes of snow reflectance because human activities (clearing of snowdrifts, for example).

Currently there aren't sensors which can solve these problems and provide satisfactory outcomes. In this case, the solution is involving data from many sources, using of ground calibration data and combination of observations with the results of generate hazards processes modeling, namely energy and mass exchange in natural systems (in our case it is snow accumulation and melting processes). These data should be calibrated and properly spatial, temporal, and energy mutually agreed.

The main issues of snow cover assessment using remote sensing traditionally associated with meteorology and climatology, hydrology, and agriculture. This range of tasks defines sets of sensors and algorithms used for snow cover analyzing. MODIS (Hall et al. 2006a) and AMSR-E (Tedesco et al. 2004) products are most helpful for most applications. So MOD10A1 (Hall et al. 2006b) product of MODIS sensor provides the daily snow cover parameters with 5-day and monthly averaging, and with 500 m spatial resolution. SWE product of AMSR-E sensor provides snow cover water equivalent with 25 km spatial resolution and 5-day time resolution (Tedesco et al. 2004). These data are well verified, the sensors are properly calibrated and algorithms allow to calculate the snow cover parameters distribution with high reliability. Reliability is limited only spatial and temporal resolutions. They can have disagreements in data distribution.

Above mentioned algorithms are most commonly used for analysis of snow cover. But their spatial resolution is big obstacle for their using for the analysis of urban areas. Furthermore, MOD10A1 and SWE products algorithm isn't robust with respect to significant changes in snow cover radiation parameters, which are typical for urban agglomerations because human activities. Therefore, we need additional sources of information with sufficient spatial resolution (comparable with spatial scale of urban areas) and with possibility to control parameters of recognition.

By analogy of (Zhang et al. 2008), considering the known trends in the radiation parameters of the snow cover at different stages of snowmelt (Dozier 1989; Ramsay 1998), and parameters of different types of snow defined by a set of ground-based measurements, we can offer snow cover normalized index:

$$\text{NDSI} = \frac{R_{[0.55-0.65]} - R_{[0.75-0.85]}}{R_{[0.55-0.65]} + R_{[0.75-0.85]}} \approx \frac{R_{\text{VIS}} - R_{\text{NIR}}}{R_{\text{VIS}} + R_{\text{NIR}}} \tag{7}$$

where R is reflection in a certain part of the spectrum.

Based on the analysis of snow reflectance distributions derived from series of observations, models, ground measurements, and data presented by other researchers (Zhang et al. 2008) we can determine the threshold value of NDSI index for areas with snow cover. It is 0.32–0.36 for open areas and 0.28–0.31 for areas covered by dense vegetation. For dense built-up areas index varies in 0.18–0.33 range. It depends on the density of buildings, anthropogenic impact level (snow cover pollution, snow compactness, etc.), and sensor type. Hence we can set a rule: if the pixel x_{ij} has NDSI index value greater than or equal to the limit value we refer this pixel to the area A_i covered by snow with maximum reliability. Ratio of NDSI value and snow cover depth is required for estimation of water equivalent and can be determined using calibration measurements on a local level.

Thus we introduce an optical index for determination of snow cover problem-oriented parameters (which determine the distribution of the relevant variables in the energy and mass exchange models) to the complex of indicators in microwave and infrared ranges that exist in the issues of integrated assessment of snow cover using remote sensing. Of course, the addition of the index does not solve the problem of information collection during sedimentation, but using of one allow us involve for analyze the data with any spatial resolution. And it contributes to solve the problem with improving spectral and spatial resolutions for analysis of snow cover.

5 Data Integration Approach

Next issue is integration of data obtained from different sources, simulation outputs, meteorological observations, and ground measurements into a single harmonized data set suitable for analysis of risks. This issue is solved using approach for harmonization of data distribution. After initial data processing from observations, measurements, and models we obtain a set of normalized distributions:

$$\xi_t = \mathbf{A}_t f(x_t) + v_t \tag{8}$$

where t is time (modeling step for a set of model data and data set measure for distributions of the meteorological observations). Henceforth we offer analyze data from modeling, observations, and measurements using the modified Ensemble Transform Kalman Filter (ETKF) (Wang and Bishop 2003). Using of Kalman Filter to identify risk parameters is well developed at the NSAU-NASU Space Research Institute. In particular, effectiveness of this approach for parameters determination of the hydrological and meteorological hazards is proved in publication (Kravchenko et al. 2008; Kussul et al. 2008].

We assume that the true system state vector $\mathbf{x}$ at time k are defined by general law:

$$\mathbf{x}_t = \mathbf{F}_t \mathbf{x}_{t-1} + \mathbf{B}_t u_t + w_t \tag{9}$$

where $\mathbf{F}_t$—matrix of the system evolution, i.e., simulated impacts on vector x_{t-1} at time $t-1$; $\mathbf{B}_t$—matrix of control measured effects u_t the vector $\mathbf{x}$; w_t—random process with covariance matrix $\mathbf{Q}_t$. Thus we enter the description of model distributions F_x and observational data $\mathbf{B}_t$.

Let's define the extrapolation value of true system state vector using state vector from the previous step:

$$\widehat{\mathbf{x}}\Big|_{t\,|\,t-1} = \mathbf{F}_t \widehat{\mathbf{x}}\Big|_{t-1\,|\,t-1} + \mathbf{B}_t u_{t-1} \tag{10}$$

General view of covariance matrix for this extrapolation value is:

$$\mathbf{P}\Big|_{t|t-1} = \mathbf{F}_t \mathbf{P}\Big|_{t-1|t-1} \mathbf{F}_t^T + \mathbf{Q}_{t-1} \tag{11}$$

The difference between the estimated (extrapolation) value of the true system state vector and obtained one at the appropriate step of modeling can be estimated as:

$$\Delta \widehat{\mathbf{x}}_t = \xi_t - \mathbf{A}_t \widehat{\mathbf{x}}\Big|_{t|t-1} \tag{12}$$

and covariance matrix of deviation is:

$$\mathbf{S}_t = \mathbf{A}_t \mathbf{P}\Big|_{t|t-1} \mathbf{A}_t^T + \mathbf{R}_t \tag{13}$$

Next, let's enter the matrix of optimal factors of Kalman strengthening based on the covariance matrixes of the extrapolation vector of state and measurements:

$$\mathbf{K}_t = \mathbf{P}\Big|_{t|t-1} \mathbf{A}_t^T \mathbf{S}_t^{-1} \tag{14}$$

Now let's correct extrapolation vector values of the true system state:

$$\widehat{\mathbf{x}}\Big|_{t|t} = \widehat{\mathbf{x}}\Big|_{t|t-1} + \mathbf{K}_t \Delta \widehat{\mathbf{x}}_t \tag{15}$$

Herewith we enter georeferenced filter for distribution of vector x_{ij}, which depend on geographically referenced coordinates (i,j) and doesn't depend on time t:

$$\left(\mathbf{x}_{ij}\right)_t = \left(\mathbf{x}_{ij}\right)_t^\alpha = \left(\mathbf{x}_{ij}\right)_t \alpha_{ij} \tag{16}$$

The coefficients α are defined by KPCA algorithm (Lee et al. 2004) using rule of assess of optimal balance for mutual validation function:

$$C^F v = \frac{1}{N} \sum_{j=1}^{N} \Phi\left(x_j\right) \Phi\left(x_j\right)^T \cdot \sum_{i=1}^{N} \alpha_i \Phi\left(x_i\right) \tag{17}$$

where nonlinear function of input data distribution Φ satisfies the conditions:

$$\sum_{k=1}^{N} \Phi\left(x_k\right) = 0 \tag{18}$$

According to (Scheolkopf et al. 1998), and $\tilde{k}_t$—averaged values of Kernel matrix $\mathbf{K} \in R^N$ (where $[\mathbf{K}]_{ij} = [k(\mathbf{x}_i, \mathbf{x}_j)]$). This matrix consists from the Kernel vectors $\mathbf{k}_t \in R^N$, while $[\mathbf{k}_i]_j = [k_t(\mathbf{x}_t, \mathbf{x}_j)]$. Kernel matrix is calculated according to the modified rule (Christianini and Shawe-Taylor 2000):

$$\mathbf{k}_t(\mathbf{x}_i, \mathbf{x}_t) = \left\langle \rho_{j,t}^{x_j} \left(1 - \rho_{j,i} \right)^{x_j} \right\rangle \tag{19}$$

where ρ—empirical parameters chosen according to the model of the phenomena (Villez et al. 2008).

If the filter is used with mutual coordination of data sets, we can propose a matrix $\mathbf{P}^a$ for the analysis of actual errors based on a errors matrix of extrapolated value of the system state vector $\mathbf{P}^f$ and covariance matrix of observational data $\mathbf{R}$:

$$\mathbf{P}^a = \mathbf{P}^f - \mathbf{P}^f \mathbf{A}^T \left(\mathbf{A} \mathbf{P}^f \mathbf{A}^T + \mathbf{R} \right)^{-1} \mathbf{A} \mathbf{P}^f \tag{20}$$

So we get a tool for optimized calculation of Kalman strengthening optimal coefficients matrix and correction of extrapolation values of the true system state vector based on aggregate modeling and observations data.

After implementation of data integration procedure we recalculate the parameters using algorithm:

$$x_t^{(ij)} = \sum_{m=1}^{n} w_{ij} \left(\widehat{x}_t^m \right) x_t^m \tag{21}$$

where $w_{ij} \left(\tilde{x}_t^m \right)$—weighting factor is determined by the minimum (Cowpertwait 1995):

$$\min \left\{ \sum_{m=1}^{n} \sum_{x_t^m \in R^m} w_{ij} \left(\tilde{x}_t^m \right) \left(1 - \frac{x_t^m}{\widehat{x}_t^m} \right)^2 \right\} \tag{22}$$

where m—the number of the experiments; n—number of data sources; x_t^m—distribution of observations; R^m—data set; $\widehat{x}_t^m$—corrected extrapolation values of the true system state vector based on aggregate modeling and observations data.

Thereby we get a regular spatial distribution of measured parameters over a local area based on remote sensing, the model simulation and regional meteorological and ground calibration measurements with a grid that corresponds to the distribution of measured data and have significantly better resolution than the original grid.

6 Risk Assessment Method

Indicators of the spatial distribution of snow and ice mass accumulated within the urban area, their water equivalent, indicators of transport infrastructure burden, and

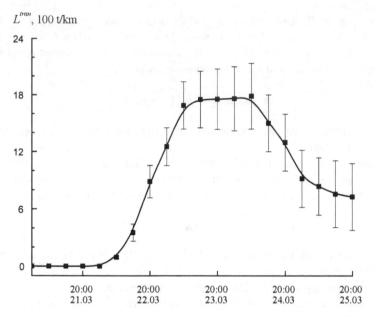

Fig. 1 Dynamic of transport infrastructure burden (L^{tran})

their temporal dynamics can be considered as a measures of risk and which can be quantitatively assess by remote sensing.

In the proposed approach, infrastructure burden, for example, transport infrastructure L_i^{tran}, can be defined as the ratio of the snow mass M_{snow}^{ij} accumulated within area during a time interval to the transport network density in a given area (km/km^2) R_j:

$$L_i^{\text{tran}} = M_{ij}/R_j \qquad (23)$$

The dynamic of transport infrastructure burden is presented in Fig. 1.

Thus, we transform the burden indicators to the spatial and temporal distributions of parameters which can be determined by remote sensing with ground validation and meteorological measurements.

Spatial distribution of snow cover and its integrated density and depth can be assessed using remote sensing. Similarly, we can use remote sensing for clarify the distribution of urban infrastructure, in particular the spatial distributions of R_j, which determines the density of the transport network.

According to the algorithms we calculated set of indicators and estimated total snow amount that created obstacles for the Kyiv city infrastructure. This amount is about 61.7 million tons. Considering total length of transport infrastructure (1675 km) (excluding yard passages and pedestrian areas) and its area (868 km^2) with significant heterogeneity (transport infrastructure density ranges from 1.43 to 2.76 km/km^2 and average is about 1.85 km/km^2), transport infrastructure burden

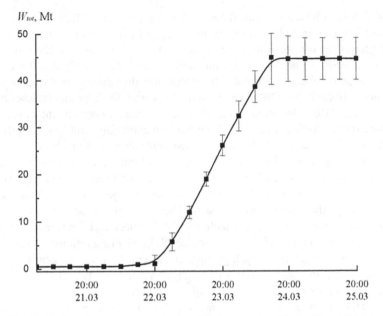

Fig. 2 The distribution of snow water equivalent (W_{tot}) accumulated during March 22–24 snowfall

was 2,742,100 tons totally or 1,700 tons of snow per 1 km (approximately 9.6 m³ of snow per meter of road including curbs, sidewalks, etc.). So we had one of the most extreme meteorological and climate emergencies in Kiev in the all observations history.

Approximately 45–50 % of transport infrastructure burden had been reduced till the second half of March 24 (i.e., within 48 h since emergence begun and 4–6 h after active phase of impact ended. Generally it is a good. But acceptable level of impact reducing (<12–15 % of the peak) (NDRF 2011; NIPP 2009] had not been achieved in 3 days. Thus necessary conditions for recovering of the normal society functioning had not been created. For example, during 4 days (till active snow melting) snow hadn't been removed outside the city in adequate quantities. It created additional threats of flooding and urban drainage systems overloading. This shows absence of adequate and/or imperfection of available mitigation strategies.

Also it should be noted that after March 22–24 snowfall it has been accumulated snow amount with approximately 44.5 million m³ water equivalent (see Fig. 2) or 51.5 thousand tons of water per km². It creates significant threats for a drainage systems and flood safety. The first problems in urban drainage systems have been fixed on March 28, i.e., on the fifth day after emergency begun and on the second day of active snow melting.

Generally the presented data agree with the data of daily observations by AMSR-E/Aqua sensor (Daily L3 Global Snow Water Equivalent) (Tedesco et al. 2004) (42.5 million tons for March 24–25) and with data of averaged 5-day distributions

AMSR-E/Aqua (5-Day L3 Global Snow Water Equivalent) (Tedesco et al. 2004) (36.5 million tons including drainage, melting, and evaporation). Thereby relative error (difference) for mean of our estimate is 4–6 % compared to daily measurements of AMSR-E and 18–20 % compared to the 5-day averaging. It has some differences with average estimates of errors and disagreements based on snow cover normalized index (NDSI) and MOD10A1 and SWE products described in (Zhang et al. 2010). According to its data, the average disagreement in estimates for China ranges from 2.5 to 15 % depending on averaging. But this disagreement in estimates (2.5 and 4.6 %) is clear because our estimates have been carried out for the urban agglomerations and estimates for China mainly for landscapes and farmland. Differences related to land cover changes are reduced because averaging multiday sets, but the overall difference related to the low resolution remains large. Thereby, the method for assessing the snow cover parameters based on complex using of remote sensing, including MOD10A1 and SWE products, snow depth normalized index NDSI, local meteorological observations, ground-based calibrations is characterized by high relative accuracy in comparison to existing methods, algorithms, and products (if MOD10A1 and SWE products have been used separately) for the urban agglomeration analysis.

Separately we should consider the methodology for assessing the overall risk of studied events. Let's consider parameters of events frequency depending on a fixed frequency, spatial and temporal heterogeneity of processes that are cause of threats and levels of dangerous impacts. The typical approach estimates event probability based on average frequency of its occurrence μ defined by the set of all available observations i:

$$\mu = \sum_i^n \frac{\lambda_i}{n} \to p \tag{24}$$

In this case, the average estimated probability will be from 0.007 to 0.01 (i.e., event recurrence ranges within once in every 100–140 years). We propose to use a more sophisticated approach based on modified Omori's law (Ogata 1983). It takes into account the complex interdependence of spatial and temporal distributions of the phenomena:

$$\lambda(t, x, y) = \mu + \sum_{j,t} \left(\frac{F_j^0(x, y)}{(t - t_j)^p} + \frac{(x - x_j; y - y_j) \cdot S_j(x - x_j; y - y_j)}{\exp(\alpha L_j)} + d \right)^q \to p \tag{25}$$

In this equation p, q, α—empirical coefficients; L—event impacts (e.g., extreme amount of precipitation caused by unpredictable untypical meteorological conditions for this season) at the point of area S_j with coordinates x, y at time t; $F_j^0(x, y)$—a stationary distribution of stress burden based on long-term observations; d—scaling factor defined by scale of risk generating processes (e.g., cyclone scale which can be determined by remote sensing). Using this approach we estimated event probability in range from 0.039 to 0.052 (i.e., event recurrence ranges within once in every 19–25 years). Also such assessments should consider climate changes not considered in our case.

7 Concluding Remarks

Satellite observation can be correct source of data for quantitative risk analysis. Spatially integrated, independent, and external data sources such as the remote sensing data allows to analyze risk parameters more effectively and so to optimize disaster mitigation and security management strategies.

Analysis of urban agglomeration requires new algorithms for correct data fusions from various sources: simulation, ground measurements and meteorological observations to obtain operational reliable situational and predictive estimations of risks with corresponding spatial and temporal resolutions.

Remote sensing data can be and should be used for control and monitoring of hazards over urban areas through improving of security management efficiency. Using problem-oriented algorithms for remote sensing data interpretation combined with meteorological observations, energy and mass exchange simulation and ground calibration measurements provide high accuracy. For risk assessments (e.g., urban infrastructure vulnerability, urban landscapes burden, moisture capacity assessment) this accuracy is higher compared with traditional approaches (Groisman and Lyalko 2012).

Emergency risks (e.g., hydrological, meteorological, and climatic emergency) and urban infrastructure vulnerability indicators currently are significantly underestimated. It leads to absence of adequate and/or imperfection of available mitigation strategies as at national and local levels. Discrepancy in preparation, risk management, and mitigation of emergency impacts with real complex risks can lead to "network" disasters (Kostyuchenko et al. 2012), which are caused by the systemic nature of the risks (Ermoliev and von Vinterfeldt 2012).

The above emergences show inefficiency of simplified decisions based on simple balance assumptions when input data and expected results have non-normal distribution (what we usually observe in extreme cases). Only an integrated modeling of entire set of deterministic and stochastic phenomena and processes relationships as risk generators can be useful for developing of effective risk management strategies. Only complex of strategic decision "ex-ante" (e.g., various engineering and structural measures) and adaptive decision "ex-post" (e.g., insurance and financial vehicles) provides flexible robust solutions (Ermoliev et al. 2000).

References

Anderson EA (1976) A point energy and mass balance model of a snow Cover. NOAA technical report NWS 19, 150

Christianini N, Shawe-Taylor J (2000) An introduction to support vector machines and other Kernel-based learning methods. Cambridge University Press, Cambridge, p 212

Cowpertwait PSP (1995) A generalized spatial–temporal model of rainfall based on a clustered point process. Proc R Soc Lond A 450(1938):163–175

Dozier J (1989) Remote sensing of snow in visible and near-infrared wavelengths. In: Asrar G (ed) Theory and applications of optical remote sensing. Wiley, New York, pp 527–547

Ermoliev YM, Ermolieva TY, Amendola A, MacDonald G, Norkin VI (2000) A system approach to management of catastrophic risks. Eur J Oper Res 122(2):452–460

Ermoliev Y, von Vinterfeldt D (2012) Systemic risk and security management. Managing safety of heterogeneous systems, Lecture notes in economics and mathematical systems. Springer, Berlin, Heidelberg, pp 19–49

Groisman P, Lyalko V (2012) Earth systems change over Eastern Europe. Akademperiodyka, Kiev, p 488. ISBN 978-966-360-195-3

Hall DK, Salomonson VV, Riggs GA (2006a) MODIS/Terra Snow Cover Monthly L3 Global 0.05Deg CMG. Version 5. National Snow and Ice Data Center, Boulder, CO

Hall DK, Salomonson VV, Riggs GA (2006b) MODIS/Terra Snow Cover Daily L3 Global 500m Grid. Version 5 [indicate subset used]. National Snow and Ice Data Center, Boulder, CO

Kostyuchenko YV, Kopachevsky I, Zlateva P, Stoyka Y, Akymenko P (2012) Role of systemic risk in regional ecological long-term threats analysis. In: Phoon KK, Beer M, Quek ST, Pang SD (eds) Proceedings of APSSRA, 23–25 May, Sustainable civil infrastructures—hazards, risk, uncertainty. Research Publishing, Singapore, pp 551–556

Kravchenko O, Kussul N, Lupian E, Savorskiy V, Hluchy L (2008) Water resource quality monitoring using heterogeneous data and high-performance computations. Cybernet Syst Anal 44(4):616–624

Kussul N, Shelestov A, Skakun S (2008) Grid system for flood extent extraction from satellite images. Earth Sci Inform 1(3):105–117

Lee J-M, Yoo CK, Choi SW, Vanrolleghem PA, Lee I-B (2004) Nonlinear process monitoring using kernel principal component analysis. Chem Eng Sci 59(1):223–234

NDRF (2011) National disaster recovery framework: strengthening disaster recovery for the nation. U.S.FEMA

NIPP (2009) IS-860.A: National infrastructure protection plan. U.S.FEMA

Ogata Y (1983) Estimation of the parameters in the modified Omori formula for aftershock frequencies by the maximum likelihood procedure. J Phys Earth 31:115–124

Ramsay B (1998) The interactive multisensor snow and ice mapping system. Hydrol Process 12(10–11):1537–1546

Scheolkopf B, Smola AJ, Muller K (1998) Nonlinear component analysis as a kernel eigenvalue problem. Neural Comput 10(5):1299–1399

Tedesco M, Kelly R, Foster JL, Chang ATC (2004) AMSR-E/Aqua 5-Day L3 Global Snow Water Equivalent EASE-Grids. Version 2. National Snow and Ice Data Center, Boulder, CO

Villez K, Ruiz M, Sin G, Colomer J, Rosen C, Vanrolleghem PA (2008) Combining multiway principal component analysis (MPCA) and clustering for efficient data mining of historical data sets of SBR processes. Water Sci Technol 57(10):1659–1666

Wang X, Bishop CH (2003) A comparison of breeding and ensemble transform Kalman filter ensemble forecast schemes. J Atmos Sci 60:1140–1158

Zhang YZ, Jin C, Yan S, Chiu LS (2008) Seasonal snow monitoring in Northeast China using space-borne sensors: preliminary results. Ann Geogr Inf Sci 14(2):113–119

Zhang Y, Yan S, Lu Y (2010) Snow cover monitoring using MODIS data in Liaoning Province, Northeastern China. Remote Sens 2(3):777–793. doi:10.3390/rs2030777

Geostatistics and Remote Sensing for Extremes Forecasting and Disaster Risk Multiscale Analysis

Yuriy V. Kostyuchenko

Abstract The method of analysis of multisource data statistics is proposed for extreme forecasting and meteorological disaster risk analysis. This method is based on nonlinear kernel-based principal component algorithm (KPCA) modified according to specific of data: socioeconomic, disaster statistics, climatic, ecological, infrastructure distribution. Using this method the set of long-term regional statistics of disasters distributions and variations of economic activity has been analyzed. On these examples the method of obtaining of the spatially and temporally normalized and regularized distributions of the parameters investigated has been demonstrated. Method of extreme distribution assessment based on analysis of meteorological measurements should be described. Analysis of regional climatic parameters distribution allows to estimate the probability of extremes (both on seasonal and annual scales) toward mean climatic values change. The way to coherent risk measures assessment based on coupled analysis of multidimensional multivariate distributions should be described. Using the method of assessment of complex risk measures on the base of coupled analysis of multidimensional multivariate distributions of data the regional risk of climatic, meteorological, and hydrological disasters were estimated basing on kernel copula semi-parametric algorithm.

Y.V. Kostyuchenko (✉)
Scientific Centre for Aerospace Research of the Earth, National Academy of Sciences of Ukraine, 55-b, O. Honchar Street, Kiev 01601, Ukraine

Department of Earth Sciences and Geomorphology, Faculty of Geography, Taras Shevchenko National University of Kiev, 2A Glushkov Avenue, 03075 Kiev, Ukraine
e-mail: yvk@casre.kiev.ua; yuriy.v.kostyuchenko@gmail.com

S. Kadry and A. El Hami (eds.), *Numerical Methods for Reliability and Safety Assessment: Multiscale and Multiphysics Systems*, DOI 10.1007/978-3-319-07167-1_16,
© Springer International Publishing Switzerland 2015

1 Introduction

Complex analysis of disaster distributions is necessary for correct assessment of its impact to vulnerability of socioeconomics and socio-ecological security.

Analysis and mapping of spatial and temporal distributions of heterogeneous disasters is very complicated problem, as well as the direct comparison of distributions is not correct approach. First, the different types of disasters have different long-term trend. Second, drivers of different disasters have different spatial and temporal scales and variability.

Problem of construction of correct techniques of complex regional risk assessment requires to estimate an integral threat of all disasters in the area studied. It requires to determination of measure of statistical distributions of observations (units of frequency of the investigated phenomena), which would be invariant toward data properties.

Problem of data analysis in context of disaster-induced socio-ecological risks is often connected with lack of reliable long-term series of catastrophic events observations, socio-ecological parameters, and natural systems state (Bartell et al. 1992; Kostyuchenko and Bilous 2009). According the general estimations based on satellite observations and statistical assessment the official data reliability on separate fields are: land-use 88–92 % (on the subregional and local scale 92–97 %), agricultural crop distribution 72–80 %, water use 92–95 %, fertilizer use 50 % (on the subregional and local scale 60–70 %), frequency and intensity of disasters 85 % (on the subregional and local scale 89–93 %). This level, and especially these variations of reliability, is not sufficient for correct integrated assessment. So, correct and regular parameter statistics is important for construction of adequate risk function and also for risk management strategies development (Kostyuchenko and Bilous 2009). Demonstration of way of observation data regularization for normalization of data reliability is the purpose of first item of this chapter.

Forecast based on multiscale analysis is another problem discussed. Main problem of local and regional climate analysis and predictions is the how the climate parameters mean values change is reflected in its extreme values distribution. We need a correct algorithm to calculate the most probable local extreme variations toward the distribution of mean values known from climate models, and based on geo-referred long-term observations. Deterministic approach based on climate models requires huge sets of heterogeneous data about climate system on regional scale. This data usually is unavailable and these types of models usually characterizes by high uncertainties. Our understanding of climate system and its local features is incomplete, so it is possible to calculate adequate only the mean values distributions with low spatial resolution (Romdhani et al. 1999).

Long- and mid-term variations of mean values of climatic parameters (first of all, the mean air temperature) we able to calculate with sufficient confidence using multiscale climate models and multidimensional sets of the observation data (meteorological measurements, satellite observations) (Romdhani et al. 1999; Villez et al. 2008). At the same time regional disaster risk depends on extreme values distribution. Therefore the analysis of stable correlations between well-calculated mean values distributions and extreme values is necessary for regional disaster risk analysis. So, regional and local analysis of behavior of extreme climatic values distributions is one of core elements of climate-related disaster risk analysis.

Multivariate character of multidimensional distributions of climate parameters generates high uncertainties, which makes impossible to use deterministic models. The system is not ergodic in rigorous sense. So the using of parametric methods is also limited.

To estimate a regional risk measure we need an approach to understand the complex systemic interrelations between distributions of mean and extreme values of climatic parameters and disasters frequency and intensity. Therefore development of alternative ways of analysis of multivariate distributions is the next core element of regional climate-related disaster risk analysis.

Here we propose to calculate the most probable distributions of extreme values of climate parameters toward the mean values change on regional scale using modified kernel-based nonlinear principal component analysis (KPCA) algorithm (Global Summary of the Day NOAA/NESDIS; Scheolkopf et al. 1998). Further, using the method of assessment of complex risk measures on the base of coupled analysis of multidimensional multivariate distributions of data, we try to estimate the regional risk of climatic, meteorological, and hydrological disasters basing on kernel copula semi-parametric algorithm.

2   Regional Risk Analysis Based on Multisource Data Statistics of Disasters

As the first step the approach to regional risk analysis based on multisource data statistics of disasters should be described. The method of analysis of multisource data statistics is proposed. This method is based on nonlinear kernel-based principal component algorithm (KPCA) modified according to specific of data: socioeconomic, disaster statistics, climatic, ecological, infrastructure distribution.

Using this method the set of long-term regional statistics of disasters distributions and variations of economic activity has been analyzed. On these examples the method of obtaining of the spatially and temporally normalized and regularized distributions of the parameters investigated has been demonstrated.

Therefore the robust technique of observation data regularization for normalization of data reliability is proposed. The technique utilize data from different sources, different nature, and with different metrics. This approach allows to calculate regularized distributions in units invariant toward data properties and quality. We can analyze simultaneously different types of disasters and driving forces, regardless of spatial and temporal scales and heterogeneities using this approach.

Multiscale analysis of regional disaster distribution trends has been done for separate areas: using approach proposed the natural and technological disaster statistic for period 1960–2010 was analyzed. On this statistics some trends were demonstrated. In particular, detected changes of frequency and intensity of natural disasters probably connected with impact of climate and environmental change both on global and regional levels. Determination of regional features of global climate impacts through disaggregation of the climate models is the direction of separate calculation using advanced stochastic approaches and algorithms.

The important vulnerability indicator is the socioeconomic reaction to increasing of dangerous impacts, which may be assessed by integrated indicative parameters. As the indicative decision-oriented parameters the indexes of damage (*IoD*) and vulnerability (*IoV*) have been proposed, calculated and analyzed. Index of damage reflects disasters losses related to regional economic parameters. Index of vulnerability includes also population distribution change. Distribution of these indexes shows that sustainable economic growth and implementation of adequate risk assessment and management strategies allow to reduce vulnerability of society toward natural catastrophes even with increasing of its frequency, intensity, and direct losses. Conclusions on efficiency of risk management regional strategies have been done. In particular, the case study demonstrates urgent necessity of implementation of systemic strategies of assessment and management of disaster risks.

2.1 Methodological Remarks

Correct statistical analysis requires the set of data x_i with controlled reliability, which reflects distribution of investigated parameters over study area during whole observation period (taking into account variances of reliability of observation and archive data x_t). Set of observation data x_t ($\mathbf{x}_t \in R^m$) consists of multi-source data: historical records, archives, observations, measurements, etc., including data with sufficient reliability x_j ($\mathbf{x}_j \in R^m$), where $j = 1, \ldots, N$. The set x_t includes also satellite observed and detected indicators. Problem of determination of controlled quality and reliability spatial–temporal distribution of investigated parameters x_i might be solved in framework of tasks of multivariate random processes analysis and multidimensional processes regularization (Raiffa and Schlaifer 1968).

2.2 Data Regularization Algorithm

Required regularization may be provided by different ways. If we able to formulate stable hypothesis on distribution of reliability of regional archives data in the framework of defined problem we may to propose relatively simple way to determine investigated parameters distributions $x_t^{(x,y)}$ towards distributions on measured sites x_t^m basing on (Fowler et al. 2003):

$$x_t^{(x,y)} = \sum_{m=1}^{n} w_{x,y} \left(\tilde{x}_t^m \right) x_t^m \tag{1}$$

where weighting coefficients $w_{x,y} \left(\tilde{x}_t^m \right)$ determined as:

$$\min \left\{ \sum_{m=1}^{n} \sum_{x_t^m \in R^m} w_{x,y} \left(\tilde{x}_t^m \right) \left(1 - \frac{x_t^m}{\tilde{x}_t^m} \right)^2 \right\} \tag{2}$$

according to (Cowpertwait 1995). Here m—number of records/points of measurements or observations; n—number of observation series; x_t^m—distribution of observations data; R^m—set (aggregate collection) of observations; $\tilde{x}_t^m$—mean distribution of measured parameters.

This is the simple way to obtain a regular spatial distribution of analyzed parameters over the study area, on which we can apply further analysis, in particular temporal regularization.

Further regularization should take into account both observation distribution temporal nonlinearity (caused by imperfection of available statistics) and features of temporal–spatial heterogeneity of data distribution caused by systemic complexity of studied phenomena—natural and technological disasters. According to Mudelsee et al. (2001), Lee et al. (2004), and Villez et al. (2008) the kernel-based nonlinear approaches are quite effective for analysis of such types of distributions.

Proposed method is based on modified kernel principal component analysis (KPCA) (Scheolkopf et al. 1998; Mika et al. 1999; Romdhani et al. 1999). In the framework of this approach the algorithm of nonlinear regularization might be described as following rule:

$$x_i = \sum_{i=1}^{N} \alpha_i^k \tilde{k}_t \left(x_i, x_t \right) \tag{3}$$

In Eq. (3) the coefficients α selected according to optimal balance of relative validation function and covariance matrix, for example, as (Lee et al. 2004):

$$C^F v = \frac{1}{N} \sum_{j=1}^{N} \Phi \left(x_j \right) \Phi \left(x_j \right)^{\mathrm{T}} \cdot \sum_{i=1}^{N} \alpha_i \Phi \left(x_i \right) \tag{4}$$

where nonlinear mapping function of input data distribution Φ determined as (Scheolkopf et al. 1998):

$$\sum_{k=1}^{N} \Phi \left(x_k \right) = 0 \tag{5}$$

and $\tilde{k}_t$—is mean values of kernel-matrix $\mathbf{K} \in R^N$ ($[\mathbf{K}]_{ij} = [k(\mathbf{x}_i, \mathbf{x}_j)]$). Vector components of matrix determined as $\mathbf{k}_t \in R^N$; $[\mathbf{k}_i]_j = [k_t(\mathbf{x}_t, \mathbf{x}_j)]$. Matrix calculated according to modified rule of (Christianini and Shawe-Taylor 2000) as: $\mathbf{k}_t (\mathbf{x}_i, \mathbf{x}_t) = \left\langle \rho_{j,t}^{x_j} (1 - \rho_{j,i})^{x_j} \right\rangle$, where ρ—empirical parameters, selected according to the classification model of study phenomena (Villez et al. 2008).

Using described algorithm it is possible to obtain regularized spatial–temporal distribution of investigated parameters over whole observation period with rectified reliability (Mudelsee et al. 2001).

2.3   Regularized Data: Analysis of Distributions

Proposed regularization algorithm has been applied to analysis of disaster statistics and obtaining of smoothed distributions of frequency of disasters, climatic and socioeconomic parameters for period 1960–2010 over study area (Ukraine) during the observation period (1960–2010).

It makes possible to calculate spatially and temporally regularized distributions of the parameters studied, to analyze interlinks and correlations between disaster distributions and, therefore, to assist a regional security and to calculate corresponding risk parameters.

For the area studied for period 1960–2010 has been analyzed wide group of natural disasters. Was used international classification of disasters according to (Guha-Sapir et al. 2011) with some minor variations caused by national classification features and data availability. For analysis were used data from international surveys and national reports (Global Summary of the Day NOAA/NESDIS; USSR National Economy 1970, 1981; State Budget of USSR 1987, 1989; National report 2004, 2006, 2010; Guha-Sapir et al. 2011).

On the Figures 1 and 2 presented calculated according the described algorithm resulting distributions of mean probability of various types of disasters per year per 1,000 km^2 during observation period.

The distributions presented may be analyzed as the multidimensional multivariate statistical distributions, and the relevant correlations might be calculated.

From viewpoint of risk assessment it is important to determinate the losses caused by disasters. Analysis of time-series of regularized geo-referred data makes possible to calculate few interesting distributions. First of all it is the distribution of direct losses from natural disasters in Ukraine, World and Europe (EU-27 region[1]), calculated in USD per square km (Fig. 3).

---

[1]EU-27: Austria, Belgium, Bulgaria, Cyprus, the Czech Republic, Denmark, Estonia, Finland, France, Germany, Greece, Hungary, Ireland, Italy, Latvia, Lithuania, Luxembourg, Malta, the Netherlands, Poland, Portugal, Romania, Slovakia, Slovenia, Spain, Sweden, and the United Kingdom.

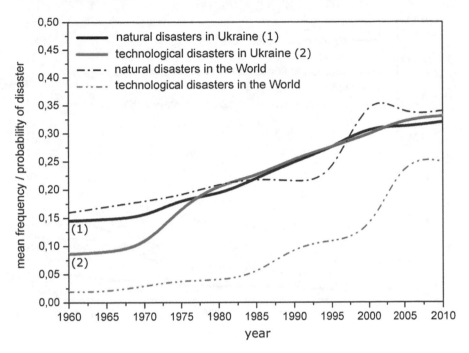

Fig. 1 Natural and technological disasters probability distribution over study area compared with average world distributions

The distribution presented demonstrates essential increasing of the losses, which is connected with registered increasing of frequency and intensity of disasters, as well as with increasing of the damaged infrastructure cost. Aiming to risks assessment and to include economic values, it is interesting to calculate relative indexes (Fig. 4).

This index (Fig. 4) was calculated using evident algorithm:

$$(IoD)_i = \langle L_{d_n} \rangle_i / (pCGDP)_i \tag{6}$$

where i—time step (in our case—calculated year); $<L_{dn}>$—estimated direct losses form disaster d of separate type n; $pCGDP$—per capita GDP (World Bank 2013). The distribution presented demonstrates that relative natural disasters damage (calculated per 1,000 km^2) during 1990 is slightly increasing, which is probably connected with impact of climate change. Common trend in world and Europe demonstrates decreasing of IoD, which is connected with economic grows (increasing of economic sustainability toward catastrophic events) and successful implementation of risk management strategies. At the same time on the territory of

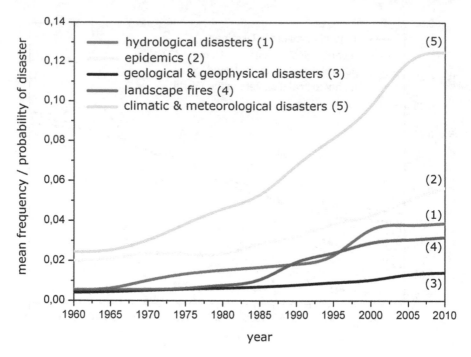

Fig. 2 Distribution of probability of separate types of natural disasters

Ukraine since 1980s and especially since 1990s *IoD* is increasing dramatically. It is connected with economical degradation and absence of adequate systemic strategies of risk management.

Also interesting to analyze separately the dimensionless index of vulnerability, which reflects losses related to GDP and population changes. This index can be calculated by simple formula:

$$(IoV)_i = \langle L_{d_n} \rangle_i / (pCGDP)_i (P_S)_i \qquad (7)$$

where *i*—time step (in our case—calculated year); $<L_{dn}>$—estimated direct losses form disaster *d* of separate type *n*; *pCGDP*—per capita GDP (World Bank 2013); P_S—population density (per square *S*) (State Statistics Service of Ukraine 2007).

The distribution presented is more evidently reflects the fact that sustainable economic growth and implementation of adequate risk assessment and management strategies allow decrease vulnerability of society toward natural catastrophes even with increasing of its frequency, intensity, and direct losses. Distribution of *IoV* for Ukraine demonstrates the necessity of implementation of systemic strategies of risk assessment and management (Fig. 5).

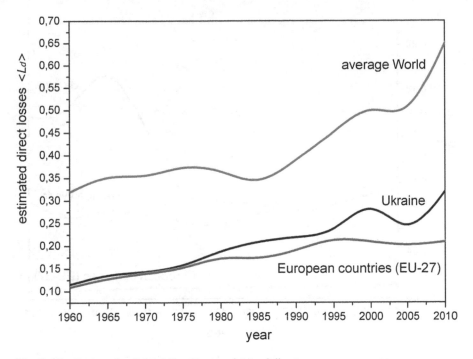

Fig. 3 Distribution of estimated direct losses of natural disasters

2.4 Discussion and Analysis

Approach proposed allows to calculate regularized distributions in units invariant toward data properties and quality. Using this approach it is possible to analyze simultaneously different types of disasters and driving forces, regardless of spatial and temporal scales and heterogeneities.

Number of natural disasters both average in the world and in separate regions is increasing. Mean probability of natural disaster calculated for square is increased during last 60 years to two times, and during last 20 years approximately to 60 %. Mean world losses is increased to two times, and to 70 % during last 20 years. In Ukraine the disaster losses is increased during last 20 years to 68 %. So intensity of natural disasters impacts is increased as over the world as well over Ukraine. Detected changes of frequency and intensity of natural disasters probably connected with impact of climate and environmental change both on global and regional levels (Groisman and Lyalko 2012). Determination of regional features of global climate impacts through disaggregation of the climate models is the direction of further investigations using advanced stochastic approaches and algorithms (Warga 1972; Ermoliev and Hordijk 2006; Guha-Sapir et al. 2011; Ermoliev and Winterfeldt 2012).

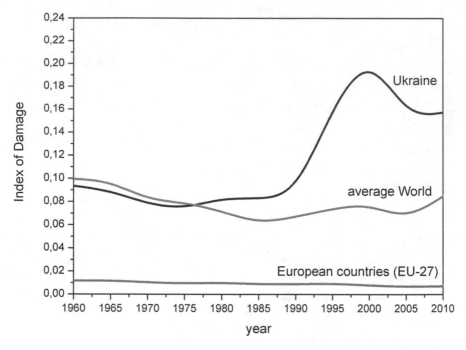

Fig. 4 *IoD* (index of damage): distribution of the estimated losses related to per capita GDP

The important indicator is the reaction to increasing of dangerous impacts, which may be assessed by indexes of damage and vulnerability. Index of damage reflects disasters losses related to regional economic parameters. Index of vulnerability includes also population distribution change. Distribution of these indexes shows that sustainable economic growth and implementation of adequate risk assessment and management strategies allow to reduce vulnerability of society toward natural catastrophes even with increasing of its frequency, intensity and direct losses. Ukrainian case demonstrates urgent necessity of implementation of systemic strategies of assessment and management of disaster risks.

3 Coherent Risk Measures Assessment Based on the Coupled Analysis of Multivariate Distributions of Multisource Observation Data

As the next step of analysis the coherent risk measures assessment based on the coupled analysis of multivariate distributions of multisource observation data we propose to describe.

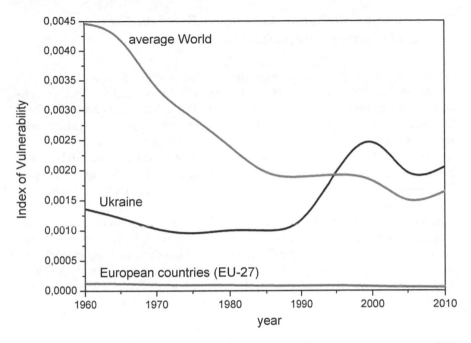

Fig. 5 *IoV* (index of vulnerability): distribution of the estimated losses related to per capita GDP and population density

First of all the method of extreme distribution assessment based on analysis of meteorological measurements should be described. Basing on KPCA the method of long-term regional analysis of statistics of meteorological measurements and climatic parameters has been demonstrated. The spatially and temporally normalized and regularized distributions of the parameters investigated have been obtained.

Further analysis of regional climatic parameters distribution allows to estimate the probability of extremes (both on seasonal and annual scales) toward mean climatic values change. Therefore the most probable distributions of extreme values of climate parameters toward the mean values change have been calculated on regional scale.

Next the way to coherent risk measures assessment based on coupled analysis of multidimensional multivariate distributions should be described. Using the method of assessment of complex risk measures on the base of coupled analysis of multidimensional multivariate distributions of data the regional risk of climatic, meteorological, and hydrological disasters were estimated basing on kernel copula semi-parametric algorithm.

3.1 Extreme Distribution Assessment Based on Analysis of Meteorological Measurements

Existing climate models, including reanalysis, has a spatial resolution 300–500 km (Kalnay et al. 1996; Parry et al. 2007). However for regional and local risk analysis we need resolution less than 100 km: about 40–70 km (Pflug and Roemisch 2007). Downscaling algorithms allow to obtain correct mean values distribution with necessary spatial grid, but not extreme values distributions. At the same time the density of meteorological stations and measurement points is about 30–50 km in developed regions and populated areas. So we have enough data for correct analysis. The problem is to construct a correct approach directed not to global but to regional and local analysis of data.

So this consideration directed to determination of explicit form of corresponding between known mean and studied extreme values of climatic parameters. In this case we should analyze probability distribution of set of data of meteorological measurements. So for every interval $[a, b]$ should be assessed probability $\Pr[a \leq X \leq b]$ of random value X will be belong to $[a, b]$. Let use the non-descending probability function $F(x)$ of simple event $p(x_i)$:

$$F(x) = \Pr[X \leq x] = \sum_{x_i \leq x} p(x_i) \qquad (8)$$

$$\lim_{x \to -\infty} F(x) = 0, \quad \lim_{x \to \infty} F(x) = 1 \qquad (9)$$

The task in this case may be formulated as determination of probability distribution:

$$P(x) = \Pr(X > x) \qquad (10)$$

And the corresponding probability distribution function $F(x)$, with $x \to \infty$.

For this purpose the distributions of meteorological measurements have been analyzed using the KPCA algorithm (Kostyuchenko et al. 2013b). Analysis was directed to determination of relationships between mean and extreme values distributions.

The area studied includes 15 meteorological stations in the site 250×250 km with center on 50.5N, 26E (Northern-West part of Ukraine, Ukrainian Polissya: Prypiat River basin), for the period 1979–2010. Mean max and min detected values of daily air temperature have been analyzed, as well as the monthly distributions of precipitation.

As the analysis demonstrates, over the whole 30-year period average annual distribution of extremes toward mean temperatures is close to normal. This is obvious result, which is interesting for strategic planning of adaptation, but is not useful for local disaster risk analysis. Climate-related disaster drivers have a seasonal nature, so extremes should be analyzed on the seasonal scale. The results

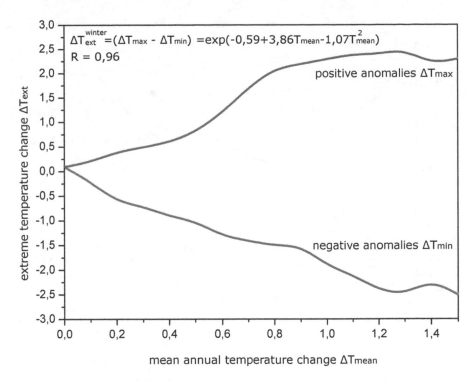

Fig. 6 Distribution of changes of max and min winter (December–February) air temperatures toward the change of mean air temperature in the study area 1990–2010

obtained (Figs. 6, 7, 8, 9) demonstrate significant deviation of seasonal distributions from the normal low. Most part of extremes distributions could be described by exponential distributions, and separate cases (for example, in spring season) are close to Pareto distribution.

Therefore we obtain a relation for determination of distribution of most probable values of temperature extremes toward known mean values. So it makes possible to estimate corresponding risks more correctly.

3.2 Coherent Risk Measures Assessment Based on Coupled Analysis of Multidimensional Multivariate Distributions

For assessment of regional climate-related disaster risk measures we propose to use the analysis of statistics of climate mean and extreme variations and multisource disasters records.

The main issue of such type of analysis is the quantitative estimation of risk measure in multidimensional multivariate case. It requires the correct assessment of every components of loss function distribution (Venter 2002). But risks in

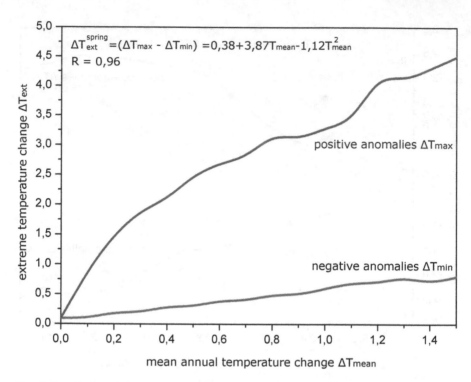

Fig. 7 Distribution of changes of max and min spring (March–May) air temperatures toward the change of mean air temperature in the study area 1990–2010

complex multi-component systems could not be described by linear superposition of scalar functions on the quite long time intervals (Ermoliev and Hordijk 2006). The complex temporal–spatial heterogeneities and significant uncertainties should be analyzed (Ermoliev and Hordijk 2006).

For analysis of the studied phenomena on intervals, in which its behavior differs essentially from normal, we propose to use a following copula (Genest et al. 1998):

$$C\,(u_1, u_2) = \exp\left(-V\left(-\frac{1}{\log u_1}, -\frac{1}{\log u_2}\right)\right) \tag{11}$$

$$V\,(x, y) = \int_0^1 \max\left(\frac{\omega}{x}, \frac{1-\omega}{y}\right) dH\,(\omega) \tag{12}$$

where

$$H\,(\omega) = \begin{cases} 0 & \omega < 0 \\ 1/2(\omega\,(1-\omega))^{-1-\alpha}\,(\omega^{-\alpha}(1-\omega)^{-\alpha})\,\frac{1}{\alpha-2}\,d\omega & 0 \le \omega < 1 \\ 1 & \omega \ge 1 \end{cases} \tag{13}$$

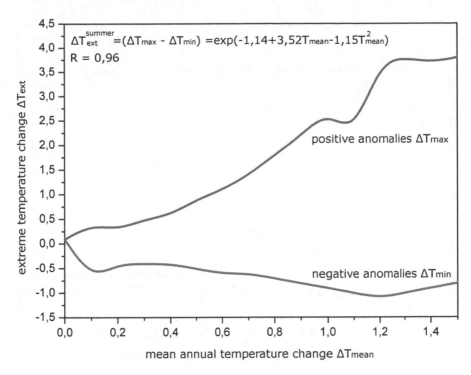

Fig. 8 Distribution of changes of max and min summer (June–August) air temperatures toward the change of mean air temperature in the study area 1990–2010

For analysis of interdependent (or weak dependent) phenomenon, for example, hydrological disasters, we can use form $0 \leq \omega < 1$.

This formalization allows better understand interdependencies between climatic parameters and disaster distribution on regional scale, and additionally allows to integrate regularization algorithms for uncertainty reducing (Juri and Wuthrich 2002).

For further analysis of behavior of risk measure dependent of number of climatic, ecological, etc., independent heterogeneous parameters we propose other algorithm. This method based on approach to coupled nonparametric analysis of multidimensional multivariate distributions by kernel copulas (Chen and Huang 2010). Using this approach it is possible to reduce uncertainties and errors connected with differences of measurement intervals, and to smooth gaps in data distributions (Embrechts et al. 2003).

If $K_{u,h}(x)$ is kernel-vector for $u \in [0; 1]$ on interval $h > 0$ we can propose according (Chen and Huang 2010):

$$K_{u,h}(x) = \frac{K(x)\,(a_2\,(u, h) - a_1\,(u, h)\,x)}{a_0\,(u, h)\,a_2\,(u, h) - a_1^2\,(u, h)} \qquad (14)$$

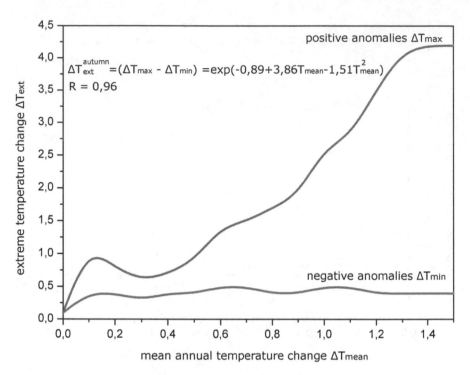

Fig. 9 Distribution of changes of max and min autumn (September–November) air temperatures toward the change of mean air temperature in the study area 1990–2010

$$a_l(u, h) = \int_{\frac{u-1}{h}}^{\frac{u}{h}} t^l K(t)dt, \quad l = 0, \ 1, \ 2 \tag{15}$$

Also in this case can be defined functions $G_{u,h}(t)$ and $T_{u,h}$:

$$G_{u,h}(t) = \int_{-\infty}^{t} K_{u,h}(x)dx \tag{16}$$

$$T_{u,h} = G_{u,h}\left(\frac{u-1}{h}\right) \tag{17}$$

Distribution function of the complex parameter will be determined by distribution functions of studied parameters $X_1, X_2, \ldots, X_n$ using copula C:

$$F(x_1, x_2, \ldots, x_n) = C(F_1(X_1), F_2(X_2), \ldots, F_n(X_n)) \tag{18}$$

Table 1 Correlation between number of disasters N_d and climatic parameters: mean air temperature T_{mean}, max detected air temperature T_{max} and the "reduced max temperature" T_{red} on different time intervals (Kostyuchenko et al. 2013a)

Observation periods	Climate parameters		
	T_{mean}	T_{max}	T_{red}
1960–1990	0.7	0.88	0.95
1990–2010	0.73	0.9	0.98
1960–2010	0.69	0.85	0.95

Distribution of extremes of studied parameters will be described by distribution functions $F_i(x)$ corresponding to threshold $x_i > u_i$ as:

$$\widehat{F}_i(x) = 1 - \frac{N_{u_i}}{n}\left(1 + \widehat{\xi}_i \frac{x - u_i}{\widehat{\beta}_i}\right)^{-\frac{1}{\xi_i}}, \quad i = 1, 2 \tag{19}$$

where ξ—smoothing parameter, β—interdependence parameter ($\beta \in [0, 1]$; $\beta = 0$ for independent distributions, and $\beta = 1$ for absolutely dependent distributions).

In this case the optimal kernel copula estimator may be presented as (Kalnay et al. 1996; Kostyuchenko et al. 2013a):

$$\widehat{C}(u, v) = n^{-1}\sum_{i=1}^{n} G_{u,h}\left(\frac{u - \widehat{F}_1(X_{i1})}{h}\right) G_{u,h}\left(\frac{v - \widehat{F}_2(X_{i2})}{h}\right)$$

$$- (uT_{u,h} + vT_{u,h} + T_{u,h}T_{v,h}) \tag{20}$$

For the area studied on the base of multi-year statistics it was determined an "optimal correlator" between air temperature and disaster frequency: "reduced max temperature" (Kostyuchenko et al. 2013a):

$$T_{red} = \left(1 - \frac{\frac{1}{N}\sum_{n=1}^{N} T_n}{T^{max}}\right)\left(1 - \frac{1}{T^{max} - \frac{1}{N}\sum_{n=1}^{N} T^{max}}\right) \tag{21}$$

Here N—number of meteorological measurements, T_n—measured air temperature, T_{max}—max registered air temperature.

Average correlation coefficients of T_{red} with quantity of disasters lie in interval 0.95–0.98, and is higher than correlation with mean temperature (0.69–0.73), and max temperature (0.85–0.9) for the period 1960–2010. Correlation coefficients are presented in Table 1.

Therefore the approach proposed is more correct relative to analysis with traditional values. Depending on time interval the multi-component correlation obtained allows increase accuracy of assessment of disasters frequency up to 22 % (11–34 %). This is essential value for mid- and long-term regional forecasting.

4 Concluding Remarks

Using multiscale approach it may be proposed an algorithm to calculate regularized distributions in units invariant toward data properties and quality. Using this approach it is possible to analyze simultaneously different types of disasters and driving forces, regardless of spatial and temporal scales and heterogeneities.

The results obtained are demonstrate the possibility of determination of explicit form of extremes distributions (which could be interpreted in terms of probability) on the base of spatial–temporal analysis of meteorological data. Basing on the results of climate modeling and reanalysis, and using the formalizations proposed it is possible to analyze disaster drivers and calculate multiscale regional risks.

Basing on existing ensemble of observation data it is possible to suppose that extremes distributions could be described by exponential distributions (Schmidt and Makalic 2009). In separate cases (for example, in spring season) this distribution is degenerates (Lawless and Fredette 2005) to Pareto distribution (Arnold 1983).

Such form of long-term approximations nonetheless not allows to conclude that observed processes are ergodic. It amount that capability of parametric methods for disaster analysis and forecasting is essentially limited, and we should focusing on non-parametric and semi-parametric approaches (Buhlmann 1970; Goovaerts et al. 2003; Kostyuchenko et al. 2013a).

The studied shifts of extreme values distribution toward mean values change is not linear and non-normal on regional scale. For example, increasing of mean air temperature to 1 °C leads to increasing of max temperature to 2.5–4 °C correspondingly. This is essential driver for disasters (Parry et al. 2007). Besides, this is important factor of environmental and socio-ecological security (Schmidt and Makalic 2009; Grigorieva and Matzarakis 2011; Kostyuchenko et al. 2013a).

Also important to note that in view of current regional temperature change about 0.91 ± 0.27 °C, we entering to zone of increasing of risks: we still in period of high risk of spring season, entering into high risks of autumn and winter seasons, and closely to zone of max risk of summer season. It should be considered in policy making.

References

Arnold BC (1983) Pareto distributions. International Co-operative Publishing House, Fairland, MD, p 216. ISBN 0-89974-012
Bartell SM, Gardner RH, O'Neill RV (1992) Ecological risk estimation. Lewis, Boca Raton, FL
Buhlmann H (1970) Mathematical methods in risk theory. Springer, Berlin, p 214

Chen SX, Huang T (2010) Nonparametric estimation of Copula functions for dependence modeling. Technical report, Department of Statistics, Iowa State University, Ames, USA. p 20

Christianini N, Shawe-Taylor J (2000) An introduction to support vector machines and other Kernel-based learning methods. Cambridge University Press, Cambridge, p 212

Cowpertwait PSP (1995) A generalized spatial–temporal model of rainfall based on a clustered point process. Proc R Soc Lond A 450:163–175

Embrechts P, Lindskog F, McNeil A (2003) Modeling dependence with copulas and applications to risk management. In: Rachev S (ed) Handbook of heavy tailed distributions in finance. Elsevier, Amsterdam, pp 329–384

Ermoliev Y, Hordijk L (2006) Global changes: facets of robust decisions. In: Marti K, Ermoliev Y, Makowski M, Pflug G (eds) Coping with uncertainty, modeling and policy issues. Springer, Berlin, pp 4–28

Ermoliev Y, Winterfeldt D (2012) Systemic risk and security management. In: Ermoliev Y et al (eds) Managing safety of heterogeneous systems, vol 658, Lecture notes in economics and mathematical systems. Springer, Berlin, pp 19–49. doi:10.1007/978-3-642-22884-1

Fowler HJ, Kilsby CG, O'Connell PE (2003) Modeling the impacts of climatic change and variability on the reliability, resilience and vulnerability of a water resource system. Water Resour Res 39:1222

Genest C, Ghoudi K, Rivest L-P (1998) Discussion of "Understanding relationships using copulas" by Edward Frees and Emiliano Valdez. North Am Actuarial J 2(3):143–149

Goovaerts MJ, Kaas RJ, Tang Q (2003) A unified approach to generate risk measures. ASTIN Bull 33(2):173–191

Grigorieva E, Matzarakis A (2011) Physiologically equivalent temperature as a factor for tourism in extreme climate regions in the Russian Far East: preliminary results. Eur J Tourism Hospitality Recreation 2:127–142

Grois man P, Lyalko V, eds (2012) Earth systems change over eastern Europe. Akademperiodyka, Kiev. 488 p, 17 p. il. ISBN 978-966-360-195-3

GSOD (Global Summary of the Day). National Climatic Data Center of U.S. Department of Commerce, Courtesy of NOAA Satellite and Information Service of National Environmental Satellite, Data, and Information Service (NESDIS). www.ncdc.noaa.gov

Guha-Sapir D, Vos F, Below R, Ponserre S (2011) Annual disaster statistical review 2010: the numbers and trends. CRED, Brussels, p 50

Juri A, Wuthrich MV (2002) Copula convergence theorems for tail events. Insur Math Econ 30:405–420

Kalnay E, Kanamitsu M, Kistler R, Collins W, Deaven D, Gandin L, Iredell M, Saha S, White G, Woollen J, Zhu Y, Leetmaa A, Reynolds R, Chelliah M, Ebisuzaki W, Higgins W, Janowiak J, Mo KC, Ropelewski C, Wang J, Roy J, Dennis J (1996) The NCEP/NCAR 40-year reanalysis project. Bull Am Meteorol Soc 77:437–470

Kostyuchenko YV, Bilous Y (2009) Long-term forecasting of natural disasters under projected climate changes in Ukraine. In: Groisman PY, Ivanov SV (eds) Regional aspects of climate–terrestrial–hydrologic interactions in non-boreal Eastern Europe. Springer in cooperation with NATO Public Diplomacy Division, Dordrecht, pp 95–102

Kostyuchenko YuV, Bilous Yu, Kopachevsky I, Solovyov D (2013a) Coherent risk measures assessment based on the coupled analysis of multivariate distributions of multisource observation data. In: Proceedings of 11th international probabilistic workshop, Brno, pp 183–192

Kostyuchenko YuV, Yuschenko M, Movchan D (2013b) Regional risk analysis based on multi-source data statistics of natural disasters. In: Zagorodny AG, Yermoliev YuM (eds) Integrated modeling of food, energy and water security management for sustainable social, economic and environmental developments. NAU, Kyiv. pp 229–238. ISBN 978-966-02-6824-1

Lawless JF, Fredette M (2005) Frequentist predictions intervals and predictive distributions. Biometrika 92(3):529–542

Lee J-M, Yoo CK, Choi SW, Vanrolleghem PA, Lee I-B (2004) Nonlinear process monitoring using kernel principal component analysis. Chem Eng Sci 59:223–234

Mika S, Scheolkopf B, Smola AJ, MJuller K-R, Scholz M, Ratsch G (1999) Kernel PCA and de-noising in feature spaces. Adv Neural Inf Process Syst 11:536–542

Mudelsee M, Börngen M, Tetzlaff G (2001) On the estimation of trends in the frequency of extreme weather and climate events. In: Raabe A, Arnold K (eds) Wissenschaftliche Mitteilungen, vol 22. Institut für Meteorologie der Universität Leipzig, Institut für Troposphärenforschung e. V. Leipzig, Leipzig, pp 78–88

National report "On Technogenic and Natural Security in Ukraine in 2003" (2004) Kiev, 435 p

National report "On Technogenic and Natural Security in Ukraine in 2005" (2006) Kiev, 375 p

National report "On Technogenic and Natural Security in Ukraine in 2009" (2010) Kiev, 252 p

Parry ML, Canziani OF, Palutikof JP, van der Linden PJ, Hanson CE (2007) In: Contribution of working group II to the fourth assessment report of the intergovernmental panel on climate change, IPCC. Cambridge University Press, Cambridge. p 987

Pflug G, Roemisch W (2007) Modeling, measuring, and managing risk. World Scientific, Singapore, p 303

Raiffa H, Schlaifer R (1968) Applied statistical decision theory. MIT Press, Massachusetts Institute of Technology, Cambridge, MA, p 356

Romdhani S, Gong S, Psarrou A (1999) A multi-view nonlinear active shape model using kernel PCA. In: Proceedings of BMVC, Nottingham, UK. pp 483–492

Scheolkopf B, Smola AJ, Muller K (1998) Nonlinear component analysis as a kernel eigenvalue problem. Neural Comput 10(5):1299–1399

Schmidt DF, Makalic E (2009) Universal models for the exponential distribution. IEEE Trans Inf Theory 55(7):3087–3090. doi:10.1109/TIT.2009.2018331

State Budget of USSR in 1981–1985. The statistical book. (1987) URSS Ministry of Finances. Finances and Statistics, Moscow, 215 p

State Budget of USSR in 1989. Brief Statistical Book (1989) URSS Ministry of Finances, Central Department of State Budget. Finances and Statistics, Moscow, 161 p

State Statistics Service of Ukraine. Demography of Ukraine (2007) http://www.ukrstat.gov.ua/operativ/operativ2007/ds/nas_rik/nas_u/nas_rik_u.html

USSR National Economy in 1969. The statistical book (1970) Central Administration of Statistics at Cabinet of Ministries of USSR. Statistics, Moscow, 840 p

USSR National Economy in 1980. The statistical book (1981) Central Administration of Statistics at Cabinet of Ministries of USSR. Finances and Statistics, Moscow, 577 p

Venter GG (2002) Tails of Copulas. Proc Casualty Actuarial Soc 89:68–113

Villez K, Ruiz M, Sin G, Colomer J, Rosen C, Vanrolleghem PA (2008) Combining multiway principal component analysis (MPCA) and clustering for efficient data mining of historical data sets of SBR processes. Water Sci Technol 57(10):1659–1666

Warga J (1972) Optimal control of differential and functional equations. Academic, New York, p 531

World Bank (2013) Global Economics Prospects, Volume 6, January 2013: Assuring Growth Over the medium Term, Washington, DC. http://databank.worldbank.org/ddp/home.do?Step=12&id=4&CNO=999

Time-Dependent Reliability Analysis
of Corrosion Affected Structures

Mojtaba Mahmoodian and Amir Alani

Abstract Incorporating the effect of corrosion into the reliability analysis of a structure is of paramount importance. Since deterioration due to corrosion is uncertain over time, it should ideally be represented as a time-dependent probabilistic (i.e. stochastic) process. For a more accurate reliability analysis and failure assessment, the multi-failure mode analysis of structures is explained in this chapter.

Sensitivity analysis is also discussed as a key part of reliability analysis from which the effect of different variables on service life of the structure can be investigated.

Worked examples of some structures such as cast iron water mains, concrete sewers and post-tension concrete bridges will also be presented and the results are discussed.

1 Introduction

Structures are required to withstand particular environmental hazards in addition to structural loads. Material corrosion in concrete and steel structures is the most common form of deterioration and should be considered in both strength and serviceability analyses of structures. To that effect, incorporating the effect of corrosion into the reliability analysis of a structure is of paramount importance. There are several parameters which may affect corrosion rate, and hence reliability. In considering the uncertainties and data scarcity associated with these parameters, various studies on the probabilistic assessment of corrosion affected structures have been undertaken (Kleiner et al. 2005; Davis et al. 2008; Guo et al. 2011; Salman and Salem 2012; Mahmoodian et al. 2012; Mahmoodian and Alani 2013a).

M. Mahmoodian (✉) • A. Alani
School of Engineering, University of Greenwich, Central Avenue,
Chatham Maritime, Kent ME44TB, UK
e-mail: m.mahmoodian@gre.ac.uk

S. Kadry and A. El Hami (eds.), *Numerical Methods for Reliability and Safety Assessment: Multiscale and Multiphysics Systems*, DOI 10.1007/978-3-319-07167-1_17, © Springer International Publishing Switzerland 2015

Since deterioration due to corrosion is uncertain over time, it should ideally be represented as a stochastic process. A stochastic process can be defined as a random function of time in which for any given point in time the value of the stochastic process is a random variable depending on other basic random variables. Therefore a perfect method for the reliability analysis and service life prediction of corrosion affected structures is a time-dependent probabilistic (i.e. stochastic) method which considers the randomness of variables involving uncertainties in a period of time.

In most of the literature, failure and reliability assessment has been carried out by considering one failure mode (Davis et al. 2005; De Silva et al. 2006; Moglia et al. 2008; Yamini 2009; Zhou 2011). However in reality, even in simple cases composed of just one element, various failure modes such as flexural failure, shear failure, buckling and deflection may exist. To familiarise ourselves with a more accurate reliability analysis and failure assessment, the multi-failure mode analysis of structures is also discussed in this chapter.

For a comprehensive reliability analysis, evaluation of the contributions of various uncertain parameters associated with reliability can be carried out by using sensitivity analysis techniques. Sensitivity analysis is conducted as a key part of reliability analysis from which the effect of different variables on service life of the structure can be investigated. Sensitivity analysis is the study of how variation in the output of a model (numerical or otherwise) can be apportioned, qualitatively or quantitatively, to different sources of variation (Saltelli et al. 2004).

Worked examples of some structures such as cast iron water mains, concrete sewers and post-tension concrete bridges will be presented and the results are discussed.

2   Corrosion in Structures and Infrastructure

2.1   Corrosion in Iron-Based Elements

Aging and deterioration of iron-based structures and infrastructure are major problems facing asset managers. Steel pipes in the oil and gas industry and cast iron pipes in water networks are good examples of corrosion affected iron-based infrastructure which are suffering from significant corrosion worldwide.

Over 50 % of the USA's oil and gas pipeline system is over 40 years old. Some 20 % of Russia's oil and gas system is near the end of its design life and in 15 years' time, 50 % of their pipeline will be at the end of its design life. In the USA, corrosion has caused 23 % of failures of oil pipelines and 39 % of gas pipelines. The consequence of the failure of these pipes can be socially, economically and environmentally devastating, causing, e.g. enormous disruption of daily life, massive costs for repair, widespread pollution and even human injuries (Anon 2002).

Iron-based pipelines have also been used in water distribution systems for many years. Data from different countries in the world shows that cast iron pipes were widely used in water distribution networks before the 1960s (Rajani and Kleiner 2004). The American Water Works Association (AWWA) estimated, based on a survey of 337 water utilities, that in the USA about 40 % of water mains are cast iron (Lillie et al. 2004). More than 60 % of water mains in the UK are estimated to comprise cast iron pipes. This corresponds to a length of 200,000 km of pipelines (Water UK 2007). The average age of cast iron pipes in existing networks is around 50 years (Misiunas 2005). Due to their long-term use, aging and deterioration of the pipes are inevitable and indeed many failures have been reported worldwide (Misiunas 2005; Rajani and Tesfamariam 2007; EPA/600 2012). It has been established (e.g. Yamini and Lence 2009; EPA/600 2012) that corrosion is the most common form of deterioration of pipes, which is a matter of concern for both the safety and serviceability of the pipes.

The predominant deterioration mechanism of iron-based pipes is electrochemical corrosion with damage occurring in the form of corrosion pits. The damage to iron is often identified by the presence of graphitisation, a result of iron being leached away by corrosion. Either form of metal loss represents a corrosion pit that grows with time and reduces the thickness and mechanical resistance of the pipe wall. This process eventually can lead to the collapse of the element. Typically, internal and external corrosion pits are observed to occur in many irregular shapes and sizes with characteristic depths, diameters (or widths) and lengths (Rajani and Kleiner 2013). They can develop randomly along any segment of pipe and tend to grow with time at a rate that depends on environmental conditions in the immediate vicinity of the pipeline (Rajani and Makar 2000).

The corrosion rate of in-service cast iron pipes is believed to be higher in the beginning and then to slow down over time, as corrosion products have an inhibiting effect (Shreir et al. 1994). Furthermore, due to the variation of service environments it is rare that the corrosion occurs uniformly along the pipe but more likely locally in the form of a corrosion pit.

A number of models for corrosion of steel and cast iron pipes have been proposed to estimate the depth of corrosion pit (e.g. Sheikh et al. 1990; Ahammed and Melchers 1997; Kucera and Mattsson 1987; Rajani et al. 2000; Sadiq et al. 2004). For example, Sheikh et al. (1990) suggested a linear model for corrosion growth in predicting the strength of cast iron pipes.

There are debates in the research community as to whether the corrosion rate can be assumed linear or otherwise. A widely used model of corrosion as measured by the depth of corrosion pit is of a power law that was first postulated for atmospheric corrosion by Kucera and Mattsson (1987) and expressed in the following form:

$$a = kt^n \tag{1}$$

where t is the exposure time and k and n are empirical coefficients which in practice are obtained by fitting the model to experimental data.

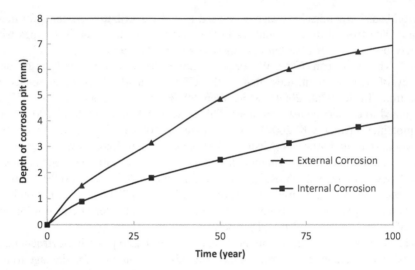

Fig. 1 Progress of corrosion depth in cast iron pipes (reproduced from Marshall 2001)

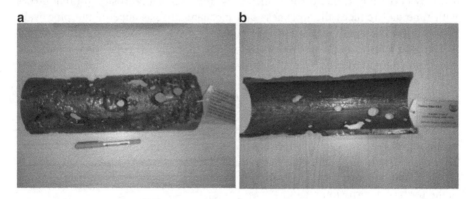

Fig. 2 A section of one of London's Victorian water mains: (**a**) external corrosion, (**b**) internal corrosion

For underground corrosion, the coefficients typically are functions of local conditions including soil type, availability of oxygen and moisture and properties of pipeline material. In many cases it may be possible to use past experience to derive estimates for the two constants in Eq. (1), but with somewhat more effort than would be necessary to estimate a constant corrosion rate as used conventionally (Ahammed and Melchers 1997).

An example of field data that shows the rate of internal and external corrosion for cast iron pipes has been illustrated in Fig. 1 (Marshall 2001). It can be concluded from this data that external corrosion progresses more speedily than internal corrosion especially during the early stages. In Fig. 2, a sample of a cast iron pipe taken from London's water mains in Victorian times (i.e. 1800–1900) also shows the severity of external corrosion compared with internal corrosion.

As the regression of available data fits a power law very well, the corrosion can be modelled, for both external and internal corrosion, in a power law format as follows (Li and Mahmoodian 2013):

$$a = 2.54t^{0.32} \quad \text{for external corrosion} \qquad (2a)$$

$$a = 0.92t^{0.4} \quad \text{for internal corrosion} \qquad (2b)$$

2.2 Reinforcement Corrosion in Concrete Structures

Reinforcement corrosion is the predominant factor in the premature degradation of reinforced concrete (RC) structures, leading to ultimate structural failure (Chaker 1992). The ingress of chloride ions into concrete may cause corrosion in RC bridges. The risk of this mode of corrosion is quite high when RC bridges are located in coastal regions and exposed to aggressive environmental conditions. Because of the penetration of chloride ions into structural members, the chloride content of concrete increases gradually, and as the concentration of chloride ions in the pore solution within the vicinity of reinforcing bars reaches a threshold value, chloride-induced corrosion is initiated (Mahmoodian and Li 2012).

Deterioration of concrete bridges is one of the most important concerns of infrastructure managers. There are thousands of in-service concrete bridges in the world (over 350,000 in the USA) which encounter corrosion as a predominant form of deterioration. Corrosion may significantly influence the long-term performance of pre-stressed concrete bridges, particularly in aggressive environments.

The rate of bridge deterioration appears to be increasing. In the northern USA the use of de-icing salts has increased from less than one million tons per year in the early 1950s to approximately 15 million tons per year in the 1990s (Baboian 1995). An aggressive chloride environment exists also for bridges sited in a marine environment within 1–2 km from the sea (e.g. Fitzpatrick 1996).

Corrosion of tendons in pre-stressed elements is a complex phenomenon that consists of several different (but interrelated) mechanisms, such as uniform corrosion and pitting corrosion. The effects of corrosion on structural behaviour in reinforced and pre-stressed concrete structures are different, with the latter having less concrete cracking but more serious structural collapses. Pitting corrosion is the main form of pre-stressing steel corrosion in aggressive environments. The corrosion rate of pre-stressing steel tendons in concrete rises with the increase of the level of stresses applied. Therefore the corrosion of pre-stressing steel in concrete structures poses a higher risk to the structure than that of reinforcing steel in terms of structural collapse (Li et al. 2011).

Reliability of the corrosion models is essential to predict the service life of corrosion affected RC and/or pre-stressed bridges and to instigating maintenance and repairs for the structures.

Numerous studies with regard to corrosion models have been made and will not be repeated here. Enright and Frangopol (1998) studied the resistance degradation of RC bridge beams under uniform corrosion, and the corrosion initiation time was predicted with a probabilistic model. Val and Robert (1997) developed a time-dependent corrosion model for the reliability analysis of RC slab bridges, including localised corrosion (pitting corrosion). This model was improved in research by Stewart (2004). Accelerated pitting corrosion tests have been used to obtain spatial and temporal maximum pit-depth data for pre-stressing strands (Darmawan and Stewart 2007). Further studies carried out by Stewart (2009) examined the mechanical behaviour of pitting corrosion of flexural and shear reinforcement and its effect on structural reliability.

A direct consequence of reinforcement corrosion is the cross-sectional area loss of reinforcement. The diameter of a corroding reinforcing bar $D(t)$ and its net cross-sectional area, $A_r(t)$ at time t can be presented as (Val and Robert 1997):

$$D(t) = D_0 - 0.0232\,(t - t_i)\,i_{corr} \tag{3}$$

$$A_r(t) = \frac{\pi}{4}\left[D(t)^2\right] \tag{4}$$

where D_0 is the initial diameter of the reinforcing bar (cm), i_{corr} is corrosion current density ($\mu A/cm^2$) and t_i is time of corrosion initiation (year).

The process of uniform corrosion, which is generally a macrocell corrosion process, is slower than pitting corrosion in a chloride environment. The maximum penetration of pitting is about four to eight times that associated with uniform corrosion (González et al. 1995). In the case of pitting corrosion, the reinforcement is more susceptible to failure due to stress concentration. For high pre-stressed stranded wires, the pitting corrosion process may be accelerated (Vu et al. 2009) and brittle rupture of the stranded wires may occur earlier than expected (Guo et al. 2011). According to Val's work, the radius of the pit at time t can be estimated as

$$p(t) = 0.0116\,(t - t_i)\,i_{corr}R \tag{5}$$

where R is the penetration ratio between the maximum and average penetration. The time of corrosion initiation t_i can be predicted by the following equation (Enright and Frangopol 1998):

$$t_i = \frac{X^2}{4D_c}\left[\mathrm{erf}^{-1}\left(\frac{C_o - C_{cr}}{C_o}\right)\right]^{-2} \tag{6}$$

where X represents concrete cover (cm), D_c is the chloride diffusion coefficient ($cm^2/year$), C_o is the chloride concentration at the concrete surface (% weight of concrete) and C_{cr} denotes threshold chloride concentration (% weight of concrete).

Fig. 3 Corrosion pit configuration (adapted from Val and Robert 1997)

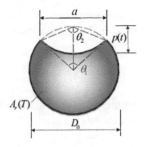

For pitting corrosion, the net cross-sectional area of a corroded rebar, $A_r(t)$ at time t is calculated by the following equations (Stewart 2009):

$$A_r(t) = \begin{cases} \frac{\pi D_0^2}{4} - A_1 - A_2 & p(t) \le \frac{\sqrt{2}}{2} D_0 \\ A_1 - A_2 & \frac{\sqrt{2}}{2} D_0 < p(t) \le D_0 \\ 0 & p(t) > D_0 \end{cases} \qquad (7a)$$

With

$$a = 2p(t) \sqrt{1 - \left[\frac{p(t)}{D_0}\right]^2} \qquad (7b)$$

$$A_1 = \frac{1}{2}\left[\theta_1 \left(\frac{D_0}{2}\right)^2 - a\left|\frac{D_0}{2} - \frac{p(t)^2}{D_0}\right|\right]; \quad A_2 = \frac{1}{2}\left[\theta_2 p(t)^2 - a\frac{p(t)^2}{D_0}\right] \qquad (7c)$$

$$\theta_1 = 2\arcsin\left(\frac{a}{D_0}\right); \quad \theta_2 = 2\arcsin\left(\frac{a}{2p(t)}\right) \qquad (7d)$$

Figure 3 illustrates the relationship between $A_r(t)$ and $p(t)$.

Experimental results have indicated that the yield stress on the net cross-sectional area is reduced by corrosion as follows (Guo et al. 2011):

$$f_y(t) = [1 - \alpha . P_{corr}(t)] . f_{y_0} \qquad (8)$$

$$P_{corr}(t) = \frac{A_o - A_r(t)}{A_o} \times 100 \qquad (9)$$

where $f_y(t)$ is deteriorated yield strength at time t, f_{y_0} corresponds to its original value and α is an empirical coefficient which has a value of 0.0054 for reinforcing bars and 0.0075 for stranded wires (based on the regression analyses of Du et al. (2005) and Vu et al. (2009)). $P_{corr}(t)$ is the percentage of corrosion loss at time t, which can be obtained from Eqs. (4) or (7a), (7b), (7c), (7d) in terms of area loss.

As mentioned earlier, maximum wire stress $(S_w(t))$ increases as the cross-sectional area decreases due to corrosion.

$$S_w(t) = \frac{T_w}{A_r(t)} \tag{10}$$

where T_w is the maximum possible tension in a wire due to the most critical loading combination.

2.3   Concrete Corrosion

The corrosion mechanism of concrete sewers: the deterioration of sewer pipes occurs at different rates depending on specific local conditions and is not determined by age alone. There have been numerous cases of severe damage to concrete pipes, where it has been necessary to replace the pipes before the desired service life has been reached. There are many cases in which sewer pipes designed to last 50–100 years have failed due to H_2S corrosion in 10–20 years. In extreme cases, concrete pipes have collapsed in as few as 3 years (Pomeroy 1976). The most corrosive agent that leads to the rapid deterioration of concrete pipelines in sewers is H_2S. Approximately 40 % of the damage in concrete sewers can be attributed to biogenous sulphuric acid attack. Sulphide corrosion, which is often called microbiologically induced corrosion, has two distinct phases as follows:

- The conversion of sulphate in wastewater to sulphide, some of which is released as gaseous hydrogen sulphide.
- The conversion of hydrogen sulphide to sulphuric acid, which subsequently attacks susceptible pipeline materials.

The surface pH of new concrete pipe is generally between 11 and 13. Cement contains calcium hydroxide, which neutralises the acids and inhibits the formation of oxidising bacteria when the concrete is new. However, as the pipe ages, this neutralising capacity is consumed, the surface pH drops, and the sulphuric acid-producing bacteria become dominant. In active corrosion areas, the surface pH can drop to 1 or even lower and can cause a very strong acid attack. The corrosion rate of the sewer pipe wall is determined by the rate of sulphuric acid generation and the properties of the cementitious materials. As sulphides are formed and sulphuric acid is produced, hydration products in the hardened concrete paste (calcium silicon, calcium carbonate and calcium hydroxide) are converted to calcium sulphate, more commonly known by its mineral name, gypsum (ASCE No. 69 1989). The chemical reactions involved in sulphide build-up can be explained as follows.

Sulphate, generally abundant in wastewater, is usually the common sulphur source, although other forms of sulphur, such as organic sulphur from animal waste, can also be reduced to sulphide. The reduction of sulphate in the presence of waste organic matter in a wastewater collection system can be described as follows:

$$SO_4{}^{-2} + \text{Organic matter} + H_2O \underset{\text{Bacteria}}{\rightarrow} 2HCO_3{}^- + H_2S \tag{11}$$

The H_2S gas in the atmosphere can be oxidised on the moist pipe surfaces above the water line by bacteria (Thiobacillus), producing sulphuric acid according to the following reaction:

$$H_2S + O_2 \underset{\text{Bacteria}}{\rightarrow} H_2SO_4 \tag{12}$$

As sulphides are formed and sulphuric acid is produced, hydration products in the hardened concrete paste (calcium silicon, calcium carbonate and calcium hydroxide) are converted to calcium sulphate. The chemical reactions involved in the corrosion of concrete are:

$$H_2SO_4 + CaSi \rightarrow CaSO_4 + 2H+ \tag{13}$$

$$H_2SO_4 + CaCO_3 \rightarrow CaSO_4 + H_2CO_3 \tag{14}$$

$$H_2SO_4 + Ca(OH)_2 \rightarrow CaSO_4 + 2H_2O \tag{15}$$

Gypsum does not provide much structural support, especially when wet. It is usually present as a pasty white mass on concrete surfaces above the water line. As the gypsum material is eroded, the concrete loses its binder and begins to spall, exposing new surfaces. This process will continue until the pipeline fails or corrective actions are taken. Sufficient moisture must be present for the sulphuric acid-producing bacteria to survive, however; if it is too dry, the bacteria will become desiccated, and corrosion will be less likely to occur. Figure 4 shows the process of sulphide build-up in a sewer system.

Concrete corrosion rate: The rate of corrosion of a concrete sewer can be calculated from the rate of production of sulphuric acid on the pipe wall, which is in turn dependent upon the rate that H_2S is released from the surface of the sewage stream. The average flux of H_2S to the exposed pipe wall is equal to the flux from the stream into the air multiplied by the ratio of the surface area of the stream to the area of the exposed pipe wall, which is the same as the ratio of the width of the stream surface (b) to the perimeter of the exposed wall (P'). The average flux of H_2S to the wall is therefore calculated as follows (Pomeroy 1976):

$$\Phi = 0.7(su)^{3/8} j \, [DS] \left(b/P' \right) \tag{16}$$

where s is pipe slope, u is velocity of stream (m/s), j is pH-dependent factor for the proportion of H_2S and [DS] is dissolved sulphide concentration (mg/L). A concrete pipe is made of cement-bonded material, or acid-susceptible substance, so the

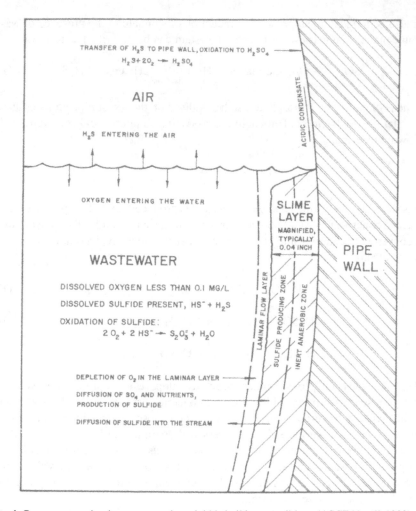

Fig. 4 Process occurring in a sewer under sulphide build-up conditions (ASCE No. 69 1989)

acid will penetrate the wall at a rate inversely proportional to the acid-consuming capability (A) of the wall material. The acid may partly or entirely react. The proportion of acid that reacts is variable (k), ranging from 100 % when the acid formation is slow, to perhaps 30–40 % when it is formed rapidly. Thus, the average rate of corrosion (mm/year) can be calculated as follows:

$$c = 11.5k\,\Phi\,(1/A) \tag{17}$$

where k is the factor representing the proportion of acid reacting, to be given a value selected by the judgement of the engineer, and A is the acid-consuming capability, alkalinity, of the pipe material, expressed as the proportion of equivalent calcium carbonate. A value for granitic aggregate concrete ranges from 0.17 to 0.24 and for calcareous aggregate concrete, A ranges from 0.9 to 1.1 (ASCE No. 60 2007). Substituting Eq. (16) into Eq. (17):

$$c = 8.05k \times (su)^{3/8} j. [DS] \times \frac{b}{P'A} \tag{18}$$

Therefore the reduction in wall thickness in elapsed time t, is:

$$d(t) = c.t = 8.05k.(su)^{3/8} j. [DS] \times \frac{b}{P'A}.t \tag{19}$$

3 Time-Dependent Reliability Analysis

3.1 Background

As the corrosion of structures and infrastructure is a time-variant phenomenon with uncertainties in affecting parameters, the most viable approach to predict a structure's reliability or its service life under future performance conditions is through probability-based techniques involving time-dependent reliability analyses.

By using these techniques, a quantitative measure of structural reliability is provided to integrate information on design requirements, material and structural degradation, damage accumulation, environmental factors, and non-destructive evaluation technology. The technique can also investigate the role of in-service inspection and maintenance strategies in enhancing reliability and extending service life. Several non-destructive test methods that detect the presence of a defect in a structure tend to be qualitative in nature in that they indicate the presence of a defect but may not provide quantitative data about the defect's size, precise location and other characteristics that would be needed to determine its impact on structural performance. None of these methods can detect a given defect with certainty. The imperfect nature of these methods can be described in statistical terms. This randomness affects the calculated reliability of a component.

Structural loads, engineering material properties, and strength-degradation mechanisms are random. The resistance, $R(t)$, of a structure and the applied loads, $S(t)$, both are stochastic functions of time. At any time, t, the safety limit state, $G(R, S, t)$, is (Melchers 1999):

$$G(R, S, t) = R(t) - S(t) \tag{20}$$

Making the customary assumption that R and S are statistically independent random variables, the probability of failure resulting from Eq. (20), $P_f(t)$, is (Melchers 1999):

$$P_f(t) = P[G(t) \leq 0] = \int_0^\infty F_R(x) f_S(x) \, dx \tag{21}$$

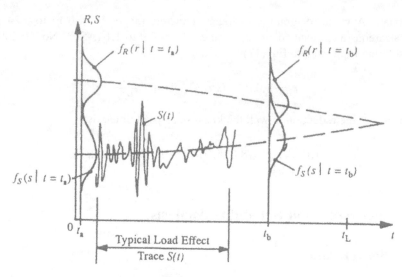

Fig. 5 Schematic time dependent reliability problem (Melchers 1999)

in which $F_R(x)$ and $f_S(x)$ are the probability distribution function of R and density function of S, respectively. Equation (21) provides a quantitative measure of structural reliability and performance, provided that P_f can be estimated and validated.

The probability that failure occurs for any one load application is the probability of limit state violation. Roughly, it may be represented by the amount of overlap of the probability density functions f_R and f_S in Fig. 5. Since this overlap may vary with time, P_f also may be a function of time.

3.2 Methods for Time-Dependent Reliability Analysis

As mentioned earlier, for the reliability and safety analysis of corrosion affected structures, probabilistic time-dependent methods need to be used. The common methods include the first passage probability method, the gamma process concept method and the Monte Carlo simulation method.

3.2.1 First Passage Probability Method

The service life of a structure is a time period at the end of which the structure stops performing the functions it is designed and built for. As already stated, to determine the service life, a limit state function $(G(t) = R(t) - S(t))$ is introduced, where $S(t)$ is the action (load) or its effect at time t and $R(t)$ is the acceptable limit (resistance) for the action or its effect. With the limit state function of Eq. (4.1), the probability of structural failure, p_f, can be determined by:

$$P_f(t) = P[G(t) \leq 0] = P[S(t) \geq R(t)] \tag{22}$$

where $P_f(t)$ is greater than the maximum acceptable risk in terms of the probability of failure, P_a, the pipe becomes unsafe or unserviceable and requires replacement or repairs. This can be determined from the following:

$$P_f(T_L) \geq P_a \tag{23}$$

where T_L is the service life for the pipe for the given assessment criterion and acceptable risk. In principle, the acceptable risk, P_a, can be determined from a risk-cost optimisation of the pipeline system during its whole service life (Li and Melchers 2005). This is beyond the scope of this chapter and will not be discussed herein but can be referred to in Mann and Frey (2011) and Dawotola et al. (2012).

Equation (22) represents a typical up-crossing problem in mathematics. Structural failure depends on the time that is expected to elapse before the first occurrence of the action process $S(t)$ up-crossing an acceptable limit (the threshold) $L(t)$ sometime during the service life of the structure [0, T_L]. Equivalently, the probability of the first occurrence of such an excursion is the probability of failure $P_f(t)$ during that time period. This is known as "first passage probability" and can be determined by Melchers (1999):

$$P_f(t) = 1 - [1 - P_f(0)] e^{-\int_0^t v \, d\tau} \tag{24}$$

where $P_f(0)$ is the probability of structural failure at time $t = 0$ and v is the mean rate for the action process $S(t)$ to up-cross the threshold $R(t)$. In many practical problems, the mean up-crossing rate is very small ($e^{-\int_0^t v \, d\tau} \cong 1 - \int_0^t v \, d\tau$), so the above equation can be approximated as follows:

$$P_f(t) = p_f(0) + \int_0^t v \, d\tau \tag{25}$$

The probability of failure due to corrosion at $t = 0$ is zero, that is $p_f(0) = 0$; therefore:

$$P_f(t) = \int_0^t v \, d\tau \tag{26}$$

The up-crossing rate in the above equation can be determined from the Rice formula (Melchers 1999):

$$\nu = \nu_R^+ = \int_R^\infty \left(\dot{S} - \dot{R} \right) f_{S\dot{S}} \left(R, \dot{S} \right) d_{\dot{S}} \tag{27}$$

where ν_R^+ is the up-crossing rate of the action process $S(t)$ relative to the threshold R, $\dot{R}$ is the slope of R with respect to time, $\dot{S}(t)$ is the time derivative process of $S(t)$ and $f_{S\dot{S}}()$ is the joint probability density function for S and $\dot{S}$.

The solution for $f_{S\dot{S}}\left(R, \dot{S} \right)$ for the special case when $S(t)$ is a stationary normal process is given by:

$$f_{S\dot{S}}\left(R, \dot{S} \right) = \frac{1}{2\pi\sigma_S\sigma_{\dot{S}}} \exp \left\{ -\frac{1}{2} \left[\left(\frac{R - \mu_S}{\sigma_S} \right)^2 + \frac{\dot{S}^2}{\sigma_{\dot{S}}^2} \right] \right\} \tag{28}$$

in which $S(t)$ is normal distributed $N(\mu_S, \sigma_S^2)$ and $\dot{S}(t)$ is $N(0, \sigma_{\dot{S}}^2)$. The mean of $\dot{S}(t)$ is zero for a stationary process. Noting that:

$$\int_0^\infty \dot{S} \, \exp \left(-\frac{\dot{S}^2}{2\sigma_{\dot{S}}^2} \right) d_{\dot{S}} = \sigma_{\dot{S}}^2 \tag{29}$$

and substituting Eq. (28) into Eq. (27) and integrating produces (Melchers 1999):

$$\nu_R^+ = \frac{1}{2\pi} \frac{\sigma_{\dot{S}}}{\sigma_S} \exp \left[-\frac{(R - \mu_S)^2}{2\sigma_S^2} \right] \tag{30}$$

For non-normal processes, the joint probability density function $f_{S\dot{S}}()$ usually will be much less amenable to definition and integration. Such processes arise, for instance, in river flows, mean hourly wind speeds and when normal processes are transformed non-linearly. It is sometimes suggested that for such processes the up-crossing rate may be approximated by Eq. (30). It should be noted that this approximation can be seriously in error (Melchers 1999).

3.2.2 Gamma Process Concept

To deal with data scarcity and uncertainties, the use of a stochastic process concept for the time-dependent reliability analysis of corrosion affected structures and infrastructures can be considered. In order to model the monotonic progression of a deterioration process, the stochastic gamma process concept can be used for modelling the corrosion progress. The gamma process is a stochastic process with independent, non-negative increments having a gamma distribution with an identical scale parameter and a time-dependent shape parameter.

A stochastic process model, such as the gamma process, incorporates the temporal uncertainty associated with the evolution of deterioration (e.g. Bogdanoff and Kozin 1985; Nicolai et al. 2004; van Noortwijk and Frangopol 2004).

The gamma process is suitable to model gradual damage monotonically accumulating over time, such as wear, fatigue, corrosion, crack growth, erosion, consumption, creep, swell and a degrading health index. For the mathematical aspects of gamma processes, see Dufresne et al. (1991), Ferguson and Klass (1972), Singpurwalla (1997) and van der Weide (1997).

To the best of the authors' knowledge, Abdel-Hameed (1975) was the first to propose the gamma process as a model for deterioration occurring randomly in time. In his chapter he called this stochastic process the "gamma wear process". An advantage of modelling deterioration processes through gamma processes is that the required mathematical calculations are relatively straightforward.

Problem formulation: The mathematical definition of the gamma process is given in Eq. (31). Given that a random quantity d has a gamma distribution with shape parameter $\alpha > 0$ and scale parameter $\lambda > 0$ if its probability density function is given by:

$$Ga\left(d\,\middle|\,\alpha, \lambda\right) = \frac{\lambda^\alpha}{\Gamma(\alpha)} d^{\alpha-1} e^{-\lambda d} \tag{31}$$

let $\alpha(t)$ be a non-decreasing, right continuous, real-valued function for $t \geq 0$, with $\alpha(0) \equiv 0$. $\Gamma(\alpha)$ denoting the gamma function of α with the mathematical definition of $\Gamma(\alpha) = (\alpha - 1)!$. The gamma process is a continuous-time stochastic process $\{d(t), t \geq 0\}$ with the following properties:

1. $d(0) = 0$ with probability one;
2. $d(\tau) - d(t)$ $Ga(\alpha(\tau) - \alpha(t), \lambda)$ for all $\tau > t \geq 0$;
3. $d(t)$ has independent increments.

Let $d(t)$ denote the deterioration at time t, $t \geq 0$, and let the probability density function of $d(t)$, in accordance with the definition of the gamma process, be given by:

$$f_{d(t)}(d) = Ga\left(d\,\middle|\,\alpha(t), \lambda\right) \tag{32}$$

with mean and variance as follows:

$$E\left(d(t)\right) = \frac{\alpha(t)}{\lambda} \tag{33}$$

$$\mathrm{Var}\left(d(t)\right) = \frac{\alpha(t)}{\lambda^2} \tag{34}$$

A corrosion affected structure is said to fail when its corrosion depth, denoted by $d(t)$, is more than a specific threshold (a_0), assuming that the threshold a_0 is

Table 1 Typical values for exponential parameter, b, in different deterioration types

Deterioration type	Exponential parameter, b	Reference
Degradation of concrete due to reinforcement corrosion	1	Ellingwood and Mori (1993)
Sulphate attack	2	Ellingwood and Mori (1993)
Diffusion-controlled aging	0.5	Ellingwood and Mori (1993)
Creep	1/8	Cinlar et al. (1977)
Expected scour-hole depth	0.4	Hoffmans and Pilarczyk (1995) and Noortwijk and Klatter (1999)

deterministic and the time at which failure occurs is denoted by the lifetime T. Due to the gamma distributed deterioration, Eq. (31), the lifetime distribution can then be written as:

$$F(t) = \Pr(T \leq t) = \Pr(d(t) \geq a_0) = \int_0^{a_0} f_{d(t)}(d)d_d = \frac{\Gamma(\alpha(t), a_o\lambda)}{\Gamma(\alpha(t))} \qquad (35)$$

where $\Gamma(v, x) = \int_{t=x}^{\infty} t^{v-1} e^{-t} \, dt$ is the incomplete gamma function for $x \geq 0$ and $v > 0$

To model corrosion in a structural element, in terms of a gamma process, the question that remains to be answered is how its expected deterioration increases over time. The expected corrosion depth at time t may be modelled empirically by a power law formulation (Ahammed and Melchers 1997):

$$\alpha(t) = ct^b \qquad (36)$$

for some physical constants $c > 0$ and $b > 0$.

Because there is often engineering knowledge available about the shape of the expected deterioration in terms of the exponential parameter b in Eq. (36), this parameter may be assumed constant. It should be noted that the gamma process is not restricted to using a power law for modelling the expected deterioration over time. As a matter of fact, any shape function $\alpha(t)$ suffices, as long as it is a non-decreasing, right continuous and real-valued function.

The typical values for b from some examples of expected deterioration according to a power law are presented in Table 1.

The reliability analysis approach, which is developed in this section by using the gamma process concept, is entitled the "Gamma Distributed Degradation, GDD" model.

In the event of expected deterioration in terms of a power law (i.e. Eq. (36)), the parameters c and λ can be estimated by using statistical estimation methods. The estimation procedure is discussed for the two scenarios, including a case where corrosion depth data is available and a case where corrosion depth data is not available.

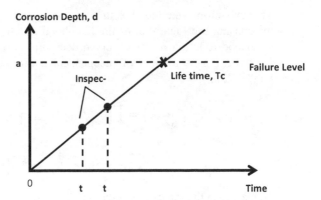

Fig. 6 Time-dependent degradation model in the cases of two inspections

(a) Gamma Distributed Degradation Model with Available Corrosion Depth Data
   This section discusses the use of the GDD model for the reliability analysis of corrosion affected structures in cases where corrosion depth data is available. The data of corrosion depth can be achieved by periodical inspections.
   To model the corrosion as a gamma process with shape function $\alpha(t) = ct^b$ and scale parameter λ, the parameters c and λ should be estimated. For this purpose, statistical methods are suggested. The two most common methods that can be used for parameter estimation are the maximum likelihood and method of moments. Both methods for deriving the estimators of c and λ were initially presented by Cinlar et al. (1977) and were developed by Noortwijk and Pandey (2003).

(i) Maximum Likelihood Estimation
   In statistics, maximum likelihood estimation (MLE) is a method of estimating the parameters of a statistical model. When applied to a dataset and given a statistical model, MLE provides estimates for the model's parameters.
   In general, for a fixed set of data and underlying statistical model, the method of maximum likelihood selects values of the model parameters to produce a distribution that gives the observed data the greatest probability (i.e. parameters that maximise the likelihood function). Given that n observations are denoted by $x_1, x_2, \ldots, x_n$, the principle of maximum likelihood assumes that the sample dataset is representative of the population. This has a probability density function of $f_x(x_1, x_2, \ldots, x_n; \theta)$, and chooses that value for θ (unknown parameter) that most likely caused the observed data to occur. In other words, once observations $x_1, x_2, \ldots, x_n$ are given, $f_x(x_1, x_2, \ldots, x_n; \theta)$ is a function of θ alone, and the value of θ that maximises the above probability density function is the most likely value for θ.
   A typical dataset consists of inspection times $t_i, i = 1, \ldots, n$ where $0 = t_0 < t_1 < t_2 < \cdots < t_n$, and corresponding observations of the cumulative amounts of deterioration $d_i, i = 1, \ldots, n$ are assumed to be given as inputs of the model. Figure 6 schematically shows a time-dependent degradation model in the case of two inspections with a deterministic path.

The maximum likelihood estimators of c and λ can be determined by maximising the logarithm of the likelihood function of the increments. The likelihood function of the observed deterioration increments $\delta_i = d_i - d_{i-1}$, $i = 1, \ldots, n$ is a product of independent gamma densities (Noortwijk and Pandey 2003):

$$l\left(\delta_1, \ldots, \delta_n \middle| c, \lambda\right) = \prod_{i=1}^{n} f_{d(t_i) - d(t_{i-1})}(\delta_i)$$

$$= \prod_{i=1}^{n} \frac{\lambda^{c[t_i^b - t_{i-1}^b]}}{\Gamma\left(c\left[t_i^b - t_{i-1}^b\right]\right)} \delta_i^{c[t_i^b - t_{i-1}^b] - 1} e^{-\lambda \delta_i} \qquad (37)$$

To maximise the logarithm of the likelihood function, its derivatives are set to zero. It follows that the maximum likelihood estimator of λ is:

$$\widehat{\lambda} = \frac{\widehat{c} t_n^b}{d_n} \qquad (38)$$

where $\widehat{c}$ must be computed iteratively from the following equation:

$$\sum_{i=1}^{n} \left[t_i^b - t_{i-1}^b\right] \left\{\psi\left(\widehat{c}\left[t_i^b - t_{i-1}^b\right]\right) - \log \delta_i\right\} = t_n^b \log\left(\frac{\widehat{c} t_n^b}{d_n}\right) \qquad (39)$$

where the function $\psi(x)$ is the derivative of the logarithm of the gamma function:

$$\psi(x) = \frac{\Gamma'(x)}{\Gamma(x)} = \frac{\partial \log \Gamma(x)}{\partial x} \qquad (40)$$

(ii) Method of Moments

In statistics, the method of moments is a method of estimation of population parameters such as mean and variance by equating sample moments with unobservable population moments and then solving those equations for the quantities to be estimated. Assuming transformed times between inspections as $w_i = t_i^b - t_{i-1}^b, i = 1, \ldots, n$, the method of moments estimates that c and λ can be found from Noortwijk and Pandey (2003):

$$\frac{\widehat{c}}{\widehat{\lambda}} = \frac{\sum_{i=1}^{n} \delta_i}{\sum_{i=1}^{n} w_i} = \frac{d_n}{t_n^b} = \overline{\delta} \qquad (41)$$

$$\frac{d_n}{\widehat{\lambda}} \left(1 - \frac{\sum_{i=1}^{n} w_i^2}{\left[\sum_{i=1}^{n} w_i\right]^2}\right) = \sum_{i=1}^{n} \left(\delta_i - \overline{\delta} w_i\right)^2 \qquad (42)$$

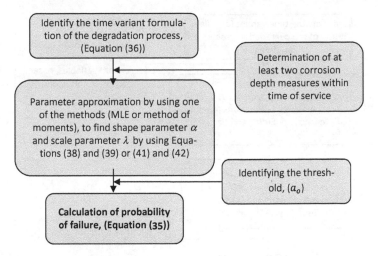

Fig. 7 Gamma distributed degradation (GDD) model in cases where corrosion depth data is available

The first equation from both methods (i.e. Eqs. (38) and (41)) are the same and the second equation in the method of moments is simpler since it does not necessarily require iterations to find the unknown parameter $(\widehat{c})$.

The flowchart in Fig. 7 illustrates the GDD model in cases where corrosion measurements are available.

(b) Developing Gamma Distributed Degradation Model in Cases Where Corrosion Depth Data is not Available

In practice, it is typical during the reliability analysis of corrosion affected structures for data such as corrosion depth to not be available. Therefore, a method should be developed for such cases to use the GDD model. As mentioned in the previous section, in order to calculate the probability of failure over elapsed time (Eq. (35)), the parameters corresponding to shape and scale parameters (α and λ) should be estimated. The steps for this purpose are:

(a) Determining the approximate moments (mean and variance).
(b) Estimating values for α and λ by using Eqs. (33) and (34).

Assuming $X_1, X_2, \ldots, X_n$ as basic random variables, moment approximation (i.e. step (a)) can be carried out by expanding the function $Y = Y(X_1, X_2, \ldots, X_n)$ in a Taylor series about the point defined by the vector of the means $(\mu_{X_1}, \mu_{X_2}, \ldots, \mu_{X_n})$. By truncating the series, the mean and variance are (Papoulis and Pillai 2002):

$$E(Y) \approx Y(\mu_{X_1}, \mu_{X_2}, \ldots, \mu_{X_n}) + \frac{1}{2}\sum_{i=1}^{n}\sum_{j=1}^{n}\frac{\partial^2 Y}{\partial X_i \partial X_j} \mathrm{cov}(X_i, X_j) \quad (43)$$

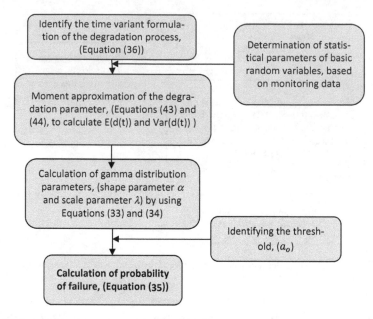

Fig. 8 GDD model in cases where corrosion depth data is not available

$$\mathrm{var}(Y) \approx \sum_{j}^{n}\sum_{j}^{n} c_i c_j \, \mathrm{cov}\left(X_i, X_j\right) \qquad (44)$$

where var and cov are variance and covariance, respectively.

The flowchart in Fig. 8 illustrates the GDD model for use in cases where corrosion measurements are not available.

3.2.3 Monte Carlo Simulation Method

Monte Carlo simulation has been successfully used for the reliability analysis of different structures and infrastructure (e.g. Camarinopoulos et al. 1999; Sadiq et al. 2004; Yamini 2009; Mahmoodian and Alani 2013b).

Monte Carlo simulation techniques involve sampling at random to artificially simulate a large number of experiments and to observe the results. To use this method in structural reliability analysis, a value for each random variable is selected randomly $(\widehat{x}_t)$ and the limit state function $(G\,(\widehat{x}))$ is checked. If the limit state function is violated (i.e. $G\,(\widehat{x}) \leq 0$), the structure or the system has failed. The experiment is repeated many times, each time with randomly chosen variables. If N trials are conducted, the probability of failure then can be estimated by dividing the number of failures to the total number of iterations:

$$P_{\mathrm{f}} \approx \frac{n\,(G\,(\widehat{x}) \leq 0)}{N} \qquad (45)$$

The accuracy of Monte Carlo simulation results depends on the sample size generated and, in cases when the probability of failure is estimated, on the value of the probability, Melchers (1999) (the smaller the probability of failure, the larger the sample size needed to ensure the same accuracy). The accuracy of the failure probability estimates can be checked by calculating their coefficient of variation (e.g. Melchers 1999).

In order to improve the accuracy of estimating the probability of ultimate strength failure, while keeping the computation time within reasonable limits, variance reduction techniques (e.g. importance sampling, Latin hypercube and directional simulation) can be employed. However, in cases where the main emphasis is on serviceability failure, which can be estimated by a crude Monte Carlo simulation with very good accuracy within a relatively short computation time, such techniques are not necessary (Val and Chernin 2009).

Importance sampling is a variance reduction technique that can be used in the Monte Carlo method (Melchers 1999). The idea behind importance sampling is that certain values of the input random variables in a simulation have more impact on the parameter being estimated than others. If these "important" values are emphasised by sampling more frequently, then the estimator variance can be reduced. Hence, the basic methodology in importance sampling is to choose a distribution that "encourages" the important values. The use of "biased" distributions will result in a biased estimator if it is applied directly to the simulation. However, the simulation outputs are weighted to the correct use of the biased distribution, and this ensures that the new importance sampling estimator is unbiased.

The fundamental issue in implementing importance sampling simulation is the choice of the biased distribution that encourages the important regions of the input variables. Choosing or designing a good biased distribution is the "art" of importance sampling. The rewards for a good distribution can be significant run-time savings; the penalty for a bad distribution can be longer run times than for a general Monte Carlo simulation without importance sampling.

Details of the Monte Carlo method including sampling techniques can be found in Ditlevsen and Madsen (1996), Melchers (1999) and Rubinstein and Kroese (2008).

4 Sensitivity Analysis

The effect of variables on the failure of a structure can be analysed by conducting a comprehensive sensitivity analysis. In view of the variables that affect the corrosion process and the failure function, it is of interest to identify those variables that affect the failure most so that more research can focus on those variables.

Variability and sensitivity analysis should be carried out to provide quantitative information necessary for classifying random variables according to their importance. These measures are essential for the failure assessment of deteriorating materials and structures.

The variability in the probability of failure is investigated by evaluating the effect of a relatively large variation in the mean of each random variable on the probability of failure. Sensitivity analysis can also be performed for the same group of random variables by estimating the relative contribution and assessing the effect of coefficient of variation of each major random variable on the failure probability.

Variability: The variability analysis is performed for the probability of failure with respect to change in the mean value of random variables. The new mean of each random variable $\overline{\mu}_{X_i}$ is:

$$\overline{\mu}_{X_i} = \mu_{X_i} + \lambda\sigma_{X_i} \tag{46}$$

where μ_{X_i} and σ_{X_i} are original mean and standard deviation of the random variable X_i, respectively; and λ is a multiplier of the standard deviation σ_{X_i}. The baseline state corresponds to $\lambda = 0$.

Relative contribution: A sensitivity index that can be used in a comprehensive failure assessment is the relative contribution of each variable in failure function. The relative contribution (α_x^2) of each random variable (x) to the variance of the failure function is introduced as follows (Ahammed and Melchers 1994):

$$\alpha_x^2 = \frac{\left(\frac{\partial G}{\partial x}\sigma_x\right)^2}{\sigma_G^2} \tag{47}$$

Variables with higher values of α_x^2 contribute more in failure function than other variables; therefore more focus and study needs to be carried out to determine the accurate values for such variables.

5 Worked Examples

5.1 Reliability Analysis of a Corrosion Affected Bridge—Ynys-y-Gwas Bridge

To practice using time-dependent reliability analysis methods for corrosion affected concrete bridges, an example adopted from Mahmoodian and Li (2012) is presented in this section. Ynys-y-Gwas bridge was a single span segmental pre-stressed structure which was built in 1953 and carried a minor road over the River Afan 6 km north-east of Port Talbot in the UK. In 1985 the bridge deck collapsed leaving only the edge beams in position. It had a simply supported segmental post-tensioned deck with a clear span of 18.3 m. The nine internal beams of the deck consisted of eight precast I-sections, with a web stiffener on one end, stressed together both longitudinally and transversely (Fig. 9).

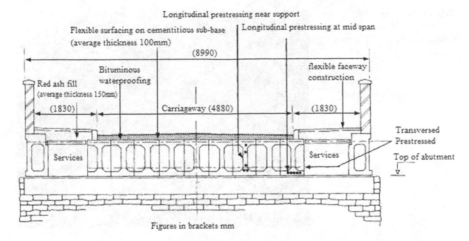

Longitudinal prestressing near support

Flexible surfacing on cementitious sub-base
(average thickness 100mm)

Longitudinal prestressing at mid span

(8990)

Bituminous
waterproofing

Red ash fill
(average thickness 150mm)

flexible faceway
construction

(1830)

Carriageway (4880)

(1830)

Services

Services

Transversed
Prestressed

Top of abutment

Figures in brackets mm

Fig. 9 Cross section of Ynys-y-Gwas bridge (Woodward and Williams 1988)

Fig. 10 Transverse collapse
line in Ynys-y-Gwas bridge
(Woodward and Williams
1988)

The deck suffered failure of all nine internal beams. A transversal line of failure evidenced the tearing off of the longitudinal tendons (Fig. 10). The edge beams and block work parapets remained standing and apparently unaffected, despite the inward and downward forces exerted by the top transverse pre-stressing tendons as the central deck failed.

It was evident from a site visit that the pre-stressing tendons were severely corroded at both longitudinal and transverse joints, indicating that this had been the cause of collapse (Fig. 11). Some wires in fully grouted ducts had lost more than 90 % of their cross-sectional area (Woodward and Williams 1988).

An investigation by Woodward and Williams (1988) on the effect of loss of tendons showed that the loss of one complete tendon within the central I-beam increased the mid-span bending moment on each beam by only 10 kNm. Similarly the loss of one complete tendon within an outer I-beam increased the bending

Fig. 11 Corrosion of
longitudinal tendons at joint
(Woodward and Williams
1988)

moment on that beam by 30 kNm and the failure of all the tendons in an outer
I-beam increased the bending moment on the adjacent beam by about 40 %. This
would overstress it and cause progressive failure of the structure.

Therefore, the criteria for the collapse of the bridge can be set as the time when
all tendons have failed. This definition confirms that for reliability analysis of the
bridge it is appropriate to consider a pre-stressed I-beam as a parallel system. Then
by using the formulation of system reliability analysis, the probability of failure of
the bridge during its service life can be estimated.

Problem formulation: The limit state function which is considered in this example
is bending moment limit state. Hence, it is assumed that failure happens when the
maximum bending moment acting on the bridge beam (M_u) exceeds the bending
capacity of the beam (M_n). The well-known limit state function $G(R, S, t)$ can be
expressed as:

$$G\left(M_n(t), M_u, t\right) = M_n(t) - M_u \qquad (48)$$

where M_u is the bending moment acting on the bridge beam (load effect) and $M_n(t)$
is the bending strength of the beam cross section at time t (structural resistance or
capacity). The tendon cross section decreases with time due to the propagation of
tendon corrosion; therefore the bending strength $(M_n(t))$ decreases. With the limit
state function presented in Eq. (48), the probability of failure due to corrosion of
tendons can be determined from:

$$P_f = P\left[G\left(M_n(t), M_u, t\right) < 0\right] = P\left[M_n(t) < M_u\right] \qquad (49)$$

The failure probability of a pre-stressed concrete I-beam can be determined using
the methods of systems reliability. There are basically two systems in the theory of
systems reliability. One is known as the series system in which the failure of one of
the system components constitutes the failure of the system. The other is known as

the parallel system in which the system fails only when all components of the system fail. For a pre-stressed concrete element, the failure can be defined when all strands in a tendon are failed or when all tendons in a section are failed. Therefore the tendon bearing capacity can be modelled as a parallel system consisting of generally n elements (n strands), in which all n elements have to fail in the specific system failure mode before the entire structure is defined to be in a state of failure.

In this example the Monte Carlo simulation method is used for reliability analysis of the pre-stressed concrete beams in Ynys-y-Gwas bridge.

Model of bending strength: According to ACI 318M 2008, the bending strength of a pre-stressed concrete element can be obtained from:

$$M_n = A_p f_{ps} \left(d_p - \frac{a}{2} \right) + A_s f_y \left(d_s - \frac{a}{2} \right) \tag{50}$$

where A_s: area of non-pre-stressed tension reinforcement, mm^2; A_p: area of pre-stressing steel, mm^2; M_n: nominal bending strength at section, N mm; a: depth of equivalent rectangular stress block, mm; d_s: distance from compression face to centroid non-pre-stressed tension reinforcement, mm; d_p: distance from compression face to centroid pre-stressing tendons, mm; f_{ps}: stress in pre-stressing steel as a portion of steel ultimate strength, MPa; f_y: yield strength of reinforcing bars, MPa.

The parameters in Eq. (50) can be calculated by using Eqs. (3), (4), (5), (6), (7a), (7b), (7c), (7d), (8) and (9). The system reliability analysis is applied to assess the possibility of an accurate prediction of the time of collapse of the bridge in the case of having required data about the condition of corroded tendons. To determine the failure probability of Ynys-y-Gwas bridge due to corrosion of tendons, the limit state function in Eq. (48) is used as the failure criteria in the Monte Carlo method. With the values of random variables given in Table 2, the probability of failure was computed and the results are shown in Fig. 12.

It is already known that the bridge failed 32 years following construction. The corresponding probability of failure determined from Fig. 12 for $t = 32$ years is $P_f = 0.017$. Considering the following definition for reliability index (Melchers 1999), this means a reliability index of $\beta = 2.12$ for the bridge at $t = 32$ years:

$$P_f = \Phi(-\beta); \quad \beta = \Phi^{-1}(P_f) \tag{51}$$

While the reliability index for designing a bridge should be more than 3.5, it is obvious that if an accurate reliability analysis had been carried out before the collapse of Ynys-y-Gwas bridge, it would have clearly shown that after 32 years, the reliability index had decreased considerably and the bridge needed special care for the maintenance and repair of corroded tendons. As mentioned earlier, this prediction would have required data about the condition of tendons, such as corrosion rate, which were not available in 1985 (when the bridge collapsed).

Table 2 Values of basic variables

Basic variables	Definition	Mean	Coefficient of variation	Sources
X	Thickness of concrete cover (cm)	4.50	0.16	Woodward and Williams (1988)
C_0	Surface chloride content (% of cement)	0.22	0.5	Woodward and Williams (1988)
C_{cr}	Threshold chloride concentration (kg/m^3)	2.0	0.20	Val and Pavel (2008)
D_c	Diffusion coefficient (cm^2/year)	0.631	0.2	Val and Pavel (2008)
i_{corr}	Corrosion current density (μA/cm^2)	$0.3686\ln(t) + 1.1305$	0.2	Li (2003) and Vu and Stewart (2002)
R	Penetration ratio	3	0.33	Stewart and Rosowsky (1998)
f_{y_0}	Yield strength of tendon (MPa)	1,600	0.028	Woodward and Williams (1988)

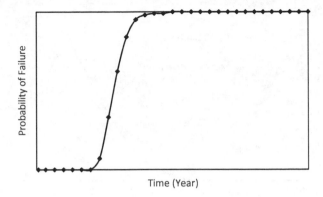

Fig. 12 Probability of failure of Ynys-y-Gwas bridge due to corrosion of tendons

Table 3 Statistical data for basic random variables	Basic variables	Units	Mean	Standard deviation
	K	–	0.8	0.1
	j	–	0.2	0.04
	DS	mg/L	1	0.5
	u	m/s	0.6	0.1
	b/P'	–	$h/D = 0.2$ $b/P' = 0.36$	0.11
			$h/D = 0.4$ $b/P' = 0.55$	0.18
			$h/D = 0.6$ $b/P' = 0.71$	0.23
	A	–	0.2	0.06

5.2 Reliability Analysis of Corrosion Affected Concrete Sewers

In this example, the time-dependent reliability analysis of a concrete sewer pipeline with 500 mm diameter and 25 mm of internal concrete cover was conducted. To incorporate the effect of increments in populations of residents and essentially the increase in the flow rate during the system's lifetime, the modelling assumes flow rates corresponding to relative depths (depth/diameter) of 0.2, 0.4 and 0.6, each occurring over a period of 25 years. Values for basic random variables which were used in the model are presented in Table 3.

According to ASCE manual No. 69 (1989), one of the performance criteria related to the stability of concrete sewers is control of the wall thickness reduction under an acceptable limit (normally concrete cover). In the theory of structural reliability this criterion can be expressed in the form of a limit state function as follows:

$$G(d_{\max}, d, t) = d_{\max}(t) - d(t) \tag{52}$$

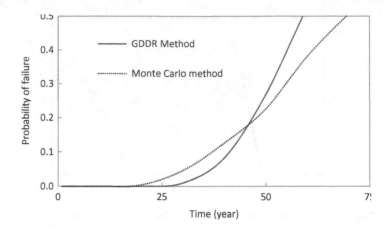

Fig. 13 Probability of failure from the gamma degradation method and Monte Carlo simulation method

where: d: reduction in wall thickness due to corrosion, mm; d_{max}: maximum permissible reduction in wall thickness (structural resistance or limit), mm; t: elapsed time (year).

d_{max} may change over time although in most practical cases it has a constant value prescribed in design codes and manuals.

With the above limit state function, the probability of failure of the concrete pipe due to the reduction of its wall thickness can be determined by:

$$P_f(t) = P\left[G\left(d_{max}, d, t\right) \leq 0\right] = P\left[d(t) \geq d_{max}(t)\right] \tag{53}$$

To calculate $P_f(t)$, the gamma distribution concept is employed.

Figure 13 shows the results obtained for the probability of pipe failure by using method (b) in Sect. 3.2.2 presented by the flowchart in Fig. 8. To compare the results with other common methods of reliability analysis, the figure also includes the results of using the Monte Carlo simulation method applied in Eq. (53).

For repair and rehabilitation planning analysis of the pipe, the time for the pipe to be unserviceable, i.e. T_c, can be determined for a given acceptable probability of failure, P_a. For example, using the graph for the gamma degradation method in Fig. 13, it can be obtained that $T_c = 42$ years for $P_a = 0.1$. If there is no intervention during the service period of (0,42) years for the pipe, such as maintenance and repairs, T_c represents the time for the failure of the pipe, based on the reliability analysis. The information of T_c (i.e. time for interventions) is of practical importance to structural engineers and infrastructure managers with regard to planning for repairs and/or rehabilitation of the pipeline. An optimum funding allocation for the pipeline system can be concluded by conducting a cost analysis for the repair and replacement of those corroded pipes with a higher risk of failure.

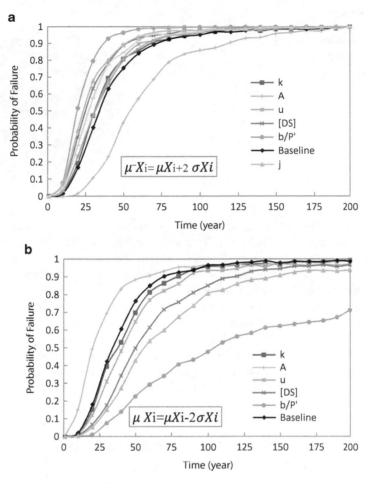

Fig. 14 Variability in the probability of failure due to the (**a**) increasing and (**b**) decreasing mean of each random variable by two standard deviations

Applying the method explained in Sect. 3, Fig. 14a, b show the variability in the probability of failure with respect to the six random variables considering λ takes the values 2.0 and -2.0, respectively. Assuming a Gaussian distribution for the random variables, this range corresponds to 95.4 % of the possible values of the variable.

It is concluded that alkalinity A and the ratio of surface width of the stream to the perimeter of the exposed wall (b/P') are the most influential variables. It should be noted that for a given cross section, (b/P') can geometrically be related to relative depth of the stream, h/D.

Figure 15 illustrates sensitivities (α^2) or the relative contribution of variables to the variance of the failure function calculated based on the definition in Eq. (47). It is obvious from the results that the relative contribution of some of the variables is

Fig. 15 Relative contributions of random variables in the probability of failure of the pipe

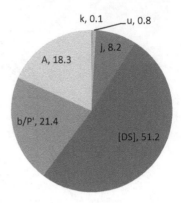

considerably lower than other variables. These variables include the acid reaction factor (*k*) and stream velocity (*u*). This indicates that the inclusion of these as variables has little influence on the probability of failure of the pipeline system. Therefore, in any future analysis, it would not be inaccurate to treat these variables as deterministic variables with constant quantities.

Among the variables, the analysis shows the particular significance of the concentration of dissolved sulphide ([DS]) in the probability of failure of the pipeline, indicating that this variable is the most significant variable on the failure of concrete sewers. Consequently, in order to achieve a more accurate failure assessment and service life prediction of sewers, infrastructure managers must note the importance of monitoring data surrounding this particular parameter.

5.3 Reliability Analysis of Corrosion Affected Cast Iron Water Pipes

In this example multi-failure mode assessment is practised through time-dependent reliability analysis methods for a corrosion affected cast iron water pipe. Experimental data from Marshall (2001) is used for the preparation of statistical properties of basic random variables. The values presented in Table 4 are the means and standard deviations of an example of a cast iron pipeline in the UK with a diameter of 305 mm and a wall thickness of 9.5 mm. The pipe is made of cast iron with the fracture toughness of $K_{IC} = 11.4$ MPa/m$^{0.5}$.

Different operational and environmental conditions can cause the failure of the pipe. The failure can be due to interaction of the different failure modes. Each mode of failure can be represented by a failure function, and the probability of a pipe system failure can be calculated by using the methods of system reliability analysis. For a pipe, the occurrence of one failure mode will incorporate its total failure. Therefore a series system is appropriate for the failure assessment of pipes.

Table 4 Statistical data for the basic random variable in the stress intensity factor model

		Multiplying constant, k		Exponential constant, n		Stress		Geometry functions, f_i, f_{bg}	
		Mean	Standard deviation	Mean	Standard deviation	Mean	Standard deviation	Mean	Standard deviation
Under hoop stress	Internal corrosion pit	0.20	0.04	0.64	0.13	216	32	$f_i = 0.658$	0.11
	External corrosion pit	0.35	0.07	0.65	0.11	216	32	$f_i = 0.727$	0.12
Under axial stress	Internal corrosion pit	0.20	0.04	0.64	0.13	113	11	$f_{bg} = 0.795$	0.13
	External corrosion pit	0.35	0.07	0.65	0.11	113	11	$f_{bg} = 0.672$	0.11

According to the theory of systems reliability, the failure probability of a series system (P_{f_s}) can be determined from (Melchers 1999):

$$P_{f_s} = 1 - \prod_{t=1}^{n} [1 - P_{f_i}] \tag{54}$$

where P_{f_i} is the probability of failure due to the ith failure mode of the pipe (can be determined by Eq. (35)) and n is the number of failure modes considered in the system.

For an underground cast iron pipe subjected to corrosion, the action process is the progression of the stress intensity factor at crack pits. For fracture assessment, the failure function of most interest to pipe engineers and infrastructure managers is the high stress intensity in the pipe wall under a combination of stress and corrosion.

The stress intensity factor of a corrosion affected cast iron pipe is a very random phenomenon, depending on time-dependent factors such as depth of corrosion pit and stress condition. It is justifiable to model stress intensity factor as a time-dependent process. If K_c is defined as the critical stress intensity factor, known as fracture toughness, beyond which the pipe cannot sustain the pit crack, the probability of pipe collapse (failure) can be determined from the following equation:

$$P_f(t) = P[K_c < K_I(t)] \tag{55}$$

In the practical application of Eq. (55) to the service life prediction of cast iron pipes, the main effort lies in developing a model of stress intensity factor K_I, i.e. the action process, since the fracture toughness K_c is a material property (ASTM-E1820-01 2001).

According to the theory of fracture mechanics, stress intensity factor K_I is a function of far-field stress level σ and the size of the pit a. This relationship can be expressed as follows (Hertzberg 1996):

$$K = f(\sigma, a) \tag{56}$$

where the functionality depends on the configuration of the crack pit and the manner in which the loads are applied. According to Laham (1999), the stress intensity factor for a crack pit in a pipe under hoop stress is as follows:

$$K_{I-h} = \sqrt{\pi a} \sum_{i=0}^{3} \sigma_i f_i(a, c, d, R) \tag{57}$$

and the stress intensity factor for a crack pit in a pipe under axial stress:

$$K_{I-a} = \sqrt{\pi a} \left(\sum_{i=0}^{3} \sigma_i f_i(a, c, d, R) + \sigma_{bg} f_{bg}(a, c, d, R) \right) \tag{58}$$

where

K_{I-h} = stress intensity factor for longitudinal cracks in Mode I, caused by hoop stress; K_{I-a} = stress intensity factor for circumferential cracks in Mode I, caused by axial stress; a = depth of the crack, i.e. corrosion pit; σ_i = stress normal to the crack plane; f_i and f_{bg} = geometry functions, depend on α, c (half-length of crack), d and R (inner radius of pipe); σ_{bg} = the global bending stress, i.e. the maximum outer fibre bending stress.

For internal and/or external crack pits, the difference in formulations of stress intensity factor (Eqs. (57) and (58)) lies in geometry functions (i.e. f_i and f_{bg}), which have been presented in different tables by Laham (1999). Due to the propagation of corrosion, a changes with time so the stress intensity factors are time variant.

Two typical scenarios of corrosion pits exist: (1) external corrosion; and (2) internal corrosion. For a comprehensive assessment, the two types of stresses (hoop and axial) that lead to Mode I fracture need to be considered in the study. Therefore four failure functions (i.e. failure modes) will be considered in determining the stress intensity factor of cast iron pipes.

Failure Mode 1, internal surface pit under hoop stress: The internal surface corrosion pit can be assumed to be semi-elliptical with length $2c$ and radial depth a as shown in Fig. 16a. f_i, the geometry functions for the ith stress distribution on a crack (pit) surface, depend on the geometry of the pits and the pipe as represented by a, c, d and R. The discrete values of f_i for a semi-elliptical external surface pit with given geometric parameters can be obtained from tables presented in Laham (1999). The probability of failure due to internal surface pit caused by hoop stress within life time would be:

$$P_{f_1} = P\left[K_c < K_{(I-h)_{\text{int.}}}(t)\right]$$
$$= P\left[K_c < \left(\sqrt{\pi a(t)}\sum_{i=0}^{3}\sigma_i(t)f_i(a(t),c,d,R)\right)\right] \quad (59)$$

Failure Mode 2, external surface pit under hoop stress: Figure 16b shows an idealised section of a cast iron pipe with external corrosion pit which is assumed to be semi-elliptical. In Fig. 2b, $2c$ is the length of the semi-elliptical pit, a is the radial depth of the pit through the pipe wall, R is the inner radius, d is the thickness of the pipe wall and the angle θ is used to describe the position around the semi-elliptical pit which varies between $0 \le \theta \le \pi$. The stress intensity factor for the Mode I fracture of a semi-elliptical surface pit K_I can also be determined by using Eq. (57). Similar to the situation in failure Mode 1, the probability of failure due to external surface pit caused by hoop stress within time is:

$$P_{f_2} = P\left[K_c < K_{(I-h)_{\text{ext.}}}(t)\right]$$
$$= P\left[K_c < \left(\sqrt{\pi a(t)}\sum_{i=0}^{3}\sigma_i(t)f_i(a(t),c,d,R)\right)\right] \quad (60)$$

Failure Mode 3, internal surface pit under axial stress: The internal surface corrosion pit can be assumed to be semi-elliptical with length $2c$ and radial depth

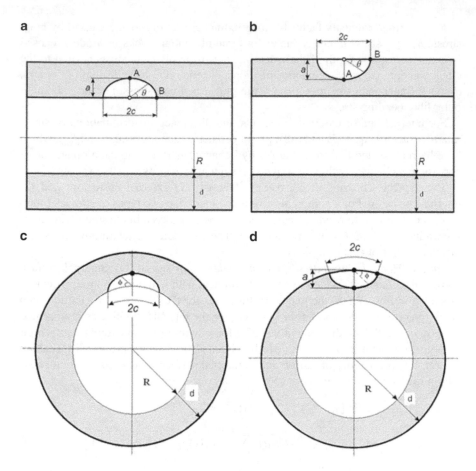

Fig. 16 (**a–d**) Geometry of all possible semi-elliptical crack pits in a pipe wall surface pit of a pipe

a as shown in Fig. 16c. In this case, the stress intensity factor for an internal semi-elliptical surface pit can be determined from Eq. (58). Considering the time dependency of the stress intensity factor, the probability of failure due to internal surface pit caused by axial stress within life time is:

$$P_{f_3} = P\left[K_c < K_{(1-a)_{int.}}(t)\right]$$

$$= P\left[K_c < \sqrt{\pi a \left(\sum_{i=0}^{3} \sigma_i(t) f_i(a(t), c, d, R) + \sigma_{bg}(t) f_{bg}(a(t), c, d, R)\right)}\right]$$

$$(61)$$

Failure Mode 4, external surface pit under axial stress: A typical example of this case is a pipe under internal pressure (e.g. water) and subjected to external corrosion. Figure 16d shows an idealised section of a cast iron pipe with external corrosion pit which is assumed to be semi-elliptical. In this figure, $2c$ is the length of the semi-elliptical pit, a is the radial depth of the pit through the pipe wall, R is the inner radius, d is the thickness of the pipe wall and the angle θ is used to describe the position around the semi-elliptical pit which varies between $0 \le \theta \le \pi$. In this case, the stress intensity factor for a Mode I fracture of a semi-elliptical surface pit K_{11} can also be determined by Eq. (58). Similar to failure Mode 3, the probability of failure due to external surface pit caused by axial stress within time:

$$
\begin{aligned}
P_{f_4} &= P\left[K_c < K_{(I-a)_{\text{ext.}}}(t)\right] \\
&= P\left[K_c < \sqrt{\pi a}\left(\sum_{i=0}^{3}\sigma_i(t)f_i\left(a(t),c,d,R\right) + \sigma_{bg}(t)f_{bg}\left(a(t),c,d,R\right)\right)\right]
\end{aligned}
\tag{62}
$$

In a comprehensive reliability analysis of a cast iron pipe, all four modes of failure should be considered by using Eq. (63):

$$
P_{f,s} = 1 - \prod_{i=1}^{4}\left[1 - P_{f,i}\right]
\tag{63}
$$

To estimate the mean and standard deviation of each stress intensity factor, the Monte Carlo simulation method is used. For this case study, the simulation is run for 200 years ($t = 1, 2, 3, \ldots, 200$). In each time step, random values for basic random variables are selected and stress intensity factors are calculated. This is repeated 10,000 times and at the end, the mean and standard deviation of the simulated values is calculated for that year.

By having the trend of the mean and standard deviation of each stress intensity factor with respect to time, the time-dependent format of the statistics of the stress intensity factors can be presented in Table 5. Shape and scale parameters in Table 5 for each type of stress intensity factor are also calculated by using Eqs. (33) and (34).

With this preparation, the probability of failure for the four different cases of failure can be calculated by Eqs. (59)–(62). Finally, the time-dependent probability of pipe collapse considering all four modes of failure can be calculated using Eq. (63). An illustration of the estimated probabilities of failure has been presented in Fig. 17.

From this figure the probability of failure for pipes with external corrosion can be compared to those with internal corrosion. It is concluded that the probability of failure for pipes with external corrosion is much higher than that for pipes with internal corrosion. This is due to the faster progression of external corrosion compared to internal corrosion.

As an example, while for an acceptable probability of failure of 50 % ($P_a = 0.5$), the service life of a pipe under hoop stress and subjected to internal corrosion is 48 years, for external corrosion, it reduces to 14 years (more than three times less).

It can also be seen from Fig. 17 that the probability of pipe failure is greater when multi-failure modes are considered as a system than when each failure mode is considered individually. It is also seen that the probability of pipe failure due

Table 5 Estimated parameters for all four types of failure

Type of crack pit	Type of stress intensity factor	Mean value of the stress intensity factor	Standard deviation of the stress intensity factor	Shape parameter, α	Scale parameter, λ
Internal surface pit under hoop stress	$K_{(l-h)_{\text{int.}}}(t)$	$\mu_{K_{(l-h)_{\text{int.}}}}(t) = 3.443t^{0.336}$	$\sigma_{K_{(l-h)_{\text{int.}}}}(t) = 0.747t^{0.470}$	$21.240t^{-0.268}$	$6.170t^{-0.604}$
External surface pit under hoop stress	$K_{(l-h)_{\text{ext.}}}(t)$	$\mu_{K_{(l-h)_{\text{ext.}}}}(t) = 5.066t^{0.337}$	$\sigma_{K_{(l-h)_{\text{ext.}}}}(t) = 1.103t^{0.448}$	$21.105t^{-0.222}$	$4.166t^{-0.559}$
Internal surface pit under axial stress	$K_{(l-a)_{\text{int.}}}(t)$	$\mu_{K_{(l-a)_{\text{int.}}}}(t) = 2.176t^{0.336}$	$\sigma_{K_{(l-a)_{\text{int.}}}}(t) = 0.415t^{0.487}$	$27.528t^{-0.302}$	$12.651t^{-0.638}$
External surface pit under axial stress	$K_{(l-a)_{\text{ext.}}}(t)$	$\mu_{K_{(l-b)_{\text{ext.}}}}(t) = 2.453t^{0.336}$	$\sigma_{K_{(l-b)_{\text{ext.}}}}(t) = 0.467t^{0.464}$	$27.601t^{-0.256}$	$11.253t^{-0.592}$

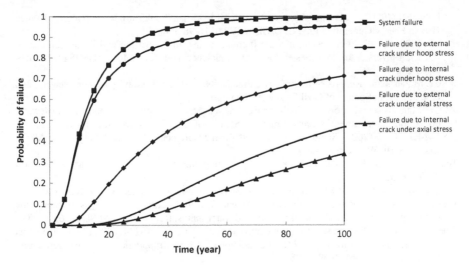

Fig. 17 Probability of pipe failure for individual failure modes and system failure mode

to hoop stress is much larger than that due to axial stress. It is clear from Fig. 17 that considering each failure mode individually in the reliability analysis of pipes is unjustifiable which may pose undue risks to the public and cause subsequent disastrous consequences.

The result obtained in Fig. 17 can also be used for estimation of the safe service life of the pipe (T). It can be determined for the given assessment criterion and acceptable probability of failure P_a. For example, using the criteria of system failure, it can be obtained from Fig. 17 that $T = 13$ years for $P_a = 0.5$. If there is no intervention during the service period of (0, 13) years for the pipe, such as maintenance and repairs, based on the criteria considered in system reliability analysis $(K_I > K_c)$, the pipe will not be serviceable after $t > 13$ years. Having knowledge of T (i.e. time for intervention or service life) is of practical importance to structural engineers and asset managers with regard to planning for repairs and/or rehabilitation of the pipe networks. It can also help to achieve an optimum funding allocation for the repair and replacement of corroded pipes with higher risk of failure.

References

Abdel-Hameed M (1975) A gamma wear process. IEEE Trans Reliab 24(2):152–153
ACI 318M (2008) Building code requirements for reinforced concrete. ACI, Farmington Hills
Ahammed M, Melchers RE (1994) Reliability of underground pipelines subject to corrosion. J Transport Eng 120(6):989–1002
Ahammed M, Melchers RE (1997) Probabilistic analysis of underground pipelines subject to combined stress and corrosion. Eng Struct 19(12):988–994

Al Laham S (1999) Stress intensity factor and limit load handbook. Structural Integrity Branch British Energy Generation Ltd., EPD/GEN/REP/0316/98, issue 2

Anon (2002) Localized corrosion, module 0 course notes. In: Corrosion and Protection Centre (CPC), University of Manchester Institute of Science and Technology (UMIST), Manchester, UK

ASCE Manuals and Reports of Engineering Practice No. 60 (2007) Gravity sanitary sewers, 2nd edn. American Society of Civil Engineers, New York

ASCE Manuals and Reports of Engineering Practice—No. 69 (1989) Sulphide in wastewater collection and treatment systems. American Society of Civil Engineers, New York

ASTM-E1820-01 (2001) Standard test method for measurement of fracture toughness, ASTM International, West Conshohocken, PA

Baboian R (1995) Environmental conditions affecting transport infrastructure. Mater Perform 34(9):48–52

Bogdanoff JL, Kozin F (1985) Probabilistic models of cumulative damage. Wiley, New York

Camarinopoulos L, Chatzoulis A, Frontistou-Yannas S, Kallidromitis V (1999) Assessment of the time-dependent structural reliability of buried water mains. Reliab Eng Syst Safety 65:41–53

Chaker V (Ed) (1992) Corrosion forms & control for infrastructure. ASTM STP 1137, Philadelphia

Cinlar E, Bazant ZP, Osman E (1977) Stochastic process for extrapolating concrete creep. J Eng Mech Div 103(EM6): 1069–1088

Darmawan MS, Stewart MG (2007) Spatial time-dependent reliability analysis of corroding pretensioned prestressed concrete bridge girders. Struct Safety 29(1):16–31

Davis P, De Silva D, Gould S, Burn S (2005) Condition assessment and failure prediction for asbestos cement sewer mains. In: Pipes Wagga Wagga Conference, Charles Sturt University, Wagga Wagga, New South Wales

Davis P, De Silva D, Marlow D, Moglia M, Gould S, Burn S (2008) Failure prediction and optimal scheduling of replacements in asbestos cement water pipes. J Water Supply Res Technol 57(4):239–252

Dawotola AW, Trafalis TB, Mustaffa Z, van Gelder PHAJM, Vrijling JK (2013) Risk based maintenance of a cross-country petroleum pipeline system. J Pipeline Syst Eng Pract. 4(3):141–148

De Silva D, Moglia M, Davis P, Burn S (2006) Condition assessment and probabilistic analysis to estimate failure rates in buried metallic pipelines. J Water Supply Res Technol 55(3):179–191

Ditlevsen O, Madsen HO (1996) Structural reliability methods. Wiley, New York

Du YG, Clark LA, Chan AHC (2005) Residual capacity of corroded reinforcing bars. Mag Concrete Res 57(3):135–147

Dufresne F, Gerber HU, Shiu ESW (1991) Risk theory with the gamma process. ASTIN Bull 21(2):177–192

Ellingwood BR, Mori Y (1993) Probabilistic methods for condition assessment and life prediction of concrete structures in nuclear power plants. Nucl Eng Des 142:155–166

Enright MP, Frangopol DM (1998) Probabilistic analysis of resistance degradation of reinforced concrete bridge beams under corrosion. Eng Struct 20(11):960–971

EPA/600/R-12/017 (2012) Condition assessment technologies for water transmission and distribution systems. United States, Environmental Protection Agency

Ferguson TS, Klass MJ (1972) A representation of independent increment processes without Gaussian components. Ann Math Stat 43(5):1634–1643

Fitzpatrick RA (1996) Corrosion of concrete structures in coastal areas. Concrete Int 18(7):46–47

González JA, Andrade C, Alonso C, Feliú S (1995) Comparison of rates of general corrosion and maximum pitting penetration on concrete embedded steel reinforcement. Cem Concr Res 25(2):257–264

Guo T, Sause R, Frangopol DM, Li A (2011) Time-dependent reliability of PSC box-girder bridge considering creep, shrinkage, and corrosion. J Bridge Eng 16(1):29–43

Hertzberg RW (1996) Deformation and fracture mechanics of engineering materials. Wiley, Chichester

Hoffmans GJCM, Pilarczyk KW (1995) Local scour downstream of hydraulic structures. J Hydraul Eng 121(4):326–340

Kleiner Y, Rajani B, Sadiq R (2005) Risk management of large-diameter water transmission mains. American Water Works Association Research Foundation, Denver, CO

Kucera V, Mattsson E (1987) Atmospheric corrosion. In: Mansfeld F (ed) Corrosion mechanics. Marcel Dekker Inc., New York

Li CQ (2003) Life cycle modelling of corrosion affected concrete structures-propagation. J Struct Eng 129(6):753–761

Li CQ, Mahmoodian M (2013) Risk based service life prediction of underground cast iron pipes subjected to corrosion. J Reliab Eng Syst Safe 119:102–108

Li CQ, Melchers RE (2005) Time-dependent reliability analysis of corrosion-induced concrete cracking. ACI Struct J 102(4):543–547

Li F, Yuan Y, Li CQ (2011) Corrosion propagation of pre-stressing steel strands in concrete subject to chloride attack. J Constr Build Mater, 25(10):3878–3885

Lillie K, Reed C, Rodgers M (2004) Workshop on condition assessment inspection devices for water transmission mains. AWWA Research Foundation, Denver, CO

Mahmoodian M, Alani A (2013a) Modelling deterioration in concrete pipes as a stochastic gamma process for time dependent reliability analysis, ASCE. J Pipeline Syst Eng Pract 5(1):04013008

Mahmoodian M, Alani A (2013b) Multi failure mode assessment of buried concrete pipes subjected to time dependent deterioration using system reliability analysis. J Fail Anal Prev 13(5):634–642

Mahmoodian M, Li CQ (2012) Failure assessment of a pre-stressed concrete bridge using time dependent system reliability method. In: 29th international bridge conference, Pittsburgh, PA, 10–13 June

Mahmoodian M, Alani A, Tee KF (2012) Stochastic failure analysis of the gusset plates in the Mississippi River Bridge. Int J Forensic Eng 1(2):153–166

Mann E, Frey J (2011) Optimized pipe renewal programs ensure cost-effective asset management. Pipelines 2011:44–54

Marshall P (2001) The residual structural properties of cast iron pipes—structural and design criteria for linings for water mains. UK Water Industry Research

Melchers RE (1999) Structural reliability analysis and prediction, 2nd edn. Wiley, Chichester

Misiunas D (2005) Failure monitoring and asset condition assessment in water supply systems. PhD dissertation, Department of Industrial Electrical Engineering and Automation Lund University

Moglia M, Davis P, Burn S (2008) Strong exploration of a cast iron pipe failure model. J Reliab Eng Syst Safety 93:863–874

Nicolai RP, Budai G, Dekker R, Vreijling M (2004) Modeling the deterioration of the coating on steel structures: a comparison of methods. In: Proceedings of the IEEE conference on systems, man and cybernetics, IEEE, Danvers, pp 4177–4182

Noortwijk JM, Frangopol DM (2004) Two probabilistic life-cycle maintenance models for deteriorating civil infrastructures. Probab Eng Mech 19(4):345–359, p 77–83

Noortwijk JM, Klatter HE (1999) Optimal inspection decisions for the block mats of the Eastern-Scheldt barrier. Reliab Eng Syst Safety 65(3):203–211

Noortwijk JM, Pandey MD (2003) A stochastic deterioration process for time-dependent reliability analysis. In: Proceedings of the eleventh IFIP WG 7.5 working conference on reliability and optimization of structural systems, 2–5 November 2003, Banff, Canada, pp 259–265

Papoulis A, Pillai SU (2002) Probability, random variables, and stochastic processes, 4th edn. McGraw-Hill, New York, 852p

Pomeroy RD (1976) The problem of hydrogen sulphide in sewers. Clay Pipe Development Association

Rajani B, Kleiner Y (2004) Non-destructive inspection techniques to determine structural distress indicators in water mains. In: Evaluation and control of water loss in urban water networks, Valencia, Spain, 21–25 June 2004, p 1–20

Rajani B, Kleiner Y (2013) External and internal corrosion of large-diameter cast iron mains. J Infrastruct Syst 19(4):486–495

Rajani B, Makar J (2000) A methodology to estimate remaining service life of grey cast iron water mains. Can J Civil Eng 27(6):1259–1272

Rajani B, Tesfamariam S (2007) Estimating time to failure of cast iron water mains. Water Manag 160(WM2):83–88

Rajani B, Makar J, McDonald S, Zhan C, Kuraoka S, Jen CK, Viens M (2000) Investigation of grey cast iron water mains to develop a methodology for estimating service life. American Water Works Association Research Foundation, Denver, CO

Rubinstein RY, Kroese DP (2008) Simulation and the Monte Carlo method, 2nd edn. Wiley, New York

Sadiq R, Rajani B, Kleiner Y (2004) Probabilistic risk analysis of corrosion associated failures in cast iron water mains. Reliab Eng Syst Safety 86(1):1–10

Salman B, Salem O (2012) Modeling failure of wastewater collection lines using various section-level regression models. ASCE J Infrastruct Syst 18(2):146–154

Saltelli A, Tarantola S, Campolongo F, Ratto M (2004) Sensitivity analysis in practice: a guide to assessing scientific models. Wiley, New York

Sheikh AK, Boah JK, Hansen DA (1990) Statistical modelling of pitting corrosion and pipeline reliability. Corrosion-NACE 46(3):190–197

Shreir LL, Jarman RA, Burstein GT (1994) Corrosion. Butterworth Heinemann, Oxford

Singpurwalla N (1997) Gamma processes and their generalizations: an overview. In: Cooke R, Mendel M, Vrijling H (eds) Engineering probabilistic design and maintenance for flood protection. Kluwer, Dordrecht, pp 67–75

Stewart MG (2004) Spatial variability of pitting corrosion and its influence on structural fragility and reliability of RC beams in flexure. Struct Safety 26(4):453–470

Stewart MG (2009) Mechanical behaviour of pitting corrosion of flexural and shear reinforcement and its effect on structural reliability of corroding RC beams. Struct Safety 31(1):19–30

Stewart MG, Rosowsky DV (1998) Time-dependent reliability of deteriorating reinforced concrete bridge decks. J Struct Safe 20(1):91–109

Val DV, Chernin L (2009) Serviceability reliability of reinforced concrete beams with corroded reinforcement. J Struct Eng 135(8):896–905

Val DV, Pavel AT (2008) Probabilistic evaluation of initiation time of chloride-induced corrosion. Reliab Eng Syst Safety 93(3):364–372

Val DV, Robert EM (1997) Reliability of deteriorating RC slab bridges. J Struct Eng 123(12): 1638–1644

van der Weide H (1997) Gamma processes. In: Cooke R, Mendel M, Vrijling H (eds) Engineering probabilistic design and maintenance for flood protection. Kluwer, Dordrecht, pp 77–83

Vu KAT, Stewart MG (2002) Spatial variability of structural deterioration and service life prediction of reinforced concrete bridges. In: Proceedings of international on bridge maintenance, safety and management—IABMAS 2002, Barcelona, CD-ROM

Vu NA, Castel A, François R (2009) Effect of stress corrosion cracking on stress-strain response of steel wires used in prestressed concrete beams. Corros Sci 51(6):1453–1459

Water UK (2007) Working on behalf of the water industry for a sustainable future. Water UK. http://www.water.org.uk/home/resources-and-links/waterfacts/resources

Woodward RJ, Williams FW (1988) Collapse of Ynys-y-Gwas bridge, West Glamorgan. In: Proceedings of the Institution of Civil Engineers, part 1, design and construction, vol 84, p 635–669

Yamini H (2009) Probability of failure analysis and condition assessment of cast iron pipes due to internal and external corrosion in water distribution systems. PhD dissertation, University of British Colombia

Yamini H, Lence BJ (2009) Probability of failure analysis due to internal corrosion in cast-iron pipes. J Infrastruct Syst 16(1):73–80

Zhou W (2011) Reliability evaluation of corroding pipelines considering multiple failure modes and time-dependent internal pressure. J Infrastruct Syst 17(4):216–224

Multicut-High Dimensional Model Representation for Reliability Bounds Estimation

A.S. Balu and B.N. Rao

Abstract The structural reliability analysis in presence of mixed uncertain variables demands more computation as the entire configuration of fuzzy variables needs to be explored. The existence of multiple design points plays an important role in the accuracy of results as the optimization algorithms may converge to a local design point by neglecting the main contribution from the global design point. Therefore, in this chapter, a method for estimating the reliability bounds of structural systems involving multiple design points in presence of mixed uncertain variables is presented. The proposed method involves weight function to identify multiple design points, multicut-high dimensional model representation for the limit state function approximation, transformation technique to obtain the contribution of the fuzzy variables and fast Fourier transform for solving the convolution integral.

1 Introduction

The Reliability analysis taking into account the uncertainties involved in a structural system plays an important role in the analysis and design of structures. Due to the complexity of structural systems the information about the functioning of various structural components has different sources and the failure of systems is usually governed by various uncertainties, all of which are to be taken into consideration for reliability estimation. Uncertainties present in a structural system can be classified

A.S. Balu (✉)
Department of Civil Engineering, National Institute of Technology Karnataka, Surathkal, Mangalore, India
e-mail: arunsbalu@gmail.com

B.N. Rao
Department of Civil Engineering, Indian Institute of Technology Madras, Chennai, Tamil Nadu, India
e-mail: bnrao@iitm.ac.in

S. Kadry and A. El Hami (eds.), *Numerical Methods for Reliability and Safety Assessment: Multiscale and Multiphysics Systems*, DOI 10.1007/978-3-319-07167-1_18, © Springer International Publishing Switzerland 2015

as aleatory uncertainty and epistemic uncertainty. Aleatory uncertainty information can be obtained as a result of statistical experiments and has a probabilistic or random character. Epistemic uncertainty information can be obtained by the estimation of the experts and in most cases has an interval or fuzzy character (Balu and Rao 2012a). When aleatory uncertainty is only present in a structural system, then the reliability estimation involves determination of the probability that a structural response exceeds a threshold limit, defined by a limit state function influenced by several random parameters. Structural reliability can be computed adopting probabilistic method involving the evaluation of multidimensional integral (Breitung 1984; Rackwitz 2001).

In first-order or second-order reliability method (FORM/SORM), the limit state functions need to be specified explicitly. Alternatively the simulation-based methods such as Monte Carlo techniques require more computational effort for simulating the actual limit state function repeated times. The response surface concept was adopted to get separable and closed form expression of the implicit limit state function in order to use fast Fourier transform (FFT) to estimate the failure probability (Sakamoto et al. 1997). The high dimensional model representation (HDMR) concepts were applied for the approximation of limit state function at the most probable point (MPP) and FFT technique to evaluate the convolution integral for estimation of failure probability (Rao and Chowdhury 2008). In this method, efforts are required in evaluating conditional responses at a selected input determined by sample points, as compared to full scale simulation methods.

Further, the main contribution to the reliability integral comes from the neighborhood of design points. When multiple design points exist, available optimization algorithms may converge to a local design point and thus erroneously neglect the main contribution to the value of the reliability integral from the global design point(s). Moreover, even if a global design point is obtained, there are cases for which the contribution from other local or global design points may be significant (Au et al. 1999). In that case, multipoint FORM/SORM is required for improving the reliability analysis (Der Kiureghian and Dakessian 1998). In the presence of only epistemic uncertainty in a structural system, possibilistic approaches to evaluate the minimum and maximum values of the response are available (Briabant et al. 1999; Penmetsa and Grandhi 2003).

All the reliability models discussed above are based on only one kind of uncertain information; either random variables or fuzzy input, but do not accommodate a combination of both types of variables. However, in some engineering problems with mixed uncertain parameters, using one kind of reliability model cannot obtain the best results. To determine the failure probability bounds of a structural system involving both random and fuzzy variables, the entire configuration of the fuzzy variables needs to be explored (Adduri and Penmetsa 2008; Balu and Rao 2014). Hence, the computational effort involved in estimating the bounds of the failure probability increases tremendously in the presence of multiple design points and mixed uncertain variables (Balu and Rao 2012b).

The HDMR techniques are efficient to estimate the failure probability if the limit state function exhibits a single design point. For the functions with multiple design points, the concepts of HDMR is extended in this paper by exploring the potential of coupled Multicut-HDMR (MHDMR)-FFT technique to evaluate the reliability of a structural system in presence of mixed uncertain variables. Comparisons of numerical results have been made with direct MCS method to evaluate the accuracy and computational efficiency of the present method.

2 Multicut-HDMR

HDMR is a general set of quantitative model assessment and analysis tools for capturing the high dimensional relationships between sets of input and output model variables (Rabitz et al. 1999; Rao and Chowdhury 2008). Since the influence of the input variables on the response function can be independent and/or cooperative, HDMR expresses the response as a hierarchical correlated function expansion in terms of the input variables. The expansion functions are determined by evaluating the input–output responses of the system relative to the defined reference point along associated lines, surfaces, subvolumes, etc. in the input variable space.

The main limitation of truncated cut-HDMR expansion is that depending on the order chosen sometimes it is unable to accurately approximate the response, when multiple design points exist on the limit state function or when the problem domain is large. In this section, a new technique based on multicut-HDMR (MHDMR) is presented for approximation of the original implicit limit state function, when multiple design points exist. The basic principles of cut-HDMR may be extended to more general cases. The MHDMR is one extension where several cut-HDMR expansions at different reference points are constructed, and the original implicit limit state function $g(\mathbf{x})$ is approximately represented not by one, but by all cut-HDMR expansions. In the present work, weight function is adopted for identification of multiple reference points closer to the limit surface.

Let $d^1, d^2, \ldots, d^{m_d}$ be the m_d identified reference points closer to the limit state function based on the weight function. MHDMR approximation of the original implicit limit state function is based on the principles of cut-HDMR expansion, where individual cut-HDMR expansions are constructed at different reference points $d^1, d^2, \ldots, d^{m_d}$ by taking one at a time as follows:

$$
g^k(\mathbf{x}) = g_0^k + \sum_{i=1}^{N} g_i^k(x_i) + \sum_{1 \le i_1 < i_2 \le N} g_{i_1 i_2}^k(x_{i_1}, x_{i_2}) + \ldots
$$

$$
+ \sum_{1 \le i_1 < \cdots < i_l \le N} g_{i_1 i_2 \ldots i_l}^k(x_{i_1}, x_{i_2}, \ldots, x_{i_l}) + \cdots + g_{12\ldots N}^k(x_1, x_2, \ldots, x_N)
$$

$$(1)$$

where $k = 1, 2, \ldots, m_d$. The original implicit limit state function $g(\mathbf{x})$ is approximately represented by blending all locally constructed m_d individual cut-HDMR expansions as follows:

$$g(x) \cong \sum_{k=1}^{m_d} \lambda_k(x) \left[g_0^k + \sum_{i=1}^{N} g_i^k(x_i) + \cdots + g_{1,2\cdots N}^k(x_1, x_2, \ldots, x_N) \right] \quad (2)$$

There are a variety of choices to define $\lambda_k(x)$. In the present study, the metric distance $\alpha_k(x)$ from any sample point to the reference point d^k; $k = 1, 2, \ldots, m_d$

$$\alpha_k(x) = \left[\sum_{i=1}^{N} \left(x_i - d_i^k \right)^2 \right]^{\frac{1}{2}} \quad (3)$$

is used to define

$$\lambda_k(x) = \frac{\overline{\lambda}_k(x)}{\sum_{s=1}^{m_d} \overline{\lambda}_s(x)} \quad (4)$$

where

$$\overline{\lambda}_k(x) = \prod_{s=1; s \neq k}^{m_d} \alpha_s(x) \quad (5)$$

The coefficients $\lambda_k(x)$ determine the contribution of each locally approximated function to the global function. The properties of the coefficients imply that the contribution of all other cut-HDMR expansions vanish except one when x is located on any cut line, plane, or higher dimensional ($\leq l$) subvolumes through that reference point, and then the MHDMR expansion reduces to single point cut-HDMR expansion. As mentioned above, the l-th order cut-HDMR approximation does not have error when x is located on these subvolumes. When m_d cut-HDMR expansions are used to construct a MHDMR expansion, the error free region in input x space is m_d times that for a single reference point cut-HDMR expansion, hence the accuracy will be improved. Therefore, first-order MHDMR approximations of the original implicit limit state function with m_d reference points can be expressed as

$$\tilde{g}(x) \cong \sum_{k=1}^{m_d} \lambda_k(x) \left[\sum_{i=1}^{N} g^k \left(d_1^k, \ldots, d_{i-1}^k, x_i, d_{i+1}^k, \ldots, d_N^k \right) - (N-1) g^k \left(d^k \right) \right]$$
$$(6)$$

The most important part of MHDMR approximation of the original implicit limit state function is identification of multiple reference points closer to the limit state function. Among the limit state function responses at all sample points, the most likelihood point is selected based on closeness to zero value, which indicates that particular sample point is close to the limit state function. The weight corresponding to each sample point is evaluated using the following weight function.

$$w^l = \exp\left(-\frac{g(c_1, \ldots, c_{i-1}, x_i, c_{i+1}, \ldots, c_N) - g(x)|_{\min}}{|g(x)|_{\min}|} \right) \quad (7)$$

3 Reliability Bounds

Let the N-dimensional input variables vector $x = \{x_1, x_2, \ldots, x_N\}$, which comprises of r number of random variables and f number of fuzzy variables be divided as, $x = \{x_1, x_2, \ldots, x_r, x_{r+1}, x_{r+2}, \ldots, x_{r+f}\}$ where the subvectors $\{x_1, x_2, \ldots, x_r\}$ and $\{x_{r+1}, x_{r+2}, \ldots, x_{r+f}\}$ respectively group the random variables and the fuzzy variables, with $N = r + f$. Then the first-order approximation of $\tilde{g}(x)$ can be divided into three parts, the first part with only the random variables, the second part with only the fuzzy variables and the third part is a constant which is the output response of the system evaluated at the reference point c, as follows

$$\tilde{g}(x) = \sum_{i=1}^{r} g\left(x_i, c^i\right) + \sum_{i=r+1}^{N} g\left(x_i, c^i\right) - (N-1) g(c) \tag{8}$$

The joint membership function of the fuzzy variables part is obtained using suitable transformation of the fuzzy variables and interval arithmetic algorithm. The fuzzy variables part of the nonlinear limit state function is expressed as a linear combination of intervening variables by the use of first-order HDMR approximation in order to apply an interval arithmetic algorithm, as follows

$$\sum_{i=r+1}^{N} g\left(x_i, c^i\right) = z_1 + z_2 + \cdots + z_f \tag{9}$$

where $z_i = (\beta_i x_i + \gamma_i)^\kappa$ is the relation between the intervening and the original variables with κ being order of approximation taking values $\kappa = 1$ for linear approximation, $\kappa = 2$ for quadratic approximation, $\kappa = 3$ for cubic approximation, and so on. The steps involved in the proposed method for failure probability estimation is as follows:

1. If $u = \{u_1, u_2, \ldots, u_r\}^T \in \Re^r$ is the standard Gaussian variable, let $u^{k*} = \{u_1^{k*}, u_2^{k*}, \ldots, u_r^{k*}\}^T$ be the MPP or design point, determined by a standard nonlinear constrained optimization. The MPP has a distance β_{HL}, which is commonly referred to as the Hasofer–Lind reliability index. Construct an orthogonal matrix $R \in \Re^{r \times r}$ whose rth column is $\alpha^{k*} = u^{k*}/\beta_{\text{HL}}$, i.e., $R = [R_1 | \alpha^{k*}]$ where $R_1 \in \Re^{r \times r-1}$ satisfies $\alpha^{k*T} R_1 = 0 \in \Re^{1 \times r-1}$. The matrix R can be obtained, for example, by Gram–Schmidt orthogonalization. For an orthogonal transformation $u = R v$.

2. Let $v = \{v_1, v_2, \ldots, v_r\}^T \in \Re^r$ be the rotated Gaussian space with the associated MPP $v^{k*} = \{v_1^{k*}, v_2^{k*}, \ldots, v_r^{k*}\}^T$. The transformed limit state function $g(v)$ therefore maps the random variables along with the values of the constant part and the fuzzy variables part at each α-cut, into rotated Gaussian space v.

First-order HDMR approximation of $g(v)$ in rotated Gaussian space v with $v^{k*} = \{v_1^{k*}, v_2^{k*}, \ldots, v_r^{k*}\}^T$ as reference point can be represented as follows:

$$
\tilde{g}^k(v) \equiv g^k(v_1, v_2, \ldots, v_r)
$$

$$
= \sum_{i=1}^{r} g^k\left(v_1^{k*}, \ldots, v_{i-1}^{k*}, v_i, v_{i+1}^{k*}, \ldots, v_r^{k*}\right) - (r-1) g\left(v^{k*}\right) \qquad (10)
$$

3. In addition to the MPP as the chosen reference point, the accuracy of first-order HDMR approximation in Eq. (10) may depend on the orientation of the first $r-1$ axes. In the present work, the orientation is defined by the matrix $\mathbf{R}$. In Eq. (10), the terms $g^k(v_1^{k*}, \ldots, v_{i-1}^{k*}, v_i, v_{i+1}^{k*}, \ldots, v_r^{k*})$ are the individual component functions and are independent of each other. Equation (10) can be rewritten as,

$$
\tilde{g}^k(v) = a^k + \sum_{i=1}^{r} g^k\left(v_i, v^{k*i}\right) \qquad (11)
$$

where $a^k = -(r-1)g(v^{k*})$.

4. New intermediate variables are defined as

$$
y_i^k = g^k\left(v_i, v^{k*i}\right) \qquad (12)
$$

The purpose of these new variables is to transform the approximate function into the following form

$$
\tilde{g}^k(v) = a^k + y_1^k + y_2^k + \cdots + y_r^k \qquad (13)
$$

5. Due to rotational transformation in v-space, component functions y_i^k in Eq. (13) are expected to be linear or weakly nonlinear function of random variables v_i. In this work both linear and quadratic approximations of y_i^k are considered.

6. Let $y_i^k = b_i + c_i\, v_i$ and $y_i^k = b_i + c_i\, v_i + e_i\, v_i^2$ be the linear and quadratic approximations, where coefficients $b_i, c_i, e_i \in \Re$ are obtained by least-squares approximation from exact or numerically simulated conditional responses $\left\{g^k\left(v_i^1, v^{k*i}\right), g^k\left(v_i^2, v^{k*i}\right), \cdots, g^k\left(v_i^n, v^{k*i}\right)\right\}^T$ at n sample points along the variable axis v_i. Then Eq. (13) results in

$$
\tilde{g}^k(v) \equiv a^k + y_1^k + y_2^k + \cdots + y_r^k
$$

$$
= a^k + \sum_{i=1}^{r} (b_i + c_i\, v_i) \qquad (14)
$$

and

$$\tilde{g}^k(v) \equiv a^k + y_1^k + y_2^k + \cdots + y_r^k$$

$$= a^k + \sum_{i=1}^{r} \left(b_i + c_i \, v_i + e_i \, v_i^2 \right) \tag{15}$$

7. The global approximation is formed by blending of locally constructed individual first-order HDMR approximations in the rotated Gaussian space at different identified reference points using the coefficients λ_k.

$$\tilde{g}(v) = \sum_{k=1}^{m_d} \lambda_k \tilde{g}^k(v) \tag{16}$$

8. Since v_i follows standard Gaussian distribution, marginal density of the intermediate variables y_i can be easily obtained by simple transformation.

$$p_{Y_i}(y_i) = p_{V_i}(v_i) |dv_i/dy_i| \tag{17}$$

9. Now the approximation is a linear combination of the intermediate variables y_i. Therefore, the joint density of $\tilde{g}(v)$, which is the convolution of the marginal density of the intervening variables y_i, is expressed as follows:

$$p_{\tilde{G}}(\tilde{g}) = p_{Y_1}(y_1) * p_{Y_2}(y_2) * \cdots * p_{Y_r}(y_r) \tag{18}$$

10. Applying FFT on both sides of Eq. (18), leads to

$$\text{FFT}\left[p_{\tilde{G}}(\tilde{g})\right] = \text{FFT}\left[p_{Y_1}(y_1)\right] \cdot \text{FFT}\left[p_{Y_2}(y_2)\right] \ldots \text{FFT}\left[p_{Y_r}(y_r)\right] \tag{19}$$

11. By applying inverse FFT on both sides of Eq. (19), joint density of $\tilde{g}(v)$ is obtained.
12. The probability of failure is given by the following equation

$$P_F = \int_{-\infty}^{0} p_{\tilde{G}}(\tilde{g}) \, d\tilde{g} \tag{20}$$

13. The membership function of failure probability can be obtained by repeating the above procedure at all confidence levels of the fuzzy variables part.

4 Numerical Examples

To evaluate the accuracy and the efficiency of the present method, comparisons of the estimated failure probability bounds, both by performing the convolution using FFT in conjunction with linear and quadratic approximations and MCS on the global approximation, have been made with that obtained using direct MCS. When comparing computational efforts by various methods in evaluating the failure probability, the number of original limit state function evaluations is chosen as the primary comparison tool in this chapter. This is because of the fact that, the number of function evaluations indirectly indicates the CPU time usage. For direct MCS, the number of original function evaluations is same as the sampling size. While evaluating the failure probability through direct MCS, CPU time is more because it involves a number of repeated actual finite-element analyses.

4.1 Parabolic Performance Function

The limit state function considered is a parabola of the form:

$$g(x) = -x_1^2 - x_2 + x_3 \tag{21}$$

where x_1 and x_2 are assumed to be independent standard normal variables, and x_3 is a fuzzy variable with triangular membership function [5.0, 7.0, 9.0]. The initial reference point c is taken as respectively the mean values and nominal values of the random and fuzzy variables.

The first-order HDMR approximation, which is constructed over the initial reference point, is divided into two parts, one with only the random variables, and the other with the fuzzy variables. The joint membership function of the fuzzy part of limit state function is obtained using suitable transformation of the fuzzy variables. In this example, the joint membership function is same as the membership function of the fuzzy variable x_3. As shown in Fig. 1, the limit state function is symmetric about x_2 for given value of x_3, and has two design points. The two actual design points of the limit state function shown in Fig. 1, obtained using recursive quadratic programming (RQP) algorithm are (2.54, 0.49) and (−2.54, 0.49) with reliability indices $\beta_1 = \beta_2 = 2.588$.

After identification of the two reference points (2, 0) and (−2, 0), local individual first-order HDMR approximations of the original limit state function are constructed at the two reference points by deploying $n = 5$ sample points along each of the variable axis. Local approximations of the original limit state function are blended together to form global approximation. The bounds of the failure probability are obtained both by performing the convolution using FFT in conjunction with linear and quadratic approximations, and MCS on the global approximation.

Fig. 1 Limit state function

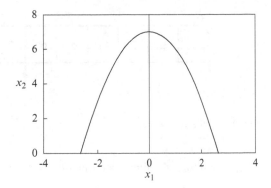

Fig. 2 Membership function of failure probability

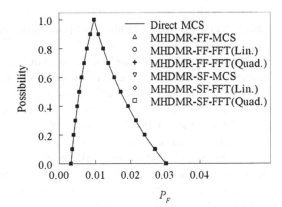

Figure 2 shows the membership function of the failure probability estimated both by performing the convolution using FFT, and MCS on the global approximation, as well as that obtained using direct MCS. In addition, effect of SF sampling scheme on the estimated membership function of the failure probability is studied. The effect of the number of sample points is studied by varying n from 3 to 9. It is observed that $n = 7$ provides the optimum number of function calls with acceptable accuracy in evaluating the failure probability with the present method.

4.2 Cantilever Steel Beam

A cantilever steel beam of 1.0 m with cross-sectional dimensions of (0.1×0.01) m is considered as shown in Fig. 3, to examine the accuracy and efficiency of the proposed method for the membership function of failure probability estimation. The beam is subjected to an in-plane moment at the free end and a concentrated load at 0.4 m from the free end. The structure is assumed to have failed if the square of the von Mises stress at the support (at A in Fig. 3) exceeds specified threshold

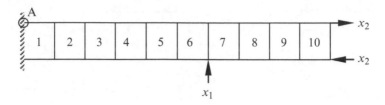

Fig. 3 Cantilever steel beam

$V_{\max}$. Therefore, the limit state function is defined as

$$g(x) = V_{\max} - V(x) \tag{22}$$

where $V(x)$ is the square of the von Mises stress, expressed as a quadratic operator on the stress vector.

In this example, loads x_1 and x_2, modulus of elasticity of the beam E and threshold quantity $V_{\max}$ are taken as uncertain variables. The variations of E and $V_{\max}$ are expressed as $E = E_0(1 + \varepsilon x_3)$ and $V_{\max} = V_{\max 0}(1 + \varepsilon x_4)$. Here, ε is small deterministic quantity representing the coefficient of variation of the random variables and are taken to equal to 0.05, $E_0 = 2 \times 10^5$ N/m^2 denotes the deterministic component of modulus of elasticity and $V_{\max 0} = 6.15 \times 10^9$ N/m^2 denotes the deterministic component of threshold quantity. All variables are assumed to be independent. The mean values of random variables x_1 and x_2 are 1 and 0 respectively, with the standard deviation of 1. The variables x_3 and x_4 are triangular fuzzy numbers with [0.0 2.0 4.0] and [0.0, 0.1, 0.2] respectively.

The limit state function given in Eq. (22) is approximated using first-order HDMR by deploying $n = 5$ sample points along each of the variable axis and taking respectively the mean values and nominal values of the random and fuzzy variables as initial reference points (1.0, 0.0, 2.0, 0.1). The approximated limit state function is divided into two parts, one with only the random variables along with the value of the constant part, and the other with the fuzzy variables. The joint membership function of the fuzzy part of approximated limit state function is obtained using suitable transformation of the fuzzy variables. Using FF sampling scheme the sample point $d = (1, -2)$ is identified as reference point closer to the limit state function producing maximum weight.

Figure 4 shows the membership function of the failure probability estimated both by performing the convolution using FFT in conjunction with linear and quadratic approximations, and MCS on the global approximation, as well as that obtained using direct MCS. It is observed that $n = 7$ provides the optimum number of function calls with acceptable accuracy in evaluating the failure probability with the present method.

Fig. 4 Membership function of failure probability

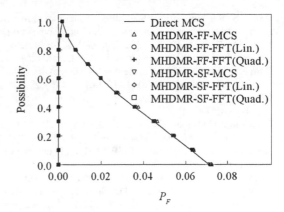

5 Summary and Conclusions

A novel uncertain analysis for estimating the membership function of the reliability of structural systems involving multiple design points in the presence of mixed uncertain variables is presented in this chapter. The method involves MHDMR technique for the limit state function approximation, transformation technique to obtain the contribution of the fuzzy variables to the convolution integral and FFT for solving the convolution integral. Weight function is adopted for identification of multiple reference points closer to the limit surface. Using the bounds of the fuzzy variables part at each confidence level along with the constant part and the random variables part, the joint density functions are obtained by (i) identifying the reference points closer to the limit state function and (ii) blending of locally constructed individual first-order HDMR approximations in the rotated Gaussian space at different identified reference points to form global approximation, and (iii) performing the convolution using FFT, which upon integration yields the bounds of the failure probability. As an alternative the bounds of the failure probability are estimated by performing MCS on the global approximation in the original space, obtained by blending of locally constructed individual first-order HDMR approximations of the original limit state function at different identified reference points. An optimum number of sample points must be chosen in approximation of the original limit state function.

References

Adduri PR, Penmetsa RC (2008) Confidence bounds on component reliability in the presence of mixed uncertain variables. Int J Mech Sci 50(3):481–489

Au SK, Papadimitriou C, Beck JL (1999) Reliability of uncertain dynamical systems with multiple design points. Struct Saf 21:113–133

Balu AS, Rao BN (2012a) High dimensional model representation based formulations for fuzzy finite element analysis of structures. Finite Elem Anal Des 50:217–230

Balu AS, Rao BN (2012b) Multicut-HDMR for structural reliability bounds estimation under mixed uncertainties. Comput Aided Civ Infrastruct Eng 27(6):419–438

Balu AS, Rao BN (2014) Efficient assessment of structural reliability in presence of random and fuzzy uncertainties. ASME J Mech Des 136(5):051008. doi:10.1115/1.4026650

Breitung K (1984) Asymptotic approximations for multinormal integrals. ASCE J Eng Mech 110(3):357–366

Briabant V, Oudshoorn A, Boyer C, Delcroix F (1999) Nondeterministic possibilistic approaches for structural analysis and optimal design. AIAA J 37(10):1298–1303

Der Kiureghian A, Dakessian T (1998) Multiple design points in first and second order reliability. Struct Saf 20(1):37–49

Penmetsa RC, Grandhi RV (2003) Uncertainty propagation using possibility theory and function approximations. Mech Base Des Struct Mach 81(15):1567–1582

Rabitz H, Alis OF, Shorter J, Shim K (1999) Efficient input-output model representations. Comput Phys Commun 117(1–2):11–20

Rackwitz R (2001) Reliability analysis—a review and some perspectives. Structural Safety 23(4):365–395

Rao BN, Chowdhury R (2008) Probabilistic analysis using high dimensional model representation and fast Fourier transform. Int J Comput Meth Eng Sci Mech 9(6):342–357

Sakamoto J, Mori Y, Sekioka T (1997) Probability analysis method using fast Fourier transform and its application. Structural Safety 19(1):21–36

Approximate Probability Density Function Solution of Multi-Degree-of-Freedom Coupled Systems Under Poisson Impulses

H.T. Zhu

Abstract A solution procedure is proposed to approximate the probability density function (PDF) solution of high-dimensional non-linear systems under Poisson impulses. The PDF solution yields the generalized Fokker–Planck–Kolmogorov (FPK) equation. First a state-space-split method is proposed to reduce the high-dimensional generalized FPK equation to a low dimensional equation. After that, the exponential–polynomial closure method is further adopted to solve the reduced FPK equation for the PDF solution. In order to show the effectiveness of the proposed solution procedure, a two-degree-of-freedom coupled pitch–roll ship motion system and a 10-degree-of-freedom mass–spring–damper system are investigated, respectively. Compared to the simulated results, the proposed solution procedure is effective to obtain the PDF solution, especially in the tail region which is very important for reliability analysis.

1 Introduction

The accurate estimation on the random response of systems is a critical issue in the field of science and engineering. In this field, the excitation model for wind, sea wave, earthquake ground motion is mostly treated as either Gaussian white noise or Poisson white noise (i.e., Poisson impulses). Under such random excitations, the probability density function (PDF) and statistical moment of the system have to be obtained in reliability analysis. However, the PDF solution is much more difficult to be accessed and much work is devoted to the evaluation on the response of

H.T. Zhu (✉)
State Key Laboratory of Hydraulic Engineering Simulation and Safety,
Tianjin University, Tianjin 300072, China
e-mail: htzhu@tju.edu.cn

S. Kadry and A. El Hami (eds.), *Numerical Methods for Reliability and Safety Assessment: Multiscale and Multiphysics Systems*, DOI 10.1007/978-3-319-07167-1_19,

single-degree-of-freedom (SDOF) systems. Even in the case of the SDOF systems, only a few stationary PDF solutions have been obtained in some special cases by solving the associated Fokker–Planck–Kolmogorov (FPK) equation (Caughey and Ma 1982; Dimentberg 1982; Lin and Cai 1988; Vasta 1995; Proppe 2002a, 2003). Most problems need some approximation methods, such as perturbation method (Roberts 1972; Cai and Lin 1992), finite element method (Langley 1985), Petrov–Galerkin method (Köylüoğlu et al. 1994), cell-to-cell mapping (path integration) technique (Köylüoğlu et al. 1995; Di Paola and Santoro 2008), and finite difference approach (Wojtkiewicz et al. 1999). Comparatively, the relevant problem of high-dimensional systems is more challenging (Cai and Lin 1996). Furthermore, if some type of non-linearity is considered, the problem becomes much more complicated. For the problem of multi-degree-of-freedom (MDOF) systems, equivalent linearization (EQL) method was widely employed for obtaining statistical moments (Caughey 1963; Atalik and Utku 1976; Spanos 1981; Proppe 2002b; Roberts and Spanos 2003; Socha 2008). The accuracy of the EQL method heavily is limited to the non-linearity degree and excitation type. Another conventional technique is Monte Carlo simulation which is a versatile method for MDOF systems (Shinozuka 1972; Muscolino et al. 2003; Proppe et al. 2003). When the tail of the PDF is considered, a high computational effort is needed with Monte Carlo simulation. Besides, a C-type Gram-Charlier series expansion method was also developed for approximating the response PDF of MDOF structural systems under Poisson white noise (Muscolino and Ricciardi 1999). In this method, the evaluation of the response cumulants up to a given order is needed and the method was applied to the case of linear systems. Recently, the PDF of MDOF dissipated Hamiltonian systems under Poisson white noises were investigated with a perturbation method (Wu and Zhu 2008a, 2008b). The numerical examples on two coupled non-linear oscillators were studied in numerical analysis.

From the above description, the PDF solution of high-dimensional non-linear systems under random excitation is less addressed, especially for the tail of the PDF solution in the problem of high-dimensional systems. This chapter adopts a solution procedure to approximate the stationary PDF solution of high-dimensional non-linear systems under Poisson impulses. First Sect. 2 introduces the state-space-split (SSS) method to reduce the high-dimensional generalized FPK equation to a low-dimensional equation (Er 2011; Er and Iu 2011). After that, the exponential–polynomial closure (EPC) method is further adopted to solve the reduced FPK equation for the PDF solution in Sect. 3 (Er 1998; Zhu et al. 2009). In order to evaluate the effectiveness of the solution procedure, a two-degree-of-freedom coupled pitch–roll ship motion system and a 10-degree-of-freedom non-linear system are investigated in Sect. 4. The applications presented in Sect. 4 are gradual. The first one is more theoretical while the second one is more practical. Comparison with the simulated results is made in the same section. According to the numerical analysis, some conclusions are drawn in the last section.

2   State-Space-Split Method

A non-linear MDOF system under Poisson impulses can be expressed as

$$\frac{d}{dt}X_i = f_i(\mathbf{X}) + W_i(t), \quad (i = 1, 2, \dots, n_\mathbf{x}) \tag{1}$$

where $n_\mathbf{x}$ is the number of the components of $\mathbf{X}$; $\mathbf{X}$ is a vector containing X_i and $\dot{X}_i$, the responses of the non-linear system; $\dot{X}_i$ is the first derivative with respect to time t. $f_i(\mathbf{X})$ is generally a non-linear function of $\mathbf{X}$, and it is assumed to be deterministic and of polynomial type. $W_i(t)$ is the ith Poisson impulse process which is formulated as follows

$$W_i(t) = \sum_{k=1}^{N(T)} Y_k \delta(t - \tau_k) \tag{2}$$

where $N(T)$ is the total number of impulses that arrive in the time interval $(-\infty, T]$. Y_k is the amplitude of the kth impulse arriving at time τ_k for $W_i(t)$. $\delta(\cdot)$ is the Dirac delta function. In this chapter, $N(T)$ is a count process yielding the Poisson law with a constant impulse arrival rate λ_i. The impulse amplitudes Y_k are independent identically distributed (i.i.d.) random variables with zero mean for $W_i(t)$, and also independent of the impulse arrival time τ_k.

Subjected to Poisson impulses, the PDF solution $p(\mathbf{x}, t)$ is governed by the generalized FPK equation. In this chapter, only the stationary response is considered and the stationary PDF solution $p(\mathbf{x})$ is determined by the following generalized FPK equation (Wu and Zhu 2008a, 2008b)

$$0 = -\frac{\partial}{\partial x_i} \{f_i(\mathbf{x}) p(\mathbf{x})\} + \frac{1}{2!} \lambda_i \delta_{ij} E[Y_i Y_j] \frac{\partial^2 p(\mathbf{x})}{\partial x_i \partial x_j}$$

$$-\frac{1}{3!} \lambda_i \delta_{ij} \delta_{ik} E[Y_i Y_j Y_k] \frac{\partial^3 p(\mathbf{x})}{\partial x_i \partial x_j \partial x_k}$$

$$+\frac{1}{4!} \lambda_i \delta_{ij} \delta_{ik} \delta_{il} E[Y_i Y_j Y_k Y_l] \frac{\partial^4 p(\mathbf{x})}{\partial x_i \partial x_j \partial x_k \partial x_l} + \cdots \tag{3}$$

where Y_i is a random amplitude of Poisson impulses associated with $W_i(t)$; δ_{ij} is the Kronecker delta. Furthermore, $p(\mathbf{x})$ is assumed to satisfy the below boundary conditions

$$\lim_{x_i \to \pm\infty} p(\mathbf{x}) = 0 \tag{4}$$

and

$$\lim_{x_i \to \pm\infty} \left\{ - f_i\left(\mathbf{x}\right) p\left(\mathbf{x}\right) + \frac{1}{2!}\lambda_i \delta_{ij} E\left[Y_i Y_j\right] \frac{\partial p\left(\mathbf{x}\right)}{\partial x_j} \right.$$

$$- \frac{1}{3!}\lambda_i \delta_{ij}\delta_{ik} E\left[Y_i Y_j Y_k\right] \frac{\partial^2 p\left(\mathbf{x}\right)}{\partial x_j \partial x_k}$$

$$\left. + \frac{1}{4!}\lambda_i \delta_{ij}\delta_{ik}\delta_{il} E\left[Y_i Y_j Y_k Y_l\right] \frac{\partial^3 p\left(\mathbf{x}\right)}{\partial x_j \partial x_k \partial x_l} + \cdots \right\} = 0 \qquad (5)$$

It means that the probability and probability flow vanish at infinite boundary.

According to the state-space-split (SSS) method (Er 2011; Er and Iu 2011), the state variables $\mathbf{x}$ are divided into two subspaces $\mathbf{x}_1 \in R^{n_{x1}}$ and $\mathbf{x}_2 \in R^{n_{x2}}$. $\mathbf{x} = \{\mathbf{x}_1, \mathbf{x}_2\}$. In analysis of a second-order dynamical stochastic system, for instance, $\mathbf{x}_1$ contains pairs of displacement and its first derivative (i.e., the corresponding velocity).

The PDF of $\mathbf{x}_1$ is defined as $p_1(\mathbf{x}_1)$, which is obtained by integrating (3) over the subspace of $R^{n_{x2}}$ as

$$\int_{R^{n_{x2}}} \left\{ -\frac{\partial}{\partial x_i}\left[f_i\left(\mathbf{x}\right) p\left(\mathbf{x}\right)\right] + \frac{1}{2!}\lambda_i \delta_{ij} E\left[Y_i Y_j\right] \frac{\partial^2 p\left(\mathbf{x}\right)}{\partial x_j \partial x_j} \right.$$

$$- \frac{1}{3!}\lambda_i \delta_{ij}\delta_{ik} E\left[Y_i Y_j Y_k\right] \frac{\partial^3 p\left(\mathbf{x}\right)}{\partial x_i \partial x_j \partial x_k}$$

$$\left. + \frac{1}{4!}\lambda_i \delta_{ij}\delta_{ik}\delta_{il} E\left[Y_i Y_j Y_k Y_l\right] \frac{\partial^4 p\left(\mathbf{x}\right)}{\partial x_i \partial x_j \partial x_k \partial x_l} + \cdots \right\} d\mathbf{x}_2 = 0 \qquad (6)$$

where $x_i, x_j, x_k, x_l, \cdots \in R^{n_x}$.

Considering (4), (5), and (6) can be further reduced as

$$\int_{R^{n_{x2}}} \left\{ -\frac{\partial}{\partial x_i}\left[f_i\left(\mathbf{x}\right) p\left(\mathbf{x}\right)\right] + \frac{1}{2!}\lambda_i \delta_{ij} E\left[Y_i Y_j\right] \frac{\partial^2 p\left(\mathbf{x}\right)}{\partial x_j \partial x_j} \right.$$

$$- \frac{1}{3!}\lambda_i \delta_{ij}\delta_{ik} E\left[Y_i Y_j Y_k\right] \frac{\partial^3 p\left(\mathbf{x}\right)}{\partial x_i \partial x_j \partial x_k}$$

$$\left. + \frac{1}{4!}\lambda_i \delta_{ij}\delta_{ik}\delta_{il} E\left[Y_i Y_j Y_k Y_l\right] \frac{\partial^4 p\left(\mathbf{x}\right)}{\partial x_i \partial x_j \partial x_k \partial x_l} + \cdots \right\} d\mathbf{x}_2 = 0 \qquad (7)$$

where $x_i, x_j, x_k, x_l, \cdots \in R^{n_{x1}}$.

In (7), only $x_i, x_j, x_k, x_l, \cdots \in R^{n \times 1}$ are retained from (6). Consequently,

$$-\frac{\partial}{\partial x_i} \int_{R^{n \times 2}} [f_i(\mathbf{x}) p(\mathbf{x})] d\mathbf{x}_2 + \frac{1}{2!} \lambda_i \delta_{ij} E[Y_i Y_j] \frac{\partial^2}{\partial x_j \partial x_j} \int_{R^{n \times 2}} p(\mathbf{x}) d\mathbf{x}_2$$

$$-\frac{1}{3!} \lambda_i \delta_{ij} \delta_{ik} E[Y_i Y_j Y_k] \frac{\partial^3}{\partial x_i \partial x_j \partial x_k} \int_{R^{n \times 2}} p(\mathbf{x}) d\mathbf{x}_2$$

$$+\frac{1}{4!} \lambda_i \delta_{ij} \delta_{ik} \delta_{il} E[Y_i Y_j Y_k Y_l] \frac{\partial^4}{\partial x_i \partial x_j \partial x_k \partial x_l} \int_{R^{n \times 2}} p(\mathbf{x}) d\mathbf{x}_2 + \cdots = 0 \quad (8)$$

where $x_i, x_j, x_k, x_l, \cdots \in R^{n \times 1}$.
and

$$\int_{R^{n \times 2}} p(\mathbf{x}) d\mathbf{x}_2 = p_1(\mathbf{x}_1) \quad (9)$$

The non-linear function in (1) can be further expressed as

$$f_i(\mathbf{x}) = f_i^{\mathrm{I}}(\mathbf{x}_1) + f_i^{\mathrm{II}}(\mathbf{x}) \quad (10)$$

Equation (8) is reformulated as

$$-\frac{\partial}{\partial x_i} \left[f_i^{\mathrm{I}}(\mathbf{x}_1) p_1(\mathbf{x}_1) + \int_{R^{n \times 2}} f_i^{\mathrm{II}}(\mathbf{x}) p(\mathbf{x}) d\mathbf{x}_2 \right] + \frac{1}{2!} \lambda_i \delta_{ij} E[Y_i Y_j] \frac{\partial^2 p_1(\mathbf{x}_1)}{\partial x_j \partial x_j}$$

$$-\frac{1}{3!} \lambda_i \delta_{ij} \delta_{ik} E[Y_i Y_j Y_k] \frac{\partial^3 p_1(\mathbf{x}_1)}{\partial x_i \partial x_j \partial x_k}$$

$$+\frac{1}{4!} \lambda_i \delta_{ij} \delta_{ik} \delta_{il} E[Y_i Y_j Y_k Y_l] \frac{\partial^4 p_1(\mathbf{x}_1)}{\partial x_i \partial x_j \partial x_k \partial x_l} + \cdots = 0$$

$$(11)$$

where $x_i, x_j, x_k, x_l, \cdots \in R^{n \times 1}$.

In general, $f_i^{\mathrm{II}}(\mathbf{x})$ is a function of a few state variables and it can be expressed as $f_i^{\mathrm{II}}(\mathbf{x}_1, \mathbf{z}_k), \mathbf{z}_k \in R^{n_{z_k}} \subset R^{n \times 2}$

$$-\frac{\partial}{\partial x_i} \left[f_i^{\mathrm{I}}(\mathbf{x}_1) p_1(\mathbf{x}_1) + \int_{R^{n_{z_k}}} f_i^{\mathrm{II}}(\mathbf{x}_1, \mathbf{z}_k) p_k(\mathbf{x}_1, \mathbf{z}_k) dz_k \right]$$

$$+\frac{1}{2!} \lambda_i \delta_{ij} E[Y_i Y_j] \frac{\partial^2 p_1(\mathbf{x}_1)}{\partial x_j \partial x_j}$$

$$-\frac{1}{3!} \lambda_i \delta_{ij} \delta_{ik} E[Y_i Y_j Y_k] \frac{\partial^3 p_1(\mathbf{x}_1)}{\partial x_i \partial x_j \partial x_k}$$

$$+\frac{1}{4!} \lambda_i \delta_{ij} \delta_{ik} \delta_{il} E[Y_i Y_j Y_k Y_l] \frac{\partial^4 p_1(\mathbf{x}_1)}{\partial x_i \partial x_j \partial x_k \partial x_l} + \cdots = 0 \quad (12)$$

where $x_i, x_j, x_k, x_l, \cdots \in R^{n \times 1}$.

$p_k(\mathbf{x}_1, \mathbf{z}_k)$ is the joint PDF of $\mathbf{x}_1$ and $\mathbf{z}_k$. It can be expressed as

$$p_k\left(\mathbf{x}_1, \mathbf{z}_k\right) = p_1\left(\mathbf{x}_1\right) q_k\left(\mathbf{z}_k \middle| \mathbf{x}_1\right) \tag{13}$$

$q_k(\mathbf{z}_k|\mathbf{x}_1)$ is the conditional PDF of $\mathbf{z}_k$ given with $\mathbf{x}_1$.
Accordingly,

$$-\frac{\partial}{\partial x_i}\left\{\left[f_i^{\mathrm{I}}\left(\mathbf{x}_1\right) + \int_{R^{n_{z_k}}} f_i^{\mathrm{II}}\left(\mathbf{x}_1, \mathbf{z}_k\right) q_k\left(\mathbf{z}_k \middle| \mathbf{x}_1\right) dz_k\right] p_1\left(\mathbf{x}_1\right)\right\}$$

$$+ \frac{1}{2!}\lambda_i \delta_{ij} E\left[Y_i Y_j\right] \frac{\partial^2 p_1\left(\mathbf{x}_1\right)}{\partial x_j \partial x_j}$$

$$- \frac{1}{3!}\lambda_i \delta_{ij}\delta_{ik} E\left[Y_i Y_j Y_k\right] \frac{\partial^3 p_1\left(\mathbf{x}_1\right)}{\partial x_i \partial x_j \partial x_k}$$

$$+ \frac{1}{4!}\lambda_i \delta_{ij}\delta_{ik}\delta_{il} E\left[Y_i Y_j Y_k Y_l\right] \frac{\partial^4 p_1\left(\mathbf{x}_1\right)}{\partial x_i \partial x_j \partial x_k \partial x_l} + \cdots = 0 \tag{14}$$

where $x_i, x_j, x_k, x_l, \cdots \in R^{n_{x_1}}$.

$q_k(\mathbf{z}_k|\mathbf{x}_1)$ is unknown and it can be approximated by the result of EQL method.

$$-\frac{\partial}{\partial x_i}\left\{\left[f_i^{\mathrm{I}}\left(\mathbf{x}_1\right) + \int_{R^{n_{z_k}}} f_i^{\mathrm{II}}\left(\mathbf{x}_1, \mathbf{z}_k\right) \overline{q}_k\left(\mathbf{z}_k \middle| \mathbf{x}_1\right) dz_k\right] \tilde{p}_1\left(\mathbf{x}_1\right)\right\}$$

$$+ \frac{1}{2!}\lambda_i \delta_{ij} E\left[Y_i Y_j\right] \frac{\partial^2 \tilde{p}_1\left(\mathbf{x}_1\right)}{\partial x_j \partial x_j}$$

$$- \frac{1}{3!}\lambda_i \delta_{ij}\delta_{ik} E\left[Y_i Y_j Y_k\right] \frac{\partial^3 \tilde{p}_1\left(\mathbf{x}_1\right)}{\partial x_i \partial x_j \partial x_k}$$

$$+ \frac{1}{4!}\lambda_i \delta_{ij}\delta_{ik}\delta_{il} E\left[Y_i Y_j Y_k Y_l\right] \frac{\partial^4 \tilde{p}_1\left(\mathbf{x}_1\right)}{\partial x_i \partial x_j \partial x_k \partial x_l} + \cdots = 0 \tag{15}$$

where $x_i, x_j, x_k, x_l, \infty \cdots \in R^{n_{x_1}}$ and $\overline{q}_k\left(\mathbf{z}_k \middle| \mathbf{x}_1\right)$ is the conditional PDF given by EQL method and $\tilde{p}_1\left(\mathbf{x}_1\right)$ is the approximate PDF solution of $\mathbf{x}_1$.

Let

$$\tilde{f}_i\left(\mathbf{x}_1\right) = f_i^{\mathrm{I}}\left(\mathbf{x}_1\right) + \int_{R^{n_{z_k}}} f_i^{\mathrm{II}}\left(\mathbf{x}_1, \mathbf{z}_k\right) \overline{q}_k\left(\mathbf{z}_k \middle| \mathbf{x}_1\right) d\mathbf{z}_k \tag{16}$$

Finally, the high-dimensional generalized FPK equation can be approximated by a low-dimensional FPK equation as follows

$$- \frac{\partial}{\partial x_i} \left\{ \tilde{f}_i \left(\mathbf{x}_1 \right) p_1 \left(\mathbf{x}_1 \right) \right\} + \frac{1}{2!} \lambda_i \delta_{ij} E \left[Y_i Y_j \right] \frac{\partial^2 \tilde{p}_1 \left(\mathbf{x}_1 \right)}{\partial x_j \partial x_j}$$

$$- \frac{1}{3!} \lambda_i \delta_{ij} \delta_{ik} E \left[Y_i Y_j Y_k \right] \frac{\partial^3 \tilde{p}_1 \left(\mathbf{x}_1 \right)}{\partial x_i \partial x_j \partial x_k}$$

$$+ \frac{1}{4!} \lambda_i \delta_{ij} \delta_{ik} \delta_{il} E \left[Y_i Y_j Y_k Y_l \right] \frac{\partial^4 \tilde{p}_1 \left(\mathbf{x}_1 \right)}{\partial x_i \partial x_j \partial x_k \partial x_l} + \cdots = 0 \qquad (17)$$

where $x_i, x_j, x_k, x_l, \cdots \in R^{n_{x_1}}$.

If $\mathbf{x}_1$ only includes one or two state variables, e.g., displacement and velocity in some degree of freedom, the approximate FPK equation is two-dimensional FPK equation for a single-degree-of-freedom system. According to Zhu et al. (2009), the exponential–polynomial closure (EPC) method can be adopted to solve the approximate FPK equation.

3 Exponential–Polynomial Closure Method

When $\mathbf{x}_1$ contains a pair of displacement and velocity of some degree of freedom, i.e., $\mathbf{x}_1 = \{x, \dot{x}\}^{\mathrm{T}}$, the reduced FPK equation is two-dimensional FPK equation for a single-degree-of-freedom system.

Letting $x = x_1$, $\dot{x} = x_2$ and considering the presence of one external Poisson white noise in that degree of freedom, the above Eq. (17) is formulated as

$$- x_2 \frac{\partial \tilde{p}}{\partial x_1} - \frac{\partial}{\partial x_2} \left\{ \tilde{f}_2 \left(x_1, x_2 \right) \tilde{p} \right\} + \frac{1}{2!} \lambda E \left[Y^2 \right] \frac{\partial^2 \tilde{p}}{\partial x_2^2}$$

$$- \frac{1}{3!} \lambda E \left[Y^3 \right] \frac{\partial^3 \tilde{p}}{\partial x_2^3} + \frac{1}{4!} \lambda E \left[Y^4 \right] \frac{\partial^4 \tilde{p}}{\partial x_2^4} = 0 \qquad (18)$$

where $\tilde{f}_1 \left(x_1, x_2 \right) = x_2$; λ is the impulse arrival rate of the Poisson white noise; Y is the amplitude of the kth impulse of the Poisson white noise. It should be noted that x_1 and x_2 are two new redefined variables denoting x and $\dot{x}$, respectively. They do not represent the variables in (1) any more in the following content. Furthermore, only the terms up to fourth-order derivative is retained for (17) for analysis on assumption that the contribution of high order terms is small to the whole equation.

The approximate PDF $\tilde{p} \left(x_1, x_2; \mathbf{a} \right)$ solution to (18) is assumed to be

$$\tilde{p} \left(x_1, x_2; \mathbf{a} \right) = C e^{Q_n \left(x_1, x_2; \mathbf{a} \right)} \qquad (19)$$

where C is a normalization constant. $\mathbf{a}$ is an unknown parameter vector containing N_p entries. The polynomial $Q_n(x_1, x_2; \mathbf{a})$ is formulated as

$$Q_n(x_1, x_2; \mathbf{a}) = \sum_{i=1}^{n} \sum_{j=0}^{i} a_{ij} x_1^{i-j} x_2^{j} \tag{20}$$

which is a complete nth degree polynomial of x_1 and x_2.

Because $\tilde{p}(x_1, x_2; \mathbf{a})$ is only an approximation, substituting $\tilde{p}(x_1, x_2; \mathbf{a})$ into (18) leads to the following residual error

$$\begin{aligned}
\Delta(x_1, x_2; \mathbf{a}) = &-x_2 \frac{\partial \tilde{p}}{\partial x_1} - \frac{\partial}{\partial x_2} \left\{ \tilde{f}_2(x_1, x_2) \tilde{p} \right\} \\
&+ \frac{1}{2!} \lambda E\left[Y^2\right] \frac{\partial^2}{\partial x_2^2} \tilde{p} \\
&- \frac{1}{3!} \lambda E\left[Y^3\right] \frac{\partial^3}{\partial x_2^3} \tilde{p} + \frac{1}{4!} \lambda E\left[Y^4\right] \frac{\partial^4}{\partial x_2^4} \tilde{p}
\end{aligned} \tag{21}$$

Substituting (19) into (21) leads to

$$\begin{aligned}
\Delta(x_1, x_2; \mathbf{a}) = &-x_2 \frac{\partial \tilde{p}}{\partial x_1} - \frac{\partial}{\partial x_2} \left\{ \tilde{f}_2(x_1, x_2) \tilde{p} \right\} + \frac{1}{2!} \lambda E\left[Y^2\right] \frac{\partial^2}{\partial x_2^2} \tilde{p} \\
&- \frac{1}{3!} \lambda E\left[Y^3\right] \frac{\partial^3}{\partial x_2^3} \tilde{p} + \frac{1}{4!} \lambda E\left[Y^4\right] \frac{\partial^4}{\partial x_2^4} \tilde{p} \\
= &\; F(x_1, x_2; \mathbf{a}) \, \tilde{p}(x_1, x_2; \mathbf{a})
\end{aligned} \tag{22}$$

If Y is Gaussian with zero mean,

$$\begin{aligned}
F(x_1, x_2; \mathbf{a}) = &-x_2 \frac{\partial Q_n}{\partial x_1} - \tilde{f}_2(x_1, x_2) \frac{\partial Q_n}{\partial x_2} \\
&+ \frac{1}{2!} \lambda E\left[Y^2\right] \left[\frac{\partial^2 Q_n}{\partial x_2^2} + \left(\frac{\partial Q_n}{\partial x_2}\right)^2 \right] + \frac{1}{4!} \lambda E\left[Y^4\right] \\
&\times \left[\frac{\partial^4 Q_n}{\partial x_2^4} + 4 \frac{\partial Q_n}{\partial x_2} \frac{\partial^3 Q_n}{\partial x_2^3} + 3 \left(\frac{\partial^2 Q_n}{\partial x_2^2}\right)^2 \right. \\
&\left. + 6 \left(\frac{\partial Q_n}{\partial x_2}\right)^2 \frac{\partial^2 Q_n}{\partial x_2^2} + \left(\frac{\partial Q_n}{\partial x_2}\right)^4 \right] - \frac{\partial \tilde{f}_2(x_1, x_2)}{\partial x_2}
\end{aligned} \tag{23}$$

Because $\tilde{p}(x_1, x_2; \mathbf{a}) \neq 0$ and usually $F(x_1, x_2; \mathbf{a}) \neq 0$, another set of mutually independent functions $H_{\overline{k}}(x_1, x_2)$ that span space R^{N_p} can be introduced to make the projection of $F(x_1, x_2; \mathbf{a})$ on R^{N_p} vanish, which leads to

$$\int_{-\infty}^{+\infty} \int_{-\infty}^{+\infty} F(x_1, x_2; \mathbf{a}) H_s(x_1, x_2) \, dx_1 dx_2 = 0 \quad s = 1, 2, \ldots, N_p \tag{24}$$

Selecting $H_s(x_1, x_2)$ as:

$$H_s(x_1, x_2) = x_1^{k-l} x_2^l f_1(x_1) f_2(x_2) \tag{25}$$

where $k = 1, 2, \ldots, n; l = 0, 1, 2, \ldots, k;$ and $s = \frac{1}{2}(k+2)(k-1) + l + 1; N_p$ non-linear algebraic equations in terms of N_p unknown parameters can be formulated. Numerical experience shows that a convenient and effective choice for $f_1(x_1)$ and $f_2(x_2)$ are the PDFs obtained with equivalent linearization or Gaussian closure under Gaussian excitation with spectral intensity $\lambda E[Y^2]$.

4 Illustrative Examples

In order to show the effectiveness of the proposed solution procedure, a two-degree-of-freedom coupled pitch–roll ship motion system and a 10-degree-of-freedom mass–spring–damper system are investigated, respectively. The results given by Monte Carlo simulation, the EQL method and the SSS-EPC method are presented and compared. The sample size of Monte Carlo simulation is 2×10^7. In the presented figures, the result given by Monte Carlo simulation is denoted as MCS and the one given by the EQL method is denoted as EQL. The result given with the SSS-EPC solution procedure is denoted as EPC ($n = 6$) when the polynomial order equals 6 in the EPC method. As well known, the EQL method is based on an important assumption that the response of systems is Gaussian. Therefore, the result of the EQL method has a Gaussian PDF distribution. Comparison with EQL shows that how the response of the above examined systems differs from being Gaussian.

4.1 Two-Degree-of-Freedom Coupled Pitch–Roll Ship Motion System

Non-linear coupled pitch–roll ship motion under excitations is given as

$$\ddot{u}_1 + c_1 \dot{u}_1 + k_1 u_1 + \alpha_1 u_1 u_2 = W_1(t) \tag{26}$$

$$\ddot{u}_2 + c_2 \dot{u}_2 + k_2 u_2 + \alpha_2 u_1^2 = W_2(t) \tag{27}$$

where u_1 and u_2 are the roll and pitch modal amplitudes; c_1 and c_2 are the modal damping coefficients; k_1 and k_2 are the linear stiffnesses of the roll and pitch modes; α_1 and α_2 are non-linear coefficients; $W_1(t)$ and $W_2(t)$ are external Poisson impulses.

According to (1), the generalized FPK equation can be formulated governing the PDF of pitch and roll amplitudes and velocities. In this chapter, only stationary PDF solutions are considered. First, u_1 and $\dot{u}_1$ are considered, the generalized FPK equation is integrated with u_2 and $\dot{u}_2$ by the SSS method. After that, the high-dimensional generalized FPK equation to a low-dimensional equation only about u_1 and $\dot{u}_1$. In terms of the reduced generalized FPK equation, the EPC method is employed to solve the low-dimensional equation for the PDF solution of u_1 and $\dot{u}_1$.

For u_2 and $\dot{u}_2$, the solution procedure is conducted in a similar manner. The original generalized FPK equation is integrated with u_1 and $\dot{u}_1$ using the SSS method. After that, the high-dimensional generalized FPK equation to a low-dimensional equation only about u_2 and $\dot{u}_2$. In terms of the reduced generalized FPK equation, the EPC method is employed to solve the low-dimensional equation for the PDF solution of u_2 and $\dot{u}_2$. It also should be noted that u_2 has a non-zero mean due to the presence of the even order term in (27) while other variables have zero means. The proposed solution procedure is also applicable in the case of non-zero mean PDF solution.

The parameters in (26) and (27) are given as $c_1 = 0.1$, $c_2 = 0.05$, $k_1 = 2$, $k_2 = 1$, $\alpha_1 = 0.2$, $\alpha_2 = 0.3$, $\lambda_1 = 2$, $\lambda_2 = 1$, $\lambda_1 E[Y^2] = 0.1$, and $\lambda_2 E[Y'^2] = 0.05$. Both Y and Y' are Gaussian with zero mean. In Monte Carlo simulation, a sample size of 2×10^7 is adopted.

The results obtained from Monte Carlo simulation (MCS), equivalent linearization method (EQL), and the proposed SSS-EPC method (EPC) are compared in Fig. 1. The numerical results show that the result of the EPC method with $n = 2$ is same as that given by equivalent linearization method. For the roll amplitude u_1, the results of MCS, EQL, and EPC ($n = 6$) are close to each other as shown in Fig. 1a, b. However, there is a significant difference in the tail region of the PDF of u_1 as shown in Fig. 1b. Similar observation is also found in the case of $\dot{u}_1$ as shown in Fig. 1c, d.

For the pitch amplitude u_2, Fig. 2a, b show that EQL and EPC ($n = 6$) are both close to MCS. Because EQL denotes a Gaussian PDF distribution, this indicates that the pitch amplitude is nearly Gaussian. Furthermore, EPC ($n = 6$) can provide an improved result for the tail region as shown in Fig. 2(b). This improvement exhibits that the pitch amplitude is not Gaussian in the tail region.

In the case of $\dot{u}_2$, Fig. 2c, d exhibit EQL and EPC ($n = 6$) are in agreement with the simulated results. EPC ($n = 6$) is much close to the simulated result for the tail region as shown in Fig. 2d. This shows that the PDF of $\dot{u}_2$ is nearly Gaussian but the tail of its PDF becomes non-Gaussian. Comparison shows that the proposed SSS-EPC method is effective for the PDF solution of non-linear coupled pitch–roll ship motion.

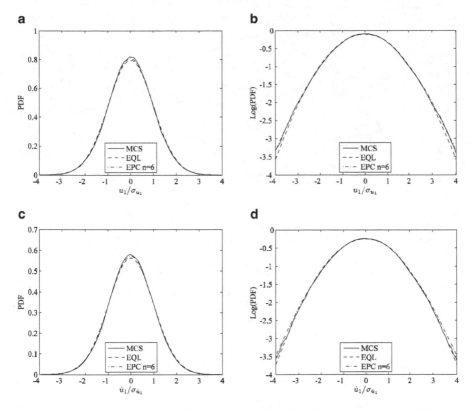

Fig. 1 Comparison of PDFs of $(u_1, \dot{u}_1)$ of the coupled pitch–roll ship motion system. (**a**) PDFs of u_1. (**b**) Logarithmic PDFs of u_1. (**c**) PDFs of $\dot{u}_1$. (**d**) Logarithmic PDFs of $\dot{u}_1$

4.2 Ten-Degree-of-Freedom Mass–Spring–Damper System

Consider a 10-degree-of-freedom non-linear system with cubic non-linearity in displacement given by Fig. 3. The non-linear system can be mathematically expressed as

$$
\begin{cases}
m_1 \ddot{y}_1 + c_1 \dot{y}_1 + k_1 y_1 + k_2 (y_1 - y_2) + v_1 y_1^3 + v_2 (y_1 - y_2)^3 = W_1(t) \\
m_i \ddot{y}_i + c_i \dot{y}_i + k_i (y_i - y_{i-1}) + k_{i+1} (y_i - y_{i+1}) \\
\quad + v_i (y_i - y_{i-1})^3 + v_{i+1} (y_i - y_{i+1})^3 = W_i(t), \quad (i = 2, \ldots, 9) \\
m_{10} \ddot{y}_{10} + c_{10} \dot{y}_{10} + k_{10} (y_{10} - y_9) + v_{10} (y_{10} - y_9)^3 = W_{10}(t)
\end{cases}
\tag{28}
$$

The parameters of the system are given as follows: $m_i = 1$, $c_i = 0.1$, $k_i = 1$, $v_i = 2$, $\lambda_i = 1$, $\lambda_i E[Y_i^2] = 2$, $i = 1, 2, \ldots, 10$. Furthermore, $W_i(t)$ are independent

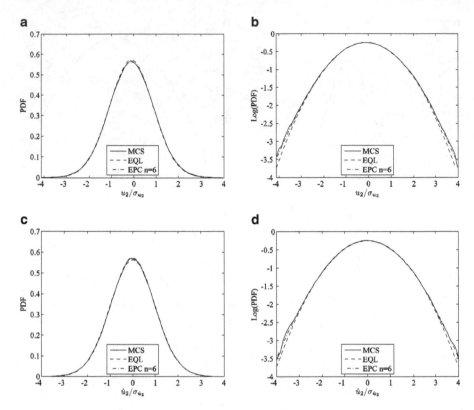

Fig. 2 Comparison of PDFs of $(u_2, \dot{u}_2)$ of the coupled pitch–roll ship motion system. (**a**) PDFs of u_2. (**b**) Logarithmic PDFs of u_2. (**c**) PDFs of $\dot{u}_2$. (**d**) Logarithmic PDFs of $\dot{u}_2$

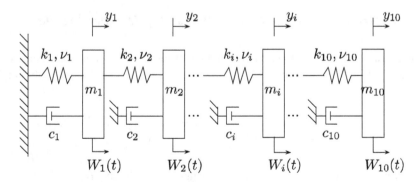

Fig. 3 A 10-degree-of-freedom mass–spring–damper system

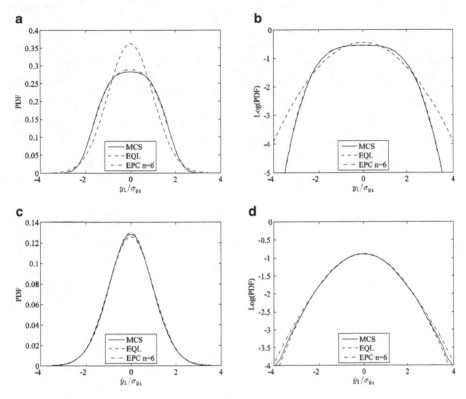

Fig. 4 Comparison of PDFs of $(y_1, \dot{y}_1)$ of the 10-degree-of-freedom mass–spring–damper system. (**a**) PDFs of y_1. (**b**) Logarithmic PDFs of y_1. (**c**) PDFs of $\dot{y}_1$. (**d**) Logarithmic PDFs of $\dot{y}_1$

identically distributed with zero mean, and they are also independent of the impulse arrival time τ_k. In this chapter, the PDF solutions of the first-, fifth-, and tenth-degree-of-freedom vibration are presented to representatively show the effectiveness of the solution procedure. It is because the first-degree-of-freedom mass is located at the bottom. The fifth-degree-of-freedom mass is located in the middle and tenth-degree-of-freedom is located at the top. The PDFs of y_1, $\dot{y}_1$, y_5, $\dot{y}_5$, y_{10}, and $\dot{y}_{10}$ are given with each method, which is shown in the following figures. σ_{y_i} and $\sigma_{\dot{y}_i}$ denote the standard deviations of y_i and $\dot{y}_i$ given by EQL method, respectively.

Figure 4 compares the PDFs of y_1 and $\dot{y}_1$ in Example 2. For the PDF of y_1, Fig. 4a shows that EQL differs significantly from the result of Monte Carlo simulation. The difference is more significant in the tail region as shown in Fig. 4b. This indicates that the PDF distribution of displacement is highly non-Gaussian. Comparatively, EPC ($n = 6$) presents good agreement with the simulated result. The non-Gaussian behavior is formulated due to the fact that the first-degree-of-freedom

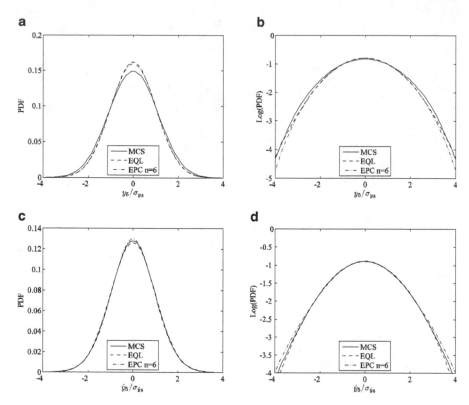

Fig. 5 Comparison of PDFs of $(y_5, \dot{y}_5)$ of the 10-degree-of-freedom mass–spring–damper system. (**a**) PDFs of y_5. (**b**) Logarithmic PDFs of y_5. (**c**) PDFs of $\dot{y}_5$. (**d**) Logarithmic PDFs of $\dot{y}_5$

spring frequently enters the non-linear region. This is because the overall external excitations have to be endured by this spring leading it to show a highly non-linear behavior.

For the velocity, both EQL and EPC ($n = 6$) are close to MCS as shown in Fig. 4c, d. A little difference exists in the tail region. Therefore, the PDF distribution of velocity is very close to being Gaussian.

Figure 5 presents the comparison on the PDFs of y_5 and $\dot{y}_5$ in Example 2. For displacement, EQL and EPC ($n = 6$) are close to MCS as shown in Fig. 5a, b. This means the PDF distribution of displacement approaches to be Gaussian. The displacement behavior is obviously different from the case of y_1 and $\dot{y}_1$. For the fifth-degree-of-freedom spring, the external excitations below and above it are all random, the total resultant force is much smaller than that of the first-degree-of-freedom spring. Therefore, the fifth-degree-of-freedom spring scarcely enters the non-linear region, which fifth-degree-of-freedom spring mostly maintains a linear behavior.

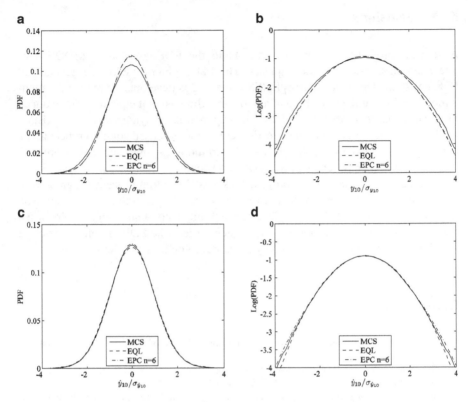

Fig. 6 Comparison of PDFs of $(y_{10}, \dot{y}_{10})$ of the 10-degree-of-freedom mass–spring–damper system. (**a**) PDFs of y_{10}. (**b**) Logarithmic PDFs of y_{10}. (**c**) PDFs of y_{10}. (**d**) Logarithmic PDFs of y_{10}

In the case of velocity, both EQL and EPC ($n = 6$) agree with the simulated result in Fig. 5c, d with a little difference in the tail region. Therefore, the PDF distribution of velocity is very close to being Gaussian.

Last Fig. 6 gives the PDFs of y_{10} and $\dot{y}_{10}$ in Example 2. The similar conclusions are drawn as the case of y_5 and $\dot{y}_5$. EQL and EPC ($n = 6$) are close to MCS in the case of displacement as Fig. 6a, b show. For the 10th-degree-of-freedom spring, only one external force acts on it. This leads the spring to mostly reside in the linear region showing a nearly Gaussian distribution.

In the case of velocity, both EQL and EPC ($n = 6$) are in agreement with the simulated result in Fig. 6c, d. An improvement is made by EPC ($n = 6$) showing that the PDF distribution of velocity is becoming a little non-Gaussian. Comparison shows that the proposed SSS-EPC method is also effective for the PDF solution of the 10-degree-of-freedom mass–spring–damper system.

5 Conclusions

A solution procedure is developed to obtain the PDF solution of MDOF non-linear systems under Poisson impulses. The PDF solution yields the generalized FPK equation. The state-space-split (SSS) method is developed to reduce the high-dimensional generalized FPK equation to a low dimensional equation. The reduced FPK equation is further solved by the exponential–polynomial closure (EPC) method. In order to evaluate the effectiveness of the proposed solution procedure, a two-degree-of-freedom coupled pitch–roll ship motion system and a 10-degree-of-freedom non-linear system are investigated, respectively. Numerical analysis shows that the PDF obtained with the proposed SSS-EPC solution procedure agrees with the simulated result, even in the tail region of the PDF solution. The comparison further shows that the PDF of displacement exhibits a non-Gaussian behavior when the restoring force with displacement is much larger. The PDF of velocity is close to being Gaussian in the case that only linear damping exists in the systems.

References

Atalik TS, Utku S (1976) Stochastic linearization of multi-degree-of-freedom non-linear systems. Earthquake Eng Struct Dyn 4:411–420

Cai GQ, Lin YK (1992) Response distribution of non-linear systems excited by non-Gaussian impulsive noise. Int J Nonlinear Mech 27:955–967

Cai GQ, Lin YK (1996) Exact and approximate solutions for randomly excited MDOF non-linear systems. Int J Nonlinear Mech 31:647–655

Caughey TK (1963) Equivalent linearization techniques. J Acoust Soc Am 35:1706–1711

Caughey TK, Ma F (1982) The exact steady-state solution of a class of nonlinear stochastic systems. Int J Nonlinear Mech 17:137–142

Di Paola M, Santoro R (2008) Non-linear systems under Poisson white noise handled by path integral solution. J Vib Control 14:35–49

Dimentberg MF (1982) An exact solution to a certain non-linear random vibration problem. Int J Nonlinear Mech 17:231–236

Er GK (1998) An improved closure method for analysis of nonlinear stochastic systems. Nonlinear Dyn 17:285–297

Er GK (2011) Methodology for the solutions of some reduced Fokker–Planck equations in high dimensions. Ann Phys (Berlin) 523:247–258

Er GK, Iu VP (2011) A new method for the probabilistic solutions of largescale nonlinear stochastic dynamic systems. In: Zhu WQ, Lin YK, Cai GQ (eds) IUTAM symposium on nonlinear stochastic dynamics and control, vol 29, IUTAM book series. Springer, New York, pp 25–34

Köylüoğlu HU, Nielsen SRK, Iwankiewicz R (1994) Reliability of nonlinear oscillators subject to Poisson driven impulses. J Sound Vib 176:19–33

Köylüoğlu HU, Nielsen SRK, Çakmak AŞ (1995) Fast cell-to-cell mapping (path integration) for nonlinear white noise and Poisson driven systems. Struct Saf 17:151–165

Langley RS (1985) A finite element method for the statistics of non-linear random vibration. J Sound Vib 101:41–54

Lin YK, Cai GQ (1988) Exact stationary response solution for second order nonlinear systems under parametric and external white noise excitations: Part II. ASME J Appl Mech 55:702–705

Muscolino G, Ricciardi G (1999) Probability density function of MDOF structural systems under non-normal delta-correlated inputs. Comput Methods Appl Mech Eng 168:121–133

Muscolino G, Ricciardi G, Cacciola P (2003) Monte Carlo simulation in the stochastic analysis of non-linear systems under external stationary Poisson white noise input. Int J Nonlinear Mech 38:1269–1283

Proppe C (2002a) The Wong-Zakai theorem for dynamical systems with parametric Poisson white noise excitation. Int J Eng Sci 40:1165–1178

Proppe C (2002b) Equivalent linearization of MDOF systems under external Poisson white noise excitation. Prob Eng Mech 17:393–399

Proppe C (2003) Exact stationary probability density functions for non-linear systems under Poisson white noise excitation. Int J Nonlinear Mech 38:557–564

Proppe C, Pradlwarter HJ, Schuëller GI (2003) Equivalent linearization and Monte Carlo simulation in stochastic dynamics. Prob Eng Mech 18:1–5

Roberts JB (1972) System response to random impulses. J Sound Vib 24:23–34

Roberts JB, Spanos PD (2003) Random vibration and statistical linearization. Dover, Mineola, NY

Shinozuka M (1972) Monte Carlo solution of structural dynamics. Comput Struct 2:855–874

Socha L (2008) Linearization methods for stochastic dynamic systems. Springer, Berlin

Spanos PD (1981) Stochastic linearization in structural dynamics. ASME Appl Mech Rev 34:1–8

Vasta M (1995) Exact stationary solution for a class of non-linear systems driven by a non-normal delta-correlated process. Int J Nonlinear Mech 30:407–418

Wojtkiewicz SF, Johnson EA, Bergman LA, Grigoriu M, Spencer BF Jr (1999) Response of stochastic dynamical systems driven by additive Gaussian and Poisson white noise: solution of a forward generalized Kolmogorov equation by a spectral finite difference method. Comput Methods Appl Mech Eng 168:73–89

Wu Y, Zhu WQ (2008a) Stationary response of MDOF dissipated Hamiltonian systems to Poisson white noises. ASME J Appl Mech 75:044502

Wu Y, Zhu WQ (2008b) Stationary response of multi-degree-of-freedom vibroimpact systems to Poisson white noises. Phys Lett A 372:623–630

Zhu HT, Er GK, Iu VP, Kou KP (2009) EPC procedure for PDF solution of non-linear oscillators excited by Poisson white noise. Int J Nonlinear Mech 44:304–310

Evaluate Reliability of Morgenstern–Price Method in Vertical Excavations

Shaham Atashband

Abstract Considering force and moment equilibrium equations simultaneously, Morgenstern–Price's method is more well mannered than other computational algorithms in slope stability field. On the other hand, because of its simplification, Rankine's theory has its particular fans in handy calculations and pre-estimations of safe depth in vertical self-stable excavations, classically and academically. To have a comparison of abovementioned methods' results, in this study, a variety of analyses have been performed using the Morgenstern–Price's algorithm with SLIDE software in which the cohesion factor of soil changes over a range between 5 and 95 kPa and the inner friction angle is applied less than 40°. The analyses results including self-stable excavation depths between 1 and 8 m and related factors of safety (1–3) are derived and collected in a 90° slope. Moreover, using Rankine's formula in the same geometry, the determination of safe vertical self-stable excavation depth is performed in various factors of safety. Finally, a relation between two methods is presented as a correlation as well as a reliability evaluation.

1 Introduction

Nowadays, the importance of a safe excavation is an undeniable fact in every earthy construction project.

As a primary simple economic solution, a self-stable slope is proposed for an excavation in the absence of neighborhood limits. As a matter of fact, a vertical slope (approximately 90°) is much optimum in comparing with other angles due to preparing much space in the excavation. However, reaching a vertical excavation

S. Atashband (✉)
Civil Engineering Department, Islamic Azad University, Arak Branch, Markazi, Iran
e-mail: sh-atashband92@iau-arak.ac.ir; shaham1_at@yahoo.com

S. Kadry and A. El Hami (eds.), *Numerical Methods for Reliability and Safety Assessment: Multiscale and Multiphysics Systems*, DOI 10.1007/978-3-319-07167-1_20,
© Springer International Publishing Switzerland 2015

needs enough strength and stability in the soil (e.g., cohesion and somewhat friction) and it limits the depth of the excavation, consequently.

In order to ensure enough safety in every excavation, various methods have been proposed for slope stability analyses (e.g., Sarma, Bishop, Ordinary, Janbu, Spencer, Morgenstern–Price) and controlling the lateral pressure of soil using or without retaining structures (e.g., Rankine, Coulomb) by extending various theories which have presented suitable results in practical experiences.

In this chapter, first of all, a review of slope stability analysis basics using limit equilibrium is presented and then a focus on Rankine theory and Morgenstern–Price method is done. Secondly, effective parameters in modeling are defined. Thirdly, collected analysis output is demonstrated and summarized in graphs. Finally, a comparison between Rankine's and Morgenstern–Price's method is presented as well as defining a correlation function of the factors of safety to evaluate the reliability of the methods.

2 Slope Stability Basics

2.1 General Definition of Safety Factor

The task of the engineer charged with analyzing slope stability is to determine the factor of safety. Generally, the factor of safety (*FS*) is defined as Eq. (1) (Das 2010).

$$FS = \frac{\tau_f}{\tau_d} \tag{1}$$

where τ_f is average shear strength of the soil and τ_d is average shear stress developed along the potential failure surface.

The shear strength of a soil consists of two components, cohesion and friction, and may be written as Eq. (2) (Das 2010).

$$\tau_f = C' + \sigma' \tan \varphi' \tag{2}$$

where C' is [effective] cohesion, φ' is [effective] friction angle of the soil, and σ' is [effective] normal stress on the potential failure surface.

In a similar manner, the average shear stress developed along the potential failure surface (τ_d) can be defined as Eq. (3) (Das 2010).

$$\tau_d = C'_d + \sigma' \tan \varphi'_d \tag{3}$$

where C'_d and φ'_d are, respectively, the [effective] cohesion and the [effective] angle of friction that develop along the potential failure surface.

Fig. 1 Two principle types of slope stability analyses: (**a**) infinite slope; (**b**) finite slope

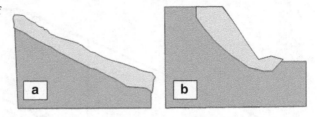

Moreover, two more factors of safety are defined in the literature, with respect to cohesion (Fc') and the factor of safety with respect to friction $(F\varphi')$.

2.2   Classification of Vertical Excavation Analyses

In general, slope stability analyses include two principle types depending on the slope geometry: infinite slopes and finite slopes (see Fig. 1).

Infinite slope analyses are useful when a thin layer of soil overlies a much harder strata or bedrock, and also may be used to evaluate the potential for shallow flowslides which are sometimes called surficial slumps (Coduto 2008).

A vertical excavation problem can be categorized as a finite analysis.

2.3   Morgenstern–Price Method Principles

The methods of slices have become the most common methods due to their ability to accommodate complex geometrics and variable soil and water pressure conditions (Tekzaghki and Peck 1967).

Morgenstern–Price is a general method of slices developed on the basis of limit equilibrium. It requires satisfying equilibrium of forces and moments acting on individual blocks. The blocks are created by dividing the soil above the slip surface by dividing planes.

Forces acting on individual blocks are displayed in Fig. 2.

The following assumptions are introduced in the (Morgenstern-Price) method to calculate the limit equilibrium of forces and moment on individual blocks (Morgenstern 1965):

- Dividing planes between blocks are always vertical (see Fig. 2).
- The line of action of weight of block W_i passes through the center of the ith segment of slip surface represented by point M.

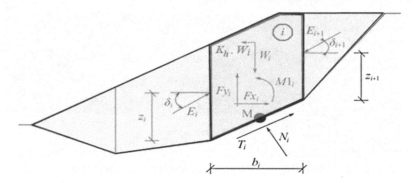

Fig. 2 Static diagram of a schematic slope in Morgenstern–Price method (Finesoftware 2010)

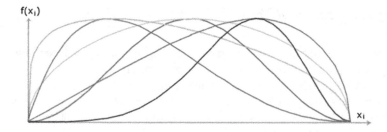

Fig. 3 Half-sine function used in Morgenstern–Price method (Finesoftware 2010)

- The normal force N_i is acting in the center of the ith segment of slip surface, at point M.
- Inclination of forces E_i acting between blocks is different on each block (δ_i) at slip surface end points is $\delta = 0$.

The only difference between Spencer and Morgenstern–Price method is shown in the above list of assumptions. Choice of inclination angles δ_i of forces E_i acting between the blocks is realized with the help of Half-sine function—one of the functions in Fig. 3 is automatically chosen. This choice of the shape of function has a minor influence on final results, but suitable choice can improve the convergency of method. Functional value of Half-sine function $f(x_i)$ at boundary point x_i multiplied by parameter λ results the value of inclination angle δ_i (Morgenstern 1967).

The factor of safety FS is determined by employing the following iteration process (Zhu et al. 2005):

Step 1: The initial value of angles δ_i is set according to Half-sine function ($\delta_i = \lambda * f(x_i)$).

Step 2: The factor of safety FS for a given value of δ_i follows from Eq. (4), while assuming the value of $E_{n+1} = 0$ at the end of the slip surface

$$E_{i+1} = \frac{\begin{array}{c} [(W_i - Fy_i) \cdot \cos\alpha_i - (K_h \cdot W_i - Fx_i) \cdot \sin\alpha_i - U_i + E_i \\ \cdot \sin(\alpha_i - \delta_i)] \cdot \frac{\tan\varphi_i}{FS} + \frac{c_i}{FS} \cdot \frac{b_i}{\cos\alpha_i} - (W_i - Fy_i) \\ \cdot \sin\alpha_i - (K_h \cdot W_i - Fx_i) \cdot \cos\alpha_i + E_i \cdot \cos(\alpha_i - \delta_i) \end{array}}{\sin(\alpha_i - \delta_{i+1}) \cdot \frac{\tan\varphi_i}{FS} + \cos(\alpha_i - \delta_{i+1})} \tag{4}$$

Step 3: The value of δ_i is provided by Eq. (5) using the values of E_i determined in the previous step with the requirement of having the moment on the last block equal to zero. Functional values $f(x_i)$ are same all the time during the iteration, only parameter λ is iterated. Equation (5) does not provide the value of z_{n+1} as it is equal to zero. For this value the moment equation of Eq. (6) must be satisfied.

$$Z_{i+1} = \frac{\begin{array}{c} \frac{b_i}{2} \cdot [E_{i+1}(\sin\delta_{i+1} - \cos\delta_{i+1} \cdot \tan\alpha_i) \\ + E_i \cdot (\sin\delta_i - \cos\delta_i \cdot \tan\alpha_i)] + E_i \cdot z_i \cdot \cos\delta_i \\ - M1_i + K_h \cdot W_i \cdot (y_M - y_{gi}) \end{array}}{E_{i+1} \cdot \cos\delta_{i+1}} \tag{5}$$

$$E_{i+1} \cdot \cos\delta_{i+1} \left(z_{i+1} - \frac{b_i}{2} \tan\alpha_i \right) - E_{i+1} \cdot \sin\delta_{i+1} \cdot \frac{b_i}{2}$$

$$- E_i \cdot \cos\delta_i \left(z_i - \frac{b_i}{2} \tan\alpha_i \right) - E_i \cdot \sin\delta_i \cdot \frac{b_i}{2}$$

$$+ M1_i - K_h \cdot W_i \left(y_M - y_{gi} \right) = 0 \tag{6}$$

Step 4: Steps 2 and 3 are then repeated until the value of δ_i (resp. parameter λ) does not change.

Moreover, it is necessary to avoid unstable solutions for successful iteration process. Such instabilities occur at points where division by zero in Eqs. (4) and (5) takes place. In Eq. (5), division by zero is encountered for $\delta_i = \pi/2$ or $\delta_i = -\pi/2$. Therefore, the value of angle $\ddot{a}_i$ must be found in the interval $(-\pi/2; \pi/2)$.

Division by zero in Eq. (4) appears when:

$$FS = \tan\varphi_i \cdot \tan(\delta_{i+1} - \alpha_i) \tag{7}$$

Another check preventing numerical instability is verification of parameter m_α— following condition must be satisfied:

$$m_\alpha = \cos\alpha_i + \frac{\sin\alpha_i \cdot \tan\varphi_i}{FS} > 0.2 \tag{8}$$

Therefore before iteration run it is required to find the highest of critical values (FS_{min}) satisfying abovementioned conditions. Values below this critical value FS_{min}

are in area of unstable solution, therefore iteration begins by setting FS to a value "just" above FS_{min} and all result values of FS from iteration runs are higher than FS_{min}.

3   Classical Theories on Lateral Earth Pressure

3.1   Rankine's Theory of Active Pressure

The phrase plastic equilibrium in soil refers to the condition where every point in a soil mass is on the verge of failure. The Scotsman W.J.M. Rankine (1857) investigated the stress conditions in soil at a state of plastic equilibrium (Das 2010). Actually, he approached the lateral earth pressure problem with the following assumptions (Coduto 2008):

- The soil is homogeneous and isotropic.
- The most critical shear surface is a plane. In reality, it is slightly concave up, but this is a reasonable assumption (especially for the active case) and it simplifies the analysis.
- The ground surface is a plane (although it does not necessarily need to be leveled).
- The wall moves sufficiently to develop the active or passive condition (in this case, active).
- The resultant of the normal and shear forces that act on the back of the wall is inclined at an angle parallel to the ground surface (Coulomb's theory provides a more accurate model of shear forces acting on the wall).

For preliminary theoretical analysis, let us consider a frictionless retaining wall represented by a plane AB as shown in Fig. 4. If the wall AB rotates sufficiently about its bottom to a position A'B, then a triangular soil mass ABC' adjacent to the wall will reach Rankine's active state (Das 2010).

Ultimately, a state will be reached when the stress condition in the soil element can be represented by the Mohr's circle b (Fig. 5), the state of plastic equilibrium, and failure of the soil will occur. This situation represents Rankine's active state, and the effective pressure on the vertical plane (which is a principal plane) is Rankine's active earth pressure (Das 2010).

Because the slip planes in Rankine's active state make angles of $\pm(45 + \varphi'/2)°$ with the major principal plane, the soil mass in the state of plastic equilibrium is bounded by the plane BC', which makes an angle of $(45 + \varphi'/2)°$ with the horizontal. The soil inside the zone ABC' undergoes the same unit deformation in the horizontal direction everywhere, which is equal to $\Delta La/La$. The lateral earth pressure on the wall at any depth z from the ground surface can be calculated by using Eq. (9) (Das 2010).

Because the slip planes in Rankine's active state make angles of $\pm(45 + \varphi'/2)°$ with the major principal plane, the soil mass in the state of plastic equilibrium

Fig. 4 Rotation of frictionless wall about the bottom based on Rankine's theory (Das 2010)

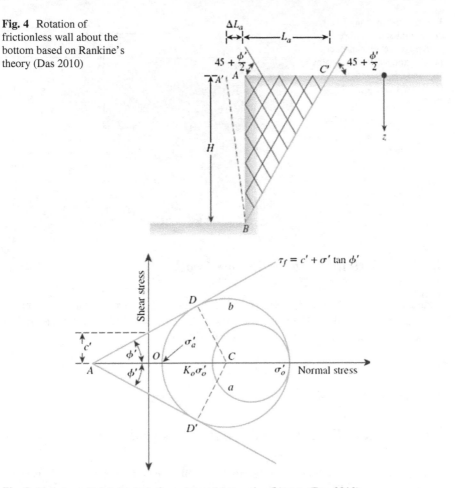

Fig. 5 Variation of Mohr's circle from Atrest (**a**) to active (**b**) state (Das 2010)

is bounded by the plane BC′, which makes an angle of $(45 + \varphi'/2)°$ with the horizontal. The soil inside the zone ABC′ undergoes the same unit deformation in the horizontal direction everywhere, which is equal to $\Delta La/La$. The lateral earth pressure on the wall at any depth z from the ground surface can be calculated by using Eq. (9) (Das 2010).

$$\sigma'_a = \gamma \cdot z \cdot \tan^2\left(45 - \frac{\varphi'}{2}\right) - 2C' \cdot \tan^2\left(45 - \frac{\varphi'}{2}\right) \qquad (9)$$

$$H_C = \frac{2C'}{\gamma \sqrt{K_a}} \qquad (10)$$

where K_a is defined as $\tan^2(45 - \varphi/2)$.

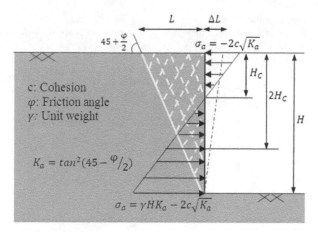

Fig. 6 Distribution of active horizontal soil pressure (σ_h) along excavation depth, crack height, and self-stable depths (H_c and $2H_c$) based on Rankine's theory

3.2 Crack Height and Self-Stable Depth

Based on Rankine's theory (Rankine 1857), the soil crack height (H_c) can be calculated as Eq. (10) and $2H_c$ applied with a factor of safety (FS_R) is considered as the self-stable depth of a vertical excavation (Fig. 6)

4 Analyses Details

4.1 Modeling Equilibrium Limit Analysis with SIDE

In this study, several models were made by SLIDE software in which various excavation depths (i.e., 1–8 m) was considered made in soil with Mohr–Coulomb behavior (linear strength type). The strength parameters vary according to Table 1, because no water table is considered, $C = C'$ and φ equal to φ'. In addition, the unit weight of soil is constant in all analysis ($\gamma = 18$ kN/m^3).

A center point grid with interval number 40 (one interval is equal to 0.5 m) in both directions (i.e., x and y) was made to assess various radiuses of sliding surfaces (Fig. 7). In addition, no surcharge was considered in the models.

SLIDE software is a user-friendly program which makes easy to check the equilibrium limit stability using circular wedges with various radiuses in Morgenstern–Price method. It also shows the value of general safety factor against overturning in the slope using colors to show the variation to help user to finding the convergence (Fig. 8).

Based on above noted considerations, several combinations of c and φ have applied to find relative factor of safety against overturning in excavation depth 1–8 m in the absence of water table. Consequently, the results could be presented as Figs. 9, 10, 11, 12, 13, 14, 15, and 16 in which the factor of safety curves are moved along friction angle and cohesion axis.

Table 1 Soil strength parameters values and ranges considered in the model

Strength parameter	Value
C: Cohesion (kPa)	5–95
φ: Friction angle (°)	0, 10, 20, 30, 40

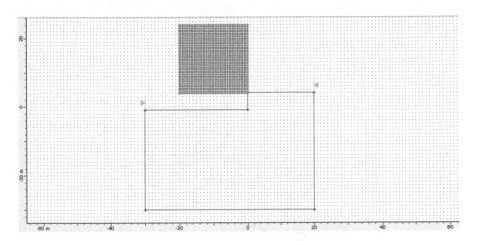

Fig. 7 A vertical excavation model using SLIDE software

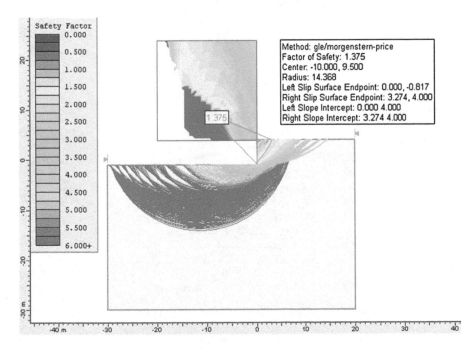

Method: gle/morgenstern-price
Factor of Safety: 1.375
Center: -10.000, 9.500
Radius: 14.368
Left Slip Surface Endpoint: 0.000, -0.817
Right Slip Surface Endpoint: 3.274, 4.000
Left Slope Intercept: 0.000 4.000
Right Slope Intercept: 3.274 4.000

Fig. 8 A slope stability analysis output with Slide software

Fig. 9 Contours of equal
factors of safety for depths
$H = 1$ m

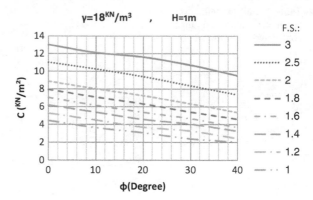

Fig. 10 Contours of equal
factors of safety for depths
$H = 2$ m

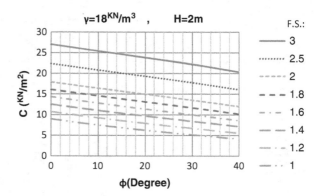

Fig. 11 Contours of equal
factors of safety for depths
$H = 3$ m

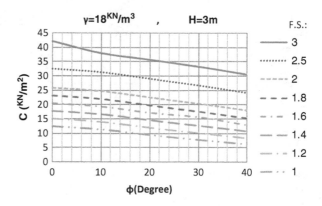

Fig. 12 Contours of equal factors of safety for depths $H = 4$ m

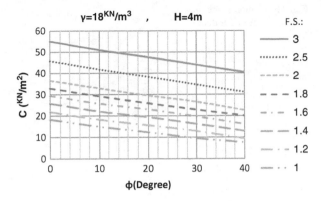

Fig. 13 Contours of equal factors of safety for depths $H = 5$ m

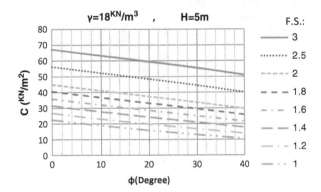

Fig. 14 Contours of equal factors of safety for depths $H = 6$ m

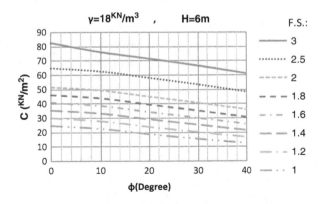

Fig. 15 Contours of equal factors of safety for depths $H = 7$ m

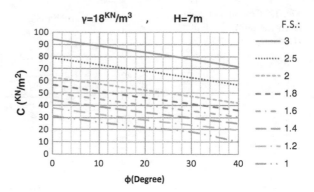

Fig. 16 Contours of equal factors of safety for depths $H = 8$ m

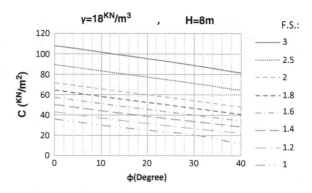

Fig. 17 Normalized contours of factor of safety for depths $H = 1$ m

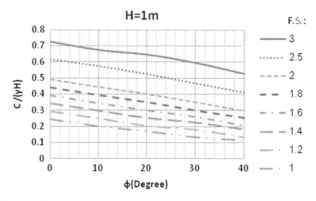

To normalize the effect of unit weight, cohesion and height of excavation in equilibrium limit analyses output, the $c\gamma/H$ phrase is used in Figs. 17, 18, 19, 20, 21, 22, 23, and 24.

To illustrate the difference between analyses output in different depths, Fig. 25 is prepared.

Fig. 18 Normalized contours of factor of safety for depths $H = 2$ m

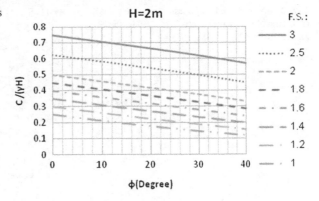

Fig. 19 Normalized contours of factor of safety for depths $H = 3$ m

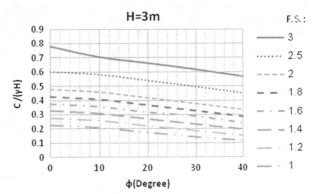

Fig. 20 Normalized contours of factor of safety for depths $H = 4$ m

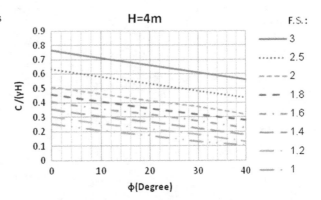

To summarize abovementioned graphs, some best-fit trend lines were searched which formulas are presented in Table 2. Although two types of trend/regression are compared in the table (i.e., linear and polynomial with order 6), the linear one is selected to apply in the rest of this study (the dark lines in Fig. 26).

Fig. 21 Normalized contours
of factor of safety for depths
$H = 5$ m

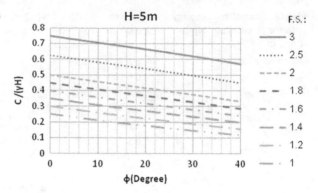

Fig. 22 Normalized contours
of factor of safety for depths
$H = 6$ m

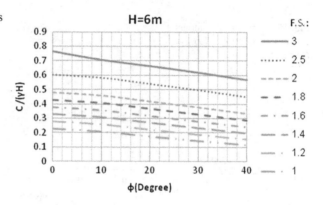

Fig. 23 Normalized contours
of factor of safety for depths
$H = 7$ m

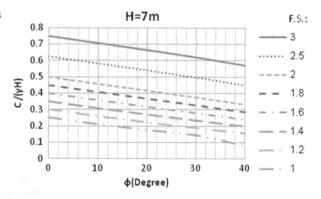

4.2 Handy Calculation Using Rankine's Equation

To calculate the factor of safety in stable depth for vertical excavation (H), the
abovementioned equation of Rankine was used. The results are presented in Fig. 26
to reach $2H_c$ according to Eq. (11).

Fig. 24 Normalized contours of factor of safety for depths $H = 8$ m

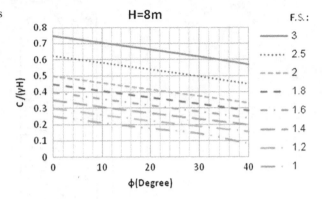

Fig. 25 The summarized SLIDE's output for normalized contours of friction angle in $H = 1$–8 m

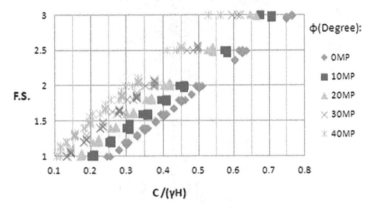

$$FS_R = \frac{2H_c}{H} = \frac{4C'}{\gamma H \sqrt{K_a}} \qquad (11)$$

The variation of FS against $c\gamma/H$ in different φ (and relevant K_a) is shown by light lines and are compared with Morgenstern–Price's ones in Fig. 26 (i.e., the dark lines).

In Fig. 26, a significant difference between Morgenstern–Price's output and Rankine's is obvious, especially for greater φ. To have a comparison between two methods, a correlation coefficient (i.e., $SA_{MP/R}$) is defined by the author in this chapter as Eq. (12) which makes easy transferring FS values.

$$FS_{MP} = SA_{MP/R} \cdot FS_R \qquad (12)$$

Table 2 Soil strength parameters values and ranges considered in the model

φ (°)	Trend/regression formula			
	Linear		Polynomial	
	FS	R^2	FS	R^2
0	$3.9733x - 0.0071$	0.998	$-498x^6 + 1346.5x^5 - 1465.9x^4 + 822.73x^3 - 251.33x^2 + 43.664x - 2.5435$	0.998
10	$3.9878x + 0.1712$	0.998	$-2584.1x^6 + 6564.3x^5 - 6712.9x^4 + 3534.9x^3 - 1010.8x^2 + 152.83x - 8.6517$	0.999
20	$4.0666x + 0.3193$	0.999	$-1548.9x^6 + 3699.9x^5 - 3552.2x^4 + 1751.1x^3 - 466.83x^2 + 67.839x - 3.168$	0.999
30	$4.173x + 0.4676$	0.997	$-990.99x^6 + 2043x^5 - 1668.8x^4 + 688.18x^3 - 150.79x^2 + 20.922x - 0.2823$	0.998
40	$4.3061x + 0.6235$	0.995	$-1765.7x^6 + 3180.2x^5 - 2264.1x^4 + 819.23x^3 - 161.36x^2 + 21.177x - 0.1182$	0.997

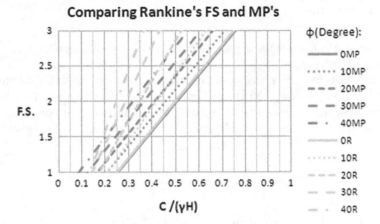

Fig. 26 Summarized SLIDE's output for normalized contours of friction angle in $H = 1\text{--}8$ m

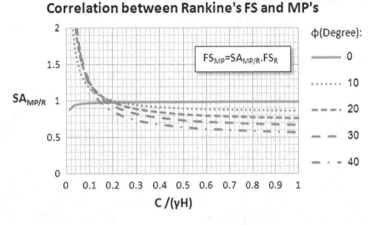

Fig. 27 Variation of $SA_{MP/R}$ (the correlation coefficient between Rankine and Morgenstern–Price results) against $c\gamma/H$ for $\varphi = 0\text{--}40°$

where FS_{MP} and FS_R are the general stability factor of safety derived from Morgenstern–Price and Rankine analysis or calculations, respectively. It is useful to have a correlation/transferring coefficient between two methods results when a calculation using Rankine is performed and it is asked to have an estimation of the computational method (i.e., Morgenstern–Price). This process can be performed using a graph like Fig. 27. In the next section, there is an example which illustrates it.

5   An Example

In this section, an example is presented to clarify the situation of transferring procedure with $SA_{MP/R}$.

The request is to calculate factor of safety against overturning for a soil (with unit weight 20 kN/m^3, cohesion 50 kPa and friction angle 30°) using both Rankine and Morgenstern–Price methods in the absence of any equilibrium limit software usage.

Answer: We know that $K_a = 0.33$ for $\varphi = 30°$ so we can use Eq. (11) to calculate FS_R:

$$FS_R = \frac{4C'}{\gamma H \sqrt{K_a}} = \frac{4 \times 50}{20 \times 5 \times \sqrt{0.33}} \approx 3.5$$

Now, we use Fig. 27 and derive $SA_{MP/R} = 0.73$ approximately, for $\varphi = 30$ and $c\gamma/H = 0.5$. Therefore $FS_{MP} = 2.55$ using Eq. (12). This estimate has less than 0.5% error ($FS_{MP} = 2.56$ using SLIDE software). Fortunately, this difference does not significantly affect the final engineering decision.

6   Conclusions

The following conclusions may be drawn from the study reported in this chapter.

- According to the final graph, in soils with $\varphi = 0$, there is a negligible difference between factors of safety obtained from two described methods (i.e., Rankine, Morgenstern–Price) which results $SA_{MP/R} = 1$ approximately.
- In soils with $\varphi \neq 0$, this chapter recommends using $SA_{MP/R}$ presented in final graph to estimate the factor of safety obtained from computational solution (e.g., Morgenstern–Price) using the simple classical handy approach (e.g., Rankine).
- In a constant $c\gamma/H$, soils with greater φ cause much difference between factors of safety obtained from two described methods (i.e., Rankine, Morgenstern–Price). It means, in such cases, $SA_{MP/R}$ is more significant.
- In the range $c\gamma/H = 0.15$–0.25, the difference between factors of safety obtained from two mentioned methods is negligible so it results $SA_{MP/R}$ near to 1. But in order to transfer the factor of safety of Morgenstern–Price from Rankine in the ranges $c\gamma/H < 0.15$–0.25 and $c\gamma/H > 0.15$–0.25, a coefficient ($SA_{MP/R}$) greater and less than 1 is needed, respectively.

For more studies, it is recommended the following

- To evaluate effects of having a surcharge at top of the excavation, a similar study using both computational and handy solutions may produce a useful correlation coefficient.

- It is possible to model ground water table and have a comparison between various methods' results to assess the pore water pressure and effective strength parameters on the formulas and equations.
- Moreover, numerical modeling (e.g., finite element) can be used to evaluate the reliability of available methods and offer suitable correlation coefficients.

Acknowledgements The author wants to proffer special thanks to Dr. Mehran Esfahanizadeh who implanted the idea of searching for differences between Morgenstern–Price's and Rankine's method results in the author's mind and encouraged him to have a research about this interesting and practical topic.

References

Coduto PD (2008) Geotechnical engineering—principles and practices. PHI Learning Private Limited, New Delhi. ISBN 978-81-203-2137-3

Das BM (2010) Principles of geotechnical engineering, 7th edn. Cengage Learning, Boston. ISBN 978-0-495-41130-7

Finesoftware (2010) http://www.finesoftware.eu/geotechnical-software/help/redi-rock-wall/morgenstern-price-01/

Morgenstern NR, Price VE (1965) The analysis stability of general slip surfaces. Geotechnique 15(1):79–93

Morgenstern NR, Price VE (1967) A numerical method for solving the equations of stability of general slip surfaces. Comput J 9:388–393

Zhu DY, Lee CF, Qian QH, Chen GR (2005) A concise algorithm for computing the factor of safety using the Morgenstern–Price method. Can Geotech J 42(1):272–278

Rankine WJM (1857) On the stability of loose earth. Philos Trans R Soc 147:9–27

Tekzaghki K, Peck RB (1967) Soil mechanics in engineering practice, 2nd edn. John Wiley and Sons, Inc., New York

Probabilistic Approach of Safety Factor from Failure Assessment Diagram

Guy Pluvinage and Christian Schmitt

Abstract Global and partial deterministic safety factors are defined through a failure assessment diagram to evaluate the failure risk of a structure or a component.

The use of a failure assessment diagram is extended to evaluate the safety factor and the security factor of a conventional probability of failure. The safety factor is defined from the failure assessment curve which has a probability of failure equal to unity. The security factor is defined from an isoprobability failure curve. Three examples of this method are described concerning gas and water pipes and boiler tubes.

1 Introduction

The earliest information about the use of the safety factor is given in the code of Hammurabi (Codex Hammurabi). The best preserved ancient law code was created in 1760 BC (middle chronology) in ancient Babylon. It was enacted by the sixth Babylonian king, Hammurabi. At the top of a basalt stele is a bas-relief image of a Babylonian god (either Marduk or Shamash), with the king of Babylon presenting himself to the god, with his right hand raised to his mouth as a mark of respect. The text covers the bottom portion with the laws written in cuneiform script. It contains a list of crimes and their various punishments, as well as settlements for common disputes and guidelines for citizen conduct. It is mentioned that an architect, who built a house which collapsed on its occupants and caused their deaths, is condemned to capital punishment (Fig. 1).

In addition, it is noted that "when a stone is necessary to build a palace, the architect has to plan to use two stones."

G. Pluvinage (✉) • C. Schmitt
Ecole Nationale d'Ingénieurs de Metz, 1 route d'Ars Laquenexy, CS 65820,
57078 METZ Cedex 03, France
e-mail: pluvinage@cegetel.net; schmitt@enim.fr

S. Kadry and A. El Hami (eds.), *Numerical Methods for Reliability and Safety Assessment: Multiscale and Multiphysics Systems*, DOI 10.1007/978-3-319-07167-1_21, © Springer International Publishing Switzerland 2015

Fig. 1 View of the stele of
the code of Hammurabi
(Louvre Museum Paris
France)

Until the nineteenth century, all constructions were mainly conceived and carried out empirically. The invention of steel construction involved developing the strength of materials. The principle of safety initially adopted consisted of making sure that the maximum effort in the most critical section of the construction remained lower than the service load obtained by dividing the material resistance by the safety factor, conventionally fixed and usually noted f_s. The method is usually determined by ensuring that permissible stresses σ_{ad} remain within the limits through the use of a safety factor and the strength limit is generally the yield stress Re for conservative reasons.

$$\sigma_{ad} = \frac{Re}{f_s} \tag{1}$$

Global safety factor (f_s) can mean either the fraction of structural capability over that required, or a multiplier applied to the maximum expected load (force, torque, bending moment, or a combination) to which a component or assembly will be subjected. The two senses of the term are completely different in that the first is a measure of the reliability of a particular design, while the second is a requirement imposed by law, standard, specification, contract, or custom. Careful engineers refer to the first sense as a safety factor, or, to be explicit, a realized factor of safety, and the second sense as a design factor (DF), but usage is inconsistent and confusing,

Table 1 Typical safety factors

Safety factor (f_S)	Applied structure loads	Structure stresses	Material behavior
$1 \leq f_S \leq 2$	Regular and well known	Known	Known after test
$2 \leq f_S \leq 3$	Regular and known	Relatively known	Known
$3 \leq f_S \leq 4$	Not well known	Not well known	No tests
	Uncertain	Unknown and uncertain	Not very well known

so engineers need to be aware of both. The safety factor is given to the engineer as a requirement. The DF is calculated by the engineer. The safety factors are defined by the "state of the art" for each field, possibly codified in standards. It is equal to or higher than 1, and if it is as much higher as the system is badly defined then the service loads are badly controlled (Pluvinage 2007). Typical values of the safety factor are given in Table 1.

Appropriate safety factors are based on several considerations. Prime considerations are the accuracy of the load, strength, and wear estimates, the consequences of engineering failure, and the cost of over engineering the component to achieve that safety factor. For example, components whose failure could result in substantial financial loss, serious injury or death can usually use a safety factor of four or higher (often ten) (Pluvinage 2007).

Buildings commonly use a safety factor of 2.0 for each structural member. The value for buildings is relatively low because the loads are well understood and most structures are redundant. Pressure vessels use 3.5–4.0, cars use 3.0, and aircraft and spacecraft use 1.4–3.0 depending on the materials. Ductile metallic materials use the lower value while brittle materials use the higher values. The field of aerospace engineering uses generally lower DFs because the costs associated with structural weight are high. This low DF is why aerospace parts and materials are subject to more stringent quality controls.

Many codes require the use of a margin of safety (MS) to describe the ratio of the strength of the structure to the requirements.

$$\text{Design factor} = \text{Failure load}/\text{Design load}$$

$$\text{Margin of safety} = (\text{Failure load}/(\text{Design load} * f_s)) - 1$$

For a successful design, the DF must always equal or exceed the required safety factor and the safety margin is greater than zero.

2 Safety Factor from Fracture Mechanics

Fracture mechanics assumes that fracture is induced by a precursor (material, defect, crack, or notch). Fracture mechanics theories therefore give a relationship between the critical gross stress $\sigma_{g,c}$ and defect size a. The well-known Griffith (1921) linear elastic fracture mechanics predict that the product of the critical gross stress and the square root of the critical gross stress are constant.

Fig. 2 Definition of global safety factor on fracture diagram

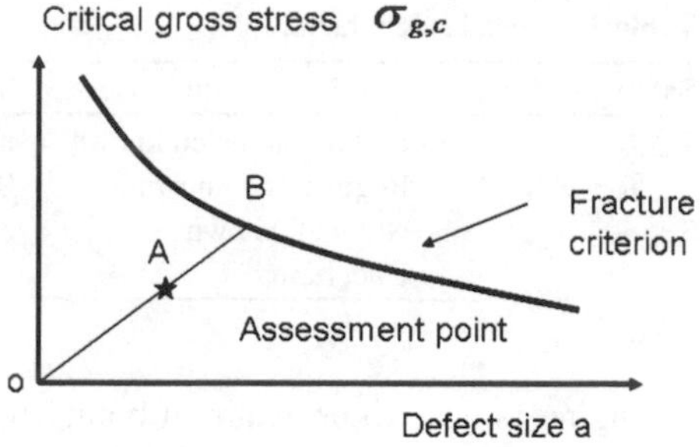

Critical gross stress $\sigma_{g,c}$

B

A

Fracture criterion

Assessment point

O

Defect size a

Table 2 Safety factors used for determination of admissible defect size a_{ad} for brittle materials using linear fracture mechanics

Partial safety factors	f_s^K	f_s^a	f_s^σ
Case 1	1.0	2.0	1.0
Case 2	1.0	1.1	1.2
Case 3	1.2	1.4	1.4

The definition of global safety factor is given directly in the diagram of critical gross stress versus defect size as the ratio of the distance from the origin to the assessment point A and the distance from the origin to the intercept point B. This definition will be used later in the failure assessment diagram (FAD) method (Fig. 2).

$$f_{g,s} = \frac{OB}{OA} \tag{2}$$

The use of a partial safety factor has been extended for the determination of the admissible defect size a_{ad} for brittle materials using linear fracture mechanics. For this, three partial safety factors are used:

f_s^K the partial safety factor on fracture toughness
f_s^a the partial safety factor on applied stress
f_s^a the partial safety factor on defect size

The admissible defect size is then given by the following relationship:

$$f_s^a a_{ad} = \frac{1}{\pi} \left(\frac{K_{Ic}}{f_s^K} \right)^2 \left(\frac{1}{f_s^\sigma \, \sigma_{g,ap}} \right)^2 \tag{3}$$

where K_{Ic} is the material fracture toughness and $\sigma_{g,ap}$ is the applied gross stress.

According to the degree of knowledge of the material properties and applied loads, the following safety factors are used (see Table 2):

Partial safety factors for fracture toughness, defect size, and applied loads are also defined in the FAD as we will see later.

3 Probabilistic Safety Factor

Engineers gradually realized the insufficiencies of this safety design, and this awakening brought about the development of the concept of reliability from a probabilistic angle. According to the probabilistic approach, a structure is considered sure if its probability of failure is lower than a conventional value, a value which depends on many factors like the expected life of the structure, the consequences generated by its ruin, the risks of obsolescence and certain economic criteria such as the value of replacement, maintenance costs, and so forth. Generally, a failure probability of 10^{-4} is used. When there is risk to human life (gas pipes, nuclear pressure vessels, etc.), a failure probability of 10^{-6} is recommended. The choice of a very low failure probability is limited by economic considerations and sometimes by the weight of the structure. One says: *if the failure probability decreases by one order of magnitude, the price increases by two orders.* The probabilistic approach introduces safety factors such as the quantitative criterion of a weak probability of rupture.

When using a probabilistic design approach, the designer no longer thinks of each variable as a single value or number. Instead, each variable is viewed as a probability distribution. The main characteristics of these distributions are the mean and standard deviation of a random variable x denoted $\overline{x}$ and s_x, respectively.

Generally, after statistical data processing, it appears that the mean and the standard deviation of a random variable x are constant and independent of the distribution shape.

The coefficient of variation is an excellent indicator of the homogeneity of the sample. This one will be declared homogeneous if $CV < 1/3$. Concerning the properties of the materials, if the mechanical tests were carried out carefully, the coefficient of variation is an excellent indicator of the process quality. According to Haugen (1980), the coefficients of variation of material mechanical properties have the range given in Table 3.

Thus, the production of a low carbon steel led to a coefficient of variation $c_{V,x} = 0.1$ with Rm the ultimate resistance. From a general point of view, one can estimate that for structural components the concept of a continuous medium is hardly applicable for $c_{V,x} < 0.2$.

For the Weibull distribution, Weibull modulus can be estimated by the following empirical relationship (Sapounov et al. 1996):

$$m = c_{V,x}^{-1.09} \tag{4}$$

Table 3 Coefficient of variation of material mechanical properties (Mankovsky et al. 1999)	Range of variation	Recommended adoptive values
Ultimate strength	0.05–0.10	0.05
Yield stress	0.05–0.10	0.07
Young's modulus	0.03	0.03
Fracture toughness	0.05–0.13	0.07

From this perspective, probabilistic design predicts the flow of variability through a system. By considering this flow, a designer can make adjustments to reduce the flow of random variability and improve quality. Proponents of the approach contend that many quality problems can be predicted and rectified during the early design stages and at a much reduced cost and weight.

The safety factor is then defined as the ratio of the ultimate strength which corresponds to the mean value of the strength distribution over the admissible stress. The admissible stress is the failure stress associated with a low and conventional probability of failure PF*(10^{-4} or 10^{-6} if there is risk to human life).

If the ultimate strength follows Weibull distribution, the probability of failure is given by the following relationship:

$$p_f^* = \exp - \left[\frac{\Gamma\left(1 + 1/m\right)}{f_s} \right]^m \tag{5}$$

and the safety factor is:

$$f_s = \frac{\Gamma\left(1 + 1/m\right)}{\left[\operatorname{Ln}\dfrac{1}{P_s^*} \right]^{1/m}} \tag{6}$$

We can see that the safety factor increases considerably when Weibull modulus decreases, that is, the scatter of the material strength increases.

4 Failure Assessment Diagram

The basic fracture mechanics relationship associates three parameters, defect size a, applied gross stress σ_g, and fracture toughness R, into a failure criterion expressed by the following equation.

$$F\left(\sigma_g, a, R\right) = 0 \tag{7}$$

R is fracture toughness in general and, according to the used fracture criterion, it can be the critical stress intensity K_{Ic}, the critical J, integral J_{Ic}, or the critical crack opening displacement δ_c (Pluvinage 1989).

Another presentation of the failure criterion can be made using a two-parameter relationship:

$$k_{r,c} = f\left(L_r\right) \tag{8}$$

where $k_{r,c}$ is the non-dimensional critical crack driving force and L_r the non-dimensional load. Equation (8) represents the failure curve.

Any loaded structure with a defect is represented by an assessment point in the plane k_r, L_r.

$$k_r = \frac{K_{ap}}{K_{Ic}} = \sqrt{\frac{J_{ap}}{J_{Ic}}} = \sqrt{\frac{\delta_{ap}}{\delta_c}}; \quad L_r = \frac{F}{F_c} \tag{9}$$

where K_{ap}, J_{ap}, and δ_{ap} are the applied stress intensity factor, the J integral or the COD. KIC, J_{Ic}, and δ_c are the fracture toughness for given conditions of constraint, F applied load and F_C critical load.

Initially, the failure curve $k_{r,c} = f(L_r)$ was obtained from a plasticity correction given by Dugdale's model (1960). This approach considers that any kind of failure from a purely brittle fracture to plastic collapse is derived from the brittle fracture by a plasticity correction. L_r is also defined as a non-dimensional stress described as the ratio of the gross stress σ_g over flow stress (chosen as the yield stress σ_Y, ultimate stress σ_U, or classical flow stress $R_c = (\sigma_Y + \sigma_U)/2$).

$$L_r = \frac{\sigma_g}{R_c} \tag{10}$$

Several failure assessment curves have been proposed in the literature. Currently, as with engineering tools, this failure curve has been established from full scale tests and using the lower bound of results and is different according to the codes. Currently, the following methods are used: EPRI in the USA (Kumar et al. 1981), R6 in the UK (R6 1998), RCC-MR in France (Moulin et al. 1993), and SINTAP in the EU (SINTAP 1999).

The failure curves are similar for these different methods but not identical. The SINTAP procedure can be generally simplified to several distinct levels according to the no yield point elongation assumptions. The mathematical expressions of the SINTAP procedure used with the aforementioned assumption for "level 1" is presented below where $f(L_r)$, L_r, L_r^{max}, σ_Y, μ, E, σ_U, N, ε_{ref} and σ_{ref} are the interpolating function, non-dimensional loading or stress-based parameter, maximum value of non-dimensional loading or stress-based parameter, yield stress, first correction factor, modulus of elasticity, ultimate stress, second correction factor, reference strain, and reference stress, respectively.

$$f(L_r) = \left[1 + \frac{L_r^2}{2}\right]^{\frac{-1}{2}} \left[0.3 + 0.7 \times e^{(-0.6 \times L_r^6)}\right],$$

$$\text{for} \quad 0 \leq L_r \leq 1 \quad \text{where} \quad L_r^{max} = 1 + \left(\frac{150}{\sigma_Y}\right)^{2.5} \tag{11}$$

In the FAD method, a failure curve is used to assess the failure zone, the safety zone, and the security zone. In Fig. 3, a typical FAD is illustrated. In this FAD, the failure curve $f(L_r)$, interpolating brittle fracture to plastic collapse, separates the safety zone from the failure zone. The security curve is obtained by dividing the failure curve by the global safety factor. It separates the safety zone from the security zone.

Fig. 3 Typical presentation of the failure assessment diagram (FAD)

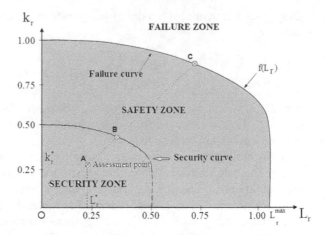

As a consequence of the definitions of L_r and K_r, the loading path OC is linear when the load increases from 0 to the critical load F_c. Under service conditions, a structure or component with a defect under service load and for a given material is represented by an assessment point A of the coordinates L_r^* and k_r^*. If this assessment point is inside the safety zone, no failure occurs, if the assessment point is on the assessment curves or above, critical conditions are reached.

The equation of the failure curve divided by the safety factor defines a security curve where the structure is safe with a safety factor greater than the conventional design and ensures a proper design according to the codes.

The safety factor associated with the structure situation is simply defined by the relationship:

$$f_s = \frac{OC}{OA} \tag{12}$$

The proper design is generally evaluated by comparing the obtained safety factor with the conventional value (often $f_s = 2$).

In a probabilistic approach, two isoprobability failure curves ($P_r = 1$ and $P_r = $ level 1 or 2) divide the FAD into three zones: the unsafe zone above the failure curve ($P_r = 1$); the safe zone with maintenance $P_r^* > 10^{-4}$ or 10^{-6}; and the safe zone without maintenance $P_r^* < 10^{-4}$ or 10^{-6}, see Fig. 4.

Any assessment point of the coordinates ($L_r^* - k_r^*$) is situated on an isoprobability curve P_r^* and the safety factor keeps the same definition as for a deterministic FAD (Adib et al. 2007).

The criticality of the situation of a structure is evaluated with better accuracy by introducing the real scatter of the parameters defining the crack extension force on the defect (load, defect size, fracture toughness). It is not based on an empirical and unique safety factor for all materials and situations but on the acceptable risk from an economic and societal definition of the risk.

Fig. 4 Typical presentation
of a probabilistic failure
assessment diagram (*PFAD*)

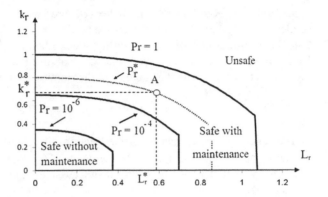

5 Monte Carlo and FORM/SORM Methods

Monte Carlo (MC) is a simple method that uses the fact that the failure probability integral can be interpreted as a mean value in a stochastic experiment (Rubinstein 1981). An estimate is therefore given by averaging a suitably large number of independent outcomes (simulations) of this experiment. The basic building block of this sampling is the generation of random numbers from a uniform distribution (between 0 and 1). This random number can be used to generate a value of the desired random variable with a given distribution. Using the cumulative distribution function $F(X)$, the random variable would then be given as:

$$X = F_X^{-1}(u) \qquad (13)$$

Then, to calculate the probability of failure, a multidimensional integral has to be evaluated:

$$P_F = P_r[g(X) < 0] = \int_{g(X)>0} f_x(x)dx \qquad (14)$$

where $g(X)$ is a limit state function and $f_x(x)$ is a known joint probability density function of the random vector X (Madsen et al. 1986).

To calculate the failure probability, one performs N deterministic simulations and determines whether the component analyzed has failed (i.e., if $g(X) < 0$) after every simulation. For a count of the number of failures, N_F, an estimate of the mean probability of failure is:

$$P_F = \frac{N_F}{N} \qquad (15)$$

The FORM (First Order Method) and SORM (Second Order Method) methods are based on the notion of reliability index. Many indexes are defined: the Rjanitzyne-Cornell index (Cornell 1969), the reliability index of first order (Madsen et al. 1986), the Hasofer and Lind index (Hasofer and Lind 1974), and Ditlevsen's (Ditlevsen and Madsen 1996) generalized index. Hasofer and Lind (1974) showed that the measurement of the reliability index should be taken in a normalized Gaussian variable space. For uncorrelated variables of any law, the principle of this transformation is to write the equality of distribution functions:

$$\Phi_u = F_x(x) \Rightarrow x \to u = \Phi^{-1} F_x(x). \tag{16}$$

The resulting transformation of the relationship is called isoprobabilistic transformation and is denoted by T in the following. The change in the variable is made in a new space of statistically independent Gaussian variables of mean zero and unit of standard deviations.

Jallouf (2006) extended this T transformation with other distributions.

First- and second-order reliability methods are general methods of structural reliability theory (Pluvinage and Sapounov 2007). The first- and second-order probabilities of failure estimates $P_{F,1}$ and $P_{F,2}$ are given by:

$$P_{F,1} = \Phi(-\beta)$$

$$P_{F,2} \approx \Phi(-\beta) \prod_{i=1}^{N-1} (1 - \kappa_i \beta)^{-1/2} \tag{17}$$

κ_i's are the principal curvatures of the limit state surface at the design point but other cumulative distributions can be used. FORM/SORM are analytical probability computation methods. Each input random variable and the performance function $g(x)$ must be continuous.

6 Example 1: Steel Pipe With a Semielliptical Defect. Surface Failure Probability Versus Internal Service Pressure

In this example, the failure probability of a gas pipe subjected to internal pressure is computed using probabilistic fracture mechanics coupled with the SINTAP procedure. Two methods are used: MC and FORM/SORM. These methods are compared; their advantages, inconveniencies, and limitations are presented.

Figure 5 shows an example of failure of a pipe under internal pressure.

6.1 Parameters Introduced in Probabilistic Fracture Mechanics

In probabilistic fracture mechanics, load, defect size, and fracture toughness are randomly distributed. Distributions are chosen according to the best fit of the experimental data associated with a confidence test such as Kolmogorov–Smirnov

Fig. 5 In spiral failure of a
gas pipe

and therefore they are not the same for each parameter. These random parameters are treated as not correlated. They can follow a normal, log normal, Weibull, or exponential distribution.

The manufacture of low carbon steel leads to an upper bound of the coefficient of variation $CV = 0.1$, for ultimate strength, yield stress, and fracture toughness (Pluvinage and Sapounov 2007). The pressure distribution obeys the same CV value. We note that for exponential distribution the coefficient of variation is necessarily taken as a unit. The presentation of the method will be arrived at with the value of the coefficient of variation.

Defect size a distribution is taken as an exponential distribution and therefore the coefficient of variation is necessarily taken as a unit. The fracture toughness KIC is assumed as Weibull distribution.

Yield stress σ_y, ultimate strength, and load or pressure are assumed to follow a normal distribution. For yield stress, ultimate strength and fracture toughness of steel, an upper bound of the coefficient of variation $CV = 0.1$ is used. The pressure distribution obeys the same CV value.

6.2 Pipe Under Internal Pressure

In a pipe, stresses occur in two directions: along the circumference (the so-called circumferential stress) and longitudinally in the axis (referred to as longitudinal or axial stress). Cracks initiate from a surface or metallurgical defect and generally grow along the axial direction and perpendicularly to the direction of principal tensile stress, which is the circumferential stress. This direction of growing obeys the principle of less energy. For conservative reason, a defect which promotes pipe failure is assumed to be a semielliptical or semispherical surface crack of depth a and length $2c$, see Fig. 6.

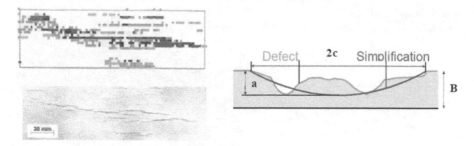

Fig. 6 Crack detection results in a pipeline from an ultrasonic intelligent pig and the corresponding crack pattern in the pipeline

For such a defect, the applied stress intensity factor is given by the following
relationship:

$$K_{ap} = \frac{pR_m}{B} \sqrt{\pi a}\, F\left(\frac{R_i}{B};\ \frac{a}{B};\ \frac{2c}{a}\right) \tag{18}$$

where p is the internal pressure; R_m and R_i are, respectively, the mean internal radius;
and B the thickness of the pipe wall. F is a geometry correction which is given in
(R6 1998).

Figure 7 shows a real defect distribution provided by the French gas company Gaz
de France. This distribution is described by three types of distribution (exponential,
normal, and beta). Figure 7 indicates that the best fit is obtained by exponential
distribution which is used further.

In the following, the crack aspect ratio is taken as $a/c = 1$ (semispherical defect).
The internal pressure in a gas pipe fluctuates continuously. It may vary depending
on the rate of gas injection into the network and the service of delivery points
downstream. Pipeline operators often cannot control these flows.

To characterize the pressure of a gas pipeline, one must consider three factors:

- The maximum pressure applied.
- The range of fluctuation of the pressure and the minimum pressure.
- The rate of pressure change (change almost instantly in some cases, over several
  days in others).

These fluctuations are commonly expressed by the R ratio which is the ratio of
the minimum pressure to the maximum. In this study, the R ratio has been given by
the gas company and is equal to 4/7. The lower limit pressure is 40 bar and an upper
limit pressure is about 70 bar.

This ratio is kept constant when the maximum pressure fluctuates with a
coefficient of variation of $CV = 0.1$. The fluctuations are normally distributed.

The Weibull distribution describes the scatter of fracture toughness and the
normal distribution the scatter of yield stress and ultimate strength. For conservative
reason, a lower bound of the coefficient of variation has been used ($CV = 0.1$).

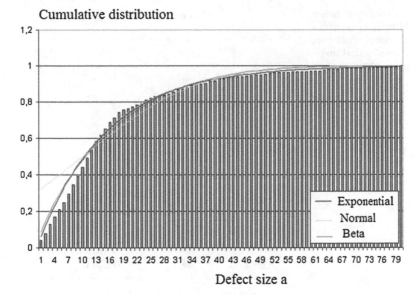

Fig. 7 Cumulative distribution of real defect depth in a natural gas network

Table 4 Parameters introduced in the probabilistic fracture mechanics analysis of a gas pipe

Parameters	Distribution	Variational coefficient	Mean
Fracture toughness	Weibull	0.1	116 MPa$\sqrt{m}$
Yield stress	Normal	0.1	410 MPa
Ultimate strength	Normal	0.1	528 MPa
Defect size	Exponential	1	3 mm
Internal pressure	Normal	0.1	55 bar

6.3 Data

A summary of distribution, mean, and coefficient of variation of the five parameters introduced in probabilistic fracture mechanics are given in Table 4.

These randomly distributed parameters are introduced in the MC and FORM/SORM methods associated with the failure criterion given by the SINTAP procedure.

Figure 8 gives the results of the failure probability given by the MC or FORM/SORM methods for a given maximum pressure converted into circumferential stress $\sigma_{\theta\theta}$. Two security levels are associated: level 2 associated with a conventional failure probability 10^{-6} and risk to human life and level 1 associated with a conventional failure probability 10^{-4} and no risk to human life. One notes that for the assessment point associated with a maximum service pressure of 70 bar or a circumferential stress of 125 MPa, the failure probability is less than 10^{-6} and the pipe with a surface defect is working in the security zone.

Fig. 8 Evolution of failure
probability of a pipe with a
surface semispherical defect
with circumferential stress
(Al Laham 1999)

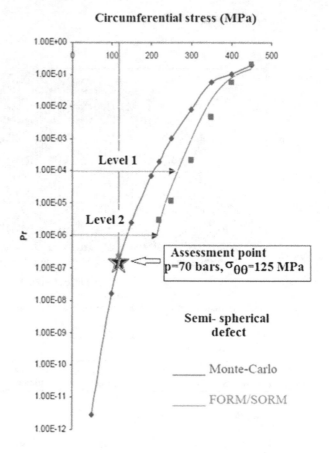

The two methods give similar results for high values of failure probability but for
low values the difference in the results is of two orders of magnitude.

We note that the MC method:

- Is generally applicable to all types of distributions method.
- Does not require special conditions on the failure functions.
- Is accurate for a number of simulations N which tends to $+\infty$ and then the
  method converges to the exact result.
- Is an effective and easy to implement method.

From the point of view of efficiency, the time associated with the MC increases
with the decrease in the probability of failure and the increase in the dimension of
the space variables.

This method is not economic in computational time for a failure probability
greater than 10^{-6}

The FORM/SORM methods have the following characteristics:

- The effectiveness of these methods is especially great in cases of small probabilities.
- The calculation time is independent of the level of probability.
- The errors in the results of these methods are difficult to estimate.

7   Example 2: Steel Pipe With a Semielliptical Surface Defect and Subjected to Internal Service Pressure. Influence of Temperature on Probabilistic Security Factor

In this example, the failure probability of a boiler pipe subjected to internal pressure is computed using probabilistic fracture mechanics coupled with the SINTAP procedure. The hoop failure stress $\sigma_{\theta\theta,c}$ for a probability of failure of 10^{-6} (level 2) is computed using the FORM/SORM method and its evolution with service temperature is determined.

Values of critical hoop stress $\sigma_{\theta\theta,c}$ are used to determine assessment points. These assessment points are plotted into an SINTAP diagram associated with two isoprobability failure curves ($P_F = 1$ and $P_F = 10^{-6}$). This latter iso probabilistic failure curve is used to determine the security factor.

7.1   Failure of Boiler Tubes

Boilers, gas turbine engines, and ovens are some of the systems which have components exposed to creep. Creep occurs under load at high temperature. An understanding of the behavior of high temperature materials is beneficial in evaluating systems failures in these systems. Failures involving creep are usually easy to identify due to the large deformation that occurs. Failures may appear to be ductile or brittle. Cracking may be either transgranular or intergranular. While creep testing is done at a constant temperature and constant load, system components may be damaged at various temperatures and loading conditions.

Figure 9 shows an example of boiler and Fig. 10 shows a failure in a boiler pipe.

The first Boiler and Pressure Vessel Code (1914 Edition) was published in 1915; it consisted of a 114-page book, measuring 5 × 8 in. Today there are 28 books, including a dozen dedicated to the construction and in-service inspection of nuclear power plant components, and two Code Case Books.

Fig. 9 Example of a boiler
with the different tubes

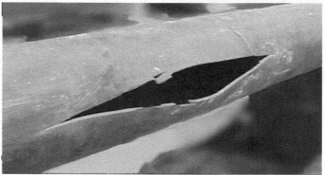

Fig. 10 Longitudinal failure in a boiler pipe

Table 5 Parameters
introduced in probabilistic
fracture mechanics analysis
of a boiler tube pipe

Parameters	Distribution	CV	Mean
Fracture toughness	Weibull	0.1	$F(T)$
Yield stress	Normal	0.1	$F(T)$
Ultimate strength	Normal	0.1	$F(T)$
Defect size	Exponential	1	3 mm
Hoop stress	Normal	0.1	55 bar

7.2 Data

The studied boiler tubes are made of steel, for which the tensile properties (yield
strength σ_y and ultimate tensile strength σ_u) and plane strain fracture toughness K_{Ic}
at five different temperatures are given in Table 5.

Boiler tubes have an external diameter of 273 mm and a wall thickness of
24 mm. The tube is butt welded and it is assumed that the welded joint contains
a longitudinal semielliptic surface crack (depth $a = 2.25$ mm, length $2c = 15$ mm,
and aspect ratio $a/2c = 0.15$).

Fig. 11 Evolution of the failure probability with the applied hoop stress. Influence of temperature. Boiler tube made in steel (Jallouf et al. 2005)

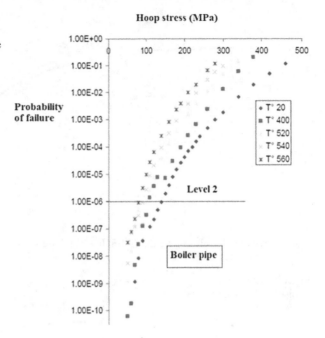

Under the effect of internal pressure, a hoop stress of 77 MPa is produced in the pipe at a normal service temperature of 400 °C. Three emergency situations are considered where the service temperature reaches 520, 540, or 560 °C. The stress intensity factor for a semielliptic surface crack is given in the SINTAP code and is given by Eq. (18)

The value of the geometrical correction for this defect is equal to $F\left(\frac{R_i}{B}, \frac{2c}{a}, \frac{a}{B}\right) = 1.445$. A small change in a produces only a small change in F, which is therefore considered as a constant in the present study.

Failure probability for five different temperatures (20 °C; 400 °C; 520 °C; 540 °C; 560 °C) has been computed using the FORM/SORM methods. Distribution, mean, and coefficient of variation values for each of the five parameters used in this analysis are reported in Table 5.

Yield strength, ultimate tensile strength, and plane strain fracture toughness are temperature dependent; hoop stress and defect size are independent of temperature.

7.3 Results and Security Factor

Figure 11 shows that the failure probability decreases with increasing temperature at constant hoop stress.

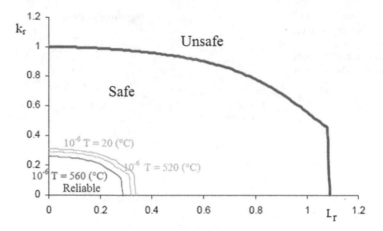

Fig. 12 Security domains where the probability of failure is less than the conventional value of 10^{-6}. Boiler tube made in steel

Fig. 13 Definition of (global) and partial security factors in the security domain of a FAD

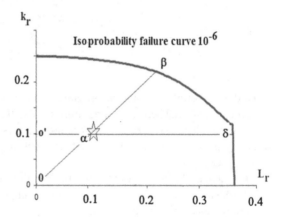

The isoprobability failure curve defines the inside of the plane (L_r, k_r), a domain where the failure probability PF is less than a conventional value P_F^*, $P_F < P_F^*$. This domain is called the security domain, except for $PF^* = 1$ which determines the safety domain.

Isoprobability failure curves are sensitive to temperature because k_r is sensitive to temperature through its denominator KIC, as is also L_r through its denominator R_c.

Figure 12 indicates security domains where the probability of failure is less than the conventional value of 10^{-6} for three different temperatures (20 °C, 520 °C, and 560 °C). One notes that the size of the security domain decreases with temperature.

The security domain defines the security factor associates where the failure probability is less than a conventional value PF^* and it is written F_{S,P_F^*}.

When the assessment point is inside the security domain, the (global) security factor is given by (Fig. 13):

$$F_{S,P^*} = \frac{0\beta}{0\alpha}. \tag{19}$$

Table 6 Safety factor F_s, determinist security factor $F_{S,2}$, the security factor F_{S,P_{fF}^*}, and the partial security factor on applied stress $F_{S,\sigma,f}$ for the five studied temperatures

Temperature (°C)	20	400	520	540	560
F_s	6.22	5.29	4.9	4.63	4.39
$F_{S,2}$	3.11	2.6	2.4	2.3	2.20
F_{S,P_F^*}	1.94	1.55	1.43	1.3	1.16
F_{S,σ,P_F^*}	1.94	1.42	1.29	1.23	1.16

The partial security factor on applied stress by (Fig. 13):

$$F_{S,\sigma,P_f^*} = \frac{0'\delta}{0\alpha}. \tag{20}$$

The security factor is also determined in a deterministic way by dividing the failure assessment curve by a conventional safety factor (generally 2) and is denoted as $F_{S,2}$.

In Table 6, the safety factor F_s, the determinist security factor associated with a safety factor of 2, $F_{S,2}$, the security factor F_{S,P_f^*} and the partial security factor on applied stress, F_{S,σ,P_f^*}, are reported for the five studied temperatures.

One notes that the safety factor is relatively high but decreases by 30 % when the temperature increases from 20 to 560 °C. The deterministic security factor decreases more for the same range of temperature (41 %) but remains higher than 2.

The probabilistic security factor is less than the determinist and decreases also with temperature according to:

$$F_{S,10^{-6}} = 2.789 \; T^{-0.1167} \tag{21}$$

The partial probabilistic security factor is less than the global and decreases similarly with temperature.

8 Example 3: Water Pipes With a Semielliptical Surface Defect and Subjected to Water Hammer. Influence of Material on Probabilistic Security Factor

In this example, the failure probability of a water pipe submitted to a water hammer is computed using probabilistic fracture mechanics coupled with the SINTAP procedure. The MC method is used.

Stochastic assessment points are introduced to the SINTAP diagram which is modified by the introduction of failure domains defined by their failure angles. The influence of the material on the statistical distribution of the safety factor is studied.

8.1 Water Hammer

Water hammer is a pressure surge or wave resulting when a fluid in motion is forced to stop or to change direction suddenly (Carlier 1972). Water hammer commonly occurs when a valve is instantaneously closed at one end of a pipeline system. Then, a pressure wave is formed and propagates in the pipe with a celerity C. The wave undergoes reflections at the other end of the pipe (a reservoir, for example) and at the valve. The time period for a complete cycle is $T = 4L/C$, where L is the pipe length. The time $T_0 = 2L/C$ is called the critical time (the propagation time for a pressure wave to travel the length of the pipe and return back). Water hammer can be analyzed by two different approaches, the column mass oscillation theory which ignores fluid compressibility and pipe wall elasticity, or transient wave analysis including compressibility and elasticity. The second approximation is appropriate when the time of closure t_c of the valve is such that $t_c < T_0$. In this case, considering elasticity is necessary and the pressure at the valve will be the same as that for instantaneous closure.

The maximum magnitude of the water hammer pulse, assuming a valve that closes instantaneously, can be estimated from the Joukowsky equation:

$$\Delta p = \rho C \Delta V \tag{22}$$

where Δp is the magnitude of the pressure wave (Pa), ρ is the density of the fluid (kgm^{-3}), C is the pressure wave celerity in the pipe (ms^{-1}), and ΔV is the change in the fluid's velocity (ms^{-1}). The pulse comes about due to Newton's law and the continuity equation applied to the deceleration of a fluid element.

More explanations about water hammer can be found in Wylie et al. (1993).

8.2 Characteristics of the Studied Water Network

The present study concerns a simple water network which consists of five pipe sections, a reservoir, and a pump. Each pipe section is connected at two nodes. The reservoir is located at node 6, the pump at node 1.

Each pipe section has the same diameter ($D = 575.6$ mm) and the same thickness ($w = 20.7$ mm). The pipe characteristics and discharges are shown in Table 7.

The schema of the water network is given in Fig. 14 with pump, reservoir, and pipe section locations. The pump is located at node # 1 and its characteristics are given in Table 8. The reservoir is located at node # 6 and its characteristics are given in Table 9.

Table 7 Characteristics of each pipe section

Pipe	Length L (m)	Diameter D (mm)	Flow rate Q (m³/s)	Wall thickness T (mm)
1	834.3	575.6	0.306	20.7
2	626.2	575.6	0.306	20.7
3	3,986.8	575.6	0.306	20.7
4	1,732.7	575.6	0.306	20.7
5	831.7	575.6	0.306	20.7

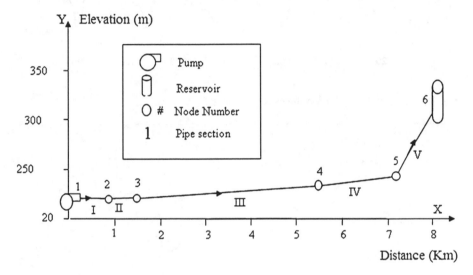

Fig. 14 Schema of the studied network with five pipe sections, one pump and one reservoir

Table 8 Pump characteristics

Node #	1
Flow rate (m³/s)	0.306
Head rate (m)	121.84
Rated pump speed (rpm)	1,450
Rated efficiency %	90
Coefficient of variation	0.1
Standard deviation (m)	12.1

Table 9 Reservoir characteristics

Node #	6
Head level (m)	326.69
Coefficient of variation	0.2
Standard deviation (m)	65.2

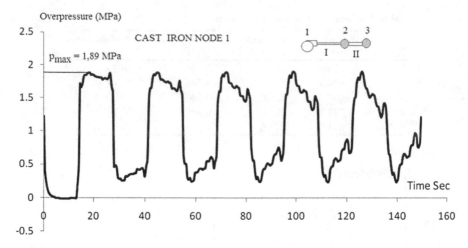

Fig. 15 Example of a head height-times curve at node 1

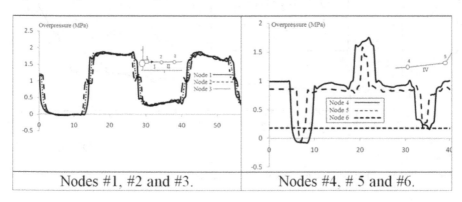

Fig. 16 Head height-times curves for different node locations

8.3 *Water Hammer Maximum Pressure*

For each node, the software provides the head height-times curve. An example of such a curve is given in Fig. 15 for a steel pipe and for node 1. We note a series of pressure pulse of about 18 s which increases with time and a maximum pressure of 1.89 MPa, see Fig. 15. The results were obtained for a head level of 121 m for the pump and 326.9 m for the reservoir.

The head height-times curves are different according to the node location: they are very similar for nodes #1, #2, and #3 but differ strongly for nodes #4, #5, and #6, as can be seen in Fig. 16.

The water hammer maximum pressure p_{max} is defined on the first pulse, after its amplitude decreases due to damping.

Table 10 Input and output coefficients of variation in the water network

	Rotated pump rate	Reservoir head level	Circumferential stress
Distribution	Normal	Normal	Normal
Coefficient of variation (CV)	0.1	0.2	0.438 (cast iron)
			0.38 (polyethylène)

The pipe diameter, rated pump speed, and head level are identical for each case studied, the friction coefficient is 0.0153 for cast iron and maximum pressure is 2.01 MPa.

Using an MC method, we generated several coupled values of reservoir head level and rotated pump rate; two parameters distributed according to a normal distribution with a coefficient of variation $CV = 0.1$. This induces an overpressure due to the random distribution of the water hammer and consequently a random distribution of circumferential maximum stress $\sigma_{\theta\theta,max}$ according to the following relationship:

$$\sigma_{\theta\theta,max} = \frac{p_{max} D}{2w} \tag{23}$$

We can note in the table that the resulting distribution of maximum circumferential stress has a considerably increasing coefficient of variation which is close to 0.4 for the two pipe materials (Table 10).

The $\sigma_{\theta\theta,max}$ distribution parameters have a mean of 41.53 MPa and a standard deviation of 18.21 (MPa). These results are introduced later to establish the FAD.

8.4 Data

Failure is assumed to be initiated by a longitudinal semielliptical crack-like defect with a notch angle of $\Psi = 0°$ and a notch radius of $\rho = 0.25$ mm. The stress intensity factor was calculated by introducing the circumferential maximum stress and geometry of the defect in Eq. (18). The obtained value of the stress intensity factor is low compared to the notch fracture toughness and consequently we are in elastic loading conditions. Pipe and defect geometries are given in Fig. 17.

In the past, water pipes were generally made of cast iron. Its mechanical properties and their distributions are shown in Table 11.

The current trend is to replace cast iron pipes which are relatively brittle. Polyethylene is an interesting material with a high performance index K_c/E ($\sqrt{m}$) and the possibility of using long sections without connections. The mechanical properties and their distributions are shown in Table 12.

The parameters of the maximal pressure distribution are extracted from Table 13. Coefficient of variation values is relatively high for both materials

Fig. 17 Defect geometries: the defect depth is $a = 2$ mm for cast iron with an aspect ratio c/a. It is 4.5 mm for polyethylene with the same aspect ratio

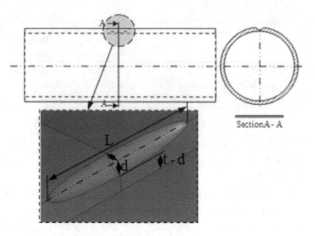

Table 11 Mechanical properties of cast iron and their distribution

Mechanical properties	Yield strength	Ultimate strength	Hoop stress	Fracture toughness	Defect
Mean	300 MPa	420	41.53 MPa	14.9 MPa$\sqrt{m}$	2 mm
CV	0.1	0.1	0.44	0.1	
Distribution	Normal	Normal	Normal	Weibull	Exponential

Table 12 Mechanical properties of polyethylene pipe and their distribution

Mechanical properties	Yield strength Re	Ultimate strength	Hoop stress	Fracture toughness	Defect
Mean	30 MPa	30 MPa	25.37 MPa	4 MPa$\sqrt{m}$	0.5 mm
CV	0.1	0.1	0.384	0.1	
Distribution	Normal	Normal	Normal	Weibull	Exponential

Table 13 Parameters of the circumferential maximum stress distribution for the two pipe materials

	Cast iron	PVC
Mean μ (MPa)	41.53	25.37
Standard deviation σ (MPa)	18.21	9.76
Coefficient of variation	0.438	0.38

9   Results

Using the MC method, many assessment points (40–50) were generated using the characteristic parameters of the distribution. The different FADs obtained are presented in Fig. 18. For these FADs, an average assessment point $\overline{A}$ is calculated from the mean values of all the variable parameters. Coordinates of the assessment point $\overline{A}$ for each pipe material and associated isofailure probability are given in Table 14. One notes that isofailure probability for polyethylene is higher. This is due to the low fracture toughness of this material.

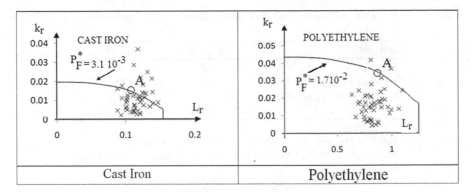

Fig. 18 Probabilistic FADs and assessment points generated by the MC method

Table 14 Coordinates of the assessment point for each of the two pipe materials and associated failure probability

Material	k_r	L_r	P_F^*
Cast iron	0.0168	0.1139	3×10^{-3}
Polyethylene	0.0377	0.833	1.7×10^{-2}

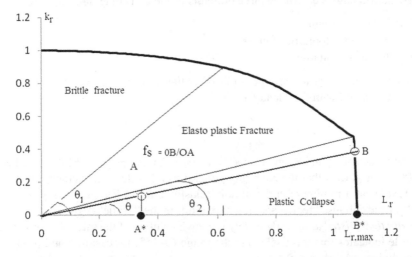

Fig. 19 Definition of brittle fracture, elastoplastic failure, and plastic collapse zones in the FAD and definition of domain angles θ_1 and θ_2

The assessment point associated with mean values of k_r and L_r, and P_F^*, the associated probability of failure, are reported in this figure and in Table 14. One notices that the probability of failure is higher for polyethylene.

Domain Angle The FAD can be presented in polar coordinates (L_r, θ) see Fig. 19. Two particular values are noted in this polar diagram: θ_1 and θ_2. The first polar angle θ_1 corresponds to the angle of the intercept of the failure curve at abscissa $L_r = 0.62$ and is associated with a conventional value of gross failure stress of 62 % of the yield stress. The second polar angle θ_2 corresponds to the intercept of the vertical line of $L_{r,max}$ abscissa.

Table 15 Mean values and standard deviation for the safety factor f_S* of the two materials

Material	Cast iron	Polyethylene
Mean (μ)	10.40	1.62
Standard deviation (σ)	1.58	0.19
Coefficient of variation (CV)	0.12	0.12
$f_{s,L}$	8.92	1.2

Table 16 Kolmogorov–Smirnov parameter values for Weibull, log normal, and normal safety factor distributions of safety factors in a cast iron pipe

	Weibull	Log normal	Normal
Cast iron	0.093354	0.053378	0.075910
Steel	0.12948	0.096424	0.11876
Polyethylene	0.060350	0.095256	0.072409

These two angles determine three domains in the FAD diagram:

If $\theta < \theta_1$ brittle fracture
If $\theta_1 < \theta < \theta_2$ elastoplastic fracture
If $\theta > \theta_2$ plastic collapse

If we consider that we have pure plastic collapse for $\theta = 0$, we define the safety factor associated with limit load $f_{s,L}$

$$f_{s,L} = OB^*/OA^* \tag{24}$$

The values of θ_1 and θ_2 are, respectively, $\theta_1 = 55$ and $\theta_2 = 22$. Muhammed et al. (2000) have shown that the emerging general trend is that, on average, the MS in the FAD is at a minimum in the middle (elastic–plastic) region, slightly higher in the "plastic collapse" region and at a maximum in the "brittle" region.

Statistical Distribution of the Safety Factor. Evolution of the safety factor with the θ angle indicates that this angle is in the range (0–15°) for cast iron and (0–4°) for polyethylene. All data are in a narrow scatter band of the range $[\mu - 3\sigma; \mu + 3\sigma]$ and also reported in the region of plastic collapse indicated by the safety factor $f_{s,L}$ computed from the ultimate pressure achieved by code ASME B31G (American National Standard Institute (ANSI)/American Society of Mechanical Engineers (ASME) 1984) (see Table 15). The safety factor distribution is well represented by a Weibull distribution as indicated by a Kolmogorov–Smirnov test (see Table 16) which is significant at 74 % for cast iron and 93 % for polyethylene.

We note that polyethylene has the lowest safety factor and that the coefficients of variation CV are similar for the two materials. The f_s mean value for polyethylene is not acceptable for safety because $f_s < 2$. The safety factor computed for fully plastic collapse is less than the mean value of f_s for each material. This can be explained by the fact that f_s is a global safety factor and $f_{s,L}$ is a partial one.

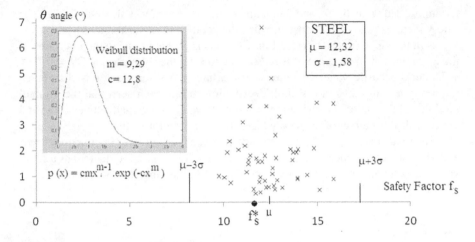

Fig. 20 Distribution of the safety factor with the θ domain angle for a cast iron pipe

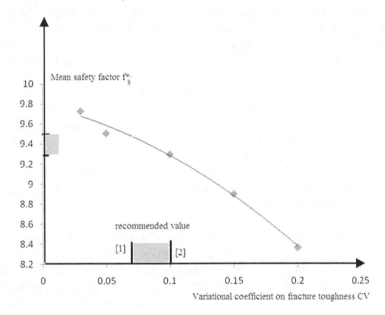

Fig. 21 Evolution of the mean safety factor with coefficient of variation

The scatter on the probabilistic safety factor is relatively moderate as can be seen in Fig. 20. One notes that the majority of the data is in the range $[\mu - 3\sigma; \mu + 3\sigma]$.

Influence on Coefficient of Variation of Safety Factor. The mean safety factor has been computed for the fracture toughness coefficient of variation which varies in the range (0.03–0.2). It can be noted that the safety factor varies from 9.8 to 8.4, that is, a relative decrease of 15 % (Fig. 21). Failure probability increases from 6.10^{-3} to $1.8.10^{-2}$. The difference in safety factor computed with Haugen's (Haugen 1980) recommended value (0.07) and Sapounov et al.'s (Sapounov et al. 1996) is 2 %. The safety factor is practically constant when the coefficient of variation on yield stress varies in the range (0.03–0.2) (see Fig. 18). This is probably due to the low value of the S_r parameter. In this case, it is reasonable to keep the simple value CV = 0.1.

Under stochastic water hammer overpressure, the safety factor is distributed randomly according to the Weibull distribution. It has been seen that the Weibull modulus of the material distribution is about 10 which is a value relatively high and the confidence interval of the safety factor has a satisfactory value.

A recommended value of 0.1 for the coefficient of variation for fracture toughness and yield stress is proposed. It has been noted that the mean safety factor has a small variation when the coefficient of variation varies in the range (0.03–0.2).

10 Conclusions

The classical deterministic global safety factor has been used for a long time, particularly its traditional value of two. Few engineers know that it was introduced 39 centuries ago. For designs against brittle fracture, it is necessary to introduce partial safety factors for each of the involved parameters, fracture toughness, load and defect size.

It has been seen that global and local safety factors for the design of a structure or a component can be easily determined using the FAD.

Due to the progress in material science and the consecutive reduction of scattering in mechanical properties, it appears that the use of the deterministic safety factor induces conservatism and increases in weight and costs. Therefore, a conventional low failure probability is targeted. The current value of a failure risk of 10^{-6} is used and it seems difficult for economic reasons to reduce this value. The MC and FORM/SORM methods are powerful tools in evaluating the security factor associated with this conventional value.

Design codes currently maintain the classical deterministic safety factor, generally by means of design curves or design values. The probabilistic approach of failure risk is only admitted as an additional method at the highest level of design (level 3). One notes that in Eurocode 5, the design stress is defined in a probabilistic way as the 5th percentile of the material strength distribution. This can be considered as a first step in introducing the probabilistic approach of the safety factor.

References

Kumar V, German MD, Shih CF (1981) An engineering approach of elastic plastic fracture mechanics. *EPRI*, NP1931, Res. Pr 1237–1

Adib H, Jallouf S, Schmitt C, Carmasol A, Pluvinage G (2007) Evaluation of the effect of corrosion defects on the structural integrity of X52 gas pipelines using the SINTAP procedure and notch theory. Int J Pres Ves Pip 84(3):123–131

Al Laham S (1999) Structural integrity branch, stress intensity factor and limit load handbook. Issue 2. British Energy Generation Ltd., Gloucester

American National Standard Institute (ANSI)/American Society of Mechanical Engineers (ASME) (1984) Manual for determining strength of corroded pipelines, ASME B31G

Carlier M (1972) Hydraulique générale et appliquée. Edition Eyrolles-Paris

Cornell CA (1969) A probability-based structured code. J Am Concr Inst 66(12):974–985

Ditlevsen O, Madsen HO (1996) Structural reliability methods. Wiley, New York

Dugdale DS (1960) Yielding of steel sheets containing slits. J Mech Phys Solid 8(2):100–104

Griffith AA (1921) The phenomena of rupture and flow in solids. Phil Trans Roy Soc Lond A 221:163–198

Hasofer AM, Lind NC (1974) An exact and invariant first-order reliability format. J Eng Mech 100:111–121

Haugen ED (1980) Probabilistic mechanical design. Wiley, New York

Jallouf S (2006) Approche probabiliste du dimensionnement contre le risque de rupture, These Metz

Jallouf S, Milović LJ, Pluvinage G, Carmasol A, Sedmak S (2005) Determination of safety margin and reliability factor of boiler tube with surface crack. Struct Integr Life 5(3):131–142

Madsen HO, Krenk S, Lind NC (1986) Methods of structural safety. Prentice Hall, Upper Saddle River

Mankovsky VA, Sapounov VT, Grebenthikova IN (1999) Le choix de la loi de distribution des caractéristiques mécaniques de courte durée des composites. Zavodsk Lab 65(4):45–52 (en russe)

Moulin D, Drubray B, Nedelec M (1993) Practical method of calculating J, in the appendix A16 of the RCC-MR: j_S method Société française des mécaniciens SFM. Journées, Courbevoie, FRANCE 1:55–65

Muhammed A, Pisarski HG, Sanderson RM (2000) Calibration of probability of failure estimates made from probabilistic fracture mechanics analysis. TWI Ltd, Offshore Technology report, 2000/021

Pluvinage G (1989) Mécanique Elastoplastique de la Rupture. Editeur Cepadues. Toulouse

Pluvinage G (2007) General approaches of pipeline defect assessment. In: Pluvinage G, Elwany M (eds) Safety, reliability and risks associated with water, oil and gas pipelines. Springer, New York, pp 1–22

Pluvinage G, Sapounov V (2007) Conception fiabiliste de la sécurité des matériaux composites, Revue des Sciences et de la technologie, No 16, pp 6–15

R6 (1998) Assessment of the integrity of structures containing defects. Nuclear electric procedure R/H/R6, Revision 3

Rubinstein RY (1981) Simulations and Monte-Carlo method. Wiley series in probability and mathematical statistics. Wiley, New York

Sapounov VT, Mankovsky VA, Jodin PH, Pluvinage G (1996) l'optimisation des caractéristiques de résistance des composites en fonction des paramètres des procédés technologiques de fabrication. Matériaux et Techniques. No 3–4, pp 21–25

SINTAP (1999) Structural integrity assessment procedure, final report EU project BE95-1462. Brite Euram Programme, Brussels

Wylie EB, Streeter VL, Suo L (1993) Fluid transients, in system. Prentice Hall, Upper Saddle River

Assessing the Complex Interaction and Variations in Human Performance Using Nonmetrical Scaling Methods

Oliver Straeter and Marcus Arenius

Abstract Human reliability and human performance in safety-critical systems is driven by various influences of complex interactions with the situational conditions under which the human needs to operate. Human reliability assessment methods (HRA methods) often do only consider these interrelations insufficiently, because they are missing a clear mathematical treatment of such interactions. Only simplified mathematical calculations like addition of the effects or simple dependency models are used to represent such interrelations. Better treatment of such interrelationships would require a multidimensional approach which also meets the mathematical constraints of data available for human reliability assessment. Typical constraints are for instance: adaptive change of weights and parameters relevant for human interactions in the course of a safety-critical interaction, multi actor environments, or individual styles of decision making.

At the department of human and organizational engineering and systemic design of the University Kassel/Germany, a method is being developed to overcome the constraints of HRA and herewith make human reliability assessment a key for the success of a resilient system. The method is built on the approach of the mathematical algorithm of NMDS (nonmetrical multidimensional scaling) and is applied to HRA issues in various industries such as nuclear, rail, aviation, or air traffic management. The section will outline the use of the method to enhance HRA for resilience engineering and will show examples of application in the industrial settings. It will conclude with an outline towards better inclusion of human contributions into reliability and safety.

O. Straeter (✉) • M. Arenius
Fachbereich Maschinenbau, Arbeits- und Organisationspsychologie, Universität Kassel,
Heinrich-Plett-Strasse 40, D-34132 Kassel, Germany
e-mail: straeter@uni-kassel.de

S. Kadry and A. El Hami (eds.), *Numerical Methods for Reliability and Safety Assessment: Multiscale and Multiphysics Systems*, DOI 10.1007/978-3-319-07167-1_22,
© Springer International Publishing Switzerland 2015

1 Importance to Model Multi-causal Relationships for Human Reliability and Safety

Human reliability is an essential part of system reliability assessment. A variety of methods exist, often distinguished into first generation HRA methods and second generation HRA methods. While first generation HRA methods focus on the task related behavior of operating crews, second generation HRA methods also take into account the goal-conflicts in safety relevant scenarios and intentions of operating crews that might deviate from achieving a safe state of the system, so called errors of commission. Critical are three difficulties, which call for a multidimensional approach to solve them. These difficulties will be described in the following. Section 2 then describes the method to address them and Sect. 3 will provide examples of applications.

1.1 Factoring PSF Dependencies

The correct assessment a human error probability (HEP) depends on the correct modeling of the PSF (performance shaping factors) which modify the human error potential. Typical PSF formulated in literature can be categorized according to Park (2014) into environmental influences, HMI, organization, procedures, and task complexity. Obviously, the influences are not independent from each other (e.g., a complex task needs to be accomplished under noisy environmental conditions with poor procedures).

The question for such complex interrelations is how much this will influence the HEP. Typically a monotone increasing function is assumed to increase the error potential given a certain impact of a PSF. As an example let us assume the basic HEP for a task might be P1. Then the HEP given P1|PSF1 is increased by certain factor f (e.g., $f1 = 5$ if PSF1 has a criticality of 5). The resulting HEP is then $P1|PSF1 = P1 * 5$. If two PSFs are valid and $f2 = 6$, then $P1|\{PSF1; PSF2\} = P1 * (5 + 6)$.

As the example shows, PSFs are considered to have an independent and linear causal relationship to increase a HEP. Some methods make extensive use of additive factored assessments like HEART-related methods (HEART, NARA, etc., Kirwan et al. 2004). In such methods one easily comes to mathematically nonsense-results (e.g., if P1 = 0, 1 in the above example then $P1|\{PSF1; PSF2\} > 1$. Such methods therefore request that the assessor needs to correct the additive calculation by own implicit experience in order to come to a reasonable result (Sträter et al. 2006).

In particular for second-generation HRA methods, a mathematically correct representation of PSF interrelations is required because of the manifold multi-causal relationships to be assessed (Reason 1990; Hollnagel 2006). This needs to be reflected in the factors and interrelationships of PSF modeling.

For a resilient assessment and design therefore the multi-relational nature of the common impact of PSFs needs to be mathematically described and modeled (Hollnagel 2006). A resilient method has to be able to assess conditional HEP given multiple PSFs and their interrelations.

The role of the NMDS for the calculation of PSF interrelationships will be demonstrated in Sect. 3.1.

1.2   Functional Fit of Automated Functions and Normal Behavior

In order to lower the (human) error probability of the system, replacement of human work by automation is often seen as the preferred solution. Main reasoning behind this conclusion is that the error probability of machines usually outmatches the error probability of human operators.

A large body of research has shown the flaws and pitfalls associated with this simplistic approach to reducing errors (Bainbridge 1983; Parasuraman and Wickens 2008; Parasuraman and Riley 1997; Sarter et al. 1997). One of the main challenges stems from the issue that the replacement of the human by a technical component is associated with an elimination of both the negative contribution as well as the (invisible) positive contribution for system safety. The latter of which ensures the flexibility of the system. Due to the elimination of the positive contribution of the human, new risks are introduced to the system that may compensate or even outweigh the gains from the elimination of the negative contribution (Hollnagel 2009). Thus, a key challenge for the design of safe automation is to ensure that the automation both eliminates the negative contribution of the human operator and ensures that the positive contribution is kept intact.

To achieve this, methods are needed that show how human operators contribute both positively to the creation of safety and how the same behavior may be associated with breakdowns in performance. Research has suggested, that this can only be achieved if models of *normal behavior* are created, that represent the *functions* that humans actually perform in order to conduct work (de Carvalho 2011; Dekker and Nyce 2012; Hollnagel 2012).

A proper human and automation strategy needs to fit the automated functions with the normal behavior of the Human, to stay safe and reliable. A misfit between the normal behavior and the functional allocation usually ends up in an error of commission due to decision errors, which is a typical second generation HRA issue.

In order to create models of normal performance, it is necessary to extract the functions that people define during work. This is done under the premise, that the functions that people perform during work are subjectively plausible to them. Thus, this subjective view is always rational from the point of view of the involved humans, given the resource constraints in terms of available time for actions and the amount of information that can be processed in that timeframe (Woods et al. 2010).

The role of the NMDS for the identification of user concepts will be demonstrated in Sect. 3.2.

1.3   Identification of Dynamic Influences of PSF on Normal Behavior

The key for a resilient system is its ability to continuously maintain operations when facing a variety of threats and disturbances (Hollnagel 2006). Human decisions and actions change, triggered by the change of performance shaping factors in the working environment and depending on the fit between their approximate adaptations and the contextual conditions that are present.

Safety assessments regarding human performance are often only made based as a static representation of such triggers in a scenario. The dynamics are usually covered in a simple manner using event-trees describing the accident development. This approach is suitable for first generation HRA methods, which deal mainly with errors of omission. In second generation HRA methods, errors of commission are modeled (e.g., due to a wrong decision, pilots switch off the wrong engine in case of a turbine failure). Errors of commission do modify the event-tree of the scenario and hence are difficult to be reflected in the event-tree technique. Ideally these require a dynamic risk modeling approach (Mosleh and Chang 2004).

A misfit between a situational context (i.e., a set of PSFs) and the user concepts at the specific time represents the potential for an error of commission. Prerequisite for a proper dynamic risk model is therefore the analysis of performance data over time to detect the potential trigger for the dynamic risk model.

The role of the NMDS representation for the representation of dynamic user concepts will be demonstrated in Sect. 3.3.

2   Modeling Multiple Interrelations

2.1   Statistical Requirements to Model Interrelations

All three issues outlined above require a method that enables to depict interrelations between situational conditions (PSF), user concepts, and the interaction of user concepts under certain situational conditions. Several statistical methods exist to deal with interrelations of parameters. The most known one is probably the factor analysis, which depicts the loads between different variables and expresses the interrelation as a correlation between the parameters.

Unfortunately factor analyses will not be appropriate for the calculation of the interrelations described above as one would need a statistical basis where correlations can be calculated. That means one needs several datasets where the parameters were systematically extracted, e.g. using an experimental setting, to then calculate higher relationships using cluster analyses or factor analysis, which are then tested using an ANOVA.

In HRA often the required data to perform such calculations is missing due to two aspects. (1) In HRA situations need to be assessed which are rare, unique, and non-repetitive. The correlation between the parameters is often not stable (e.g., an operator making an error due to bad ergonomic design during time stress at the beginning of a simulator session will adapt to the situation and be able to work error free after some trials. (2) Usually the interrelations are observable in events, where an accident came into existence. Hence most data on relevant PSF-interactions comes from event reports. However these only provide frequencies of common appearance but cannot be used to calculate correlations (event 1 is dissimilar to event 2 and hence both cannot be treated as one statistical variable which is needed for a factor analysis).

As correlations cannot be calculated for most interrelations of PSFs, potential other methods for analyzing interrelations of parameters are the multidimensional scaling methods (MDS) and in further development the nonmetrical multidimensional scaling methods (NMDS). See Borg and Groenen (2005) for an overview.

Both, MDS and NMDS perform an iterative algorithm to come to a picture about the interrelations. The MDS and NMDS both generate a 2- or n-dimensional graph showing the interrelations as distances in a scatterplot (NMDS graph). The closer the relation of two parameters is, the closer the points are located in the graph. The calculation is generally performed in the following steps:

First a matrix C representing the coherencies is required.

$$C = \begin{pmatrix} c_{11} & \cdots & c_{1j} \\ \vdots & \cdots & \vdots \\ c_{i1} & \cdots & c_{ij} \end{pmatrix} \tag{1}$$

Second a matrix of randomly spread points P is generated.

$$P = \begin{pmatrix} r_{11} & \cdots & r_{1j} \\ \vdots & \cdots & \vdots \\ r_{i1} & \cdots & r_{ij} \end{pmatrix} \tag{2}$$

From both a distance-matrix D is calculated.

$$D = C - P = \begin{pmatrix} d_{11} & \cdots & d_{1j} \\ \vdots & \cdots & \vdots \\ d_{i1} & \cdots & d_{ij} \end{pmatrix} \tag{3}$$

Given these distances, a new matrix P' is generated which moves the points of matrix C towards minimizing D. Based on this new P' a new D' is generated and so on. This iteration proceeds until the sum of the absolute values of all d_{ij} reaches a minimum or a solid baseline. This baseline depicts the uncertainty in the data and is usually called "stress" (s) or sometimes alienus. If $s = 0$ then the ideal fit of R to C is found. The greater the s is, the less the solution of matrix R fits to the set of matrix C (s is comparable to the standard deviation in parametric statistics).

While the MDS needs data on a relative scale (e.g., PSF1 is twice as important as PSF2), the NMDS only needs a rank of the PSFs (e.g., PSF1 is more important as PSF2). Therefore the data requirement for an NMDS is much simpler and hence more robust for the practice. Experts can for instance much easily rank the importance of PSF rather than assess the distance between them. Despite using ranks, an NMDS is able to generate data on the relative scale level or on the level of factor analyses (Borg and Staufenbiel 1989).

As the NMDS has the least requirements on the data quality and reveals equal results than the other multidimensional methods, it is the preferable approach for assessing interrelations of PSFs.

2.2   NonMetrical Multidimensional Scaling with Thermal Variation

MDS and NMDS have one disadvantage in using an iterative approach to find the solution. The solution is reconstructed from the randomly generated matrix which is then optimized regarding the distances known between the different parameters.

This approach bears the risk that finding a solution which is better than the original randomly chosen matrix for P is not necessarily the real optimum for P. These states are called local minima solutions.

To tackle the problem of local minima, the proposed solution is a temperature-dependent variation of a NMDS. Such algorithms for avoiding ending up in local minima are known from physics or biology. The idea is to overcome the local minima by randomly introducing an impacting energy (temperature). In physics this is known in the "potential gradient theory" for instance. Figure 1 illustrates the idea of the temperature effect on local minima.

Arrows symbolize the energy required (in the form of temperature) to overcome a minimum. Due to the energy, the ball "jumps" in all possible directions, the direction itself is random. Slowly the system cools down (i.e. from iteration to iteration) and the probability that the ball "lands" in absolute minima when cooled down completely is maximized. A local minimum can be overcome.

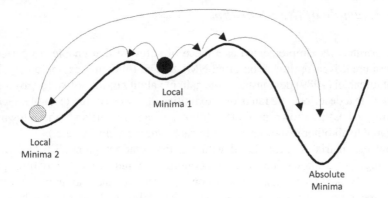

Fig. 1 The temperature effect to avoid local minima

A mathematical solution to the problem of local minima in NMDS is to define a random factor r that consists of a random number rnd calculated for each iteration i divided by the number of iterations multiplied by a weighting factor (0.005).

$$r = \left(\frac{rnd_i}{i * 0.005} \right) \qquad (4)$$

This random factor r can be used to increase the distortions in matrix D in the first iterations and then to cool down the distortions with an increasing number of iterations.

The distortion can further be amended by introducing a direction in which the points of matrix P will be moved. Best results could have been achieved introducing the direction rd which is applied to all columns of matrix P:

$$rd = \left\{ \begin{array}{l} 100|\, rnd - 0.5 > 0 \\ -100|\, rnd - 0.5 \leq 0 \end{array} \right\} \qquad (5)$$

$$p_{\bullet j}' = rd * r + p_{\bullet j} \qquad (6)$$

with $p_{\bullet j}$ is the column j of matrix P, $p_{\bullet j}'$ the revised value in column j of matrix P, i the number of Iterations of the NMDS, rnd the random number between 0 and 1, rd the random direction of the move, and r the amount of the move.

A further optimization is to weight matrix C by the relative frequency of observations of the interrelation between two parameters (e.g., one knows from incident analyses that the factor "poor interface design" was mentioned 10 times in 100 events ($h_i = 0, 1$) and the factor "time pressure" 20 times in 100 events ($h_i = 0, 2$). Weighting strategies are:

$$p_{\bullet j}'' = p_{\bullet j}' * w \quad \text{and} \quad w \in \left\{ \frac{1}{h_i + h_j}; \frac{1}{h_i * h_j}; \frac{1}{\min\left(h_i; h_j\right)}; \frac{1}{\max\left(h_i; h_j\right)} \right\}$$
$$(7)$$

2.3   Example of the Algorithm

The algorithm was implemented as a piece of software that enables to perform a temperature driven optimization of an NMDS. The following example from Borg and Staufenbiel (1989) demonstrates the gain in validity of the software: given is the matrix C that describes the ranks of proximities of towns (in the example: German speaking towns) as shown in Table 1 (rank 1 means for instance these towns are compared to all other distances closest to each other; 2 s closest, etc.).

Now the matrix R is produced with the first random point distribution. The following Fig. 2 shows how the temperature-triggered algorithm optimized the results in the next step. First in Fig. 2a one can see the classical approach without any temperature algorithm used. Dots move along a line and order themselves to the optimal setting of points. However, the algorithm ended up in a local minimum and placed Hamburg and Munich both in the lower right corner. In fact both towns are far more remote to each other as the final result in Fig. 2d shows. Now temperature is introduced to distort the points (Fig. 2b); one can see that dots move almost across the entire graph and are slowly cooled down (Fig. 2c) to then come to the valid result of the landscape of German speaking towns. The distances between the towns are equidistant. That means if one knows one true distance, one can calculate the remaining true distances proportionally.

Figure 3 shows the stress combined with the different states. Figure 2a is at the first minimum around iteration 10; Fig. 2b is about the iteration step 100 and Fig. 2c at 200 and the final result of Fig. 2d at 400+ iterations. The "mountain" crossed between iteration 50 and 150 is a typical behavior of the temperature curve showing that the algorithm achieved to overcome local minima and to find the absolute minimum.

3   Application of NMDS in Safety and Reliability Engineering

3.1   Factoring PSF Dependencies Based on Incident Analysis

3.1.1   NMDS of PSFs Based on Incidents

Due to the fact, that events are multi-causal they show different PSFs as well as their common appearance. This information can be used to calculate interrelations. Table 2 shows for instance, a table reflecting coinciding performance shaping factors in 143 railway events (from Arenius et al. 2013). The table shows the frequency of PSF interrelations. A high frequency signifies a high similarity of the PSFs in terms of their negative effect on human performance.

These frequencies can be directly used to calculate interrelations using an NMDS. Figure 4 illustrates the NMDS representation of the coinciding PSF from Table 2.

Table 1 Matrix C of proximities of German speaking towns

Town	Basel	Berlin	Frankfurt	Hamburg	Hannover	Kassel	Köln	München	Nürnberg	Stuttgart
Basel	–									
Berlin	45	–								
Frankfurt	20	33	–							
Harnburg	44	16	31	–						
Hannover	42	13	17	2	–					
Kassel	34	21	3	15	1	–				
Köln	28	37	4	27	14	8	–			
München	22	39	23	43	38	30	35	–		
Nürnberg	24	29	11	36	26	12	25	5	–	
Stuttgart	9	40	6	41	32	18	19	10	7	–

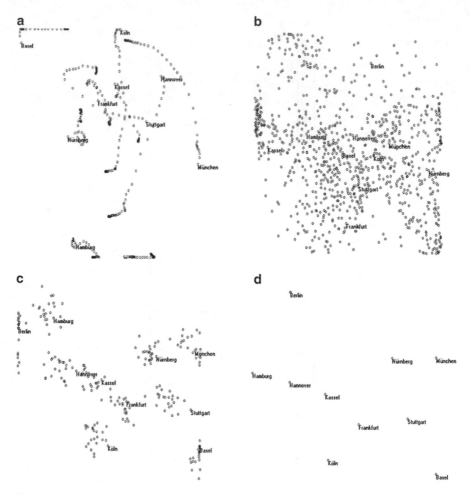

Fig. 2 Point calculation of matrix *P* (**a**) without temperature, (**b**) on temperature peak, (**c**) while cooling down, and (**d**) final result

The closer the PSFs are located, the more they tend to coincide in the event data and commonly to contribute to human errors. The more distant the points are, the more independent the PSFs are.

As evident from the illustration, human errors were associated with the coincidence of deficiencies in signal monitoring, the effect of distractions in the driver cabin and irregularities regarding other actors in the work environment (other delayed trains, unexpected change in tracks, etc.). The cluster of PSFs indicated by the dashed line in the figure therefore shows the vulnerabilities of the system that are associated with breakdowns in human performance.

Based on this NMDS representation, the analyst is able to explore and uncover the narratives associated with specific sets of coinciding PSF and therefore to detect the nature of the vulnerabilities of the system regarding the psf.

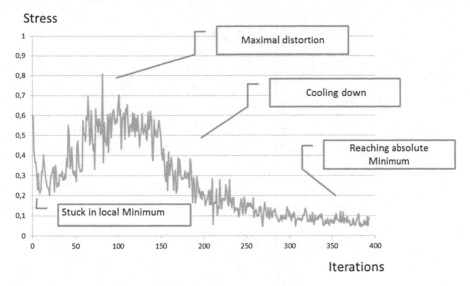

Fig. 3 Stress (alienus) curve for the temperature driven NMDS

3.1.2 Dependency-Measure Between PSFs

As discussed in Sect. 1, a dependency model is required if a combination of different PSFs is to be assessed. The NMDS graph can easily be used to calculate the dependencies between PSFs.

Because the NMDS graph delivers equidistant information, the factor of dependency is directly dependent of the distance in the graph. Two fixed points are known. If two dots lie on exactly the same position, then the dependency is 1 and the two most remote dots have the dependency of zero. Given this the dependency between two points can be expressed as:

$$f_{ij} = \frac{d_{\max} - d_{ij}}{d_{\max}} \tag{8}$$

with f_{ij} is the dependency-factor between PSF i and PSF j, d_{ij} the distance between PSF i and PSF j, $d_{\max}$ the maximal diameter of the NMDS graph.

In the above NMDS "signal monitoring" and "distraction" are very close and "time pressure" is comparable remote from "signal monitoring". Given $d_{\max} = 16$, $d_{\text{signal monitoring distraction}} = 0.8$ and $d_{\text{signal monitoring time pressure}} = 8$, the following factors can be calculated:

$$f_{\text{signal monitoring distraction}} = 0.95 \quad \text{and} \quad f_{\text{signal monitoring time pressure}} = 0.5$$

Table 2 Matrix *C* of proximities of PSFs in railway incidents

PSF	Total frequency	Authority	Job experience	Distraction	Stress	Time pressure	Signal monitoring	Irregularities	Cooperation
Authority	5	–							
Job experience	22	0	–						
Distraction	45	2	10	–					
Stress	7	1	1	5	–				
Time pressure	5	0	1	3	2	–			
Signal monitoring	55	4	11	41	5	2	–		
Irregularities	44	4	10	25	4	0	30	–	
Cooperation	9	3	1	4	1	1	5	8	–

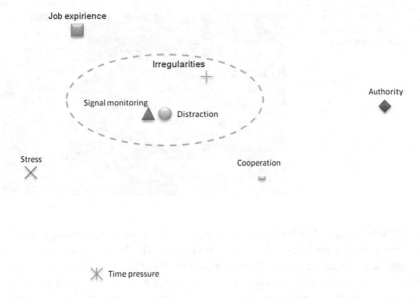

Fig. 4 NMDS of most important performance shaping factors for 143 events in railways (from Arenius et al. 2013)

Given one knows the HEP in a Task 1 given a PSF i then the additional effect of another HEP given PSF j can be calculated as outlined in the following example:

$$HEP_{Task1}|\{PSF_i; PSF_j\} = HEP_{Task1}|PSF_i + (1 - f_{ij}) * HEP_{Task1}|PSF_j \qquad (9)$$

with PSF_i is the leading PSF for assessment (e.g., "signal monitoring"), PSF_j the additional PSF (e.g., "distraction" or "time pressure").

3.2 Functional Fit of Automated Functions and Normal Behavior in a Process Control Environment

3.2.1 User Concepts and System Safety

The NMDS yields a representation that allows the analyst to look at the structure or underlying dimensions behind humans perception as reflected in a given metric (Borg and Groenen 2005). This means that, given the appropriate set of data for an NMDS analysis, the user concept of the human operators can be uncovered in order to generate models of normal performance.

A study conducted for the mining industry demonstrates how a NMDS can be applied to eye-tracking data of human bucket-wheel excavator operators in order to

Fig. 5 Areas of interest
(AoIs) of the bucket-wheel
excavator control room

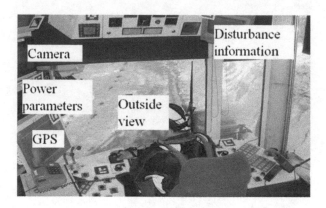

extract user concepts. Bucket-wheel excavators are large scale and highly automated machines used for surface coal mining operations. The human operators of these machines are located in a control room attached to excavator with a good view on the bucket-wheel that excavates the coal. Eye-tracking data is analyzed in terms on gaze location in the work place. These gaze locations are termed areas of interest (AoI) and denote objects of interest for the tasks to perform; an AoI can be anything from windows, to display and controls, etc. The control room and attached AoIs are presented in Fig. 5.

As a next step, the distribution of the gaze on the different AoIs can be represented by means of their relative cumulative proportion over the recording time. Figure 6 shows the result of such calculation.

When rotating the cumulative graphs of the figure clockwise 45°, the graphs show a vertical flow related to the *x*-axis. This flow then can be used to correlate the AoI-graphs. The resulting matrix shows positive and negative correlations between the areas of interest which serve as an input to the NMDS algorithm (Table 3).

Based on this matrix, a NMDS graph of the AoIs can be calculated as shown in Fig. 7a. By applying the NMDS to the correlation of the scanning of information sources (the AoIs) it was revealed, that there was a misfit between the way that the information displays were arranged in the control room (*shaded boxes* in Fig. 7b) and the way that the operators actually used them during normal work. Or, in more technical terms, the *structure behind the perceived similarity* of displays by the human operators did not match the actual physical placement of the displays in the control room. This hidden structure behind the perceived similarities of display represents the *user concept* (*white boxes* in Fig. 7b).

The NMDS produced by this study was based on data of the gaze distribution on displays. However, the same approach could be applied for the analysis of gaze distributions on individual elements of graphical displays, in order to reveal the best reconfigurations that fit the user concepts for a given scenario.

Cummulative AoIs allocation

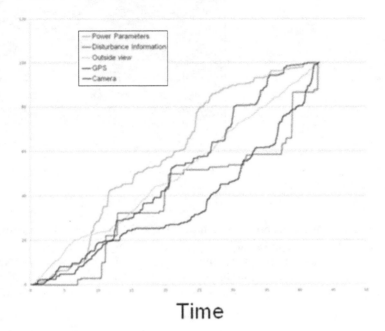

Time

Fig. 6 Cumulative gaze distribution on areas of interest over time in control room

Table 3 Matrix *C* of proximities of AoIs of mining operators

Correlation	Disturbance information	Camera	Power parameters	GPS	Outside view
Disturbance information	–				
Camera	0.06	–			
Power parameters	−0.06	−0.77	–		
GPS	−0.35	−0.45	0.41	–	
Outside view	−0.41	0.41	−0.66	−0.31	–

3.2.2 Calculating the Fit of User Concepts to Functional Arrangements

Based on the results of the NMDS representation, the information sources in the control room were rearranged in order to ensure a better mapping between how the operators visually scanned the display (the user concept) and the way that they were placed in the control room. Based on the results, it was possible to

- See what is happening: reveal the user concept, the hidden structure behind the distribution of the gaze on the information sources (areas of interest, AoI). The user concept shows how the information sources should be arranged in order to minimize transition times and ensure that linked information is found as efficiently as possible by the operators.

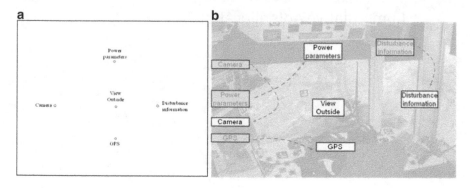

Fig. 7 Using the NMDS representation to fit control rooms to user concepts (design before—*shaded*, after NMDS redesign—*white boxes*)

- Implement a direct mapping and quick fixes: Map the NMDS representation directly into the physical working environment. Furthermore, due to the application of isotonic transformation (i.e. scaling and rotating of the NMDS representation), solutions can be found that fit the engineering and space constraint of the information-dense control room of the bucket-wheel excavator.

Based on this approach, the NMDS representation of performance data can be used in order to improve the fit between user concept and actual working environment. The misfit mf is simply expressed by the total distance of all elements of the AoI matrix of the workplace arrangements (in the NMDS Matrix C) and the AoI matrix of the user concept (in the NMDS matrix P).

$$mf = \sum \left| d_{ij} \right| \qquad (10)$$

Using an NMDS this is the absolute distance in the residual matrix D. The higher the misfit, the less fit is between the user concept and the working environment and the higher is the potential for a safety-critical impact of user concepts.

3.3 Identification of Dynamic Influences of PSF on Normal Behavior

3.3.1 Change of User Concepts Due to Trigger Events

The following NMDS shows changes in user concepts during the interaction of anesthetists with technology and team members in a highly critical and dynamic situation during a simulated operation in a high fidelity simulator.

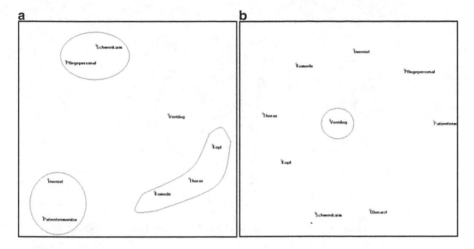

Fig. 8 NMDS representation of focused (*left*) and unfocused (*right*) user concepts in anesthesia

The left-hand NMDS-representation (Fig. 8a) shows the user concept of the anesthetist before the operating team was complemented with a senior team member and while she was being stressed by the team to diagnose the problem. Clusters of information sources are clearly visible (highlighted in *red*). These clusters reflect the user concept of the anesthetist during that particular time period: as one member of the operating team was pushing the anesthetist to actively diagnose the reason behind a dramatic decline in O_2-saturation of the patient, the gaze of the anesthetist often shifted between this stress-inducing team member and the display showing the vitality parameters of the patient. Furthermore, the gaze of the anesthetist shifted regularly between the body parts of the patient that can be used to determine signs of respiration (head and torso) and the health record of the patient, where the medical history and information of the patient is located which is critical for diagnosis.

However, this active role changed radically when the senior medical staff member entered the operating theater and took control over the scenario (Fig. 8b). The anesthetist adapted to this change by altering her monitoring strategy to a more heterogeneous and unfocused global monitoring of information sources, as evident from the approximate equidistance of the information displays in the NMDS representation. Thus, she exhibited a passive user concept that supplemented the senior medical staff member with information when requested.

Thus, the NMDS representation provides a very parsimonious representation of coping strategies that human operators make during work. If several measurements are done, gradually models of normal performance can be constructed and set into relation to the outcome of the applied user concepts, that is, which user concepts that prove efficient and which user concepts that lead to adverse outcomes.

This can also be done for scenarios in any given system where PSFs have been collected. For event data in railways e.g. PSFs could be analyzed in terms of their

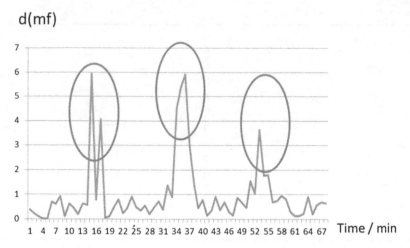

Fig. 9 Identification of dynamic changes of user concepts triggered by change of situational conditions

NMDS representation. If clusters of safety-critical PSF emerge, then these can in turn be tested empirically in terms of their effect on the user concepts and coping strategies of the human operators. Based on this, measures can be derived that target the specific coping strategies uncovered in the NMDS analysis of performance data.

3.3.2 Detection of Changes in User Concepts

The analysis of the eye-tracking data revealed that there was a difference in the NMDS representation depending on the presence or absence of additional expertise in the staff and regarding the influence of stressful interaction (PSF: bad team composition and stress).

The eye-tracking data was analyzed according to the approach outline in Sect. 3.2. The areas of interest (AoI) were defined and the gaze distribution was calculated as cumulative proportion of the total gaze distribution. Based on this representation, the correlations between the AoI were calculated. In order to detect the change of user concepts given a change in situational conditions, the distribution of the misfit *mf* over time can be used.

$$d(mf) = mf_t - mf_{t-1} \tag{11}$$

If one draws this function over time, a distribution as outlined in Fig. 9 will be generated. One can see three peaks in the example that represents a change in user concepts (highlighted with *circles*).

4   Outlook

Human performance is variable, as the adaptive and approximate process of coping with uncertainty forces people to "fill in the gaps" by relying on heuristics, trade-offs, and shortcuts that mostly produce the desired results, but in very rare cases lead to adverse outcomes (Hollnagel 2009). This adaptive capacity of the human is a necessity for system functioning as it guarantees the flexibility of the system and therefore constitutes an important contribution in keeping the system safe (e.g. Hollnagel et al. N 2005).

The section outlined the use of multidimensional methods for better modeling human adaptations for a better safety and reliability of systems. Human contributions are dependent of contextual conditions and of time dynamics under which a system operates. The fit of user concepts to the functions provided by a system determine to a great extent the human error potential. In particular the error potential for decision based, errors is determined by this fit. The modeling of interrelationships of PSF using a NMDS representation make it possible to adequately capture the interrelationships and to represent them in ways that make it possible for the analyst to identify the fit and vulnerabilities of the system.

The dynamic NMDS provides information regarding the alignment of human cognition/heuristics with the technical working environment. Thus, a new foundation for the design of human–automation interaction can be achieved in the sense that the HMI can be fitted to the structures that human regard as relevant in interaction for certain tasks and in specific contextual conditions.

With the approach to represent approximate adaptations and change of user concepts of human operators under the influence of PSFs, dynamic aspects of system safety can be enriched by human safety contributions.

This approach allows identifying where working systems are underspecified regarding human aspects. This also holds true for the humans working in the system and who will eventually face conditions for which neither they nor the system designers prepared the system. Using a multidimensional approach toward human performance herewith allows better assessment of the safety contribution and also to identify and assess the positive aspects of human behavior for safety.

References

Arenius M, Milius B, Schwencke D, Lindner T, Grippenkoven J, Sträter O (2013) System Mensch-Sicherheit modellieren (SMSmod): Sichere und robuste Systemgestaltung durch Modellierung ergonomischer Bedingungen für eine hohe menschliche Zuverlässigkeit (Forschungsbericht). Universität Kassel, Kassel, p 52

Bainbridge L (1983) Ironies of automation. Automatica 19(6):775–779. doi:10.1016/0005-1098(83)90046-8

Borg I, Groenen PJF (2005) Modern multidimensional scaling. Springer, New York

Borg I, Staufenbiel T (1989) Theorien und Methoden der Skalierung—Eine Einführung. Hans Huber, Bern

De Carvalho PVR (2011) The use of functional resonance analysis method (FRAM) in a mid-air collision to understand some characteristics of the air traffic management system resilience. Reliab Eng Syst Saf 96(11):1482–1498. doi:10.1016/j.ress.2011.05.009

Dekker SWA, Nyce JM (2012) Cognitive engineering and the moral theology and witchcraft of cause. Cogn Technol Work 14(3):207–212. doi:10.1007/s10111-011-0203-6

Hollnagel E (2006) Resilience—the challenge of the unstable. In: Hollnagel E, Woods D, Leveson N (eds) Resilience engineering: concepts and precepts. Ashgate, Aldershot, pp 9–17

Hollnagel E (2009) The ETTO principle: efficiency-thoroughness trade-off: why things that go right sometimes go wrong. Ashgate, Surrey

Hollnagel E (2012) FRAM: the functional resonance analysis method: modelling complex socio-technical systems. Ashgate, Farnham

Hollnagel E, Woods D, Leveson, N (2005) Resilience engineering - concepts and precepts. Ashgate. Aldershot. (ISBN 0754646416)

Kirwan B, Gibson H, Kennedy R, Edmunds J, Cooksley G, Umbers I (2004) Nuclear action reliability assessment (NARA)—a data-based HRA tool. PSAM 7 2004, Berlin

Mosleh A, Chang Y (2004) Model-based human reliability analysis: prospects and requirements. In: Apostolakis G, Soares CG, Kondo S, Sträter O (eds) Human reliability data issues and errors of commission. Special edition of the reliability engineering and system safety. Elsevier, Amsterdam

Parasuraman R, Riley V (1997) Humans and automation: use, misuse, disuse, abuse. Hum Factors J Hum Factors Ergon Soc 39(2):230–253. doi:10.1518/001872097778543886

Parasuraman R, Wickens CD (2008) Humans: still vital after all these years of automation. Hum Factors J Hum Factors Ergon Soc 50(3):511–520. doi:10.1518/001872008X312198

Park J (2014) A guideline to collect HRA data in the simulator of nuclear power plants. KAERI, South Korea. KAERI/TR-5206/2013

Reason J (1990) Human error. Cambridge University Press, Cambridge

Sarter N, Woods D, Billings CE (1997) Automation surprises. In: Salvendy G (ed) Handbook of human factors & ergonomics, 2nd edn. Wiley, New York

Sträter O, Leonhardt J, Durrett D, Hartung J (2006) The dilemma of ill-defining the safety performance of systems if using a non-resilient safety assessment approach—experiences based on the design of the future air traffic management system. In: 2nd Symposium on resilience engineering, Nice/France

Woods D, Dekker S, Cook R, Johannesen L, Sarter N (2010) Behind human error, 2nd edn. Ashgate, Surrey

Markov Modeling for Reliability Analysis Using Hypoexponential Distribution

Therrar Kadri, Khaled Smaili, and Seifedine Kadry

Abstract Reliability is the probability that the system will perform its intended function under specified working condition for a specified period of time. It is the analysis of failures, their causes and consequences. Reliability analysis is the most important characteristic of product quality as things have to be working satisfactorily before considering other quality attributes. Usually, specific performance measures can be embedded into reliability analysis by the fact that if the performance is below a certain level, a failure can be said to have occurred. Markov model is widely used technique in reliability analysis. The hypoexponential distribution is used in modeling multiple exponential stages in series. This distribution can be used in many domains of application. In this chapter we find a modified and simple form of the probability density function for the general case of the hypoexponential distribution. The new form obtained is used in an application of Markovian systems to find its reliability.

1 Introduction

The sum of random variable plays an important role in modeling many events (Ross 2011; Feller 1971). In particular the sum of exponential random variable has important applications in the modeling in many domains such as communications

T. Kadri (✉)
Beirut Arab University, Beirut, Lebanon
e-mail: therrar@hotmail.com

K. Smaili
Lebanese University, Zahle, Lebanon
e-mail: ksmeily@hotmail.com

S. Kadry
American University of the Middle East, Eguaila, Kuwait
e-mail: skadry@gmail.com

S. Kadry and A. El Hami (eds.), *Numerical Methods for Reliability and Safety Assessment: Multiscale and Multiphysics Systems*, DOI 10.1007/978-3-319-07167-1_23, © Springer International Publishing Switzerland 2015

and computer science (Trivedi 2002; Anjum and Perros 2011), Markov process (Jasiulewicz and Kordecki 2003; Mathai 1982), insurance (Willmot and Woo 2007; Minkova 2010) and reliability and performance evaluation (Trivedi 2002; Jasiulewicz and Kordecki 2003; Bolch et al. 2006; Amari and Misra 1997). Nadarajah (2008) presented a review of some results on the sum of random variables.

Many processes in nature can be divided into sequential phases. If the time of a process spends in each phase is independent and exponentially distributed, then the overall time is hypoexponentially distributed, see Trivedi (2002).

The hypoexponential distribution is the distribution of the sum of $m \geq 2$ independent exponential random variables. The general case of this distribution is when the m parameters do not have to be distinct. This case was examined by Smaili et al. (2013a). However, the case when the parameters are identical the hypoexponential distribution is the Erlang distribution (Anjum and Perros 2011). Also the case when the parameters are distinct, it is known to be the hypoexponential distribution with different parameters discussed by Smaili et al. (2013b).

In this chapter we find a modified and simple form of the probability density function (PDF) of the general case of the hypoexponential distribution. This modified form is a linear combination of the PDF of the known Erlang distribution. In the same way, the cumulative distribution function (CDF), moment generating function (MGF), reliability function, hazard function, and moment of order k for the general case of the hypoexponential distribution is written as a linear combination of CDF, MGF, reliability function, hazard function, and moment of order k of the Erlang distribution, respectively. Therefore, in order to obtain a comprehensive study of the hypoexponential random variable is from the known Erlang random variable. Moreover, we propose a recursive method to determine the coefficient of the linear combinations, A_{ij}, for the expressions obtained. At the end of first section an application of Markovian systems is given alongside the reliability analysis. Next, we examine the case of the hypoexponential RV with different parameters. Also, in the domain of reliability and performance evaluation of systems and software many authors used the geometric and arithmetic parameters such as (Gaudoin and Ledoux 2007; Jelinski and Moranda 1972; Moranda 1979). We study these two particular cases. We illustrate two examples for each case.

2   The Hypoexponential Random Variable: General Case

In this section we examine the general case of the hypoexponential random variable. We find a modified and simple form of the PDF for the general case of the hypoexponential distribution when its parameters do not have to be distinct. This modified form is found by writing the PDF of this distribution as a linear combination of the PDF of the known Erlang distribution. Also, this modified form shall generate a simple form of the CDF, reliability function, hazard function, MGF, and moment of order k for the general case of the hypoexponential distribution. Moreover, we give a recursive method to determine the coefficient of the linear combinations, A_{ij}, for the expressions obtained. At the end an application of Markovian systems is given alongside the reliability analysis.

Let X_{ij}, $1 \leq i \leq n$ and $1 \leq j \leq k_i$, be $m \geq 2$ independent exponential random variables with parameter α_i, written as $X_{ij} \sim \text{Exp}(\alpha_i)$, where $m = \sum_{i=1}^{n} k_i$. We define the random variable

$$S_m = \sum_{i=1}^{n} \sum_{j=1}^{k_i} X_{ij}$$

to be the hypoexponential random variable with parameters $\overrightarrow{\alpha} = (\alpha_1, \alpha_2, \ldots, \alpha_n)$ and $\overrightarrow{k} = (k_1, k_2, \ldots, k_n)$, where $\alpha_i > 0$ is repeated k_i times, $i = 1, 2, \ldots, n$. We write $S_m \sim Hypoexp\left(\overrightarrow{\alpha}, \overrightarrow{k}\right)$, see Smaili et al. (2013a).

However, the case when $n = 1$ the m parameters are identical and the hypoexponential distribution is the Erlang distribution (Anjum and Perros 2011). Written as $S_m \sim Erl(m, \alpha_i)$

In the following proposition, we state the PDF, CDF, reliability function, hazard function, MGF, and moment of order k for the Erlang RV, see Abdelkader (2003) and Zukerman (2012).

Proposition 1 Let Y be an Erlang RV with parameter $\alpha > 0$ and positive integer n. Then we have for Y

1. PDF is $f_Y(t) = \frac{(\alpha t)^{n-1} \alpha e^{-\alpha t}}{(n-1)!} I_{(0,\infty)}(t)$
2. CDF is $F_Y(x) = \frac{\gamma(n,\alpha x)}{(n-1)!} = 1 - \frac{\Gamma(n,\alpha x)}{(n-1)!}$
3. Reliability function is $R_Y(x) = \frac{\Gamma(n,\alpha x)}{(n-1)!}$
4. Hazard function is $h_Y(x) = \frac{(\alpha x)^{n-1} \alpha e^{-\alpha x}}{\Gamma(n,\alpha x)} I_{(0,\infty)}(t)$
5. MGF is $\Phi_Y(t) = \frac{\alpha^n}{(\alpha-t)^n}$ and
6. Moment of order k is $E\left[Y^k\right] = \frac{\Gamma(k+n)}{\alpha^k \Gamma(n)}$.

where $\gamma(n, \alpha x)$ is the lower incomplete gamma function and $\Gamma(n, \alpha x)$ is the upper incomplete gamma function defined as $\Gamma(n, \alpha x) = (n-1)! e^{-\alpha x} \sum_{k=0}^{n-1} \frac{(\alpha x)^k}{k!} = (n-1)! - \gamma(n, \alpha x)$ and $\Gamma(n)$ is the gamma function.

2.1 The PDF for the Hypoexponential RV

We start by stating in the following theorem which gives the PDF for the general case of the hypoexponential distribution given by Jasiulewicz and Kordecki (2003).

Theorem 1 Let $m \geq 2$ and $S_m \sim Hypoexp\left(\overrightarrow{\alpha}, \overrightarrow{k}\right)$ where $\overrightarrow{\alpha} = (\alpha_1, \alpha_2, \ldots, \alpha_n)$ and $\overrightarrow{k} = (k_1, k_2, \ldots, k_n)$. Then the PDF of S_m is given by,

$$f_{S_m}(t) = \left(\prod_{i=1}^{n} \alpha_i^{k_i}\right) \sum_{i=1}^{n} \sum_{j=1}^{k_i} c_{ij} \frac{t^{j-1}e^{-\alpha_i t}}{(j-1)!} I_{(0,\infty)}(t)$$

$$\text{where} \quad c_{ij} = \frac{1}{(k_i - j)!} \lim_{s \to -\alpha_i} \frac{d^{(k_i-j)}}{ds^{(k_i-j)}} \left(\prod_{j=1, j \neq i}^{n} \frac{1}{(s + \alpha_j)^{k_j}}\right) \tag{1}$$

Smaili et al. (2013a) gave a modified and simple form of the PDF of this distribution as a linear combination of the probability density function of the known Erlang distribution, given in the following theorem.

Theorem 2 Let $m \geq 2$ and $S_m \sim Hypoexp\left(\overrightarrow{\alpha}, \overrightarrow{k}\right)$ where $\overrightarrow{\alpha} = (\alpha_1, \alpha_2, \ldots, \alpha_n)$ and $\overrightarrow{k} = (k_1, k_2, \ldots, k_n)$. Then the PDF of S_m is

$$f_{S_m}(t) = \sum_{i=1}^{n} \sum_{j=1}^{k_i} A_{ij} f_{Y_{ij}}(t), \tag{2}$$

where

$$A_{ij} = \left(\prod_{i=1}^{n} \alpha_i^{k_i}\right) \frac{c_{ij}}{\alpha_i^j} \tag{3}$$

and c_{ij} is defined in Eq. (1) and $Y_{ij} \sim Erl(j, \alpha_i)$, where $1 \leq i \leq n$ and $1 \leq j \leq k_i$.

Proof From Theorem 1, the PDF of S_m can be written as

$$f_{S_m}(t) = \left(\prod_{i=1}^{n} \alpha_i^{k_i}\right) \sum_{i=1}^{n} \sum_{j=1}^{k_i} \frac{c_{ij}}{a_i^j} \frac{(\alpha_i t)^{j-1} \alpha_i e^{-\alpha_i t}}{(j-1)!} I_{(0,\infty)}(t) \tag{4}$$

where c_{ij} given in Eq. (1). Now, we define that $A_{ij} = \left(\prod_{i=1}^{n} \alpha_i^{k_i}\right) \frac{c_{ij}}{\alpha_i^j}$ and the $f_{Y_{ij}}(t) = \frac{(\alpha_i t)^{j-1} \alpha_i e^{-\alpha_i t}}{(j-1)!} I_{(0,\infty)}(t)$, for $1 \leq i \leq n$ and $1 \leq j \leq k_i$. However $f_{Y_{ij}}(t)$ is the PDF of the Erlang RV the so-called Y_{ij} having the parameters j and α_i. Therefore, we can rewrite Eq. (4) in the following form $f_{S_m}(t) = \sum_{i=1}^{n} \sum_{j=1}^{k_i} A_{ij} f_{Y_{ij}}(t)$. $\qquad \square$

2.2 The CDF for the Hypoexponential RV

In this part, we give a modified form of the CDF for the general case of the hypoexponential RV. Moreover, an equality for the coefficients A_{ij} in Eq. (3) of the linear combination is obtained, which shall be used later.

Proposition 2 Let $m \geq 2$ and $Y_{ij} \sim Erl(j, \alpha_i)$ for $1 \leq i \leq n$ and $1 \leq j \leq k_i$. Then the CDF of the hypoexponential RV $S_m \sim Hypoexp\left(\overrightarrow{\alpha}, \overrightarrow{k}\right)$, $\overrightarrow{\alpha} = (\alpha_1, \alpha_2, \ldots, \alpha_n)$ and $\overrightarrow{k} = (k_1, k_2, \ldots, k_n)$ is

$$F_{S_m}(x) = \sum_{i=1}^{n} \sum_{j=1}^{k_i} A_{ij} F_{Y_{ij}}(x) \tag{5}$$

where $F_{Y_{ij}}(x)$ is the CDF of Y_{ij} and A_{ij} is defined in Eq. (3)

Proof We have from Theorem 2, $f_{S_m}(t) = \sum_{i=1}^{n} \sum_{j=1}^{k_i} A_{ij} F_{y_{ij}}(t)$. Then the CDF of S_m is given as $F_{S_m}(x) = \int_{-\infty}^{x} f_{S_m}(t) dt = \sum_{i=1}^{n} \sum_{j=1}^{k_i} A_{ij} \int_{-\infty}^{x} f_{Y_{ij}}(t) dt = \sum_{i=1}^{n} \sum_{j=1}^{k_i} A_{ij} F_{Y_{ij}}(x)$. $\qquad\square$

In the next proposition, we determine an identity for the coefficients of the linear combination stated in Eq. (2) using the above proposition.

Proposition 3 $\sum_{i=1}^{n} \sum_{j=1}^{k_i} A_{ij} = 1$, where is the expression defined in Eq. (3).

Proof We have the CDF of any random variable at ∞ is 1. Then $\lim_{x \to \infty} F_{S_m}(x) = \lim_{x \to \infty} F_{Y_{ij}}(x) = 1$ where $F_{S_m}(x)$ and $F_{Y_{ij}}(x)$ are the CDF of $S_m \sim Hypoexp\left(\overrightarrow{\alpha}, \overrightarrow{k}\right)$ and $Y_{ij} \sim Erl(j, \alpha_i)$ for $1 \leq i \leq n$ and $1 \leq j \leq k_i$, respectively. Now, Eq. (4) at ∞ gives that $\sum_{i=1}^{n} \sum_{j=1}^{k_i} A_{ij} = 1$. $\qquad\square$

In the next, theorem we use Propositions 1 and 2 to find a new form of CDF for the general case of hypoexponential RV.

Theorem 3 Let $m \geq 2$. Then the CDF of hypoexponential RV $S_m \sim Hypoexp\left(\overrightarrow{\alpha}, \overrightarrow{k}\right)$, $\overrightarrow{\alpha} = (\alpha_1, \alpha_2, \ldots, \alpha_n)$ and $\overrightarrow{k} = (k_1, k_2, \ldots, k_n)$ is

$$F_{S_m}(x) = \sum_{i=1}^{n} \sum_{j=1}^{k_i} A_{ij} \frac{\gamma(j, \alpha_i x)}{(j-1)!} I_{(0,\infty)}(x)$$

$$= 1 - \sum_{i=1}^{n} \sum_{j=1}^{k_i} A_{ij} \frac{\Gamma(j, \alpha_i x)}{(j-1)!} I_{(0,\infty)}(x) \tag{6}$$

where A_{ij} is defined in Eq. (3) and $\gamma(j, \alpha_i x)$ and $\Gamma(j, \alpha_i x)$ are the lower incomplete gamma function and the upper incomplete gamma function.

Proof We have from Proposition 2, $F_{S_m}(x) = \sum_{i=1}^{n} \sum_{j=1}^{k_i} A_{ij} F_{Y_{ij}}(x)$. Moreover, it is known that the CDF of the Erlang distribution from Proposition 1 we have $F_Y(x) = \frac{\gamma(n, \alpha x)}{(n-1)!}$. Then $F_{S_m}(x) = \sum_{i=1}^{n} \sum_{j=1}^{k_i} \frac{\gamma(j, \alpha_i x)}{(j-1)!} I_{(0,\infty)}(x)$.

Now, we have from Proposition 1, $\frac{\gamma(n,\alpha x)}{(n-1)!} = 1 - \frac{\Gamma(j,\alpha_i x)}{(j-1)!}$. Thus $F_{S_m}(x) =$
$\sum_{i=1}^{n}\sum_{j=1}^{k_i} A_{ij}\left(1 - \frac{\Gamma(j,\alpha_i x)}{(j-1)!}\right) I_{(0,\infty)}(x) = \sum_{i=1}^{n}\sum_{j=1}^{k_i} A_{ij} - \sum_{i=1}^{n}\sum_{j=1}^{k_i}$
$A_{ij}\frac{\Gamma(j,\alpha_i x)}{(j-1)!}$. But from Proposition 3 $\sum_{i=1}^{n}\sum_{j=1}^{k_i} A_{ij} = 1$. Therefore, we obtain
$$F_{S_m}(x) = 1 - \sum_{i=1}^{n}\sum_{j=1}^{k_i} A_{ij}\frac{\Gamma(j,\alpha_i x)}{j-1} I_{(0,\infty)}(x).$$ □

2.3   The Reliability and Hazard Functions for the Hypoexponential RV

Mathematically, the reliability function $R(t)$ is the probability that a system will be successfully operating without failure in the interval from time 0 to t. The reliability function is given by $R(t) = P(T > t) = 1 - P(T < t) = 1 - F(T)$, $t \geq 0$, where T is a random variable representing the failure time or time-to-failure and $F(T)$ is the cumulative density function. The failure rate function, or hazard function, is very important in reliability analysis because it specifies the rate of the system aging. The hazard function is defined by $h(t) = f(t)/R(t)$ where $f(t)$ is the probability density function and $R(t)$ is the reliability function (Trivedi 2002).

In this part we give a modified form of the reliability (survivor) and hazard (failure) functions.

Proposition 4 The reliability function of the hypoexponential RV $S_m \sim Hypoexp$ $\left(\overrightarrow{\alpha}, \overrightarrow{k}\right)$, $\overrightarrow{\alpha} = (\alpha_1, \alpha_2, \ldots, \alpha_n)$, and $\overrightarrow{k} = (k_1, k_2, \ldots, k_n)$ is given by

$$R_{S_m}(t) = \sum_{i=1}^{n}\sum_{j=1}^{k_i} A_{ij} R_{Y_{ij}}(t) = \sum_{i=1}^{n}\sum_{j=1}^{k_i} A_{ij}\frac{\Gamma(j,\alpha_i t)}{(j-1)!} I_{(0,\infty)}(t)$$

where $R_{Y_{ij}}(t)$ is the reliability function of $Y_{ij} \sim Erl(j, \alpha_i)$, $1 \leq i \leq n$, and $1 \leq j \leq k_i$ and $\Gamma(j, \alpha_i t)$ is the upper incomplete gamma function.

Proof We have reliability function of S_m by $R_{S_m}(t) = 1 - F_{S_m}(t)$, and by Proposition 2, we get $R_{S_m}(t) = 1 - \sum_{i=1}^{n}\sum_{j=1}^{k_i} A_{ij} F_{y_{ij}}(x)$. Now, let $R_{Y_{ij}}(t)$, be reliability function of $Y_{ij} \sim Erl(j, \alpha_i)$, $1 \leq i \leq n$, and $1 \leq j \leq k_i$, since $F_{Y_{ij}}(t) = 1 - R_{Y_{ij}}(t)$, then $R_{S_m}(t) = 1 - \sum_{i=1}^{n}\sum_{j=1}^{k_i} A_{ij}\left(1 - R_{Y_{ij}}(t)\right) = 1 - \sum_{i=1}^{n}\sum_{j=1}^{k_i} A_{ij} +$ $\sum_{i=1}^{n}\sum_{j=1}^{k_i} A_{ij} R_{Y_{ij}}(t)$ and $\sum_{i=1}^{n}\sum_{j=1}^{k_i} A_{ij} = 1$ from Proposition 3. Then $R_{S_m}(t) = \sum_{i=1}^{n}\sum_{j=1}^{k_i} A_{ij} R_{Y_{ij}}(t)$. However, the second formula of $R_{S_m}(t)$ is obtained directly from Proposition 1, where $R_{Y_{ij}}(t) = \frac{\Gamma(j,\alpha_i t)}{(j-1)!}$.   □

Proposition 5 The hazard (failure) function of the hypoexponential RV $S_m \sim$ $Hypoexp\left(\overrightarrow{\alpha}, \overrightarrow{k}\right)$, $\overrightarrow{\alpha} = (\alpha_1, \alpha_2, \ldots, \alpha_n)$, and $\overrightarrow{k} = (k_1, k_2, \ldots, k_n)$ is given by

$$h_{S_m}(t) = \frac{\sum_{i=1}^{n}\sum_{j=1}^{k_i} A_{ij} h_{Y_{ij}}(t) R_{Y_{ij}}(t)}{\sum_{i=1}^{n}\sum_{j=1}^{k_i} A_{ij} R_{Y_{ij}}(t)}$$

where $h_{Y_{ij}}(t)$ and $R_{Y_{ij}}(t)$ are the hazard and reliability function of $Y_{ij} \sim Erl(j, \alpha_i)$ for $1 \leq i \leq n$ and $1 \leq j \leq k_i$, and A_{ij} is defined in Eq. (3.3).

Proof We have the hazard function of S_m is given by $h_{S_m}(t) = \frac{f_{S_m}(t)}{R_{S_m}(t)}$, where $f_{S_m}(t)$ and $R_{S_m}(t)$ are the PDF and reliability function of $S_m \sim Hypoexp\left(\overrightarrow{\alpha}, \overrightarrow{k}\right)$. By substituting the formulas of $f_{S_m}(t)$ and $R_{S_m}(t)$ in Theorem 2 and Proposition 4,

we obtain that $h_{S_m}(t) = \dfrac{\sum_{i=1}^{n}\sum_{j=1}^{k_i} A_{ij} f_{Y_{ij}}(t)}{\sum_{i=1}^{n}\sum_{j=1}^{k_i} A_{ij} R_{Y_{ij}}(t)}$ where $Y_{ij} \sim Erl(j, \alpha_i)$. However $F_{Y_{ij}}(t) = h_{Y_{ij}}(t) R_{Y_{ij}}(t)$ Thus we obtain the result. □

2.4 The MGF for the Hypoexponential RV

In this part, we introduce a modified form of the MGF for the general case of the hypoexponential RV. Next, we give two new forms of the moment of S_m of order k of the general case of the hypoexponential RV. These new forms are compared to determine a generalized equality. Our results are applied on the two particular cases $k = 1$ and $k = 2$.

Proposition 6 Let $m \geq 2$ and $Y_{ij} \sim Erl(j, \alpha_i)$ for $1 \leq i \leq n$ and $1 \leq j \leq k_i$. Then the MGF of the hypoexponential RV $S_m \sim Hypoexp\left(\overrightarrow{\alpha}, \overrightarrow{k}\right)$, $\overrightarrow{\alpha} = (\alpha_1, \alpha_2, \ldots, \alpha_n)$ and $\overrightarrow{k} = (k_1, k_2, \ldots, k_n)$ is

$$\Phi_{S_m}(t) = \sum_{i=1}^{n}\sum_{j=1}^{k_i} A_{ij} \Phi_{Y_{ij}}(t) \tag{7}$$

where A_{ij} is defined in Eq. (3) and $t < min\left\{\overrightarrow{\alpha}\right\}$.

Proof Let $S_m \sim Hypoexp\left(\overrightarrow{\alpha}, \overrightarrow{k}\right)$ and $Y_{ij} \sim Erl(j, \alpha_i)$ for $1 \leq i \leq n$ and $1 \leq j \leq k_i$. Then we have from Theorem 2, $f_{S_m}(t)\sum_{i=1}^{n}\sum_{j=1}^{k_i} A_{ij} f_{Y_{ij}}(t)$. However the MGF of S_m and Y_{ij} are give as

$$\Phi_{S_m}(t) = \int_{-\infty}^{+\infty} e^{tx} \left(\sum_{i=1}^{n} \sum_{j=1}^{k_i} A_{ij} \, f_{Y_{ij}}(x) \right)$$

$$dx = \sum_{i=1}^{n} \sum_{j=1}^{k_i} A_{ij} \left(\int_{-\infty}^{+\infty} e^{tx} f_{Y_{ij}}(x) dx \right) = \sum_{i=1}^{n} \sum_{j=1}^{k_i} A_{ij} \, \Phi_{Y_{ij}}(t)$$

$$\Phi_{S_m}(t) = \int_{-\infty}^{+\infty} e^{tx} f_{S_m}(x) dx \text{ and } \Phi_{Y_{ij}}(t) = \int_{-\infty}^{+\infty} e^{tx} f_{Y_{ij}}(x) dx, \text{ respectively.}$$

Therefore, we obtain that. $\qquad\qquad\qquad\qquad\qquad\qquad\qquad\qquad\qquad\qquad$ □

Proposition 7 The MGF of S_m is $\Phi_{S_m}(t) = \sum_{i=1}^{n} \sum_{j=1}^{k_i} A_{ij} \frac{\alpha_i^j}{(\alpha_i - t)^j}$ for $t <$ min $\left\{ \overrightarrow{\alpha} \right\}$, $\overrightarrow{\alpha} = (\alpha_1, \alpha_2, \dots, \alpha_n)$ where A_{ij} is defined in Eq. (3).

Proof From Eq. (7), $\Phi_{S_m}(t) = \sum_{i=1}^{n} \sum_{j=1}^{k_i} A_{ij} \, \Phi_{Y_{ij}}(t)$. However, from Proposition 1, the MGF of the Erlang Distribution $Y_{ij} \to Erl(j, \alpha_i)$ is given by $\Phi_{Y_{ij}}(t) = \frac{\alpha_i^j}{(\alpha_i - t)^j}$ for $t < \alpha_i$, $i = 1, 2, \dots, n$, see Abdelkader (2003) and Minkova (2010). Hence $t < \min \left\{ \overrightarrow{\alpha} \right\}$, $\overrightarrow{\alpha} = (\alpha_1, \alpha_2, \dots, \alpha_n)$. Thus, we obtain the result. $\qquad$ □

Corollary 1 Let $m \geq 2$ and $Y_{ij} \to Erl(j, \alpha_i)$ for $1 \leq i \leq n$ and $1 \leq j \leq k_i$. Then the moment of order k of $S_m \sim Hypoexp\left(\overrightarrow{\alpha}, \overrightarrow{k} \right)$, $\overrightarrow{\alpha} = (\alpha_1, \alpha_2, \dots, \alpha_n)$ and $\overrightarrow{k} = (k_1, k_2, \dots, k_n)$ is

$$E\left[S_m^k \right] = \sum_{i=1}^{n} \sum_{j=1}^{k_i} A_{ij} E\left[Y_{ij}^k \right] \qquad\qquad (8)$$

where A_{ij} is defined in Eq. (3).

Proof From Proposition 6, the MGF of $S_m \sim Hypoexp\left(\overrightarrow{\alpha}, \overrightarrow{k} \right)$ is given in Eq. (6) as $\Phi_{S_m}(t) = \sum_{i=1}^{n} \sum_{j=1}^{k_i} A_{ij} \, \Phi_{Y_{ij}}(t)$. But we have the moment of order k of S_m $E\left[S_m^k \right] = \frac{d^k \Phi_{S_m}(t)}{dt^k} \Big|_{t=0}$ and the moment of order k of Y_{ij}, $E\left[Y_{ij}^k \right] = \frac{d^k \Phi_{Y_{ij}}(t)}{dt^k} \Big|_{t=0}$. Then applying the kth differentiation on Eq. (6), then finding the value at $t = 0$, we obtain that $E\left[S_m^k \right] = \sum_{i=1}^{n} \sum_{j=1}^{k_i} A_{ij} E\left[Y_{ij}^k \right]$. $\qquad$ □

Next we give another expression of moment of S_m of order k.

Proposition 8 Let $m \geq 2$ and k be a positive integer. Then the moment of S_m of order k is

$$E\left[S_m^k \right] = \sum_{i=1}^{n} \sum_{j=1}^{k_i} A_{ij} \frac{j(j+1)\dots(j+k-1)}{\alpha_i^k}. \qquad\qquad □$$

Proof From Corollary 1, $E\left[S_m^k\right] = \sum_{i=1}^{n}\sum_{j=1}^{k_i} A_{ij} E\left[Y_{ij}^k\right]$. However, $Y_{ij} \sim Erl(j, \alpha_i)$ and from Proposition 1 the moment of Y_{ij} of order k is given by

$$E\left[Y_{ij}^k\right] = \frac{\Gamma(k+j)}{\alpha_i^k \Gamma(j)}$$

Also $\frac{\Gamma(k+j)}{\Gamma(j)} = j(j+1)\ldots(j+k-1)$. Therefore, we obtain the result. $\square$

In the next proposition, we use the idea of writing S_m as a sum of Erlang RV to write the moment of S_n of order k in another new form.

Proposition 9 Let $m \geq 2$ and k be a positive integer. Then

$$E\left[S_m^k\right] = \sum_{E_k} k! \binom{k_1 + l_1 - 1}{l_1}\binom{k_2 + l_2 - 1}{l_2}\cdots\binom{k_n + l_n - 1}{l_n}$$

$$\cdot \frac{1}{\alpha_1^{l_1}} \cdot \frac{1}{\alpha_2^{l_2}} \cdots \frac{1}{\alpha_n^{l_n}}$$

where $E_k = \{(l_1, \ldots, l_n)/0 \leq l_i \leq k; \sum_{i=1}^{n} l_i = k; 1 \leq i \leq n\}$.

Proof We have the general case of $S_m \sim Hypoexp\left(\vec{\alpha}, \vec{k}\right)$ where $\vec{\alpha} = (\alpha_1, \alpha_2, \ldots, \alpha_n)$ and $\vec{k} = (k_1, k_2, \ldots, k_n)$ can be written as a sum of n Erlang RV, where $Y_i \sim Erl(k_i, \alpha_i)$, $i = 1, 2, \ldots, n$. That is $S_m = \sum_{j=1}^{n} Y_i$. Then $E[S_m^k] = E[(Y_1 + Y_2 + \cdots + Y_n)^k]$. Now using the multinomial expansion formula, we obtain that $E\left[S_m^k\right] = E\left[\sum_{E_k} \frac{k!}{l_1! l_2! \ldots l_n!}\left(Y_1^{l_1} Y_2^{l_2} \ldots Y_n^{l_n}\right)\right]$.

Now since expectation is linear then we may write $E\left[S_m^k\right] = \sum_{E_k} E\left[\frac{k!}{l_1! l_2! \ldots l_n!}\right.$ $\left(Y_1^{l_1} Y_2^{l_2} \ldots Y_n^{l_n}\right)]$, also Y_i, $i = 1, 2, \ldots, n$. Then $E\left[S_m^k\right] = \sum_{E_k} \frac{k!}{l_1! l_2! \ldots l_n!} E\left[Y_1^{l_1}\right] \cdot$ $E\left[Y_2^{l_2}\right] \ldots E\left[Y_n^{l_n}\right]$. Moreover, the moment of Erlang $Y_i \sim Erl(k_i, \alpha_i)$ is given in Proposition 1 as $E\left[Y_i^{l_i}\right] = \frac{\Gamma(l_i + k_i)}{\alpha_i^{l_i}\Gamma(k_i)} = \binom{k_i + l_i - 1}{l_i}\frac{l_i!}{\alpha_i^{l_i}}$ $i = 1, 2, \ldots, n$

Therefore, $E\left[S_m^k\right] = \sum_{E_k} k! \binom{k_1 + l_1 - 1}{l_1}\binom{k_2 + l_2 - 1}{l_2}\cdots\binom{k_n + l_n - 1}{l_n} \cdot$ $\frac{1}{\alpha_1^{l_1}} \cdot \frac{1}{\alpha_2^{l_2}} \cdots \frac{1}{\alpha_n^{l_n}}$. $\square$

The two forms of the moment of order k obtained in Propositions 8 and 9, gives an important identity concerning the coefficient A_{ij}, given in the following proposition.

Proposition 10 Let $m \geq 2$ and k be a positive integer

$$\sum_{i=1}^{n}\sum_{j=1}^{k_i} A_{ij} \frac{j(j+1)\ldots(j+k-1)}{\alpha_i^k} \sum_{E_k} k! \binom{k_1+l_1-1}{l_1}\binom{k_2+l_2-1}{l_2}$$

$$\ldots \binom{k_n+l_2-1}{l_n} \cdot \frac{1}{\alpha_1^{l_1}} \cdot \frac{1}{\alpha_2^{l_2}} \cdots \frac{1}{\alpha_n^{l_n}}$$

$$(9)$$

where $E_k = \{(l_1, \ldots, l_n)/0 \leq l_i \leq k; \ \sum_{i=1}^{n} l_i = k; 1 \leq i \leq n\}$.
Note that we may write

$$\sum_{E_k} \frac{1}{\alpha_1^{l_1}\alpha_2^{l_2}\ldots\alpha_n^{l_n}} = \sum_{I_k} \frac{1}{\alpha_{i_1}\alpha_{i_2}\alpha_{i_3}\ldots\alpha_{i_k}}$$

$$\text{where} \quad I_k = \{(i_1,\ldots,i_k)/1 \leq i_1 \leq i_2 \leq \cdots \leq i_k \leq n\}$$

However E_k and I_k are equivalent representing a set of combination with repetition having $\binom{n+k-1}{k}$ possibilities and $E_0 = I_0 = 0$, thus the above summation (9) shall be 1. This verifies Proposition 3.

In the following two corollaries, we use Propositions 3 and 4 to formulate two important special cases: $k = 1$, Expectation, and $k = 2$. However, some Identities concerning A_{ij} have been established.

Corollary 2 Let $m \geq 2$ and $S_m \sim Hypoexp\left(\overrightarrow{\alpha}, \overrightarrow{k}\right)$ where $\overrightarrow{\alpha} = (\alpha_1, \alpha_2, \ldots, \alpha_n)$
and $\overrightarrow{k} = (k_1, k_2, \ldots, k_n)$. Then $E[S_m] = \sum_{i=1}^{n}\sum_{j=1}^{k_i} A_{ij}\frac{j}{\alpha_i} = \sum_{i=1}^{n}\frac{k_i}{\alpha_i}$, where A_{ij} is defined in Eq. (3).

Proof From Proposition 8, $E[S_m^k] = \sum_{i=1}^{n}\sum_{j=1}^{k_i} A_{ij} \frac{j(j+1)\ldots(j+k-1)}{\alpha_i^k}$. Set $k = 1$,
then $E[S_m] = \sum_{i=1}^{n}\sum_{j=1}^{k_i} A_{ij}\frac{j}{\alpha_i}$. From Definition 1, $S_m = \sum_{i=1}^{n}\sum_{j=1}^{k_i} X_{ij}$,
$X_{ij} \sim Exp(\alpha_i)$, for $1 \leq i \leq n$ and $1 \leq j \leq k_i$. Thus $[S_m] = E\left[\sum_{i=1}^{n}\sum_{j=1}^{k_i} X_{ij}\right] = \sum_{i=1}^{n}\sum_{j=1}^{k_i} E[X_{ij}] = \sum_{i=1}^{n}\sum_{j=1}^{k_i}\frac{1}{\alpha_i} = \sum_{i=1}^{n}\frac{k_i}{\alpha_i}$. $\square$

Corollary 3 Let $m \geq 2$ and $S_m \sim Hypoexp\left(\overrightarrow{\alpha}, \overrightarrow{k}\right)$ where $\overrightarrow{\alpha} = (\alpha_1, \alpha_2, \ldots, \alpha_n)$
and $\overrightarrow{k} = (k_1, k_2, \ldots, k_n)$. Then $E[S_m^2] = \sum_{i=1}^{n}\sum_{j=1}^{k_i} A_{ij}\frac{j(j+1)}{\alpha_i^2} = \sum_{i=1}^{n}\frac{k_i(k_i+1)}{\alpha_i^2} + 2\sum_{1 \leq j \leq n}\frac{k_i k_j}{\alpha_i \alpha_j}$, where A_{ij} is defined in Eq. (3).

Proof From Proposition 8, take $k = 2$, we obtain that $E\left[S_m^2\right] = \sum_{i=1}^{n}\sum_{j=1}^{k_i}$ $A_{ij}\frac{j(j+1)}{\alpha_i^2}$. Moreover, from Proposition 9, we have $E\left[S_m^k\right] = \sum_{E_k} k!\begin{pmatrix} k_1 + l_1 - 1 \\ l_1 \end{pmatrix}$

$\begin{pmatrix} k_2 + l_2 - 1 \\ l_2 \end{pmatrix}\cdots\begin{pmatrix} k_n + l_n - 1 \\ l_n \end{pmatrix}\cdot\frac{1}{\alpha_1^{l_1}}\cdot\frac{1}{\alpha_2^{l_2}}\cdots\frac{1}{\alpha_n^{l_n}}$. Set $k = 2$ we have $E_2 = \{(l_1, \ldots,$ $l_n)/0 \le l_i \le 2; \sum_{i=1}^{n} l_i = 2; 1 \le i \le n\}$. Then we have two possible cases either $l_i = 2$ and $l_j = 0$ for $i \ne j$ for $1 \le i \le n$ or $l_i = l_j = 1$ for all $1 \le i < j \le n$. Therefore, we obtain that $\left[S_m^2\right] = \sum_{i=1}^{n} 2!\begin{pmatrix} k_i + 1 \\ 2 \end{pmatrix}\frac{1}{\alpha_i^2} + \sum_{1 \le i < j \le n} 2!\begin{pmatrix} k_i \\ 1 \end{pmatrix}\begin{pmatrix} k_j \\ 1 \end{pmatrix}\frac{1}{\alpha_i\alpha_j} =$ $\sum_{i=1}^{n}\frac{k_i(k_i + 1)}{\alpha_i^2} + 2\sum_{1 \le i < j \le n}\frac{k_i k_j}{\alpha_i\alpha_j}.$ $\qquad\square$

2.5 Algorithm for Finding A_{ij}

In the previous section we gave the PDF, CDF, MGF, reliability function, hazard function, and moment of order k of and $S_m \sim Hypoexp\left(\overrightarrow{\alpha}, \overrightarrow{k}\right)$ where $\overrightarrow{\alpha} = (\alpha_1, \alpha_2, \ldots, \alpha_n)$ and $\overrightarrow{k} = (k_1, k_2, \ldots, k_n)$. In all expression, the coefficient A_{ij}, defined in Eq. (3) exists. For the importance of this coefficient, we present a method in finding A_{ij}. The method is a recursive algorithm. This method uses logarithmic properties and the Leibnitz's mth derivative.

Proposition 11 Let $1 \le i \le n$, $1 \le j \le k_i$. Then

$$A_{ij} = \frac{\prod_{i=1}^{n}\alpha_i^{k_i}}{\alpha_i^j (k_i - j)!} \lim_{s \to -\alpha_i} g_i^{(k_i - j)}(s) \qquad (10)$$

where $g_i^{(q)}(s)$ is the qth derivative of the function

$$g_i(s) = \prod_{j=1, j \ne i}^{n} \frac{1}{(s + \alpha_j)^{k_j}} \qquad (11)$$

Proof We have from Eqs. (1) and (3), $A_{ij} = \left(\prod_{i=1}^{n}\alpha_i^{k_i}\right)\frac{c_{ij}}{\alpha_i^j}$ and $c_{ij} = \frac{1}{(k_i - j)!}\lim_{s \to -\alpha_i} g_i^{(k_i - j)}(s)$, respectively. Thus, we may write A_{ij} in the given form. $\qquad\square$

In the next theorem, we relate the qth derivative of the function $g_i(s)$ defined in Eq. (11) to the less order derivatives of $g_i(s)$. This gives a recursive sequence of A_{ij}.

Theorem 4 Let $1 \le i \le n$, $1 \le j \le k_i$. Then

$$g_i^{(q)}(s) = \sum_{l=0}^{q-1} \left[\binom{q-1}{l} (-1)^{l+1} \left(\sum_{j=1, j \neq i}^{n} \frac{(l)! k_j}{(s + \alpha_j)^{l+1}} \right) g_i^{(q-l-1)}(s) \right] \quad \text{for} \quad q \ge 1$$

(12)

Proof We shall find the $(k_i - j)$th derivative of $g_i(s)$ in a recursive way. First we take the logarithm of both sides of $g_i(s) = \prod_{j=1, j \neq i}^{n} \frac{1}{(s+\alpha_j)^{k_j}}$, we obtain that $\ln g_i(s) =$

$$- \sum_{j=1, j \neq i}^{n} k_j \ln (s + \alpha_j).$$

Second we differentiate the above equation with s, leads to $\frac{g_i'(s)}{g_i(s)} = - \sum_{j=1, j \neq i}^{n} \frac{k_j}{s + \alpha_j}$

and $g_i'(s) = -g_i(s) \sum_{j=1, j \neq i}^{n} \frac{k_j}{s+\alpha_j}$. Now, let $u_i = - \sum_{j=1, j \neq i}^{n} \frac{k_j}{s+\alpha_j}$ and $v_i = g_i(s)$, thus $g_i'(s) = u_i v_i$. By applying Leibnitz's qth derivative to $u_i v_i$, we obtain

$$(g_i'(s))^{(q)} = \sum_{l=0}^{q} \left[\binom{q}{l} u_i^{(l)}(s) v_i^{(q-l)}(s) \right]$$

However, $u_i^{(l)}(s) = (-1)^{l+1} \sum_{j=1, j \neq i}^{n} \frac{(l)! k_j}{(s+\alpha_j)^{l+1}}$ and $v_i^{(q-l)}(s) = g_i^{(q-l)}(s)$.

Therefore,

$$g_i^{(q+1)}(s) = \sum_{l=0}^{q} \left[\binom{q}{l} (-1)^{l+1} \left(\sum_{j=1, j \neq i}^{n} \frac{(l)! k_j}{(s+\alpha_j)^{l+1}} \right) g_i^{(q-l)}(s) \right] \qquad \square$$

2.6 Application

Markov chains are mathematical models which have several applications in computer science, particularly in performance and reliability modeling. In this application, we consider a system that can go through five degraded stages before it completely failed (D6), at which time a repair will take place. When the system is still operating, periodic maintenance (1) activities are triggered with constant

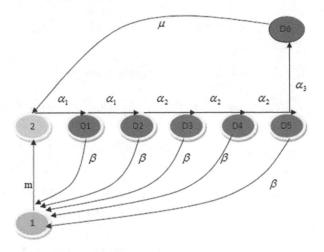

Fig. 1 Stochastic system with hypoexponential lifetime

rate β. The rate for repair is μ. Both repair and maintenance take the system back to the original good state (2). A Markov chain to represent this system is shown in Fig. 1. The initial state is running (2), where the system is fully operational. With the degradation rate $\alpha_i (i = 1, 2)$, the system transits from D1 to D5, which are degraded operational states. With rate α_3, the system can move from state D5 to state D6, which is the failed state. The repairing rate is μ will take the system back to the running (2), and all maintenance actions take the system back to state (2). The system is available when it is in states (2) or D1–D5. While the system is available, periodic maintenance will take place with rate β. The system is not available when it is in maintenance state (1) or in state D6. The rate to finish maintenance is denoted as m.

Figure 1 corresponds to a system that has a hypoexponential failure time distribution, with parameters α_i $(i = 1 \ldots 3)$. Let X be the random variable representing the failure times using hypoexponential distribution. Then X follows an hypoexponential distribution, $S_6 \sim Hypoexp\left(\vec{\alpha}, \vec{k}\right)$, $\vec{\alpha} = (1, 2, 6)$ and $\vec{k} = (2, 3, 1)$. We shall start by finding the coefficients A_{ij}. Now, since $1 \leq i \leq 3$ and $1 \leq j \leq k_i$, we have the following coefficients $A_{11}e, A_{12}, A_{21}, A_{22}, A_{23}$ and A_{31}. Using Eqs. (10), (11), and (12), we shall find these coefficients.

- For $i = 1$, $\alpha_1 = 1$, $k_1 = 2$, Eq. (11) gives that $g_1(s) = \prod_{j=2,3} \dfrac{1}{(s + \alpha_j)^{k_j}}$

$= \dfrac{1}{(s + 2)^3} \times \dfrac{1}{(s + 6)}$ and $g_1(-1) = \frac{1}{5}$. Then Eq. (10) gives $A_{12} = \dfrac{1^2 \cdot 2^3 \cdot 6^1}{1^2} g_1(-1) = \frac{48}{5}$. Now from Eq. (3.12), $g_1^{(1)}(s) = (-1)\left(\sum_{j=2,3} \dfrac{k_j}{(s + \alpha_j)}\right)$

$g_1^{(0)}(s) \, l \; = \; -\left(\frac{3}{(s+2)} + \frac{1}{(s+6)}\right) g_1^{(0)}(s), \; g_1^{(1)}(-1) \; = \; -\frac{16}{25}.$ Using Eq. (10)
$A_{11} = 48 g_1^{(1)}(-1) = -\frac{768}{25}.$

- For $i = 2$, $\alpha_2 = 2$, $k_2 = 3$, Eq. (11) gives that $g_2(s) = \prod_{j=1,3} \dfrac{1}{(s + \alpha_j)^{k_j}} =$

$\dfrac{1}{(s+1)^2} \times \dfrac{1}{(s+6)},$ and $g_2^{(0)}(-2) \; = \; \frac{1}{4},$ Then Eq. (10) gives $A_{23} \; =$

$\frac{28}{2^3} g_2^{(0)}(-2) \; = \; \frac{3}{2}.$ Again from Eq. (2), $g_2^{(1)}(s) \; = \; (-1) \left(\displaystyle\sum_{j=1,j=3} \dfrac{k_j}{(s + \alpha_j)} \right)$

$g_2^{(0)}(s) \; = \; -\left(\frac{2}{(s+1)} + \frac{1}{(s+6)}\right) g_2^{(0)}(s)\frac{1}{4}, \; g_2^{(1)}(-2) \; = \; \frac{7}{16}$ and from Eq. (10) we

obtain $A_{22} \; = \; \frac{48}{2^2(1)!} g_2^{(1)}(-2) \; = \; \frac{21}{4}.$ Now, in order to find A_{21}, we shall find
$g_2^{(2)}(-2).$ Eq. (12) we have

$g_2^{(2)}(s) \; = \; \displaystyle\sum_{l=0}^{1} \left[\binom{1}{l} (-1)^{l+1} \left(\sum_{j=1,j=3} \dfrac{(l)! k_j}{(s + \alpha_j)^{l+1}} \right) g_2^{(1-l)}(s) \right] \; =$

$\binom{1}{0} (-1) \left(\displaystyle\sum_{j=1,j=3} \dfrac{k_j}{(s + \alpha_j)} \right) g_2^{(1)}(s) \; + \; \binom{1}{1} \left(\sum_{j=1,j=3} \dfrac{(1)! k_j}{(s + \alpha_j)^2} \right)$

$g_2^{(0)}(s) = -\left(\dfrac{2}{(s+1)} + \dfrac{1}{(s+6)}\right) g_2^{(1)}(s) + \left(\dfrac{2}{(s+1)^2} + \dfrac{1}{(s+6)^2}\right) g_2^{(0)}(s).$

Thus, $g_2^{(2)}(-2) \; = \; \frac{41}{32}$ and $A_{21} \; = \; \frac{48}{2^1(2)!} g_2^{(2)}(-2) \; = \; \frac{123}{8}.$

- For $i = 3$, $\alpha_3 = 6$, $k_3 = 1$, we have from Eq. (11) $g_3(s) = \prod_{j=1,2} \dfrac{1}{(s + \alpha_j)^{k_j}} =$

$\dfrac{1}{(s+1)^2} \times \dfrac{1}{(s+2)^3}$ and $g_3^{(0)}(-6) \; = \; -\frac{1}{1600}$ and Eq. (10) leads to $A_{31} \; =$
$\frac{48}{6(0)!} g_3^{(0)}(-6) = -\frac{1}{200}.$

So $A_{11} \; = \; -\frac{768}{25}, \; A_{12} \; = \; \frac{48}{5}, \; A_{21} \; = \; \frac{123}{8}, \; A_{22} \; = \; \frac{21}{4}, \; A_{23} \; = \; \frac{3}{2}$ and $A_{31} \; = \;$
$-\frac{1}{200}.$ For verification, Proposition 3 states that $\sum_{i=1}^{3} \sum_{j=1}^{k_i} A_{ij} = 1.$ So we may

conclude that the PDF of S_6 from Theorem 2 is $f_{S_6}(t) = \sum_{i=1}^{3} \sum_{j=1}^{k_i} A_{ij} f_{Y_{ij}}(t) =$
$-\frac{768}{25}\left(e^{-t}\right) + \frac{48}{5}\left(te^{-t}\right) + \frac{123}{8}\left(2e^{-2t}\right) + \frac{21}{4}\left(4te^{-2t}\right) + \frac{3}{2}\left(4t^2 e^{-2t}\right) - \frac{1}{200}\left(6e^{-6t}\right)$ when
$t > 0$ and $f_{S_6}(t) \; = \; 0.$ Moreover, the CDF, MGF, reliability function, hazard
function, and moment of order k of S_6 can be obtained.

Below we graph the PDF (Fig. 2), reliability function (Fig. 3) and the hazard
function (Fig. 4) of S_6.

Fig. 2 PDF of S_6

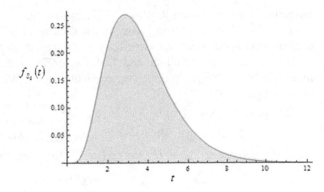

Fig. 3 Reliability function of S_6

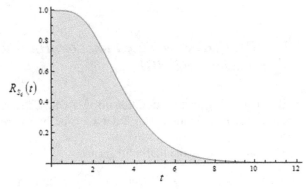

Fig. 4 Hazard function of S_6

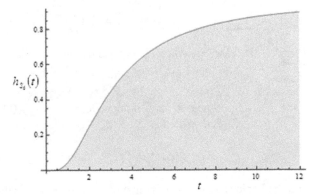

3 The Hypoexponential Random Variable with Different Parameters

In this section, we examine the case of the hypoexponential RV with different parameters. This case is a particular case of the general case of Sect. 2. A given form of PDF and CDF can be found in Akkouchi (2008), Amari and Misra (1997), and Kordecki (1997). We give in this section a modified form of the PDF, CDF, reliability function, hazard function, MGF, and moment of order k of the particular case, see Smaili et al. (2013b). Also, in the domain of reliability and performance evaluation of systems and software many authors used the geometric and arithmetic parameters such as (Gaudoin and Ledoux 2007; Jelinski and Moranda 1972; Moranda 1979). We study these two particular cases. We illustrate two examples of such a case.

The general case of this RV is $S_m = \sum_{i=1}^{n} \sum_{j=1}^{k_i} X_{ij}$. Now, taking $k_i = 1, i = 1, 2, \ldots, n$, we get $m = n$ and $S_n = \sum_{i=1}^{n} X_i$, where $X_i \to \exp(\alpha_i)$ and we write this case as $S_n \to Hypoexp\left(\vec{\alpha}\right)$. Moreover, the PDF in Eq. (2) has the expression $f_{S_n}(t) = \sum_{i=1}^{n} A_i f_{X_i}(t)$, where $A_i = A_{ij}$.

3.1 PDF, CDF, Reliability Function, Hazard Function, and MGF

In this part we give the modified form PDF, CDF, reliability function, hazard function, MGF, and moment of order k of the hypoexponential RV with different parameters.

We start by finding the PDF of this case.

Theorem 5 Let $S_n \to Hypoexp\left(\vec{\alpha}\right)$, where $\vec{\alpha} = (\alpha_1, \alpha_2, \ldots, \alpha_n)$. Then

$$f_{S_n}(t) = \sum_{i=1}^{n} A_i f_{X_i}(t), \text{ where } A_i = \prod_{j=1, j \neq i}^{n} \left(\frac{\alpha_j}{\alpha_j - \alpha_i}\right).$$

Proof This case is when $k_i = 1$, thus $m = n$. From Eq. (3.2), $f_{S_n}(t) = \sum_{i=1}^{n} A_{i1} f_{Y_{i1}}(t)$ and $Y_{i1} \to Erl(1, \alpha_i) = \exp(\alpha_i)$. Let $Y_{i1} = X_i$. We have the Laplace transform of $f_{X_i}(t)$ is given by $\mathscr{L}\{f_{X_i}(t)\} = \frac{\alpha_i}{\alpha_i + s}$, where $s > max\{-\alpha_i\}$ for $i = 1, 2, \ldots, n$. Since X_i are independent then $f_{S_n}(t)$ is the convolutions of $f_{X_i}, i = 1, 2, \ldots, n$ written as

$$f_{S_n}(t) = (f_{X_1} * f_{X_2} * \cdots * f_{X_n})(t)$$

and the Laplace transform of convolution of functions is the product of their Laplace transform, thus

$$\mathscr{L}\{f_{S_n}(t)\} = \prod_{i=1}^{n} \mathscr{L}\{f_{X_i}(t)\} = \prod_{i=1}^{n} \frac{\alpha_i}{\alpha_i + s} = \prod_{i=1}^{n} (\alpha_i) \prod_{i=1}^{n} \frac{1}{\alpha_i + s} \tag{13}$$

However, by Heaviside Expansion theorem (Spiegel 1965), for distinct poles gives that $\mathscr{L}\{f_{S_n}(t)\} = \left(\prod_{i=1}^{n} \alpha_i\right) \sum_{i=1}^{n} \frac{P_i}{s + \alpha_i}$ where $P_i = \dfrac{1}{\prod_{j=1, j \neq i}^{n} (\alpha_j - \alpha_i)}$.

Therefore,

$$f_{S_n}(t) = \left(\prod_{i=1}^{n}\alpha_i\right)\mathscr{L}^{-1}\left\{\frac{P_i}{s+\alpha_i}\right\} = \left(\prod_{i=1}^{n}\alpha_i\right)\sum_{i=1}^{n}P_i\mathrm{e}^{-\alpha_i t}I_{(0,\infty)}(t)$$

But $(\prod_{i=1}^{n}\alpha_i)P_i = A_i\alpha_i$. Thus $f_{S_m}(t) = \sum_{i=1}^{m}A_i f_{X_i}(t)$.
Next, we give the CDF of the hypoexponential RV with different parameters.   □

Corollary 4 Let $S_n \sim Hypoexp\left(\overrightarrow{\alpha}\right)$. Then the CDF of S_n is given by $F_{S_n}(x) =$
$1 - \sum_{i=1}^{n}A_i\mathrm{e}^{-\alpha_i t}I_{(0,\infty)}(t)$
where $F_{X_i}(x)$ is the CDF of the exponential random variable, $X_i \sim \mathrm{Exp}(\alpha_i)$,
$i = 1, 2, \ldots, n$.

Proof The CDF of the general case of the hypoexponential RV is given in Eq. (6)
as $F_{S_m}(x) = 1 - \sum_{i=1}^{n}\sum_{j=1}^{k_i}A_{ij}\frac{\Gamma(j, \alpha_i x)}{(j-1)!}I_{(0,\infty)}(x)$.
Taking the particular case of the hypexponential RV with different parameters,
thus $k_i = 1$ and $A_{ij} = A_i$, then $F_{S_m}(x) = 1 - \sum_{i=1}^{n}A_i\Gamma(1, \alpha_i x)I_{(0,\infty)}(x) = 1 - \sum_{i=1}^{n}A_i\mathrm{e}^{-\alpha_i t}I_{(0,\infty)}(x)$ □

Corollary 5 Let $S_n \sim Hypoexp\left(\overrightarrow{\alpha}\right)$. Then the reliability function of S_n is given by

$$R_{S_n}(x) = \sum_{i=1}^{n}A_i\mathrm{e}^{-\alpha_i t}I_{(0,\infty)}(t)$$

Proof The proof is a direct consequence of Corollary 4 knowing that $R_{S_n}(x) = 1 - F_{S_n}(x)$ □

Proposition 12 The hazard (failure) function for the hypoexponential distribution
Hypoexp$\left(\overrightarrow{\alpha}\right)$, $\overrightarrow{\alpha} = (\alpha_1, \alpha_2, \ldots, \alpha_n)$ is given by

$$h_{S_n}(t) = \frac{\sum_{i=1}^{n}A_i\alpha_i\mathrm{e}^{-\alpha_i t}I_{(0,\infty)}(t)}{\sum_{i=1}^{n}A_i\mathrm{e}^{-\alpha_i t}I_{(0,\infty)}(t)}$$

Proof We have the hazard function given by $h_{S_n}(t) = \frac{f_{S_n}(t)}{R_{S_n}(t)}$ thus we obtain the result.   □

Corollary 6 Let $n \geq 2$. Then
$$\Phi_{S_n}(t) = \sum_{i=1}^{n}A_i\Phi_{X_i}(t) \quad \text{and} \quad E\left[S_n^k\right] = \sum_{i=1}^{n}\frac{A_i k!}{\alpha_i^k}$$

Proof From Proposition 6 the moment generating function is given for the general case as $\Phi_{S_m}(t) = \sum_{i=1}^{n} \sum_{j=1}^{k_i} A_{ij} \Phi_{Y_{ij}}(t)$. Taking the particular case $k_i = 1$, the case of the hypoexponential with different parameters we obtain the result. In a similar manner, Proposition 2 in this particular case gives the moment of order k of S_n as $E\left[S_n^k\right] = \sum_{i=1}^{n} \dfrac{A_i k!}{\alpha_i^k}$. $\qquad\qquad\square$

3.2   Case of Arithmetic and Geometric Parameters

The study of reliability and performance evaluation of systems and software use in general sum of independent exponential RV with distinct parameters. The model of (Jelinski and Moranda 1972) considered that the parameters change in an arithmetic sequence $\alpha_i = \alpha_{i-1} + d$. Moreover, (Moranda 1979), considered the model when α_i changes in a geometric sequence $\alpha_i = \alpha_{i-1} r$. In this section, we study the hypoexponential distribution in these two cases when the parameters are arithmetic and geometric, and we present their PDFs.

3.2.1   The Case of Arithmetic Parameters

Let $S_n \sim Hypoexp\left(\overrightarrow{\alpha}\right)$, where $\overrightarrow{\alpha} = (\alpha_1, \alpha_2, \ldots, \alpha_n)$. Where α_i, $i = 1, 2, \ldots, n$ form an arithmetic sequence of common difference d.

Lemma 1. Let $\gamma_i = \prod_{j=1, j \neq i}^{n}(\alpha_j - \alpha_i)$. Then for all $1 \le i \le n$. $\gamma_i = (-1)^{i-1} \dfrac{(n-1)!}{\binom{n-1}{i-1}} d^{n-1}$.

Proof Suppose that α_i form an arithmetic sequence of common difference d. Then $\alpha_j - \alpha_i = (j - i)d$. We have $\gamma_i = \prod_{j=1, j \neq i}^{n}(\alpha_j - \alpha_i)$. Hence,

$$\gamma_i = (\alpha_1 - \alpha_i)(\alpha_2 - \alpha_i) \cdots (\alpha_{i-1} - \alpha_i)(\alpha_{i+1} - \alpha_i) \cdots (\alpha_n - \alpha_i)$$
$$= (-(i-1)d) \cdots (-2d)(d)(2d) \cdots ((n-i)d)$$
$$= (-1)^{i-1}(i-1)!(n-i)! d^{n-1}$$

However, $(i-1)!(n-i)! = \dfrac{n!}{i\binom{n}{i}} = \dfrac{(n-1)!}{\binom{n-1}{i-1}}$. Then $\gamma_i = (-1)^{i-1} \dfrac{(n-1)!}{\binom{n-1}{i-1}} d^{n-1}$. $\qquad\qquad\square$

Lemma 2. For all $1 \le i \le n$. $\gamma_i = (-1)^{n-1} \gamma_{n-(i-1)}$.

Proof We have from Lemma 1, $\gamma_i = (-1)^{i-1} \dfrac{(n-1)!}{\binom{n-1}{i-1}} d^{n-1}$ for all $1 \leq i \leq n$.

Replace i by $n - (i - 1)$, we obtain

$$\gamma_{n-(i-1)} = (-1)^{n-i} \frac{(n-1)!}{\binom{n-1}{n-(i-1)}} d^{n-1} = (-1)^{n-i} \frac{(n-1)!}{\binom{n-1}{i-1}} d^{n-1} =$$

$(-1)^{n-1} \gamma_i$.

Thus we obtain the result. $\qquad\square$

Proposition 13 Let $n \geq 2$. Then the PDF of S_n is $f_{S_n}(t) = \left(\prod_{i=1}^{n} \alpha_i \right) \sum_{i=1}^{n} \dfrac{e^{-\alpha_i t}}{\gamma_i}$ $I_{(0,\infty)}(t)$.

Proof We have from Theorem 5 $f_{S_n}(t) = \sum_{i=1}^{n} A_i f_{X_i}(t)$, where $X_i \to \exp(\alpha_i)$

and $A_i = \prod_{j=1, j \neq i}^{n} \left(\dfrac{\alpha_j}{\alpha_j - \alpha_i} \right)$. We have $A_i = \prod_{j=1, j \neq i}^{n} \left(\dfrac{\alpha_j}{\alpha_j - \alpha_i} \right) =$

$\dfrac{1}{\prod_{j=1, j \neq i}^{n} \left(1 - \frac{\alpha_i}{\alpha_j} \right)}$ Then

$$f_{S_n}(t) = \sum_{i=1}^{n} \frac{\alpha_i e^{-\alpha_i t}}{\prod_{j=1, j \neq i}^{n} \left(1 - \dfrac{\alpha_i}{\alpha_j} \right)} I_{(0,\infty)}(t)$$

$$= \sum_{i=1}^{n} \frac{\alpha_i e^{-\alpha_i t} \prod_{j=1, j \neq i}^{n} (\alpha_j)}{\prod_{j=1, j \neq i}^{n} (\alpha_j - \alpha_i)} I_{(0,\infty)}(t)$$

$$= \left(\prod_{i=1}^{n} \alpha_i \right) \sum_{i=1}^{n} \frac{e^{-\alpha_i t}}{\gamma_i} I_{(0,\infty)}(t) \qquad (14)$$

where $\gamma_i = \prod_{j=1, j \neq i}^{n} (\alpha_j - \alpha_i)$ and by the Lemmas 1 and 2 we obtain the result. $\quad\square$

3.2.2 The Case of Geometric Parameters

Next, we consider the case when α_i, $i = 1, 2, \ldots, n$ form a geometric sequence of common ratio r.

Proposition 14 Let $n \geq 2$. Then

$$f_{S_n}(t) = \sum_{i=1}^{n} \frac{\alpha_i e^{-\alpha_i t}}{\prod_{j=1, j \neq i}^{n} (1 - r^{i-j})} I_{(0,\infty)}(t)$$

Fig. 5 Reliability function

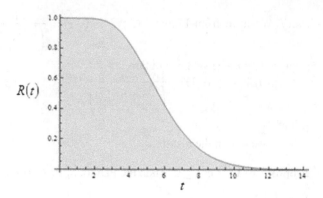

$R(t)$

t

Fig. 6 Hazard function

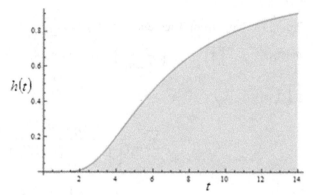

$h(t)$

t

Proof We have from Eq. (14), $f_{S_n}(t) = \sum_{i=1}^{n} \dfrac{\alpha_i e^{-\alpha_i t}}{\prod_{j=1, j \neq i}^{n} \left(1 - \dfrac{\alpha_i}{\alpha_j}\right)} I_{(0,\infty)}(t).$

Suppose now the parameter α_i forms geometric sequence of common ratio r, then
$\alpha_i = \alpha_j r^{i-j}$ and $\frac{1}{A_i} = \prod_{j=1, j \neq i}^{n} \left(1 - \dfrac{\alpha_i}{\alpha_j}\right) = \prod_{j=1, j \neq i}^{n} \left(1 - r^{i-j}\right).$ □

3.2.3 Application

The model of Jelinski and Moranda (Jelinski and Moranda 1972) considered that the
parameters change α_i (the rate of defective of the product) in an arithmetic sequence
$\alpha_i = \alpha_{i-1} + d$. We take $\alpha_1 = 1$ and $d = 0.2$ and $m = 5$ Applying Corollary 5 and
Propositions 11 and 12, we find the reliability function (Fig. 5) and Hazard function
(Fig. 6) represented in the following figures.

Fig. 7 Reliability function

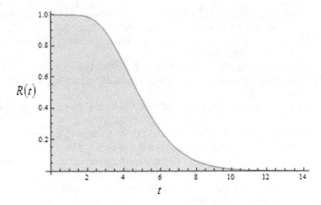

Fig. 8 Hazard function

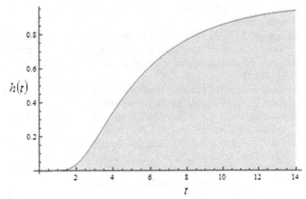

In our second application is a model given by Moranda in (Moranda 1979), considered the model when α_i changes in a geometric sequence $\alpha_i = \alpha_{i-1} r$. We take $\alpha_1 = 1$, $r = 1.2$, and $m = 10$. Using Corollary 5 and Propositions 11 and 19, we find the reliability function (Fig. 7) and Hazard function (Fig. 8) represented in the following figures.

References

Ross SM (2011) Introduction to probability models, 10th edn. Academic, San Diego

Feller W (1971) An introduction to probability theory and its applications, vol II. Wiley, New York

Trivedi KS (2002) Probability and statistics with reliability, queuing and computer science applications, 2nd edn. Wiley, Hoboken

Anjum B, Perros HG (2011) Adding percentiles of erlangian distributions. IEEE Commun Lett 15(3):346–348

Jasiulewicz H, Kordecki W (2003) Convolutions of Erlang and of Pascal distributions with applications to reliability. Demonstratio Math 36(1):231–238

Mathai AM (1982) Storage capacity of a dam with gamma type inputs. Ann Inst Stat Math 34(1):591–597. doi:10.1007/BF02481056

Willmot GE, Woo JK (2007) On the class of Erlang mixtures with risk theoretic applications. N Am Actuarial J 11(2):99–115. doi:10.1080/10920277.2007.10597450

Minkova LD (2010) Insurance risk theory. Lecture notes, TEMPUS Project SEE doctoral studies in mathematical sciences

Bolch G, Greiner S, Meer H, Trivedi K (2006) Queuing networks and Markov chains: modeling and performance evaluation with computer science applications, 2nd edn. Wiley-Interscience, New York

Nadarajah S (2008) A review of results on sums of random variables. Acta Appl Math 103(2): 131–140. doi:10.1007/s10440-008-9224-4

Smaili K, Kadri T, Kadry S (2013a) A modified-form expressions for the hypoexponential distribution. Brit J Math Comput Sci 4(3):322–332

Smaili K, Kadri T, Kadry S (2013b) Hypoexponential distribution with different parameters. Appl Math 4(4):624–631. doi:10.4236/am.2013.44087

Amari SV, Misra RB (1997) Closed-form expression for distribution of the sum of independent exponential random variables. IEEE Trans Reliab 46(4):519–522. doi:10.1109/24.693785

Abdelkader YH (2003) Erlang distributed activity times in stochastic activity networks. Kybernetika 39:347–358

Zukerman M (2012) Introduction to queueing theory and stochastic teletraffic model. Teaching textbook manuscript. http://www.ee.cityu.edu.hk/126zukerman/classnotes.pdf

Akkouchi M (2008) On the convolution of exponential distributions. Chungcheong Math Soc 21(4):501–510

Kordecki W (1997) Reliability bounds for multistage structure with independent components. Stat Probab Lett 34:43–51. doi:10.1016/S0167-7152(96)00164-2

Gaudoin O, Ledoux J (2007) Modélisation aléatoire en fiabilité des logiciels. Hermès Science Publications, Paris

Jelinski Z, Moranda PB (1972) Software reliability research, Statistical computer performance evaluation. Academic, New York, pp 465–484

Moranda PB (1979) Event-altered rate models for general reliability analysis. IEEE Trans Reliab R-28:376–381. doi:10.1109/TR.1979.5220648

Spiegel MR (1965) Schaum's outline of theory and problems of Laplace transforms. Schaum, New York

Part IV
Decision Making Under Uncertainty

Reliability-Based Design Optimization and Its Applications to Interaction Fluid Structure Problems

Abderahman Makhloufi and Abdelkhalak El Hami

Abstract The objectives of this work are to quantify the influence of material and operational uncertainties on the performance of structures coupled with fluid, and to develop a reliability-based design and optimization (RBDO) methodology for this type of the structures. Such a problem requires a very high computation cost, which is mainly due to the calculation of gradients, especially when a finite element model is used. To simplify the optimization problem and to find at least a local optimum solution, two new methods based on semi-numerical solution are proposed in this chapter. The results demonstrate the viability of the proposed reliability-based design and optimization methodology relative to the classical methods, and demonstrate that a probabilistic approach is more appropriate than a deterministic approach for the design and optimization of structures coupled with fluid.

1 Introduction

The objective of the RBDO model is to design structures which should be both economic and reliable where the solution reduces the structural weight in uncritical regions. It does not only provide an improved design but also a higher level of confidence in the design. The classical approach (Feng and Moses 1986) can be carried out in two separate spaces: the physical space and the normalized space. Since very many repeated searches are needed in the above two spaces, the computational time for such an optimization is a big problem. To overcome these difficulties, two points of view have been considered. From reliability view point, RBDO involves the evaluation of probabilistic constraints, which can be executed in two different ways: either using the Reliability Index Approach (RIA), or the Performance Measure

A. Makhloufi • A. El Hami (✉)
Laboratoire d'Optimisation et Fiabilité en Mécanique des Structures, INSA de Rouen,
Avenue de l'Université BP 08, 6800 Saint Etienne de Rouvray, France
e-mail: abdelkhalak.elhami@insa-rouen.fr

S. Kadry and A. El Hami (eds.), *Numerical Methods for Reliability and Safety Assessment: Multiscale and Multiphysics Systems*, DOI 10.1007/978-3-319-07167-1_24,
© Springer International Publishing Switzerland 2015

Approach (PMA) (Tu et al. 1999; Youn et al. 2003). Recently, the enhanced hybrid mean value (HMV+) method is proposed by Youn et al. 2005, to improve numerical stability and efficiency in the most probable point (MPP) search. The major difficulty lies in the evaluation of the probabilistic constraints, which is prohibitively expensive and even diverges with many applications. However, from optimization view point, an efficient method called the hybrid method (HM) has been elaborated by (Kharmanda et al. 2002) where the optimization process is carried out in a Hybrid Design Space (HDS). This method has been shown to verify the optimality conditions relative to the classical RBDO method. The advantage of the HM allows us to satisfy a required reliability level for different cases (static, dynamic, ...), but the application of the hybrid RBDO to engineering design problems is often complicated by multi-physics phenomena, such as fluid–structure interaction (FSI) for naval and aeronautical structures. In these cases, the stochastic coupled multi-physics response needs to be accounted for in the design optimization process. To simplify the optimization problem that rely on fluid–structure interaction for performance improvement and aims to find at least a local optimum solution, two new methods based on semi-numerical solution are proposed: the first one is called optimum safety factor (Kharmanda et al. 2004a), developed for structures in static and dynamic. The second method called safest point (SP) has been next proposed by (Kharmanda et al. 2007), developed for a special case, when a failure interval $[f_a, f_b]$ is given. This method can be considered as a conjoint of the OSF method in order to solve the freely vibrating structures. In this chapter, we will underline the different methods of the RBDO analysis and we highlight the advantage of new semi-numerical approaches proposed. To demonstrate the methodology, results are shown for a marine propeller and cavity acoustics, but the methodology is generally applicable to other adaptive structures that undergo fluid–structure interaction.

2   Reliability Analysis

The design of structures and the prediction of their good functioning lead to the verification of a certain number of rules resulting from the knowledge of physical and mechanical experience of designers and constructors. These rules traduce the necessity to limit the loading effects such as stresses and displacements. Each rule represents an elementary event and the occurrence of several events leads to a failure scenario. The objective is then to evaluate the failure probability corresponding to the occurrence of critical failure modes.

2.1   Importance of Safety Criteria

In deterministic structural optimization, the designer aims to reduce the construction cost without caring about the effects of uncertainties concerning materials,

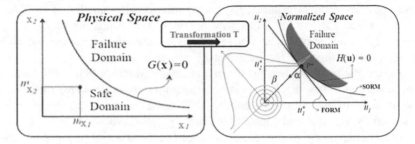

Fig. 1 The transformation between the physical space and normalized one

geometry, and loading. In this way, the resulting optimal configuration may present a lower reliability level and then, leads to higher failure rate. The equilibrium between the cost minimization and the reliability maximization is a great challenge for the designer. In general design problems, we distinguish between two kinds of variables:

- The design variables X which are deterministic variables to be defined in order to optimize the design. They represent the control parameters of the mechanical system (e.g. dimensions, materials, loads) and of the probabilistic model (e.g. means and standard deviations of the random variables).
- The random variables Y which represent the structural uncertainties, identified by probabilistic distributions. These variables can be geometrical dimensions, material characteristics, or applied external loading.

2.2 Failure Probability

In addition to the vector of deterministic variables X to be used in the system design and optimization, the uncertainties are modelled by a vector of stochastic physical variables affecting the failure scenario. The knowledge of these variables is not, at best, more than statistical information and we admit a representation in the form of random variables. For a given design rule, the basic random variables are defined by their joint probability distribution associated with some expected parameters; the vector of random variables is denoted herein Y whose realizations are written y. The safety is the state in which the structure is able to fulfill all the functioning requirements (e.g. strength and serviceability) for which it is designed. To evaluate the failure probability with respect to a chosen failure scenario, a limit state function $G(x, y)$ is defined by the condition of good functioning of the structure. In Fig. 1, the limit between the state of failure $G(x, y) < 0$ and the state of safety $G(x, y) > 0$ is known as the limit state surface $G(x, y) = 0$. The failure probability is then calculated by:

$$P_{\mathrm{f}} = P_r \left[G\left(x, y\right) \leq 0 \right] = \int_{G(x,y) \leq 0} f_Y(y) \, dy_1 \cdots dy_n \tag{1}$$

where P_f is the failure probability, $f_Y(y)$ is the joint density function of the random variables Y and $P_r[\cdot]$ is the probability operator. The evaluation of the integral in (Eq. (1)) is not easy, because it represents a very small quantity and all the necessary information for the joint density function are not available. For these reasons, the First and the Second Order Reliability Methods FORM/SORM have been developed. They are based on the reliability index concept, followed by an estimation of the failure probability. The invariant reliability index β was introduced by Hasofer and Lind (1974), who proposed to work in the space of standard independent gaussian variables instead of the space of physical variables. The transformation from the physical variables y to the normalized variables u is given by:

$$u = T(x, y) \quad \text{and} \quad y = T^{-1}(x, u) \tag{2}$$

This operator $T(\cdot)$ is called the probabilistic transformation. In this standard space, the limit state function takes the form:

$$H(x, u) \equiv G(x, y) = 0 \tag{3}$$

In the FORM approximation, the failure probability is simply evaluated by:

$$P_f \approx \Phi(-\beta) \tag{4}$$

where $\Phi(\cdot)$ is the standard Gaussian cumulated function. For practical engineering, (Eq. (4)) gives sufficiently accurate estimation of the failure probability.

2.3   Reliability Evaluation

For a given failure scenario, the reliability index β is evaluated by solving a constrained optimization problem (Fig. 1). The calculation of the reliability index can be realized by the following form:

$$\beta = \min\left(\sqrt{u^{\mathsf{T}} u}\right) \quad \text{subject to} \quad H(x, u) \leq 0 \tag{5}$$

The solution of this problem is called the design point P^*, as illustrated in Fig. 1. When the mechanical model is defined by numerical methods, such as the finite element method, the evaluation of the reliability implies a special coupling procedure between both reliability and mechanical models.

3 Reliability-Based Design Optimization

Using the Deterministic Design Optimization (DDO) procedure by a reliability analysis (see Kharmanda et al. 2004a), we can distinguish between two cases:

- Case 1: High reliability level: when choosing high values of safety factors for certain parameters, the structural cost (or weight) will be significantly increased because the reliability level becomes much higher than the required level for the structure. So, the design is safe but very expensive.
- Case 2: Low reliability level: when choosing small values of safety factors or bad distribution of these factors, the structural reliability level may be too low to be appropriate. For example, (Grandhi and Wang 1998) found that the resulting reliability index of the optimum deterministic design of a gas turbine blade is $\beta = 0.0053$ under some uncertainties. This result indicated that the reliability at the deterministic optimum is quite low and needs to be improved by probabilistic design.

For both cases, we can find that there is a strong need to integrate the reliability analysis in the optimization process in order to control the reliability level and to minimize the structural cost or weight in the non-critical regions of the structure.

3.1 Classical Method (CM)

Traditionally, for the reliability-based optimization procedure we use two spaces: the physical space and the normalized space (Fig. 1). Therefore, the reliability-based optimization is performed by nesting the two following problems:

Optimization problem:

$$\min : f(\mathbf{x}) \quad \text{subject to} \quad g_k(\mathbf{x}) \leq 0 \quad \text{and} \quad \beta(\mathbf{x}, \mathbf{u}) \geq \beta_t \tag{6}$$

where $f(\mathbf{x})$ is the objective function, $g_k(\mathbf{x}) \leq 0$ are the associated constraints, $\beta(\mathbf{x}, \mathbf{u})$ is the reliability index of the structure, and β_t is the target reliability.

Reliability analysis:

The reliability index $\beta(\mathbf{x}, \mathbf{u})$ is determined by solving the minimization problem:

$$\beta = \min \ dis(\mathbf{u}) = \sqrt{\sum_1^m u_j^2} \quad \text{subject to} \ : \ H(\mathbf{x}, \mathbf{u}) \leq 0 \tag{7}$$

where $dis(\mathbf{u})$ is the distance in the normalized random space and $H(\mathbf{x}, \mathbf{u})$ is the performance function (or limit state function) in the normalized space, defined such that $H(\mathbf{x}, \mathbf{u}) < 0$ implies failure, see Fig. 1. In the physical space, the image of $H(\mathbf{x}, \mathbf{u})$ is the limit state function $G(\mathbf{x}, \mathbf{y})$, see Fig. 1. The solution of these nested problems leads to very large computational time, especially for large-scale structures.

Fig. 2 Hybrid Design Space
for normal distribution

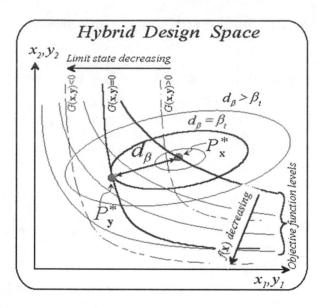

3.2   Hybrid Method (HM)

In order to improve the numerical performance, the hybrid approach consists in minimizing a new form of the objective function $F(\mathbf{x}, \mathbf{y})$ subject to a limit state and to deterministic as well as to reliability constraints, as:

$$
\begin{aligned}
\min_{\mathbf{x}, \mathbf{y}} \; : \; & F(\mathbf{x}, \mathbf{y}) = f(\mathbf{x}) \cdot d_\beta(\mathbf{x}, \mathbf{y}) \\
\text{subject to} \; : \; & G(\mathbf{x}, \mathbf{y}) \leq 0, \quad g_k(\mathbf{x}) \leq 0 \quad \text{and} \quad d_\beta(\mathbf{x}, \mathbf{y}) \geq \beta_t
\end{aligned}
\tag{8}
$$

The minimization of the objective function $F(\mathbf{x}, \mathbf{y})$ is carried out simultaneously with respect to $\mathbf{x}$ and $\mathbf{y}$ in the HDS of deterministic variables $\mathbf{x}$ and random variables $\mathbf{y}$. Here $d_\beta(\mathbf{x}, \mathbf{y})$ is the distance in the hybrid space between the optimum point and the design point. Since the random variables and the deterministic ones are treated in the same space (HDS), it is important to know the types of the used random variables (continuous and/or discrete) and the distribution law that has been applied. In engineering design, there are many standard distributions or continuous and discrete random variables. In this chapter, the normal distribution is used to evaluate the reliability index β. Other distributions can be used, such as the lognormal and uniform distributions (Kharmanda et al. 2004b). Using anyone of these distributions, we can determine the failure probability and then the reliability index corresponding. An example of this HDS is given in Fig. 2, containing design and random variables, where the reliability levels d_β can be represented by ellipses in case of normal distribution, the objective function levels are given by solid curves and the limit state function is represented by dashed level lines except for

Fig. 3 Bracket to be designed

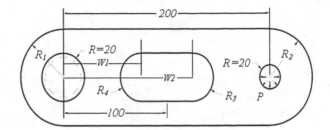

Table 1 Data of the problem

Variables	R1	R2	R3	R4	W	R	L
Dimension (mm)	45	45	20	20	50	10	200

Table 2 Material parameters

Variables	E (MPa)	ν	ρ (kg/m^3)
Values	71,018	0.33	8250

Table 3 RBDO variables

Variable **y**	Mean **x**	Standard deviation	Initial design
$R1$ (mm)	m_{R1} (mm)	4.5	45
$R2$ (mm)	m_{R2} (mm)	4.5	45

$G(\mathbf{x}, \mathbf{y}) = 0$. We can see two important points: the optimal solution P_x^* and the reliability solution P_y^* (i.e. the design point found on the curves $G(\mathbf{x}, \mathbf{y}) = 0$ and $d_\beta = \beta_t$). In the paper of Kharmanda et al. (2002), we had demonstrated that the HM reduced the computational time almost 80 % relative to the classical RBDO approach. Using the HM, the optimization process is carried out in the HDS where all numerical information about the optimization process can be modeled. Furthermore, the classical method (CM) has weak convergence stability because it is carried out in two spaces (physical and normalized spaces).

A bracket as shown in Fig. 3 is made of *7075-T651* Aluminum ($E = 71,018$ MPa, $\nu = 0.33$), the yield strength of the material is assumed to be $S_Y = 524$ MPa. An initial geometry of the bracket is given in Fig. 1 (dimensions in mm). The bracket is clamped at the left hub and carries a downward load at the right hub. The load is modeled as a uniform pressure $P = 50$ N/mm^2 as shown. Material properties and geometric dimensioning used in this model are shown respectively in Tables 1 and 2. In this study, the objective is to minimize the weight design of the structure under the design constraints that's the reliability constraint. To optimize the structure, the mean values of the dimensions m_{R1} and m_{R2} are the design variables. The physical dimensions $R1$ and $R2$ are the random variables y, which are supposed to be normally distributed. Table 3 gives the RBDO variables, as well as the corresponding standard deviations and initial values. In this problem, we have four optimization variables: two random variables y and two design variables x.

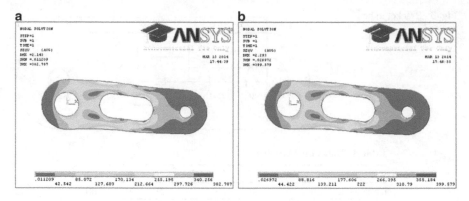

Fig. 4 Stress distribution with RBDO procedure: (**a**) classical method and (**b**) hybrid method

For this design, the target reliability level is $\beta_c = 3$ with convergence tolerance equal to 1 %. The equivalent maximum failure probability is $P_f = 4 \times 10^{-4}$.

Classical RBDO approach: Using the classical model, the optimization problem can be written in two sub-problems:

1. optimization problem subject to reliability constraint:

$$
\begin{aligned}
\min : \quad & \text{volume} (R1, R2) \\
\text{subject to} : \quad & \beta (\{x\}, \{y\}) \geq \beta_c
\end{aligned}
$$

2. calculation of the reliability index:

$$
\min : \quad \left(d_1 (\{u\}) = \sqrt{\sum_i^m u_i^2} \right) \quad i = 1, 2
$$

$$
\begin{aligned}
\text{subject to} : \quad & \sigma_{\text{eqv}} (R1, R2) \leq \sigma_y \\
\text{with} \quad & \sigma_y = 524 \text{ MPa}, \quad \beta_c = 3 \quad \text{and} \quad u_i = \frac{x_i - m_{xi}}{\sigma_{xi}}
\end{aligned}
$$

Hybrid RBDO approach: Using the hybrid reliability-based design model, we can simplify the two last sub-problems into one problem:

$$
\begin{aligned}
\min \quad : \quad & F (\{x\}, \{y\}) = \text{volume} (\{x\}) \times d_\beta (\{x\}, \{y\}) \\
\text{subject to} : \quad & \sigma_{\text{eqv}} (R1, R2) \leq \sigma_y \quad \text{and} \quad \beta (\{x\}, \{y\}) \geq \beta_c
\end{aligned}
$$

Figure 4 shows the stress distribution after the application of the reliability-based optimization procedure, and Table 4 shows the results of the classical and HMs.

Table 4 Classical and hybrid RBDO results

Parameter	Initial point	Classical RBDO	Hybrid RBDO
$R1$ (mm)	45	41.05	39.56
$R2$ (mm)	45	35.95	36.75
Volume (mm^3)	19,534.31	15,254.65	15,016.45
β	–	3.2	3.00
Von Mises stress (MPa)	242	383	399

Table 5 Efficiency comparison

Model	Classical RBDO	Hybrid RBDO
Volume (mm^2)	15,254.65	15,016.45
β	3.2	3.00
n_{det}	7	8
n_{rel}	$5 + (7*3)$	0
n_{calls}	80	32

The classical RBDO approach requires 80 Finite Element Analyses (FEA) to reach the minimal volume $V^* = 15{,}254.65$ mm^3 and to satisfy the target reliability level $\beta = 3.2 \geq \beta_c$. However, the HM needs only 32 evaluations to reach the minimal volume $V^* = 15{,}016.45$ mm^3 and to satisfy the target reliability level $\beta = 3.00$.

At each deterministic iteration, the CM needs a complete reliability analysis in order to calculate the reliability index. Furthermore, for each reliability iteration, we need two FEA (equal to the random variables number $m = 2$) that leads to a very high FEA (for this example: five reliability iterations for the first deterministic iteration and 3 ones for the following optimization iterations). By comparing their results, the HM gives a computational time clearly reduced with respect to the classical approach. In addition, for each deterministic iteration, we need a $n + 1 = 3$ FEA, n is the design variables number and one FEA for evaluating the stresses. In the hybrid RBDO procedure, a gradient calculation for the design variables ($n + 1 = 3$ FEA) and two FEA (one for the design variables and the other for the random ones) are necessary for each iteration. Table 5 gives the comparison between the two methods, where n_{det} and n_{rel} are the numbers of deterministic and reliability iterations, respectively, and n_{calls} is the number of FEA.

In fact, when using the HM, we have a complex optimization problem with many variables. Solving this problem, we get a local optimum. When changing the starting point, we may get another local optimum. This way the designer has to repeat the optimization process to get several local optima. However, to overcome these drawbacks, two semi-numerical RBDO methods have been proposed (Kharmanda et al. 2007).

4 Semi-numerical RBDO Method

4.1 Optimum Safety Factor Method (OSF)

In this section, we select the normal distribution, to demonstrate the efficiency of the OSF method relative to the hybrid RBDO method. Other distributions such as the lognormal and uniform distributions are developed by Kharmanda et al. (2004b) to compute safety factors satisfying a required reliability level to a trimaterial structure.

In general, when considering the normal distribution law, the normalized variable u_i is given by:

$$u_i = \frac{y_i - m_i}{\sigma_i}, \quad i = 1, \ldots, n \tag{9}$$

The standard deviation σ_i can be related to the mean value m_i by:

$$\sigma_i = \gamma_i \cdot m_i \quad \text{or} \quad \sigma_i = \gamma_i \cdot x_i, \quad i = 1, \ldots, n \tag{10}$$

This way we introduce the safety factors S_{f_i} corresponding to the design variables x_i. The design point can be expressed by:

$$y_i = S_{f_i} \cdot x_i, \quad i = 1, \ldots, n \tag{11}$$

For an assumed failure scenario $G(y) < 0$, the equation of the optimum safety factor for a single limit state case can be written in the following form (see the developments in Kharmanda et al. 2004a):

$$S_{f_i} = 1 \pm \gamma_i \cdot \beta_t \sqrt{\frac{\left|\frac{\partial G}{\partial y_i}\right|}{\sum_{i=1}^{n} \left|\frac{\partial G}{\partial y_i}\right|}}, \quad i = 1, \ldots, n \tag{12}$$

Here, the sign $\pm$ depends on the sign of the derivative, i.e.

$$\frac{\partial G}{\partial y_i} > 0 \iff S_{f_i} > 1 \quad \text{or} \quad \frac{\partial G}{\partial y_i} < 0 \iff S_{f_i} < 1, \quad i = 1, \ldots, n \tag{13}$$

Implementation of the OSF approach: The algorithm of optimum safety factor approach consists of a simple optimization problem to find the failure point followed by calculation of the optimum safety factor and finally reevaluation of the new model using the safety factors. This optimization problem presented in Fig. 5 can be expressed according to the following steps:

1. Input the initial values of the variable vector y_0 of the studied model.
2. Evaluate the objective function $f(y)$.
3. Calculate the deterministic constraints $G(y)$.
4. Test the convergence constraints $G(y) \leq 0$, if converged, stop or update y and go to step 2.

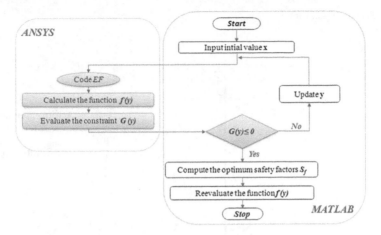

Fig. 5 The optimum safety factor algorithm

5. Compute the safety factors S_{fi} using (Eq. (6)). Here, the derivatives of the limit state function are evaluated at the design point.
6. Reevaluate the new model that presents the optimum solution.

The OSF method cannot be used in some dynamic cases of freely vibrating structures and using the HM, we can obtain local optima and designer may then select the best optimum. This is the reason for why; the Safest Point method (SP) has been proposed (Kharmanda et al. 2007).

4.2   Safest Point Method (SP)

The reliability-based optimum structure under free vibrations for a given interval of eigen-frequency is found at the safest position of this interval where the safest point has the same reliability index relative to both sides of the interval. A simple method has been proposed here to meet the safest point requirements relative to a given frequency interval.

Let consider a given interval $[f_a, f_b]$ (see Fig. 6). For the first shape mode, to get the reliability-based optimum solution for a given interval, we consider the equality of the reliability indices:

$$\beta_a = \beta_b \quad \text{or} \quad \beta_1 = \beta_2 \tag{14}$$

$$\text{with} \quad \beta_a = \sqrt{\sum_{i=1}^{n} \left(u_i^a\right)^2} \quad \text{and} \quad \beta_b = \sqrt{\sum_{i=1}^{n} \left(u_i^b\right)^2} \quad i = 1, \ldots, n \tag{15}$$

Fig. 6 The safest point at
frequency f_n

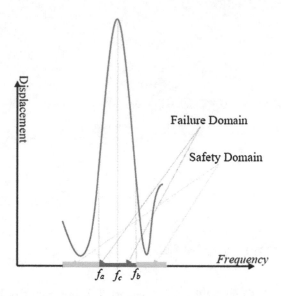

To verify the equality (14), we propose the equality of each term. So we have:

$$u_i^a = -u_i^b, \quad i = 1, \ldots, n \tag{16}$$

Normal distribution for SP method: One of the most commonly used distributions of
a random variable y_i in engineering problems is the normal or Gaussian distribution.
The mean value m_i and the standard deviation σ_i are two parameters of the
distribution, usually estimated from available data. The normalized variable u_i is
defined by:

$$u_i = \frac{y_i - m_{yi}}{\sigma_i}, \quad i = 1, \ldots, n \tag{17}$$

with : $x_i = m_{yi} \quad i = 1, \ldots, n$ and $\sigma_i = \gamma_i \times x_i, \quad i = 1, \ldots, n \tag{18}$

We consider the equality of the reliability indices

$$\frac{y_i^a - m_i}{\sigma_i} = -\frac{y_i^b - m_i}{\sigma_i} \quad \text{or} \quad \frac{y_i^a - x_i}{\sigma_i} = -\frac{y_i^b - x_i}{\sigma_i}, \quad i = 1, \ldots, n \tag{19}$$

Safest point corresponding to the frequency f_n and located in the interval $[f_a, f_b]$
is given by:

$$m_i = x_i = \frac{y_i^a + y_i^b}{2}, \quad i = 1, \ldots, n \tag{20}$$

Lognormal distribution for SP method: If a random variable has a lognormal distribution; the normalized variable u_i is thus defined as:

$$u_i = \frac{\ln(y_i) - \mu_i}{\xi_i}, \quad i = 1, \ldots, n \tag{21}$$

where μ_i and ζ_i are two parameters of the lognormal distribution that are defined as follows:

$$\mu_i = \ln\left(\frac{x_i}{\sqrt{1+\gamma_i^2}}\right), \quad \zeta_i = \sqrt{\ln(1+\gamma_i^2)} \quad \text{with} \quad \gamma_i = \frac{\sigma_i}{x_i}, \quad i = 1, \ldots, n \tag{22}$$

This distribution is valid for any value of a random variable in the interval from 0 to $+\infty$. The normalized variable u_i can then be expressed by:

$$u_i = \frac{\ln\left(\frac{y_i\sqrt{1+(\sigma_i/x_i)^2}}{x_i}\right)}{\sqrt{\ln\left(1+(\sigma_i/x_i)^2\right)}} \quad i = 1, \ldots, n \tag{23}$$

We consider the equality of the reliability indices:

$$\frac{\ln(y_i^a) - \mu_i}{\xi_i} = -\frac{\ln(y_i^b) - \mu_i}{\zeta_i} \quad \text{or} \quad \mu_i = \ln\left(\frac{x_i}{\sqrt{1+\gamma_i^2}}\right)$$

$$= \frac{\ln(y_i^a) + \ln(y_i^b)}{2} \quad i = 1, \ldots, n \tag{24}$$

Safest point corresponding to the frequency f_n. and located in the interval $[f_a, f_b]$ is given by:

$$m_i = x_i = \sqrt{1+\gamma_i^2}\exp\left(\frac{\ln(y_i^a \cdot y_i^b)}{2}\right) \quad i = 1, \ldots, n \tag{25}$$

Uniform distribution for SP method: In general, when considering the uniform distribution law, the transformation between the physical and the normalized space is expressed by:

$$y_i = a + (b-a)\,\Phi(u_i), \quad i = 1, \ldots, n \tag{26}$$

with

$$\Phi(U_i) = \frac{1}{\sqrt{2\pi}} \int_{-\infty}^{U_i} e^{-\frac{u^2}{2}} \, du, \quad i = 1, \ldots, n \tag{27}$$

where a and b in (26) are the interval boundaries, and the normalized variables are denoted by U_i. The mean value m_i (or x_i) is given by

$$m_i = x_i = \frac{a+b}{2}, \quad i = 1, \ldots, n \tag{28}$$

and the standard deviation σ_i by

$$\sigma_i = \frac{a-b}{\sqrt{12}}, \quad i = 1, \ldots, n \tag{29}$$

From (28) and (29), we easily get

$$a = x_i - \sqrt{3}\sigma_i \quad \text{and} \quad b = x_i + \sqrt{3}\sigma_i, \quad i = 1, \ldots, n \tag{30}$$

We then propose to write (26) as follows

$$u_i = \Phi^{-1}\left(\frac{y_i - a}{b - a}\right) \quad \text{or} \quad u_i = \Phi^{-1}\left(\frac{y_i - x_i + \sqrt{3}\sigma_i}{2\sqrt{3}\sigma_i}\right), \quad i = 1, \ldots, n \tag{31}$$

Safest point corresponding to the frequency f_n and located in the interval $[f_a, f_b]$ is given by:

$$m_i = x_i = \frac{y_i^a + y_i^b}{2}, \quad i = 1, \ldots, n \tag{32}$$

Implementation of the SP approach: The SP algorithm for symmetric case can be expressed by the three following steps (two sequential optimization steps and an analytical evaluation one) (see Fig. 7):

1. Compute the design point a: The first optimization problem is to minimize the objective function subject to the first bound of the frequency interval f_a. The resulting solution is considered as a most probable point A.
2. Compute the design point b: The second optimization problem is to minimize the objective function subject to the second bound of the frequency interval f_b. The resulting solution is considered as a most probable point B.
3. Compute the optimum solution: Here, we analytically determine the optimum solution of the studied structure for linear distribution case.

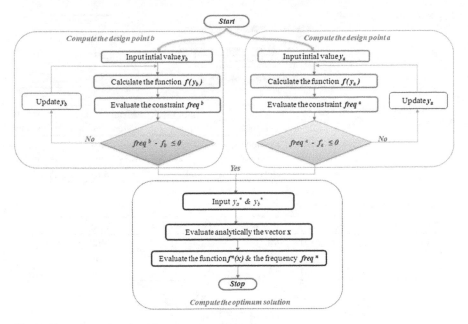

Fig. 7 The safest point algorithm

The reliability-based optimum structure under free vibrations for a given interval of eigen-frequency is found at the safest position of this interval where the safest point has the same reliability index relative to both sides of the interval. A simple method has been proposed here to meet the safest point requirements relative to a given frequency interval.

5   Numerical Applications

The following applications are carried out using ANSYS as a Finite Element Software. All optimization process is carried out using a zero order method in ANSYS optimization tools. This method uses curve fitting for all dependant variables. The gradient tool in ANSYS computes the gradient of the state variables and the objective function with respect to the design variables. A reference design set is defined as the point of evaluation for the gradient. The gradient is approximated from the following forward difference (ANSYS Guide 2013). For simplicity, we consider that all random variables follow the normal (Gauss) distribution law and the standard deviations are considered as proportional of the mean value of the random variables.

Fig. 8 Studied plate layers

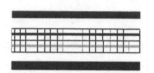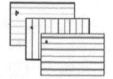

Table 6 Probabilistic model parameters

Parameter	Initial values	Mean values	Standard deviation
e_1 (mm)	0.025	me_1 (mm)	0.0025
e_2 (mm)	0.050	me_2 (mm)	0.0050
e_3 (mm)	0.025	me_3 (mm)	0.0025

Table 7 Material properties

Parameter	$E_{11} = E_{33}$	E_{22}	$G_{12} = G_{21}$	G_{13}	$\upsilon_{12} = \upsilon_{13}$	ρ (kg/m^3)	θ
Layer 1	200	1.0	40	2.0	0.3	2000	0/90
Layer 2	100	1.0	15	2.5	0.1	50	–
Layer 3	150	1.0	15	2.5	0.2	1400	90/0

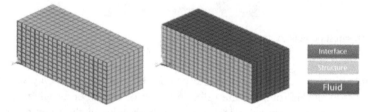

Fig. 9 Meshing model of fluid–structure interaction

5.1 Optimization of a Beam Under Fluid–Structure Interaction

In the first application, we use the OSF and HMs to integrate fluid–structure interaction phenomena into reliability-based design optimization. The studied trimaterial plate structure is excited by a harmonic force (0–500 Hz) considering the fluid-structure interaction phenomenon. The simplified model is presented in Fig. 8. A rectangular plate consists of three layers fixed on the four corners. Each layer has a thickness as: e_i, $i = 1, 2, 3$ (see Table 6). The material properties: E_{ij} (Young's modulus), ρ_i (volume mass), v_i (Poisson's ratio), and G_{ij} (shear modulus) are presented in Table 7. This rectangular plate is obscure in the fluid (air) being perfect, compressible, nonrotational, and initially in rest. Its volume mass and celerity of sonorous waves are respectively: $\rho_f = 1.2$ kg/m^3 and $c = 340$ m/s. The meshing model presented in Fig. 9 is carried out for both structure and fluid: 200 *Shell81* elements (bidimensional linear shell element) and 1600 *Fluid30* elements (tridimensional linear acoustic fluid element). Figure 10 presents the acoustic pressure inside the cavity in relation with the frequency interval [0–500] Hz.

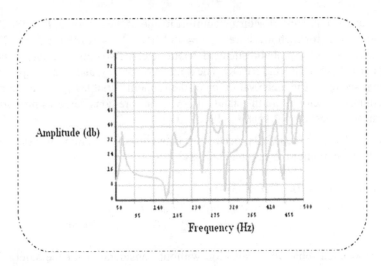

Fig. 10 Response of the acoustic pressure inside the cavity

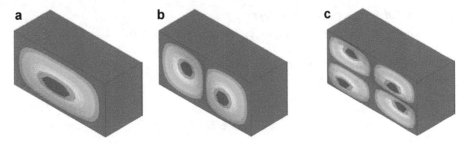

Fig. 11 Three first shape modes of the plate. (**a**) Mode $(1, 1)$, $f = 66$ Hz. (**b**) Mode$(2, 1)$, $f = 113$ Hz. (**c**) Mode$(2, 2)$, $f = 211$ Hz

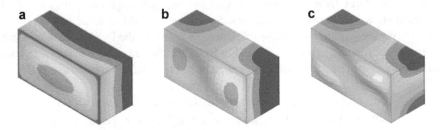

Fig. 12 Three first shape modes of the acoustic cavity. (**a**) Mode$(1, 1)$, $f = 107$ Hz. (**b**) Mode$(2, 1)$, $f = 138$ Hz. (**c**) Mode$(2, 2)$, $f = 265$ Hz

The three shape modes of the plate and of the acoustic cavity are respectively presented in Figs. 11 and 12. To optimize this structure, we consider the stress Von Mises and the interior noise level inside the acoustic cavity as constraints. The target (or allowable) constraint of acoustic comfort inside of the cavity is: $P_t = 90$ db and the yield stresses for each layer are: $\sigma_y^{M_1} = 48$ MPa, $\sigma_y^{M2} = 18$MPa, $\sigma_y^{M3} = 42$MPa.

Table 6 presents the probabilistic model parameters. Two optimization procedures can be carried out: DDO and RBDO. The dimensions e_1, e_2, and e_3 are regrouped in a random vector y. The mean values m_i of the random variables y_i are regrouped in a deterministic vector x and the standard deviations are considered as proportional of the mean values: $\sigma_i = 0.1*m$; $i = 1, \ldots, 3$.

DDO procedures: In the DDO procedure, it is the objective to minimize the volume subject to the maximum stress constraint as:

$$\text{min}: \text{ volume} (me_1, me_2, me_3)$$
$$\text{subject to}: \sigma_{max}^{Mi} (me_1, me_2, me_3) \leq \sigma_w^{Mi} = \sigma_i^y / S_f, \quad i = 1, \ldots, 3$$
$$P (me_1, me_2, me_3) - P_t \leq 0, \ldots P_t = 90 \text{ db}$$

The associated reliability evaluation without consideration of the safety factor can be written in the form

$$\text{min}: d (e_1, e_2, e_3, ue_1, ue_2, ue_3)$$
$$\text{subject to}: \sigma_y^{Mi} - \sigma_{max}^{Mi} (e_1, e_2, e_3, me_1, me_2, me_3) \leq 0, \quad i = 1, \ldots, 3$$

The objective function and the constraints are evaluated by using ANSYS Finite Element Analysis software.

$$\text{structural volume} = \sum_{i=1}^{\text{element}} V_i$$

where V_i designate the volume of the ith element

RBDO procedures (HM method): The CM implies very high computational cost and exhibits weak convergence stability. So we use the HM to satisfy the required reliability level (within admissible tolerances of 1 %). In the hybrid procedure, we minimize the product of the volume and the reliability index subject to the limit state functions and the required reliability level. The hybrid RBDO problem is written as:

$$\text{min}: \text{ volume} (me_1, me_2, me_3) \cdot d_\beta (e_1, e_2, e_3, me_1, me_2, me_3)$$
$$\text{subject to}: \sigma_{max}^{Mi} (me_1, me_2, me_3) \leq \sigma_w^{Mi}, \quad i = 1, \ldots, 3$$
$$P (me_1, me_2, me_3) - P_t \leq 0$$
$$P_t = 90 \text{ db}$$
$$d_\beta (e_1, e_2, e_3, me_1, me_2, me_3) \geq \beta_c$$

RBDO procedures (OSF method): The RBDO by OSF includes three main steps:

- The first step is to obtain the design point (the most probable point). Here, we minimize the volume subject to the design constraints without consideration of the safety factors. This way the optimization problem is simply written as:

Table 8 Sensitivities of limit state functions and optimum safety factors

Variables	$\partial G_{M1}/\partial y_i$	$\partial G_{M2}/\partial y_2$	$\partial G_{M3}/\partial y_3$	S_{fi}
e_1	−0.4159	−0.0273	−0.0895	0.8198
e_2	−0.3104	−0.2100	−0.3765	0.7662
e_3	−0.0370	−0.0403	−0.3348	0.8415

$$\min : \text{ volume} (me_1, me_2, me_3)$$
$$\text{subject to}: \ \sigma_{\max}^{Mi} (me_1, me_2, me_3) \leq \sigma_w^{Mi} = \sigma_i^y / S_f, \quad i = 1, \dots, 3$$
$$P (me_1, me_2, me_3) - P_t \leq 0$$
$$P_t = 90 \text{ db}$$

The design point is found to correspond to the maximum Von Mises stresses $\sigma_y^{M1} = 47.966$ MPa, $\sigma_y^{M2} = 27.998$, $\sigma_y^{M3} = 41.275$ MPa that is almost equivalent to the given yield stresses.

- The second step is to compute the optimum safety factors using (Eq. (12)). In this example, the number of the deterministic variables is equal to that of the random ones. During the optimization process, we obtain the sensitivity values of the limit state with respect to all variables. So there is no need for additional computational cost. Table 8 shows the results leading to the values of the safety factors, namely the sensitivity results for the different limit state functions.
- The third step is to calculate the optimum solution. This encompasses inclusion of the values of the safety factors in the values of the design variables in order evaluate the optimum solution.

Table 9 shows the results of the DDO and RBDO methods. In the DDO problem, we cannot control the required reliability levels but when using the RBDO procedures (HM and OSF), the target reliability index is satisfied. For the computational time, the solution of the hybrid problem needs a high computing time (28,670 s) because of the big number of optimization variables (deterministic and random vectors). However, using OSF, we need only a little computing time (9,236 s). The reduction of the computing time is almost 68 %. Furthermore, the RBDO using OSF does not need additional cost computing time relative to DDO (9,332 s).

5.2 Numerical Application on Marine Propeller

In the second application, we compare between the SP method and the hybrid one relative to the computational time. Scheme and data problem are presented respectively in Fig. 13 and Table 10.

The objective is to find the eigen-frequency for a given interval $[14, 20]$Hz, that is located on the safest position of this interval. So $f_a = 14$ Hz, $f_b = 20$ Hz and

Table 9 DDO and RBDO results

Variables	DDO procedure	RBDO procedure Hybrid method	OSF method
me_1 (cm)	0.5142	0.5245	0.5142
me_2 (cm)	0.2257	0.2269	0.2257
me_3 (cm)	0.74493	0.7473	0.7449
σ_y^{M1} (MPa)	47.966	47.009	47.966
σ_y^{M2} (MPa)	27.998	27.999	27.998
σ_y^{M3} (MPa)	41.275	41.075	41.275
e_1 (cm)	0.5259	0.6229	0.6273
e_2 (cm)	0.2869	0.2997	0.2945
e_3 (cm)	0.7082	0.8902	0.8852
σ_{max}^{M1} (MPa)	40.001	34.585	33.315
σ_{max}^{M2} (MPa)	23.286	23.095	21.869
σ_{max}^{M3} (MPa)	34.261	33.663	26.637
β	2.76	3.35	3.35
P_f (%)	0.3	0.04	0.04
P (db)	76	88.3	89.5
Volume (cm^3)	3,042,127	3,625,621	3,616,843
Time (s)	9,332	28,670	9,236

Fig. 13 Scheme of the marine propeller

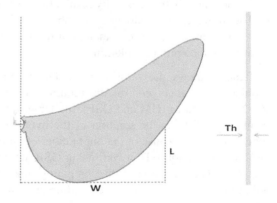

Table 10 Data of the problem

	Parameter	Value
Plate	W	4.5 m
	L	1.5 m
	Thickness	0.1 m
	Outer radius	0.50 m
	Inner radius	0.35 m
	Density	7,860 kg/m^3
	Young modulus	2.0 e^{11} Pa
	Coefficient of poisson	0.3
Water	Density	1000 kg/m^3

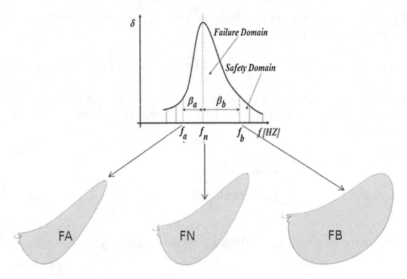

Fig. 14 Marine propeller optimization models for both cases

$f_n = ?$ Hz, where f_n must verify the equality of reliability indices: $\beta_a = \beta_b$. We can deal with three models: the first structure must be optimized subject to the first frequency value of the given f_a, the second one must be optimized at the end frequency value of the interval f_b, and the third structure must be optimized subject to a frequency value f_n that verifies the equality of reliability indices relative to both sides of the given interval (see Fig. 14).

Here, we can deal with two reliability-based design optimization methods: hybrid method and SP method. The HM simultaneously optimizes the three structures but the SP method consists in optimizing three simple problems.

HM procedure: We minimize the weight of one pale of the objective function subject to the different frequencies constraint and the reliability one as follows:

$$
\begin{aligned}
\min : \quad & \text{weight}\,(m_{Th}, m_W, m_L) \cdot d_{\beta_1}\,(Th_1, W_1, L_1, m_{Th}, m_W, m_L) \\
& \cdot d_{\beta_2}\,(Th_2, W_2, L_2, m_{Th}, m_W, m_L) \\
\text{subject to}: \quad & f_{\max}^1\,(Th_1, W_1, L_1) - f_a \le 0, \quad f_{\max}^2\,(Th_2, W_2, L_2) - f_b \le 0 \\
& : d_{\beta_1}\,(Th_1, W_1, L_1, m_{Th}, m_W, m_L) - d_{\beta_2}\,(Th_2, W_2, L_2, m_{Th}, m_W, m_L) \le 0 \\
& : u_i^a + u_i^b = 0, \quad i = 1, \ldots, I
\end{aligned}
$$

SP procedure: We have two simple optimization problems:

- The first is to minimize the objective function of the first model subject to the frequency f_a constraint as follows:

$$
\min : \quad \text{Weight}_a\,(Th_1, W_1, L_1) \quad \text{subject to} \quad : f_{\max}^1\,(Th_1, W_1, L_1) - f_a \le 0
$$

- The second is to minimize the objective function of the second model subject to the frequency f_b constraint as follows:

$$\min : \quad \text{Weight}_b \, (Th_2, W_2, L_2) \quad \text{subject to} \quad : f_{max}^2 \, (Th_2, W_2, L_2) - f_b \leq 0$$

Let consider a given interval $[f_a, f_b]$ (see Fig. 14). For the first shape mode, to get the reliability-based optimum solution for a given interval, we consider the equality of the reliability indices:

$$\beta_a = \beta_b \quad \text{or} \quad \beta_1 = \beta_2$$

$$\text{with} \quad \beta_a = \sqrt{\sum_{i=1}^{n} \left(u_i^a \right)^2} \quad \text{and} \quad \beta_b = \sqrt{\sum_{i=1}^{n} \left(u_i^b \right)^2} \quad i = 1, \ldots, n$$

To verify the equality (Eq. (14)), we propose the equality of each term. So we have:

$$u_i^a = - u_i^b, \quad i = 1, \ldots, n$$

According to the normal distribution law, the normalized variable u_i is given by (Eq. (17)), we get:

$$\frac{y_i^a - m_i}{\sigma_i} = -\frac{y_i^b - m_i}{\sigma_i} \quad \text{or} \quad \frac{y_i^a - x_i}{\sigma_i} = -\frac{y_i^b - x_i}{\sigma_i}, \quad i = 1, \ldots, n$$

To obtain equality between the reliability indices (see Eq. (14)), the mean value of variable corresponds to the structure at f_n. So the mean values of safest solution are located in the middle of the variable interval $[y_i^a, y_i^b]$ as follows:

$$m_i = x_i = \frac{y_i^a + y_i^b}{2}, \quad i = 1, \ldots, n$$

The coordinates of the third model which corresponds to f_n according to (Eq. (20)).

Table 11 shows the results of the SP method and presents the reliability-based optimum point for a given interval [14 Hz, 20 Hz]. The value of f_n presents the equality of reliability indices. The SP method reduces the computing time by 75 % relative to the HM. The advantage of the SP method is simple to be implemented on the machine and to define the eigen-frequency of a given interval and provides the designer with reliability-based optimum solution with a small tolerance relative to the HM.

Table 11 Results for the marine propeller

Variables	Initial design	Optimum design with HM	Optimum design with SP
Th (m)	0.2	0.175	0.20
L (m)	2.00	2.253	2.55
W (m)	4.00	4.151	4.50
Th1 (m)	0.1	0.115	0.120
L1 (m)	1.5	1.65	1.750
W1 (m)	3.5	3.55	3.750
Th2 (m)	0.3	0.286	0.278
L2 (m)	3.5	3.485	3.353
W2 (m)	5.5	5.175	5.246
FA (Hz)	13	14.22	14.10
FN (Hz)	21	18.29	18.90
FB (Hz)	22	19.89	19.90
Volume (m^3)	2.95	2.836	3.05
Time (s)	–	4,125	995

6 Conclusion

A RBDO solution that reduces the structural weight in uncritical regions both provides an improved design and a higher level of confidence in the design. The classical RBDO approach can be carried out in two separate spaces: the physical space and the normalized space. Since very many repeated searches are needed in the above two spaces, the computational time for such an optimization is a big problem. The structural engineers do not consider the RBDO as a practical tool for design optimization. An efficient method called the HM has been elaborated where the optimization process is carried out in a HDS. But the application of the hybrid RBDO to engineering design problems is often complicated by multi-physics phenomena, such as fluid–structure interaction (FSI), the use of HM necessitates a high computing time and a complex implementation. Two new methods based on semi-numerical solution called the OSF and SP Methods have been proposed to simplify the Reliability-based design optimization problem of structures coupled with fluid. As it is shown in the numerical applications, the SP and OSF method can reduce efficiently the computing time and aims to find at least a local optimum solution relative to the HM.

References

Ansys Guide (2013) Ansys structural analysis guide

Feng YS, Moses F (1986) A method of structural optimization based on structural system reliability. J Struct Mech 14:437–453

Grandhi RV, Wang L (1998) Reliability-based structural optimization using improved two-point adaptive nonlinear approximations. Finite Elem Anal Des 29:35–48

Hasofer AM, Lind NC (1974) An exact and invariant first order reliability format. J Eng Mech ASCE EM1 100:111–121

Kharmanda G, Mohamed A, Lemaire M (2002) Efficient reliability-based design optimization using hybrid space with application to finite element analysis. Struct Multidiscip Optim 24(2002):233–245

Kharmanda G, Mohsine A, El Hami A (2003) Efficient reliability-based design optimization for dynamic structures. In: Cinquini C, Rovati M, Venini P, Nascimbene R (eds) Proceedings of the fifth world congress of structural and multidisciplinary optimization, WCSMO-5, May 19–23, 2003, Lido di Jesolo-Venice, Italy. University of Pavia, Italy, p 6

Kharmanda G, El Hami A, Olhoff N (2004a) Global reliability-based design optimization. In: Floudas CA (ed) Frontiers on global optimization. Kluwer Academic, Boston, pp 255–275

Kharmanda G, Olhoff N, El Hami A (2004b) Optimum safety factor approach for reliability-based design optimization with extension to multiple limit state case. Struct Multidiscip Optim 26:295–307

Kharmanda G, Altonji A, El Hami A (2006) In: Safest point method for reliability-based design optimization of freely vibrating structures, CIFMA01—IFCAM01, Aleppo—Syria02-04 mai/May 2006

Kharmanda G, Souza de Cursi E, El Hami A (2007) Advanced technologies for reliability-based design optimization 2nd International Francophone congress for Advanced mechanics, 14–16 May 2007 Aleppo, Syria

Mohsine A, Kharmanda G, El Hami A (2005) Reliability-based design optimization study using normal and lognormal distributions with applications to dynamic structures. In: The fifth international conference on structural safety and reliability ICOSSAR05, June, 19–22, 2005, Rome (Italy)

Mohsine A, Kharmanda G, El Hami A (2006) Improved hybrid method as a robust tool for reliability-based design optimization. J Struct Multidiscip Optim 32:203–213

Mohsine A (2006) Contribution à l'optimisation fiabiliste en dynamique des structures mécaniques. PhD thesis, INSA de Rouen, France (French version)

Tu J, Choi KK, Park YH (1999) A new study on reliability-based design optimization. J Mech Des ASME 121(4):557–564

Youn BD, Choi KK, Park YH (2003) Hybrid analysis method for reliability-based design optimization. J Mech Des 125(2):221–232

Youn BD, Choi KK, Du L (2005) Adaptive probability analysis using an enhanced hybrid mean value method. Struct Multidiscip Optim 29:134–148

Improved Planning In-Service Inspections of Fatigued Aircraft Structures Under Parametric Uncertainty of Underlying Lifetime Models

Nicholas A. Nechval and Konstantin N. Nechval

Abstract Certain fatigued structures must be inspected in order to detect fatigue damages that would otherwise not be apparent. A technique for obtaining optimal inspection strategies is proposed for situations where it is difficult to quantify the costs associated with inspections and undetected failure. For fatigued structures for which fatigue damages are only detected at the time of inspection, it is important to be able to determine the optimal times of inspection. Fewer inspections will lead to lower fatigue reliability of the structure upon demand, and frequent inspection will lead to higher cost. When there is a fatigue reliability requirement, the problem is usually to develop an inspection strategy that meets the reliability requirements. It is assumed that only the functional form of the underlying invariant distribution of time to crack detection is specified, but some or all of its parameters are unspecified. The invariant embedding technique proposed in this paper allows one to construct an optimal inspection strategy under parametric uncertainty. Under fatigue crack growth, the damage tolerance approach is considered. The new technique proposed for planning in-service inspections of fatigued structures under crack propagation requires a quantile of the t-distribution and is conceptually simple and easy to use. The numerical examples are given.

N.A. Nechval (✉)
EVF Statistics Department, EVF Research Institute, University of Latvia,
Raina Blvd 19, Riga 1050, Latvia
e-mail: nechval@junik.lv

K.N. Nechval
Transport and Telecommunication Institute, Riga, Latvia
e-mail: konstan@tsi.lv

S. Kadry and A. El Hami (eds.), *Numerical Methods for Reliability and Safety Assessment: Multiscale and Multiphysics Systems*, DOI 10.1007/978-3-319-07167-1_25, © Springer International Publishing Switzerland 2015

1   Introduction

In spite of decades of investigation, fatigue response of materials is yet to be fully understood. This is partially due to the complexity of loading at which two or more loading axes fluctuate with time. Examples of structures experiencing such complex loadings are automobile, aircraft, off-shores, railways, and nuclear plants. Fluctuations of stress and/or strains are difficult to avoid in many practical engineering situations and are very important in design against fatigue failure. While most industrial failures involve fatigue, the assessment of the fatigue reliability of industrial components being subjected to various dynamic loading situations is one of the most difficult engineering problems. The traditional analytical method of engineering fracture mechanics (EFM) usually assumes that crack size, stress level, material property and crack growth rate, etc. are all deterministic values which will lead to conservative or very conservative outcomes. According to many experimental results and field data, even in well-controlled laboratory conditions, crack growth results usually show a considerable statistical variability (as shown in Fig. 1).

Yet more considerable statistical variability is the case under variable amplitude loading (as shown in Fig. 2).

Fatigue is one of the most important problems of aircraft arising from their nature as multiple-component structures, subjected to random dynamic loads. The analysis of fatigue crack growth is one of the most important tasks in the design and life prediction of aircraft fatigue-sensitive structures (for instance, wing, fuselage) and their components (for instance, aileron or balancing flap as part of the wing panel, stringer, etc.). An example of in-service cracking from B727 aircraft (year of manufacture 1981; flight hours not available; flight cycles 39,523) is given in Fig. 3.

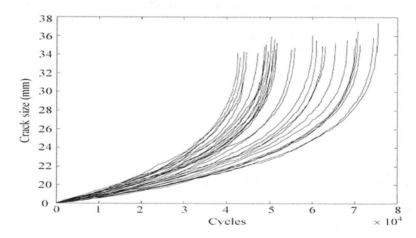

Fig. 1 Constant amplitude loading fatigue test data curves. (Reproduced from Wu and Ni 2003)

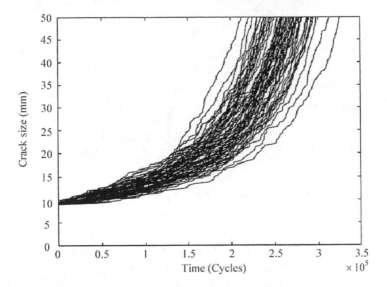

Fig. 2 Variable amplitude loading fatigue test data curves

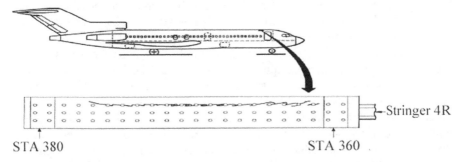

Fig. 3 Example of in-service cracking from B727 aircraft. (Reproduced from Jones et al. 1999)

From an engineering standpoint the fatigue life of a component or structure consists of two periods (this concept is shown schematically in Fig. 4):

(1) crack initiation period, which starts with the first load cycle and ends when a technically detectable crack is present, and (2) crack propagation period, which starts with a technically detectable crack and ends when the remaining cross section can no longer withstand the loads applied and fails statically. Periodic inspections of aircraft are common practice in order to maintain their reliability above a desired minimum level.

For guaranteeing safety, the structural life ceiling limits of the fleet aircraft are defined from three distinct approaches: safe-life, damage tolerance, and fail-safe approaches. The common objectives to define fleet aircraft lives by the three approaches are to ensure safety while at the same time reducing total ownership costs.

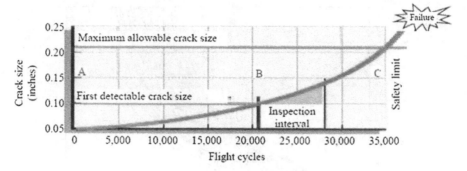

Fig. 4 Schematic fatigue crack growth curve [crack initiation period (A–B); crack propagation period (B–C)]

The safe-life approach is based on the concept that significant damage, i.e. fatigue cracking, will not develop during the service life of a component. The life is initially determined from fatigue test data (*S–N* curves) and calculations using a cumulative damage "law". Then the design safe-life is obtained by applying a safety factor. When the service life equals the design Safe-Life the component must be replaced. However, there are two major drawbacks to this approach: (1) components are taken out of service even though they may have substantial remaining lives; (2) despite all precautions, cracks sometimes occur prematurely. This latter fact led airlines to introduce the damage tolerance approach (Iyyer et al. 2007). The damage tolerance approach recognizes that damage can occur and develop during the service life of a component. Also, it assumes that cracks or flaws can be present in new structures. Safety is obtained from this approach by the requirements that either (1) any damage will be detected by routine inspection before it results in a dangerous reduction of the static strength (inspectable components), or (2) initial damage shall not grow to a dangerous size during the service life (non-inspectable components). For damage tolerance analysis to be successful it must be possible to:

1. Define either a minimum crack length that will not go undetected during routine inspections, or else an initial crack length, nominally based on preservice inspection capability.
2. Predict crack growth during the time until the next inspection or until the design service life is reached.

The fail-safe approach assumes initial damage as manufactured and its subsequent growth during service to detectable crack sizes or greater. Service life in fail-safe structures can thus be defined as the time to a service detectable damage. Inspection intervals are determined by using appropriate safety factors on calculated crack growth time interval from service detectable cracks to critical crack sizes. The prediction of crack growth is similar to that of damage tolerance approach, except that a much smaller initial crack length is used.

Many important fatigued structures (for instance, transportation systems and vehicles: aircraft, space vehicles, trains, ships; civil structures: bridges, dams, tunnels; and so on) for which extremely high reliability is required are maintained by in-service inspections to prevent the reliability degradation due to fatigue damage. However, temporal transition of the reliability is significantly affected by the inspection strategy selected. Thus, to keep structures reliable against fatigue damage by inspections, it is clearly important in engineering to examine the optimal inspection strategy. In particular, it should be noticed that periodical inspections with predetermined constant intervals are not always effective, since a fatigue crack growth rate is gradually accelerated as fatigue damage grows, i.e. the intervals between inspections should be gradually smaller in order to restrain the reliability degradation by repeated inspections. Therefore, we need to construct the inspection strategy by paying attention to this case.

Barlow et al. (1963) tackled this problem by assuming a known, fixed cost of making an inspection and a known fixed cost per unit time due to undetected failure. They then found a sequence of inspection times for which the expected cost is a minimum. Their results have been extended by various authors (Luss and Kander 1974; Sengupta 1977). Unfortunately, it is difficult to compute optimal checking procedures numerically, because the computations are repeated until the procedures are determined to the required degree by changing the first check time. To avoid this, Munford and Shahani (1972) suggested a suboptimal (or nearly optimal) but computationally easier inspection policy. This policy was used for Weibull and gamma failure distribution cases (Munford and Shahani 1973; Tadikamalla 1979). Numerical comparisons among certain inspection policies are given by Munford (1981) for the case of Weibull failure times. This case under parametric uncertainty was considered by (Nechval et al. 2008a, 2008b, 2009, 2011a, 2011b).

Most models, which are used for solving the problems of planning inspections, are developed under the assumptions that the parameter values of the models are known with certainty. When these models are applied to solve real-world problems, the parameters are estimated and then treated as if they were the true values. The risk associated with using estimates rather than the true parameters is called estimation risk and is often ignored. In this paper, we consider the case when the functional form of the underlying invariant lifetime distribution is assumed to be known, but some or all of its parameters are unspecified. To make the discussion clear, we make the following restrictions on the inspection: (1) there is only one objective structure component of the inspection; (2) if a fatigue crack is detected by the inspection, the component is immediately replaced with a new (virgin) one. To construct the optimal reliability-based inspection strategy in this case, the two criteria are proposed and the invariant embedding technique (Nechval et al. 2008a, 2011a) is used.

2 Planning Inspections for Fatigue Crack Detection Under Crack Initiation and Complete Information

In this paper we look at inspection strategies for items or structures that can be described as being in one of two states, one of which is preferable to the other. This preferred state might be described as "working" whilst the other may represent some sort of "fatigue damage". The structures are originally known to be in a working state but may subsequently fatigue damage. In other words, at $t_0 = 0$ the structure is in state S_0 (working) but at a later time, t_1, the structure will move into state S_1 (fatigue damage). We suppose that we do not know when the transition from S_0 into S_1 will occur, and that fatigue damage (crack) can only be detected through inspection. We deal with situations, where it is difficult to quantify the costs associated with inspections and undetected fatigue damage, or when these costs vary in time.

The inspection strategy defined is based on the conditional reliability of the structure. It is given as follows. Fix $0 < \gamma < 1$ and let

$$\tau_1 = \arg \left(\Pr \{ X > \tau_1 \} = \gamma \right) \tag{1}$$

$$\tau_j = \arg \left(\Pr \left\{ X > \tau_j \,\middle|\, X > \tau_{j-1} \right\} = \gamma \right), \quad j \geq 2 \tag{2}$$

where $\{\tau_j\}_{j=1,2,\dots}$ are inspection times, X is a random variable representing the lifetime of the structure. This is named as "reliability-based inspection". The above inspection strategy makes use of the information about the remaining life that is inherent in the sequence of previous inspection times. The value of γ can be seen as "minimum fatigue reliability required" during the next period when the structure was still operational at last inspection time, that is, in other words, the conditional probability that the failure (fatigue crack) occurs in the time interval (τ_{j-1}, τ_j) without failure at time τ_{j-1} is always assumed $1 - \gamma$.

It is clear that if F_θ, the structure lifetime distribution with the parameter θ (in general, vector), is continuous and strictly increasing, the definition of the inspection strategy is equivalent to

$$\tau_j = \arg \left(\overline{F}_\theta \left(\tau_j \right) = \gamma^j \right), \quad j \geq 1 \tag{3}$$

or equivalent to

$$\tau_j^* = \arg \min_{\tau_j} \left[\overline{F}_\theta \left(\tau_j \right) - \gamma^j \right]^2, \quad j \geq 1 \tag{4}$$

where

$$\overline{F}_\theta \left(\tau_j \right) = 1 - F_\theta \left(\tau_j \right) \tag{5}$$

If it is known that each inspection costs c_1 and the cost of leaving an undetected failure (fatigue crack) is c_2 per unit time, then the total expected cost per inspection cycle is given by

$$E_\theta\{C\} = \sum_{j=1}^{\infty} \int_{\tau_{j-1}}^{\tau_j} \left[jc_1 + c_2\left(\tau_j - x\right) \right] f_\theta(x) dx$$

$$= c_1 \sum_{j=1}^{\infty} j \left[F_\theta\left(\tau_j\right) - F_\theta\left(\tau_{j-1}\right) \right]$$

$$+ c_2 \sum_{j=1}^{\infty} \tau_j \left[F_\theta\left(\tau_j\right) - F_\theta\left(\tau_{j-1}\right) \right] - c_2 \int_0^{\infty} x f_\theta(x) dx$$

$$= c_1 \sum_{j=0}^{\infty} \overline{F}_\theta\left(\tau_j\right) + c_2 \sum_{j=1}^{\infty} \tau_j \left[\overline{F}_\theta\left(\tau_{j-1}\right) - \overline{F}_\theta\left(\tau_j\right) \right] - c_2 E_\theta\{X\} \quad (6)$$

where $f_\theta(x)$ is the probability density function of the structure lifetime X,

$$E_\theta\{X\} = \int_0^{\infty} x f_\theta(x) dx \quad (7)$$

Thus, we can choose γ such that $E_\theta\{C\}$ as defined in (6) is minimized.

3 Planning Inspections for Fatigue Crack Detection Under Crack Initiation and Incomplete Information

To construct the optimal reliability-based inspection strategy under parametric uncertainty, the two criteria are proposed.

The first criterion, which takes into account (3) and the past lifetime data of the structures of the same type, allows one to construct the inspection strategy given by

$$\tau_j = \arg\left(E_\theta\left\{ \overline{F}_\theta\left(\tau_j\right) \right\} = \gamma^j \right), \quad j \geq 1 \quad (8)$$

where $\tau_j \equiv \tau_j\left(\widehat{\theta}\right)$, $\widehat{\theta}$ represents either the maximum likelihood estimator of θ or sufficient statistic S for θ, i.e., $\tau_j \equiv \tau_j(S)$. This criterion is named as "unbiasedness criterion".

The second criterion (preferred), which takes into account (4) and the past lifetime data of the structures of the same type, allows one to construct the inspection strategy given by

$$\tau_j^* = \arg \min_{\tau_j} E_\theta \left\{ \left[\overline{F}_\theta (\tau_j) - \gamma^j \right]^2 \right\}, \quad j \geq 1 \tag{9}$$

This criterion is named as "minimum variance criterion".

It will be noted that in practice, under parametric uncertainty, the criterion,

$$\tau_j = \arg \left(\overline{F}_{\widehat{\theta}} (\tau_j) = \gamma^j \right), \quad j \geq 1 \tag{10}$$

is usually used. This criterion is named as "maximum likelihood criterion".

To find a sequence of inspection times, $\tau_j \equiv \tau_j \left(\widehat{\theta} \right)$ or $\tau_j \equiv \tau_j(S)$, $j \geq 1$, satisfying either (8) or (9), the invariant embedding technique (Nechval et al. 2008a, 2011a) can be used. Let us assume that each inspection costs c_1 and the cost of leaving an undetected failure (fatigue crack) is c_2 per unit time, then under parametric uncertainty we can choose γ such that $E_\theta \{ E_\theta \{ C \} \}$ is minimized.

4   Exponential Distribution of Time to Crack Detection

Theorem 1 Let $X_1, \ldots, X_n$ be the random sample of the past independent observations of time to crack detection from the fatigued structures of the same type, which follow the exponential distribution with the probability density function

$$f_\theta(x) = (1/\theta) \exp(-x/\theta), \quad x \geq 0, \quad \theta > 0 \tag{11}$$

where the parameter θ is unknown. Then the reliability-based inspection strategies for a new fatigued structure of the same type are given as follows.

The unbiased inspection strategy (UIS):

$$\tau_j = \left[\gamma^{-j/n} - 1 \right] S, \quad j \geq 1 \tag{12}$$

The minimum variance inspection strategy (MVIS):

$$\tau_j^* = \frac{1 - \gamma^{\frac{j}{n+1}}}{2\gamma^{\frac{j}{n+1}} - 1} S, \quad j \geq 1 \tag{13}$$

where $S = \sum_{i=1}^n X_i$ is the sufficient statistic for θ.

The maximum likelihood inspection strategy (MLIS):

$$\tau_j = j \ln \gamma^{-1} \frac{S}{n}, \quad j \geq 1 \tag{14}$$

Proof Using the invariant embedding technique (Nechval et al. 2008a, 2011a), we obtain the unbiased inspection strategy (UIS) from (8):

$$\tau_j = \arg \left[E_\theta \{ \overline{F}_\theta (\tau_j) \} = \gamma^j \right]$$

$$= \arg \left[E_\theta \left\{ \exp \left(-\frac{\tau_j}{\theta} \right) \right\} = \gamma^j \right] = \arg \left[E_\theta \left\{ \exp \left(-\frac{\tau_j}{S} \frac{S}{\theta} \right) \right\} = \gamma^j \right]$$

$$= \arg \left[E \left\{ \exp \left(-\eta_j V \right) \right\} = \gamma^j \right]$$

$$= \arg \left[\frac{1}{(1 + \eta_j)^n} = \gamma^j \right] = \left[\gamma^{-j/n} - 1 \right] S, \quad j \geq 1 \tag{15}$$

where

$$\eta_j = \tau_j / S \tag{16}$$

$$V = S/\theta \sim f(v) = \frac{1}{\Gamma(n)} v^{n-1} \exp(-v), \quad v \geq 0 \tag{17}$$

The minimum variance inspection strategy (MVIS) is obtained from (9):

$$\tau_j^* = \arg \min_{\tau_j} E_\theta \{ [\overline{F}_\theta (\tau_j) - \gamma^j]^2 \}$$

$$= \arg \min_{\tau_j} E_\theta \left\{ [\exp (-\tau_j/\theta) - \gamma^j]^2 \right\}$$

$$= \arg \min_{\tau_j} E \{ \exp (-2\eta_j V) - 2 \exp(-\eta_j V) \gamma^j + \gamma^{2j} \}$$

$$= \arg \min_{\tau_j} \left(\frac{1}{(1 + 2\eta_j)^n} - \frac{2\gamma^j}{(1 + \eta_j)^n} + \gamma^{2j} \right)$$

$$= \left[\left(1 - \gamma^{j/(n+1)} \right) / \left(2\gamma^{j/(n+1)} - 1 \right) \right] S, \quad j \geq 1 \tag{18}$$

The maximum likelihood inspection strategy (MLIS) follows immediately from (10):

$$\tau_j = \arg \left(\overline{F}_{\hat{\theta}} (\tau_j) = \gamma^j \right) = \arg \left[\exp \left(-\frac{\tau_j}{\hat{\theta}} \right) = \gamma^j \right]$$

$$= j \ln \gamma^{-1} \frac{S}{n}, \quad j \geq 1 \tag{19}$$

where the maximum likelihood estimator of θ is $\hat{\theta} = S/n$. This ends the proof. $\square$

5   Determination of the Optimal Value of γ

Theorem 2 Let us assume that under conditions of Theorem 1 each inspection of the UIS costs c_1 and the cost of leaving an undetected failure (fatigue crack) is c_2 per unit time, then γ minimizing $E\{E_\theta\{C\}\}$ is given by

$$\gamma^* = \arg\left(\left[\frac{1-\gamma^{(n+1)/n}}{1-\gamma}\right]^2\left[\frac{1-\gamma^{-1/n}+(\gamma^{-1}-1)/n}{\gamma^{1/n}}\right]=\frac{n+1}{n}\frac{c_1}{c_2}\frac{1}{S}\right) \quad (20)$$

Proof Taking into account (6) and (12), we have

$$\gamma^* = \arg\min_\gamma E_\theta\{E_\theta\{C\}\}$$

$$= \arg\ \min_\gamma E_\theta\left\{c_1\sum_{j=0}^\infty \overline{F}_\theta\left(\tau_j\right)+c_2\sum_{j=1}^\infty \tau_j\left[\overline{F}_\theta\left(\tau_{j-1}\right)-\overline{F}_\theta\left(\tau_j\right)\right]-c_2 E_\theta\{X\}\right\}$$

$$= \arg\min_\gamma \theta\left(\frac{c_1 n}{S}\frac{1}{1-\gamma^{(n+1)/n}}+c_2 n\frac{\gamma^{-1/n}-1}{1-\gamma}-c_2\right)$$

$$= \arg\min_\gamma\left(\frac{c_1 n}{S}\frac{1}{1-\gamma^{(n+1)/n}}+c_2 n\frac{\gamma^{-1/n}-1}{1-\gamma}-c_2\right)$$

$$= \arg\left(\left[\frac{1-\gamma^{(n+1)/n}}{1-\gamma}\right]^2\left[\frac{1-\gamma^{-1/n}+(\gamma^{-1}-1)/n}{\gamma^{1/n}}\right]=\frac{n+1}{n}\frac{c_1}{c_2}\frac{1}{S}\right)$$

$$(21)$$

This ends the proof.   □

5.1   *Numerical Example 1*

Let $X_1, \ldots, X_n$ be the random sample of the past independent observations of time to crack detection (from the same fatigued structures) which follow the exponential distribution (11), where $n=2$ and the parameter θ is unknown. The sufficient statistic for θ is $S=335$ h. In order to construct the reliability-based inspection strategy for a new fatigued structure of the same type, the unbiasedness criterion (8) will be used. Let us assume that each inspection of the UIS costs $c_1 = 1$ (in terms of money) and the cost of leaving an undetected failure (fatigue crack) is $c_2 = 2$ (in terms of money) per unit time. Then it follows from (20) that $\gamma^* = 0.95$. Figure 5 depicts the relationship between $E_\theta\{E_\theta\{C\}\}/\theta$ and γ.

Fig. 5 Relationship between
$E_\theta \{E_\theta \{C\}\}/\theta$ and γ

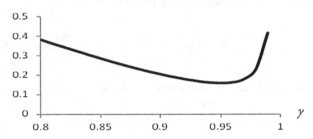

The optimal inspection times (in terms of hours) for the unbiased inspection strategy are given in Table 1.

Table 1 Optimal inspection times (in terms of hours)

j	1	2	3	4	5	6	...
τ_j	8.70	17.63	26.79	36.19	45.83	55.73	...

6 Gumbel Distribution of Time to Crack Detection

Theorem 3 Let $X_1 \leq \cdots \leq X_r$ be the first r ordered past observations of time to crack detection from the n fatigued structures of the same type, which follow the Gumbel distribution with the probability distribution function

$$\Pr\{X \leq x\} = 1 - \exp\left[-\exp\left(\frac{x-\mu}{\sigma}\right)\right], \quad -\infty < x < \infty \qquad (22)$$

where μ is the location parameter, and σ is the scale parameter ($\sigma > 0$). The shape of the Gumbel model does not depend on the distribution parameters. Then the unbiased reliability-based strategy of simultaneous inspection of the m new fatigued structures of the same type is given as follows:

$$\tau_j = \arg\left(\Pr\{Y_1 > \tau_j | \mathbf{z}\} = \gamma^j\right)$$

$$= \arg\left[\frac{\displaystyle\int_0^\infty v^{r-2} e^{v\sum_{i=1}^r z_i} \left(me^{v(\tau_j - \widehat{\mu})/\widehat{\sigma}} + \sum_{i=1}^r e^{vz_i} + (n-r)e^{vz_r}\right)^{-r} dv}{\displaystyle\int_0^\infty v^{r-2} e^{v\sum_{i=1}^r z_i} \left(\sum_{i=1}^r e^{vz_i} + (n-r)e^{vz_r}\right)^{-r} dv} = \gamma^j\right], \quad j \geq 1$$

$$(23)$$

where Y_1 is the smallest observation from a set of m future ordered observations $Y_1 \leq \cdots \leq Y_m$ also from the distribution (22),

$$\mathbf{z} = (z_1, z_2, \ldots, z_r), \quad Z_i = \frac{X_i - \widehat{\mu}}{\widehat{\sigma}}, \quad i = 1, \ldots, r \tag{24}$$

$\widehat{\mu}$ and $\widehat{\sigma}$ are the maximum likelihood estimators of μ and σ based on the first r ordered past observations ($X_1 \leq \cdots \leq X_r$) from a sample of size n from the Gumbel distribution, which can be found from solution of

$$\widehat{\mu} = \widehat{\sigma} \ln \left(\left[\sum_{i=1}^{r} e^{x_i/\widehat{\sigma}} + (n-r) e^{x_r/\widehat{\sigma}} \right] \bigg/ r \right) \tag{25}$$

and

$$\widehat{\sigma} = \left(\sum_{i=1}^{r} x_i e^{x_i/\widehat{\sigma}} + (n-r) x_r e^{x_r/\widehat{\sigma}} \right) \left(\sum_{i=1}^{r} e^{x_i/\widehat{\sigma}} + (n-r) e^{x_r/\widehat{\sigma}} \right)^{-1} - \frac{1}{r} \sum_{i=1}^{r} x_i \tag{26}$$

Y_1 is the smallest observation from a set of m future ordered observations $Y_1 \leq \cdots \leq Y_m$ also from the distribution (22).

Proof The joint density of $X_1 \leq \cdots \leq X_r$ is given by

$$f\left(x_1, \ldots, x_r \big| \mu, \sigma\right) = \frac{n!}{(n-r)!} \prod_{i=1}^{r} \frac{1}{\sigma} \exp\left(\frac{x_i - \mu}{\sigma} - \exp\left(\frac{x_i - \mu}{\sigma}\right)\right)$$

$$\times \exp\left(-(n-r)\exp\left(\frac{x_r - \mu}{\sigma}\right)\right) \tag{27}$$

Let $\widehat{\mu}$, $\widehat{\sigma}$ be the maximum likelihood estimates of μ, σ, respectively, based on $X_1 \leq \cdots \leq X_r$ from a complete sample of size n, and let

$$V_1 = \frac{\widehat{\mu} - \mu}{\widehat{\sigma}}, \quad V = \frac{\widehat{\sigma}}{\sigma} \quad \text{and} \quad Z_i = \frac{X_i - \widehat{\mu}}{\widehat{\sigma}}, \quad i = 1(1)r \tag{28}$$

Parameters μ and σ in (28) are location and scale parameters, respectively, and it is well known that if $\widehat{\mu}$ and $\widehat{\sigma}$ are estimates of μ and σ, possessing certain invariance properties, then V_1 and V are the pivotal quantities whose distributions depend only on n. Most, if not all, proposed estimates of μ and σ possess the necessary properties; these include the maximum likelihood estimates and various linear estimates. Z_i, $i = 1(1)r$, are ancillary statistics, any $r - 2$ of which form a functionally independent set. For notational convenience we include all of $z_1, \ldots, z_r$ in (24); z_{r-1} and z_r can be expressed as function of $z_1, \ldots, z_{r-2}$ only.

Using the invariant embedding technique (Nechval et al. 2008a, 2011a), we then find in a straightforward manner, that the probability element of the joint density of V_1, V, conditional on fixed $z = (z_1, z_2, \ldots, z_r)$, is

$$
f\left(v_1, v \middle| \mathbf{z}\right) dv_1 dv = \vartheta\left(\mathbf{z}\right) v^{r-2} \exp\left(v \sum_{i=1}^{r} z_i\right)
$$

$$
\times e^{rv_1} \exp\left(-e^{v_1}\left[\sum_{i=1}^{r} \exp\left(z_l v\right) + (n - r) \exp\left(z_r v\right)\right]\right) dv_1 dv,
$$

$$
v_1 \in (-\infty, \infty), \quad v \in (0, \infty) \tag{29}
$$

where

$$
\vartheta\left(\mathbf{z}\right) = \left(\int_0^\infty \Gamma(r) v^{r-2} \exp\left(v \sum_{i=1}^{z} z_i\right)\left[\sum_{i=1}^{r} \exp\left(z_i v\right) + (n - r) \exp\left(z_r v\right)\right]^{-r} dv\right)^{-1} \tag{30}
$$

is the normalizing constant. Writing

$$
\Pr\left\{Y_1 > \tau_j \middle| \mu, \sigma\right\} = \prod_{j=1}^{m} \Pr\left\{Y_i > \tau_j \middle| \mu, \sigma\right\} = \exp\left\{-m \exp\left(\frac{\tau_j - \mu}{\sigma}\right)\right\}
$$

$$
= \exp\left\{-m \exp\left(z_{\tau_j} v + v_1\right)\right\} \Pr\left\{Y_1 > \tau_j \middle| v_1, v\right\} \tag{31}
$$

where

$$
z_{\tau_j} = \left(\tau_j - \widehat{\mu}\right) / \widehat{\sigma} \tag{32}
$$

we have that

$$
\Pr\left\{Y_1 > \tau_j \middle| \mathbf{z}\right\} = \int_0^\infty \int_{-\infty}^\infty \Pr\left\{Y_1 > \tau_j \middle| v_1, v\right\} f\left(v_1, v \middle| \mathbf{z}\right) dv_1 dv \tag{33}
$$

Now v_1 can be integrated out of (33) in a straightforward way to give

$$
\Pr\left\{Y_1 > \tau_j \,\middle|\, \mathbf{z}\right\} = \frac{\displaystyle\int_0^\infty v^{r-2} e^{v\sum_{i=1}^r z_i} \left(m e^{v z_{\tau j}} + \sum_{i=1}^r e^{v z_i} + (n-r) e^{v z_r} \right)^{-r} dv}{\displaystyle\int_0^\infty v^{r-2} e^{v\sum_{i=1}^r z_i} \left(\sum_{i=1}^r e^{v z_i} + (n-r) e^{v z_r} \right)^{-r} dv} \tag{34}
$$

This completes the proof.   □

7 Weibull Distribution of Time to Crack Detection

Theorem 4 Let $X_1 \le \cdots \le X_r$ be the first r ordered past observations of time to crack detection from the n fatigued structures of the same type, which follow the Weibull distribution with the probability distribution function

$$
\Pr\{X \le x\} = 1 - \exp\left[-\left(\frac{x}{\beta}\right)^\delta \right], \quad x \ge 0 \tag{35}
$$

where both distribution parameters (δ-shape, β-scale) are positive. Then the unbiased reliability-based strategy of simultaneous inspection of the m new fatigued structures of the same type is given as follows:

$$
\tau_j = \arg\left[\Pr\left\{Y_1 > \tau_j \,\middle|\, \mathbf{w}\right\} = \gamma^j \right]
$$

$$
= \arg\left[\frac{\displaystyle\int_0^\infty v_2^{r-2} \prod_{i=1}^r w_i^{v_2} \left(m w_{\tau j}^{v_2} + \sum_{i=1}^r w_i^{v_2} + (n-r) w_r^{v_2} \right)^{-r} dv_2}{\displaystyle\int_0^\infty v_2^{r-2} \prod_{i=1}^r w_i^{v_2} \left(\sum_{i=1}^r w_i^{v_2} + (n-r) w_r^{v_2} \right)^{-r} dv_2} = \gamma^j \right] \quad j \ge 1 \tag{36}
$$

where

$$V_2 = \frac{\delta}{\widehat{\delta}}, \quad \mathbf{w} = (w_1, w_2, \ldots, w_r), \quad W_i = \left(\frac{x_i}{\widehat{\beta}}\right)^{\widehat{\delta}}, \quad i = 1, \ldots, r, \quad w_{\tau_j} = \left(\frac{\tau_j}{\widehat{\beta}}\right)^{\widehat{\delta}}$$

(37)

$\widehat{\beta}$ and $\widehat{\delta}$ are the maximum likelihood estimators of β and δ based on the first r ordered past observations $(X_1 \leq \cdots \leq X_r)$ from a sample of size n from the two-parameter Weibull distribution (35), which can be found from solution of

$$\widehat{\beta} = \left(\left[\sum_{i=1}^{r} x_i^{\widehat{\delta}} + (n-r) x_r^{\widehat{\delta}}\right]/r\right)^{1/\widehat{\delta}}$$

(38)

and

$$\widehat{\delta} = \left[\left(\sum_{i=1}^{r} x_i^{\widehat{\delta}} \ln x_i + (n-r) x_r^{\widehat{\delta}} \ln x_r\right)\left(\sum_{i=1}^{r} x_i^{\widehat{\delta}} + (n-r) x_r^{\widehat{\delta}}\right)^{-1} - \frac{1}{r}\sum_{i=1}^{r} \ln x_i\right]^{-1}$$

(39)

Proof The proof of Theorem 4 is similar to that of Theorem 3 if we consider the distribution of the logarithm of a Weibull variate ($\ln X$), which follows the Gumbel distribution. □

7.1 Numerical Example 2

Consider, for example, the data of fatigue tests on a particular type of structural components (stringer) of aircraft IL-86. The data are for a complete sample of size $r = n = 5$, with observations of time to crack initiation (in number of 10^4 flight hours): $X_1 = 5$, $X_2 = 6.25$, $X_3 = 7.5$, $X_4 = 7.9$, $X_5 = 8.1$; $m = 1$.

Goodness-of-fit Testing. It is assumed that X_i, $i = 1(1)5$, follow the two-parameter Weibull distribution (35), where the parameters β and δ are unknown. We assess the statistical significance of departures from the Weibull model by performing empirical distribution function goodness-of-fit test. We use the K statistic (Kapur and Lamberson 1977). For censoring (or complete) datasets, the K statistic is given by

Table 2 Inspection time sequence under crack initiation

Inspection time τ_j ($\times 10^4$ flight hours)	Interval $\tau_{j+1} - \tau_j$ (flight hours)
$\tau_0 = 0$	–
$\tau_1 = 2.5549$	25,549
$\tau_2 = 3.2569$	7,020
$\tau_3 = 3.6975$	4,406
$\tau_4 = 4.0212$	3,237
$\tau_5 = 4.2775$	2,563
$\tau_6 = 4.4898$	2,123
$\tau_7 = 4.6708$	1,810
$\tau_8 = 4.8287$	1,579
$\tau_9 = 4.9685$	1,398
$\vdots$	$\vdots$

Fig. 6 Graphical representation of inspection intervals under crack initiation

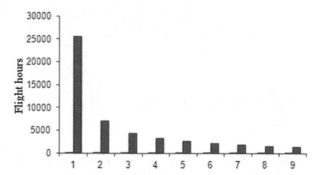

$$K = \frac{\sum_{i=[r/2]+1}^{r-1}\left(\dfrac{\ln(x_{i+1}/x_i)}{M_i}\right)}{\sum_{i=1}^{r-1}\left(\dfrac{\ln(x_{i+1}/x_i)}{M_i}\right)} = \frac{\sum_{i=3}^{4}\left(\dfrac{\ln(x_{i+1}/x_i)}{M_i}\right)}{\sum_{i=1}^{4}\left(\dfrac{\ln(x_{i+1}/x_i)}{M_i}\right)} = 0.184 \qquad (40)$$

where $[r/2]$ is a largest integer $\leq r/2$, the values of M_i are given in Table 13 (Kapur and Lamberson 1977). The rejection region for the α level of significance is $\{K > K_{n;\alpha}\}$. The percentage points for $K_{n;\alpha}$ were given by Kapur and Lamberson (1977). For this example,

$$K = 0.184 < K_{n=5;\alpha=0.05} = 0.86 \qquad (41)$$

Thus, there is not evidence to rule out the Weibull model.

Inspection Planning Under Fatigue Crack Initiation. Using (36), we have obtained the following inspection time sequence (see Table 2). Graphical representation of inspection intervals is shown in Fig. 6.

8 Stochastic Models of Fatigue Crack Propagation (Growth)

8.1 Stochastic Model Based on Paris–Erdogan Law

To capture the statistical nature of fatigue crack growth, different stochastic models have been proposed in the literature. Some of the models are purely based on direct curve fitting of the random crack growth data, including their mean value and standard deviation (Bogdanoff and Kozin 1985). These models, however, have been criticized by other researchers, because less crack growth mechanisms have been included in them. To overcome this difficulty, many probabilistic models adopted the crack growth equations proposed by fatigue experimentalists, and randomized the equations by including random factors into them (Lin and Yang 1985; Yang et al. 1985; Yang and Manning 1990; Nechval et al. 2003, 2004; Straub and Faber 2005). The random factor may be a random variable, a random process of time, or a random process of space. It then creates a random differential equation. The solution of the differential equation reveals the probabilistic nature as well as the scatter phenomenon of the fatigue crack growth. To justify the applicability of the probabilistic models mentioned above, fatigue crack growth data are needed. However, it is rather time-consuming to carry out experiments to obtain a set of statistical meaningful fatigue crack growth data. To the writers' knowledge, there are only a few datasets available so far for researchers to verify their probabilistic models. Among them, the most famous dataset perhaps is the one produced by Virkler et al. (1979) more than twenty years ago. More frequently used datasets include one reported by Ghonem and Dore (1987). Itagaki and his associates have also produced some statistically meaningful fatigue crack growth data, but have not been mentioned very often (Itagaki et al. 1993). In fact, many probabilistic fatigue crack growth models are either lack of experimental verification or just verified by only one of the above datasets. It is suspected that a model may explain a dataset well but fail to explain another dataset. The universal applicability of many probabilistic models still needs to be checked carefully by other available datasets.

Many probabilistic models of fatigue crack growth are based on the deterministic crack growth equations. The most well-known equation is

$$\frac{da(t)}{dt} = q(a(t))^b \tag{42}$$

in which q and b are constants to be evaluated from the crack growth observations. The independent variable t can be interpreted as stress cycles, flight hours, or flights depending on the applications. It is noted that the power-law form of $q(a(t))^b$ at the right hand side of (42) can be used to fit some fatigue crack growth data appropriately and is also compatible with the concept of Paris–Erdogan law

(Paris and Erdogan 1963). The service time for a crack to grow from size $a(t_0)$ to $a(t)$ (where $t > t_0$) can be found by performing the necessary integration

$$\int_{t_0}^{t} dt = \int_{a(t_0)}^{a(t)} \frac{dv}{qv^b} \tag{43}$$

to obtain

$$t - t_0 = \frac{[a(t_0)]^{-(b-1)} - [a(t)]^{-(b-1)}}{q(b-1)} \tag{44}$$

For the particular case (when $b = 1$), it can be shown, using Lopital's rule, that

$$t - t_0 = \frac{\ln[a(t)/a(t_0)]}{q} \tag{45}$$

Thus, we have obtained the Exponential model

$$a(t) = a(t_0) e^{q(t-t_0)} \tag{46}$$

In Nechval et al. (2008a, 2008b), it was considered a stochastic version of (44),

$$\frac{1}{a_0^{b-1}} - \frac{1}{a^{b-1}} = (b-1)q(t-t_0) + V_{\bullet} \tag{47}$$

where $a_0 \equiv a(t_0)$, $a \equiv a(t)$.

Stochastic Model. If $V_{\bullet} \sim N(0, [(b-1)\sigma(t-t_0)^{1/2}]^2)$, then the probability that crack size $a(t)$ will exceed any given (say, maximum allowable) crack size $a^{\bullet}$ can be derived and expressed as

$$\Pr\{a(t) \geq a^{\bullet}\} = 1 - \Phi\left(\left[\frac{\left(a_0^{-(b-1)} - (a^{\bullet})^{-(b-1)}\right) - (b-1)q(t-t_0)}{(b-1)\sigma(t-t_0)^{1/2}}\right]\right) \tag{48}$$

where $\Phi(\cdot)$ is the standard normal distribution function. In this case, the conditional probability density function of a is given by

$$f_{b,q,\sigma}\left(a|t\right) = \frac{a^{-b}}{\sigma[2\pi(t-t_0)]^{1/2}}$$

$$\times \exp\left(-\frac{1}{2}\left[\frac{\left(a_0^{-(b-1)} - a^{-(b-1)}\right) - (b-1)q(t-t_0)}{(b-1)\sigma(t-t_0)^{1/2}}\right]^2\right) \tag{49}$$

This model allows one to characterize the random properties that vary during crack growth and may be suitable for planning in-service inspections of fatigued structures under fatigue crack growth when the damage tolerance approach is used.

8.2 New Stochastic Model of Fatigue Crack Propagation

New Stochastic Model. As a result of our investigations of the experimental data of fatigue crack growth, we have found that for planning in-service inspections of fatigued structures under fatigue crack growth it may be used as the simplest of probabilistic models—the straight line model—which derives its name from the fact that the deterministic portion of the model graphs as a straight line,

$$y = \beta_0 + \beta_1 x + \varepsilon \tag{50}$$

where

$$y \equiv a \ln t(a), \quad x \equiv a \tag{51}$$

a is a crack size, *t* can be interpreted as either stress cycles, flight hours, or flights depending on the applications, β_0 and β_1 are parameters depending on loading spectra, structural/material properties, etc., ε is a stochastic factor.

Let us assume that we have a sample of n data points consisting of pairs of values of x and y, say $(x_1, y_1), (x_2, y_2), \ldots, (x_n, y_n)$, where

$$y_i = \beta_0 + \beta_1 x_i + \varepsilon_i, \quad 1 \le i \le n, \quad \varepsilon_i \overset{\text{iid}}{\sim} N\left(0, \sigma^2\right) \tag{52}$$

σ^2 is the variance.

The constant real parameters $(\beta_0, \beta_1, \sigma^2)$ are unknown. The parameters (β_0, β_1) are estimated by maximum likelihood, which in the case of normally distributed iid factors ε_i as assumed here is equivalent to choosing β_0, β_1 in terms of $\{(x_i, y_i)\}_{i=1}^n$ by least squares, i.e. to minimize $\sum_{i=1}^n (y_i - \beta_0 - \beta_1 x_i)^2$ which results in:

$$\widehat{\beta}_0 = \overline{y} - \widehat{\beta}_1 \overline{x} \tag{53}$$

$$\widehat{\beta}_1 = \frac{s_{xy}}{s_x^2} \tag{54}$$

where

$$\overline{x} = \frac{1}{n}\sum_{i=1}^n x_i, \quad \overline{y} = \frac{1}{n}\sum_{i=1}^n y_i \tag{55}$$

$$s_x^2 = \frac{1}{n-1}\sum_{i=1}^{n}(x_i - \overline{x})^2 = \sum_{i=1}^{n}x_i^2 - \frac{1}{n}\left(\sum_{i=1}^{n}x_i\right)^2 \tag{56}$$

$$s_{xy} = \frac{1}{n-1}\sum_{i=1}^{n}(x_i - \overline{x})(y_i - \overline{y}) = \sum_{i=1}^{n}x_i y_i - \frac{1}{n}\left(\sum_{i=1}^{n}x_i\right)\left(\sum_{i=1}^{n}y_i\right) \tag{57}$$

The standard (unbiased) estimator of σ^2 is the mean residual sum of squares (per degree of freedom) given by

$$\widehat{\sigma}^2 = \frac{1}{n-2}\sum_{i=1}^{n}\left(y_i - \widehat{\beta}_0 - \widehat{\beta}_1 x_i\right)^2 \tag{58}$$

It can be shown that

$$y_{n+1} - \widehat{\beta}_0 - \widehat{\beta}_1 x_{n+1} \sim N\left(0, \sigma^2\left[1 + \frac{1}{n} + \frac{(x_{n+1} - \overline{x})^2}{(n-1)s_x^2}\right]\right) \tag{59}$$

where y_{n+1} is a single future value of y (not yet observed) corresponding to a chosen value x_{n+1} of x which was not in fact part of the dataset. Then

$$T = \left(y_{n+1} - \widehat{\beta}_0 - \widehat{\beta}_1 x_{n+1}\right)\left(\widehat{\sigma}\left[1 + \frac{1}{n} + \frac{(x_{n+1} - \overline{x})^2}{(n-1)s_x^2}\right]^{1/2}\right)^{-1} \sim t_{n-2} \tag{60}$$

i.e., the statistic T follows a t-distribution with $(n-2)$ degrees of freedom. In this case, a $100(1-\alpha)\%$ lower prediction limit $h_{n+1}^{(1-\alpha)}$ for a single future value of $y = y_{n+1}$ at a chosen value of $x = x_{n+1}$ is given by

$$h_{n+1}^{(1-\alpha)} = \widehat{\beta}_0 + \widehat{\beta}_1 x_{n+1} - t_{n-2;1-\alpha}\widehat{\sigma}\left[1 + \frac{1}{n} + \frac{(x_{n+1} - \overline{x})^2}{(n-1)s_x^2}\right]^{1/2} \tag{61}$$

where $t_{n-2;1-\alpha}$ denotes the $(1-\alpha)$ quantile of the t-distribution with $(n-2)$ degrees of freedom, which corresponds to a cumulative probability of $(1-\alpha)$.

For constant amplitude fatigue tests, the typical crack growth curve is shown in Fig. 7.

Using the proposed stochastic model of crack propagation (50), the typical crack growth curve in Fig. 7 can be transformed as follows (see Fig. 8).

It is interesting to point out that the coefficient of determination R^2 is close to 1.

A variety of functions were reviewed in the search for a functional relationship which would provide a useful empirical model for consolidated fatigue data. Hyperbolic, exponential, and power functions were all investigated and found to be unsatisfactory for fitting the complete range of available data. In order to find

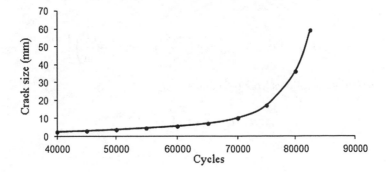

Fig. 7 Example of typical crack growth curve

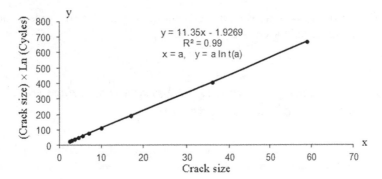

Fig. 8 Example of transformed typical crack growth curve

a reasonable model of fatigue data trends throughout the life range of interest, the following functional relationship,

$$a \ln t(a) = \beta_0 + \beta_1 a + \varepsilon \tag{62}$$

is proposed in this paper, where ε is a stochastic factor. In principle, this model of fatigue crack growth is based on the following deterministic crack growth equation,

$$\frac{dt(a)}{da} = \frac{q_1 q_0}{a^2} \exp\left(-\frac{q_0}{a}\right) \tag{63}$$

in which q_1 and q_0 are constants to be evaluated from the crack growth observations. The variable t can be interpreted as stress cycles, flight hours, or flights depending on the applications. The service time for a crack to grow from size a_0 to a (where $a > a_0$) can be found by performing the necessary integration

$$\int_{t(a_0)}^{t(a)} dt = q_1 \int_{a_0}^{a} \frac{q_0}{v^2} \exp\left(-\frac{q_0}{v}\right) dv \tag{64}$$

to obtain

$$t(a) - t(a_0) = q_1 \left[\exp\left(-\frac{q_0}{a}\right) - \exp\left(-\frac{q_0}{a_0}\right) \right] \tag{65}$$

If $a_0 = 0$, then it follows from (65) that

$$t(a) = q_1 \exp\left(-\frac{q_0}{a}\right) \tag{66}$$

or

$$a \ln t(a) = \beta_0 + \beta_1 a \tag{67}$$

where $\beta_1 = \ln q_1$, $\beta_0 = -q_0$. Including a stochastic factor ε into (67), we obtain the stochastic model of fatigue crack propagation (62).

8.3 New Equation for Planning In-Service Inspections Under Fatigue Crack Propagation

New Equation. Let us assume that we have a sample of k data points consisting of pairs of values of a and τ, say $(\tau_1, a_1), (\tau_2, a_2), \ldots, (\tau_k, \tau_k)$, where $k > 2$, τ_j is the time of the jth inspection, a_j is the crack size detected by means of the jth inspection, $j = 1(1)k$. Then the time τ_{k+1} of the next inspection is determined as

$$\tau_{k+1} = \exp\left(\frac{h_{k+1}^{(1-\alpha)}}{\tilde{a}_{k+1}}\right) \tag{68}$$

where

$$\tilde{a}_{k+1} = a_k + \Delta_k \tag{69}$$

represents a chosen value of future crack size a_{k+1} (not yet observed), Δ_k is a selected increment of a_k,

$$h_{k+1}^{(1-\alpha)} = \widehat{\beta}_0 + \widehat{\beta}_1 x_{k+1} - t_{k-2;1-\alpha} \widehat{\sigma} \left[1 + \frac{1}{k} + \frac{(x_{k+1} - \overline{x})^2}{(k-1) s_x^2} \right]^{1/2} \tag{70}$$

represents the $100(1-\alpha)\%$ lower prediction limit for a single future value of $y = y_{k+1} = a_{k+1} \ln \tau_{k+1}$ at a chosen value of $x = x_{k+1} = \tilde{a}_{k+1}$,

$$\widehat{\beta}_0 = \overline{y} - \widehat{\beta}_1 \overline{x}, \quad \widehat{\beta}_1 = \frac{s_{xy}}{s_x^2} \tag{71}$$

$$\overline{x} = \frac{1}{k}\sum_{j=1}^{k} x_j = \frac{1}{k}\sum_{j=1}^{k} a_j \tag{72}$$

$$\overline{y} = \frac{1}{k}\sum_{j=1}^{k} y_i = \frac{1}{k}\sum_{j=1}^{k} a_j \ln \tau_j \tag{73}$$

$$s_x^2 = \frac{1}{k-1}\sum_{j=1}^{k}(x_j - \overline{x})^2 = \sum_{j=1}^{k} x_j^2 - \frac{1}{k}\left(\sum_{j=1}^{k} x_j\right)^2 = \sum_{j=1}^{k} a_j^2 - \frac{1}{k}\left(\sum_{j=1}^{k} a_j\right)^2 \tag{74}$$

$$s_y^2 = \frac{1}{k-1}\sum_{j=1}^{k}(y_j - \overline{y})^2 = \sum_{j=1}^{k}(a_j \ln \tau_j)^2 - \frac{1}{k}\left(\sum_{j=1}^{k} a_j \ln \tau_j\right)^2 \tag{75}$$

$$s_{xy} = \frac{1}{k-1}\sum_{j=1}^{k}(x_j - \overline{x})(y_j - \overline{y}) = \sum_{j=1}^{k} x_j y_j - \frac{1}{k}\left(\sum_{j=1}^{k} x_j\right)\left(\sum_{j=1}^{k} y_j\right)$$

$$= \sum_{j=1}^{k} a_j^2 \ln \tau_j - \frac{1}{k}\left(\sum_{j=1}^{k} a_j\right)\left(\sum_{j=1}^{k} a_j \ln \tau_j\right) \tag{76}$$

$$\widehat{\sigma} = \left[\frac{1}{k-2}\sum_{j=1}^{k}\left(y_j - \widehat{\beta}_0 - \widehat{\beta}_1 x_j\right)^2\right]^{1/2}$$

$$= \left[\frac{1}{k-2}\sum_{j=1}^{k}\left(a_j \ln \tau_j - \widehat{\beta}_0 - \widehat{\beta}_1 a_j\right)^2\right]^{1/2} \tag{77}$$

8.4   Numerical Example 3

For illustration, the procedure of inspection planning based on Eq. (68) will be used for the upper longerons of RNLAF F-16 aircraft (Grooteman 2008), see Fig. 9. Since the longerons are not safety-critical, a baseline deterministic durability analysis was done, starting from a 0.178 mm corner crack and ending at the "functional impairment" crack length of 4.75 mm. The crack growth curve was obtained from a Lockheed Martin crack growth model and a load spectrum representing the average

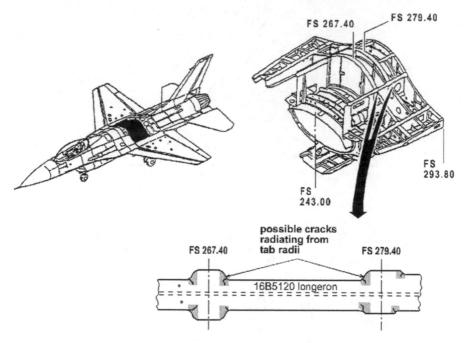

Fig. 9 Inspection points of the upper longeron of RNLAF F-16 aircraft. (Reproduced from Grooteman 2008)

usage of RNLAF F-16s. This crack growth model is based on the well-known Forman equation and the generalized Willenborg retardation model, and has been validated with data from a test of aircraft, i.e. it gives the correct trend of actual crack growth. Figure 10 shows the calculated mean crack growth curve and superimposes the damage tolerance inspection requirements (Military Specification 1974) on the crack growth curve, whereby the 90 %/95 % reliably detectable crack length, a_d, for in-service inspections was specified to be 2.54 mm. These requirements led to the following inspection scheme, which, however, has an unknown safety level:

1. Initial inspection 2,655 FH (flight hours);
2. Repeat inspection 62 FH.

Let us assume that $1 - \alpha = 0.95$, $\Delta_k = 4$ mm for all k. Using the initial data points $(\tau_1 = 2,655$ FH, $a_1 = 0.35$ mm), $(\tau_2 = 4,200$ FH, $a_2 = 0.75$ mm) and the subsequent data points taken from the calculated mean crack growth curve (see Fig. 10, where the coefficient of determination $R^2 = 1$), we obtain from (68) the following sequence of inspections and detectable crack lengths (see Table 3):

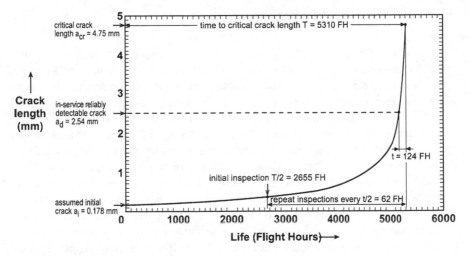

Fig. 10 Deterministic damage tolerance inspection requirements for the RNLAF longerons. (Reproduced from Grooteman 2008)

Fig. 11 Graphical representation of inspection intervals under crack propagation

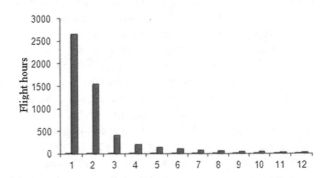

Table 3 Inspection time sequence under crack propagation

Inspection time τ_j (flight hours)	Crack length (mm)	Interval $\tau_{j+1}-\tau_j$ (flight hours)
$\tau_0 = 0$	—	—
$\tau_1 = 2,655$	$a_1 = 0.35$	$\tau_1 - \tau_0 = 2,655$
$\tau_2 = 4,200$	$a_2 = 0.75$	$\tau_2 - \tau_1 = 1,545$
$\tau_3 = 4,600$	$a_3 = 1.15$	$\tau_3 - \tau_2 = 400$
$\tau_4 = 4,800$	$a_4 = 1.55$	$\tau_4 - \tau_3 = 200$
$\tau_5 = 4,930$	$a_5 = 1.95$	$\tau_5 - \tau_4 = 130$
$\tau_6 = 5,030$	$a_6 = 2.35$	$\tau_6 - \tau_5 = 100$
$\tau_7 = 5,110$	$a_7 = 2.75$	$\tau_7 - \tau_6 = 80$
$\tau_8 = 5,170$	$a_8 = 3.15$	$\tau_8 - \tau_7 = 60$
$\tau_9 = 5,220$	$a_9 = 3.55$	$\tau_9 - \tau_8 = 50$
$\tau_{10} = 5,260$	$a_{10} = 3.95$	$\tau_{10} - \tau_9 = 40$
$\tau_{11} = 5,290$	$a_{11} = 4.35$	$\tau_{11} - \tau_{10} = 30$
$\tau_{12} = 5,310$	$a_{12} = 4.75$	$\tau_{12} - \tau_{11} = 20$

9   Conclusions and Future Work

The technique proposed in this paper represents a simple and computationally attractive statistical method based on the constructive use of the invariance principle in mathematical statistics. The main advantage of this technique consists in that it allows one to eliminate unknown parameters from the problem and to use the past lifetime data for planning future inspections as completely as possible.

The unbiasedness and minimum variance criteria, which are proposed in the paper for constructing inspection strategies under fatigue crack initiation such as the unbiased inspection strategy (UIS) and minimum variance inspection strategy (MVIS), respectively, represent the novelty of the work. It is clear that these inspection strategies, which have such properties as unbiasedness and minimum variance, are preferable as compared to the maximum likelihood inspection strategy (MLIS). We have illustrated the prediction methods for log-location-scale distributions (such as the exponential, Gumbel, or Weibull distribution). Application to other distributions could follow directly.

Under fatigue crack growth, the damage tolerance approach is considered. As a result of our investigations of the experimental data of fatigue crack growth, we have found that for planning in-service inspections of fatigued structures under crack propagation it may be used the simplest of probabilistic models—the linear regression model. Once the appropriate stochastic model is established, it can be used for the fatigue reliability prediction of structures made of the tested material. As such the model presented here provides a fast and computationally efficient way to predict the fatigue lives of realistic structures. The new technique proposed for planning in-service inspections of fatigued structures under crack propagation requires a quantile of the t-distribution and is conceptually simple and easy to use.

The results obtained in this work can be used to solve the service problems of the following important engineering structures: (1) transportation systems and vehicles—aircraft, space vehicles, trains, ships; (2) civil structures—bridges, dams, tunnels; (3) power generation—nuclear, fossil fuel, and hydroelectric plants; (4) high-value manufactured products—launch systems, satellites, semiconductor and electronic equipment; (5) industrial equipment—oil and gas exploration, production and processing equipment, chemical process facilities, pulp and paper.

Acknowledgments This research was supported in part by Grant Nos. 06.1936, 07.2036, and 09.1014 from the Latvian Council of Science and the National Institute of Mathematics and Informatics of Latvia.

References

Barlow RE, Hunter LC, Proschan F (1963) Optimum checking procedures. J Soc Ind Appl Math 11:1078–1095
Bogdanoff JL, Kozin F (1985) Probabilistic models of cumulative damage. Wiley, New York
Ghonem H, Dore S (1987) Experimental study of the constant probability crack growth curves under constant amplitude loading. Eng Fract Mech 27:1–25

Grooteman F (2008) A stochastic approach to determine lifetimes and inspection schemes for aircraft components. Int J Fatigue 30:138–149

Itagaki H, Ishizuka T, Huang PY (1993) Experimental estimation of the probability distribution of fatigue crack growth lives. Probab Eng Mech 8:25–34

Iyyer N, Sarkar S, Merrill R, Phan N (2007) Aircraft life management using crack initiation and crack growth models-P-3C Aircraft experience. Int J Fatigue 29:1584–1607

Jones R, Molent L, Pitt S (1999) Studies in multi-site damage of fuselage lap joints. J Theor Appl Fract Mech 32:18–100

Kapur KC, Lamberson LR (1977) Reliability in engineering design. Wiley, New York

Lin YK, Yang JN (1985) On statistical moments of fatigue crack propagation. Eng Fract Mech 18:243–256

Luss H, Kander Z (1974) Inspection policies when duration of checkings is non-negligible. Oper Res Quart 25:299–309

Military Specification (1974) Airplane Damage Tolerance Requirements. MIL-A-83444 (USAF)

Munford AG (1981) Comparison among certain inspection policies. Manage Sci 27:260–267

Munford AG, Shahani AK (1972) A nearly optimal inspection policy. Oper Res Quart 23:373–379

Munford AG, Shahani AK (1973) An inspection policy for the Weibull case. Oper Res Quart 24:453–458

Nechval NA, Nechval KN, Vasermanis EK (2003) Statistical models for prediction of the fatigue crack growth in aircraft service. In: Varvani-Farahani A, Brebbia CA (eds) Fatigue damage of materials 2003. WIT Press, Southampton, pp 435–445

Nechval NA, Nechval KN, Vasermanis EK (2004) Estimation of warranty period for structural components of aircraft. Aviation VIII:3–9

Nechval NA, Nechval KN, Berzins G, Purgailis M, Rozevskis U (2008a) Stochastic fatigue models for efficient planning inspections in service of aircraft structures. In: Al-Begain K, Heindl A, Telek M (eds) Analytical and stochastic modeling techniques and applications, vol 5055, Lecture notes in computer science (LNCS). Springer, Berlin, pp 114–127

Nechval NA, Berzins G, Purgailis M, Nechval KN (2008b) Improved estimation of state of stochastic systems via invariant embedding technique. WSEAS Trans Math 7:141–159

Nechval KN, Nechval NA, Berzins G, Purgailis M, Rozevskis U, Strelchonok VF (2009) Optimal adaptive inspection planning process in service of fatigued aircraft structures. In: Al-Begain K, Fiems D, Horvath G (eds) Analytical and stochastic modeling techniques and applications, vol 5513, Lecture notes in computer science (LNCS). Springer, Berlin, pp 354–369

Nechval NA, Nechval KN, Purgailis M (2011a) Inspection policies in service of fatigued aircraft structures. In: Ao SI, Gelman L (eds) Electrical engineering and applied computing, vol 90, Lecture notes in electrical engineering. Springer, Berlin, pp 459–472

Nechval NA, Nechval KN, Purgailis M (2011b) Prediction of future values of random quantities based on previously observed data. Eng Lett 9:346–359

Paris R, Erdogan F (1963) A critical analysis of crack propagation laws. J Basic Eng 85:528–534

Sengupta B (1977) Inspection procedures when failure symptoms are delayed. Oper Res Quart 28:768–776

Straub D, Faber MH (2005) Risk based inspection planning for structural systems. Struct Saf 27:335–355

Tadikamalla PR (1979) An inspection policy for the gamma failure distributions. Oper Res Quart 30:77–78

Virkler DA, Hillberry BM, Goel PK (1979) The statistic nature of fatigue crack propagation. ASME J Eng Mater Technol 101:148–153

Wu WF, Ni CC (2003) A study of stochastic fatigue crack growth modeling through experimental data. Probab Eng Mech 18:107–118

Yang JN, Manning SD (1990) Stochastic crack growth analysis methodologies for metallic structures. Eng Fract Mech 37:1105–1124

Yang JN, His WH, Manning SD (1985) Stochastic crack propagation with applications to durability and damage tolerance analyses. Technical report, Flight Dynamics Laboratory, Wright-Patterson Air Force Base, AFWAL-TR-85-3062

Diffuse Response Surface Model Based on Advancing Latin Hypercube Patterns for Reliability-Based Design Optimization

Peipei Zhang, Piotr Breitkopf, and Catherine Knopf-Lenoir-Vayssade

Abstract Since variances in the input parameters of engineering systems cause subsequent variations in the product performance, Reliability-Based Design Optimization (RBDO) is getting a lot of attention recently. However, RBDO is computationally expensive. Therefore, the Response Surface Methodology (RSM) is often used to improve the computational efficiency in the solution of problems in RBDO. In this chapter, the Diffuse Approximation (DA), a variant of the well-known Moving Least Squares (MLS) approximation based on a progressive sampling pattern is used within a variant of the First Order Reliability Method (FORM). The proposed method simultaneously uses points in the standard normal space (*U*-space) as well as the physical space (*X*-space). At last, we investigate the optimization of the process parameters for Numerical Control (NC) milling of ultrahigh strength steel. The objective functions are tool life and material removal rate. The results show that the method proposed can decrease the number of 'exact' function calculations needed and reduce the computation time. It is also helpful to adopt this new method for other engineering applications.

Since variances in the input parameters of engineering systems cause subsequent variations in the product performance, and deterministic optimum designs that are obtained without taking uncertainties into consideration could lead to unreliable designs. Reliability-based design optimization (RBDO) is getting a lot of attention. However, RBDO is computationally expensive. Therefore, the response surface methodology (RSM) is often used to improve the computational efficiency in the solution of problems in RBDO.

P. Zhang (✉)
School of Mechatronics Engineering, University of Electronic Science and Technology of China, 2006 Xiyuan Road, Gaoxin Western District, Chengdu, Sichuan 611731, China
e-mail: peipei.zhang@uestc.edu.cn

P. Breitkopf • C. Knopf-Lenoir-Vayssade
Roberval Laboratory, University of Technology of Compiegne, Compiègne, France

S. Kadry and A. El Hami (eds.), *Numerical Methods for Reliability and Safety Assessment: Multiscale and Multiphysics Systems*, DOI 10.1007/978-3-319-07167-1_26, © Springer International Publishing Switzerland 2015

RBDO is extensively applied in various fields due to its uncertainty consideration. For example, RBDO is used in the crashworthiness of vehicle analysis to consider simulation uncertainties and manufacturing imperfections (Youn et al. 2004). Allen and Maute (2004) combined RBDO and high-fidelity aero elastic simulations of structures. Behnam and Eamon (2013) obtained minimum cost of ductile hybrid fiber reinforced polymer bar configurations by RBDO. Chen et al. (2014) developed a cell evolution method for RBDO.

Several popular RSMs are applied widely in computational mechanics. RSM is used to predict the tool life in end milling Titanium Alloy Ti–6AI–4V (Ginta 2009). It was used by Bashir et al. (2010) for optimization of ammoniacal nitrogen removal from semi-aerobic landfill leachate using ion exchange resin. Kumar and Singh (2014) used RSM to optimize process parameters for catalytic pyrolysis of waste high-density polyethylene to liquid fuel.

This chapter begins with the introduction of RBDO. After that diffuse approximation is presented as main part of RSM. Then design of experiments (DOE), most important part of RSM, focuses on advancing Latin hypercube patterns. At last a case of optimizing the process parameters of numerically controlled milling of ultrahigh strength steel is investigated.

1 Reliability-Based Design Optimization

1.1 General Definition of RBDO Model

1.1.1 Deterministic Optimization Problem Statement

In the optimization process, the goal is to

$$\text{Minimize} \quad f(\mathbf{x}), \quad \mathbf{x} \in R^n$$

subject to a set of m constraints

$$g_j(\mathbf{x}) \leq 0; \quad j = 1, \ldots, m$$

with

$$L_i \leq x_i \leq U_i, \quad i = 1, \ldots, n$$

where f is the objective function, x_i are the design variables, g_j is the jth nonlinear constraint. The region of interest is defined by L_i and U_i which are the lower and upper bounds on the design variables, respectively.

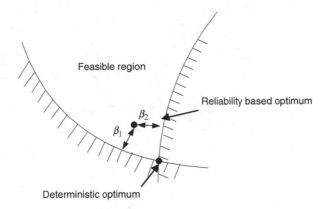

Fig. 1 Comparison between RBDO and deterministic optimization

1.1.2 RBDO Problem Definition

RBDO is similar to deterministic optimization. The main difference lies in the constraints. The constraints in RBDO are the reliability indices β which ensure that the design point is in the safe region. Figure 1 shows the role of the reliability index β.

In system parameter design, the RBDO model (Choi et al. 2007) can generally be defined as:

Minimize or maximize $f(\mathbf{x})$

$$\text{Subject to } \beta_i\,(\mathbf{u}) \geq \overline{\beta_i} \quad i = 1, 2, \cdots, n$$

where $\mathbf{x}$ is the vector of the random variables in the design space (X-space) and $\mathbf{u}$ is the associated variable in the standard normal space (U-space). So the RBDO process is performed in two different random spaces: X-space and U-space.

Figure 2 clearly shows the difference between deterministic optimization and reliability-based optimization. Some points (filled circle points) may be out of the safe region when the deterministic optimum is being considered. Reliability-based optimization on the other hand is able to ensure that all the possible points (generated by the tolerance) are within the safe region with the probability defined by the reliability index β.

The flow chart of reliability-based optimization is shown in Fig. 3.

1.2 Basic Reliability Problems

In the RBDO model, the reliability constraints are most important. This section will explain how to define the reliability.

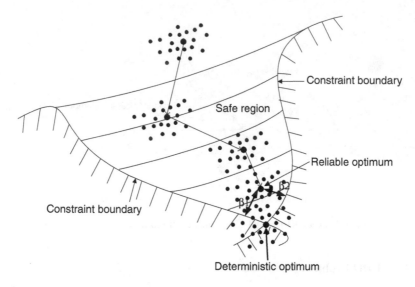

Fig. 2 Sketch of reliability-based optimization

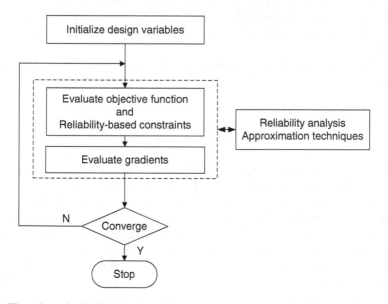

Fig. 3 Flow chart of reliability-based optimization

In effect, repeated realizations of the same physical phenomena may generate multiple outcomes. Among these multiple outcomes, some outcomes are more frequent than others. This means that the result is obtained with a certain amount of uncertainty, so satisfactory performances cannot be absolutely ensured. Instead, assurance can only be given in terms of the probability of success in satisfying

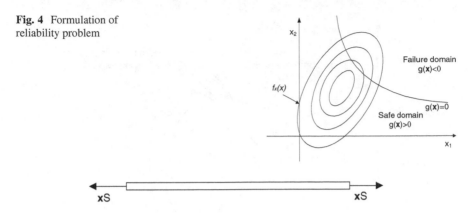

Fig. 4 Formulation of reliability problem

Fig. 5 Rod under tension load

some performance criterions. A probabilistic assurance of performance determines reliability. It can also be described as the probability of failure P_f.

A vector of basic random variables $\mathbf{x} = (x_1, x_2, \ldots, x_n)$ represents the uncertain quantities that the define state of the structure. In structural reliability analysis, the system output $Z = g(\mathbf{x})$ is usually a functional relationship between the specific performance criteria and the relevant load and resistance parameters $\mathbf{x}$. The failure surface can be given by $g(\mathbf{x}) = 0$. This is the boundary between the safe and failure regions in the random variables space. When $g(\mathbf{x}) > 0$, the structure is considered to be safe and when $g(\mathbf{x}) < 0$, the structure can no longer fulfill the design requirements (see Fig. 4).

Here, we use the example of a simple rod under tension load (Fig. 5) to illustrate the basic reliability problem. Suppose xS is the tension load and xR is the strength. xS and xR are independent random variables with probability density functions ($f_R(xR)$ and $f_S(xS)$). The failure (rod is broken) state is given by: $xR < xS$; critical state: $xR = xS$; and safe state: $xR > xS$. These three states are shown in Fig. 6.

Probability of failure P_f is

$$P_f = P\left[g\left(\mathbf{x}\right) \leq 0\right] = \int_{g(\mathbf{x}) \leq 0} f_X(x) dx \tag{1}$$

where $f_X(x)$ is Probability Density Function (PDF) of basic variables $\mathbf{x}$.

The reliability index β is related to the probability of failure P_f by

$$\beta = -\Phi^{-1}\left(P_f\right) \tag{2}$$

where Φ^{-1} is the inverse Cumulative Distribution Function (CDF) of standard normal variable. Figure 7 gives the interpretation of reliability index β.

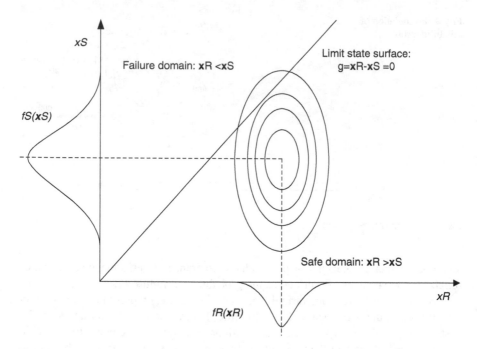

Fig. 6 Basic reliability problem of a rod under tension load

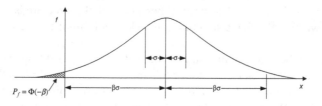

Fig. 7 Reliability index β

1.3 Transformations

The RBDO process is performed in two different spaces: the design space (X-space) and the standard normal space (U-space). It is necessary to do the transformation between two spaces at any design point to estimate the reliability index. During the transformation, the most important is that the failure probability P_f in the U-space is the same as in the X-space.

The transformation $\mathbf{T}$ between the two different spaces is given by Youn (2001):

$$\mathbf{u} = \mathbf{T}(\mathbf{x}) \quad \mathbf{x} = \mathbf{T}^{-1}(\mathbf{u}) \tag{3}$$

$$h(\mathbf{u}) = h(\mathbf{T}(\mathbf{x})) \quad g(\mathbf{x}) = g(\mathbf{T}^{-1}(\mathbf{u})) \tag{4}$$

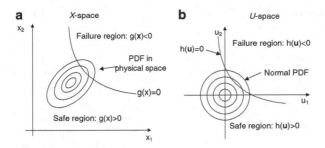

Fig. 8 Mapping from X-space to U-space

Table 1 Comparison of RIA and PMA

Approach	RIA	PMA
Main problem	Minimize function f	Minimize function f
	Subject to $\beta(\mathbf{x}) >= \beta_{\text{limit}}$	Subject to $-g(\mathbf{x}) \leq 0$
	and $g(\mathbf{x}) > 0$	
Sub-problem	Minimize $\beta(\mathbf{u})$	Minimize $h(\mathbf{u})$
	Subject to $h(\mathbf{u}) = 0$	Subject to $\beta(\mathbf{u}) = \beta_{\text{limit}}$

Figure 8 shows the mapping of design point from X-space (Fig. 8a) to U-space (Fig. 8b) by transformation.

For independent variables, the transformation may be expressed as

$$\mathbf{u}_i = \Phi^{-1}\left[F_{\mathbf{x}_i}(\mathbf{x}_i)\right] i = 1, \cdots, n \tag{5}$$

where Φ^{-1} is the inverse CDF of standard normal variable, $F_{\mathbf{x}_i}(\mathbf{x}_i)$ is CDF of $\mathbf{x}_i$.

When the variables are not independent, the transformation is more complicated. The readers are referred to Rosenblatt (1952), Der Kiureghian (1986) and Box–Cox (1964) for more information.

1.4 Methods for Solving Inner Loop

In the RBDO formulation, the double loop method has two iterative processes. The inner loop is an optimization process to search for the Most Probable Point (MPP) of each constraint.

Among several reliability approaches for locating the MPP to estimate the probability index, both performance measure approach (PMA) and reliability index approach (RIA) are used. The idea behind the PMA is to convert the probability measure to a performance measure in order to search for the MPP. RIA involves the evaluation of reliability by using a Taylor series of the limit state function to obtain the reliability index for each constraint. The two of them are compared in Table 1.

Fig. 9 MPP estimation of
PMA and RIA

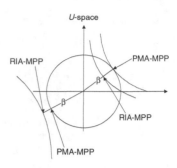

1.5   The Concept of MPP

The concept of the MPP was originally developed in structural reliability in the 1970s (Hasofer and Lind 1974). The probability of failure is maximal at the MPP. For the PMA approach, $-\nabla(g)$ at MPP is parallel to the vector from the origin to that point. The MPP lies on the β-circle for PMA approach and on the limit surface boundary for RIA approach (Fig. 9). Exact MPP calculation is an optimization problem.

1.6   Methods for the Computation of Reliability

Figure 10 shows the methods for the computation of reliability presented thus far. The First- and Second-Order Reliability Methods (left side of Fig. 10) are often used for low dimensionality problems, since their limit state surface $g(\mathbf{x}) = 0$ is often approximated (linear or quadratic) by a first- or second-order Taylor series expansion. Different approximate response surfaces $h(\mathbf{u}) = 0$ correspond to different methods for the calculation of failure probability. If the response surface is approached by a first-order approximation at the MPP, the method is termed as the First-Order Reliability Method (FORM), while the response surface is approached by a second-order approximation at the MPP, the method is termed as the Second-Order Reliability Method (SORM) and the response surface is approached by a higher order approximation at the MPP, the method is termed as the Higher-Order Reliability Method (HORM). It is actually a mathematical optimization process to find the MPP on the limit state surface which has the shortest distance from the origin to the limit state surface in the U-space.

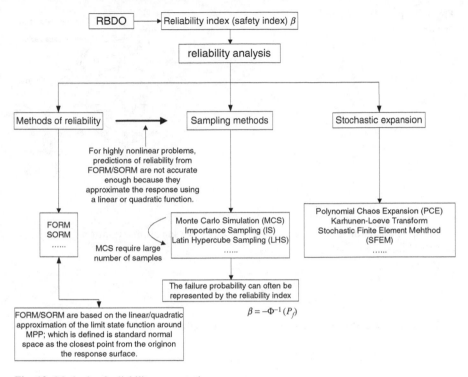

Fig. 10 Methods of reliability computation

However, for highly nonlinear problems, the FORM and SORM are not accurate enough to predict reliability because of the linear and quadratic approximations. Sampling methods (see in the middle of Fig. 10) avoid this disadvantage, as they estimate the reliability index by repeating the experiments many times. They are an inexpensive way to study the uncertainty in the given problem. Stochastic expansion is third method to calculate reliability index. More detailed information about these methods is given by Choi et al. (2007). First-Order Reliability Methods (FORM) is only introduced in detail here due to limited space.

In the FORM, the limit state surface is approximated by a tangent plane (normal direction α) at the MPP, and the FORM results are used to specify a bound based on the probability of failure.

The FORM is a mathematical optimization process to find the MPP on the limit state surface. Reliability index β is the shortest distance from the origin to the limit state surface in the U-space (Fig. 11).

Fig. 11 First-order reliability
method

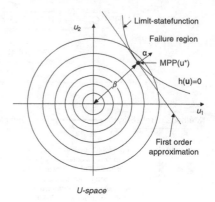

U-space

The response surface is approached by a first-order approximation at the MPP, and the limit surface state is

$$h\left(\mathbf{u}\right) = -\alpha^{\mathrm{T}}\mathbf{u} + \beta \tag{6}$$

Normal direction to the approximated limit surface state α is

$$\alpha = -\frac{\nabla h\left(\mathbf{u}\right)}{\|\nabla h\left(\mathbf{u}\right)\|}\big|_{\mathbf{u}=\mathbf{u}*} \tag{7}$$

Reliability index β is the shortest distance from the origin to the approximated limit state surface. The probability of failure P_{f} is expressed in Eq. (9).

$$\beta = \mathrm{sign}\left[h(0)\right]\|\mathbf{u}^*\| \tag{8}$$

$$P_{\mathrm{f}} = \Phi\left(-\beta\right) \tag{9}$$

Here, the Hasofer-Lind-Rackwitz-Fiessler (HLRF) algorithm is used to find $\mathbf{u}^*$ in the U-space. Find

$$\mathrm{Min.}\ \left\{Q\left(\mathbf{u}\right)\right\} = \|\mathbf{u}\|^2 = \mathbf{u}^{\mathrm{T}}\mathbf{u} \tag{10}$$

Subject to

$$h\left(\mathbf{u}\right) = 0 \tag{11}$$

It is iteratively calculated using

$$\mathbf{u}_{i+1} = \frac{1}{\nabla h^{\mathrm{T}}(\mathbf{u}_i)\nabla h(\mathbf{u}_i)} \left(\nabla h^{\mathrm{T}}(\mathbf{u}_i)\mathbf{u}_i - h(\mathbf{u}_i)\right)\nabla h(\mathbf{u}_i) \quad i = 0,\ldots,um \quad (12)$$

$\nabla h(\mathbf{u}_i)$ is the gradient vector of the limit state function $h(\mathbf{u})$ at $\mathbf{u}_i$. um is the number of iteration in U-space. Convergence criterion is

$$|\mathbf{u}_{i+1} - \mathbf{u}_i| \leq \varepsilon \tag{13}$$

$\mathbf{u}^*$ is the convergence point, the MPP.

2 Response Surface Model (RSM): Diffuse Approximation

For reliability computation, such as FORM and SORM require the computation of function values, first- and second-order derivatives. In order to apply well the formula in FORM, Response Surface Methodology is introduced in this section for approximating the function value and its gradients.

RSM is a collection of mathematical and statistical techniques useful for empirical model building. It replaces the true functional relationship by a mathematical expression that is much cheaper to evaluate. The response surface is an approximation of the relational expression of the response y predicted from n design variables x_i.

In general, for mathematical expression, a polynomial is often used because it is easy to handle; a nonlinear function that can be linearized through variable transformation (such as an exponential function) may also be used. There are many methods to make a response surface: least square method, polynomials, exponential, logarithm, neural network, spline interpolation, Lagrange interpolation etc. Here, the diffuse approximation (DA), a version of moving least squares (MLS) approximation is proposed as RSM.

The diffuse approximation (DA), a version of MLS approximation, estimates the value of a function $g: \Re^n \to \Re$ and of its derivatives based on a certain number k of samples $g(\mathbf{x}_i)$, $i = 1,2,\ldots,k$. DA is usually presented in one to three dimensional spaces, which is generally enough when applying DA as an alternative for the Finite Element approximation, but is not enough for the optimization purposes. The design spaces of interest are of higher dimension and specific issues of sampling and weighting have to be addressed for a proper implementation. Here a scalable DA framework for an arbitrary number of design variables is provided.

Fig. 12 Data point $\mathbf{x}_i$ and
evaluation point $\mathbf{x}$

The function value at a data point $\mathbf{x}_i$ (filled circles in Fig. 12) may be formulated
using the first-order Taylor expansion in terms of the function and of the gradient
values at the evaluation point $\mathbf{x}$ (hollow circle in Fig. 12) and in terms of the
distances $\Delta \mathbf{x}_i = \mathbf{x}_i\text{-}\mathbf{x}$, $i = 1 \ldots k$.

Using first-order expansion for the set of samples $g(\mathbf{x}_i)$ gives

$$g\left(\mathbf{x}_1\right) = g\left(\mathbf{x}\right) + \nabla g^{\mathrm{T}} \Delta \mathbf{x}_1 + \varepsilon_1$$
$$\ldots$$
$$g\left(\mathbf{x}_i\right) = g\left(\mathbf{x}\right) + \nabla g^{\mathrm{T}} \Delta \mathbf{x}_i + \varepsilon_i$$
$$\ldots$$
$$g\left(\mathbf{x}_k\right) = g\left(\mathbf{x}\right) + \nabla g^{\mathrm{T}} \Delta \mathbf{x}_k + \varepsilon_k \qquad (14)$$

from which we readily obtain expressions for the errors (Eq. (15)).

$$\varepsilon_1 = g\left(\mathbf{x}_1\right) - g\left(\mathbf{x}\right) - \nabla g^{\mathrm{T}} \Delta \mathbf{x}_1$$
$$\ldots$$
$$\varepsilon_i = g\left(\mathbf{x}_i\right) - g\left(\mathbf{x}\right) - \nabla g^{\mathrm{T}} \Delta \mathbf{x}_i$$
$$\ldots$$
$$\varepsilon_k = g\left(\mathbf{x}_k\right) - g\left(\mathbf{x}\right) - \nabla g^{\mathrm{T}} \Delta \mathbf{x}_k \qquad (15)$$

In matrix notation, denoting $\boldsymbol{\varepsilon}_{\mathrm{K}} = (\varepsilon_1 \ldots \varepsilon_k)^{\mathrm{T}}$

$$\boldsymbol{\varepsilon}_{\mathrm{K}} = \mathbf{g}_{\mathrm{K}} - \mathbf{Pa} \qquad (16)$$

Fig. 13 Circle domain of influence in 2D

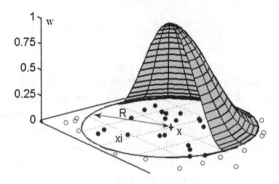

with

$$\mathbf{P}^{\mathrm{T}} = \begin{bmatrix} 1 & \cdots & 1 \\ \Delta x_1^1 & \cdots & \Delta x_k^1 \\ \vdots & & \vdots \\ \Delta x_1^n & \cdots & \Delta x_k^n \end{bmatrix} = \begin{bmatrix} 1 & \cdots & 1 \\ \Delta \mathbf{x}_1 & \cdots & \Delta \mathbf{x}_k \end{bmatrix} \quad \mathbf{a} = \begin{bmatrix} g(\mathbf{x}) \\ \nabla g(\mathbf{x}) \end{bmatrix},$$

and $\mathbf{g}_K$ stands for the sample value vector $\mathbf{g}_K = (g_1 \ \ldots \ g_k)^{\mathrm{T}}$, where $\boldsymbol{\Delta}\mathbf{x}_i = (\Delta x_i^1 \ \ldots \ \Delta x_i^n)^{\mathrm{T}}$. The approximate function $\tilde{g}(\mathbf{x})$ and the approximate gradient $\tilde{\nabla}g(\mathbf{x})$ values at the evaluation point $\mathbf{x}$ may then be found by minimizing the weighted squared error

$$J\left(\tilde{g}(\mathbf{x}), \tilde{\nabla}g(\mathbf{x})\right) = \frac{1}{2}\boldsymbol{\varepsilon}_K^{\mathrm{T}}\mathbf{W}\boldsymbol{\varepsilon}_K \tag{17}$$

From the second-order Taylor expansion, the matrix $\mathbf{P}$ has additional lines corresponding to $n + (n^2 - n)/2$ second-order terms $\Delta x_i^k \Delta x_i^l$, $k, l = 1, \ldots, n$, and one gets readily the criterion for the approximation of the function $\tilde{g}(\mathbf{x})$, of the gradient $\tilde{\nabla}g(\mathbf{x})$ and of the Hessian $\tilde{\mathbf{H}}(\mathbf{x})$

$$J\left(\tilde{g}(\mathbf{x}), \tilde{\nabla}g(\mathbf{x}), \tilde{\mathbf{H}}(\mathbf{x})\right) = \frac{1}{2}\boldsymbol{\varepsilon}_K^{\mathrm{T}}\mathbf{W}\boldsymbol{\varepsilon}_K \tag{18}$$

The diagonal weight matrix

$$\mathbf{W} = \begin{bmatrix} w(\mathbf{x}_1, \mathbf{x}) & & \\ & \ddots & \\ & & w(\mathbf{x}_k, \mathbf{x}) \end{bmatrix} \tag{19}$$

involves the weight functions $w(\mathbf{x}_i, \mathbf{x})$ which translate the influence of the ith sample point at the evaluation point $\mathbf{x}$ as illustrated in Fig. 13. R is the radius of influence domain.

$$w\left(\mathbf{x}_i, \mathbf{x}\right) = w_{\text{ref}}\left(\frac{|\mathbf{x}_i - \mathbf{x}|}{R}\right) \tag{20}$$

The influence of a data points decreases with the relative distance, according to the reference weight function $w_{\text{ref}}(d)$. The choice of weight function is critical for the quality of the approximation. Weight functions are obtained from the reference window functions by substituting the relative distance between the evaluation point and the data points. It provides three features to the approximation: the locality, the continuity, and the interpolation capacity. The locality is obtained when w_{ref} disappears outside the unit region. The C^m continuity is governed by vanishing of the m-order derivative of w_{ref} at the boundary. The interpolation property is obtained when $w(\mathbf{x}_i, \mathbf{x}_j) = \delta_i^j$ (Kronecker delta). A common choice for w_{ref}, which satisfies the first two properties, is a spline function. More other weight functions (basic hat, exponential) are presented by Belytschko et al. (1996). Here, we choose (Eq. (21)) as the weight function.

$$w_{\text{ref}}(d) = \begin{cases} 1 - 3d^2 + 2d^3 & \text{for } 0 \le d < 1 \\ 0 & \text{for } d \ge 1 \end{cases} \tag{21}$$

The weight function $w(\mathbf{x}_i, \mathbf{x})$ built from the reference weight function $w_{\text{ref}}(d)$ serves implicitly as a selection tool for determining the domain of influence of the given point. An obvious and common choice is

$$w\left(\mathbf{x}_i, \mathbf{x}\right) = w_{\text{ref}}\left(\sqrt{\Delta\mathbf{x}_i^{\text{T}}\Delta\mathbf{x}_i}/R\right) \tag{22}$$

with R to be the given size of the domain of influence, which results in an n-spherical domain. Therefore, a tensor product of the reference weight functions (Eq. (23)), calculated separately for each component of $\Delta\mathbf{x}_i$, is better suited for diffuse approximation

$$w\left(\mathbf{x}_i, \mathbf{x}\right) = w_{\text{ref}}\left(\Delta x_i^1/R_1\right) \times \cdots \times w_{\text{ref}}\left(\Delta x_i^k/R_n\right) \tag{23}$$

resulting in n-cubical shaped domains (Fig. 14) and permitting to adjust every R_j, $j = 1 \ldots n$ according to the required resolution in individual directions.

Once the weight matrix is defined, one obtains the approximation of the function and of the gradient by minimizing the criterion (Eq. (18)).

$$\begin{bmatrix} \tilde{g}\left(\mathbf{x}\right) \\ \tilde{\nabla} g\left(\mathbf{x}\right) \end{bmatrix} = \left[\mathbf{PWP}^{\text{T}}\right]^{-1}\mathbf{PWg}_{\text{K}} \tag{24}$$

The condition of the matrix $\mathbf{PWP}^{\text{T}}$ determines the quality of the approximation and depends on the pattern and on its size. The choice of the pattern, avoiding the degenerate situations is the subject of the DOE. In order to make the system

Fig. 14 Rectangular domain of influence in 2D

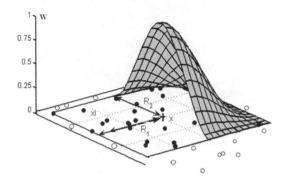

independent of the pattern size, the diagonal scaling matrix $\mathbf{D}$ is defined

$$
\mathbf{D} =
\begin{bmatrix}
1 & & 0 \\
& R_1 & \\
& & \ddots & \\
0 & & R_n
\end{bmatrix}^{-1}
\tag{25}
$$

where we can use the same radiuses of influence R_j as for the weight function computation. Finally, the system is Eq. (26)

$$
\begin{bmatrix} \tilde{g}\,(\mathbf{x}) \\ \tilde{\nabla} g\,(\mathbf{x}) \end{bmatrix} = \mathbf{D} \underbrace{\left[(\mathbf{DP})\,\mathbf{W}(\mathbf{DP})^{\mathrm{T}} \right]^{-1}}_{\mathbf{A}} \underbrace{[(\mathbf{DP})\,\mathbf{W}]}_{\mathbf{B}} \mathbf{g}_K = \mathbf{D}\mathbf{A}^{-1}\mathbf{B}\mathbf{g}_K
\tag{26}
$$

in which every $A_{ij} \in (0,1)$ is independently of the scale of the pattern.

The difficulty in the use of the formulation (Eq. (26)) is the control of the number of neighboring points k when the dimensionality n of the design space increases. This is the problem of DOE.

3　DOE: Advancing Latin Hypercube Sampling

The objective of DOE is to choose a set of appropriate sampling points with which the response surface should be evaluated. It is an important aspect of RSM. These strategies were originally developed for the model fitting of physical experiments and numerical experiments. DOE is thus how to arrange m experiments with n input variables. The variables are called factors in the DOE context. Each factor is fixed to several specified values called levels. So an experimental design is usually written in matrix form where the rows show the individual experiments (called sampling points) and the columns denote the particular factors.

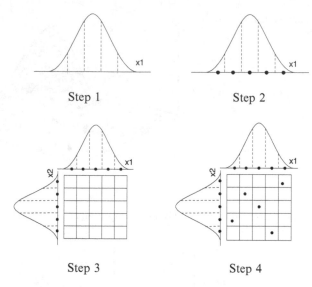

Fig. 15 Basic concept of LHS with two variables and five realizations

The choice of the DOE depends on the accuracy of the approximation, the cost of constructing the response surface, and the cost of producing physical experiments (maximum number of acceptable sampling points).

Advancing Latin hypercube sampling (LHS) is present due to its superiority on the number of samples.

3.1 Latin Hypercube Sampling Theory

In statistics, a Latin square is an $n \times n$ table filled with n different symbols in such a way that each symbol occurs exactly once in each row and exactly once in each column. The term "Hypercube" extends this concept to higher dimensions for lots of design variables. Therefore, LHS method, also known as the "stratified sampling technique," represents a multivariate sampling method without overlapping designs. LHS, which has been successfully used to generate multivariate samples of statistical distributions, was first proposed by McKay et al. (1979). In LHS, the distribution for each random variable can be subdivided into n equal probability intervals or bins. Each bin has one analysis point. There are n analysis points, randomly mixed, so each of the n bins has the distribution probability of $1/n$. Figure 15 shows the basic steps for the general LHS method. They are:

1. Divide the distribution for each variable into n non-overlapping intervals on the basis equal probability.
2. Select one value at random from each interval with respect to its probability density.

3. Repeat steps (1) and (2) until the values of all random variables, i.e., $x_1, x_2, \ldots, x_k$ are selected.
4. Associate the n values obtained for each x_i with the n values obtained for the other $x_{j \neq i}$ at random.

The regularity of probability intervals on the probability distribution function ensures that each of the input variables has all portions of its range represented, resulting in relatively a small variance in the response. At the same time, the analysis becomes much less computationally expensive.

3.2 Selection of Number of Points

LHS method can reduce the number of sampling points for calculations. However, minimal number of needed sampling points exists. How many points are enough? These sampling points are used to perform approximation by DA. So the number of sampling points depends on the approximation formulation.

For Eq. (24), firstly, the following symbols are defined:

k, Number of data points needed
n, Number of dimensions of variable (n)
col, Number of columns of matrix $\mathbf{P}$

Number of rows and columns of matrix $\mathbf{P}$ is k and col respectively: $\mathbf{P}_{k \times col}$.

- Matrix $\mathbf{P}$ is built from linear base polynomials, $col = (n + 1)$;

$$
\mathbf{P}_{k \times (n+1)}^{\mathrm{T}} =
\begin{bmatrix}
1 & \cdots & 1 \\
\Delta x_1^1 & \cdots & \Delta x_k^1 \\
\vdots & & \vdots \\
\Delta x_1^n & \cdots & \Delta x_k^n
\end{bmatrix}_{k \times (n+1)}
=
\begin{bmatrix}
1 & \cdots & 1 \\
\Delta \mathbf{x}_1 & \cdots & \Delta \mathbf{x}_k
\end{bmatrix}_{k \times (n+1)}
$$

- Matrix $\mathbf{P}$ is the case of quadratic basis functions, $col = (1 + 2n + (n^2 - n)/2)$;

$$
\mathbf{P}_{k \times col}^{\mathrm{T}} =
\begin{bmatrix}
1 & \cdots & 1 \\
\Delta x_1^1 & \cdots & \Delta x_k^1 \\
\vdots & & \vdots \\
\Delta x_1^n & \cdots & \Delta x_k^n \\
\Delta x_1^1 \Delta x_{i \neq 1}^1 & \cdots & \Delta x_k^1 \Delta x_{i \neq k}^1 \\
\vdots & & \vdots \\
\Delta x_1^n \Delta x_{i \neq 1}^n & \cdots & \Delta x_k^n \Delta x_{i \neq k}^n \\
\left(\Delta x_1^1\right)^2 & \cdots & \left(\Delta x_k^1\right)^2 \\
\vdots & & \vdots \\
\left(\Delta x_1^n\right)^2 & \cdots & \left(\Delta x_k^n\right)^2
\end{bmatrix}
=
\begin{bmatrix}
1 & \cdots & 1 \\
\Delta \mathbf{x}_1 & & \Delta \mathbf{x}_k \\
\Delta \mathbf{x}_1 \Delta \mathbf{x}_{i \neq 1} & & \Delta \mathbf{x}_k \Delta \mathbf{x}_{i \neq k} \\
\left(\Delta \mathbf{x}_1\right)^2 & & \left(\Delta \mathbf{x}_k\right)^2
\end{bmatrix}_{k \times col}
$$

Table 2 Number of points (k) needed with different dimensions

Dimensions	Full factorial designs	Linear base $\mathbf{P}$	Quadratic base $\mathbf{P}$	Bilinear base $\mathbf{P}$
1-D	3	≥ 2	≥ 3	≥ 2
2-D	9	≥ 3	≥ 6	≥ 5
3-D	27	≥ 4	≥ 9	≥ 7
4-D	81	≥ 5	≥ 15	≥ 11
5-D	243	≥ 6	≥ 21	≥ 16
n-D	3^n	$\geq (n+1)$	$\geq 1 + 2n + (n^2 - n)/2$	$\geq 1 + n + (n^2 - n)/2$

- Matrix $\mathbf{P}$ is the case of bilinear basis functions, $col = (1 + n + (n^2 - n)/2)$;

$$
\mathbf{P}^{\mathrm{T}}_{k \times col} =
\begin{bmatrix}
1 & \cdots & 1 \\
\Delta x_1^1 & \cdots & \Delta x_k^1 \\
\vdots & & \vdots \\
\Delta x_1^n & \cdots & \Delta x_k^n \\
\Delta x_1^1 \Delta x_{i \neq 1}^1 & \cdots & \Delta x_k^1 \Delta x_{i \neq k}^1 \\
\vdots & & \vdots \\
\Delta x_1^n \Delta x_{i \neq 1}^n & \cdots & \Delta x_k^n \Delta x_{i \neq k}^n
\end{bmatrix}
=
\begin{bmatrix}
1 & \cdots & 1 \\
\Delta \mathbf{x}_1 & \cdots & \Delta \mathbf{x}_k \\
\Delta \mathbf{x}_1 \Delta \mathbf{x}_{i \neq 1} & \cdots & \Delta \mathbf{x}_k \Delta \mathbf{x}_{i \neq k}
\end{bmatrix}_{k \times col}
$$

The number of rows and the number of columns of matrix $\mathbf{W}$, which depends on the number of points, are k and k respectively: $\mathbf{W}_{k \times k}$.

The number of rows and the number of columns of vector $\mathbf{g}_K$ are k and 1 respectively: $\mathbf{g}_{K(k \times 1)}$.

$$
\mathbf{a} = \begin{bmatrix} \tilde{g}(\mathbf{x}) \\ \tilde{\nabla} g(\mathbf{x}) \end{bmatrix} = [\mathbf{P}\mathbf{W}\mathbf{P}^{\mathrm{T}}]^{-1} \mathbf{P}\mathbf{W}\mathbf{g}_K \tag{27}
$$

For linear base, it is not difficult to calculate the number of rows and the number of columns of vector $\mathbf{a}$. they are $(n + 1)$ and 1: $\mathbf{a}_{(n+1) \times 1}$.

Here, we can say that number of unknown is $(n + 1)$. It means that number of points k should be at least bigger than $(n + 1)$, or else, system of equations will not get the solution. Table 2 lists the relations between k and n. Full factorial designs take three levels as the example.

The DA scheme needs an efficient and scalable DOE, limiting the number of the "exact" function evaluations. From Table 2, we can see that in n dimensions, DA requires $k > n + 1$ data points with the linear $\mathbf{P}$ (for $k = n + 1$, the approximation degenerates to the Least Squares fitting as the $\mathbf{P}$ matrix becomes square, and $[\mathbf{P}\mathbf{W}\mathbf{P}^{\mathrm{T}}]^{-1}\mathbf{P}\mathbf{W}\mathbf{g}_K = \mathbf{P}^{-\mathrm{T}}\mathbf{W}^{-1}\mathbf{P}^{-1}\mathbf{P}\mathbf{W}\mathbf{g}_K = \mathbf{P}^{-\mathrm{T}}\mathbf{g}_K$) and $k > 1 + 2n + (n^2 - n)/2$ with the quadratic case.

Fig. 16 Advancing LHS
pattern applied in FORM

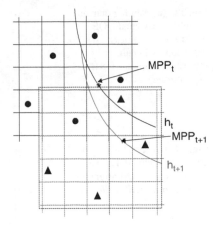
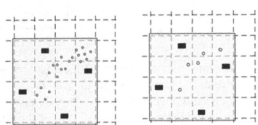

Fig. 17 Mixing points in
both spaces

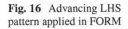

3.3 Advancing LHS Patterns

The overall idea of advancing LHS patterns is shown in Fig. 16. The pattern at the
iteration t surrounds the current approximation of the MPP. Then, Eq. (12) is used
to estimate the new MPP approximation. At this point the pattern has to be updated
in the vicinity of MPP$_{t+1}$. The goal is to minimize the number of "exact" function
evaluations. The idea is thus to reuse as many points as possible from previous
computations and to add new points. The new points are selected in the way that
together with the previous points they form a new Latin Hypercube. Four points
(triangles) are needed for iteration $(t + 1)$, while reuse a point (circle) from previous
iteration.

After mixing the sampling points from two spaces (rectangles from X-space,
circles from U-space), there are several points in the same cell. Their average is
taken instead of all the points in same cell. Now, only one point exists in one cell
(Fig. 17). Other strategies are of course possible.

The possible adaptations of the pattern are panning, expanding, and shrinking. As
the current evaluation point moves during the optimization iterations, three cases are
possible:

– The point stays inside the pattern, but the convergence stops, here, we are going to
  use the "shrink" operator, say by halving the interval of grid to refine the pattern.

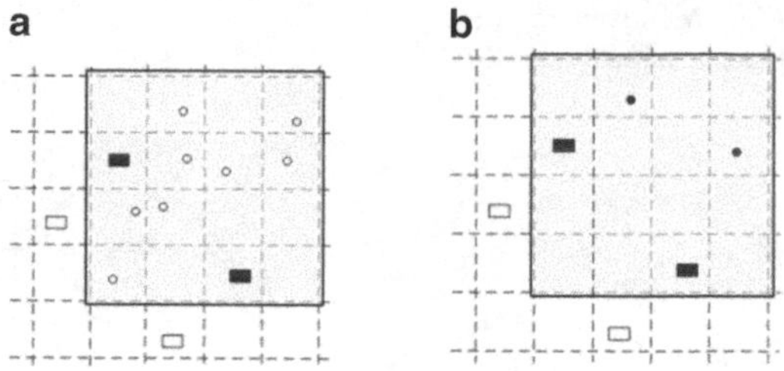

Fig. 18 (**a**) Panning domain of influence, virtual DOE (eight *hollow circles*); (**b**) Panning domain of influence, adapted LHS (two *filled circles*)

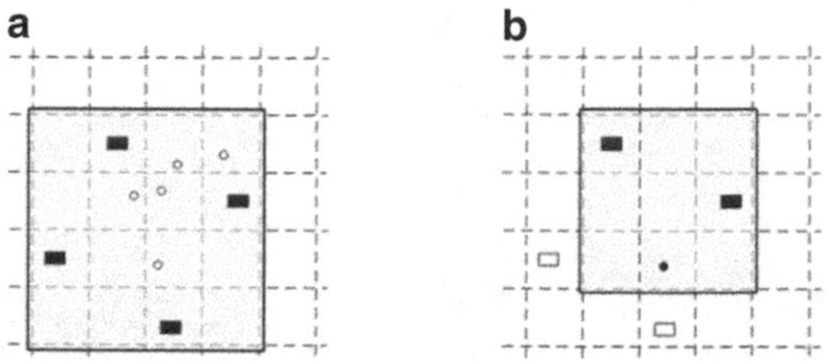

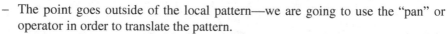

Fig. 19 (**a**) Shrinking domain, virtual DOE (five *hollow circles*); (**b**) Shrinking LHS, adapted LHS (*filled circle*)

- The point goes outside of the local pattern—we are going to use the "pan" or operator in order to translate the pattern.
- When subsequent iteration points fall outside of the pattern, that means that the size of the pattern is too small, this is the situation when we use the "expand" operator.

In Fig. 18a, sampling point increases in the interested region due to the increment of iterations in optimization process. Two new sampling points have to be chosen among the virtual DOE noted now by eight hollow circles. The goal is to select the points in the way that to obtain the new Latin square. Figure 18b shows the two filled circles chosen in the way to obtain an equilibrated Latin square together with the two points from the former rectangles.

The size of the domain of interest decreases when the optimization process converges. It can reduce the number of "exact" function evaluation without affecting quality of DOE (errors, condition number ...). In Fig. 19a, the gray domain corresponds to the reduced domain of interest from Fig. 19b. The hollow circles represent again the virtual DOE points along which now one additional point (the size of the square decreases from 4×4 to 3×3) in the way has to be chosen to preserve the Latin square. Figure 19b shows the chosen point as a filled circle.

The last situation happens when the domain of interest expands. In Fig. 20a, the gray square expands the original 4×4 domain. The virtual DOE is once again denoted by triangles and one of them has to be chosen. The four potential points verifying the LHS criterion for the updated pattern are denoted by the filled triangles in Fig. 20a.

Table 3 gives the value of the correlation coefficient, which is the usual criterion for LHS, for four possible choices of the candidate point. According to the correlation criterion, the best choice is the filled triangle No 4. The current design point advances across the pattern, for instance towards upper left, the choice of the

a

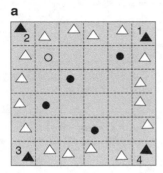

b

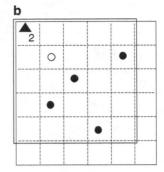

Fig. 20 (**a**) Expanding domain, virtual DOE (seven *hollow circles*); (**b**) Expanding domain, adapted LHS (*filled circle*)

Table 3 Comparison of condition number and correlation coefficient for four possible points

	1	2	3	4
Correlation coefficient	0.7796	−0.2717	0.7820	−0.2640
κ ($\kappa = \lambda_{max}/\lambda_{min}$)	132.8759	4.0642	126.7684	66.9423

fourth point would give a potentially unbalanced pattern by adding points in the direction opposite to the descent. This happens, because the correlation coefficient does not depend on the position of the evaluation point. To compensate this effect, a criterion is used based on the conditioning of the approximation at the current point **x**. The following line in Table 3 gives the condition numbers of matrix $\mathbf{A}(\mathbf{x})$ at a specific evaluation point (hollow circle in the upper left part of Fig. 20a), for the four possible choices. The pattern giving minimal value of condition number is selected. Therefore, the filled triangle No 2 point (left, top) is selected as new point for forming a new 5×5 Latin square together with the four filled circle points from the previous iteration Fig. 20b.

This choice results in a well-balanced pattern of sampling points, surrounding the current evaluation point approximation and evolving in the direction of the current RBDO iteration.

3.4 Coupling Advancing LHS Patterns With Reliability Computations

A classical RBDO ($n = 2$; $k = 3$) approach is used to discuss the situation about coupling advancing LHS patterns with reliability computations. The left and the right side of Fig. 21 illustrate the process respectively in the X-space and the U-space. At each optimization iteration, the eventually correlated and non-normally distributed design variables from the X-space are transformed into the U-space.

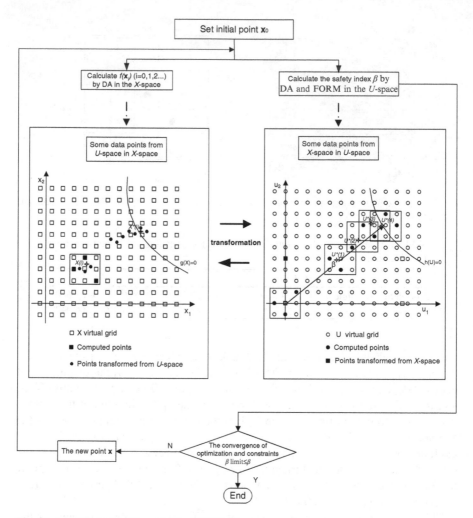

Fig. 21 The RSM with DA model for RBDO

In the U-space, the solid round points are used to calculate the derivative by the DA. The FORM is used to find the new point. At convergence, The MPP is found on the limit state surface. The distance between the original point $(0, 0)$ and MPP is the safety index β. The reliability index is then evaluated in the U-space, which involves an iterative process based on Eq. (12) and the generation of new data points when the current estimation of MPP advances. In Fig. 21, in the U-space, both the solid rounds points (generated in the U-space) and the square solid points (transformed from X-space) are used together to calculate the gradient and the function terms involved in the FORM equation Eq. (12). At convergence, the MPP on the limit state surface is located.

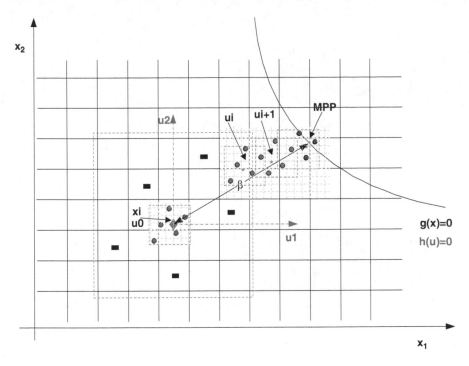

Fig. 22 Proposed model with advancing LHS pattern for RBDO

All the points evaluated in the U-space are then transformed into the X-space using the inverse transformation $\mathbf{x} = \mathbf{T}^{-1}(\mathbf{u})$ and are denoted by small circles in the left side of Fig. 21. In the X-space, the next design point is determined by standard optimization using approximated values and gradients, based again on the incremented LHS pattern.

In both spaces, the points in the vicinity of the current design are used to calculate approximated value of functions and gradients using Eq. (26). And the process of optimization continues until the optimal point, satisfying that prescribed safety indices, is found. For the iteration in U-space shown in Fig. 21, the number of evaluations with advancing LHS patterns is 10 rather than 45 which is calculated traditional LHS patterns (the number of point each iteration 9, the number of iterations 5).

The potential benefit of the proposed method is obvious when the design point $\mathbf{x}$ approaches the limit surface and the points from both spaces (X-space and U-space) are mixed. Mixing makes the points denser around $\mathbf{x}_i$ or $\mathbf{u}_i$ and provides a more precise local approximation and a faster convergence.

An example of this process involving an advancing LHS pattern, taken from an actual computation, is given in Fig. 22, with $n = 2$ and $k = 5$. From this figure, it is easy to find that 17 data points are used to calculate reliability index β. 20 data points have to be used with advancing LHS patterns.

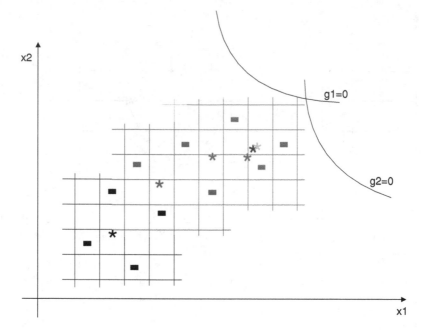

Fig. 23 Advancing LHS patterns in X-space

Figure 23 shows the application of advancing LHS pattern in X-space. A black rectangle point is reused in second iteration (red rectangle points). However, in last iteration (pink rectangle points), three points are from last iteration. It means that a new point is calculated for this iteration. For these five iterations, 10 "exact" functions are computed to do approximation, not traditional number of "exact" functions 20 (4 (data points for each iteration) × 5 (number of iteration)). Figure 24 shows flow chart of coupling advancing LHS patterns with optimization. In this flow chart, firstly, original point x_0 is calculated with DA and LHS patterns separately in two spaces. Two sets of points in both design spaces are combined to form a virtual DOE in order to adequately used previous calculated points. After that, points in the region of influence are checked to find the previous computed points. If there are no previous computed points, all new points are to be developed for forming a new LHS patterns. If there are some previous computed points, less new points need to be developed. Functions $f(\mathbf{x}_i)$ and reliability index β_i with points in region of influence. Then constraints are checked to be satisfied. If yes, stop the program, if no, develop new point and go back to first step to repeat the all the steps.

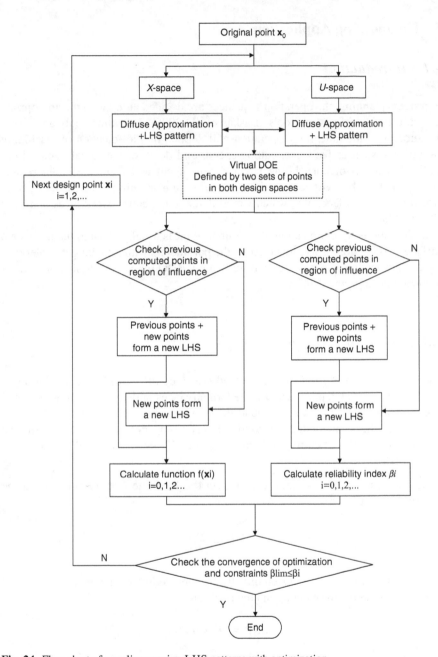

Fig. 24 Flow chart of coupling moving LHS patterns with optimization

4 Engineering Application

4.1 Introduction

In process planning, the operation sequences are decided along with the appropriate machine tools and cutting tools (Baskar et al. 2006). If cutting tools are given, the success of the machining operation will depend on the selection of machining parameters, such as feed rate, cutting speed, and depth of cut (axial depth of cut and radial depth of cut). Using larger depth of cut and feed rate could reduce machining time but will cause serious problems in machining, such as chatter, tool breakage, poor surface, etc., which would damage the cutter, work piece and even NC machine.

In deterministic optimization, the optimal point is usually close to the limit state surface; it is dangerous to adopt directly the parameters without taking the tolerance region into account. So it is useful to study the RBDO for NC machining operations of ultrahigh strength steel.

4.2 NC Milling Model

The parameters of NC machining are feed rate f, cutting speed N, and axial depth of cut d_p and radial depth of cut d_e. The tool advances in the y direction with a feed rate f (mm/min) and it rotates around the z-axis with the cutting speed N (rpm). Here seven different models are used for machining time (T), material removal rate (MRR), quality of cutting (Q), cutting force (F), power (P), tool life (T_{tl}), and torque (M).

1. The machining time T (min) depends on the feed rate f (mm/min) and on the length of cut L(mm)

$$T = \frac{L}{f}$$

2. The MRR (mm³/min) is given directly by the product of the axial depth of cut d_p, the radial depth of cut d_e and the feed rate f.

$$MRR = f \times d_p \times d_e$$

3. Quality of cutting (Q) is also called roughness of surface (R_a) of machining which is determined by material capability (c) and four machining parameters (d_p, d_e, N, f). Exponents, a_1, a_2, a_3, and a_4, are the coefficients of roughness of surface that are identified by experiments and the least square method.

$$R_a = c d_p^{a_1} d_e^{a_2} N^{a_3} f^{a_4}$$

$$R_a = 0.7298 d_p^{0.0658} d_e^{-0.0904} N^{-0.8865} f^{1.071}$$

4. Cutting force (F) is also composed of four machining parameters (d_p, d_e, N, f), coefficient of machining (C_F), and exponents $b_i^1, b_i^2, b_i^3, b_i^4$. Similarly, these coefficients should be identified by experiments and the least square method.

$$F_{i\,\max} = C_F d_p^{b_i^1} N^{b_i^2} f^{b_i^3} d_e^{b_i^4}, \quad \text{with } i = x, y, z.$$

$$\begin{cases} F_{x\,\max} = 767.1848 \times d_p^{0.7153} N^{-0.2420} f^{0.2269} d_e^{0.0606} \\ F_{y\,\max} = 35.6615 \times d_p^{0.8264} N^{-0.5024} f^{0.5993} d_e^{1.0831} \\ F_{z\,\max} = 9.8560 \times d_p^{0.7388} N^{-0.1749} f^{0.5151} d_e^{0.5358} \end{cases}$$

5. The spindle power P (Kw) developed by the cutter of diameter D depends on the tangential component of the cutting force produced in the x direction and on the longitudinal component y but does not depend on the vertical component z as there is no displacement in this direction.

$$P = \frac{(F_{x\,\max} \times \pi D N + F_{y\,\max} \times f)}{6 \times 10^4}$$

6. The tool life model T_{tl} (min) is

$$T_{tl} = \left(\frac{19650}{d_p^{0.15} d_e^{0.1} N f^{0.25}} \right)^5$$

7. The torque M (N·m) is given by

$$M = \frac{F_{x\,\max} D}{2 \times 10^3}.$$

4.3 Optimization Problem Statement

The statistical nature of design variables and functions considered in RBDO process is taken as constraints (limit state functions) involving bounds on reliability indices β. Actual variation ranges of the four parameters are:

$$d_p \in [2, 5], \qquad d_e \in [4, 15]$$
$$f \in [300, 600], \quad N \in [800, 1400]$$

A Sequential Quadratic Programming (SQP) algorithm will be used to find optimal and reliable parameters.

From experimental results, the following expressions of the possible limit state functions have been identified:

- Machining time: $g_T = 250 - T = 0$
- Material removal rate: $g_{MRR} = MRR - 8{,}000 = 0$
- Torque: $g_{torque} = 35 - M = 0$
- Cutting force: $g_{force} = 3{,}500 - F = 0$
- Tool life: $g_{toollife} = T_{tl} - 60 = 0$.

The failure domain is defined by $g \geq 0$. Other parameters of the model are specified as: $D = 32(mm)$, $L = 84{,}000(mm)$.

Based on the engineering practice and the possible state limit functions mentioned above, a test case of NC machining is considered here.

4.4   Test Case

In general, machining parameters obey normal distribution. And its variance σ is affected by the capability of machine tool and experiences of workers. The variances of machining parameters are obtained from the experiences of workers. The vector of variance σ is [0.005; 0.005; 2; 5]. $k = 8$, sampling points k are set according to the Table 3. Interval of girds (distance between evaluation point and data points) h are 0.01 (X-space) and 0.001 (U-space). The radius of influence is $R = [2.5\,h;$ $2.5\,h; 2.5\,h; 2.5\,h]$. Stopping criterion (absolute value of error) is 10^{-2} (X-space) and 0.5×10^{-3} (U-space). Initial point is given as [3.5; 7; 450; 1000].

The tool life is one of the most attended targets in practical machining. So the tool life (T_{tl}) is considered as the objective to be maximized. Minimum values of the safety indices on machining time (T) and material removal rate (MRR) are taken as constraints:

$$\text{Maximize : tool life} \quad f_{toollife}\left(d_p, d_e, f, N\right) = \left(\frac{19650}{d_p^{0.15} d_e^{0.1} N f^{0.25}}\right)^5$$

$$\text{Subject to} \quad \begin{aligned} \beta_{MRR} &\geq \overline{\beta}_{MRR} \\ \beta_T &\geq \overline{\beta}_T \end{aligned}$$

$$\text{with} \quad \begin{aligned} g_T &= 250 - \tfrac{L}{f} \geq 0 \\ g_{MRR} &= f \times d_p \times d_e - 8000 \geq 0 \end{aligned}$$

The limit values of safety indices must correspond to acceptable failure probabilities. They depend on the limitations of practical machining, and may be different in different limit state surfaces. Three situations are considered:

Table 4 Results of each iteration in X-space for $\beta_T \geq 0$ and $\beta_{MRR} \geq 0$

Ite.	NP	x	$-f_{toollife}$	β_T	β_{MRR}
1	8	[3.5, 7,450, 1000]	−208.784010	57.000000,	59.655566
2	8	[2, 7.479764, 449.7383, 998.9519]	−309.153158	56.869172,	−35.468890
3	8	[2, 8.74186, 447.9655, 994.6622]	−293.632990	55.982739,	−4.146687
4	8	[2, 8.925221, 448.009, 993.5693]	−292.167382	56.004483,	−0.069051
5	8	[2, 8.891009, 449.8863, 990.5713]	−295.636464	56.943141	−0.002770
6	8	[2, 6.526084, 569.58, 800]	−747.860864	116.789999	−17.074853
7	8	[2, 7.018623, 567.442, 800]	−724.541133	115.721020,	−0.993031
8	8	[2, 7.055865, 566.8924, 800]	−723.501844	115.446217	−0.004762
9	8	[2, 7.089607, 564.1934, 800]	−726.096996	114.096712,	−0.005142
10	8	[2, 9.954485, 336.2161, 800]	−1170.328884	0.108140,	−29.382826
11	8	[2, 11.6258, 336.1989, 800]	−1083.013832	0.099484	−3.605565
12	7	[2, 11.89959, 335.986, 800]	−1071.329326	−0.006992	−0.073595
13	4	[2, 11.90477, 335.9998, 800]	−1071.041726	0.000000,	0.000013
14	4	[2, 11.90529, 335.9852, 800]	−1071.076394	−0.007405,	0.000000
15	4	[2, 11.90476, 336, 800]	−1071.040924	0.000000	0.000018
16	4	[2, 11.90515, 335.989, 800]	−1071.067252	−0.005514	0.000000
17	0	[2, 11.90476, 336, 800]	−1071.041345	0.000000	0.000000

Ite. standard order of iteration, *NP* standard number of new points

- $\beta_T \geq 0$, $\beta_{MRR} \geq 0$: this case is equivalent to deterministic optimization
- $\beta_T \geq 2$, $\beta_{MRR} \geq 2$: corresponding to a probability of failure $P_f \leq 2.275 \times 10^{-2}$
- $\beta_T \geq 3$, $\beta_{MRR} \geq 4$: $P_f \leq 1.35 \times 10^{-3}$ and $P_f \leq 3.17 \times 10^{-5}$ respectively

The detailed results of each iteration (number of new created points, optimal points **x**, objective function $f_{toollife}$, two reliability constraints) in X-space for three situations mentioned above are shown in Table 4 ($\beta_T \geq 0$, $\beta_{MRR} \geq 0$), Table 5 ($\beta_T \geq 2$, $\beta_{MRR} \geq 2$), and Table 6 ($\beta_T \geq 3$, $\beta_{MRR} \geq 4$). Other results will not be presented here due to limited space. In these tables, the number of new created points reduces to zero at the end of iterations due to the adaptation of advancing LHS patterns. This kind of phenomenon is more sensitive in U-space.

Table 7 shows the results obtained by a deterministic formulation and by a reliability approach performed with three levels of reliability. The same solution is reached by the proposed method (with advancing LHS patterns) and by the reference one (without advancing LHS patterns), but due to the computation of safety indices, the number of functions evaluations increases from 13 in the deterministic approach to several hundred in probabilistic design. The use of approximations capable to handle previously calculated points becomes then necessary.

In Table 7, bigger radial depth of cut (d_e) and smaller feed rate (f) with minimum d_p and N can bring a longer tool life. More rigorous requirements on the T and MRR constraints ($\beta_T \geq 3$ and $\beta_{MRR} \geq 4$) imply to increase the axial depth of cut d_p and feed rate f. Both constraints are active at the optimum solution. The other important information in this table is the number of "exact" function evaluations. Comparing the results with reference solution, the proposed method leads to a significant

Table 5 Results of each iteration in X-space for $\beta_T \geq 2$ and $\beta_{MRR} \geq 2$

Ite.	NP	x	$-f_{toollife}$	β_T	β_{MRR}
1	8	[3.5, 7, 450, 1000]	−208.783989	57.000001	59.655566
2	8	[2, 7.566049, 449.762, 998.9558]	−307.359228	56.881014	−33.004245
3	8	[2, 8.82677, 448.3544, 995.1743]	−291.150266	56.177210	−2.082074
4	8	[2, 9.014734, 448.1595, 993.55]	−290.619454	56.079768	1.929792
5	8	[2, 9.030208, 447.5577, 958.8009]	−347.526902	55.778850	1.999489
6	8	[2, 9.094933, 444.3734, 800]	−863.984319	54.186676	1.989206
7	8	[2, 9.115617, 443.3951, 800]	−865.384026	53.697562,	1.998938
8	8	[2.010617, 11.20993, 339.9495, 800]	−1083.419315	1.974754	−6.847918
9	8	[2, 11.0647, 365.7266, 800]	−999.252079	14.863298	1.915340
10	8	[2, 11.40406, 354.8065, 800]	−1022.283895	9.403265	1.851303
11	8	[2, 11.89463, 340.012, 800]	−1055.716696	2.005979	1.713298
12	8	[2, 11.91707, 340, 800]	−1054.769016	1.999995	1.999411
13	0	[2, 11.91711, 340, 800]	−1054.766991	2.000000	2.000000

Ite. standard order of iteration, *NP* standard number of new points

Table 6 Results of each iteration in X-space for $\beta_T \geq 3$ and $\beta_{MRR} \geq 4$

Ite.	NP	x	$-f_{toollife}$	β_T	β_{MRR}
1	8	[3.5, 7, 450, 1000]	−208.783962	57.000000	59.655566
2	8	[2,7.652575, 449.7481, 998.9639]	−305.616150	56.874047	−30.591211
3	8	[2, 8.923739, 448.222, 994.6144]	−290.487188	56.111011	−0.008934
4	8	[2, 9.108272, 448.198, 993.8661]	−288.632768	56.099015	3.934414
5	8	[2, 9.1035, 448.5802, 993.6597]	−288.700766	56.290090	3.999946
6	8	[2, 8.965542, 455.2779, 990.2563]	−290.515130	59.638972,	3.956814
7	8	[2, 6.074225, 600, 916.3444]	−368.393980	132.000024	−22.427106
8	8	[2, 6.727603, 600, 913.6504]	−355.239418	132.000033	2.144853
9	8	[2, 6.780666, 600, 913.4239]	−354.285623	132.000029,	3.989801
10	8	[2, 6.780961, 600, 913.0783]	−354.948782	132.000031	3.999999
11	8	[2, 6.780962, 600, 906.0688]	−368.892797	132.000028,	4.000033
12	8	[2, 9.7328, 341.436, 800]	−1161.007826	2.718009	−31.011318
13	8	[2, 11.63319, 342.0275, 800]	−1059.656614	3.013776,	−0.832562
14	8	[2, 11.99065, 342.011, 800]	−1043.804345	3.005516	3.874502
15	8	[2, 12.00084, 341.9994, 800]	−1043.405327	2.999738	3.999929
16	0	[2, 12.00082, 342, 800]	−1043.404011	3.000000	4.000000

Ite. standard order of iteration, *NP* standard number of new points

reduction of the number of "exact" function evaluations: for the constraint g_T function, this number is decreased by 32, 90, and 190 respectively for the three situations. For the constraint g_{MRR} function they are reduced by 20, 38, and 27 % when using the DA. The total number of objective function evaluations is also decreased by 25, 8, and 8: this is not directly related to the use of mixed X-space and U-space points, because no reliability index is computed for the objective function, but it can be interpreted as the effect of the high quality of the gradients values due to the DA, which allows a faster convergence of SQP.

Table 7 Numerical results for test case

| | Dete. Opt. | $\beta_T \geq 0, \beta_{\mathrm{MRR}} \geq 0$ | | $\beta_T \geq 2, \beta_{\mathrm{MRR}} \geq 2$ | | $\beta_T \geq 3, \beta_{\mathrm{MRR}} \geq 4$ | |
		Ref.	Pro.	Ref.	Pro.	Ref.	Pro.
Num. of g_T evaluations	13	488	456	440	350	640	450
Num. of g_{MRR} evaluations	13	352	280	336	207	424	312
Num. of f_{toollife} evaluations	13	136	111	104	96	128	120
d_p	2	2		2		2	
d_e	11.90476	11.90476		11.91711		12.00082	
f	336	336		340		342	
N	800	800		800		800	
Value—f_{toollife}	1071	−1071.0417		1054.767		−1043.404	
Value—g_T	0	1.2158×10^{-4}		−2.9412		−4.3859	
Value—g_{MRR}	0.00	-6.7471×10^{-4}		−103.6368		−208.5614	
β_T	0	0.000000		2.000000		3.000000	
β_{MRR}	0	0.000000		2.000000		4.000000	

Ref. standard reference method, *Pro.* standard proposed method, *Dete. Opt.* standard Deterministic optimization

5 Conclusions

In this chapter, Diffuse Approximation (DA) as a RSM is described in detail. For reducing data points which are used to develop RSM in process of optimization, Advancing Latin Hypercube patterns are presented. In the application part, a case of optimizing the process parameters of milling of ultrahigh strength steel is investigated. The optimal reliable parameters are obtained within computing time and with less calculations of function.

References

Allen M, Maute K (2004) Reliability-based design optimization of aeroelastic structures. Struct Multidiscip Optim 27:228–242

Bashir MJK, Aziz HA, Yusoff MS, Adlan MN (2010) Application of response surface methodology (RSM) for optimization of ammoniacal nitrogen removal from semi-aerobic landfill leachate using ion exchange resin. Desalination 254:154–161

Baskar N, Asokan P, Saravanan R (2006) Selection of optimal machining parameters for multi-tool milling operations using a memetic algorithm. J Mater Process Technol 174:239–249

Behnam B, Eamon C (2013) Reliability-based design optimization of concrete flexural members reinforced with ductile FRP bars. Construct Build Mater 47:942–950

Belytschko T, Krongauz Y, Organ D et al (1996) Meshless methods: an overview and recent developments. Comput Methods Appl Mech Eng 139:3–47

Box GEP, Cox DR (1964) An analysis of transformations. J R Stat Soc Series B 26:211–252

Chen CT, Chen MH, Horng WT (2014) A cell evolution method for reliability-based design optimization. Appl Soft Comput 15:67–79

Choi SK, Grandhi RV, Canfield RA (2007) Reliability-based structural design. Springer, London

Der Kiureghian A, Liu PL (1986) Structural reliability under incomplete probability information. J Eng Mech ASCE 112(1):85–104

Ginta TL (2009) Tool life prediction by response surface methodology in end milling Titanium alloy Ti-6AI-4V using uncoated WC-Co inserts. Eur J Sci Res 28:533–541

Hasofer AM, Lind NC (1974) An exact and invariant first order reliability format. J Eng Mech ASCE 100(EM1):111–121

Kumar S, Singh RK (2014) Optimization of process parameters by response surface methodology (RSM) for catalytic pyrolysis of waste high-density polyethylene to liquid fuel. J Environ Chem Eng 2:115–122

McKay MD, Beckman RJ, Conover WJ (1979) A comparison of three methods for selecting values of input variables in the analysis of output form a computer code. Technometrics 21(2):239–245

Rosenblatt M (1952) Remarks on a multivariate transformation. Ann Math Stat 23(3):470–472

Youn BD (2001) Advances in reliability-based design optimization and probability analysis. Dissertation, University of Iowa

Youn BD, Choi KK, Yang RJ et al (2004) Reliability-based design optimization for crashworthiness of vehicle side impact. Struct Multidiscip Optim 26:272–283

The Stochastic Modeling of the Turning Decision by Left-Turning Vehicles at a Signalized Intersection in a University Campus

Md. Shohel Reza Amin and Ciprian Alecsandru

Abstract The turning decision of a vehicle, at a signalized intersection, depends on the characteristics of the road users (e.g., vehicle, pedestrians, bicycles) and the intersection. The objective of this chapter is to estimate the turning decision of left-turning vehicles at a signalized intersection in a university campus. The signalized intersection, at the crossing of Boulevard de Maisonneuve Ouest and Rue MacKay within the Sir George Williams campus of the Concordia University (Montreal, Canada), is considered as a case study. The traffic video data were collected from 10 a.m. to 5 p.m. during the period of July–October in the year 2010. Vehicles turn at the intersection based on the gap between those and the crossing traffic, and complete the turning maneuver accepting the adequate gap (time or distance). The mean value of accepting the gap is known as the critical gap acceptance (CGA). The stochastic modeling of the left-turning decision is implemented at two stages—the estimation of the CGA by using probabilistic approaches; and the determination of the factors' contribution by applying backpropagation neural network (BPN). The stochastic distribution functions estimate the CGA for passenger cars and other vehicles (e.g., buses, trucks, and vans) as 14.3 s and 16.5 s, respectively. The BPN models determine the bicycle distance from conflict point, platoon bicycles, existence of bicycle at conflict zone, bicycles' speed, vehicles' speed, pedestrians' speed, number of vehicles passed, and vehicle moving at conflict zone are the predominant factors of left-turning decision.

M.S.R. Amin (✉) • C. Alecsandru
Concordia University, Montreal, Canada
e-mail: md_amin@encs.concordia.ca

S. Kadry and A. El Hami (eds.), *Numerical Methods for Reliability and Safety Assessment: Multiscale and Multiphysics Systems*, DOI 10.1007/978-3-319-07167-1_27,
© Springer International Publishing Switzerland 2015

707

1 Introduction

The turning vehicles, at a signalized intersection, are frequently interacted with other road users (e.g., crossing pedestrians and bicycles, oncoming vehicles). The turning decision of a vehicle is not only dependent on the vehicles' characteristics but also on the attributes of pedestrian and bicycle, such as pedestrians' speed, bicycles' speed, and platoon bicycle. The crossing pedestrians are the major factors affecting the turning decision of vehicles at intersections, especially within a university campus. Drivers are more cautious for pedestrian safety in the university campus as the students are more frequently crossing the intersection comparing to other urban intersections. In a busy campus environment, the likelihood of interactions with platoons of pedestrians and/or bicycle is significantly higher. The additional structural safety measures of road geometric design to protect the cyclists and pedestrians also guide the vehicle's turning decision. In these scenarios, turning vehicles first yield the right-of-way (ROW) to cyclists and then to pedestrians.

The previous studies (Geruschat and Hassan 2005; Katz et al. 1975) found out that the likelihood of turning vehicles to yield the crossing pedestrians was associated with the vehicles' speed, relative position of the pedestrians to the curb, and platoon pedestrians. Some researches (Varhelyi 1998; Hamed 2001) identified that the drivers had less willingness to give the ROW to pedestrians; and they only slowed or stopped when the speed of their vehicles was low. The turning decision of vehicles is also influenced by specific bicycle attributes. For example, the gap acceptance (GA) of vehicle is significantly determined by the number and speed of bicycles in the conflict zone (Li et al. 2010) and the platoon bicycles (Zhang et al. 2009). Different drivers may react differently during the turning maneuver. A truck driver's reaction to pedestrian platoon is not same to that of car driver. The car driver usually tends to move slowly to turn, while the truck or bus driver stops the vehicle completely.

The objective of this chapter is to estimate the turning decision of left-turning vehicles at a signalized intersection in a university campus. Vehicles turn at the intersection based on the gap between those and the crossing traffic, and complete the turning maneuver accepting the adequate gap (time or distance). The mean value of accepting the gap is known as the critical gap acceptance (CGA). The signalized intersection, at the crossing of Boulevard de Maisonneuve Ouest and Rue MacKay within the Sir George Williams campus of the Concordia University (Montreal, Canada), is considered as a case study. The stochastic modeling of the left-turning decision is implemented at two stages—the estimation of the CGA by using probabilistic approaches; and the determination of the factors' contribution by applying backpropagation neural network (BPN).

2   Traffic Data and Study Area

The traffic video data were collected from 10 a.m. to 5 p.m. during the period of July–October in the year 2010. There are two conflict zones at the crossing of Boulevard de Maisonneuve Ouest and Rue Mackay for the left- and right-turning vehicles. The left-turning vehicles, on Boulevard de Maisonneuve Ouest, interfere with pedestrians and cyclists (Fig. 1a). The right-turning vehicles, on Rue Mackay, interfere with pedestrian flows only (Fig. 1b). Since this study carried out the stochastic modeling of left-turning vehicles, this study considered the conflict zone of Fig. 1a. A total of 638 traffic data were recorded during the 30 h period in the year of 2010.

a

b

Fig. 1 (**a**) Left-turning vehicles on Boulevard de Maisonneuve Ouest (source: Google Maps). (**b**) Right-turning vehicles on Rue Mackay (source: Google Maps)

A total of 606 vehicles out of 638 cases were accepted gap, while less than 5 % of turning vehicles accepted a lag during the left-turning maneuver. In this study, lag acceptance records were removed. A lag is defined as the time needed for a vehicle to reach the conflict zone. The proportions of observed vehicles are 85.48 % passenger cars, 2.31 % buses, 2.64 % trucks, and 9.24 % vans.

3 Methodology

The left-turning maneuver of vehicles is modeled as a two-step process—the decision stage and the implementation stage. In the decision stage (Sects. 3.1 and 4.1), the CGA is calculated to estimate the mean value of the gap accepted by different types of vehicles. In the implementation stage (Sects. 3.2 and 4.2), this study identifies the significant factors contributing to the turning decision.

3.1 Critical Gap Acceptance (CGA)

This study defines the GA as the gap time when a vehicle accepts to allow a pedestrian and/or bicycles to pass in front of it. For example, let's assume the gap commences at T_1 when pedestrians reach the conflict point, the left-turning vehicle reaches the conflict point at T_2, and the gap ends at T_3 when following pedestrians reach the conflict point. The accepted gap by vehicle is $T = T_3 - T_1$. More complex scenarios can be encountered, such as the case of pedestrian–bicycle–vehicle interference. For example, the gap may start at T_1, when the bicyclist reaches the conflict point, pedestrian (or platoon of pedestrians) are screened by the bicyclists and reach the conflict point at T_2, the turning vehicle reaches the conflict point at T_3, and the gap ends at T_4 when the following pedestrian(s) or bicyclist(s) reaches the conflict point, then the accepted gap by vehicle is $T = T_4 - T_1$.

The CGA, the mean value of GA calibrated for local conditions (Troutbeck and Brilon 2002), can be determined as the minimum gap duration accepted by a vehicle in a specific situation (Miller 1971). The deterministic value of CGA can be identified by measuring the mean of the GA distribution without considering the randomness and heterogeneity (Taylor and Mahmassani 1998). The GA data are randomly distributed; therefore, a simple mean value does not represent the appropriate CGA. The cumulative distribution function (CDF) of GA distribution can estimate the 50 % probability of a particular GA value, which is defined as CGA.

The stochastic approaches are mentioned in the Next-Generation micro-SIMulation (NGSIM) research effort of Federal Highway Administration (FHWA) in 2004 (Alhajyaseen et al. 2012). Several research studies (Abernethy 2004; Alhajyaseen et al. 2012) suggested that GA probability distributions could be adjusted and fitted by cumulative Weibull distribution function. This study applies

Easyfit Professional software to check the fitness of the probability distribution of GA against 49 cumulative distribution functions (CDF). The goodness-of-fit of GA distributions, compatibility of a random sample with a theoretical probability distribution function, is determined by Kolmogorov–Smirnov (K–S), Anderson Darling (A–D), and Chi-square tests. Initially, Chi-square goodness-of-fit test was used because it can be applied to any univariate distribution. However, the Chi-square test is restricted to discrete distributions. Both the K–S and A–D tests were applied to determine the goodness-of-fit of continuous distributions. It was expected that the K–S and A–D tests would yield better results because the distribution of GA by each vehicle was continuous. The K–S test quantifies a distance between the empirical and normal (theoretical) CDF. The K–S test statistic is defined by $\max_{1 \leq i \leq N}(F(Y_i) - (i - 1/N), (i/N) - F(Y_i))$ where $F(Y_i)$ is the normal cumulative distribution of GA (Chakravarti et al. 1967). The A–D test is a modification of the K–S test by giving more weight to the tails than the K–S test does. The difference between these two test stems from the fact that the K–S test is distribution free (i.e., the critical values do not depend on the specific distribution being tested), while the A–D test is more sensitive because it uses the specific distribution in calculating critical values. The A–D test is defined as $A^2 = -N - s$, where $S = \sum_{i=1}^{N}((2i - 1)/N)[\ln F(Y_i) + \ln(1 - F(Y_{N+1-i}))]$ (Chakravarti et al. 1967).

3.2   *Factors Contributing to Turning Decision*

The calibration of a simulator to replicate vehicle's turning maneuvers at a signalized intersection is not a trivial exercise because it has to account for complex interactions among all road users (i.e., pedestrians, vehicles, and bicycles) under the prevailing traffic conditions. Several factors can influence the turning decision of vehicles at signalized intersection, such as vehicles' speed, vehicle in queue, pedestrian distance from the curb, number of pedestrians in the conflict zone, pedestrians' speed, and behavior and number of preceding pedestrians. In addition, one has to evaluate the impact of specific bicycle-related parameters, such as bicycles' speed, flow, and bicyclists' behavior. The existence of bicycles may have a pedestrian screening effect and may contribute to the road crossing decision by pedestrians. Often times, turning vehicles waiting in the queue or at signal tail of the green phase, may not yield ROW to pedestrians in order to use the crossing opportunity of the current signal cycle. The contribution of these variables to vehicle's left-turning decision (accept the critical gap) can be determined by different GA models. Ben-Akiva and Lerman (1985) and Cassidy et al. (1995) proposed logit GA model, while Mahmassani and Sheffi (1981) and Madanat et al. (1994) proposed probit GA model.

This study models the CGA of left-turning vehicles by applying an artificial neural network (ANN). In this model, a certain relationship between dependent and independent variables is not a priory hypothesis. However, the ANN is capable of determining this relationship during the learning process (IBM 2010). This study used the backpropagation learning algorithm of ANN based on the mathematical derivation of the Back-Propagation Neural Networks (BPN) by Freeman and Skapura (1991). This BPN model the outputs as a binary value of CGA accepted (1) or rejected (0) by vehicles.

The fundamental concept of BPN networks for a two-phase propagate-adapt cycle is that input variables (vehicle, pedestrian and bicycle characteristics; and traffic condition) are applied as a stimulus to the input layer of network units that is propagated through each upper layer until an output is generated. This estimated output is then compared to the desired output, and an error is computed for each output unit. These errors are then transferred backward from the output layer to each unit in the intermediate layer that contributes directly to the output. Each unit in the intermediate layer receives only a portion of the total error signal, based roughly on the relative contribution the unit made to the original output. This process repeats layer-by-layer until each node in the network has received an error that represents its relative contribution to the total error. Based on the error received, connection weights are then updated by each unit to cause the network to converge toward a state that allows all the training patterns to be encoded (Freeman and Skapura 1991).

This study applies generalized delta rule (GDR) to learn the algorithm for the neural network. Suppose we have a set of P vector-pairs in the training set, $(x_1, y_1), (x_2, y_2), \ldots (x_p, y_p)$, which are examples of a functional mapping $y = \phi(x) : x \in R^N, y \in R^M$ We also assume that $(x_1, d_1), (x_2, d_2), \ldots (x_p, d_p)$ is some processing function that associates input vectors, $\mathbf{x}_k$ (vehicle, pedestrian, bicycle, and traffic attributes) with the desired output value, d_k (CGA accepted or rejected). The mean square error, or expectation value of error, is defined by Eq. (1).

$$\varepsilon_k^2 = \theta_k = (d_k - y_k)^2 = \left(d_k - \mathbf{w}^t \mathbf{X}_N\right)^2 \quad \text{where} \quad y = \mathbf{w}^t \mathbf{X} \tag{1}$$

$$\nabla \theta_k (\mathbf{w}) = \frac{\partial (\theta)}{\partial \mathbf{w}} = -2\varepsilon_k \mathbf{X}_N \tag{2}$$

Since the weight vector is an explicit function of iteration (t), the initial weight vector is denoted as $\mathbf{w}(0)$ and the weight vector at iteration t is $\mathbf{w}(t)$. At each step, the next weight vector is calculated according to Eq. (3).

$$\mathbf{w}(t + 1) = \mathbf{w}(t) + \Delta \mathbf{w}(t) = \mathbf{w}(t) - \mu \nabla \theta_k (\mathbf{w}(t))$$

$$= \mathbf{w}(t) + 2\mu \varepsilon_N \mathbf{X}_N \quad \forall \nabla \theta (\mathbf{w}(t)) \approx \nabla \theta (\mathbf{w}) \tag{3}$$

where $\Delta \mathbf{w}(t)$ is the change in $\mathbf{w}$ at the tth iteration. The error surface is assumed as a paraboloid. Initially, the error weight surface is negative gradient in the direction of steepest descent, where μ is constant of negative gradient and determines the stability and speed of convergence of the weight vector toward the minimum error value (Freeman and Skapura 1991).

The input layer of input variables distributes the values to the hidden layer units. Assuming that the activation of input node is equal to the net input, the output of this input node is given by Eq. (4).

$$I_{pj} = f_j \left(net_{pj} \right) \quad net_{pj} = \sum_{i=1}^{N} w_{ji} x_{pi} + \theta_j \tag{4}$$

where net_{pj} is the net input to the jth hidden unit, w_{ji} is the weight on the connection from the ith input unit derived from Eq. (3), and θ_j is the bias term derived from Eq. (1).

Similarly, the output of output node is given by Eq. (5), where the net output from the jth hidden unit to kth output units is net_{pk}.

$$O_{pk} - f_k \left(net_{pk} \right) \quad \forall net_{pk} - \sum_{j=1}^{L} w_{kj} I_{pj} + \theta_k \tag{5}$$

There are multiple units in a layer, a single error value (θ_k) is not suffice for the BPN. The sum of the squares of the errors for all output units is calculated by Eq. (6).

$$\theta_{pk} = \frac{1}{2} \sum_{k=1}^{M} \varepsilon_{pk}^2 = \frac{1}{2} \sum_{k=1}^{M} \left(y_{pk} - O_{pk} \right)^2$$

$$\Delta_p \theta_p \left(\mathbf{w} \right) = \frac{\partial \left(\theta_p \right)}{\partial w_{kj}} = - \left(y_{pk} - O_{pk} \right) \frac{\partial}{\partial w_{kj}} \left(O_{pk} \right)$$

$$= - \left(y_{pk} - O_{pk} \right) \frac{\partial f k}{\partial \left(net_{pk} \right)} \frac{\partial \left(net_{pk} \right)}{\partial w_{pk}} \tag{6}$$

Combining Eqs. (4), (5), and (6), change in weight of output layer can be determined by Eq. (7).

$$\frac{\partial \left(\theta_p \right)}{\partial w_{kj}} = - \left(y_{pk} - O_{pk} \right) \frac{\partial f k}{\partial \left(net_{pk} \right)} \frac{\partial}{\partial w_{kj}} \left(\sum_{j=1}^{L} w_{kj} I_{pj} + \theta_k \right)$$

$$= - \left(y_{pk} - O_{pk} \right) f_k' \left(net_{pk} \right) I_{pj} \tag{7}$$

Following Eq. (8), the weights on the output layer can be written as Eq. (8).

$$w_{kj} \left(t + 1 \right) = w_{kj} (t) + \tau \left(y_{pk} - O_{pk} \right) f_k' \left(net_{pk} \right) I_{pj} \tag{8}$$

where τ is a constant and is also known as learning-rate parameter. However, $f_k(net_{pk})$ needs to be differentiated to derive f_k'. There are two forms of output functions for paraboloid [$f_k(net_{jk}) = net_{jk}$] and sigmoid or logistic function [$f_k(net_{jk}) = (1 + e^{-net_{jk}})^{-1}$] As the output of this model is binary, sigmoid function is applied for the output function because the sigmoid is output limiting and quasi-bistable, but is also differentiable.

The weights on the output layer can be re-written as Eq. (9).

$$w_{kj}\,(t+1) = w_{kj}(t) + \tau\left(y_{pk} - O_{pk}\right) O_{pk} \left(1 - O_{pk}\right) I_{pj} = w_{kj}(t) + \tau\delta_{pk} I_{pj} \tag{9}$$

The errors from the outputs were backpropagated in order to distribute the errors into the contributing nodes of the neural network. Reconsidering Eqs. (6), (7), and (9) for backpropagation algorithm, change of weights on hidden layer is given by Eq. (10).

$$\theta_p = \frac{1}{2}\sum_{k=1}^{M}\left(y_{pk} - O_{pk}\right)^2 = \frac{1}{2}\sum_{k=1}^{M}\left(y_{pk} - f_k\left(net_{pk}\right)\right)^2$$

$$= \frac{1}{2}\sum_{k=1}^{M}\left(y_{pk} - f_k\left(\sum_{j=1}^{L}w_{kj}I_{pj} + \theta_k\right)\right)^2$$

$$\frac{\partial\theta_p}{\partial w_{ji}} = -\sum_{k=1}^{M}\left(y_{pk} - O_{pk}\right)\frac{\partial O_{pk}}{\partial w_{ji}}$$

$$= -\sum_{k=1}^{M}\left(y_{pk} - O_{pk}\right)\frac{\partial O_{pk}}{\partial\left(net_{pk}\right)}\frac{\partial\left(net_{pk}\right)}{\partial I_{pj}}\frac{\partial I_{pj}}{\partial\left(net_{pj}\right)}\frac{\partial\left(net_{pj}\right)}{\partial w_{ji}}$$

$$\frac{\partial\theta_p}{\partial w_{ji}} = -\sum_{k=1}^{M}\left(y_{pk} - O_{pk}\right)f_k'\left(net_{pk}\right)w_{kj}f_k'\left(net_{pj}\right)x_{pi}$$

$$\Delta_p w_{ji} = \frac{\partial\theta_p}{\partial w_{ji}} = \tau f_k'\left(net_{pj}\right)x_{pi}\sum_{k=1}^{M}\left(y_{pk} - O_{pk}\right)f_k'\left(net_{pk}\right)w_{kj}$$

$$= \tau f_j'\left(net_{pk}\right)x_{pi}\sum_{k=1}^{M}\partial_{pk}w_{kj} \tag{10}$$

Equation (10) explains that every weight update on hidden layer depends on all the error terms (∂_{pk}) on the output layer, which is the fundamental essence of backpropagation algorithm. By defining the hidden layer error term as $\delta_{pj} = f_j'(net_{pj})\sum_{K=1}^{M}\partial_{pk}w_{ki}$, we can update the weight Eq. (11) to become analogous to those for the output layer.

$$w_{ji}\,(t+1) = w_{ji}(t) + \tau\delta x_{pi} \tag{11}$$

Equations (9) and (11) have the same form of delta rule.

4 The Behavior Analysis of the Left-Turning Vehicles

The behavior analysis is made by comparing the passenger cars with other vehicles. In Sect. 4.1, the CGA is estimated by fitting the traffic data to different distribution functions. The CGA is evaluated by different goodness-of-fit tests. The contribution of the selected explanatory variables to the turning decision of the left-turning vehicles is examined by BPN models in Sect. 4.2.

4.1 Critical Gap Acceptance (CGA)

The stochastic analysis of the GA reveals that the CDF of gap acceptance by all vehicles is best fitted by the Weibull distribution function (Eq. (12)), where $\alpha = 1.8865$ (continuous shape), $\beta = 16.108$ (continuous scale), and $\gamma = 0.30685$ (continuous location). The best-fitted distribution function is selected based on the goodness-of-fit nonparametric tests (i.e., K–S, A–D, and Chi-square tests). The nonparametric tests evaluate the null hypothesis that there is a distance between the empirical and theoretical CDF.

The disaggregated data for passenger cars show that the CDF of GA is best fitted by the Burr distribution function (Eq. (13)), where $k = 9.4307$ (continuous shape), $\alpha = 1.9914$ (continuous shape), $\beta = 46.906$ (continuous scale), and $\gamma = 0.22807$ (continuous location). The CDF of GA for the other vehicles is fitted best by the Rayleigh distribution function (Eq. (14)) with $\sigma = 13.156$ (continuous scale).

All three non-parametric tests (i.e., K–S, A–D, and Chi-square tests) for all vehicles and passenger cars reject the null hypothesis which assumes that there is a distance between the empirical and theoretical CDF (Table 1).

$$F(x) = 1 - \exp\left(-\left(\frac{x - \gamma}{\beta}\right)^{\alpha}\right) \tag{12}$$

$$F(x) = 1 - \left(1 + \left(\frac{x - \gamma}{\beta}\right)^{\alpha}\right)^{-k} \tag{13}$$

$$F(x) = 1 - \exp\left(-\frac{1}{2}\left(\frac{x}{\sigma}\right)^{2}\right) \tag{14}$$

The CGA of all vehicles is 14.57 s, while the CGA of passenger cars and other vehicles are 7.97 s and 12.66 s, respectively. The CGA is 4.7 s higher for other vehicles comparing to that for passenger cars. This study identifies three reasons to accept higher critical gap by the other vehicles. First, the other vehicles started turning maneuver at a longer distance from the conflict point (for example, the turning maneuver was 9.44 m for passenger cars and 14.15 m for other vehicles). Second, the bicycles had a higher acceptance of risk when interfering

Table 1 Model summary of probability distribution of GA (sec)

Statistics				Goodness of fit (5 % significance level)[a]			
Criteria	All vehicles	Passenger cars	Other vehicles	Criteria	All vehicles	Passenger cars	Other vehicles
Mean	14.57	14.252	16.488	K–S test	0.0580 (0.055) 5 %	0.0575 (0.05368) 10 %	0.08934 (0.144)
Standard deviation	7.9606	7.8774	8.2344	A–D test	2.3229 (1.9286) 10 %	2.1083 (1.9286) 10 %	1.0453 (2.502)
COV	0.5464	0.5527	0.4994	Chi-square test	19.217 (16.92) 5 %	29.65 (21.666) 1 %	8.674 (8.5581) 20 %

[a]Each cell contains statistics value and critical value at a significance level

Table 2 Case processing summary of BPN models

Cases	All vehicles	Passenger cars	Other vehicles
Training	64.10 %	63.10 %	65.10 %
Testing	26.30 %	26.30 %	19.80 %
Holdout	9.60 %	10.60 %	15.10 %
Total	605	518	86

with passenger cars and other vehicles (i.e., the bicycles' speed was 1.23 m/s and 1.9 m/s, respectively). Third, other vehicles were more accommodating to bicycles and pedestrians on the conflict zone comparing to passenger cars as other vehicles provided ROW to more distanced bicycles (1.90 m for cars and 3.84 m for other vehicles) to pass ahead of them.

4.2 Factors Contributing to Turning Decision-Making Process

This study applies BPN models without a prior hypothesis of a certain relationship between the CGA and the characteristics of the road users and intersection. The BPN models divide the CGA data into training, testing, and holdout data. From the total of 605 gap-acceptance records processed for all vehicles (i.e., 518 for passenger cars and 86 for other vehicles), 64.10 % are used during the training phase, 26.30 % are assigned for cross-validation, and 9.60 % are assigned as testing data (Table 2). Training data are used to train the neural network, while the cross-validation data are used to identify errors during training in order to prevent overtraining. The testing data are used to assess the final neural network. The error for the testing data gives a true estimate of the predictive ability of the model because the testing data are not used to build the BPN model (IBM 2010).

The results of the BPN models for passenger cars, other vehicles, and all vehicles are summarized in Table 3. Table 3 explains the importance of response quantities or independent variables to CGA. The importance of an independent variable is a measure of how much the network's model-predicted value changes for different values of the independent variable. The CGA decision model for all left-turning vehicles is mainly determined by the pedestrians' speed (14.8 %), bicycle distance from the conflict point (13.9 %), number of vehicles passed (11 %), number of bicycle passed during the interference (9.8 %), and bicycles' speed (8.5 %) (Table 3). Therefore, the CGA decision by all left-turning vehicles is mainly structured by the pedestrian and bicycle attributes.

The contribution of road users' characteristics to CGA decision model is different for different types of vehicles. The CGA decision model for the left-turning passenger cars is explained by the pedestrians' speed (9 %), existence of bicycle at conflict zone during vehicles' turning maneuver (7.5 %), platoon bicycles (11 %), number of vehicles passed (11.5 %), and bicycles' speed (11.2 %) (Table 3). Similar

Table 3 Importance of response quantities or variables to CGA by BPN model

Factors	Passenger cars	Other vehicles	All vehicles
Distance from conflict point (m)	4.8	4.4	7
Vehicle in queue	7.3	2.7	2.3
Vehicles' speed (m/s)	8	4.4	1.6
Existence of bicycle at conflict point	7.5	5.7	2.2
Number of bicycle passed during the interference	3.8	9.8	3.2
Platoon bicycles when passing conflict point	11	4.4	10.3
Bicycle distance from the conflict point (m)	6.3	13.9	21.5
Bicycles' speed (m/s)	11.2	8.5	2.3
Pedestrians' speed (m/s)	9	14.8	3
Platoon pedestrians when passing conflict point	2.7	3.4	5.6
Pedestrians are in rush when crossing the intersection	6.5	1.7	2.5
Vehicle moving at conflict zone	1.1	2	8.8
Number of vehicles passed	11.5	11	15
Existence of traffic jam	4.5	6.9	6.7
Vehicle is in signal tail during turning	1.3	2.5	0.8
Vehicle turning from turning lane	3.5	3.9	7.3

to the CGA decision model for all vehicles, the CGA decision model for passenger cars is mainly determined by the pedestrian and bicycle attributes.

The CGA decision model of other vehicles identifies the bicycle distance from conflict point (21.5 %), platoon bicycles (10.3 %), number of vehicles passed (15 %), vehicle moving at conflict zone (8.8 %), and vehicle turning from turning lane (7.3 %) as the most important attributes (Table 3). It reveals that bicycle attributes have limited contribution (39.5 %) to the decision model, majority (21.5 %) of which is contributed by the bicycle distance from conflict point. This may happen because of higher turning maneuver time required by other vehicles (vehicle distance from conflict point 14.15 m). During the turning maneuvers, of 43 % cases other vehicles were in queue, of 48 % cases vehicle were in signal tail, and of 83 % cases vehicles were moving at conflict zone.

This study observes some statistical anomalies in the BPN models. For example, vehicles' speed has insignificant contribution to the CGA decision model for all vehicles; however, contribute significantly to CGA decision models both for passenger cars (8.0 %) and other vehicles (4.4 %) (Table 3). The platoon bicycles passing conflict zone have peculiar behavior for defining the CGA decision model (4.4 %, 11 %, and 0 % contributions to the CGA decision models for all vehicles, passenger cars, and other vehicles, respectively) (Table 3). This study estimates the "sum of square error" and "relative error" to define the statistical significance of the training, testing, and holdout procedures of the BPN models (Table 4).

The estimation of BPN decision models has significant difference between values implied by estimators and the true values of the outputs being estimated especially for training data. Testing data, used to track errors during training in order to prevent overtraining, also contain noteworthy expected value of squared error loss (Table 4). Error for holdout data explains less accurate predictive ability of the constructed *BPN*—Multi Layer Perceptron (MLP) network (Table 4). Moreover, relative error

Table 4 Model summary of BPN models

Cases	Statistical significance	All vehicles	Cars	Other vehicles
Training	Sum of square error	38.114	40.7	3.254
	Relative error	0.81	0.947	0.536
Testing	Sum of square error	17.548	13.253	2.148
	Relative error	0.878	0.956	0.794
Holdout	Relative error	0.856	0.926	1.48

of CGA, a ratio of the sum-of-squares error for CGA to the sum-of-squares error for "null model" (in which mean value is used as the predicted value), explains significant amount of errors in modeling CGA decision (Table 4). However, the average overall relative error of model and relative error of dependent variables are fairly constant across the training, testing, holdout data, which give some confidence that the model is not over-trained and that the error in future cases scored by the neural network will be closed to the error (Table 4).

5 Conclusion

The turning vehicles, at a signalized intersection in a university campus, are frequently interfered by the pedestrians and bicycles. The turning decision of vehicles is very crucial to ensure traffic safety. Vehicle's turning decision subjects to road user's characteristics and traffic condition at the signalized intersection. The turning decision of vehicles, at the signalized intersection in a university campus, is significantly different from that of urban road intersection. The objective of this study is to estimate the turning decision of left-turning vehicles at a signalized intersection in a university campus. The signalized intersection at the crossing of the Boulevard de Maisonneuve Ouest and Rue MacKay within the Sir George Williams campus at the Concordia University (Montreal, Canada) is considered as the case study. The traffic video data were collected from 10 a.m. to 5 p.m. during the period of July–October in the year 2010. Vehicles turn at the intersection based on the gap between those and the crossing traffic, and complete the turning maneuver accepting the adequate gap (time or distance). The mean value of accepting the gap is known as the critical gap acceptance (CGA). The stochastic distribution functions estimated the CGA for passenger and non-Passenger Cars (e.g., buses, trucks, and vans) as 14.3 s and 16.5 s, respectively.

The BPN decision model identifies that the pedestrians' speed (9 %), existence of bicycle at conflict zone (7.5 %), platoon bicycles (11 %), number of vehicles passed (11.5 %), and bicycles' speed (11.2 %) are the main decision-making factors for left-turning passenger cars. The BPN prediction model also identifies that the bicycle distance from conflict point (21.5 %), platoon bicycles (10.3 %), number of vehicles passed (15 %), vehicle moving at conflict zone (8.8 %), and vehicle turning from turning lane (7.3 %) are the most important attributes of the turning decision by other vehicles. The CGA decision by all left-turning vehicles is predominantly guided by the bicycle and vehicles attributes. The outcomes of this research may help the transportation planners to take preventive measures reducing the accident risk at the conflict zone of signalized intersection in a university campus.

Acknowledgements Amir Hossein and Hamed Shahrokhi, M.Sc. students of Civil Engineering at the Concordia University, helped this research by partially extracting traffic video data from video cassette.

References

Abernethy RB (2004) The New Weibull handbook.http://www.barringer1.com/pdf/preface-toc-5th-edition.pdf. Accessed 10 Jan 2012

Alhajyaseen WKM, Asano M, Nakamura H et al (2012) Estimation of left-turning vehicle maneuvers for the assessment of pedestrian safety at intersections. IATSS Res 36:66–74

Ben-Akiva M, Lerman SR (1985) Discrete choice analysis: theory and applications to travel demand. MIT, Cambridge

Cassidy JM, Madanat SM, Wang M et al (1995) Unsignalized intersection capacity and level of service: revisiting critical gap. Transport Res Rec 1484:16–23

Chakravarti IM, Laha RG, Roy J (1967) Handbook of methods of applied statistics. Wiley, Mishawaka

Freeman JA, Skapura DM (1991) Neural networks. Algorithms, applications, and programming techniques. Addison-Wesley, New York

Geruschat D, Hassan S (2005) Driver behavior in yielding to sighted and blind pedestrians at Roundabouts. J Vis Impair Blind 99(5):286–302

Hamed MM (2001) Analysis of pedestrian's behavior at pedestrian crossings. Saf Sci 38:63–82

IBM (2010) IBM SPSS neural networks 19. http://www.sussex.ac.uk/its/pdfs/SPSS_Neural_Network_19.pdf. Accessed 10 Jan 2012

Katz A, Zaidel D, Elgrishi A (1975) An experimental study of driver and pedestrian interaction during the crossing conflict. Hum Factors 17(5):514–527

Li S, Qian D, Li N (2010) BP simulation model and sensitivity analysis of right-turn vehicles' crossing decisions at signalized intersection. J Transport Syst Eng Inform Technol 10(2):49–56

Madanat SM, Cassidy MJ, Wang M (1994) Probabilistic delay model at stop-controlled intersection. J Transport Eng 120(1):21–36

Mahmassani H, Sheffi Y (1981) Using gap sequences to estimate gap acceptance functions. Transport Res 15(B):143–148

Miller AJ (1971) Nine estimators of gap acceptance parameters. In: Proceeding of the 5th international symposium on the theory of traffic flow and transportation, pp 215–235

Taylor DB, Mahmassani HS (1998) Bicyclist and motorist gap acceptance behavior in mixed-traffic. In: Proceeding of the 78th annual meeting of the transportation research board, Washington

Troutbeck RJ, Brilon W (2002) Unsignalized intersection theory. In: Gartner NH, Messer CJ, Rathi A (eds) Traffic flow theory. Transportation Research Board, Washington

Varhelyi A (1998) Drivers' speed behavior at a zebra crossing: a case study. Accid Anal Prev 30(6):731–743

Zhang X, Ding Q, Ye H (2009) Influence of cyclist behavior on mixed traffic flow at signal intersection. In: International conference on management and service science (MASS '09), 20–22 Sept. 2009, Beijing, China

Decision Making Behavior of Earthquake Evacuees: An Application of Discrete Choice Models

Umma Tamima and Luc Chouinard

Abstract Destination choice modeling, after an earthquake, is challenging for the moderate seismic zones due to the shortage of evacuation data. Destination choice decisions are important for emergency planners to ensure the safety of evacuees, and to estimate the demand and specify the capacity of shelters. This study proposes a model for the behavior of evacuees in the aftermath of an earthquake using households as the unit of analysis. This study also considers heterogeneous mixtures of population in terms of income and ethnicity from different parts of the city. The Stated Preference (SP) method, using various hypothetical scenarios of shelter choice game in the event of a large earthquake, is applied to collect information on destination choices. Data were collected by e-mail back surveys, door-to-door surveys, and surveys in public places (e.g., at shopping malls, public parks, and student dormitories). This study proposes an error component model and a random coefficient model. The random coefficient models capture the heterogeneous responses of respondent while the error component model counts for correlations between destination choices. The results from the proposed disaggregate method are more comprehensive than those from the HAZUS method since it accounts for factors that impact decisions on destination choices.

1 Introduction

The importance of evacuation studies in the aftermath of a disaster is recognized by the natural hazards researchers (Tamima and Chouinard 2012; Hasan et al. 2010; Mesa-Arango et al. 2013). Disasters result in displaced households and the estimation for the demand for shelters is a part of disaster preparedness studies.

U. Tamima (✉) • L. Chouinard
Department of Civil Engineering and Applied Mechanics, McGill University, 817 Sherbrooke Street West, Montreal, Quebec H3A 2K6, Canada
e-mail: umma.tamima@mail.mcgill.ca; luc.chouinard@mcgill.ca

S. Kadry and A. El Hami (eds.), *Numerical Methods for Reliability and Safety Assessment: Multiscale and Multiphysics Systems*, DOI 10.1007/978-3-319-07167-1_28,
© Springer International Publishing Switzerland 2015

For example, the recent earthquake at Balochistan, Pakistan in 2013 that killed more than five hundred people, and thousands of people were displaced and have spent the night in the open (Jillani 2013). The prediction of the number of displaced households in the aftermath of an earthquake is challenging due to the uncertainties on the level of damage and on the decisions by evacuees. Decisions by evacuees are influenced by the behavior of households and should be formulated probabilistically (Ben-Akiva and Bierlaire 1999). Modeling the decision behavior of evacuees is challenging and the complexity of this task adds dimensionality of the problem for evacuation planning. Ignoring this component in evacuation planning may lead to inaccurate estimation of the demand for shelters. Destination choice modeling is a useful tool to understand which factors influence the decisions of evacuees.

The objective of this chapter is to model destination choices in the aftermath of large earthquakes for a moderate seismic zone. The destination choices include public shelters, staying home, and other locations such as: friends' home, hotels, and rental apartments.

2 Literature Review

Numerous researchers addressed the destination choice models for hurricane hazards (Mesa-Arango et al. 2013; Hasan et al. 2010; Dash and Gladwin 2007; Fu and Wilmot 2004, Lindell et al. 2005). The age of the decision maker, presence of children or elderly persons in the household, gender, disability, race and ethnicity, income, previous experience, household size, living in a single-family dwelling unit, risk perception, geographic characteristics and location (proximity to highways and exit routes) play important roles in the evacuation decision making process (Dash and Gladwin 2007; Lindell et al. 2005; Gladwin and Peacock 1997).

Several studies have defined factors affecting destination choices after earthquakes based on empirical data from past earthquakes (MCEER 2008; Chang et al. 2009; FEMA 2003; Harrald et al. 1992). FEMA (2003) developed the HAZUS method that is widely used worldwide for seismic risk analyses. The HAZUS methodology assumes that destination choices in the aftermath of an earthquake is a function of the degree of building damage, income, ethnicity, ownership, and age.

The weighting factors for these variables were developed from data of the Northridge earthquake and expert opinion.

The HAZUS methodology uses census tracts as the basic unit of analysis; and assumes that socioeconomic characteristics of census tracts are homogenous. To address the limitation of the method, Chang et al. (2009) adopted the household as the unit of analysis; and considered both building damage and socioeconomic characteristics. Households are assumed to go through a series of "yes/no" decisions, the outcomes of which are affected by various household and neighborhood-level variables. However, the results for two earthquakes vary significantly from the outcome predicted by the HAZUS. Moreover, the actual number of people who took shelters after Northridge earthquake was about three times (2.70) higher than the prediction from HAZUS.

In addition to these limitations, the application of the HAZUS method to settings other than the United States requires a calibration/validation of the parameters and weighting factors. One of the challenges associated with destination choice modeling for no-notice disasters is the shortage of data on evacuation, especially in moderate seismic region such as Montreal. This study applies the Stated Preference (SP) method for data collection which consists in collecting responses from participants to predefined hypothetical scenarios. The objective of this study is to develop mixture logit models for decisions with error components and random coefficients. The random coefficients describe heterogeneous responses of participants while the error components depict the correlation among destination choices.

3 Methodology

Destination choice modeling in the aftermath of disasters must include human response after a disaster which depends on preferences for destinations as a function of attributes, risk perception, and coping capacity of households. Understanding human response is used to forecast demand for shelters and can also be used by emergency planners to improve the environment of shelters. There are number of techniques available to understand human response and behavior. The most widely used technique is discrete choice analysis (DCA). DCA is applicable to destination choice modeling in which the decision making agent selects from a finite set of discrete alternatives (Ben-Akiva and Lerman 1985). The choice of individuals can be explained by the principle of utility maximization assuming the alternatives are mutually exclusive and collectively exhaustive. This study calibrates a utility function representing individual preferences to choose shelters in the aftermath of earthquake in Montreal. Walker (2001) mentions that Multinomial Logit (MNL) models are robust for formulating the choice probabilities which are widely used in choice model (McFadden 1978; Ben-Akiva and Lerman 1985; Ben-Akiva and Bierlaire 1999; Walker 2001; Koppelman and Bhat 2006).

The estimation of utility to choose shelters is a function of individual characteristics and shelter attributes. The utility function for shelter i and the individual j has the following form:

$$U_{it} = V_{it} + \varepsilon_{it} \tag{1}$$

V_{it} is a deterministic term and ε_{it} is a disturbance term modeled as a random variable with the extreme value Type-1 distribution. The deterministic term is a function of a set of k factors that affect decisions,

$$V_{it} = \beta_{i0} + \beta_{i1'} X_{it1} + \beta_{i2} X_{it2} + \cdots + \beta_{ik} X_{itk} \tag{2}$$

where, V_{it}—is the total utility provided by shelter i to individual t, β_{ij} is the parameter for factor j and shelter i and x_{ikj} is the individual (or shelter) characteristics k (age, income, travel time) for shelter i and individual t, and ε_{it} = the error term associated with the shelter i and individual t.

The logit model is applied for model estimation. The expression of the logit model for probability of choosing an option (shelter), A_{it} can be derived from the following equation:

$$P_{it} = P\left(A_{it}\right) = \frac{e^{V_{it}}}{\sum_{A_j \in C_n} e^{V_{jt}}} \tag{3}$$

Where, C_n is the choice set faced by individual t. However, one of the assumptions of the standard logit model is that parameters are fixed across observations. When this assumption does not hold, inconsistent estimates of parameters and outcome probabilities will result (Hasan et al. 2010). To address the heterogeneity and flexible correlation structure, the random parameters or mixed logit models are usually considered. To allow for parameter variations across households (represented by variations in β), a mixed model is defined (i.e., a model with a mixing distribution). Mixed logit probabilities are the integrals of standard logit probabilities over a density of parameters that can be expressed in the following form:

$$P_i = \int P_{it}\left(\underline{\beta}\right) f\left(\underline{\beta}/\underline{\theta}\right) d\underline{\beta} \tag{4}$$

Where the logit probability is evaluated with the vector of parameters $\underline{\beta}$ and $f\left(\underline{\beta}/\underline{\theta}\right)$ is the joint probability density function with $\underline{\theta}$ referring to a vector of parameters of that density function (i.e., mean and variance).

The mixed logit model for the probability of individual n for choosing a destination in the aftermath of an earthquake is,

$$P_{ni} = \int \frac{e^{V_i}}{\sum_{Aj \in C_n} e^{V_j}} f\left(\underline{\beta}/\underline{\theta}\right) d\underline{\beta} \tag{5}$$

The mixed logit probability is a weighted average of the logit formula evaluated at different values of $\underline{\beta}$, with the weights given by the joint density $f(\underline{\beta})$. The values of $\underline{\beta}$ have some interpretable meaning as representing the decision criteria of individual decision makers. In the statistics literature, the weighted average of several functions is called a mixed function, and the density that provides the weights is called the mixing distribution (Train 2009). This density is a function of parameters $\underline{\theta}$ that represent, for example, the mean and covariance of the β's in the population. This specification is the same as for standard logit except that $\underline{\beta}$ varies over decision makers rather than being fixed.

Another way of representing mixed logit models is the error component model that creates correlation among utilities for different alternatives. Utility is specified as;

$$U_{nj} = a'x_{nj} + \mu'Z_{nj} + \varepsilon_{nj} \tag{6}$$

Where, x_{nj} and Z_{nj} are vectors of observed variables relating to alternative j, $\underline{\alpha}$ is a vector of fixed coefficients, $\underline{\mu'}$ is a vector of random terms with zero mean, and are iid extreme value variables.

In this study, panel data is composed of ten observations per individual. These ten observations correspond to the choices made by a single respondent for ten hypothetical destination choice situations described in the questionnaire. The specification of the utility function treats coefficients varying over population but fixed across choices. The probability that a person makes a sequence of T choices is the product of logit formula for each element of the sequence:

$$P_i = \prod_{t=1}^{T} \frac{e^{vi}}{\sum_{Aj \in C_n} e^{vj}} \tag{7}$$

The Mixed logit is well suited to estimation methods based on simulation (Train 2009, McFadden and Train 2000). A Maximum Simulation-based Maximum-Likelihood Method is used to estimate the Mixed Logit Model. Several researchers (McFadden and Ruud 1994; Geweke et al. 1994; Böersch-Supan and Hajivassiliou 1993; Stern 1997; Brownstone and Train 1998; McFadden and Train 2000) offer the details of simulation-based maximum-likelihood methods for estimating mixed logit models (Hasan et al. 2010).

The Mixed logit probabilities are approximated through simulation for a given value of θ by drawing a value of $\underline{\beta}$ from $f\left(\underline{\beta}/\theta\right)$ and then the logit formula is calculated for this draw. These steps are repeated and results are averaged to obtain the simulated probability. The simulated probabilities are inserted into the log-likelihood function to give a simulated log-likelihood which is used to calculate the likelihood function.

4 Dataset

One of the challenges associated with destination choice modeling for non-data-rich regions is the calibration of the model. The Montreal region was affected by a major ice-storm in January 1998. During the ice-storm, several people took refuge in shelters or other suitable accommodations (with family and friends) but data on evacuation was not recorded in the aftermath of the ice-storm (Tamima and Chouinard 2012). Montreal is ranked second in Canada for seismic risks and

Variables	Choice game 1	Choice game 2	Choice game 3	Choice game 4	Choice game 5	Choice game 6	Choice game 7	Choice game 8
Building damage (home)	Moderate	Moderate	Extensive	Extensive	Extensive	Extensive	Extensive	Extensive
Power outage (home)	Less than 2 days	2-4 days	Less than 2 days	2-4 days	More than 4 days	Less than 2 days	2-4 days	More than 4 days
Distance from home to shelter	Less than 4 Km	More than 4 Km	Less than 4 Km	Less than 4 Km	Less than 4 Km	More than 4 Km	More than 4 Km	More than 4 Km
Road conditions (including bridges) after EQ in Montreal	Moderate damage	Moderate damage	Extensive damage	Extensive damage	Extensive damage	Extensive damage	Extensive damage	Extensive damage
Private space, space for pet, parking space, social network capacity e.g. TV, elevator (shelter)	No	No	No	No	No	No	No	No
Backup power (generator), food and clothes in shelter	Yes	Yes	Yes	Yes	Yes	Yes	Yes	Yes
Special arrangements (beds and diets) for aged and disabled in shelter	No	No	No	No	No	No	No	No
DECISION Which option would you prefer in this case?	○ I will go to a public shelter ○ I will stay home ○ Other	○ I will go to a public shelter ○ I will stay home ○ Other	○ I will go to a public shelter ○ I will stay home ○ Other	○ I will go to a public shelter ○ I will stay home ○ Other	○ I will go to a public shelter ○ I will stay home ○ Other	○ I will go to a public shelter ○ I will stay home ○ Other	○ I will go to a public shelter ○ I will stay home ○ Other	○ I will go to a public shelter ○ I will stay home ○ Other

Fig. 1 A sample of stated preference survey

a number of small earthquakes have occurred in the Montreal area but none was strong enough to cause any significant damage or evacuation. The scarcity of data on evacuation for this region prompted the need for performing a stated preferences survey. The idea is to obtain data on behavior by studying the choice process under hypothetical scenarios designed by the researcher (Ben-Akiva and Abou-Zeid 2013).

This study uses the Stated Preference (SP) method including various hypothetical scenarios of shelter choice game in the event of a large earthquake in Montreal. A household-based questionnaire is developed to determine the preferences to shelter decisions. It includes a heterogeneous mixture of population in terms of income, ethnicity, and location. The questionnaire survey was conducted during June–August, 2012 by e-mail, door-to-door surveys, and by surveys at shopping malls, public parks, etc. A typical sample of the questionnaire is shown in Fig. 1.

In the stated preference questionnaire, an experimental setting is defined which consists of the context of the hypothetical scenarios and the alternatives or profiles that make up these scenarios. The experimental factors are building damage, power outage, distance from home to shelter, road conditions, private space, space for pets, parking space, social networking availability, elevator, backup power (generator), food and clothes in the shelter, and special arrangements (beds and diets) for aged and disabled evacuees. Each factor has multiple levels and each level of each attribute is combined to construct profiles.

There are three destination options in the questionnaire: (1) public shelters (2) staying home and (3) others such as; friends' home, apartment rental, motel, hotel, religious centers such as mosque and church. The destination options are selected based on the literature review on the ice-storm in Montreal.

Table 1 Demographic structure of the respondents

Name of the variables	Percentage	Name of the variables	Percentage
Age of the respondents		*Income*	
Less than 20 years	3.7	Less than $24,999	40.2
20–30 years	35.9	25,000–49,999	24.9
30–40 years	34.6	25,000–49,999	13.0
40–50 years	10.6	75,000–99,999	10.6
50–60 years	8.6	100,000–149,999	6.0
60–70 years	5.3	More than 150,000	5.3
More than 70 years	1.3		
Ethnicity		*Home Type*	
Canadian	28.2	College Dorm	2.7
Asian	42.2	Apartment	54.2
Latin American/Hispanic	7.0	Condominium	11.0
European	4.0	Townhouse	10.6
Arab	7.0	Single-family home	18.3
Others	11.7	Others	3.3
Education		*No of kids*	
Less than high school	1.7	No kid	70.4
High school graduate	10.6	One	14.3
College	14.0	Two or more	15.3
Associate degree	3.0		
Bachelor degree	31.6		
Graduate/Professional degree	39.6		
Decision		*Dwelling type*	
Public Shelter	58.0	Rent	64.5
Staying home	24.0	Owned	35.5
Others	18.0		

Source: Field survey, 2012

The Centre de la sécurité civile de Montréal, has sixty-two shelters in Montréal. In summary, considering all the scenarios, the frequency distribution of the collected data shows that 58 % of the respondents would choose to go to a shelter following a major earthquake, 24 % would elect to stay home and the remainder (18 %) would opt for the other options (Table 1).

Demographic data on the respondents as well as on other variables that have influence on choice decision are presented in Table 1. The majority (70.5 %) of the respondents of this study were within the age interval of 20–40 years old (Table 1). Table 1 also shows that 40.2 % respondents have annual income of less than $24,999, while 37.9 and 21.9 % respondents have annual income within the range of $25,000–$49,999, and $50,000 and above, respectively. The ethnic composition of the respondents is: 28.2 % Canadian, 42.2 % Asian, 7 % Latin American/Hispanic, 4 % European, 7 % Arab, and 11.7 % from other origins.

Most of these respondents are living at apartments and single-family homes (72.5 %). Since most of the respondents are living at apartments and single-family homes, a majority of them (70.4 %) have no kids (Table 1). The evacuation decision model may be biased if the proportion of respondents in a particular group is not sufficient. However, in this case, the number of respondents per category was judged to be sufficient in that respect.

The questionnaire survey contains a large proportion of highly educated people (71.2 %), which may have a higher awareness relative to earthquake hazards, their impacts, and what-to-do during, before, and after an earthquake (Table 1).

5 Model Estimation

The estimates for the parameters of the MNL model and the two mixed logit models are presented in Table 2. In all cases, the estimation procedure is based on the maximum-likelihood approach which maximizes the probability of a chosen alternative.

The basic test for the adequacy of the model is the examination of the values and sign of the estimates. For example, the estimates of the coefficients for building damage, power outage, and road condition, have all positive signs which indicate that households are more likely to go to shelters with an increase in building damage, electricity outage, and deteriorated road conditions (Table 2). The t-value of the coefficients for age, distance from home to shelters, ethnicity, and number of kids, income, road conditions, family situation, building damage, and power outage are all significant at the 5 % significance level. The positive sign of the coefficients for age and households with/without children indicates the increasing utility to choose public shelters with age and increasing number of children (Table 2). These findings

Table 2 Multinomial logit model for destination choice

Variables	Estimated coefficients	Standard error	t-value	p-value
ASC_Home	4.17	0.28	14.89	0
ASC_Others	−1.08	0.15	−7.07	0
ASC_Shelter	0	Fixed		
B_Age	0.02	0.005	4.61	0
B_Distance from home to shelters	−0.48	0.086	−5.65	0
B_Latin	−0.52	0.204	−2.56	0.01
B_No of kids	0.35	0.121	2.9	0
B_High income	−0.48	0.235	−2.04	0.04
B_Road conditions	0.34	0.156	2.21	0.03
B_Single family	−0.39	0.138	−2.79	0.01
B_bldg damage	1.91	0.164	11.61	0
B_power outage	0.21	0.086	2.49	0.01
Null log-likelihood	−3,306.82			
Final log-likelihood	−2,373.17			
Adjusted ρ^2	0.279			

are consistent with the findings of previous research. Dash and Gladwin (2007) concluded that factors such as age of the decision maker, presence of children or elderly persons in the household, gender, disability, ethnicity, and income play important roles in evacuation decision making.

On the other hand, the probability to choose public shelters after an earthquake decreases for Latin households, households with high income and families who live in single-family housing (Table 2). Previous earthquake and hurricane studies reveal that Hispanic populations from Central America and Mexico tend to be more concerned about reoccupying buildings than other groups (FEMA 2003). This tendency appears to be because of the fear of collapsed buildings instilled from past disastrous Latin American earthquakes (FEMA 2003). However, the results obtained from the survey data show the opposite result for Montreal and may be related to the lack of previous experience of the Latin population relative to earthquakes. Households with high income are less likely to go to shelters since that may have more destination options since they can more easily pay for hotels, motels, or rental space.

The probability to choose shelter as an option also increases with the degree of building damage and the number of days with power outage during the winter season in Montreal. During winter season households without power are forced to evacuate irrespective of the level of building damage (Table 2).

The Mixed logit model was estimated to incorporate taste heterogeneity among households (Eq. (4)). Error component model was estimated to test whether there is a correlation among utilities for different alternatives (Eq. (5)).

In the Alternative Specific Variance Model, Alternative Specific Constants (ASCs) are randomly distributed. ASC home and ASC others are randomly distributed with mean α_{home} and α_{others} standard deviation σ_{home} and σ_{others} which are both estimated (Table 3). Staying home alternative is normalized because it has minimum variance. Staying home and shelters are preferred compared to others. The standard deviation of home option is less compared to mean which indicates less variability. It may not be necessary to consider ASC home to be a random variable. On the other hand, standard deviation of ASC_{others} is greater than the mean, which indicates a great variability. This suggests that it is a good idea to consider ASC_{others} to be a random variable instead of a constant.

In the Error Component model, the assumption is that public shelters and other options are correlated; both being safe, but also that Staying home and Shelters are correlated because of ease of access to living necessities. The random terms are assumed to be normally distributed

$$\varsigma_{safety} \sim N\left(m_{safety}, \sigma^2_{safety}\right)$$

$$\varsigma_{accessibility\ stuffs} \sim N\left(m_{accessibilty\ stuffs}, \sigma^2_{accessibility\ stuffs}\right)$$

where m is the mean and σ is the standard deviation. Standard deviation of both safety and accessibility are estimated. ASC_{others} is negative, indicating a preference

Table 3 Estimation of the alternative specific variance model and error component model

Variables	Alternative specific variance model			Error component model		
	Estimated coefficients	Standard error	t-test	Estimated coefficients	Standard error	t-test
ASC_Home	4.83	0.874	5.52	4.34	0.453	9.59
ASC_HomeStandard deviation	−0.96	0.742	−1.3			
ASC_Others	−5.14	1.9	−2.71	−4.07	1.12	−3.63
ASC_OthersStandard deviation	5.69	2.09	2.73			
ASC_Shelter	0	Fixed		0	Fixed	
B_Age	0.022	0.007	3.34	0.02	0.0051	3.89
B_DistanceHome to shelters	−1.17	0.297	−3.92	−0.99	0.177	−5.6
B_Latin	−0.56	0.266	−2.09	−0.49	0.222	−2.22
B_No of kids	0.51	0.185	2.78	0.441	0.138	3.2
B_High income	−0.63	0.318	−1.98	−0.55	0.261	−2.1
B_Road conditions	0.061	0.243	0.25	0.11	0.208	0.54
B_Single family	−0.57	0.208	−2.73	−0.49	0.158	−3.11
B_bldg damage	2.71	0.595	4.55	2.33	0.293	7.95
B_powerout	0.24	0.162	1.51	0.25	0.141	1.75
SAFETY				0	Fixed	
SAFETY_Standard deviation				−0.198	1.1	−0.18
Accessibility of living necessities				0	Fixed	
Accessibility_Standard deviation				4.5	1.22	3.7
No. of observations	3,010			3,010		
Null log-likelihood	−3,306.82			−3,306.823		
Final log-likelihood	−2,352.53			−2,354.92		
Adjusted ρ^2	0.285			0.284		

toward Home over Others option, all the rest being constant (Table 3). Standard deviation of accessibility of stuffs is significantly different than zero that means home and shelter share common unobserved attribute.

In the Random Coefficient Model, random parameters are assumed to be randomly distributed over the population to capture the taste variation of individual. In this specification, unknown parameters are normally distributed. The generic coefficient for Latin, rich and building damage variables, which are normally distributed with mean m_{latin}, m_{rich}, and $m_{building\ damage}$ and standard deviation σ_{latin}, σ_{rich}, and $\sigma_{building\ damage}$, respectively. The estimation results are shown in Table 4.

ASC_{others} shows negative sign means that all the rest, remaining constant, staying home is most preferred alternative compared to other destination options. The mean of Latin and Rich variables coefficients are negative, and the standard deviation of σ_{latin} and σ_{rich} are significantly different than zero. For the standard deviation value for Latin and Rich variables are higher than the mean value. It means that different individuals in the same ethnic group and same income categories perceive the alternatives' utilities differently.

During the field survey, each individual faces ten scenarios. In order to address the sequence of choices, mixture of logit with panel data was estimated where the coefficients vary over population but fixed across choices (Eq. (7)). Python Biogeme 3.2 software is used to estimate the parameters of mixture of logit model with panel data (Table 4). Individual specific error term is added, where the standard deviation is estimated while the mean is fixed to zero. The coefficient of SIGMA_PANEL is significant which means that this model allows for capturing intrinsic correlations among the observations of the same individual.

Under all circumstances, 24 % households want to stay home. One of the probable reasons is that most of the surveyed households do not have any experience to face disasters. This is why; they could not feel the severity of the disasters and decided to stay home. Riad et al. (1999) reported evacuation experience as the single best predictor of evacuation in Hurricanes Hugo and Andrew. A study on evacuation decision in Bangladesh during the super cyclone SIDR in 2007 depicts that evacuation decision of female households depends on the decision of male. Female households are not permitted to leave the home until the return of their husband, brother, or father (Khatun, personal communication, June 11, 2007) One of the possible reasons might be the threat of insecurity for the females. In some cases, female households are reluctant to go to the shelters due to the privacy issues (Khatun, personal communication, June 11, 2007). For Montrealers, privacy is one of the major factors that decrease the percentage to choose public shelters. A shelter is a construction built specially for providing relief during disasters or a multipurpose construction taken over for temporary use in an emergency, such as a school or a gymnasium (Li et al. 2008). In case of Montreal, this is not an exception. Public schools and gymnasiums are designated as shelters. During the field survey at Cote De Negies, an anonymous respondent stated "under any circumstances I will not go to the public shelter because I will be infected by the disease carried by others".

Table 4 Estimation of the random coefficient model and mixture of logit with panel data

Variables	Random coefficient model			Mixture of logit with panel data		
	Estimated coefficients	Standard error	t-test	Estimated coefficients	Standard error	t-test
ASC_Home	4.27	0.34	12.74	4.58	0.31	14.64
ASC_Others	−1.08	0.15	−7.02	−1.08	0.15	−7.08
ASC_Shelter	0	Fixed		0	Fixed	
B_Age	0.021	0.005	4.53	0.021	0.0045	4.76
B_Distance (Home to shelter)	−0.481	0.086	−5.57	−0.49	0.086	−5.66
B_Latin	−0.48	0.24	−2.04	−0.55	0.20	−2.69
B_LatinStandard deviation	−0.751	0.65	−1.16	–	–	–
B_No of kids	0.367	0.13	2.94	0.36	0.12	2.98
B_High income	−0.424	0.28	−1.53	−0.49	0.24	−2.09
B_High incomeStandard deviation	0.817	0.74	1.10	–	–	–
B_Road conditions	0.366	0.17	2.21	0.34	0.16	2.20
B_Single family	−0.398	0.14	−2.79	0.36	0.14	2.62
B_bldg damage	1.95	0.18	10.75	1.92	0.17	11.63
B_bldg damageStandard deviation	−0.121	0.23	−0.53	–	–	–
B_powerout	0.214	0.09	2.49	0.21	0.085	2.49
SIGMA_PANEL	–			−0.13	0.052	−2.52
ZERO				0	Fixed	
No. of estimated parameters	14			12		
No. of observations	3,010			3,010		
Null log-likelihood	−3,306.82			−3,306.82		
Final log-likelihood	−2,372.63			−2,369.99		
Adjusted $\rho 2$	0.28			0.28		

Although Mosques are not defined as public shelters in Montreal, few respondents choose Mosque as a probable option to take shelters. The reason is that religious women are reluctant to share common rooms with male. Usually, in the Mosque, there are separate rooms for male and female. For the rest of the people who prefers to go to the other options, choose relatives and friends' home outside the damaged area, hotels, and motels.

6 Discussion and Conclusion

This study develops a mixture of logit model for the household decision making on the destination choice in the aftermath of large earthquake. The SP method was used for data collection. Data were collected from field survey assuming the hypothetical earthquake scenarios for the City of Montreal. Three destination options were given as choices, such as: public shelters, staying home, and others (staying in hotels, motels, renting apartments, friend's or relative's home).

The MNL model identifies age, ethnicity, income, types of accommodation, number of kids, road condition, building damage, distance from home to shelter and power outage as the significant factors influencing destination choice of households. The MNL model depicts that age, number of kids, building damage, road condition, and the number of days without power in the home, have positive effects on shelter choice in the aftermath of earthquakes. On the other hand, ethnicity (e.g., Latin), high income, type of accommodation (e.g., single family) and distance from home to shelters have negative effect on the utility to choose shelters as a destination option. It is expected that probability to choose shelter as a destination option increases with age and number of kids. People with kids are more concerned about the after effect of earthquakes and they are risk averse. The study finds out that the probability to choose shelters increases with age. Young people are less likely to go to shelters as they perceive risk differently. Senior people are more likely to leave home because of the fear of trapping. Moreover, shelters can serve some basic needs for the evacuees that may attract seniors. Latin households are less likely to go to shelters which contradict with the previous findings of FEMA (2003). This study concludes that the lack of experience is the main reason of the less likelihood of the Latin households to choose shelters. Evacuation behavior is found to be related to previous experience to similar events (Hasan et al. 2010).

Households living in a single-family house are less likely to evacuate. Usually, the single-family houses are owned by the dwellers and they are concerned about the safety of the belongings. Another reason might be the alternative arrangement of the power and the confidence about the less likelihood of collapse of the whole building during earthquakes.

Error component model depicts that there is a correlation between home and shelter that shares common unobserved attribute. In the Random Coefficient Model, random parameters are assumed to be randomly distributed over the population to capture the taste variation of individual. The model concludes that different individuals in the same ethnic groups and same income categories perceive the alternatives' utilities differently. In order to address the sequence of choices, mixture of logit with panel data was estimated. This model captures the intrinsic correlations among the observations of the same individual.

The widely used destination choice method is HAZUS-MH that is designed to produce the loss estimates for earthquake risk mitigation, emergency preparedness, response, and recovery. The shelter model provides two estimates: the number of displaced households (due to the loss of habitability), and the number of people requiring only short-term shelter. The model is calibrated based on the earthquake data in the United States. The application of HAZUS method in different settings needs calibration or validation of the parameters. The parameter values are estimated based on the subjective judgment of the experts. This study develops the mixture of logit models to address the limitations of the traditional approach of the HAZUS. This study develops a database on the behavior of evacuees for a moderate seismic region such as the City of Montreal.

This study helps emergency planners to plan for the facilities in case of natural hazards. The discrete choice model identifies factors influencing evacuation decision of the city dwellers. The emergency planners can take the factors into consideration to improve the facilities that will eventually reduce the risk of affected people. This model is applicable in any other settings vulnerable to the natural hazards.

The future improvement of this work will include collecting more data on demographic characteristics and on other decision variables of evacuation. Incorporation of hypothetical scenarios for two seasons, summer and winter time can be another extension of the future work.

References

Ben-Akiva ME, Lerman SR (1985) Discrete choice analysis: theory and application to travel demand. MIT, Cambridge

Ben-Akiva M, Bierlaire M (1999) Discrete choice methods and their applications to short-term travel decisions. In: Hall RW (ed) Handbook of transportation science. Kluwer Academic, Norwell

Ben-Akiva M, Abou-Zeid M (2013) Stated preference (SP) methods-1, Discrete choice analysis: predicting demand and market shares. Massachusetts Institute of Technology, Cambridge

Böersch-Supan A, Hajivassiliou V (1993) Smooth unbiased multivariate probability simulators for maximum likelihood estimation of limited dependent variable models. J Economet 58(3): 347–368

Brownstone D, Train K (1998) Forecasting new product penetration with flexible substitution patterns. J Economet 89(1–2):109–129

Chang SE, Pasion C, Yavari S, Elwood K (2009) Social impacts of lifeline losses: modeling displaced populations and health care functionality. Proceedings of the 2009 ASCE technical council on lifeline earthquake engineering conference, Oakland

Dash N, Gladwin H (2007) Evacuation decision making and behavioral responses: individual and household. Nat Hazards Rev 8(3):69–77

FEMA (2003) Multi-hazard loss estimation methodology earthquake model HAZUS®MH MR4 technical manual. National Institute of Building Sciences, Washington

Fu H, Wilmot CG (2004) Sequential logit dynamic travel demand model for hurricane evacuation. Trans Res B 1882:19–26

Geweke J, Keane M, Runkle D (1994) Alternative computational approaches to inference in the multinomial probit model. Rev Econ Stat 76(4):609–632

Gladwin H, Peacock WG (1997) Warning and evacuation: a night for hard houses. In: Morrow BH, Gladwin H (eds) Hurricane Andrew: gender, ethnicity and the sociology of disasters. Routledge, New York, pp 52–74

Hasan S, Ukkusuri S, Gladwin H, Murray-Tuite P (2010) Behavioral model to understand household-level hurricane evacuation decision making. J Transp Eng 137(5):341–348

Harrald JR, Fouladi B, Al-Hajj SF (1992) Estimates of demand for mass care services in future earthquakes affecting the San Francisco Bay region. Prepared by George Washington University for the American Red Cross Northern California Earthquake Relief and Preparedness Project (NCERPP), 41pp. plus appendices

Jillani S (2013) Pakistan earthquake: hundreds dead in Balochistan, *BBC*. Retrieved from http://www.bbc.co.uk/news/world-asia-24222760

Khatun A (2007). Personal interview with Haque A. June 11

Koppelman FS, Bhat C (2006) A self-instructing course in mode choice modeling: multinomial and nested logit models, http://www.ce.utexas.edu/prof/bhat/courses/lm_draft_060131final-060630.pdf

Lindell MK, Lu JC, Prater CS (2005) Household decision making and evacuation in response to hurricane Lili. Nat Hazards Rev 6(4):171–179

Li X, Claramunt C, Kung H-T, Guo Z, Wu J (2008) A decentralized and continuity-based algorithm for delineating capacitated shelters' service areas. Environ Plann B 35(4):593–608

Mesa-Arango R, Samiul H, Ukkusuri SV, Murray-Tuite P (2013) Household-level model for hurricane evacuation destination type choice using hurricane ivan data. Nat Hazards Rev 14(1):11–20

McFadden D (1978) Modeling the choice of residential location. In: Karlqvist A, Lunqvist L, Snickars F, Weibull J (eds) Spatial interaction theory and residential location. North-Holland, Amsterdam

McFadden D, Ruud P (1994) Estimation by simulation. Rev Econ Stat 76(4):591–608

McFadden D, Train K (2000) Mixed MNL models for discrete response. J Appl Econ 15(5):447–470

Multidisciplinary Centre for Earthquake Engineering and Research (MCEER) (2008) Linking lifeline infrastructure performance and community disaster resilience: models and multi-stakeholder processes. Technical report MCEER-08-0004, MCEER, Buffalo

Riad JK, Norris FH, Ruback RB (1999) Predicting evacuation in two major disasters: risk perception, social influence, and access to resources. J Appl Soc Psychol 29(5):918–934

Stern S (1997) Simulation-based estimation. J Econ Lit 35(4):2006–2039

Tamima U, Chouinard LE (2012) Framework for earthquake evacuation planning. J Leadership Manage Eng, ASCE 12(4):222–230

Train EK (2009) Discrete choice methods with simulation, 2nd edn. Cambridge University, New York

Walker JL (2001) Extended discrete choice models: integrated framework, flexible error structures, and latent variables. Doctor of Philosophy, Massachusetts Institute of Technology, Cambridge

Preventive Maintenance and Replacement Scheduling in Multi-component Systems

Seyed Ahmad Ayatollahi, Mirmehdi Seyyed-Esfahani, and Taha-Hossein Hejazi

Abstract Maintenance and replacement schedule is one of the most important issues in industrial-production systems to ensure that the system is sufficient. This chapter presents a multi-objective model to schedule preventive maintenance activities for a series system of several standby subsystems where each component has an increasing rate of occurrence of failure (ROCOF). The planning horizon divided into the same length and discrete intervals that in each period three different maintenance actions such as maintenance, replacement, and do nothing can be performed. The objectives of this model are maximizing the system reliability and minimizing the total system cost. Because of nonlinear and complex structure of the mathematical model, non-dominated sorting genetic algorithm (NSGA-II) is used to solve this model. Finally, a numerical example is illustrated to show the model's effectiveness.

Notation

N	Number of subsystems
T	Length of planning horizon
J	Number of intervals
K	Number of maintenance levels
C	Number of components in each subsystem
λ	Characteristic life (scale) parameter of component c of subsystem i
β_i^c	Shape parameter of component c of subsystem i
α_i^k	Improvement factor of subsystem i in maintenance level k

S.A. Ayatollahi • M. Seyyed-Esfahani • T.-H. Hejazi (✉)
Department of Industrial Engineering and Management Systems, Amirkabir University of Technology (Tehran Polytechnic), Tehran, Iran
e-mail: t.h.hejazi@aut.ac.ir

S. Kadry and A. El Hami (eds.), *Numerical Methods for Reliability and Safety Assessment: Multiscale and Multiphysics Systems*, DOI 10.1007/978-3-319-07167-1_29, © Springer International Publishing Switzerland 2015

$\vartheta_i^c(t)$	ROCOF of component c of subsystem i
F_i	Unexpected failure of subsystem i
$M_i^{c,k}$	Level kth maintenance cost of component c of subsystem i
R_i^c	Replacement cost of component c of subsystem i
S_i	Switching cost in subsystem i
Z	System outage cost
Cost	Total system cost
$E(N_{i,j})$	Number of expected failures in subsystem i in period j
Reliability$_{i,j}^c$	Reliability of component c of subsystem i in period j
Reliability$_{i,j}^{SS}$	Reliability of subsystem i in period j
Reliability	Total system reliability
$Re_{i,j}^{SS,c}$	Reliability of subsystem i if component c be loaded at the start of period j
$Re_{i,j}^c(t)$	Operation probability of component c of subsystem i in interval $[0,t]$ of period j
$Q_{i,j}^c(t)$	Failure probability of component c of subsystem i in interval $[0,t]$ of period j

1 Introduction

Preventive maintenance and replacement is a schedule of planned maintenance and replacement activities in order to prevent system failures. The main objective of preventive maintenance and replacement is to prevent failure occurrences earlier than in reality. This concept says that by replacing old components, the reliability could be kept or improved.

Several misconceptions exist about preventive maintenance and replacement. One of them is that preventive maintenance and replacement is very expensive. This logic shows that planned maintenances are more expensive than the time a component works till its failure and by a corrective maintenance become repaired. It may be true for some components but just costs shouldn't be compared, also long-term benefits and savings should be considered. For example, without preventive maintenance and replacement, by unplanned failure occurrences, some costs like lost production cost will impose to the system. Also, by increasing in system service effective age, some savings will be brought (http reliawiki com Preventive_Maintenance).

Long-term benefits of preventive maintenance include:

1. Improved system reliability.
2. Decreased cost of replacement.
3. Decreased system downtime.
4. Better spares inventory management.

Long-term costs and benefits comparison usually shows preventive maintenance and replacement superiority.

One of the fundamental questions is that, when preventive maintenance and replacement is effective? The answer is preventive maintenance is a logical choice if, and only if, the following two conditions are met (http reliawiki com Preventive_Maintenance):

1. The component in question has an increasing failure rate. In other words, the failure rate of the component increases with time, implying wear-out. Preventive maintenance of a component that is assumed to have an exponential distribution (which implies a constant failure rate) does not make sense
2. The overall cost of the preventive maintenance action must be less than the overall cost of a corrective action

Modern world has realized the importance of preventive maintenance. So all system types, including conveyers, vehicles, and overhead cranes, have predefined maintenance and replacement schedules to reduce system failure risk. Preventive maintenance and replacement activities usually include inspection, cleaning, lubrication, adjustment, worn components replacement, etc. In addition, in preventive maintenance and replacement, labors can record equipment failures and maintain or replace old components before their failure. An ideal maintenance plan prevents all equipment failures earlier than their occurrence in reality. Regardless of specific systems, preventive maintenance and replacement activities could be divided into two categories: component maintenance and its replacement (Usher et al. 1998).

A simple example of component maintenance is air pressure controlling in a car tires in desirable limits. It's notable that this task changes the tire age characteristics and if occurs correctly will reduce the failure rate. Replacing a tire with new one is a simple replacement example.

Standby systems are widely used in different industries. For example considering a spare tire for cars, in fact is using a standby system to start up the car after a tire failure. Using standby systems reduce system total costs and cause improvement in reliability level. Outage in steel factories imposes enormous costs to the system for its restart. So, for different parts of this factory, standby systems will be considered. As mentioned before this policy causes earlier system restart after its outage and prevents imposed costs.

In real world, factories consider multi-level maintenances for each system. It's obvious that different maintenances cause different changes in age characteristics. For example, primary maintenances can be performed on equipment that don't change its effective age so much. But by equipment overhaul, it will improve deeply and will change to a state like a new equipment.

It is known that preventive maintenance and replacement includes a trade off between costs of conducting maintenance activities and saving costs resulted from the reduction in the overall rate of system failure. Preventive maintenance and replacement scheduling designers must measure these costs separately to minimize total system operating costs. They may like to improve system reliability to highest level and maximize it according to budget constraint.

In this chapter the problem is finding the best sequence of maintenance and replacement activities for each component of the considered system in each period of a specified planning horizon, in order to minimize total cost and maximize system reliability.

Optimization problems, in terms of objective functions and optimization criteria, are dividable into two categories: single objective function problems and multi-objective optimization problems. In single objective optimization problems, a unique performance index improves, so that its minimum or maximum value shows the obtained responses quality completely. But in multi-objective models more than one objective should be optimized. In other words, in this type of problems, several objective functions or operating indexes must be defined and optimized simultaneously.

Multi-objective optimization is one of the known research fields among optimization concepts. Usually, multi-objective optimization known as multicriteria optimization and vector optimization. So far, several methods have been introduced for solving multi-objective optimization problems.

NSGA-II algorithm is one of the most popular and powerful algorithms in solving multi-objective optimization problems and its performance is proofed in solving different problems.

2 Problem Definition

Consider a system consists of N subsystems that are connected to each other in series. Each subsystem is a standby system with two or more components. It means that if the loaded component fails, one of the other standby components become active and the system continues its working. Figure 1 shows the assumed system.

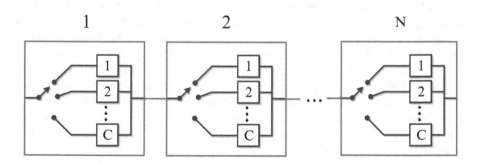

Fig. 1 Schematic view of the assumed system

The main purpose of this system modeling is finding a set of maintenance and replacement schedules for the components of each subsystem during the planning horizon to minimize the total system cost and maximize its total reliability.

For this end, it is assumed that component c of subsystem i has an increasing rate of occurrence of failure (ROCOF) $\vartheta_i^c(t)$, in which $t \cdot (t > 0)$ shows actual time. It is assumed that component failures follow well-known Non-Homogenous Poisson Process (NHPP) with ROCOF given as:

$$\vartheta_i^c(t) = \lambda_i^c \cdot \beta_i^c \cdot t^{\beta_i^c - 1} \text{ for } i = 1, \dots, N \text{ and } c = 1, \dots, C \quad (1)$$

Where λ_i^c and β_i^c are the characteristic life (scale) and the shape parameters of component i, respectively. It's notable that NHPP is similar to HPP with this difference that NHPP is a function of time.

As mentioned above, we seek to find a schedule for future maintenance and replacement activities on each component in interval $[0, T]$. For this purpose, the planning horizon is divided into J separated periods that each period length is equal to T/J. Similar to John Usher assumption, at the end of period j, one of these activities can be performed on each component: do nothing, maintenance, and replacement (Usher et al. 1998). These actions will review deeply as follows. One of the main assumptions of this model is that performing maintenance or replacement activities reduce the age of the components effectively. So that the ROCOF of the component will decrease. For simplicity, it is assumed that these activities perform instantaneously. In other words, the required time for maintenance or replacement is negligible compared to the planning horizon. Although this time is zero, but a cost associated to the maintenance and replacement have been considered. As mentioned above, one of these activities can be performed on each component at the end of a period:

1. Do nothing: In this policy, nothing performs on the component and the component age stays on the state "bad as old." So the component continues its working without any changes.
2. Maintenance: By maintaining a component, its age changes into a state between "bad as old" and "good as new." In this model, maintenance action reduces the effective age as a percentage of total lifetime. It's obvious that by a reduction in a component effective age, the component ROCOF decreases. An important point is that, the percentage of age reduction depends on the maintenance performance. So, multi-level maintenance with different improvement factors can be considered.
3. Replacement: In this case, the component replaces with a new one. So the component effective age drops to zero and the component state will be "good as new."

3 Effective Age of a Component at the Start of Each Period

3.1 Maintenance

For simplicity, an assumption is considered in which maintenance and replacement actions perform at the end of period j instantaneously. As mentioned above, according to the maintenance type, the age reduction differs. For example, three maintenance levels are defined: primary, intermediate, and moderate maintenance. When a primary maintenance performs, less age reduction occurs with less cost. But by overhauling a component, that component improves effectively with more cost. So:

$$
\begin{aligned}
&X^c_{i,j+1} = a^k_i \cdot X'^c_{i,j} \\
&\text{for } i = 1, \ldots, N; \quad j = 1, \ldots, T-1; k = 1, \ldots, K; \\
&c = 1, \ldots, C \text{ and } \left(0 \le a^k_i \le 1\right)
\end{aligned}
\tag{2}
$$

To consider instantaneous changes in system age and its failure rate, $X^c_{i,j}$ is defined as the effective age of component c of subsystem i at the start of period j and $X'^c_{i,j}$ as the effective age of component c of subsystem i at the end of period j. a^k_i displays an improvement factor. Actually, this factor shows the effectiveness of the maintenance action. When $a^k_i = 0$, the effect of the maintenance action is similar to replacement and the effective age becomes zero. But in the state $a^k_i = 1$, the maintenance effect is similar to do nothing policy with no changes in the component age. It means that, increasing the improvement factor shows the maintenance weaknesses.

As is evident in Fig. 2, maintenance action at the end of period j, creates an instantaneous drop in ROCOF of component c. So by performing a maintenance action on component c of subsystem i at the end of period j, its ROCOF will change from $\vartheta^c_i(X'^c_{i,j})$ to $\vartheta^c_i(X^c_{i,j+1})$.

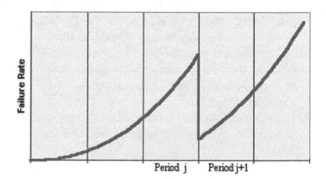

Fig. 2 Effect of period-j maintenance on component ROCOF (Usher et al. 1998)

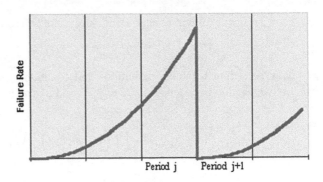

Fig. 3 Effect of period-j replacement on component ROCOF (Usher et al. 1998)

Period j   Period j+1

3.2 Replacement

If component c of subsystem i is replaced with a new one at the end of period j, its effective age at the start of period $j + 1$ will be:

$$X_{i,j+1}^c = 0$$
$$\text{for } i = 1, \ldots, N; j = 1, \ldots, T - 1; c = 1, \ldots, C \tag{3}$$

In other words, system will return to a state "good as new," in which the component effective age becomes zero like a new component. So the ROCOF of this component drops from $\vartheta_i^c(X_{i,j}'^c)$ to $\vartheta_i^c(0)$. Figure 3 shows the replacement effect on the component failure rate.

3.3 Do Nothing

If no action performs on a component in period j, that component continues its working without any changes in its effective age and ROCOF. So:

$$X_{i,j+1}^c = X_{ij}'^c$$
$$\text{for } i = 1, \ldots, N; j = 1, \ldots, T - 1; c = 1, \ldots, C \tag{4}$$

$$\vartheta_i^c\left(X_{i,j+1}^c\right) = \vartheta_i^c\left(X_{i,j}'^c\right) \quad \text{for } i = 1, \ldots, N; \ c = 1, \ldots, C \tag{5}$$

In order to construct a recursive function, which calculate $X_{i,j}^c$ according to any of above policies, two binary decision variables $m_{i,j}^{c,k}$ and $r_{i,j}^c$ are defined. These two variables represent the maintenance and replacement states on component c of subsystem i at the end of period j.

$$m_{i,j}^{c,k} = 1 \text{ if component } c \text{ of subsystem } i \text{ maintained in } k\text{th}$$
$$\text{level at the end of period } j; 0 \text{ otherwise} \tag{6}$$

$$r_{i,j}^c = 1 \text{ if component } c \text{ of subsystem } i \text{ replace at the end of}$$

$$\text{period } j; 0 \text{ otherwise} \tag{7}$$

Now according to above definitions and equations from Eq. (2) to Eq. (4), a recursive function between $X_{i,j}^c, X_{i,j}'^c, m_{i,j}^{c,k}, r_{i,j}^c$ and α_i^k can be rewritten as below:

$$X_{i,j}^c = \left(1 - r_{i,j-1}^c\right) \left[\prod_{k=1}^{K} \left(1 - m_{i,j-1}^{c,k}\right)\right] X_{i,j-1}'^c$$

$$+ \sum_{k=1}^{K} m_{i,j-1}^{c,k} \cdot \left(a_i^k \cdot X_{i,j-1}'^c\right)$$

$$\text{for } i = 1, \ldots, N; j = 2, \ldots, T \text{ and } c = 1, \ldots, C \tag{8}$$

Equation (8) presents a closed form to calculate the effective age of component c of subsystem i at the end of period j, according to maintenance and replacement actions are performed in previous period. In this recursive function if a component is replaced in the previous period, then $r_{i,j-1}^c = 1$ and $m_{i,j-1}^{c,k} = 0$, where the result will be $X_{i,j}^c = 0$. But if a component is maintained in one of maintenance levels, then $r_{i,j-1}^c = 0$ and $m_{i,j-1}^{c,k}$ for that maintenance level become one where $X_{i,j}^c = \alpha_i^k \cdot X_{i,j-1}'^c$ and the improvement factor of that maintenance will be used. Finally, if certain operation doesn't perform and the component continues its working and the equation becomes $X_{i,j}^c = X_{i,j-1}'^c$.

There is a basic assumption that the system starts its working from a completely new state. So the primary lifetime for each component at the start of first period is 0. It is clear that this assumption could be changed according to real system characteristics.

$$X_{i,1}^c = 0 \text{ for } i = 1, \ldots N \text{ and } c = 1, \ldots, C \tag{9}$$

4 Costs Related to Maintenance and Replacement Activities

In this section the costs that are necessary to consider in this model will be analyzed.

4.1 Maintenance Cost

It's obvious that performing maintenance on a component imposes a cost to the system. So when a maintenance action performs in level k on component c of susbsystem i in a period, the constant maintenance cost $M_i^{c,k}$ will add to the total system cost at the end of that period.

$$M_{i,j} = \sum_{c=1}^{C} \sum_{k=1}^{K} M_i^{c,k} \cdot m_{i,j}^{c,k} \tag{10}$$

for $i = 1, \ldots, N$ and $j = 1, \ldots, T$

4.2 Replacement Cost

Similar to the maintenance cost, when a component is replaced in period j, a constant replacement cost, R_i^c, that is equal to initial purchase price will add to total cost.

$$R_{i,j} = \sum_{c=1}^{c} R_i^c \cdot r_{i,j}^c \quad \text{for } i = 1, \ldots, N \quad \text{and} \quad j = 1, \ldots, T \tag{11}$$

4.3 Fixed Cost

Imagine a state in which all components of a subsystem is maintained or replaced at the end of a period. It's evident that in this state the total system stops working. Since the considered system is a series system of N subsystems, by failing a subsystem the total system will stop working. In this order the system will contact to a cost related to next system setup. In this purpose, fixed cost Z has been considered in a period when all components of a subsystem is maintained or replaced.

$$\text{Fixed cost} = \sum_{j=1}^{T} \left[Z \left(1 - \prod_{i=1}^{N} \left(1 - \prod_{c=1}^{c} \left(r_{i,j}^c + \sum_{k=1}^{K} m_{i,j}^{c,k} \right) \right) \right) \right] \tag{12}$$

As is obvious in Eq. (12), when all components of a subsystem are maintained or replaced in a period, the expression $\prod_{c=1}^{c} (r_{i,j}^c + \sum_{k=1}^{K} m_{i,j}^{c,k})$ becomes equal to one and the multiply operation on subsystems becomes 0. Then the fixed cost Z will add to total cost.

Considering this cost has two benefits. First, this cost prevents system stops. Second, when a system stops working at the end of a period, other subsystem's components can be maintained or replaced also without any changes in fixed cost. Actually, considering this cost helps centralization in maintenance and replacement activities.

So far it has been found that this model is a mixed integer nonlinear programming model (MINLP). For solving this model using meta-heuristic algorithms is undeniable. So because of wide using of genetic algorithm in preventive maintenance and replacement scheduling, an edition of this algorithm is used which is suitable for multi-objective models. This algorithm is non-dominated sorting genetic algorithm (NSGA-II). This algorithm will be checked in the following.

Up to now the problem is defined, the system is analyzed and primary assumptions are set. Now for system modeling three different system approaches for standby systems with different fundamental assumptions are considered.

- Preventive maintenance and replacement scheduling with failure impossibility assumption
- Preventive maintenance and replacement scheduling with failure possibility assumption

  Non-optional switching
  Optional switching

5 System Modeling

In this section the system that is defined in the second section and is shown in Fig. 1, will be modeled according to above assumptions.

5.1 Preventive Maintenance and Replacement Scheduling with Failure Impossibility Assumption

In this model there is no possibility for components failure during the periods. It means that when a component is loaded at the start of a period must continue its working to the end of that period without any failure. In Fig. 4, the system function is illustrated. As shown in this figure the switching operations perform at the end of the periods not during them. Actually in this approach switching operations perform to maintain or replace the components that are not under load.

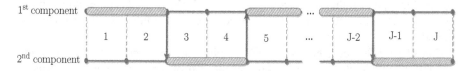

Fig. 4 Schematic view of a subsystem operation in preventive maintenance and replacement scheduling with failure impossibility assumption

5.1.1 System Configuration

It's obvious that if a component is under load in a period, its effective age will increase in amount of T/J but if the component wouldn't be under load in a period, no changes will happen to its age, because each subsystem is cold standby. In the other words, in each period if a component be under load its age increases otherwise no changes will happen to the component age on that period. So for calculating the age of components a binary variable $l^c_{i,j}$ is defined which will be equal to one if component c of subsystem i is under load at the start of period j. So:

$$X'^c_{i,j} = X^c_{i,j} + l^c_{i,j} \cdot T/J$$

$$\text{for } i = 1, \ \ldots \ , N; \quad j = 1, \ \ldots \ , T; \quad c = 1, \ \ldots \ , C \qquad (13)$$

Another important point is that, in each period and in each subsystem, just one component must be under load. Therefore:

$$\sum_{c=1}^{C} l^c_{i,j} = 1 \text{ for } i = 1, \ \ldots \ , N \text{ and } j = 1, \ \ldots \ , T \qquad (14)$$

For simplicity it is assumed that at the start of first period, first component of each subsystem is under load.

$$l^1_{i,1} = 1 \text{ for } i = 1, \ \ldots \ , N \qquad (15)$$

5.1.2 System Costs

Failure Cost

Unplanned failures impose a cost to the system. An important point is that at the start of first period, failures occurrence time is unknown. However, it's known that if the component has a high ROCOF in a period, the probability of failure occurrence is more, so higher cost must be considered and vice versa when ROCOF in a period is low, less failure cost will yield. Since in each period just one component of each subsystem is under load and switching operation can be done just at the end of period, so failure of loaded component causes system failure. For this reason, the expected number of failures in each period for each subsystem will be calculated. In this chapter J.S. Usher et al. methodology is used where average of failure rate with a fixed cost has been used (Usher et al. 1998). The expected number of failures in component c of subsystem i in period j can be calculated as below:

$$E\left(N_{i,j}\right) = \sum_{c=1}^{C} l^c_{i,j} \cdot \int_{X^c_{i,j}}^{X'^c_{i,j}} \vartheta^c_i(t)dt$$

$$\text{for } i = 1, \ \ldots \ , N; j = 1, \ \ldots \ , T \qquad (16)$$

According to Sect. 2 and by the assumption NHPP for rate of occurrence of failure, expected number of failures in component c of subsystem i in period j will be:

$$E\left(N_{i,j}\right) = \sum_{c=1}^{C} l_{i,j}^{c} \cdot \int_{X_{i,j}^{c}}^{X_{i,j}^{\prime c}} \lambda_i^c \cdot \beta_i^c \cdot t^{\beta_i^c - 1} dt$$

$$= \sum_{c=1}^{C} l_{i,j}^{c} \cdot \left[\lambda_i^c \left(X_{i,j}^{\prime c} \right)^{\beta_i^c} - \lambda_i^c \left(X_{i,j}^{c} \right)^{\beta_i^c} \right]$$

$$\text{for } i = 1, \ldots, N; \quad j = 1, \ldots, T \qquad (17)$$

It is assumed that cost of each failure is F_i (a dollar per failure occurrence) which allows the mode to calculate $F_{i,j}$, where $F_{i,j}$ is failure cost of subsystem i in period j.

$$F_{i,j} = F_i \cdot \sum_{c=1}^{C} l_{i,j}^{c} \cdot \left[\lambda_i^c \left(\left(X_{i,j}^{\prime c} \right)^{\beta_i^c} - \left(X_{i,j}^{c} \right)^{\beta_i^c} \right) \right]$$

$$\text{for } i = 1, \ldots, N; j = 1, \ldots, T \qquad (18)$$

So regardless of maintenance or replacement action (that is assumed to occur at the end of period) in period j, there is a cost related to failures which may occur in each period.

Switching Cost

If a switching action occurs at the end of period j in subsystem i, then a cost equal to S_i would be imposed to the system. Considering this cost is essential to avoid unnecessary switching. Switching operation can be recognized from the changes in binary variable $l_{i,j}^{c}$. When $l_{i,j}^{c}$ value changes from one to zero or from zero to one means that a switching operation has occurred. So the expression $|l_{i,j+1}^{c} - l_{i,j}^{c}|$ will be equal to one if a switching operation occurs. This expression will be equal to one for two components, one the component that becomes loaded and the component that removes from loading state. So using $1/2 S_i$ is inevitable.

$$S_{i,j} = \frac{1}{2} S_i \cdot \left| l_{i,j+1}^{c} - l_{i,j}^{c} \right| \quad \text{for } i = 1, \ldots, N; j = 1, \ldots, T \qquad (19)$$

Total Cost

At the start of period $j = 1$, a set of maintenance, replacement, and do nothing actions for all components of each subsystem in each period must be specified to minimize the total cost. According to different costs definition, the total cost can be written as a simple sum on the costs.

$$
\text{Total Cost} = \sum_{i=1}^{N} \sum_{j=1}^{T} \left[F_i \cdot \sum_{c=1}^{C} l_{i,j}^{c} \cdot \left[\lambda_i^{c} \left(\left(X_{i,j}^{\prime c} \right)^{\beta_i^c} - \left(X_{i,j}^{c} \right)^{\beta_i^c} \right) \right] \right.
$$
$$
\left. + \sum_{c=1}^{C} \left(\sum_{k=1}^{K} \left(M_i^{c,k} \cdot m_{i,j}^{c,k} \right) + R_i^{c} \cdot r_{i,j}^{c} + \frac{1}{2} S_i \cdot \left| l_{i,j+1}^{c} - l_{i,j}^{c} \right| \right) \right]
$$
$$
+ \sum_{j=1}^{T} \left[Z \left(1 - \prod_{i=1}^{N} \left(1 - \prod_{c=1}^{C} \left(r_{i,j}^{c} + \sum_{k=1}^{K} m_{i,j}^{c,k} \right) \right) \right) \right]
$$

$$(20)$$

5.1.3 System Reliability

For considering the reliability objective function in this model, the reliability function of subsystem i in period j is defined as Eq. (20) and since the subsystems are series, total reliability could be calculated by multiplying all subsystems in all periods as shown in Eq. (21).

$$
R_{i,j} = e^{-\left[\sum_{c=1}^{C} l_{i,j}^{c} \cdot \int_{X_{i,j}^{c}}^{X_{i,j}^{\prime c}} \vartheta_i^{c}(t) dt \right]} = e^{-\left[\sum_{c=1}^{C} l_{i,j}^{c} \cdot \left[\lambda_i^{c} \left(\left(X_{i,j}^{\prime c} \right)^{\beta_i^c} - \left(X_{i,j}^{c} \right)^{\beta_i^c} \right) \right] \right]}
$$

for $i = 1, \ \ldots \ N; j = 1, \ \ldots \ , T$

$$(21)$$

$$
\text{Reliability} = \prod_{i=1}^{N} \prod_{j=1}^{T} e^{-\left[\sum_{c=1}^{C} l_{i,j}^{c} \cdot \left[\lambda_i^{c} \left(\left(X_{i,j}^{\prime c} \right)^{\beta_i^c} - \left(X_{i,j}^{c} \right)^{\beta_i^c} \right) \right] \right]} \qquad (22)
$$

5.2 Preventive Maintenance and Replacement Scheduling with Failure Possibility Assumption

In this model two approaches can be supposed. In first approach, optional switching is not applicable and by a failure in the loaded component, switching operation performs. In second approach, optional switching is considered. It means that two kinds of switching are conceivable, non-optional switching and optional one according to failures and maintenance-replacement plan, respectively. These two approaches will be surveyed in details.

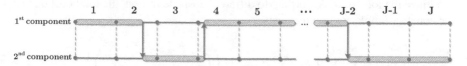

Fig. 5 Schematic view of a subsystem operation in preventive maintenance and replacement scheduling with failure possibility assumption and non-optional switching

5.2.1 Non-optional Switching

In this approach, when the loaded component breaks down, the switching operation performs and the other component of subsystem has to be loaded. In other words, the component that is under load at the start of a period wouldn't be under load at the end of that period necessarily. It is obvious that the time of failure occurrence is unknown. It means that we don't know when a failure occurs so this model deals to a probability which is related to the component failure time. For simplicity just two components are considered for each subsystem as shown in Fig. 5, the switching operations are performed just when a failure occurs. It's obvious that these failures can occur during a period and its time is unknown.

Components Effective Age

Because of this probability approach, the expected effective age of the components must be calculated. It means that in each period a fraction of each period length, T/J, will add to the component effective age. For controlling the model parameters and proper modeling, a variable is defined, $p_{i,j}^c$, which indicates the probability of component c of subsystem i loading status at the start of period j. It's clear that sum of these probabilities in each subsystem must be equal to one.

$$\sum_{i=1}^{n} p_{i,j}^c = 1 \quad \text{for} \quad i = 1, \ \ldots \ , N \quad \text{and} \quad j = 1, \ \ldots \ , T \qquad (23)$$

By considering two components for each subsystem, Eq. (23) could be rewritten as below:

$$p_{i,j}^1 + p_{i,j}^2 = 1 \quad \text{for} \quad i = 1, \ \ldots \ , N \quad \text{and} \quad j = 2, \ \ldots \ , T \qquad (24)$$

An assumption is considered that at the start of first period, the first component of each subsystem is under load. So:

$$p_{i,1}^1 = 1 \quad \text{for} \quad i = 1, \ \ldots \ , N \qquad (25)$$

$$p_{i,1}^2 = 0 \quad \text{for} \quad i = 1, \ \ldots \ , N \qquad (26)$$

Now, according to the above definitions, the expected effective age of each component at the end of each period is computable. By considering two components for each subsystem, the first component will be under load at the end of a period in two states. These two states are:

1. First component operates from start to end of a period without any failure.
2. Second component is loaded at the start of a period, but by a failure in that component the related subsystem switches to first component.

It's evident that reliability of a component shows the probability of its operation without any failure during a period and the complementary of that reliability expresses the failure probability of that component in a period. Now, the expected effective age of that component at the end of period j can be rewritten as a sum of expected effective age of that component at the start of period j and during that period. So the expected effective age of first component of each subsystem can be formulated as below:

$$X_{i,j}'^1 = X_{i,j}^1 + \left[p_{i,j}^1 \cdot \text{Reliabilty}_{i,j}^1 + p_{i,j}^2 \cdot \left(1 - \text{Reliabilty}_{i,j}^2 \right) \right] \cdot T/J$$

$$\text{for } i = 1, \ldots, N \text{ and } j = 1, \ldots, T \qquad (27)$$

Similar to the first component two states of computing the expected effective age for second component of each subsystem are:

1. Second component operates from start to end of a period without any failure.
2. First component is loaded at the start of a period, but by a failure in that component the related subsystem switches to second component.

According to above states the expected effective of the second component of each subsystem can be rewritten as below:

$$X_{i,j}'^2 = X_{i,j}^2 + \left[p_{i,j}^2 \cdot \text{Reliabilty}_{i,j}^2 + p_{i,j}^1 \cdot \left(1 - \text{Reliabilty}_{i,j}^1 \right) \right] \cdot T/J$$

$$\text{for } i = 1, \ldots, N \text{ and } j = 1, \ldots, T \qquad (28)$$

It should be noted that just one failure is considered for each subsystem in a period. It's evident that more than one failure in a period in each subsystem causes subsystem and whole system failure finally because two components is considered in each subsystem and maintenance and replacement activities occur at the end of periods.

One of the parameters that should be updated in each period is $p_{i,j}^c$. It's obvious that in the subsystems with two components, in two states a component will be under load at the start of period $j + 1$, which are:

1. The component is under load at the start of period j and stays loaded during that period without any failure.
2. The other component is under load at the start of period j, but by a failure, a switching operation performs.

According to expressed points, the probability of first component loading status at the end of period $j + 1$ will be:

$$p^1_{i,j+1} = p^1_{i,j} \cdot \text{Reliabilty}^1_{i,j} + p^2_{i,j} \cdot \left(1 - \text{Reliabilty}^2_{i,j}\right)$$

$$\text{for } i = 1, \ldots, N \text{ and } j = 1, \ldots, T - 1 \qquad (29)$$

It is clear that because of being two components in each subsystem, and according to Eq. (24), $p^2_{i,j+1}$ is complementary of $p^1_{i,j+1}$. It means that:

$$p^2_{i,j+1} = 1 - p^1_{i,j+1}$$

$$\text{for } i = 1, \ldots, N \text{ and } j = 1, \ldots, T - 1 \qquad (30)$$

System Costs

Failure Cost

By looking at the foregoing operation periods, the unplanned components failures should be considered. As mentioned before two components are considered for each subsystem in which by a component failure the other one will be loaded. Since the maintenance and replacement operations perform just at the end of periods, if both components of a subsystem fail in a period, the related subsystem and the whole system will fail.

Two approaches can be considered for failure cost. First is calculating the expected failure cost. For this purpose, the subsystem failure probability that is complementary of subsystem reliability is multiplied to constant failure cost F_i.

$$F_{i,j} = F_i \cdot \left(1 - \text{Reliability}^{SS}_{i,j}\right)$$

$$\text{for } i = 1, \ldots, N \text{ and } j = 1, \ldots, T \qquad (31)$$

In the above equation, $\text{Reliability}^{SS}_{i,j}$ shows the reliability of subsystem i in period j. So, its complementary, $(1 - \text{Reliability}^{SS}_{i,j})$, expresses the failure probability of that subsystem.

Second approach is considering expected failure numbers in a subsystem. It's obvious that because of being just two components in each subsystem, at most one failure in each subsystem during a period is reasonable. So, when expected failure number is more than one, subsystem fails. At first the expected number of failures in a component should be calculated:

$$E\left(N^c_{i,j}\right) = \int_{X^c_{i,j}}^{X'^c_{i,j}} \vartheta^c_i(t) dt$$

$$\text{for } i = 1, \ldots, N \text{ and } j = 1, \ldots, T \text{ and } c = 1, \ldots, C \qquad (32)$$

$E(N_{i,j}^c)$ shows the expected number of failures in component c of subsystem i in period j. According to Eq. (1) and by the assumption NHPP for ROCOF, expected number of failures in component c of subsystem i in period j will be:

$$E\left(N_{i,j}^c\right) = \int_{X_{i,j}^c}^{X_{i,j}^{\prime c}} \lambda_i^c \cdot \beta_i^c \cdot t^{\beta_i^c - 1} dt = \left[\lambda_i^c \left(X_{i,j}^{\prime c}\right)^{\beta_i^c} - \lambda_i^c \left(X_{i,j}^c\right)^{\beta_i^c}\right]$$

$$\text{for } i = 1, \ \ldots, \ N \ \text{ and } \ j = 1, \ldots, \ T \ \text{ and } \ c = 1, \ \ldots, C \tag{33}$$

According to component c loading probability, the expected failure numbers of subsystem i will be:

$$E\left(N_{i,j}\right) = \sum_{c=1}^{C} p_{i,j}^c \cdot E\left(N_{i,j}^c\right) = \sum_{c=1}^{C} p_{i,j}^c \cdot \left[\lambda_i^c \left(X_{i,j}^{\prime c}\right)^{\beta_i^c} - \lambda_i^c \left(X_{i,j}^c\right)^{\beta_i^c}\right]$$

$$\text{for } i = 1, \ \ldots, N \ \text{ and } \ j = 1, \ \ldots, T \tag{34}$$

As expressed before just one failure in each subsystem is acceptable. So, if the expected failure number is greater than one, failure cost exists. So failure cost of subsystem i in period j will be:

$$F_{i,j} = F_i \cdot \max\left\{0, E\left(N_{i,j}\right) - 1\right\}$$

$$\text{for } i = 1, \ \ldots \ , N \ \text{ and } \ j = 1, \ \ldots \ , T \tag{35}$$

Total Cost

At the start of period $j = 1$, a set of maintenance and replacement activities should be specified for each component in following periods which minimizes the total cost. According to different costs definitions, the total cost function can be rewritten as below:

$$\text{Total Cost} = \sum_{i=1}^{N} \sum_{j=1}^{T} \left[F_i \cdot \left(1 - \text{Reliabilty}_{i,j}^{SS}\right) \right.$$

$$+ \sum_{c=1}^{C} \left(\sum_{k=1}^{K} \left(M_i^{c,k} \cdot m_{i,j}^{c,k}\right) + R_i^c \cdot r_{i,j}^c\right) \right]$$

$$+ \sum_{j=1}^{T} \left[z \left(1 - \prod_{i=1}^{N} \left(1 - \prod_{c=1}^{C} \left(r_{i,j}^c + \sum_{k=1}^{K} m_{i,j}^{c,k}\right)\right)\right)\right] \tag{36}$$

Above function calculates the total cost as a sum of component costs in each period according to maintenance and replacement cost, down tie fixed cost and expected system failure cost.

System Reliability

In order to calculate the system reliability, at first the reliability of subsystem i in period j should be calculated. It's notable that each subsystem is a cold standby system. In other words, in this subsystems when a component is in standby mode no changes will happen in its age and failure rate. Charles O. Smith defined a general function for reliability of a standby system with two components (Smith 1976).

$$R_{SS} = R_1(t) + Q_1(t_1) \cdot R_2(t - t_1) \quad 0 \leq t_1 \leq t \tag{37}$$

Above expression shows that the reliability of a standby system is composed of two parts. First part, $R_1(t)$, is related to a case in which the first component is active from time 0 to t without any failure. But the second part, $Q_1(t_1) \cdot R_2(t-t_1)$, represents the state in which first component breaks down in time t_1 and the second component continues its operation from t_1 to t. In the following, each subsystem reliability will be surveyed in details.

In the above equation, $R_c(t)$ and $Q_c(t)$ represent the reliability of component c from 0 to t and the failure probability of component C at time t, respectively. In a system with C components in each subsystem, $C! \sum_{r=0}^{C-1} \frac{1}{[(C-1)-r]!}$ states can occur for each subsystem in each period. So in this model by considering two components in each subsystem, four states can occur in each period. These states are:

State 1:
In this state, the first component is under load at the start of period j, and will work till the end of this period without any failure (Table 1).

State 2:
In this state, the first component cannot finish the period and because of a failure at time t in first component a switching operation performs. So, the second component will be loaded and the system continues its working from time t to T/J by second component (Table 2).

State 3:
This state is similar to first state, but the second component is under load at the start of period j (Table 3).

Table 1 First state in a standby subsystem

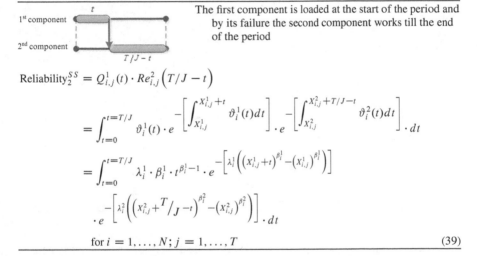

T/J	First component works during the period without
1st component	any failure
2nd component	

$$\text{Reliability}_1^{SS} = Re_{i,j}^1(T/J) = e^{-\left[\int_{X_{i,j}^1}^{X_{i,j}^1+T/J} \vartheta_i^1(t)dt\right]}$$

$$= e^{-\left[\left(\lambda_i^1\left(X_{i,j}^1+T/J\right)^{\beta_i^1} - \left(X_{i,j}^1\right)^{\beta_i^1}\right)\right]}$$

$$\text{for } i = 1,\ldots,N; j = 1,\ldots,T \tag{38}$$

Table 2 Second state in a standby subsystem

t	The first component is loaded at the start of the period and
1st component	by its failure the second component works till the end
2nd component	of the period
$T/J-t$	

$$\text{Reliability}_2^{SS} = Q_{i,j}^1(t) \cdot Re_{i,j}^2\left(T/J - t\right)$$

$$= \int_{t=0}^{t=T/J} \vartheta_i^1(t) \cdot e^{-\left[\int_{X_{i,j}^1}^{X_{i,j}^1+t} \vartheta_i^1(t)dt\right]} \cdot e^{-\left[\int_{X_{i,j}^2}^{X_{i,j}^2+T/J-t} \vartheta_i^2(t)dt\right]} \cdot dt$$

$$= \int_{t=0}^{t=T/J} \lambda_i^1 \cdot \beta_i^1 \cdot t^{\beta_i^1-1} \cdot e^{-\left[\lambda_i^1\left(\left(X_{i,j}^1+t\right)^{\beta_i^1} - \left(X_{i,j}^1\right)^{\beta_i^1}\right)\right]}$$

$$\cdot e^{-\left[\lambda_i^2\left(\left(X_{i,j}^2+T/J-t\right)^{\beta_i^2} - \left(X_{i,j}^2\right)^{\beta_i^2}\right)\right]} \cdot dt$$

$$\text{for } i = 1,\ldots,N; j = 1,\ldots,T \tag{39}$$

State 4:
In this state, second component is loaded at the start of period j, but similar to second state, component failure causes a switching operation and the first component become loaded (Table 4).

Total Reliability

In this section, the above functions for total reliability computation will be combined. Since, at the start of each period, the components loading condition is

Table 3 Third state in a standby subsystem

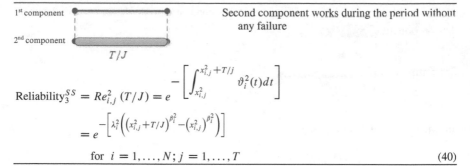

$$\text{Reliability}_3^{SS} = Re_{i,j}^2 (T/J) = e^{-\left[\int_{x_{i,j}^2}^{x_{i,j}^2 + T/j} \vartheta_i^2(t)dt\right]}$$

$$= e^{-\left[\lambda_i^2\left(\left(x_{i,j}^2 + T/J\right)^{\beta_i^2} - \left(x_{i,j}^2\right)^{\beta_i^2}\right)\right]}$$

$$\text{for } i = 1,\dots,N; j = 1,\dots,T \tag{40}$$

Table 4 Fourth state in a standby subsystem

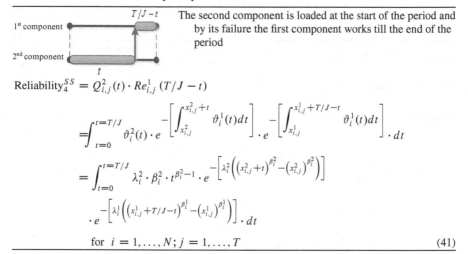

$$\text{Reliability}_4^{SS} = Q_{i,j}^2(t) \cdot Re_{i,j}^1 (T/J - t)$$

$$= \int_{t=0}^{t=T/J} \vartheta_i^2(t) \cdot e^{-\left[\int_{x_{i,j}^2}^{x_{i,j}^2 + t} \vartheta_i^1(t)dt\right]} \cdot e^{-\left[\int_{x_{i,j}^1}^{x_{i,j}^1 + T/J - t} \vartheta_i^1(t)dt\right]} \cdot dt$$

$$= \int_{t=0}^{t=T/J} \lambda_i^2 \cdot \beta_i^2 \cdot t^{\beta_i^2 - 1} \cdot e^{-\left[\lambda_i^2\left(\left(x_{i,j}^2 + t\right)^{\beta_i^2} - \left(x_{i,j}^2\right)^{\beta_i^2}\right)\right]}$$

$$\cdot e^{-\left[\lambda_i^1\left(\left(x_{i,j}^1 + T/J - t\right)^{\beta_i^1} - \left(x_{i,j}^1\right)^{\beta_i^1}\right)\right]} \cdot dt$$

$$\text{for } i = 1,\dots,N; j = 1,\dots,T \tag{41}$$

unknown, calculating the expected reliability in each period for each subsystem is inevitable. So according to law of total probability, expected reliability will be:

$$\text{Reliability}_{i,j}^{SS} = p_{i,j}^1 \cdot Re_{i,j}^{SS,1} + p_{i,j}^2 \cdot Re_{i,j}^{SS,2}$$

$$= p_{i,j}^1 \cdot \left(R_{i,j}^1 (T/J) + Q_{i,j}^1(t) \cdot R_{i,j}^2 (T/J - t)\right)$$

$$+ p_{i,j}^2 \cdot \left(R_{i,j}^2 (T/J) + Q_{i,j}^2(t) \cdot R_{i,j}^1 (T/J - t)\right)$$

$$\text{for } i = 1,\dots,N; j = 1,\dots,T \tag{42}$$

In this equation, $Re_{i,j}^{SS,1}$ illustrates ith subsystem reliability in which the first component is under load at the start of period j and its measure is equal to sum of first and second states reliability as follows:

$$Re_{i,j}^{SS,1} = e^{-\left[\int_{x_{i,j}^1}^{x_{i,j}^1+T/J} \vartheta_i^1(t)dt\right]}$$

$$+ \int_{t=0}^{t=T/J} \vartheta_i^1(t) \cdot e^{-\left[\int_{x_{i,j}^1}^{x_{i,j}^1+t} \vartheta_i^1(t)dt\right]} \cdot e^{-\left[\int_{x_{i,j}^2}^{x_{i,j}^2+T/J-t} \vartheta_i^2(t)dt\right]} \cdot dt$$

$$= e^{-\left[\lambda_i^1\left(\left(x_{i,j}^1+T/J\right)^{\beta_i^1} - \left(x_{i,j}^1\right)^{\beta_i^1}\right)\right]}$$

$$+ \int_{t=0}^{t=T/J} \lambda_i^1 \cdot \beta_i^1 \cdot t^{\beta_i^1-1} \cdot e^{-\left[\lambda_i^1\left(\left(x_{i,j}^1+t\right)^{\beta_i^1} - \left(x_{i,j}^1\right)^{\beta_i^1}\right)\right]}$$

$$\cdot e^{-\left[\lambda_i^2\left(\left(x_{i,j}^2+T/J-t\right)^{\beta_i^2} - \left(x_{i,j}^2\right)^{\beta_i^2}\right)\right]} \cdot dt$$

$$\text{for } i = 1, \ldots, N; j = 1, \ldots, T \tag{43}$$

Similarly, $Re_{i,j}^{SS,2}$ is referred to the states that the second component is loaded at the start of period j. Sum of third and fourth states reliability equations will be:

$$Re_{i,j}^{SS,2} = e^{-\left[\int_{x_{i,j}^2}^{x_{i,j}^2+T/J} \vartheta_i^2(t)dt\right]}$$

$$+ \int_{t=0}^{t=T/J} \vartheta_i^2(t) \cdot e^{-\left[\int_{x_{i,j}^2}^{x_{i,j}^2+t} \vartheta_i^2(t)dt\right]} \cdot e^{-\left[\int_{x_{i,j}^1}^{x_{i,j}^1+T/J-t} \vartheta_i^1(t)dt\right]} \cdot dt$$

$$= e^{-\left[\lambda_i^2\left(\left(x_{i,j}^2+T/J\right)^{\beta_i^2} - \left(x_{i,j}^2\right)^{\beta_i^2}\right)\right]}$$

$$+ \int_{t=0}^{t=T/J} \lambda_i^2 \cdot \beta_i^2 \cdot t^{\beta_i^2-1} \cdot e^{-\left[\lambda_i^2\left(\left(x_{i,j}^2+t\right)^{\beta_i^2} - \left(x_{i,j}^2\right)^{\beta_i^2}\right)\right]}$$

$$\cdot e^{-\left[\lambda_i^1\left(\left(x_{i,j}^1+T/J-t\right)^{\beta_i^1} - \left(x_{i,j}^1\right)^{\beta_i^1}\right)\right]} \cdot dt$$

$$\text{for } i = 1, \ldots, N; j = 1, \ldots, T \tag{44}$$

So, according to above descriptions, the expected total reliability for subsystem i in period j is equal to:

$$\text{Reliability}_{i,j}^{SS} = p_{i,j}^1 \cdot Re_{i,j}^{SS,1} + p_{i,j}^2 \cdot Re_{i,j}^{SS,2}$$

$$= p_{i,j}^1 \cdot \Bigg[e^{-\left[\lambda_i^1 \left(\left(x_{i,j}^1 + T/J \right)^{\beta_i^1} - \left(x_{i,j}^1 \right)^{\beta_i^1} \right) \right]}$$

$$+ \int_{t=0}^{t=T/J} \lambda_i^1 \cdot \beta_i^1 \cdot t^{\beta_i^1 - 1} \cdot e^{-\left[\lambda_i^1 \left(\left(x_{i,j}^1 + t \right)^{\beta_i^1} - \left(x_{i,j}^1 \right)^{\beta_i^1} \right) \right]}$$

$$\cdot e^{-\left[\lambda_i^2 \left(\left(x_{i,j}^2 + T/J - t \right)^{\beta_i^2} - \left(x_{i,j}^2 \right)^{\beta_i^2} \right) \right]} \cdot dt$$

$$+ p_{i,j}^2 \cdot \Bigg[e^{-\left[\lambda_i^2 \left(\left(x_{i,j}^2 + T/J \right)^{\beta_i^2} - \left(x_{i,j}^2 \right)^{\beta_i^2} \right) \right]}$$

$$+ \int_{t=0}^{t=T/J} \lambda_i^2 \cdot \beta_i^2 \cdot t^{\beta_i^2 - 1} \cdot e^{-\left[\lambda_i^2 \left(\left(x_{i,j}^2 + t \right)^{\beta_i^2} - \left(x_{i,j}^2 \right)^{\beta_i^2} \right) \right]}$$

$$\cdot e^{-\left[\lambda_i^1 \left(\left(x_{i,j}^1 + T/J - t \right)^{\beta_i^1} - \left(x_{i,j}^1 \right)^{\beta_i^1} \right) \right]} \cdot dt$$

$$\text{for} \quad i = 1, \ldots, N; \, j = 1, \ldots, T \tag{45}$$

With regard to considering a series system, the total reliability will be computable by a simple multiply operation on N subsystems and J periods.

$$\text{Reliability} = \prod_{i=1}^{N} \prod_{j=1}^{T} \text{Reliability}_{i,j}^{SS}$$

$$= \prod_{i=1}^{N} \prod_{j=1}^{T} \Bigg(p_{i,j}^1 \cdot \Bigg[e^{-\left[\lambda_i^1 \left(\left(x_{i,j}^1 + T/J \right)^{\beta_i^1} - \left(x_{i,j}^1 \right)^{\beta_i^1} \right) \right]}$$

$$+ \int_{t=0}^{t=T/J} \lambda_i^1 \cdot \beta_i^1 \cdot t^{\beta_i^1 - 1} \cdot e^{-\left[\lambda_i^1 \left(\left(x_{i,j}^1 + t \right)^{\beta_i^1} - \left(x_{i,j}^1 \right)^{\beta_i^1} \right) \right]}$$

$$\cdot e^{-\left[\lambda_i^2 \left(\left(x_{i,j}^2 + T/J - t \right)^{\beta_i^2} - \left(x_{i,j}^2 \right)^{\beta_i^2} \right) \right]} \cdot dt$$

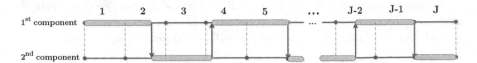

Fig. 6 Schematic view of a subsystem operation in third approach

$$
+ p_{i,j}^2 \cdot \left[e^{-\left[\lambda_i^2 \left(\left(x_{i,j}^2 + T/J \right)^{\beta_i^2} - \left(x_{i,j}^2 \right)^{\beta_i^2} \right) \right]} \right]
$$

$$
+ \int_{t=0}^{t=T/J} \lambda_i^2 \cdot \beta_i^2 \cdot t^{\beta_i^2 - 1} \cdot e^{-\left[\lambda_i^2 \left(\left(x_{i,j}^2 + t \right)^{\beta_i^2} - \left(x_{i,j}^2 \right)^{\beta_i^2} \right) \right]}
$$

$$
\cdot e^{-\left[\lambda_i^1 \left(\left(x_{i,j}^1 + T/J - t \right)^{\beta_i^1} - \left(x_{i,j}^1 \right)^{\beta_i^1} \right) \right]} \cdot dt \tag{46}
$$

5.2.2 Optional Switching

In this section, optional switching possibility is considered to previous model. It means that two types of switching can perform in this system, optional and non-optional. When a failure occurs non-optional switching performs but, at the end of some periods, optional switching performs to maintain or replace the component that is under load. In Fig. 6, these two types of switching are illustrated.

In Fig. 6, black arrows represent non-optional switching and red arrows show optional switching. By assuming optional switching, it is assumed that by maintaining or replacing a component at the end of period j that component cannot be loaded at the start of period $j + 1$. By this assumption, when $\sum_{k=1}^{K} m_{i,j-1}^{c,k} + r_{i,j-1}^c$ is equal to one, $p_{i,j}^c$ become 0. So:

$$
p_{i,j}^c \cdot \left(\sum_{k=1}^{K} m_{i,j-1}^{c,k} + r_{i,j-1}^c \right) = 0
$$

$$
\text{for} \quad i = 1, \ldots, N; j = 2, \ldots, T \quad \text{and} \quad c = 1, 2 \tag{47}
$$

Another assumption is that at the start of each period at least one component in each subsystem must be ready to become loaded. In other words, at the end of each period there is at least one component that will not maintain or replace. So:

$$\sum_{c=1}^{c} \left(\sum_{k=1}^{K} m_{i,j}^{c,k} + r_{i,j}^{c} \right) \leq 1$$

for $i = 1, \ldots, N$; $j = 1, \ldots, T - 1$ and $c = 1, 2$ \hfill (48)

By this assumption fixed cost become equal to 0. Because at the end of a period there is no possibility to perform maintenance or replacement activities on all components of a subsystem and at least one component in each subsystem works. So, the total system never stops working for maintenance or replacement.

6 Solution Approach

NSGA-II is one of the most useful and powerful existing algorithms for solving various multi-objective optimization problems that its performance has been proved (Deb et al. 2000). This algorithm combines GA algorithm with dominancy concept to form Pareto front. Deb and his colleague designed first version of this algorithm in 1995 and completed it in 2000 (Srinivas and Deb 1995; Deb et al. 2000). These algorithm parameters are defined as follow:

- Crossover procedure
  In this research, according to specific nature of this problem, two new methods are considered. These two crossover procedures are:

  1. Reverse three points crossover: In this method, at first three elements of $N \times C \times T$ matrix will be selected randomly. By this selection the parent chromosomes will divide into two parts. The procedure of children chromosomes construction is that the arrays out of these three points will select from the first parent and the other arrays will copy from the second parent reversely. By this method, if the parents have been selected similarly, the children will create differently.
  2. NCT points crossover: In this type of crossover, even genes will select from the first parent and the odd genes from the second one.

  So, if the selected solutions be similar, the algorithm uses the reverse three points crossover, otherwise the NCT points crossover will be used.

- Mutation procedure
  It's clear that mutation operator changes the coding design of chromosomes for diversity in solutions. According to the problem construction, if a component is maintained or replaced at the end of a period, the total system contacts to a cost. So a specific mutation procedure is defined for this problem. In this procedure, a number will be generated between one and $N \times C \times T$ randomly. Then if this is a non-zero number, the algorithm changes it to zero and reversely if it is equal to zero changes to a number belongs to $\{1,2,3,4\}$ randomly.

7   Numerical Example

In this section, three mentioned optimization models will be solved with a set of data (considered in Table 5) to be compared in application area and examining their strength and weaknesses. In addition to considered data set, for each subsystem three similar components is assumed and 24 month is defined for planning horizon. Fixed cost will be 800$. In Table 5 α^1, α^2, and α^3 show three levels of maintenance from overhaul to elementary maintenance and M^1, M^2, and M^3 represent the costs associated to maintenance levels. It's notable that MATLAB R2013a environment is used for model solving.

Table 5 Numerical example parameters

Subsystem	λ	β	α^1	α^2	α^3	Failure cost ($)	Maintenance cost ($)			Replacement Cost ($)	Switching Cost ($)
							M^1	M^2	M^3		
1	0.00022	2.20	0.27	0.62	0.86	250	47	35	22	200	150
2	0.00035	2.00	0.31	0.58	0.82	240	43	32	24	210	142
3	0.00038	2.05	0.22	0.55	0.91	270	78	65	48	245	167
4	0.00034	1.90	0.24	0.50	0.78	210	59	42	28	180	112
5	0.00032	1.75	0.29	0.48	0.83	220	67	50	37	205	187
6	0.00028	2.10	0.33	0.65	0.89	280	51	38	23	235	145
7	0.00015	2.25	0.37	0.75	0.93	200	58	45	31	175	138
8	0.00012	1.80	0.37	0.68	0.90	225	43	30	21	215	129
9	0.00025	1.85	0.26	0.52	0.85	215	56	48	40	210	162
10	0.00020	2.15	0.38	0.67	0.84	255	69	55	43	250	178

In this section, these three models will be checked deeply and compared with each other. In many literature, optimization model is designed as a single objective model by using one limitation formulation, like, Moghaddam and Usher (2011) recent research in which two limitation models have been presented, where the first model minimizes the total cost for a given system reliability and the second one maximizes the total reliability with a budgetary constraint. Of course they completed their research by considering a multi-objective model for a system of several components that are connected in series configuration (Moghaddam and Usher 2011). This chapter presented models are also multi-objective models which are maximizing the total reliability by minimizing the total cost for a series system of several standby subsystems.

7.1   First Model

As surveyed before in this model, preventive maintenance and replacement scheduling with failure impossibility assumption is considered. Final optimization model is:

$$
\text{Min Total Cost} = \sum_{i=1}^{N}\sum_{j=1}^{T}\left[F_i \cdot \sum_{c=1}^{C} l_{i,j}^c \cdot \left[\lambda_i^c\left(\left(X_{i,j}^{\prime c}\right)^{\beta_i^c} - \left(X_{i,j}^c\right)^{\beta_i^c}\right)\right]\right.
$$

$$
+ \sum_{c=1}^{C}\left(\sum_{k=1}^{K}\left(M_i^{c,k}\cdot m_{i,j}^{c,k}\right) + R_i^c \cdot r_{i,j}^c + \frac{1}{2}S_i \cdot \left|l_{i,j+1}^c - l_{i,j}^c\right|\right)\right]
$$

$$
+ \sum_{j=1}^{T}\left[z\left(1 - \prod_{i=1}^{N}\left(1 - \prod_{c=1}^{C}\left(r_{i,j}^c + \sum_{k=1}^{K} m_{i,j}^{c,k}\right)\right)\right)\right]
$$

$$
\text{Max Reliability} = \prod_{i=1}^{N}\prod_{j=1}^{T} e^{-\left[\sum_{c=1}^{C} l_{i,j}^c \cdot \left[\lambda_i^c\left(\left(X_{i,j}^{\prime c}\right)^{\beta_1^c} - \left(X_{i,j}^c\right)^{\beta_1^c}\right)\right]\right]}
$$

Subject to:

$$
X_{i,1}^c = 0 \text{ for } i = 1,\ldots N \text{ and } c = 1,\ldots,C
$$

$$
X_{i,j}^c = \left(1 - r_{i,j-1}^c\right)\left[\prod_{k=1}^{K}\left(1 - m_{i,j-1}^{c,k}\right)\right] X_{i,j-1}^{\prime c}
$$

$$
+ \sum_{k=1}^{K} m_{i,j-1}^{c,k}\cdot\left(\alpha_i^k \cdot X_{i,j-1}^{\prime c}\right)
$$

$$
\text{for } i = 2,\ldots,N; j = 1,\ldots,T \text{ and } c = 1,\ldots,C
$$

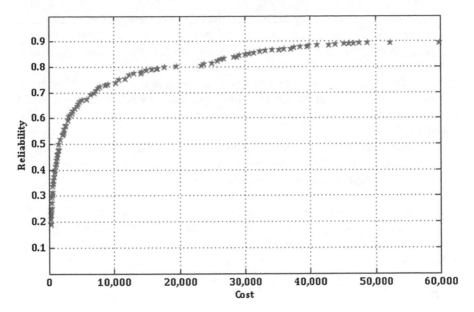

Fig. 7 First model Parto front

$$X_{i,j}^{\prime c} = X_{i,j}^{c} + l_{i,j}^{c} \cdot T/J \text{ for } i = 1, \ldots, N; j = 1, \ldots, T \text{ and } c = 1, \ldots, C$$

$$\sum_{k=1}^{K} m_{i,j}^{c,k} + r_{i,j}^{c} \leq 1 \text{ for } i = 1, \ldots, N; j = 1, \ldots, T \text{ and } c = 1, \ldots, C$$

$$\sum_{c=1}^{C} l_{i,j}^{c} = 1 \text{ for } i = 1, \ldots, N \text{ and } j = 1, \ldots, T$$

$$l_{i,j}^{c}, m_{i,j}^{c,k}, r_{i,j}^{c}, = 0 \text{ } or \text{ } 1$$

$$\text{for } i = 1, \ldots, N; j = 1, \ldots, T; c = 1, \ldots, C \text{ and } k = 1, \ldots, K$$

$$X_{i,j}^{\prime c}, X_{i,j}^{c} \geq 0 \text{ for } i = 1, \ldots, N \text{ and } j = 1, \ldots, T \tag{49}$$

The proposed model solved by the considered parameters. The Pareto front related to this problem is shown in Fig. 7. An important point is that all the points in this front provide an approximation of the optimal values and none of them dominate the others. It means that in different strategies different policies can be chosen. In other words the primary points of Pareto front by low costs have less reliability level and reversely by increasing the costs, the reliability levels are also increasing. So according to designers and managers strategies different policies can be implemented.

As is obvious in the components effective age diagrams, effective age of a component can increase, decrease or have no changes during the planning horizon. Increasing in effective age diagram of a component expresses that the component is loaded and operated during the period. No changes in a component effective age shows that the component wouldn't be under load in that period and when an instantaneous drop or decrease occurs in the diagram illustrates the maintenance and replacement activities happened on the component that its measure depends on the maintenance level.

It's notable that in the maintenance and replacement schedule tables, M1 shows overhaul maintenance, M2, intermediate maintenance, M3, primary maintenance, and R illustrates replacement activity on a component.

An obvious point about components effective age diagrams is that, in low reliability levels, system prefers to maintain or replace the first component and the second component isn't active so much. But by increases in reliability levels and costs, the second component starts to working (Figs. 8, 9, 10, 11, and 12; Tables 6, 7, 8, and 9).

7.2 Second Model

This model prepared with failure possibility assumption but non-optional switching (Fig. 13). Total mathematical model, its reliability, costs, and components effective age diagrams are shown in the following:

$$
\text{Min Total Cost} = \sum_{i=1}^{N} \sum_{j=1}^{T} \left[F_i \cdot \left(1 - \text{Reliability}_{i,j}^{SS} \right) \right.
$$

$$
+ \sum_{c=1}^{C} \left(\sum_{k=1}^{K} \left(M_i^{c,k} \cdot m_{i,j}^{c,k} \right) + R_i^c \cdot r_{i,j}^c \right) \right]
$$

$$
+ \sum_{i=1}^{T} \left[z \left(1 - \prod_{i=1}^{N} \left(1 - \prod_{c=1}^{C} \left(r_{i,j}^c + \sum_{k=1}^{K} m_{i,j}^{c,k} \right) \right) \right) \right]
$$

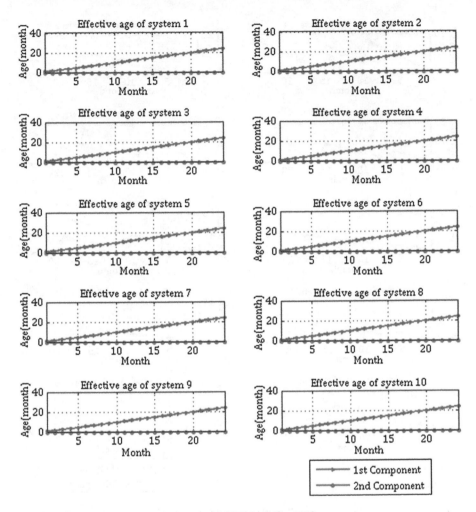

Fig. 8 Components effective age by $R = 19.25~\%$ and $C = 401\$$

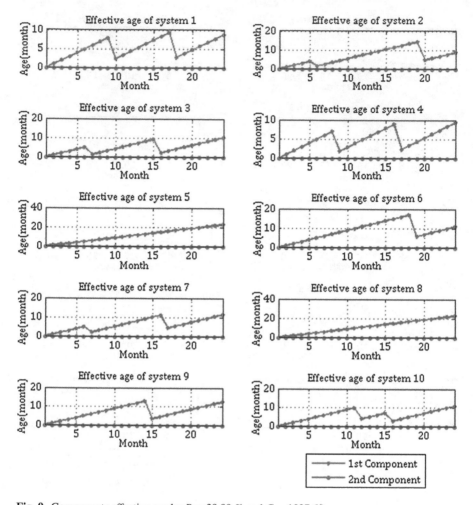

Fig. 9 Components effective age by $R = 39.99$ % and $C = 1037.6$$

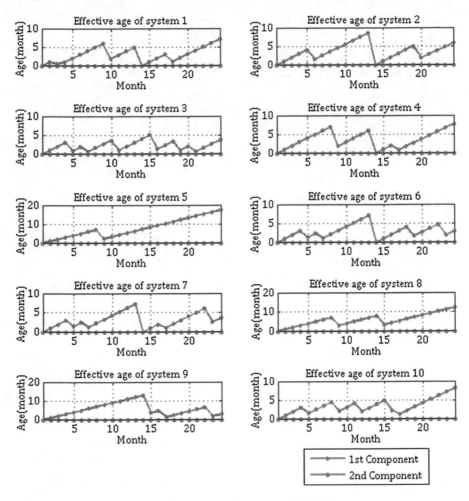

Fig. 10 Components effective age by $R = 60.55\%$ and $C = 3116.3\$$

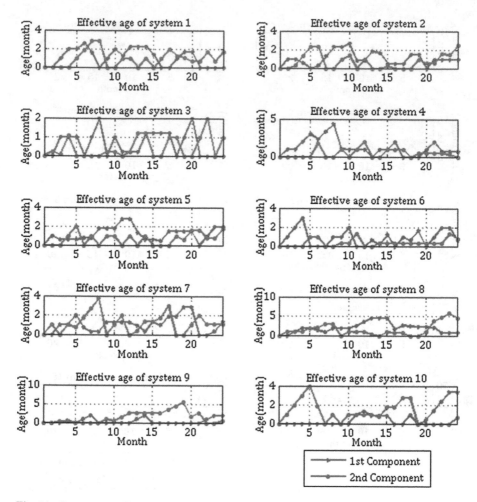

Fig. 11 Components effective age by $R = 85.16\%$ and $C = 30700\$$

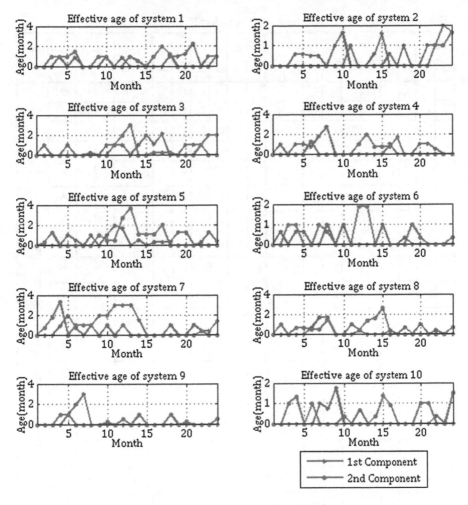

Fig. 12 Components effective age by $R = 89.49\%$ and $C = 59930\$$

Table 6 Maintenance and replacement schedule by $R = 39.99$ % and $C = 1037.6$\$

Subsystem	10		9		8		7		6		5		4		3		2		1		Component
Component	2	1	2	1	2	1	2	1	2	1	2	1	2	1	2	1	2	1	2	1	
	I	I	I	I	I	I	I	I	I	I	I	I	I	I	I	I	I	I	I	I	1
	I	I	I	I	I	I	I	I	I	I	I	I	I	I	I	I	I	I	I	I	2
	I	I	I	I	I	I	I	I	I	I	I	I	I	I	I	I	I	I	I	I	3
	I	I	I	I	I	I	I	I	I	I	I	I	I	I	I	I	I	I	I	I	4
	I	I	I	I	I	I	I	I	I	I	I	I	I	I	I	I	I	M1	I	I	5
	I	I	I	I	I	I	I	M1	I	I	I	I	I	I	I	M1	I	I	I	I	6
	I	I	I	I	I	I	I	I	I	I	I	I	I	I	I	I	I	I	I	I	7
	I	I	I	I	I	I	I	I	I	I	I	I	I	M1	I	I	I	I	I	I	8
	I	I	I	I	I	I	I	I	I	I	I	I	I	I	I	I	I	I	I	M1	9
	I	I	I	I	I	I	I	I	I	I	I	I	I	I	I	I	I	I	I	I	10
	I	M1	I	I	I	I	I	I	I	I	I	I	I	I	I	I	I	I	I	I	11
	I	I	I	I	I	I	I	I	I	I	I	I	I	I	I	I	I	I	I	I	12
	I	I	I	I	I	I	I	I	I	I	I	I	I	I	I	I	I	I	I	I	13
	I	I	I	M1	I	I	I	I	I	I	I	I	I	I	I	I	I	I	I	I	14
	I	M1	I	I	I	I	I	I	I	I	I	I	I	I	I	M1	I	I	I	I	15
	I	I	I	I	I	I	I	M1	I	I	I	I	I	M1	I	I	I	I	I	I	16
	I	I	I	I	I	I	I	I	I	I	I	I	I	I	I	I	I	I	I	M1	17
	I	I	I	I	I	I	I	I	I	M1	I	I	I	I	I	I	I	I	I	I	18
	I	I	I	I	I	I	I	I	I	I	I	I	I	I	I	I	I	M1	I	I	19
	I	I	I	I	I	I	I	I	I	I	I	I	I	I	I	I	I	I	I	I	20
	I	I	I	I	I	I	I	I	I	I	I	I	I	I	I	I	I	I	I	I	21
	I	I	I	I	I	I	I	I	I	I	I	I	I	I	I	I	I	I	I	I	22
	I	I	I	I	I	I	I	I	I	I	I	I	I	I	I	I	I	I	I	I	23
	I	I	I	I	I	I	I	I	I	I	I	I	I	I	I	I	I	I	I	I	24

Maintenance and replacement schedule

Table 7 Maintenance and replacement schedule by $R = 60.55\%$ and $C = 3116.3\$$

10		9		8		7		6		5		4		3		2		1		Subsystem
2	1	2	1	2	1	2	1	2	1	2	1	2	1	2	1	2	1	2	1	Component
I	I	I	I	I	I	I	I	I	I	I	I	I	I	I	I	I	I	I	I	1
I	I	I	I	I	I	I	I	I	I	I	I	I	I	I	I	I	I	I	M1	2
I	I	I	I	I	I	I	I	I	I	I	I	I	I	I	I	I	I	I	M2	3
I	M1	I	I	I	I	I	M1	I	M1	I	I	I	I	I	M1	I	I	I	I	4
I	I	I	I	I	I	I	I	I	I	I	I	I	I	I	I	M1	I	I	I	5
I	I	I	I	I	I	I	M1	I	M1	I	I	I	I	I	M1	I	I	I	I	6
I	I	I	I	I	I	I	I	I	I	I	I	I	I	I	I	I	I	I	I	7
I	M1	I	I	I	M1	I	I	I	I	I	M1	I	M1	I	I	I	I	M1	I	8
I	I	I	I	I	I	I	I	I	I	I	I	I	I	I	I	I	I	M1	I	9
I	I	I	I	I	I	I	I	I	I	I	I	I	I	I	M1	I	I	I	I	10
I	M1	I	I	I	I	I	I	I	I	I	I	I	I	I	I	I	I	I	I	11
I	I	I	I	I	I	I	I	I	I	I	I	I	I	I	I	I	I	I	I	12
I	I	I	I	I	I	I	R	I	R	I	I	I	R	I	R	I	I	I	R	13
I	I	I	M1	I	M1	I	I	I	I	I	I	I	I	I	I	I	I	I	I	14
I	M1	I	I	I	I	I	I	I	I	I	I	I	I	I	M1	I	I	I	I	15
M1	M1	I	M1	I	I	I	M1	I	I	I	I	I	M1	I	I	I	I	I	I	16
I	I	I	I	I	I	I	I	I	I	I	I	I	I	I	I	I	I	M1	I	17
I	I	I	I	I	I	I	I	M1	I	I	I	I	I	I	M1	I	I	I	I	18
I	I	I	I	I	I	I	I	I	I	I	I	I	I	I	I	M1	I	I	I	19
I	I	I	I	I	I	I	I	I	I	I	I	I	I	I	M1	I	I	I	I	20
I	I	I	I	I	I	I	I	I	I	I	I	I	I	I	I	I	I	I	I	21
I	I	I	M1	I	I	I	M1	I	M1	I	I	I	I	I	I	I	I	I	I	22
I	I	I	I	I	I	I	I	I	I	I	I	I	I	I	I	I	I	I	I	23
I	I	I	I	I	I	I	I	I	I	I	I	I	I	I	I	I	I	I	I	24

Maintenance and replacement schedule

Table 8 Maintenance and replacement schedule by $R = 85.16$ % and $C = 30700$$

10		9		8		7		6		5		4		3		2		1		Subsystem
2	1	2	1	2	1	2	1	2	1	2	1	2	1	2	1	2	1	2	1	Component
–	–	M3	–	–	–	–	–	–	–	M1	–	M3	–	M1	–	–	–	–	R	1
–	–	M1	M1	–	–	R	–	–	–	–	M1	R	–	M1	–	M1	–	–	–	2
–	–	–	M2	–	–	–	–	–	–	–	–	–	–	–	–	–	M2	–	–	3
–	M2	M3	R	–	–	M2	–	–	R	–	–	–	–	M3	–	–	R	–	–	4
M1	–	R	–	M2	–	M2	–	–	M1	–	R	M3	M2	–	M3	M3	R	–	M1	5
M1	M3	–	–	–	M2	–	M1	R	–	M3	–	M3	–	M3	–	–	–	–	M2	6
–	–	–	R	–	–	–	–	M2	–	M3	–	–	–	–	–	–	–	–	R	7
R	R	–	–	M3	–	R	–	M1	–	–	M1	M1	R	–	–	M3	–	M3	–	8
–	–	M1	–	–	–	–	–	–	–	M2	–	M3	M2	–	–	M2	M3	–	–	9
–	–	–	M2	M3	M3	–	–	–	R	R	–	–	–	M3	M1	M3	M1	–	M2	10
–	M2	–	–	–	–	R	–	M3	–	–	–	M1	–	–	–	–	–	–	–	11
–	M1	–	–	M1	–	M1	M2	R	M2	M3	M2	R	R	–	–	R	–	M3	–	12
M1	–	–	–	M3	–	–	M1	M1	M1	–	M2	–	–	–	–	–	M2	–	–	13
–	–	–	R	–	–	–	–	–	–	R	–	–	–	M3	–	M1	M1	M3	M2	14
–	R	–	R	M1	M1	M2	M2	–	R	–	M1	–	–	M3	–	M3	–	–	R	15
–	M1	–	–	–	–	M2	M3	–	–	–	–	–	R	–	–	R	–	–	R	16
–	–	–	M2	R	M2	M3	–	–	M1	–	–	M3	M1	–	R	M1	–	M2	–	17
M3	R	–	–	M1	M2	–	–	–	M1	M1	–	M2	–	M3	–	–	–	R	M3	18
–	M1	M1	–	–	–	–	–	–	R	–	–	–	–	M2	–	M2	R	M2	M2	19
M2	–	–	–	–	M3	–	R	–	–	–	R	–	R	–	R	–	–	–	R	20
–	–	R	–	–	M1	M1	M1	–	–	M2	–	–	M1	–	–	–	–	–	–	21
–	–	–	–	–	–	–	M1	–	–	–	–	M1	–	M3	–	M2	–	M1	–	22
M2	–	M2	–	M2	–	–	–	M1	M1	–	–	M3	–	–	–	–	–	–	–	23
–	–	R	–	M2	M1	–	–	–	–	M1	–	M1	–	M2	–	M3	–	–	M3	24

Maintenance and replacement schedule

Table 9 Maintenance and replacement schedule by $R = 89.49\ \%$ and $C = 59930\$$

Maintenance and replacement schedule

Subsystem	10		9		8		7		6		5		4		3		2		1		
Component	2	1	2	1	2	1	2	1	2	1	2	1	2	1	2	1	2	1	2	1	
	M3	M2	M3	–	–	–	M2	–	M1	M2	M1	–	M3	–	M1	–	–	R	M2	R	1
	–	R	R	–	R	R	–	R	–	R	–	R	–	R	R	R	–	R	–	R	2
	M2	R	–	M2	M2	M2	–	M3	–	M2	R	–	–	–	M3	–	–	M2	–	–	3
	R	R	–	–	–	R	R	–	R	–	–	–	R	–	R	–	R	–	–	R	4
	–	M2	R	–	M2	M2	–	M1	R	R	M3	M2	M2	M3	M3	R	–	M1	M2	M3	5
	R	–	R	–	–	–	–	R	–	R	–	R	R	–	R	–	R	–	R	R	6
	M1	M1	R	R	–	–	–	–	M2	–	M3	–	–	–	M2	M1	–	R	–	R	7
	R	–	R	R	R	R	R	–	R	R	–	R	R	R	R	R	–	–	–	R	8
	M1	R	M1	R	R	M1	–	–	–	–	M2	–	M3	M2	R	–	M2	M3	–	–	9
	R	R	R	–	–	R	R	–	R	R	–	–	R	–	–	–	–	R	R	R	10
	R	M2	M2	–	M1	M1	–	–	–	M3	–	M3	–	–	–	–	R	–	M3	M3	11
	R	R	R	R	–	R	R	–	–	R	–	R	–	R	R	R	–	R	R	–	12
	M1	M2	–	M2	M2	M2	M2	M1	R	M1	M1	M2	M1	R	M3	–	–	M2	M1	M2	13
	–	–	R	–	–	–	R	R	R	–	–	R	–	R	–	–	–	–	R	R	14
	M1	R	R	R	M3	M1	M3	M2	R	R	–	M1	–	–	M1	M2	M2	R	–	R	15
	R	R	R	R	R	R	R	R	R	R	–	–	–	R	–	–	R	R	–	–	16
	R	–	–	M2	R	M2	M3	–	–	M1	R	–	M3	M1	–	R	M1	–	M2	–	17
	R	R	R	–	R	R	R	R	–	R	–	–	R	R	R	R	–	R	R	–	18
	–	M1	M1	–	M3	–	M1	R	M1	R	R	–	M1	–	M2	–	M2	R	M2	M2	19
	–	R	R	R	R	R	–	–	R	R	R	R	R	–	R	–	R	–	–	–	20
	M3	M1	R	–	M2	M1	M1	M1	–	R	M2	M1	–	M1	–	–	–	–	R	R	21
	R	R	R	R	–	R	R	–	–	R	–	–	R	R	R	–	–	–	R	–	22
	M2	M3	M2	M1	M2	–	M2	–	M1	M1	M3	M1	M3	R	R	–	M3	M3	–	–	23
	–	–	–	–	R	–	R	–	–	–	–	–	–	R	–	R	R	–	R	–	24

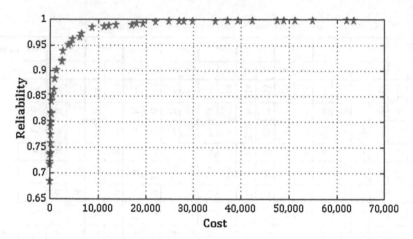

Fig. 13 Second model Parto front

$$
\text{Max Reliability} = \prod_{i=1}^{N} \prod_{j=1}^{T} \left(p_{i,j}^1 \cdot \left[e^{-\left[\lambda_i^1 \left(\left(X_{i,j}^1 + T/J \right)^{\beta_i^1} - \left(X_{i,j}^1 \right)^{\beta_i^1} \right) \right]} \right. \right.
$$

$$
+ \int_{t=0}^{t=T/J} \lambda_i^1 \cdot \beta_i^1 \cdot t^{\beta_i^1 - 1} \cdot e^{-\left[\lambda_i^1 \left(\left(X_{i,j}^1 + t \right)^{\beta_i^1} - \left(X_{i,j}^1 \right)^{\beta_i^1} \right) \right]}
$$

$$
\cdot e^{-\left[\lambda_i^2 \left(\left(X_{i,j}^2 + T/J - t \right)^{\beta_i^2} - \left(X_{i,j}^2 \right)^{\beta_i^2} \right) \right]} \cdot dt \Bigg]
$$

$$
+ p_{i,j}^2 \cdot \left[e^{-\left[\lambda_i^2 \left(\left(X_{i,j}^2 + T/J \right)^{\beta_i^2} - \left(X_{i,j}^2 \right)^{\beta_i^2} \right) \right]} \right.
$$

$$
+ \int_{t=0}^{t=T/J} \lambda_i^2 \cdot \beta_i^2 \cdot t^{\beta_i^2 - 1} \cdot e^{-\left[\lambda_i^2 \left(\left(X_{i,j}^2 + t \right)^{\beta_i^2} - \left(X_{i,j}^2 \right)^{\beta_i^2} \right) \right]}
$$

$$
\cdot e^{-\left[\lambda_i^1 \left(\left(X_{i,j}^1 + T/J - t \right)^{\beta_i^1} - \left(X_{i,j}^1 \right)^{\beta_i^1} \right) \right]} \cdot dt \Bigg] \Bigg)
$$

subject to:

$$X_{i,1}^c = 0 \text{ for } i = 1,\ldots N \text{ and } c = 1,2$$

$$X_{i,j}^c = \left(1 - r_{i,j-1}^c\right)\left[\prod_{k=1}^{K}\left(1 - m_{i,j-1}^{c,k}\right)\right]X_{i,j-1}^{\prime c} + \sum_{k=1}^{K} m_{i,j-1}^{c,k} \cdot \left(\alpha_i^k \cdot X_{i,j-1}^{\prime c}\right)$$

for $i = 1,\ldots,N; j = 2,\ldots,T$ and $c = 1,2$

$$X_{i,j}^{\prime 1} = X_{i,j}^1 + \left[p_{i,j}^1 \cdot \text{Reliability}_{i,j}^1 + \left(1 - p_{i,j}^1\right)\left(1 - \text{Reliability}_{i,j}^2\right)\right].T/J$$

for $i = 1,\ldots,N$ and $j = 1,\ldots,T$

$$X_{i,j}^{\prime 2} = X_{i,j}^2 + \left[\left(1 - p_{i,j}^1\right) \cdot \text{Reliability}_{i,j}^2 + p_{i,j}^1 \cdot \left(1 - \text{Reliability}_{i,j}^1\right)\right] \cdot T/J$$

for $i = 1,\ldots,N$ and $j = 1,\ldots,T$

$$p_{i,1}^1 = 1 \text{ for } i = 1,\ldots,N$$

$$p_{i,1}^2 = 0 \text{ for } i = 1,\ldots,N$$

$$p_{i,j+1}^1 = p_{i,j}^1 \cdot \text{Reliability}_{i,j}^1 + \cdot \left(1 - p_{i,j}^1\right) \cdot \left(1 - \text{Reliability}_{i,j}^2\right)$$

for $i = 1,\ldots,N$ and $j = 1,\ldots,T - 1$

$$p_{i,j}^2 = 1 - p_{i,j}^1 \text{ for } i = 1,\ldots,N \text{ and } j = 1,\ldots,T$$

$$\sum_{k=1}^{K} m_{i,j}^{c,k} + r_{i,j}^c \le 1 \text{ for } i = 1,\ldots,N; j = 1,\ldots,T \text{ and } c = 1,2$$

$$m_{i,j}^{c,k}, r_{i,j}^c = 0 \text{ or } 1 \text{ for } i = 1,\ldots,N; j = 1,\ldots,T; c = 1,2 \text{ and } k = 1,\ldots,K$$

$$0 \le p_{i,j}^1, p_{i,j}^2 \le 1 \text{ for } i = 1,\ldots,N \text{ and } j = 1,\ldots,T$$

$$X_{i,j}^{\prime c}, X_{i,j}^c \ge 0 \text{ for } i = 1,\ldots,N \text{ and } j = 1,\ldots,T \qquad (50)$$

This point is notable that second component effective age doesn't change any more and its reason that optional switching isn't possible in this model. Second component effective age changes obviously in the next model (Figs. 14, 15, 16, and 17; Tables 10, 11, and 12).

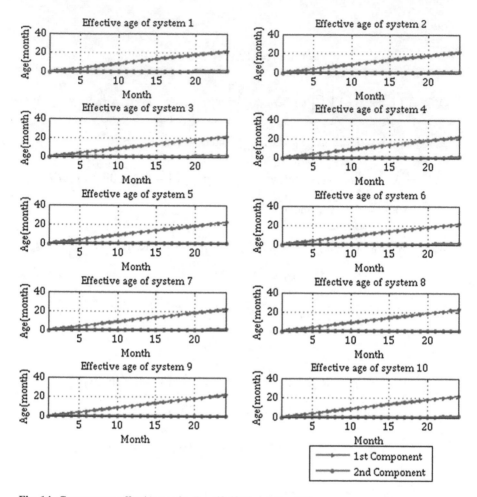

Fig. 14 Components effective age by $R = 68.48\%$ and $C = 93\$$

7.3   Third Model

This model is designed for a case in which both failure and optional switching is possible (Figs. 18, 19, 20, 21, and 22; Tables 13, 14, and 15). This model formulation and its diagrams are shown as below:

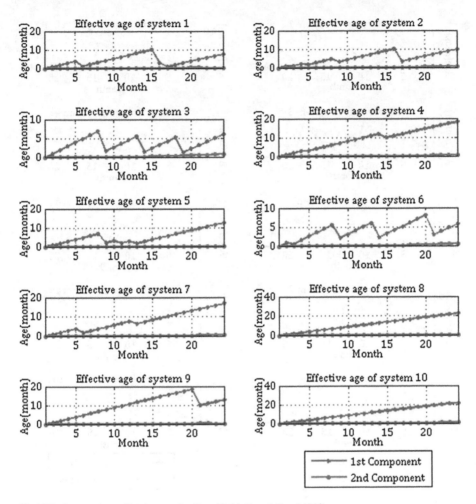

Fig. 15 Components effective age by $R = 90.11\%$ and $C = 1481\$$

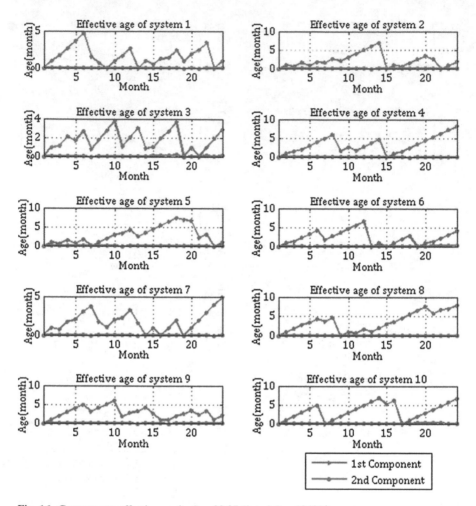

Fig. 16 Components effective age by $R = 99.37\%$ and $C = 19431\$$

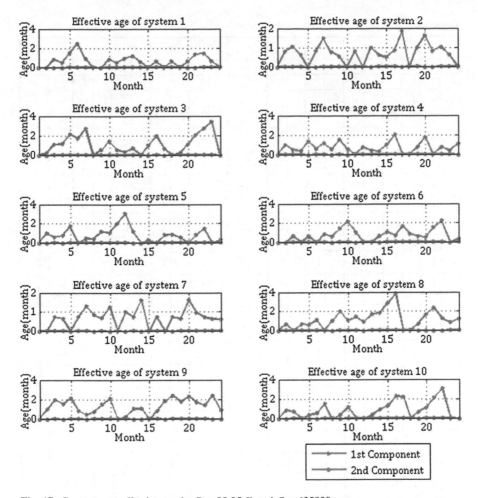

Fig. 17 Components effective age by $R = 99.95\%$ and $C = 63583\$$

Table 10 Maintenance and replacement schedule by $R = 90.11\%$ and $C = 1481\$$

Subsystem	10		9		8		7		6		5		4		3		2		1		
Component	2	1	2	1	2	1	2	1	2	1	2	1	2	1	2	1	2	1	2	1	
	–	–	–	–	–	–	–	–	–	–	–	–	–	–	–	–	–	–	–	–	1
	–	–	–	–	–	–	M2	–	–	M1	–	–	–	–	–	–	–	M2	–	–	2
	–	–	–	–	–	–	–	–	–	–	–	–	–	–	–	–	–	–	–	–	3
	–	–	–	–	–	–	–	–	–	–	–	–	–	M3	–	–	–	M2	–	–	4
	–	–	–	–	–	–	–	M1	–	–	–	–	–	–	–	–	–	–	–	M1	5
	–	–	–	–	–	–	–	–	–	–	–	–	M1	–	–	–	–	–	–	–	6
	–	–	–	–	–	–	–	–	–	–	–	–	–	–	–	–	–	–	–	–	7
	–	–	–	–	–	–	–	M1	–	M1	–	–	M2	M1	–	M2	–	–	–	–	8
	–	–	–	–	–	–	–	–	–	–	–	–	–	–	–	–	–	–	–	–	9
	–	–	–	–	–	–	–	M1	–	M2	–	–	–	–	–	–	–	–	–	–	10
	–	–	–	–	–	–	–	–	–	–	–	–	–	–	–	–	–	–	–	–	11
	–	–	–	–	–	–	–	M2	–	M2	–	–	–	–	–	–	–	–	–	–	12
	–	–	–	–	–	–	–	–	–	M1	–	–	–	–	–	M1	–	–	–	–	13
	–	–	–	–	–	–	–	–	–	–	–	–	M2	M3	–	–	–	–	–	–	14
	–	–	–	–	–	–	–	–	–	–	–	–	–	–	–	–	–	–	M1	–	15
	–	–	–	–	–	–	–	–	–	–	–	–	–	–	–	–	–	M1	M1	–	16
	–	–	–	–	–	–	–	–	–	–	–	–	–	–	–	–	–	–	–	–	17
	–	–	–	–	–	–	–	–	–	–	–	–	–	–	–	M1	–	–	–	–	18
	–	–	–	–	–	–	–	–	–	–	–	–	–	–	–	–	–	–	–	–	19
	–	–	–	M2	–	–	–	–	–	M1	–	–	–	–	–	–	–	–	–	–	20
	–	–	–	–	–	–	–	–	–	–	–	–	–	–	–	–	–	–	M1	–	21
	–	–	M1	–	–	–	–	–	–	–	–	–	–	–	–	–	–	–	–	–	22
	–	–	–	–	–	–	–	–	–	–	–	–	–	–	–	–	–	–	–	–	23
	–	–	–	–	–	–	–	–	–	–	–	–	–	–	–	–	–	–	–	–	24

Maintenance and replacement schedule

Table 11 Maintenance and replacement schedule by $R = 99.37\ \%$ and $C = 19431\$$

| Subsystem | 10 | | 9 | | 8 | | 7 | | 6 | | 5 | | 4 | | 3 | | 2 | | 1 | | |
Component	2	1	2	1	2	1	2	1	2	1	2	1	2	1	2	1	2	1	2	1	
–	–	–	–	–	–	–	–	–	–	–	–	M3	–	–	–	–	–	–	–	1
–	–	M2	–	–	M3	–	M1	M2	M2	M3	M1	M2	M3	–	M2	M2	M1	–	M3	2
–	–	M3	–	–	–	–	–	–	–	–	–	M3	M3	–	–	–	–	–	–	3
M2	–	–	–	M1	M3	–	M2	R	–	–	M1	–	–	M1	M2	M3	M1	–	–	4
–	–	–	–	–	–	–	–	–	–	–	–	–	–	–	–	–	–	–	–	5
–	R	–	M2	M1	M2	R	M3	M2	M1	–	R	M1	–	R	M1	M1	M2	–	M1	6
–	–	–	–	–	–	–	M1	–	–	–	–	–	–	–	–	–	–	–	M1	7
–	–	M2	–	M3	R	M3	M1	M1	–	M2	–	M1	M1	–	–	M1	M2	M1	R	8
–	–	–	–	–	–	–	–	–	–	–	–	–	–	–	–	–	–	–	–	9
M2	–	–	M1	M3	M1	–	M2	M2	–	R	M3	–	M2	M3	M1	–	–	–	M3	10
–	–	–	–	–	–	–	–	–	–	–	–	–	–	–	–	–	–	–	–	11
–	–	R	M3	M1	M1	R	M1	–	R	M1	M2	M3	–	M2	–	R	–	M3	R	12
–	–	–	–	–	–	–	R	–	–	–	–	–	–	–	M1	–	–	–	–	13
M3	M2	M3	M2	–	–	–	–	M1	R	M3	–	M2	R	–	M2	–	R	–	M1	14
–	–	–	M1	–	M3	–	R	–	–	–	–	–	–	–	–	M3	–	–	M3	15
–	R	R	M2	M2	–	–	–	–	–	M3	–	M2	M3	–	M3	M1	M1	M1	M2	16
–	–	–	–	–	–	–	–	–	–	–	–	R	–	–	–	M1	–	–	–	17
–	–	–	M3	M3	–	R	R	–	R	–	M3	–	–	R	R	M3	–	R	M1	18
–	–	–	–	–	–	–	–	–	–	–	M3	–	–	–	–	M1	–	–	–	19
–	–	–	M2	M2	M2	–	–	–	M2	M3	M1	M2	–	–	R	M1	M2	R	M3	20
–	–	–	–	–	–	–	–	–	–	–	–	–	–	–	–	–	R	–	–	21
M1	–	–	M1	–	M3	R	–	–	–	R	R	–	–	M2	–	M3	–	M1	R	22
–	–	–	–	–	–	–	–	–	–	–	–	–	–	–	–	–	–	–	–	23
–	M1	M2	–	–	R	R	–	M2	–	–	–	–	–	R	–	–	–	M1	–	24

Maintenance and replacement schedule

Table 12 Maintenance and replacement schedule by $R = 99.95\,\%$ and $C = 63583\$$

Subsystem	10		9		8		7		6		5		4		3		2		1	
Component	2	1	2	1	2	1	2	1	2	1	2	1	2	1	2	1	2	1	2	1
1	R	M3	R	–	M1	M2	M2	R	R	R	R	–	–	–	R	M1	M2	M3	R	R
2	–	M1	M3	–	M1	R	M1	M2	M2	M2	M2	M1	–	M1	–	M3	M3	M2	R	M3
3	M1	R	R	M2	M2	M2	–	M1	M1	R	R	M2	–	M1	M2	M2	M3	M1	R	M1
4	R	M1	M1	M3	M3	M1	R	R	–	M2	M3	–	R	–	M3	–	R	R	M2	–
5	R	M1	M2	M1	–	M2	M2	M2	–	R	M2	R	M3	M1	–	M2	R	M3	M1	–
6	M2	–	R	M1	–	R	–	M2	M2	M3	M2	M2	M3	M3	M3	–	–	M3	M2	M1
7	M2	M3	M3	M2	R	–	R	M1	R	M1	M2	M1	R	M1	M1	R	M1	M1	M2	R
8	R	M1	–	M3	M3	–	M1	M1	M2	M3	–	M3	–	–	M2	M2	M1	M1	R	R
9	M2	M3	M2	M3	–	M1	R	M2	M1	M3	M3	M2	R	M1	M2	M3	M2	R	M3	M3
10	M2	R	–	R	M2	M2	M2	R	R	M3	M1	–	M2	R	M1	M3	M1	R	M1	M1
11	M2	R	M3	M1	R	M1	R	–	M1	R	M1	–	M2	M3	M3	R	M2	M2	M2	M2
12	–	M1	R	M3	–	M3	M1	M1	–	R	R	M1	M2	M1	M1	M2	–	M2	M2	M2
13	–	M2	R	M2	M1	M2	R	M3	R	M2	R	R	M3	M1	–	R	M3	M1	M2	M1
14	–	M2	M3	R	–	–	–	R	M3	M2	M1	M1	M3	M3	R	–	–	M1	M3	R
15	–	–	M1	M3	M1	–	M2	M2	–	M1	M3	R	–	–	M1	–	M1	M2	M3	M2
16	M2	M2	R	–	M3	R	M1	R	–	–	–	M3	–	R	R	M1	R	–	M3	R
17	–	M2	M1	M3	M2	R	R	M2	–	M1	M2	–	–	R	M2	R	M1	R	–	M2
18	R	M2	–	M2	R	M2	–	M1	–	M1	–	M2	–	M3	M2	M1	–	–	–	R
19	–	M2	M3	M3	R	–	M3	–	–	M1	–	R	–	–	M2	M3	R	M3	R	M2
20	M1	–	M1	M2	R	M3	M3	M1	M3	–	M2	M3	M2	R	M3	–	R	M1	R	M3
21	–	–	M3	M2	R	M1	M2	M1	M3	M3	M2	M2	–	M3	M3	M3	M1	M2	M1	M2
22	–	R	R	–	M1	M1	M1	M1	R	R	–	R	M2	M1	M3	M3	M1	M1	M1	M1
23	R	R	R	M1	–	M2	M2	M1	M3	M1	M3	M1	R	M3	R	R	M2	R	–	R
24	M2	M2	M2	M2	M1	–	M2	–	M2	M1	M2	M3	M1	R	R	M1	M1	–	M1	M3

Maintenance and replacement schedule

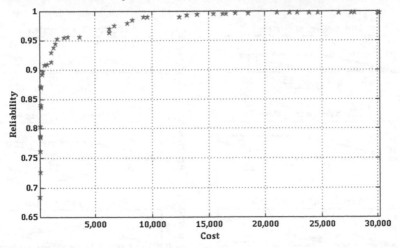

Fig. 18 Third model Parto front

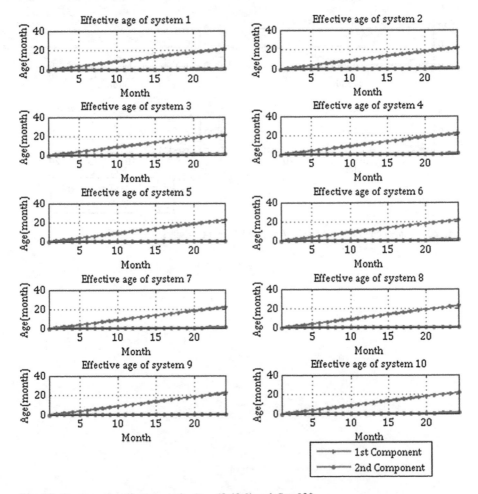

Fig. 19 Components effective age by $R = 68.48\%$ and $C = 93\$$

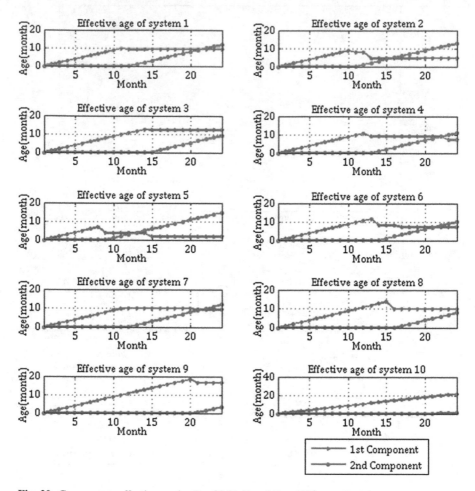

Fig. 20 Components effective age by $R = 90.81\ \%$ and $C = 499\$$

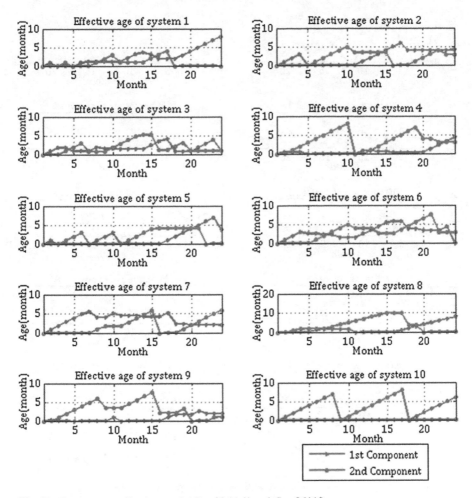

Fig. 21 Components effective age by $R = 99.11 \%$ and $C = 9611\$$

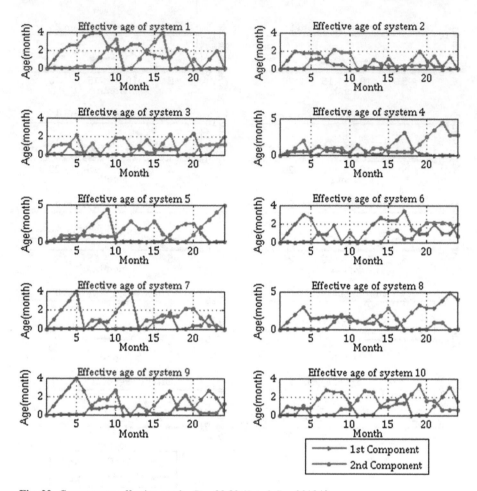

Fig. 22 Components effective age by $R = 99.89\%$ and $C = 30184\$$

Table 13 Maintenance and replacement schedule by $R = 90.81$ % and $C = 499$\$

Overall label (far right): Maintenance and replacement schedule

Period	S10-2	S10-1	S9-2	S9-1	S8-2	S8-1	S7-2	S7-1	S6-2	S6-1	S5-2	S5-1	S4-2	S4-1	S3-2	S3-1	S2-2	S2-1	S1-2	S1-1
1																				
2																				
3																				
4																				
5																				
6																				
7																				
8												M2								
9																				
10																				M3
11									M3											
12																M3			M2	
13											M2									
14													M2					M3		
15								M2												
16												M3								
17																				
18																				
19																				
20					M3					M3										
21																				
22																M3				
23																				
24																				

Table 14 Maintenance and replacement schedule by $R = 99.11\ \%$ and $C = 9611\$$

10		9		8		7		6		5		4		3		2		1		Subsystem
2	1	2	1	2	1	2	1	2	1	2	1	2	1	2	1	2	1	2	1	Component
–	–	–	M1	–	M1	–	–	–	–	M1	–	–	M2	–	–	M3	–	M3	–	1
M2	–	–	M3	–	M3	–	–	M2	–	–	R	–	–	–	M3	–	–	–	R	2
–	–	–	–	–	–	–	–	–	–	–	–	–	–	–	–	–	–	–	–	3
–	–	–	R	M2	–	M2	–	–	M2	–	–	–	R	–	M2	–	R	R	–	4
–	–	–	–	–	–	–	–	–	–	–	–	–	–	–	–	–	–	–	–	5
–	–	–	–	–	M2	–	M3	–	M3	R	–	–	M1	M1	–	–	–	–	M2	6
–	–	–	–	–	–	–	M2	–	–	–	–	–	–	–	–	–	–	–	–	7
–	R	M2	–	–	–	M3	–	–	M2	M3	–	–	–	–	M2	–	–	–	–	8
–	–	–	–	–	–	–	–	–	–	–	–	–	–	–	–	–	–	–	–	9
–	–	–	R	–	R	–	M2	M2	–	–	R	R	–	–	M3	M2	–	M1	–	10
–	–	–	–	–	–	–	–	–	–	–	–	–	M3	–	–	–	–	–	–	11
–	–	–	–	–	–	–	–	–	–	–	–	–	–	–	–	–	–	–	–	12
–	–	–	–	–	–	–	M3	M2	–	–	–	–	M3	M3	–	–	–	–	M3	13
–	–	–	–	–	–	–	–	–	–	–	–	–	–	–	–	–	–	M3	M3	14
–	–	M1	–	M3	–	R	–	–	M3	M3	–	–	M2	M1	–	–	R	M2	–	15
–	–	–	–	–	–	–	–	–	–	–	–	–	–	–	M3	–	–	–	–	16
R	–	–	M3	M1	–	M1	–	M1	–	–	M2	–	–	–	M1	–	M2	R	–	17
–	–	–	–	–	–	–	–	–	–	–	–	–	–	–	–	–	–	–	–	18
–	–	R	–	R	–	–	–	M3	–	M3	–	M2	–	M1	–	–	–	–	–	19
–	–	–	–	–	–	–	–	–	–	–	–	–	–	–	–	–	–	–	–	20
–	–	–	M2	M1	–	–	–	–	M1	R	–	M3	–	–	–	–	–	M3	–	21
–	–	–	–	–	–	–	–	–	–	–	–	–	–	–	–	M2	–	–	–	22
–	–	M2	–	–	–	–	M3	–	R	–	M2	–	–	–	M1	–	M3	R	–	23
–	–	–	–	–	–	–	–	–	–	–	–	–	–	–	–	–	–	–	–	24

Maintenance and replacement schedule

Table 15 Maintenance and replacement schedule by $R = 99.89\ \%$ and $C = 30184\$$

Maintenance and replacement schedule

Subsystem	10		9		8		7		6		5		4		3		2		1		
Component	2	1	2	1	2	1	2	1	2	1	2	1	2	1	2	1	2	1	2	1	
	–	–	R	–	–	–	M1	–	–	–	–	M1	–	M2	M2	–	M3	–	M1	–	1
	–	M1	M1	–	M3	–	M3	–	R	–	M3	–	–	–	–	M2	–	–	M2	–	2
	–	–	M1	–	–	–	M1	–	M1	–	M1	–	–	–	M3	–	–	M2	–	M3	3
	R	–	M2	–	–	M1	–	–	–	M2	M2	–	R	–	M1	–	–	–	M1	–	4
	R	–	–	M2	R	–	–	R	–	M1	–	–	–	M3	–	R	M2	–	M3	–	5
	–	–	–	M1	–	M2	M3	–	R	–	–	–	–	M2	–	–	–	M1	–	M3	6
	–	M2	M3	–	–	–	–	M2	M1	–	M3	–	M2	–	R	–	–	M2	–	–	7
	M2	–	–	M2	–	–	R	–	–	R	–	–	–	M1	R	–	M2	–	–	M2	8
	–	M1	–	–	M1	–	M2	–	–	M1	–	R	R	–	–	M3	–	R	–	M3	9
	–	R	R	–	–	M1	–	–	R	–	–	–	M2	–	R	–	R	–	R	–	10
	–	–	–	R	M1	–	–	–	–	–	–	–	–	M1	–	M1	–	M1	–	M3	11
	M2	–	–	M3	–	M3	–	R	M1	–	M2	–	–	M1	M3	–	M2	–	–	–	12
	M1	–	M1	–	–	M1	M3	–	–	M3	–	R	M2	–	M1	–	M1	–	M2	–	13
	–	M3	M1	–	–	M1	–	M2	–	M3	–	M3	M2	–	–	M1	M2	–	M3	M3	14
	M2	–	–	–	M1	–	M2	–	M2	–	M1	–	–	–	–	–	–	M1	R	–	15
	–	M3	–	M3	R	–	M3	–	M1	–	M1	–	–	–	–	–	–	M1	R	–	16
	–	R	–	M1	R	–	–	R	–	M1	R	–	M1	–	R	–	–	–	M3	–	17
	–	M3	–	–	M1	–	M3	–	–	M2	–	M3	M1	–	M3	–	–	–	M2	–	18
	M1	–	M1	–	–	M2	–	M1	M3	–	–	–	M3	–	M3	–	M1	–	–	R	19
	–	–	M1	–	M3	–	M1	–	–	–	–	M2	–	R	–	R	M1	–	R	–	20
	M1	–	–	–	M1	–	–	–	M1	–	–	R	–	–	M2	–	R	–	M2	–	21
	–	–	–	M2	R	–	–	R	M2	–	–	M1	M2	–	–	–	–	–	M1	–	22
	–	M1	–	M1	–	M2	R	–	M1	–	M2	–	R	–	–	–	R	–	R	R	23
	–	–	M2	–	–	–	–	M3	M1	–	–	M1	–	R	–	–	R	R	M2	–	24

$$\text{Min Total Cost} = \sum_{i=1}^{N} \sum_{j=1}^{T} \left[F_i \cdot \left(1 - \text{Reliability}_{i,j}^{SS} \right) \right.$$

$$+ \sum_{c=1}^{C} \left(\sum_{k=1}^{K} \left(M_i^{c,k} \cdot m_{i,j}^{c,k} \right) + R_i^c \cdot r_{i,j}^c \right) \right]$$

$$+ \sum_{j=1}^{T} \left[Z \left(1 - \prod_{i=1}^{N} \left(1 - \sum_{c=1}^{C} \left(r_{i,j}^c + \sum_{k=1}^{K} m_{i,j}^{c,k} \right) \right) \right) \right]$$

$$\text{Max Reliability} = \prod_{i=1}^{N} \prod_{j=1}^{T} \left(p_{i,j}^1 \cdot \left[e^{-\left[\lambda_i^1 \left(\left(X_{i,j}^1 + T/J \right)^{\beta_i^1} - \left(X_{i,j}^1 \right)^{\beta_i^1} \right) \right]} \right. \right.$$

$$+ \int_{t=0}^{t=T/J} \lambda_i^1 \cdot \beta_i^1 \cdot t^{\beta_i^1 - 1} \cdot e^{-\left[\lambda_i^1 \left(\left(X_{i,j}^1 + t \right)^{\beta_i^1} - \left(X_{i,j}^1 \right)^{\beta_i^1} \right) \right]}$$

$$\cdot e^{-\left[\lambda_i^2 \left(\left(X_{i,j}^2 + T/J - t \right)^{\beta_i^2} - \left(X_{i,j}^2 \right)^{\beta_i^2} \right) \right]} \cdot dt \right]$$

$$+ p_{i,j}^2 \cdot \left[e^{-\left[\lambda_i^2 \left(\left(X_{i,j}^2 + T/J \right)^{\beta_i^2} - \left(X_{i,j}^2 \right)^{\beta_i^2} \right) \right]} \right.$$

$$+ \int_{t=0}^{t=T/J} \lambda_i^2 \cdot \beta_i^2 \cdot t^{\beta_i^2 - 1} \cdot e^{-\left[\lambda_i^2 \left(\left(X_{i,j}^2 + t \right)^{\beta_i^2} - \left(X_{i,j}^2 \right)^{\beta_i^2} \right) \right]}$$

$$\cdot e^{-\left[\lambda_i^1 \left(\left(X_{i,j}^1 + T/J - t \right)^{\beta_i^1} - \left(X_{i,j}^1 \right)^{\beta_i^1} \right) \right]} \cdot dt \right] \right)$$

subject to:

$$X_{i,j}^c = 0 \text{ for } i = 1, \ldots N \text{ and } c = 1, 2$$

$$X_{i,j}^c = \left(1 - r_{i,j-1}^c \right) \left[\prod_{k=1}^{K} \left(1 - m_{i,j-1}^{c,k} \right) \right] X_{i,j-1}^{/c}$$

$$+ \sum_{k=1}^{K} m_{i,j-1}^{c,k} \cdot \left(\alpha_i^k \cdot X_{i,j-1}^{/c} \right)$$

$$\text{for } i = 1, \ldots N; j = 2, \ldots, T \text{ and } c = 1, 2$$

$$X_{i,j}'^1 = X_{i,j}^1 + \left[p_{i,j}^1 \cdot \text{Reliability}_{i,j}^1 \right.$$

$$\left. + \left(1 - p_{i,j}^1 \right) \left(1 - \text{Reliability}_{i,j}^2 \right) \right] \cdot T/J$$

$$\text{for } i = 1, \ldots, N \text{ and } j = 1, \ldots, T$$

$$X_{i,j}'^2 = X_{i,j}^2 + \left[\left(1 - p_{i,j}^1 \right) \cdot \text{Reliability}_{i,j}^2 \right.$$

$$\left. + p_{i,j}^1 \cdot \left(1 - \text{Reliability}_{i,j}^1 \right) \right] \cdot T/J$$

$$\text{for } i = 1, \ldots, N \text{ and } j = 1, \ldots, T$$

$$p_{i,1}^1 = 1 \text{ for } i = 1, \ldots, N$$

$$p_{i,1}^2 = 0 \text{ for } i = 1, \ldots, N$$

$$p_{i,j+1}^1 = p_{i,j}^1 \cdot \text{Reliability}_{i,j}^1 + \cdot \left(1 - p_{i,j}^1 \right)$$

$$\cdot \left(1 - \text{Reliability}_{i,j}^2 \right)$$

$$\text{for } i = 1, \ldots, N \text{ and } j = 1, \ldots, T - 1$$

$$p_{i,j}^2 = 1 - p_{i,j}^1 \text{ for } i = 1, \ldots, N \text{ and } j = 1, \ldots, T$$

$$\sum_{k=1}^{K} m_{i,j}^{c,k} + r_{i,j}^c \leq 1 \text{ for } i = 1, \ldots, N; j = 1, \ldots, T \text{ and } c = 1, 2$$

$$\sum_{c=1}^{c} \left(\sum_{k=1}^{K} m_{i,j}^{c,k} + r_{i,j}^c \right) \leq 1 \text{ for } i = 1, \ldots, N; j = 1, \ldots, T - 1 \text{ and } c = 1, 2$$

$$p_{i,j}^c \cdot \left(\sum_{k=1}^{K} m_{i,j-1}^{c,k} + r_{i,j-1}^c \right) = 0 \text{ for } i = 1, \ldots, N; j = 2, \ldots, T \text{ and } c = 1, 2$$

$$m_{i,j}^{c,k}, r_{i,j}^c = 0 \text{ or } 1 \text{ for } i = 1, \ldots, N; j = 1, \ldots, T;$$

$$c = 1, 2 \text{ and } k = 1, \ldots, K$$

$$0 \leq p_{i,j}^1, p_{i,j}^2 \leq 1 \text{ for } i = 1, \ldots, N \text{ and } j = 1, \ldots, T$$

$$X_{i,j}'^c, X_{i,j}^c \geq 0 \text{ for } i = 1, \ldots, N \text{ and } j = 1, \ldots, T \qquad (51)$$

Table 16 NSGA-II parameters

NSGA-II parameters	1st model	2nd & 3rd model
Number of Generations	1000	100
Population Size	100	50
Probability of Crossover	0.7	0.7
Probability of Mutation	0.4	0.4

NSGA-II parameters are similar to GA parameters that are set as shown in Table 16.

8 Conclusion

For better conclusion, these three models will be compared. As illustrated in Fig. 23, second and third models with failure possibility are more powerful than first model without failure possibility. As shown in Fig. 24, a brief focus on second and third models shows third model strength. For example, by a simple comparison between points (1) and (2), although the solution reliabilities are similar, the third model cost is 33 % less than the other. Reversely, in the plot primary points (points (8) and (9)), when the costs of these models are close together, the third model reliability is 10 % more than second one. It means that by considering optional switching in a standby system, the total cost reduces effectively without any changes in reliability or in the same costs, optional switching increases total reliability.

Table 17 shows a comparison between one schedule of second and third models, non-optional and optional switching in a same reliability. As is obvious in Table 17 optional switching causes less cost and more applicable scheduling than non-optional switching without any changes in reliability. So it can be found out that, in reality, the third model is more effective and powerful with failure possibility and optional switching.

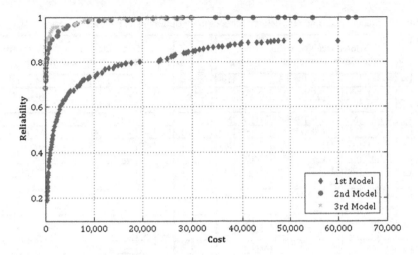

Fig. 23 Models Pareto front comparison

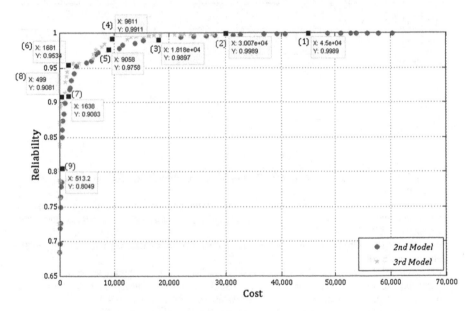

Fig. 24 Second and third model Pareto front comparison

Table 17 Maintenance and replacement scheduling in optional and non-optional switching cases

Maintenance and replacement schedule in non-optional switching:

Subsystem	Component	1	2	3	4	5	6	7	8	9	10
1	1	M3	M2	M1	M2	M2	--	M1	R	M2	--
1	2	R	R	M1	M1	--	--	M3	M3	R	M2
2	1	M3	M1	M2	M1	R	--	M1	M1	R	M1
2	2	M3	--	M2	--	R	M3	M1	--	M1	M3
3	1	R	M3	--	M1	R	--	M1	--	--	M1
3	2	M2	M2	M3	M2	--	M3	M3	M2	M1	--
4	1	R	--	M3	M1	M1	--	R	M1	M3	M1
4	2	M3	--	R	R	M1	M2	--	--	--	--
5	1	M1	M2	M3	--	M2	M3	--	M1	R	M2
5	2	M3	--	M2	--	M1	M2	--	--	R	--
6	1	M1	M1	R	--	M1	--	M2	M1	M1	R
6	2	--	--	--	M1	M1	--	--	M2	--	--
7	1	M2	M2	R	M3	M3	M3	--	M1	R	--
7	2	R	--	--	M3	R	M1	--	M1	M1	M1
8	1	M1	R	R	M2	M3	R	R	R	M3	M2
8	2	--	--	--	--	R	M3	M2	--	R	--
9	1	R	M3	M2	M1	M3	--	--	M3	R	R
9	2	M1	--	M3	M3	M2	--	--	--	M2	M1
10	1	--	--	--	--	--	--	--	R	M2	M3
10	2	--	--	M3	M1	M2	--	R	--	M3	--

Reliability= 99.89%   Cost= 45002$

Maintenance and replacement schedule in optional switching:

Subsystem	Component	1	2	3	4	5	6	7	8	9	10
1	1	--	--	M3	--	--	M3	--	M2	M3	--
1	2	M1	M2	--	M1	M3	--	--	--	--	R
2	1	--	--	M2	--	--	M1	M2	--	R	--
2	2	M3	--	--	--	M2	--	--	M2	--	R
3	1	--	M2	--	--	R	--	--	--	M3	--
3	2	M2	--	M3	M1	--	--	R	R	--	R
4	1	M2	--	--	--	M3	M2	--	M1	--	--
4	2	--	--	--	R	--	--	M2	--	R	M2
5	1	M1	--	M1	--	--	--	--	--	R	--
5	2	--	M3	--	M2	--	--	M3	--	--	--
6	1	--	--	--	M2	M1	--	--	R	M1	--
6	2	--	R	M1	--	--	R	M1	--	--	R
7	1	--	--	--	--	R	--	M2	--	--	--
7	2	M1	M3	M1	--	--	M3	--	R	M2	--
8	1	--	--	--	M1	--	M2	--	--	--	M1
8	2	--	M3	--	--	R	--	--	--	M1	--
9	1	--	--	--	--	M2	M1	--	M2	--	--
9	2	R	M1	M1	M2	--	--	M3	--	--	R
10	1	--	M1	--	--	--	--	M2	--	M1	R
10	2	--	--	--	R	R	--	M2	--	--	

Reliability= 99.89%   Cost= 30184$

References

http://reliawiki.com/Preventive_Maintenance

Usher JS, Kamal AH, Hashmi SW (1998) Cost optimal preventive maintenance and replacement scheduling. IIE Transactions 30(12):1121–1128

Moghaddam KS, Usher JS (2011) Preventive maintenance and replacement scheduling for repairable and maintainable systems using dynamic programming. Computers & Industrial Engineering 60:654–665

Srinivas N, Deb K (1995) Multiobjective function optimization using nondominated sorting genetic algorithms. Evol Comput 2(3):221–248

Deb K, Agrawal S, Pratap A, Meyarivan T (2000). A fast elitist non-dominated sorting genetic algorithm for multi-objective optimization: NSGA-II, vol 5. In: Proceedings of the parallel problem solving from nature VI conference, 16–20 September, Paris, pp 849–858

Smith CO (1976) Introduction to reliability in design. Illustrated, McGraw-Hill

Index

S. Kadry and A. El Hami (eds.), *Numerical Methods for Reliability and Safety
Assessment: Multiscale and Multiphysics Systems*, DOI 10.1007/978-3-319-07167-1,
© Springer International Publishing Switzerland 2015

CPSIA information can be obtained
at www.ICGtesting.com
Printed in the USA
LVHW02*1740150918
590267LV00001B/112/P

9 783319 071664